Y0-DKL-473

Tricarboxylic acid cycle and respiratory chain

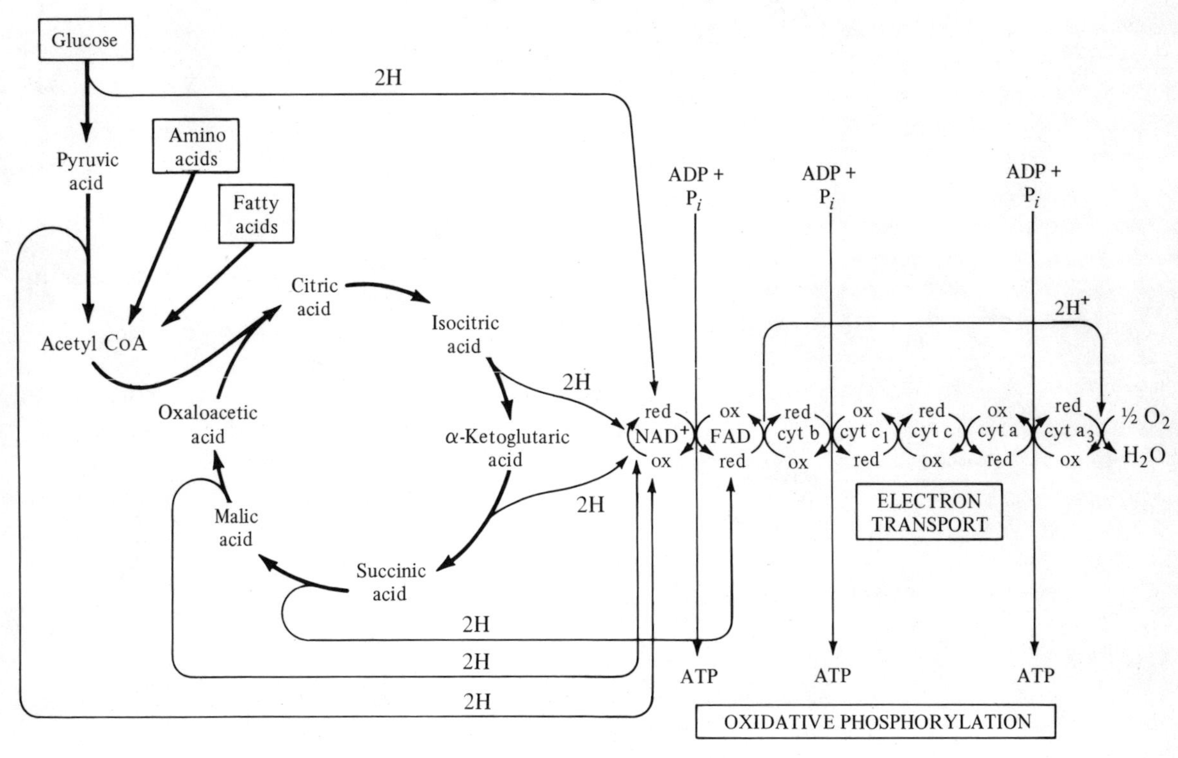

The Chemistry of Living Systems

The Chemistry of Living Systems

Robert F. Steiner
University of Maryland Baltimore County

Seymour Pomerantz
University of Maryland School of Medicine

D. VAN NOSTRAND COMPANY
New York Cincinnati Toronto London Melbourne

D. Van Nostrand Company Regional Offices:
New York Cincinnati

D. Van Nostrand Company International Offices:
London Toronto Melbourne

Copyright © 1981 by Litton Educational Publishing, Inc.

Library of Congress Catalog Card Number: 80-51095
ISBN: 0-442-28128-5

All rights reserved. No part of this work covered by the copyright hereon may be reproduced or used in any form or by any means—graphic, electronic, or mechanical, including photocopying, recording, taping, or information storage and retrieval systems—without written permission of the publisher. Manufactured in the United States of America.

Published by D. Van Nostrand Company
135 West 50th Street, New York, N.Y. 10020

10 9 8 7 6 5 4 3 2 1

Preface

The Chemistry of Living Systems is intended for a one-semester course in biochemistry at the junior or senior level, given in chemistry, biochemistry, or biology departments for premedical, predental, preveterinary, health science, or agronomy students, as well as for other nonchemistry majors. The book can be used most effectively by students with one year of basic chemistry and biology and at least one semester of organic chemistry. Although calculus notation appears in one chapter, its use is minimal and students lacking a knowledge of calculus will not be significantly impeded.

The Chemistry of Living Systems is unique in several respects. It stresses methods, including physical methods, and techniques of structural determination. Some topics, such as muscle contraction and the properties of plasma proteins, have been given more extensive coverage than is usually found in elementary texts. A historical approach has been adopted where appropriate, with stress upon classical experiments. A concluding chapter summary elucidates the primary points covered in each chapter. The problems presented at the end of each chapter vary in complexity and have been coded with respect to difficulty. A few problems do not have obvious answers and would be appropriate for future research. An Instructor's Manual, available from the publisher, for the text provides commentary and background reading suggestions for each chapter, a test for each chapter, and the rest of the answers to the text problems.

The book begins with a brief description of the living cell (Chapter 1), including a discussion of the two kinds of cells and of the principal organelles observed within cells. Chapter 2 deals with the properties of water, the ionization of weak electrolytes, and the nature of buffers. Chapters 3–5 consider the structure and general properties of proteins, including the properties of their amino acid constituents and the methods employed to determine their amino acid sequence and three-dimensional structure. Chapter 6 is concerned with the properties of enzymes, while Chapter 7 describes the major coenzymes. Chapter 8, which deals with nutrition, describes the classification of species with respect to external sources of mass and energy.

Chapter 9 is concerned with the structures and properties of the major carbohydrates, including the simple sugars and the polysaccharides. Chapter 10 deals with the metabolism of carbohydrates, including their degradation by glycolysis and the biosynthesis of polysaccharides from monomeric sugars. Chapter 11 discusses the mechanism of photosynthesis, including the nature of the primary photochemical events and the subsequent synthesis of carbohydrates. Chapter 12 describes the terminal stages of oxidative metabolism, which involve the tricarboxylic acid cycle and the linked processes of respiration and oxidative phosphorylation. Chapter 13 is concerned with the structure, metabolic degradation, and biosynthesis of the triacylglycerol fats, while Chapter 14 deals with the phospholipids. Chapter 15 describes the properties of natural membranes. Chapter 16 discusses the structures and biological roles of the terpenes and steroids, which include a number of important vitamins and hormones. Chapter 17 outlines the properties of hemes and heme-containing proteins, including hemoglobin and the cytochromes. Chapters 18 and 19 are concerned with nucleic acids. Chapter 18 describes the structure, biological role, and biosynthesis of DNA, as well as its mediation of the directed biosynthesis of RNA. Chapter 19 is focused upon the mechanism of protein synthesis and the genetic code. Chapters 20 and 21 deal with the degradation and biosynthesis of the amino acids and nucleotides, respectively. Chapter 22 discusses the structures and biological activities of the major hormones. Chapter 23 describes the properties of the principal constituents of blood plasma, while chapter 24 is concerned with several specialized tissues, including connective tissue, muscle, and bone.

I would like to thank the following reviewers for their helpful comments: Paul W. Chun, University of Florida; Edward D. Harris, Texas A&M University; Peter H. von Hippel, University of Oregon; and Delano V. Young, Boston University.

Robert F. Steiner
Seymour Pomerantz

Contents

1 The Cell 1

1-1 The Central Importance of the Cell 1
1-2 Biological Systems 1
1-3 The Nature of the Living Cell 2
1-4 Chromosomes and Genes 9
1-5 Cell Replication 11

2 The Chemical Nature of Biological Systems 15

2-1 A Chemical Prelude 15
2-2 The Elemental Composition of Living Cells 15
2-3 Biological Significance of Water 16
2-4 Hydrogen Ion Concentration of Biological Fluids 18

3 Amino Acids and Peptides 26

3-1 Amino Acids 26
3-2 Characteristics of the Amino Acids Occurring in Natural Proteins 28
3-3 Less Common Amino Acids 33
3-4 Reactions of the Amino Acids 33
3-5 Optical Activity of the Amino Acids 35
3-6 Ionization of the Amino Acids 37
3-7 Separation and Analysis of the Amino Acids and Peptides 39

4 The Chemistry of Polypeptides and Proteins 50

4-1 Chemical Structure and Biological Function 50
4-2 Purification and Criteria for Homogeneity of Proteins 52
4-3 Determination of the Primary Structure of Proteins 52

4-4 Proteins of Known Primary Structure 57
4-5 Genetically Controlled Variations in Primary Structure 58

5 Size, Shape, and Conformation of Proteins 68

5-1 Spatial Organization of Proteins 68
5-2 General Aspects of Protein Conformation 68
5-3 Origins of the Stability of Protein Struture 69
5-4 Hydrodynamic Methods for Estimating Size and Shape 73
5-5 Alternative Techniques for Molecular Weight Determination 84
5-6 Secondary Structure of Proteins 86
5-7 X-Ray Crystallography 87
5-8 An Example of a Protein of Known Structure 92
5-9 General Conclusions about Protein Structure 93
5-10 The Amino Acid Sequence of a Protein Dictates Its Three-Dimensional Structure 95

6 The Enzymes and Biocatalysis 100

6-1 General Properties of Enzymes 100
6-2 Types of Enzymic Reactions 102
6-3 Enzyme Kinetics 104
6-4 Thermodynamics of Biochemical Reactions 109
6-5 Proteolytic Enzymes 113
6-6 The Structural Basis for the Activity of Chymotrypsin 115
6-7 The Chymotrypsin Family of Enzymes 120
6-8 Other Proteolytic Enzymes 120
6-9 Allosteric Enzymes 121

7 Coenzymes and Biological Oxidations 129

7-1 The Function of Coenzymes 129
7-2 Coenzymes That Transfer Groups Other Than Hydrogen 130
7-3 Coenzymes That Transfer Hydrogen 136
7-4 Oxidation and Reduction 141
7-5 The Redox Potential 143

8 Nutrition 151

8-1 Requirements of Living Systems 151
8-2 Transfer of Mass and Energy between Living Cells and Their Environment 151
8-3 Energy Sources for Biological Organisms 152
8-4 Amino Acid Requirements 156
8-5 Requirements for Specialized Compounds 156

9 The Carbohydrates 162

9-1 The Biological Role of the Carbohydrates 162
9-2 Monosaccharides 162
9-3 Chemical Properties and Derivatives of the Monosaccharides 168
9-4 Oligosaccharides 171
9-5 Polysaccharides 173
9-6 Glycoproteins 181
9-7 Enzymic Hydrolysis of Polysaccharides 181

10 Carbohydrate Metabolism 186

10-1 Carbohydrates as Biological Fuels 186
10-2 Digestion of Carbohydrates 187
10-3 Biochemical Conversions of Glucose 188
10-4 Glycolysis 190
10-5 Enzymes of Glycolysis 196
10-6 The Phosphogluconate Pathway 200
10-7 Hexose Transformations 203
10-8 Synthesis of Oligo- and Polysaccharides 204
10-9 Synthesis and Degradation of Glycogen 206
10-10 Control of Glycogen Degradation 209

11 Photosynthesis 214

11-1 The Photosynthetic Apparatus of Plants 214
11-2 Photosynthetic Pigments 216
11-3 The Primary Photosynthetic Events 219
11-4 Biosynthesis of Carbohydrates 222

12 The Tricarboxylic Acid Cycle and Oxidative Phosphorylation 231

12-1 Biological Significance 231
12-2 Oxidative Metabolism 232
12-3 Mitochondria 233
12-4 The TCA Cycle 235
12-5 Reactions of the TCA Cycle 236
12-6 Metabolic Pathways Linked to the TCA Cycle 242
12-7 The Respiratory Chain and Oxidative Phosphorylation 245

13 The Triacylglycerol Fats 257

13-1 Biological Roles 257
13-2 Lipids 257
13-3 Fatty Acids 258

13-4 Digestion of the Triacylglycerol Fats 262
13-5 Distribution of Triacylglycerol Fats 263
13-6 Degradation of Fatty Acids 264
13-7 Biosynthesis of Fatty Acids 268
13-8 Synthesis of Triacylglycerol Fats 273
13-9 Ketone Bodies 275

14 Phosphatides and Related Compounds 279

14-1 Phospholipids 279
14-2 Glycerophosphatides 283
14-3 Sphingolipids 285
14-4 Metabolism of the Phosphatides 288

15 Biological Membranes 295

15-1 Structure and Composition of Membranes 295
15-2 Physical Properties of Membranes 297
15-3 Active Transport 298

16 Terpenes and Steroids 304

16-1 Biological Functions 304
16-2 Terpenes 305
16-3 Steroids 309
16-4 Sterols 310
16-5 Bile Salts 315
16-6 Biosynthesis of Cholesterol 317
16-7 Steroid Hormones 321
16-8 Biosynthesis of the Steroid Hormones 323

17 Porphyrins, Hemes, and Heme Proteins 327

17-1 General Properties of the Porphyrins 327
17-2 Hemes 328
17-3 Hemoglobin 329
17-4 Myoglobin 339
17-5 Other Hemoproteins 340
17-6 The Metabolism of Heme Groups 342

18 Nucleotides and Nucleic Acids 347

18-1 The Biological Function of the Nucleic Acids 347
18-2 Nucleotides 348
18-3 Chemical Structure of the Nucleic Acids 352

18-4 Physical Structure of the Nucleic Acids 354
18-5 Molecular Genetics 360
18-6 Replication of DNA 366
18-7 Restriction and Modification of DNA 373
18-8 Replication of a Viral DNA 374
18-9 Nucleoproteins 375
18-10 Different Kinds of RNA 375
18-11 DNA Transcription 379

19 The Biosynthesis of Proteins 387

19-1 Origins of the Currently Accepted Model for Protein Synthesis 387
19-2 The Protein-Synthesizing System of the Cell 391
19-3 The Genetic Code 397
19-4 Control of Protein Synthesis 399
19-5 Viruses as Model Systems 401
19-6 Mutations and the Genetic Code 405

20 Amino Acid Metabolism 409

20-1 Metabolic Conversions of Amino Acids 409
20-2 Essential and Nonessential Amino Acids 410
20-3 Digestion of Protein 410
20-4 Pathways of Amino Acid Conversion 411
20-5 The Urea Cycle 414
20-6 Metabolic Degradation of Individual Amino Acids 419
20-7 Biosynthetic Reactions of the Amino Acids 432
20-8 Biosynthesis of the Nonessential Amino Acids by Mammals 436

21 The Metabolism of Nucleotides and Their Components 444

21-1 Biosynthesis of Pyrimidine Ribonucleotides 444
21-2 Biosynthesis of Purine Ribonucleotides 448
21-3 Formation of the Deoxyribonucleotides 455
21-4 Degradation of the Nucleotide Bases 456
21-5 Biosynthesis of the Nucleotide Coenzymes 459

22 Hormonal Regulation 464

22-1 General Aspects 464
22-2 Thyroid Hormones 466
22-3 Pancreatic Hormones 467
22-4 Hormones of the Adrenal Medulla 469
22-5 Pituitary Hormones 472
22-6 Parathyroid Hormones 475
22-7 Control of Blood Glucose 476

23 Plasma 479

23-1 Multiple Functions of Blood Plasma 479
23-2 Composition of Plasma 479
23-3 The Blood Clotting System 480
23-4 Plasma Albumin 484
23-5 Immunoglobulins 485
23-6 Other Plasma Proteins 491

24 Specialized Tissues 494

24-1 Connective Tissue 494
24-2 Myofibril 499
24-3 Bone Tissue 507

Glossary 511

Selected Answers to Review Questions 529

Index 533

The Chemistry of Living Systems

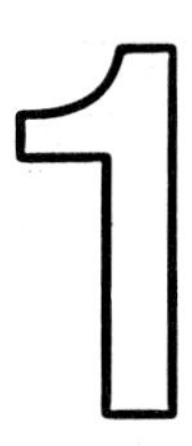

The Cell

1-1 THE CENTRAL IMPORTANCE OF THE CELL

The orientation of this initial chapter will be entirely biological. We will be considering the properties of the fundamental unit of living systems, namely, the **cell**. This is the simplest organized structure to which can be attributed the characteristics commonly associated with life. The molecules comprising living cells do not possess these characteristics and differ only in size and complexity from the smaller molecules familiar to the organic chemist. The molecular components of cells must be assembled into an **organized structure** to exhibit the properties of living systems. These properties do not survive the loss of this organized structure.

We are beginning with the cell, whose structure, composition, and replication will be described here, because it is the natural medium within which occur the processes that comprise biochemistry and outside of which these processes cannot operate to any collective purpose.

1-2 BIOLOGICAL SYSTEMS

The division of the visible biological world into **plants** and **animals** is sufficiently obvious to be almost forced upon human consciousness. This distinction, which probably dates from long before the civilized era, has been formalized by biologists in the classification of living systems into the two kingdoms of Plantae and Animalia.

The differentiation between the two kingdoms is sharp and unambiguous for the higher forms, but is a little blurred for the more primitive species. Animals are capable of active movement; plants are not. Animals acquire their essential chemical energy by the processing of organic materials ingested from their surroundings; plants obtain their energy by the utilization of the radiant energy of sunlight in the process called **photosynthesis**. Animals tend to grow only to a limiting size characteristic of the species; the limits of growth for plants are less rigid. The differences persist to the microscopic level of structural organization.

The biologists of the nineteenth century introduced a third kingdom, Protista, to include the **microorganisms**. The classes of living systems traditionally placed in this category include **algae, protozoa, fungi**, and **bacteria**, all of which share a structural organization that is relatively uncomplicated in comparison to those of the plants and animals. Many protists consist of a single cell and the protists that are multicellular do not exhibit the extensive differentiation into tissues characteristic of plants and animals. Some modern classifications place the bacteria and blue-green algae, which differ in several fundamental respects from the other groups, in the separate kingdom of Monera.

The **viruses** lie on the threshold of the living world. They possess some, but by no means all, of the characteristics of true living systems and are best regarded as biological *objects* rather than biological *organisms.*

We will not pursue further the formal classification of biological systems. The frame of reference of this book will inevitably shift according to the relative availability of information about a given topic. Thus the discussion of molecular genetics will be largely centered about bacteria and viruses, for which the most definitive information is available, while the treatment of metabolic pathways will be focused on mammalian systems.

The existence of biochemistry as a unified scientific discipline is made possible by an underlying basic similarity of living systems that is somewhat masked by their seemingly endless external diversity. Biological systems resemble each other in many more ways than they differ. All are endowed with **biological continuity**; alterations of species do not occur in the ordinary processes of reproduction. This continuity is achieved in basically similar ways by all living forms.

A second common feature is the occurrence of parallel chemical processes in all living systems. These are designated collectively as **metabolism**. Although the details of metabolism may differ considerably from system to system, the basic purpose is always the synthesis of the essential components of the organism from chemically simpler precursors and the extraction from the surroundings of energy in a biologically usable form.

1-3 THE NATURE OF THE LIVING CELL

Among the few generalizations in biology that seem established beyond dispute is the generalization that the structural organization of all true living systems is based upon a single kind of fundamental unit—the **cell**. The overwhelming superficial variety of the biological world can be understood in terms of the operation of the evolutionary process upon a primitive prototype of the living cell.

All cells have a rather similar chemical nature. In particular, two kinds of giant molecules are of universal occurrence and seem inextricably associated with life as we know it. The **proteins** serve structural and transport functions and include a class of biochemical catalysts called **enzymes**, which make possible the multitude of chemical processes occurring in the cell. The **nucleic acids** function as the cellular repository of the genetic information that directs the synthesis of the proteins and ultimately governs the structure and activity of the cell.

Certain structural features are common to all cells. They are all enclosed by a membrane or **plasmalemma** of selective permeability. All cells show some degree of localization of certain specific functions within particular structural elements that recur for cells of very different origins.

The extensive application of the electron microscope to studies of the architecture of living cells has resulted in a genuine breakthrough in our knowledge of cellular organization. Among the most important findings is the discovery that there are two basic kinds of cells, whose differences are such as to suggest that they reflect a major event in evolutionary time.

The **eukaryotic** cell, which has the more complex internal structure, occurs in all plants and animals as well as fungi, protozoa, and most algae (Fig. 1-1). The simpler **prokaryotic** cell is confined to the bacteria and to the blue-green algae.

Prokaryotic cells

The bacteria and blue-green algae, which comprise the group of prokaryotic organisms and which have been placed in the Monera kingdom according to several modern systems of classification, have a distinctive cellular organization. The prokaryotic cell is enclosed by a membrane or plasmalemma, plus an external cell wall, which confers strength and structural support. The interior of the cell is differentiated into two distinguishable regions: the **cytoplasm** and the **nucleus**. In contrast to eukaryotic cells, the nucleus is not separated from the cytoplasm by a membrane.

The plasmalemma, whose integrity is essential for the survival of the cell, is about 10 nm thick. In some organisms it is folded inward at certain locations. In blue-green algae these folded regions contain photosynthetic pigments. Apart from the plasmalemma, prokaryotic cells contain no permanent membrane structures. While localized regions that are differentiated from the balance of the cytoplasm with respect to composition or activity may exist, these regions are not membrane bounded.

The only particulate entities, other than the nucleus, that occur in all prokaryotic cells are the **ribosomes**. These particles, which are about 20 nm in diameter, are present in high concentrations in the cytoplasmic region. Their composition, by mass, corresponds to roughly equal proportions of protein and **ribonucleic acid** (RNA), one of the two types of nucleic acids. The ribosomes are of central biological importance as the sites of protein synthesis.

In prokaryotes the nucleus is visualized by electron microscopy as an irregular region in the interior of the cytoplasm, which is seen at high resolution to consist of tangled filaments of **deoxyribonucleic acid** (DNA). These are not permanently associated with protein. In both prokaryotes and eukaryotes the nuclear DNA serves as the cellular repository of the genetic information that directs the synthesis of the proteins and ultimately governs the structure and function of the cell. Prokaryotic cells contain only a single copy of this genetic information, which is encoded into a single DNA molecule.

Cell division in prokaryotes usually occurs by binary fission. However, the complex mitotic patterns characteristic of cell division in eukaryotes are never observed. Differentiation into tissues does not occur.

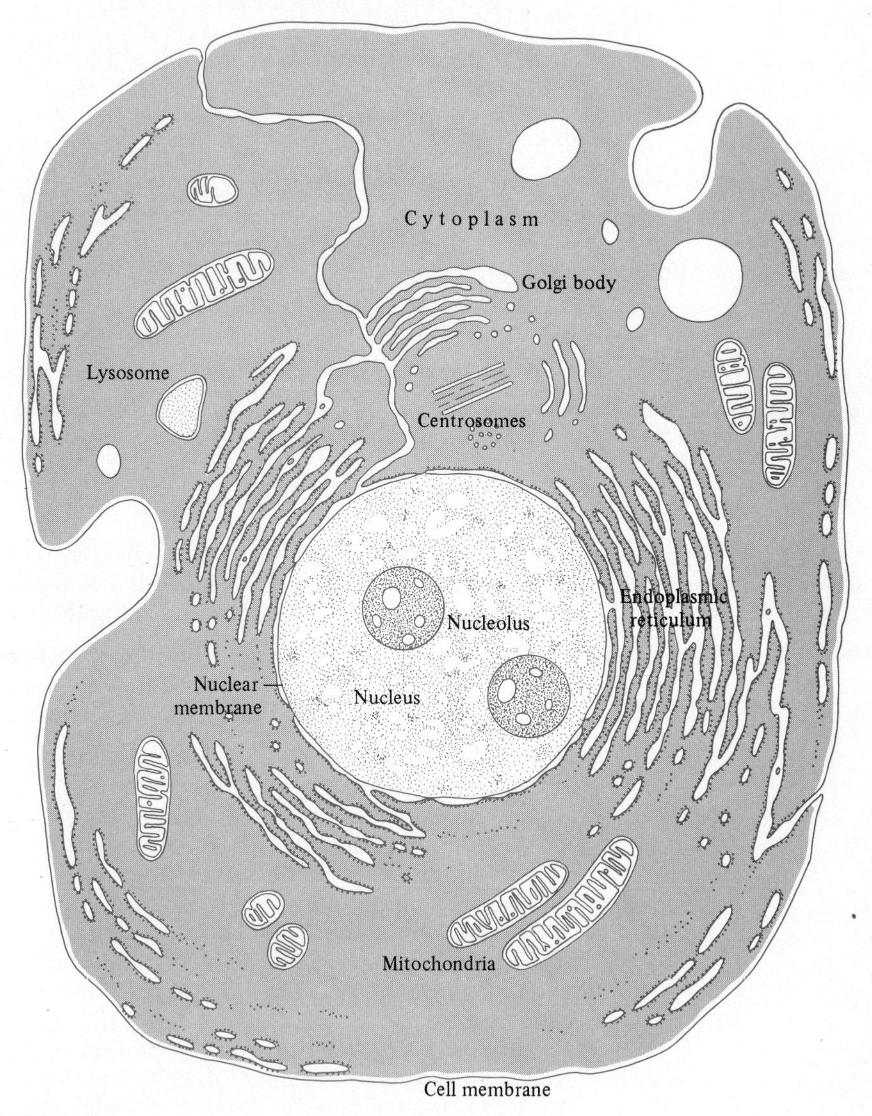

Figure 1-1
Generalized diagram of a eukaryotic cell showing the various particulate bodies. The diagram is not meant to represent any specific cell, but rather to depict structural features shared by cells of this kind.

The **genome**, or the unit of DNA that carries a complete set of genetic instructions for an organism, is, in bacteria, a single chromosome consisting of one DNA molecule. The DNA of *Escherichia coli* (*E. coli*) is about five times larger (molecular weight 2.5×10^9) than that of *Mycoplasma arthritides* (molecular weight 5×10^8). The DNA of higher animals is much (up to 10^5 times) larger (Table 1-1).

Eukaryotic cells

The principal characteristic of the eukaryotic cell that distinguishes it from the prokaryotic cell is the presence of membrane-enclosed compartments, which are not extensions of the plasmalemma. The most prominent of these compartments is the nucleus, whose DNA exists as organized DNA-protein complexes called **chromosomes**. The number of chromosomes is never less than two, and ranges into the hundreds in some species. The eukaryotic nucleus is enclosed by the nuclear membrane, which contains numerous openings or **pores**. In addition to the chromosomes, the eukaryotic nucleus always contains one or more **nucleoli**. These are globular structures that are formed at particular regions of certain chromosomes and within which ribosome precursors are synthesized.

The chromosomes, nucleoli, and nuclear membranes are the only recognizable organized structures occurring in all eukaryotic nuclei. The rest of the nucleus is structureless **nucleoplasm**. The chromosomes are transmitted between cell generations.

Between the plasmalemma and the nuclear membrane is the cytoplasm, which contains a complex system of membranes and particles. Among the most prominent features is an extensive network of membranous channels called the **endoplasmic reticulum** or ER. This consists of a set of flattened sacs called **cisternae**, which arise by folding of the enclosing membrane. Numerous ribosomes are attached to certain regions of the ER, as well as to the nuclear membrane (Fig. 1-2).

One differentiated region of the ER is the **Golgi apparatus**. This region processes many proteins that are excreted by the cell for use elsewhere and also encloses proteins in membranous compartments or **vesicles**, which persist and function within the original cell. The latter include the **lysosomes**, which contain high concentrations of digestive enzymes capable of catalyzing the degradation of biopolymers into smaller fragments. The lysosomes function as intracellular disposal units which fuse with and eliminate material foreign to the cell.

Table 1-1
The Approximate Sizes of Some Cells

Cell	*Diameter* (μm)	*Volume* (μm^3)
Liver cell	20	4000
Thymus cell	6	120
Yeast cell	10	500
E. coli	1	120

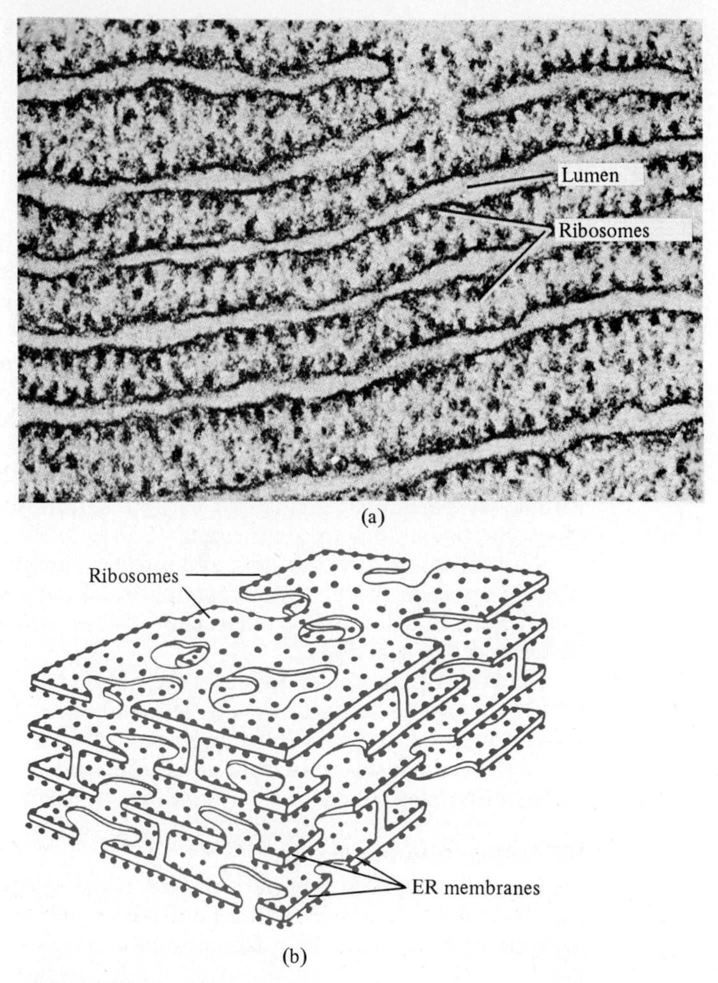

Figure 1-2
(a) Electron micrograph (× 85,000) of the endoplasmic reticulum, showing ribosomes. (b) A three-dimensional diagram of the endoplasmic reticulum system. (Courtesy of K. R. Porter.)

All eukaryotic cells contain cytoplasmic compartments called **mitochondria** (Fig. 1-3). Each mitochondrion is enclosed by a double membrane, of which the inner membrane is folded into projections called **cristae**. The mitochondria are the primary sites of **oxidative metabolism**, which is the terminal phase of the processes whereby the synthesis of the energy-storing compounds essential for life is accomplished. Each mitochondrion contains a complex system of enzymes required for oxidative metabolism. Some of these occur in the interior; others are integrated into the membrane structure. Mitochondria possess their own DNA.

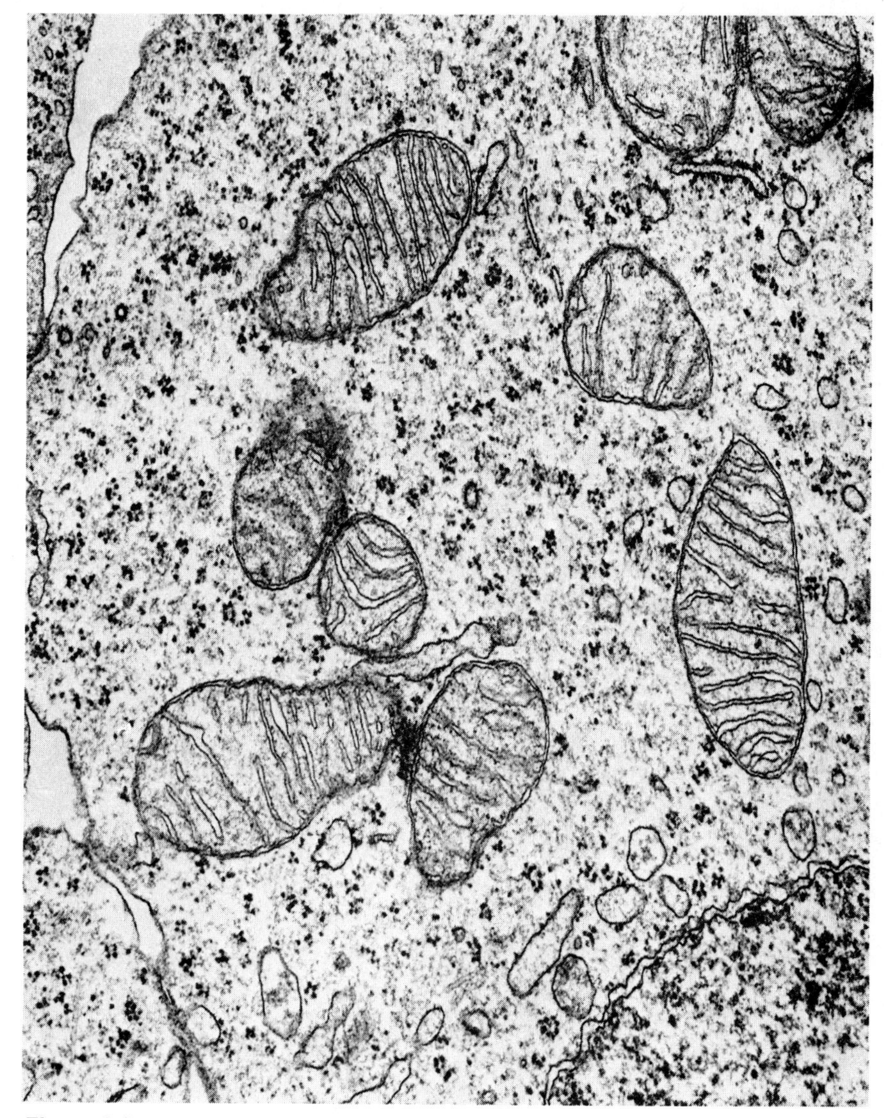

Figure 1-3
Electron micrograph (× 65,000) of the mitochondria of a cell of rat testis. Note the internal extensions or **cristae** of the inner membrane (Courtesy of H. H. Mollenhauer).

The **centrioles** are small granules that appear to have a prominent function in cell replication and are associated with the formation of the spindle during mitosis. The centrioles are also involved in the independent process of the formation of **cilia** and **flagella**, which are external filaments responsible for the mobility of many cells.

Some cytoplasmic particles occur conspicuously in certain eukaryotic cells and are absent in others. The cells of tissue-forming plants and of eukaryotic algae contain one or more types of **plastids**, of which the **chloroplasts** of photosynthetically active green plants are an example. The chloroplasts, which are the sites of photosynthesis, are enclosed by two membranes. The photosynthetic processes occurring within them are essential for the maintenance of life on this planet because they are responsible for the replenishment of the oxygen supply and the utilization of the radiant energy of sunlight in the production of carbohydrates.

The cells of plants, as well as those of algae and fungi, share several characteristics that endow them with a distinctive appearance upon microscopic examination (Fig. 1-4). These characteristics include a very prominent and rigid cell wall, which may survive the death of the organism,

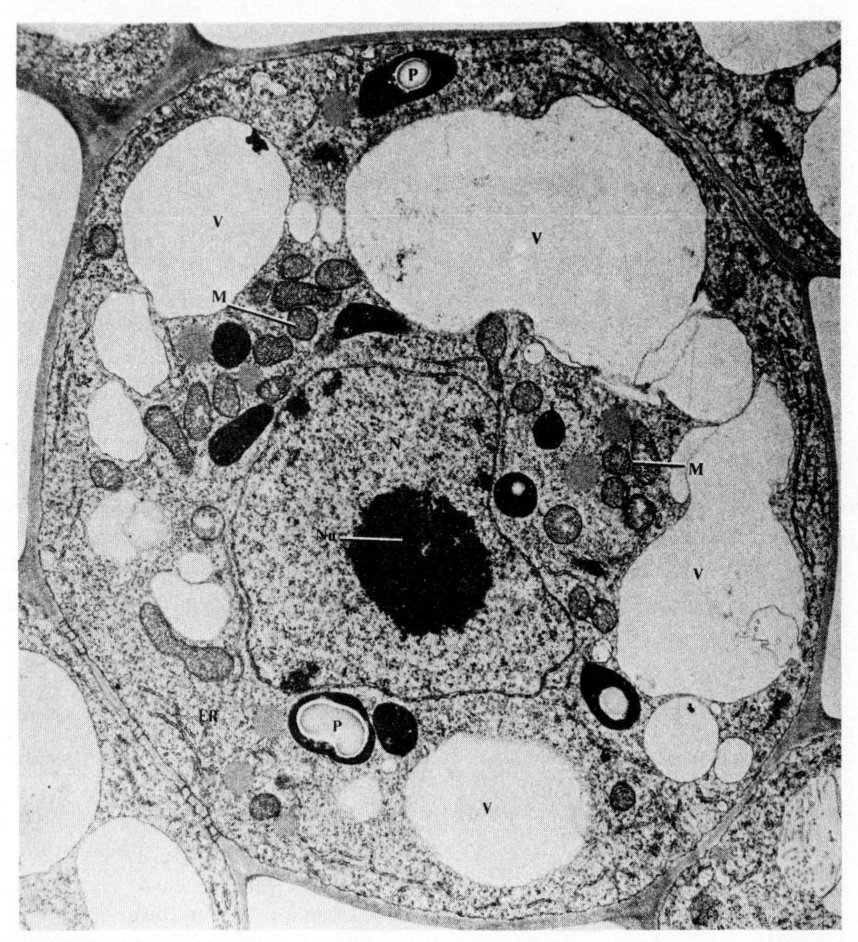

Figure 1-4
Electron micrograph of a mature plant cell (*Potamogeton natans* root). The large vacuoles (V) are a prominent feature. Note also the starch-filled plastids (P), the mitochondria (M), and the nucleolus (Nu), as well as the nuclear membrane separating the nucleus (N) from the cytoplasm. The endoplasmic reticulum (ER) is also visible. (Courtesy M. C. Ledbetter, Brookhaven National Laboratory.)

and the presence of large cavities called **vacuoles**, which are filled with solutions or suspensions of diverse types of molecules. The vacuoles may occupy a major fraction of the cell volume, confining the nucleus and other organelles to the periphery. In contrast, the vacuoles occurring in animal cells are generally quite small.

1-4 CHROMOSOMES AND GENES

In higher organisms, including animals and flowering plants, there is a distinction between the germ cells (**gametes**) specializing in reproduction and the ordinary functional cells (**somatic cells**) of the organism. The somatic cells contain a double number of chromosomes equal to $2n$, where n is an integer that depends upon the species. Each somatic cell thus possesses two of each type of chromosome and is said to be **diploid**. In contrast, the gametes each contain a single set of n different chromosomes and are said to be **haploid**. Prokaryotic cells are invariably haploid, possessing a single chromosome.

The chromosomes are the principal carriers of genetic information and, as such, have a directive influence upon the development of a new cell or individual. The genetic information preserved within each chromosome is carried in coded form by its DNA component. The functional units of heredity are chromosomal subelements called **genes**. Operationally, a chromosome may be regarded for many purposes as a linear sequence of genes. Each gene is associated with a particular biological function which is usually realized ultimately by the directed synthesis of a specific protein molecule. Since the latter is often an enzyme, this model has sometimes been summed up by the phrase: "One gene—one enzyme." As diploid cells contain two sets of equivalent chromosomes, they also possess a double copy of each gene, while haploid cells have only a single copy.

A particular gene may undergo a chemical alteration or **mutation** that forms a new gene with significantly different properties. Mutant genes may arise through spontaneous chemical events or through the action of physical agents, such as ionizing radiation, or of chemical mutagenic reagents. If a mutant form of a gene is consistent with survival and replication of the organism, it may be transmitted to subsequent generations.

An organism with two identical copies of a particular gene is said to be **homozygous** with respect to that gene. If two different forms of the gene are present, the organism is **heterozygous**.

By quantitative studies of the distribution of progeny that result from crosses between individuals of differing genetic complements, it has been possible in many cases to construct **linear genetic** maps for particular chromosomes that show the sequence of genes. Further details about the technique and results of genetic mapping may be found in the references at the end of this chapter.

The association of a gene with a particular chromosome is not irrevocable. Rupture of a chromosome may occur and may be followed by an interchange of chromosomal segments between chromosomes of the same type. The transfer of chromosomal segments is known to occur for both eukaryotic and prokaryotic organisms. An interesting example of the latter case is the chromosomal transfer or **conjugation**, which occurs between two different cells of *E. coli.* In this case, a fraction of a chromosome is transferred during the transient contact of two bacterial cells. All, or part, of the transferred chromosomal segment may subsequently be incorporated into the genetic complement of the recipient cell. Since, as for many prokaryotic organisms, the *E. coli* chromosome is circular, chromosomal transfer is preceded by a rupture of the chromosome at a particular point.

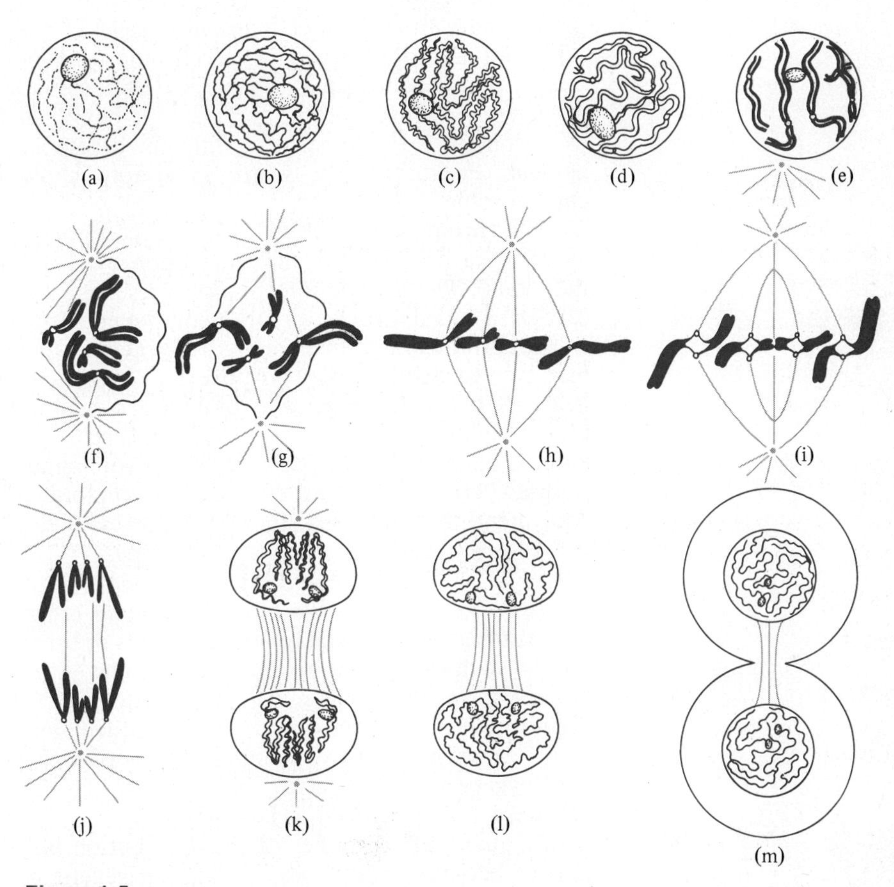

Figure 1-5
Schematic generalized version of mitosis, showing only the nucleus, for a somatic cell containing four chromosomes. Stage (a) corresponds to **interphase**, the interval between divisions. In **prophase**, (b)–(e), the chromosomes become visible; each is composed of two chromatids, which share a common centromere [clear circles (e)]. In **prometaphase**, (f) and (g), the spindle (dotted lines) forms and the nuclear membrane disappears. In **metaphase**, (h) and (i), the chromatids begin to separate; in **anaphase** (j) separation is completed. Finally, in telophase (k)–(m), the process culminates in the formation of two complete cells.

1-5 CELL REPLICATION

Mitosis

Biological continuity, or the persistence in successive generations of characteristics associated with a particular species, is achieved at the cellular level and is a consequence of the nature of cell replication. In eukaryotic organisms the dominant mechanism for the replication of somatic cells is by a form of cell division called **mitosis** (Fig. 1-5). This results in the formation of two new cells, each of which is equivalent to the single original cell and, like the latter, contains $2n$ chromosomes. In species whose reproduction occurs by a sexual mechanism, the role of mitosis is confined to the formation of new somatic cells within an individual; in primitive organisms, which reproduce asexually, it is responsible for the formation of new individuals.

During mitosis the chromosomes, which are indistinct in the interval between cell divisions (termed interphase), acquire a compact rodlike shape and become arranged into characteristic patterns or **mitotic figures**, which are readily observed with the microscope. Each chromosome undergoes longitudinal division to give rise to a pair of daughter chromosomes, which separate to become incorporated into different cells.

Mitotic replication assures that each daughter cell receives a complete set of genetic instructions. The multitude of somatic cells in a mammal all have the same genetic content. The wide divergence in the appearance and function of cells from different tissues arises from the action of regulatory systems that affect gene activity and from other internal and external influences. The ultimate expression of genetic information stored in the chromosomes in terms of specialized cells and tissues is the result of the directive influence of these other factors.

The replication of prokaryotic cells also occurs by cell division, but the characteristic mitotic patterns do not appear. The chromosome is a single naked DNA molecule. DNA replication and subsequent cell division proceed without a major change in cell appearance.

Meiosis

Gametes are formed by a special kind of cell division called **meiosis** and consequently they appear as haploid cells with a single set of chromosomes (Fig. 1-6). In contrast to mitosis, meiosis is a very restricted process with respect to both cell type and time of occurrence. Only the specialized gamete cells of sexually reproducing species undergo meiosis at specific times in the life cycle. Basically, meiosis consists of two consecutive nuclear divisions, while the chromosomes divide only once. In this way a diploid cell of the reproductive tract gives rise to four haploid gametes, with n chromosomes each.

DNA replication occurs during the interphase prior to the initial meiotic division; no further substantial increase in DNA content occurs during meiosis. The first meiotic division proceeds through a series of stages analogous to those of mitosis. However, the initial prophase is substantially more complex than that occurring during mitosis. A close pairing of homologous replicated chromosomes occurs during this phase, with the frequent occurrence of exchange between homologous chromosomal segments. The latter

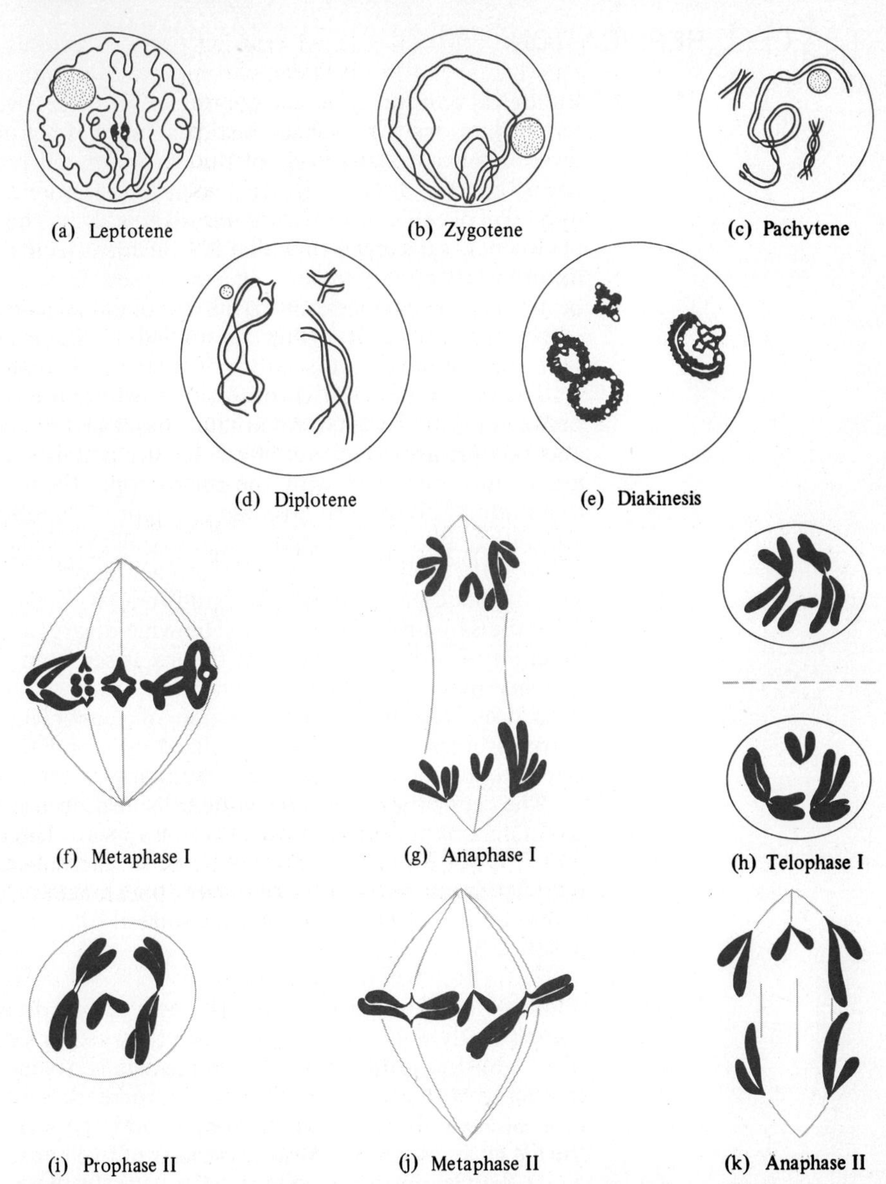

Figure 1-6
Generalized diagram of the stages of meiosis for a nucleus containing three pairs of chromosomes. Stages (a)–(e) are the consecutive phases of the first prophase (prophase I). During the first meiotic division, stages (a)–(h), chromosomal pairs persist. Formation of haploid nuclei, each containing only a single set of chromosomes, occurs during the second meiotic division.

process, which is termed **crossing over**, is responsible for the transfer of genetic information between chromosomes of maternal and paternal origin and is of profound genetic importance.

During this and subsequent phases of the first meiotic division the chromosomes remain as chromatid pairs; haploid nuclei are not formed at this stage. During the first anaphase each replicated chromosome separates from the homologous chromatid pair. Depending upon the species, the anaphase nuclei may enter directly into the second meiotic division, or a telophase stage may occur.

During the second meiotic division the daughter nuclei pass through the consecutive mitotic stages in a relatively conventional manner. Separation of chromatid pairs and the formation of haploid nuclei occur at this stage. The net achievement of meiosis is the formation of haploid gametes from a diploid cell and the exchange of genetic material between the homologous chromosomes occurring in the original diploid cell.

Sexual reproduction

The first stage of the sexual reproduction of eukaryotic organisms is the fusion of two gametes (each parent contributes one gamete). The product of this fusion, which is called a **zygote**, possesses a double set of genetic determinants. The new individual develops by subsequent cell replications.

SUMMARY

The structure and functioning of all living systems is based upon the **cell**. Despite the vast diversity of living cells, certain structural features recur in most cells. These common features include some form of **cell wall**, which surrounds and protects the cell contents, a cell **membrane** of selective permeability, a set of particulate **ribosomes** that mediate protein synthesis, and a **nuclear** region containing **chromosomal** DNA, the carrier of genetic programming.

There are two kinds of cells that differ fundamentally in properties. **Prokaryotic cells,** which occur in bacteria, have a relatively simple structure. No membrane other than the cell membrane is present, therefore, the nuclear region and the organelles are not bounded by a membrane. Only one chromosome is present, which consists of DNA only.

Eukaryotic cells, which occur in higher organisms, have a more complex organization. A **nuclear membrane** separates the **nucleus**, which contains the chromosomes, from the **cytoplasm**. The cytoplasm contains various membrane-enclosed organelles, each of which is specialized in function. These organelles include the **mitochondria**, which are important in the oxidative metabolism of fats, carbohydrates, and amino acids; the **endoplasmic reticulum**, which is associated with ribosomes; the **Golgi apparatus**, which functions in the secretion of cell products to the exterior; and the **lysosomes**, which contain enzymes capable of digesting the macromolecules present in the cell.

In contrast to prokaryotic cells, the cells of eukaryotic organisms, other than those involved in reproduction, contain two of each type of chromosome. These cells, which are termed **somatic**, replicate by a process called **mitosis**, in which cell division occurs, accompanied by a doubling of the total number of chromosomes.

The cells of eukaryotic systems which are directly concerned with reproduction are formed by a different process termed **meiosis**, which produces **haploid** cells with only a single copy of each chromosome.

The nucleus of a eukaryotic cell also contains a **nucleolus** which is rich in RNA.

REFERENCES

The following code is used to classify references. I: particularly useful as an introduction to the subject; R: useful primarily as a reference text; A: an advanced account of the material; H: a publication of historical importance.

General

A. L. Lehninger, *Biochemistry*, Worth, New York (1975). (A)

D. Metzler, *Biochemistry*, Academic, New York (1979). (R)

A. Roller, *Discovering the Basis of Life: An Introduction to Molecular Biology*, McGraw-Hill, New York (1974). (I)

L. Stryer, *Biochemistry*, Freeman, San Francisco (1976). (I)

J.D. Watson, *Molecular Biology of the Gene*, Benjamin, New York (1965). (I)

The Cell

C. J. Avers, *Cell Biology*, 2nd ed., Van Nostrand, New York (1981). (A)

J. Brachet, "The Living Cell," *Scientific American* 205, 50 (1961). (I)

K. R. Porter and M. A. Bonneville, *Fine Structure of Cells and Tissues*, Lea and Febiger, Philadelphia (1968). (A)

Replication

D. Mazia, "The Cell Cycle," *Scientific American* 230, 54 (1974). (I)

2

The Chemical Nature of Biological Systems

2-1 A CHEMICAL PRELUDE

Our point of view will now shift abruptly to some basic *chemical* considerations. The molecules present in living cells are necessarily composed of the same atoms that are found elsewhere in the earth's crust and atmosphere, although, as we shall see, their relative abundance is very different.

However, the role of one simple molecule, H_2O, is quite as pervasive within biological systems as on the earth's surface and it is important to grasp this fact early. After an initial survey of the relative importance of different elements in living cells, this chapter will be concerned with the dominant role of the *aqueous* environment of living cells in governing the structural organization of cellular elements and the ionization of electrolytes.

2-2 THE ELEMENTAL COMPOSITION OF LIVING CELLS

The relative occurence of the chemical elements in living systems is very different from their distribution in the earth's crust and atmosphere. For example, silicon, which is one of the most abundant elements, is of significance in only a few biological species. Of the 100 or so elements occurring in the earth's crust, only 27 have been identified as essential components of any organism and only 16 are invariably present.

Most of the mass of biological organisms is accounted for by four elements—carbon, hydrogen, nitrogen, and oxygen. All of these occur in the first row of the periodic table (atomic numbers 1–9). Of the other elements of the first row, boron and fluorine have significant, though specialized, functions; the others are probably not essential (Table 2-1).

In the second row of the periodic table (atomic numbers 10–17), sodium, magnesium, phosphorus, sulfur, and chlorine are essential and invariably present in living cells. Among the elements of higher atomic number, calcium and potassium are of first importance, while manganese, iron, cobalt, copper, and zinc have essential functions. Only a few elements of atomic number greater than 30, including molybdenum and iodine, have been definitely shown to be of biological importance. However, the list of elements for which a biological function has been established tends to expand with time and should not be regarded as necessarily complete.

2-3 BIOLOGICAL SIGNIFICANCE OF WATER

Quantitatively, water is by far the most important constituent of living tissues, often accounting for three-fourths or more of their mass. The only exceptions are some relatively inert tissues, such as bone and hair, as well as plant and bacterial spores.

Water, which is both the most commonplace and the most mysterious of liquids, appears to provide an essential milieu for the structural organization and biochemical activities of living cells. Indeed its function is of such central importance as to be often taken for granted and passed over with little discussion in many biology texts. The biochemical processes to be discussed in the chapters to follow should always be understood as occurring in an aqueous environment, by which they are profoundly influenced.

Table 2-1
Approximate Relative Abundance of Natural Elements in All Organisms in Atoms (per 100 Atoms)

Element	*Abundance*
H	49
C	25
O	25
N	0.27
Ca	0.073
K	0.046
S	0.033
Mg	0.031
P	0.030
Na	0.015
Others	Traces*

*Trace elements occurring in all organisms include Mn, Fe, Co, Cu, and Zn. Trace elements occurring in some organisms include B, Al, V, Mo, I, Br, As, Se, and Sn.

The water present in living tissues always contains substantial quantities of dissolved inorganic ions, especially Na^+, K^+, Ca^{2+}, Mg^{2+}, Cl^-, HPO_4^{2-}, HCO_3^-, and SO_4^{2-}. The relative levels of these ions vary considerably. In animal tissues, potassium and magnesium tend to be concentrated within cells, while sodium and chloride occur primarily in extracellular fluids such as plasma and lymph.

Water is however very far from being merely an inert carrier of biological materials. As a consequence of its unique physical characteristics, it has a controlling influence, not only upon biochemical reactions but also on the molecular form assumed by proteins and other biopolymers, as well as upon the architecture of cells.

The unusual properties of water arise largely from the powerful attractive forces between its molecules. Such quantities as the heat of vaporization, the specific heat, and the surface tension are all elevated relative to the other common liquids, reflecting this high degree of internal cohesion.

The strong intermolecular forces in liquid water arise from the high degree of *polarity* of the O—H bond, which has about one-third ionic character. Because of the resultant separation of charge, each of the hydrogen atoms has effectively a fractional positive charge, while the oxygen bears a fractional negative charge. As a consequence, there is a significant electrostatic attraction between the oxygen atom and the hydrogen atom of an adjacent water molecule. This contributes to the stability of the type of linkage called **hydrogen bonding**. Because of its geometry, each water molecule tends to hydrogen bond with four adjacent water molecules (Fig. 2-1).

Liquid water exists largely as molecular clusters linked by hydrogen bonds. As a result it may be regarded as possessing a high degree of order. While this is short range and mobile, rather than static, it has more in common with crystalline order than with the chaos of gases or the relatively random molecular arrangement of simple liquids. Any substance dissolved

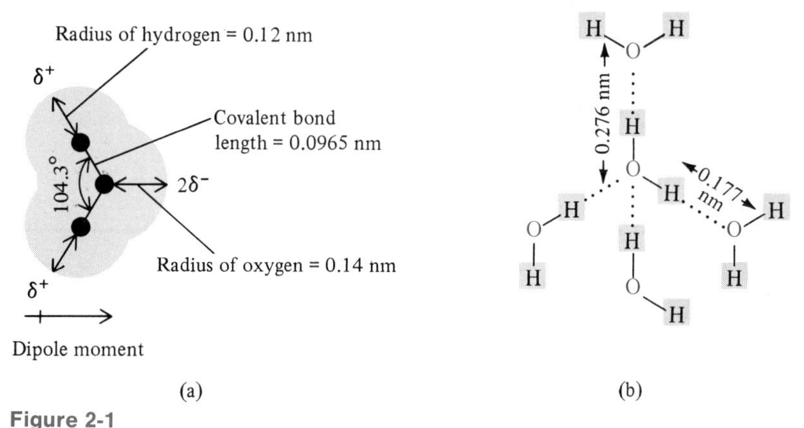

Figure 2-1
(a) The polarity of the H_2O molecule. The electrons of the covalent O—H bond are attracted by the more electronegative oxygen nucleus. (b) Hydrogen bonding between H_2O molecules.

Table 2-2
Dielectric Constants of Some Liquids

Liquid	*Dielectric Constant*
Water	80
Methanol	33
Ethanol	24
Carbon tetrachloride	2

in water, or in contact with it, will have a perturbing effect upon this quasi-crystalline lattice. Molecules which contain groups of atoms that are not compatible with the water lattice will tend to cluster together in solution. Such groups are termed **hydrophobic**. As subsequent chapters will make clear, this phenomenon is a dominant factor in biology.

Another important property of water is its unusually high **dielectric constant*** ($\sim$ 80) whose magnitude is much larger (Table 2-2) than for most other liquids and reflects the polarity of the water molecule discussed earlier. As a consequence of this high dielectric constant and of the tendency of water molecules to form complexes with charged species, water favors the **dissociation of electrolytes**, which tend to exist as charged ions in aqueous solution.

Ions will exist only if the solvent in which they are formed prevents their natural recombination. Solvents of high dielectric constant reduce the mutual electrostatic attraction of oppositely charged ions; the solvent dipoles tend to become aligned in opposition to the electric field associated with each ion and thereby to attenuate it.

2-4 HYDROGEN ION CONCENTRATION OF BIOLOGICAL FLUIDS

Ionization of Water

As a consequence of the highly polar character of the O—H bond, there exists a significant tendency for a water molecule to dissociate a proton, leaving a hydroxyl (OH^-) ion. The dissociated proton becomes linked with the oxygen atom of a second water molecule to form a hydronium (H_3O^+) ion. Although the actual products of the ionization of water are H_3O^+ and OH^- ions, the process is conventionally written

$$H_2O \rightleftarrows H^+ + OH^- \qquad \textbf{(2-1)}$$

Like any other reversible process it corresponds to an equilibrium constant, which in this case has the form

$$K = [H^+][OH^-]/[H_2O] \qquad \textbf{(2-2)}$$

where the quantities in brackets are concentrations in moles per liter.

* The dielectric constant is defined by the equation $F = e_1e_2/Dr^2$ where F is the force between two particles of charge e_1 and e_2, r is their separation, and D is the dielectric constant.

The value of $[H_2O]$, the molar concentration of water, is very high relative to $[H^+]$ and $[OH^-]$ and is virtually unaffected by the ionization process. If one substitutes its numerical value, 55.5, in Equation (2-2), there results

$$\begin{aligned} 55.5\,K &= [H^+][OH^-] \\ &= K_w = 1.0 \times 10^{-14} \quad \text{(at 25°C)} \end{aligned} \tag{2-3}$$

In Equation (2-3) the numerical value of $[H_2O]$ has been combined with K to define K_w, the **ion product** of water. The values of $[H^+]$ and $[OH^-]$ are thus related by

$$[H^+] = K_w/[OH^-] = 1.0 \times 10^{-14}/[OH^-] \tag{2-4}$$

Alternatively, we may rewrite Equation (2-4) in terms of logarithms:

$$\log_{10}[H^+] = -\log_{10}[OH^-] - 14 \tag{2-5}$$

Definition of the pH Scale

The hydrogen ion concentrations of interest to the biochemist usually range from 10^{-1} to 10^{-13}. The inconvenience of expressing concentrations in terms of powers of ten which are large and negative has led to the introduction of the **pH scale**. The pH of a solution is defined by

$$\text{pH} = -\log_{10}[H^+] \tag{2-6}$$

At **neutral pH** the concentrations of hydrogen and hydroxyl ions are equal. We then have

$$\begin{aligned} [H^+][OH^-] &= [H^+]^2 \\ &= 1.0 \times 10^{-14} \end{aligned} \tag{2-7}$$

and

$$\begin{aligned} [H^+] &= [OH^-] \\ &= 1.0 \times 10^{-7}. \end{aligned} \tag{2-8}$$

Also,

$$\begin{aligned} \text{pH} &= -\log_{10} 1.0 \times 10^{-7} \\ &= 7.0 \end{aligned} \tag{2-9}$$

The assignment of a value of 7.0 to the pH of an exactly neutral solution thus arises directly from the absolute value of the ion product of water at 25°. It should be stressed that the pH scale is logarithmic in nature. If two solutions differ in pH by 2.0 pH units, the solution of lower pH has a hydrogen ion concentration that is higher by a factor of 10^2, or 100, than the other.

The primary standard for the measurement of pH is the **hydrogen electrode**, which measures the difference in electromotive force between a reference electrode of known electromotive force (emf) and a platinum electrode immersed in the solution to be measured and equilibrated with hydrogen gas at a known pressure and temperature. The electromotive force generated corresponds to the equilibrium

$$H_2 \rightleftarrows 2H^+ + 2e^- \tag{2-10}$$

In practice, pH is usually determined by means of the **glass electrode**, which responds to H^+-ion level in the absence of H_2 gas. It requires calibration against buffers of accurately known pH.

Acids and Bases

Acids and bases may be defined in various ways; the usefulness of a particular definition depends upon the context. For the purposes of the biochemist, perhaps the most useful definition is that which designates an acid as a **donor** of protons and a base as an **acceptor** of protons. In generalized terms, the ionization of an acid, HA, to yield a proton and the **conjugate base**, A^-, may be written

$$HA \rightleftarrows H^+ + A^- \quad \textbf{(2-11)}$$

HA may be, for example, an undissociated carboxylic acid (—COOH), a protonated amine ($—NH_3^+$), or a phenolic hydroxyl (—OH).

Most acids of biochemical interest belong to the class of **weak electrolytes**, whose extent of ionization depends upon concentration and other conditions. The ionization of a weak acid may be characterized by its **dissociation constant**, which is an equilibrium constant defined by

$$K = \frac{[H^+][A^-]}{[HA]} \quad \textbf{(2-12)}$$

Equation (2-12) may be rewritten

$$\log_{10}K = \log_{10}[H^+] + \log_{10}[A^-] - \log_{10}[HA] \quad \textbf{(2-13)}$$

or

$$-\log_{10}K = -\log_{10}[H^+] + \log_{10}\frac{[HA]}{[A^-]}$$

Since $[HA]/[A^-]$ is equal to $(1 - \alpha)/\alpha$, where α is the fraction of molecules which have dissociated, we have, introducing the notation $pK = -\log_{10}K$,

$$pK = pH + \log_{10}\frac{1 - \alpha}{\alpha} \quad \textbf{(2-14)}$$

When one-half of the molecules are dissociated, $\alpha = \frac{1}{2}$, $[A^-] = [HA]$ and $pK = pH$. The pK is thus numerically equal to the pH at which dissociation is half complete. Equation (2-14) is often called the **Henderson-Hasselbalch** equation.

The above discussion applies explicitly to the dissociation of a collection of similar sites, which are *isolated* and *independent*. The behavior of such a system can be described in terms of a single dissociation constant. If two or more dissociable sites occur in the same molecule, the situation cannot be described so simply. Because of electrostatic interactions the dissociation of each site will depend upon the state of dissociation of the other, so that a complicated set of equations is required to describe the overall behavior. The algebraic complexity escalates sharply with each additional dissociable site. The dissociation of a large molecule with many dissociable sites, such as a protein, can be treated quantitatively only with the aid of drastic approximations.

Buffers

Equation (2-14) is particularly useful for the computation of pH values for mixtures of a weak acid and the corresponding salt. Mixtures of this kind are termed **buffer solutions.**

Buffers have the property of absorbing OH^- or H^+ ions and thereby minimize the pH changes which would otherwise result from the addition of acid or base. The added OH^- or H^+ is largely consumed by titration of the weak acid or its conjugate base. Buffering capacity is maximal at the pH equal to the pK of the acid and becomes negligible at pH's one or two units removed from the pK.

More quantitatively, one may define a **buffer value** by $dX/d\text{pH}$, where X is the number of moles per liter of strong base, such as NaOH, which is added to the given buffer solution. The derivative $dX/d\text{pH}$ is equivalent to the limit approached by the ratio of the increments of base and pH as the latter become very small.

$$\frac{dX}{d\text{pH}} = \lim_{x \to 0} \frac{\Delta X}{\Delta \text{pH}} \equiv B. \tag{2-15}$$

The larger the value of B, the more acid (or base) is required to alter the pH to a given extent and hence the more resistant is the solution to alterations in pH (Fig. 2-2).

It may also be shown that

$$\begin{aligned} B &= m \frac{d\alpha}{d\text{pH}} \\ &= 2.303 m\alpha(1 - \alpha) \end{aligned} \tag{2-16}$$

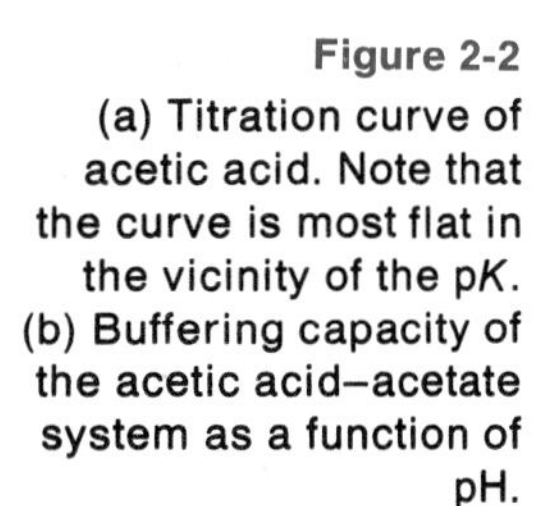

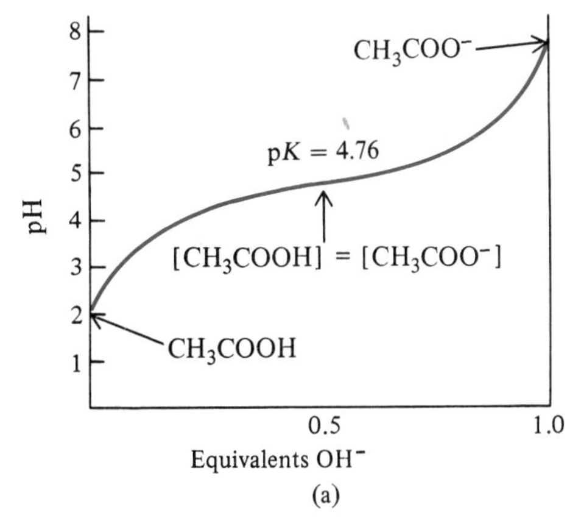

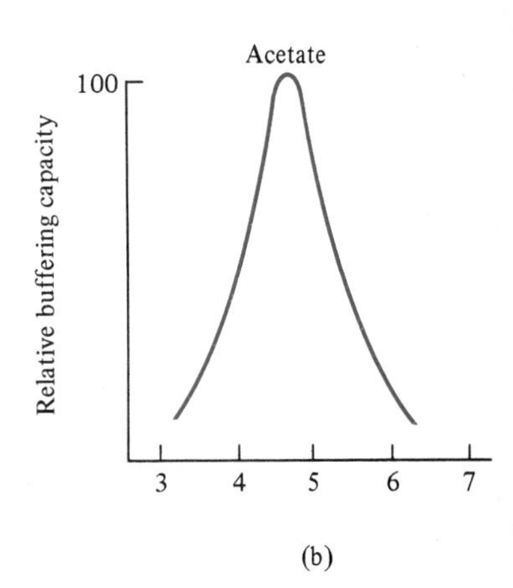

Figure 2-2
(a) Titration curve of acetic acid. Note that the curve is most flat in the vicinity of the pK. (b) Buffering capacity of the acetic acid–acetate system as a function of pH.

where m is the *total* molarity of buffer (weak acid plus conjugate base). The buffer value is thus largest at the midpoint of titration of the weak acid, when $\alpha = 0.5$ and the pH is numerically equal to the pK; it approaches zero at pH's far removed from the pK, where α or $1 - \alpha$ approach zero (pH $\ll$ pK or pH $\gg$ pK).

Biological systems are very sensitive to alterations in pH and both intracellular and extracellular biological fluids generally have closely controlled pH values. The major *intracellular* buffer is the $H_2PO_4^- - HPO_4^{2-}$ system with p$K = 7.2$. The chief *extracellular* buffer, which controls the pH of plasma, is the $H_2CO_3 - HCO_3^-$ (bicarbonate) system.

Buffer System of Blood

The pH of blood is maintained nearly constant at 7.4 primarily by the bicarbonate buffer system (H_2CO_3–HCO_3^-). At first glance it might appear surprising that H_2CO_3, a relatively strong acid with a pK near 3.8, can serve as a buffer at a pH near 7.0. The explanation lies in the reversible equilibrium existing between H_2CO_3 and dissolved CO_2:

$$H^+ + HCO_3^- \rightleftarrows H_2CO_3 \rightleftarrows CO_2 \text{ (dissolved)} + H_2O$$

$$CO_2 \text{ (dissolved)} \rightleftarrows CO_2 \text{ (gaseous)}$$

The dissolved CO_2 is itself in equilibrium with gaseous CO_2 from the large pool available in the lungs. Under conditions where the blood must absorb excess OH^-, the H_2CO_3 is quickly replaced by drawing on the gaseous reservoir of CO_2. Conversely, the addition of H^+ results in a loss of dissolved CO_2 to the gaseous reservoir.

The buffering action of this system is remarkably effective. The addition of 10 ml of 1-M HCl to one liter of blood plasma reduces its pH only to 7.2; if it is added to the same volume of 0.15-M NaCl, the pH falls to 2.0.

SUMMARY

The naturally occurring elements are very unevenly represented in living cells; their relative abundance is unrelated to the distribution of elements in the earth's crust. Nearly all of the mass of most cells is accounted for by hydrogen, oxygen, carbon, nitrogen, phosphorus, and sulfur. The most important monoatomic ions are Na^+, K^+, Mg^{2+}, Ca^{2+}, and Cl^-. Other elements are present in small quantities, including Fe, Mn, Cu, Co, Zn, B, V, Mo, Sn, F, Cr, I, and Se.

A dominant factor in determining thc propcrtics of living systems is the high proportion of water that they contain. The structure and polarity of liquid water favor ionization of dissolved salts and the mutual association of hydrocarbon groups. The latter effect, which is responsible for **hydrophobic bonding**, is an important determinant of protein structure.

Many molecules occurring in biological systems are **weak acids**, whose extent of combination with protons depends upon conditions. The dissociation of such a weak acid is characterized by its dissociation constant, K, which is often tabulated as pK ($= -\log_{10}K$).

An **acid**, or proton donor, is associated with a conjugate **base**, or proton acceptor. If HA is a weak acid, which ionizes to form H^+ and A^-, then A^- is the corresponding base.

The hydrogen ion concentration of a solution is usually characterized by its pH, which is equal to the negative logarithm of its hydrogen ion concentration. The pH of pure water is 7.0.

A mixture of a weak acid and the corresponding base may function to stabilize the pH, so that its value changes relatively slowly upon the addition of increments of H^+ or OH^-. Such mixtures are termed **buffers**, and are very important in controlling the pH of living cells, which are quite sensitive to changes in H^+ level.

REFERENCES

The following code is used to classify references. I: particularly useful as an introduction to the subject; R: useful primarily as a reference text; A: an advanced account of the material; H: a publication of historical importance.

General

A. L. Lehninger, *Biochemistry*, Worth, New York (1975). (A)

D. Metzler, *Biochemistry*, Academic, New York (1979). (R)

L. Stryer, *Biochemistry*, Freeman, San Francisco (1976). (I)

Elemental Composition of Living Systems

E. S. Devery, *Scientific American*, 149, September (1970). (I)

Properties of Water

D. Eisenberg and W. Kauzmann, *The Structure and Properties of Water*, Oxford University, Fair Lawn, New Jersey (1969). (A)

Acid-Base and Buffer Calculations

R. Montgomery and C. A. Swenson, *Quantitative Problems in Biochemical Sciences*, Freeman, San Francisco (1969). (I)

W. B. Wood, J. H. Wilson, R. M. Benbow, and L. E. Hood, *Biochemistry: A Problems Approach*, Benjamin, Menlo Park (1974). (I)

REVIEW QUESTIONS

Questions marked with an asterisk are of a high level of difficulty.

2-1 Living cells, when compared to the inanimate matter in the earth's crust: (Choose one or more)
(a) have about the same chemical composition.
(b) have about the same degree of order.
(c) obey the same chemical and physical laws.
(d) are in thermodynamic equilibrium with their surroundings.

2-2 Choose *one* explanation from the right-hand column that best describes the property of water listed in the left-hand column.

(a) dissolves ions well
(b) dissolves alcohols and sugars well
(c) Liquid water contains icelike structures.
(d) dissolves hydrocarbons poorly

(1) good hydrogen bond formation
(2) charge separation, resulting in positive and negative ends of water molecule
(3) Pure water dissociates slightly into ions.
(4) Energy (E) of interaction between water and nonpolar molecules is unfavorable.
(5) None of the above

2-3 (a) Indicate which of the properties of carbon listed below help make it a very important element of biomolecules.
(1) very high atomic mass
(2) can bond to important electronegative elements like O, S, and N
(3) common oxide is insoluble in H_2O
(4) single atoms can easily form bonds in five different directions
(5) can readily form double and triple bonds

(b) Indicate similarly the properties of H_2O which are important in its function as a biomolecule.
(1) ability to hydrogen bond to other water molecules
(2) low dielectric constant
(3) strong hydrophobic interactions with ions
(4) high boiling temperature
(5) increase in density when liquid water freezes

*2-4 Consider the general situation of a buffer system consisting of an acid HA (with dissociation constant K) at concentration C_{HA}, plus the salt of its conjugate base, NaA, at concentration C_{NaA}.

(a) Write the five equations in five unknowns required to solve the system exactly. Hint: Write (1) and (2) as equilibrium equations, (3) as conservation of charge, (4) and (5) as conservation of mass.

(b) Use (3), (4), and (5) to obtain equations for $[A^-]$ and $[HA]$ in terms of C_{HA}, C_{NaA}, $[H^+]$, and $[OH^-]$.

(c) Use the results of (a) and (b), and make the appropriate approximations to obtain a simple formula for $[H^+]$ in terms of C_{HA}, C_{NaA}, and K.

(d) Under which of the following conditions is the simple formula (c) likely to be a good approximation (accurate to within $\pm 5\%$). Hint: You do not need to solve the whole problem. Look at part (b).

(1) $C_{HA} = C_{NaA} = 10^{-3}M, [H^+] = 3 \times 10^{-4}M$
(2) $C_{HA} = C_{NaA} = 10^{-3}M, [H^+] = 3 \times 10^{-6}M$
(3) $C_{HA} = C_{NaA} = 10^{-3}M, [H^+] = 3 \times 10^{-9}M$
(4) $C_{HA} = C_{NaA} = 10^{-3}M, [H^+] = 3 \times 10^{-12}M$

2-5 What is the degree of dissociation of a weak acid of $pK = 4.7$ when the pH is 5.0, 7.5, or 3.6?

3

Amino Acids and Peptides

3-1 AMINO ACIDS

In 1953, Miller found that the discharge of an electric arc, meant to simulate lightning, through a mixture of the gases NH_3, H_2O, CH_4, and H_2, which were then believed to constitute the earth's primeval atmosphere, resulted after several days in the formation of significant quantities of compounds called **amino acids**. The experiment aroused considerable interest, which has continued to the present day, and provided the first direct evidence for a plausible initial step in the **chemical evolution** of living systems.

The excitement generated by this finding arose from the fact that the amino acids are the basic constituents of the biopolymers termed **proteins**, which have a central function in the structure and activity of living cells. While we shall not pursue further the topic of chemical evolution, the Miller experiment provides a fitting point of entry into a discussion of proteins with which we enter the domain of biochemistry proper.

The most remarkable feature of the proteins as a class is their incredible versatility of function. Even the relatively simple organism *E. coli* contains 500–1000 different proteins, each with its characteristic biological activity. While some proteins, such as **collagen**, have passive and purely structural function, the majority have a more active role. In particular, the **enzymes**, which are proteins with **catalytic** properties, are indispensable for mediating the multitude of metabolic reactions essential for the survival and replication of the cell.

Like the other major biopolymers, the proteins are **linear combinations** of their basic repeating units, the amino acids, having the form *A-B-C-D-E-F-*, where the letters stand for amino acids. Twenty different amino acids are commonly found in natural proteins. As we shall see in chapter 4, the three-dimensional architecture of each protein, and hence its biological activity, is essentially dictated by the specific **sequences** of amino acids in the linear polymeric combinations of amino acids comprising the protein. A protein may be thought of as possessing an **informational content** that is specified by its amino acid sequence, somewhat as the information contained in a message transmitted in Morse code is specified by its sequence of dots and dashes. In the case of the protein, the information encoded into its amino acid sequence determines its structure and function. Since there are 20 different amino acids and both the amino acid composition and the linear arrangement of amino acids may vary, the number of possible sequences, and hence of possible protein structures, is enormous.

The Chemical Nature of Amino Acids

As the name implies, the amino acids possess an amino ($—NH_2$) and a carboxyl (—COOH) group. A generalized structure for an amino acid may be written

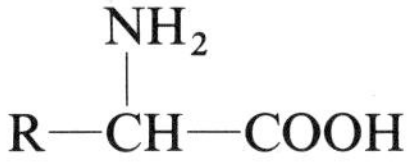

$$\begin{array}{c} NH_2 \\ | \\ R—CH—COOH \end{array}$$

The carbon atom to which the amino and carboxyl groups are covalently attached is known as the **α-carbon**, so that amino acids with this kind of structure are often termed **α-amino acids**. The symbol R represents the **side-chain**, to which the amino acids owe their chemical individuality.

Under physiological conditions, which normally correspond to a pH close to neutrality, the amino groups of the free amino acids are protonated and the carboxyl groups ionized, so that they exist as dipolar **zwitterions**:

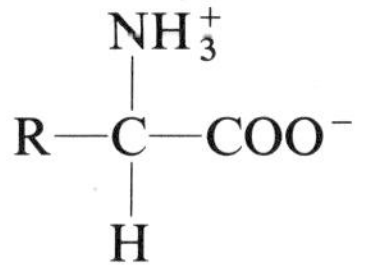

$$\begin{array}{c} NH_3^+ \\ | \\ R—C—COO^- \\ | \\ H \end{array}$$

The natural proteins consist of one or more chains of amino acids, chemically joined in linear combination and folded into a specific three-dimensional structure. These linear polymers are called **polypeptides**, and the chemical linkage joining the amino acids is termed a **peptide** bond. The polypeptide chains occurring in natural proteins may contain as many as 1000 or more amino acids linked by peptide bonds.

The peptide linkage is of the amide type and may be formally thought of as arising from the elimination of water between the **α-amino** and the **α-carboxyl** groups of adjacent amino acids:

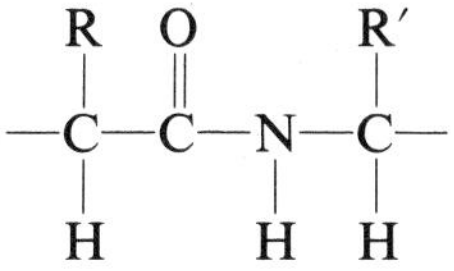

$$\begin{array}{ccccccc} & R & & O & & R' & \\ & | & & \| & & | & \\ — & C & — & C & —N— & C & — \\ & | & & & | & | & \\ & H & & & H & H & \end{array}$$

Because of the nature of the peptide linkage, each polypeptide chain has a free α-amino group at one terminus, called the **NH_2-terminal** end, and a free α-carboxyl at the other, called the **COOH-terminal** end:

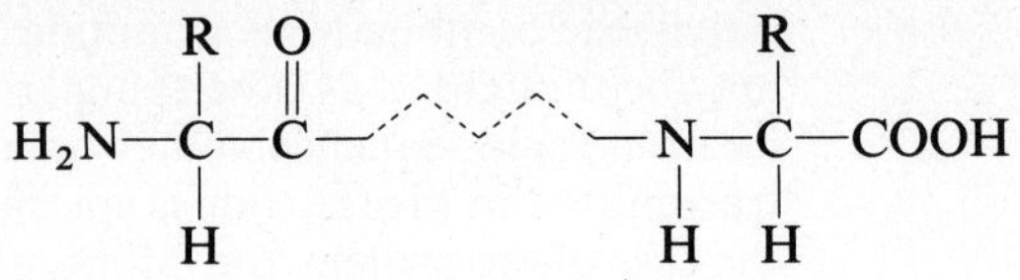

$$H_2N-\underset{H}{\overset{R}{C}}-\overset{O}{\overset{\|}{C}}- \cdots -\underset{H}{N}-\underset{H}{\overset{R}{C}}-COOH$$

3-2 CHARACTERISTICS OF THE AMINO ACIDS OCCURRING IN NATURAL PROTEINS

Amino Acids with Saturated Hydrocarbon Side-Chains

Of the 20 amino acids that commonly occur in natural proteins (Fig. 3-1) four have side-chains that are saturated aliphatic hydrocarbons. Residues of this class, which includes alanine, valine, leucine, and isoleucine, are chemically rather inert when incorporated into a polypeptide. The nonpolar character of their side-chains renders water an energetically unfavorable solvent, with important consequences for protein structure (chapter 5).

Amino Acids with Hydroxyl-Containing Side-Chains

Two amino acids, serine and threonine, have side-chains containing an **aliphatic** hydroxyl group. These hydroxyls have no acidic properties and remain un-ionized over the entire range of pH.

In contrast, the **aromatic** hydroxyl groups of tyrosine and the related iodinated compounds, diiodotyrosine and thyroxine, are weakly acidic and ionize to form a phenoxide ion at alkaline pH:

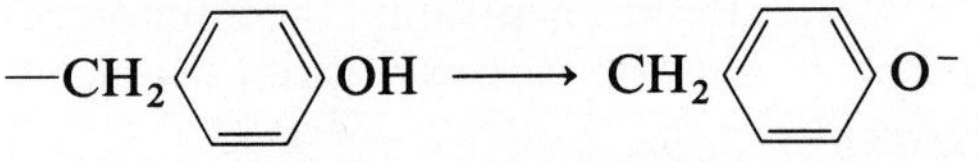

$$-CH_2-C_6H_4-OH \longrightarrow CH_2-C_6H_4-O^-$$

Other Aromatic Amino Acids

Diiodotyrosine and thyroxine are of very restricted occurrence in proteins. They are found in the thyroid protein thyroglobulin.

Phenylalanine and tryptophan complete the list of amino acids with aromatic side-chains. Tyrosine, phenylalanine, and tryptophan all possess strong absorption bands in the near ultraviolet, with maxima near 280 nm (2800 Å), which account for the absorption of proteins at wavelengths above 230 nm. The ionization of tyrosine is accompanied by a shift in position of the absorption band to longer wavelengths (Fig. 3-2).

All three aromatic amino acids display ultraviolet fluorescence. In proteins which contain tryptophan, the emission spectrum is usually dominated by this residue.

Acidic Amino Acids

Two amino acids, aspartic and glutamic acids, have side-chains containing carboxyl groups, which can dissociate

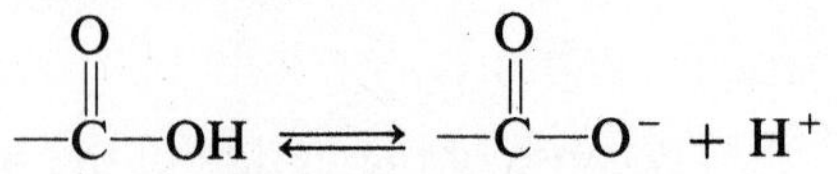

$$-\overset{O}{\overset{\|}{C}}-OH \rightleftharpoons -\overset{O}{\overset{\|}{C}}-O^- + H^+$$

The corresponding amide derivatives, asparagine and glutamine, are also of common occurrence in proteins:

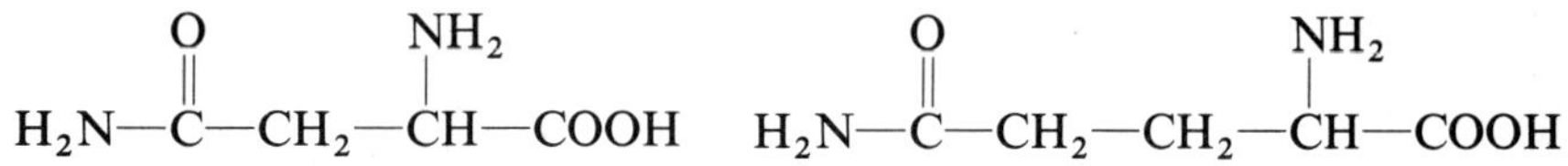

$$H_2N-\overset{\displaystyle O}{\overset{\|}{C}}-CH_2-\overset{\displaystyle NH_2}{\overset{|}{C}H}-COOH$$

Asparagine

$$H_2N-\overset{\displaystyle O}{\overset{\|}{C}}-CH_2-CH_2-\overset{\displaystyle NH_2}{\overset{|}{C}H}-COOH$$

Glutamine

Basic Amino Acids

Three basic amino acids—lysine, arginine, and histidine—are found frequently in proteins. A fourth, hydroxylysine, has been identified only in the structural protein collagen.

Lysine and histidine are weak electrolytes whose state of ionization depends upon the pH. The guanidinium group of arginine is essentially a strong electrolyte, being completely protonated at pH's below 13.

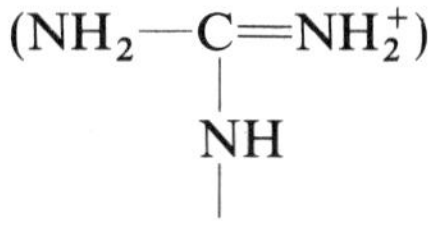

$$(NH_2-\underset{\underset{\displaystyle \underset{|}{NH}}{|}}{C}=NH_2^+)$$

Sulfur-Containing Amino Acids

Of the three sulfur-containing amino acids—methionine, cysteine, and cystine—the latter two are unique in being interconvertible. Cystine may be converted into two molecules of cysteine by reduction; cystine is reformed by oxidation:

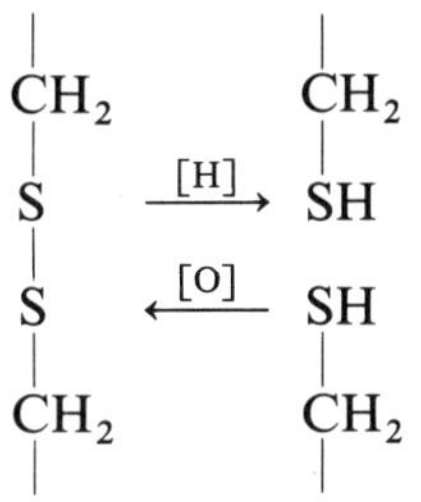

$$\begin{array}{c} | \\ CH_2 \\ | \\ S \\ | \\ S \\ | \\ CH_2 \\ | \end{array} \quad \underset{[O]}{\overset{[H]}{\rightleftarrows}} \quad \begin{array}{c} | \\ CH_2 \\ | \\ SH \\ \\ SH \\ | \\ CH_2 \\ | \end{array}$$

Because of its Siamese-twin character cysteine may be shared by two different polypeptide chains, or two different parts of the same chain, serving to cross-link them. An essential preliminary to the determination of the amino acid sequence of a protein is the separation of the different polypeptide chains. This requires the splitting of all cystine (disulfide) bridges and is usually accomplished by reduction with a sulfhydryl reagent, such as β-mercaptoethanol ($HSCH_2CH_2OH$):

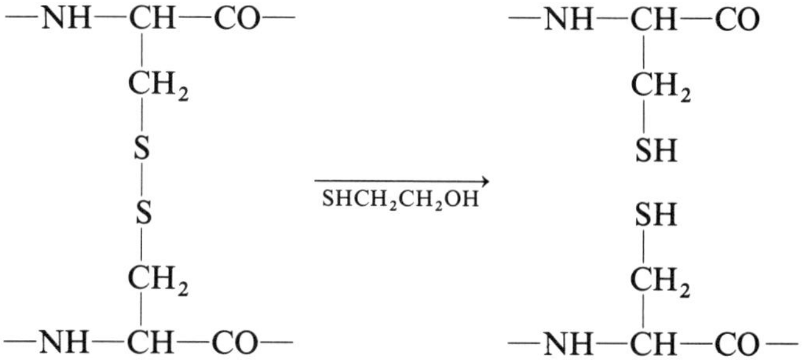

$$\begin{array}{c} -NH-CH-CO- \\ | \\ CH_2 \\ | \\ S \\ | \\ S \\ | \\ CH_2 \\ | \\ -NH-CH-CO- \end{array} \quad \xrightarrow[SHCH_2CH_2OH]{} \quad \begin{array}{c} -NH-CH-CO \\ | \\ CH_2 \\ | \\ SH \\ \\ SH \\ | \\ CH_2 \\ | \\ -NH-CH-CO- \end{array}$$

Free amino acid

Peptide chain

Group I. Nonpolar aliphatic R groups

Glycine (Gly) · Alanine (Ala) · Valine (Val) · Leucine (Leu) · Isoleucine (Ile)

Group II. Hydroxyl-containing R groups, aliphatic and aromatic

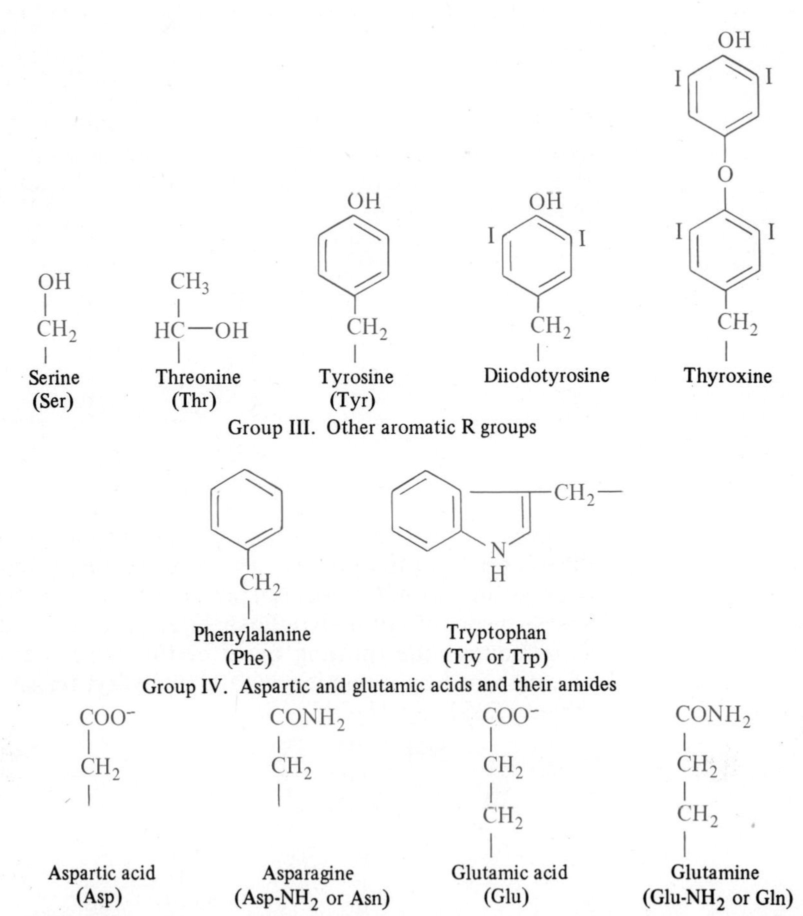

Figure 3-1

Structures of the amino acids which occur in natural proteins. In the cases of proline and hydroxyproline (Group VII) the complete structures are shown. For the other amino acids only the structures of the side-chains are shown. Asparagine and glutamine are alternatively abbreviated Asn and Gln, respectively.

Group V. Basic amino R groups

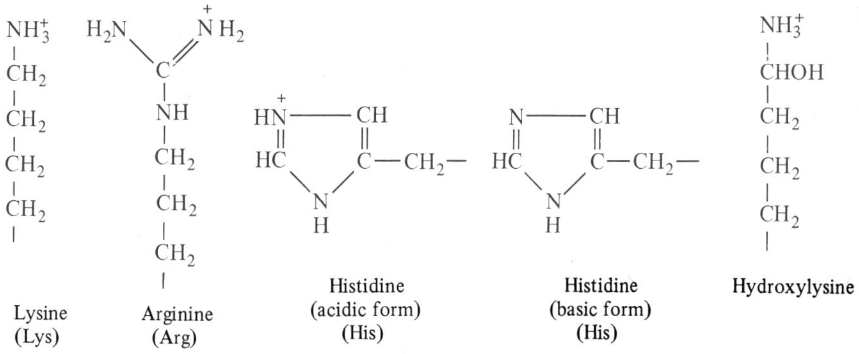

Lysine (Lys): NH_3^+—CH_2—CH_2—CH_2—CH_2—

Arginine (Arg): H_2N—C(=$\overset{+}{N}H_2$)—NH—CH_2—CH_2—CH_2—

Histidine (acidic form) (His)

Histidine (basic form) (His)

Hydroxylysine: NH_3^+—CHOH—CH_2—CH_2—

Group VI. Sulfur-containing R groups

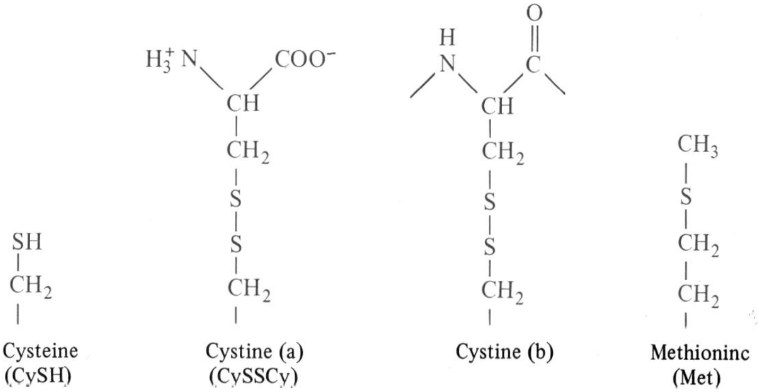

Cysteine (CySH): SH—CH_2—

Cystine (a) (CySSCy): H_3^+N—CH(COO^-)—CH_2—S—S—CH_2—

Cystine (b)

Methioninc (Met): CH_3—S—CH_2—CH_2—

Group VII. Imino acid R groups

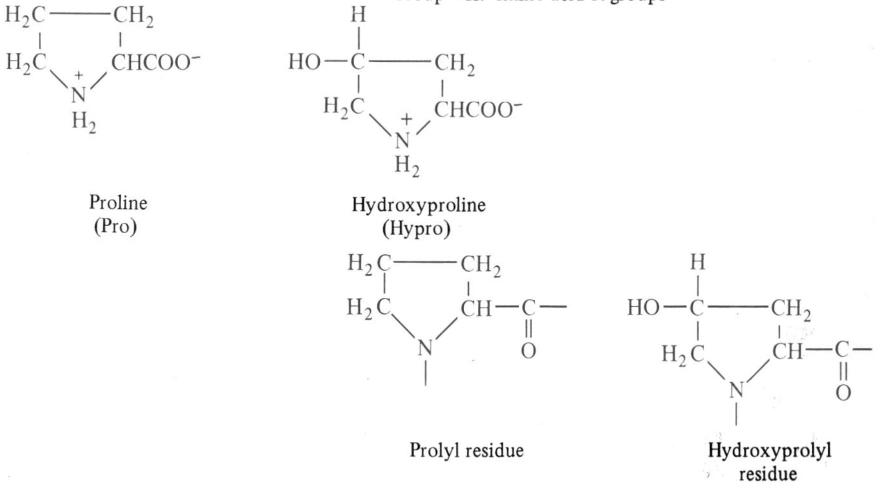

Proline (Pro)

Hydroxyproline (Hypro)

Prolyl residue

Hydroxyprolyl residue

Figure 3-1 *(continued)*

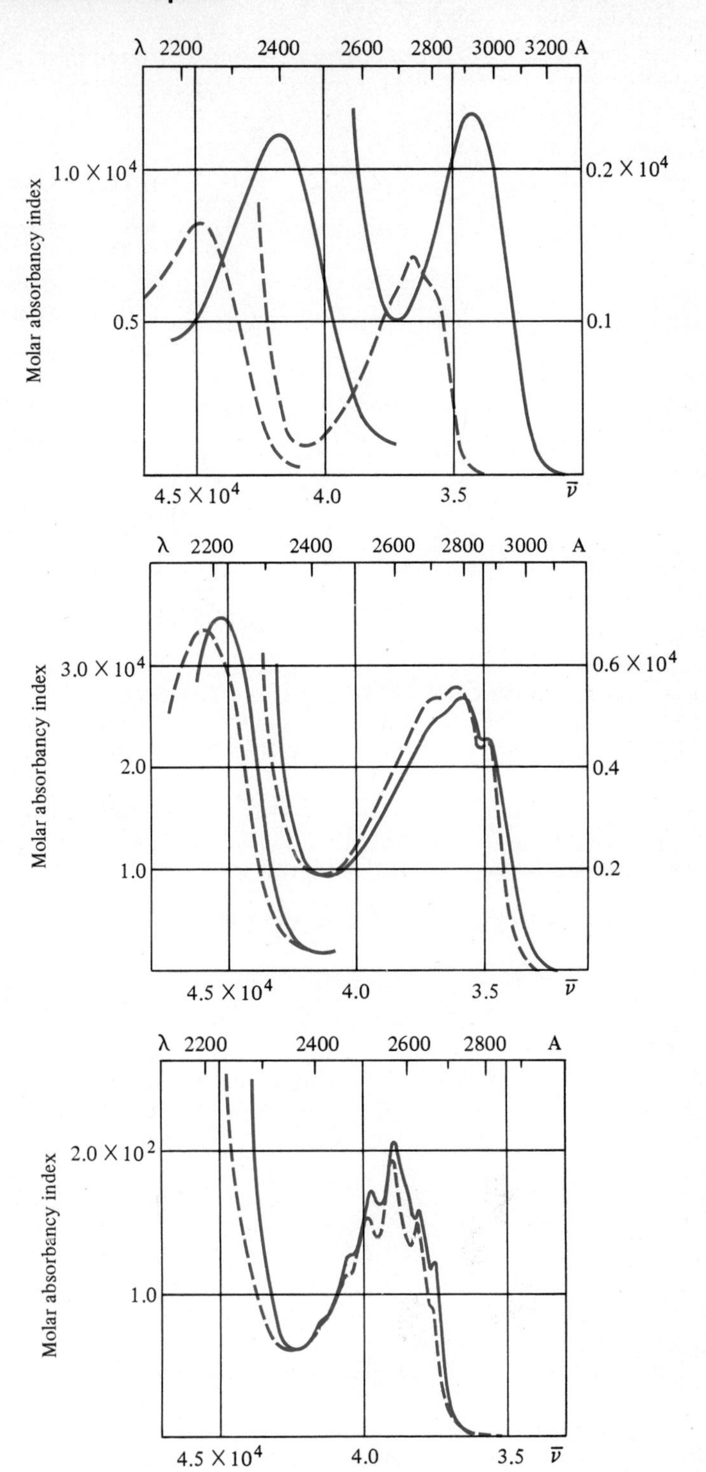

Figure 3-2
The ultraviolet absorption spectra of the aromatic amino acids in 0.1-*M* HCL (---) and in 0.1-*M* NaOH (—). The upper abscissa is the wavelength in angstroms; the lower is the frequency in wave numbers. Upper curve: tyrosine. Middle curve: tryptophan. In both cases the right ordinate refers to the curves on the right, for wavelengths above 2,300 angstroms; the left ordinate refers to the curves on the left for lower wavelengths. The two scales are necessary to display the entire absorption spectrum on one graph. Lower curve: phenylalanine. The molar absorbancy, ε, is defined by $\varepsilon = -(1/c)\log_{10} T$, where c is the molar concentration and T is the fraction of the light intensity at the given wavelength which is transmitted by 1 cm of solution.

The cysteine groups formed in this way can be stabilized against reoxidation by reaction with iodoacetate:

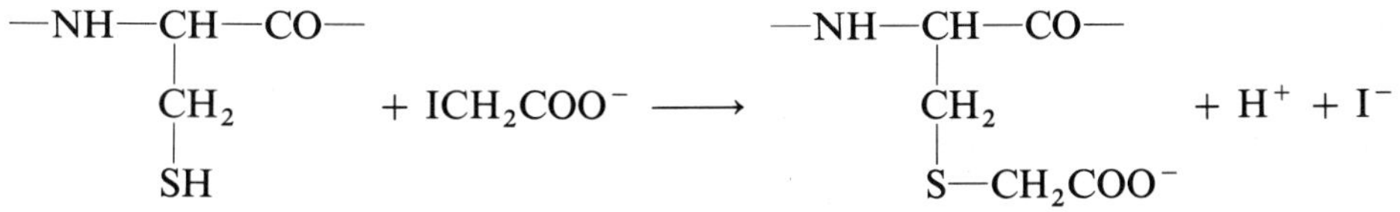

$$\begin{array}{c} -NH-CH-CO- \\ | \\ CH_2 \\ | \\ SH \end{array} + ICH_2COO^- \longrightarrow \begin{array}{c} -NH-CH-CO- \\ | \\ CH_2 \\ | \\ S-CH_2COO^- \end{array} + H^+ + I^-$$

Alternatively, the cystine bridges may be ruptured by oxidation with performic acid to two residues each of cysteic acid:

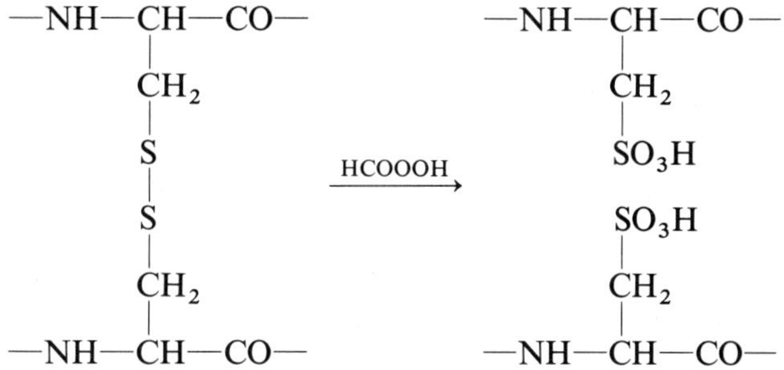

$$\begin{array}{c} -NH-CH-CO- \\ | \\ CH_2 \\ | \\ S \\ | \\ S \\ | \\ CH_2 \\ | \\ -NH-CH-CO- \end{array} \xrightarrow{HCOOOH} \begin{array}{c} -NH-CH-CO- \\ | \\ CH_2 \\ | \\ SO_3H \\ \\ SO_3H \\ | \\ CH_2 \\ | \\ -NH-CH-CO- \end{array}$$

α-Imino Acids

Proline and hydroxyproline differ from all other residues found in proteins in that their side-chains are closed rings which include the α-nitrogen. They are thus **α-imino** rather than **α-amino** acids. Hydroxyproline is of restricted occurrence, having been identified only in collagen and its derivatives.

The formation of a peptide bond by an α-imino acid leaves the α-nitrogen with no attached hydrogen (Fig. 3-1). This is an important factor in the structure of many proteins.

3-3 LESS COMMON AMINO ACIDS

In addition to the amino acids already listed, a number of others exist which are not ordinarily incorporated into proteins. Some are found in the free state; some occur in small peptides or specialized proteins; others serve as metabolic intermediates. Among the most important of these are ornithine, an intermediate in urea synthesis; β-alanine, a component of the vitamin pantothenic acid; and taurine, a decarboxylated derivative of cysteine (Fig. 3-3).

Lanthionine has been isolated from wool. It is equivalent to cystine with one sulfur atom removed. Homocysteine, whose side-chain is $-CH_2-CH_2-SH$, corresponds to demethylated methionine. γ-Aminobutyric acid occurs in the free state in the brain.

3-4 REACTIONS OF THE AMINO ACIDS

Reactions of the Amino Group

Several reagents have been introduced which combine with α-amino groups to form colored or fluorescent derivatives which are useful in structural

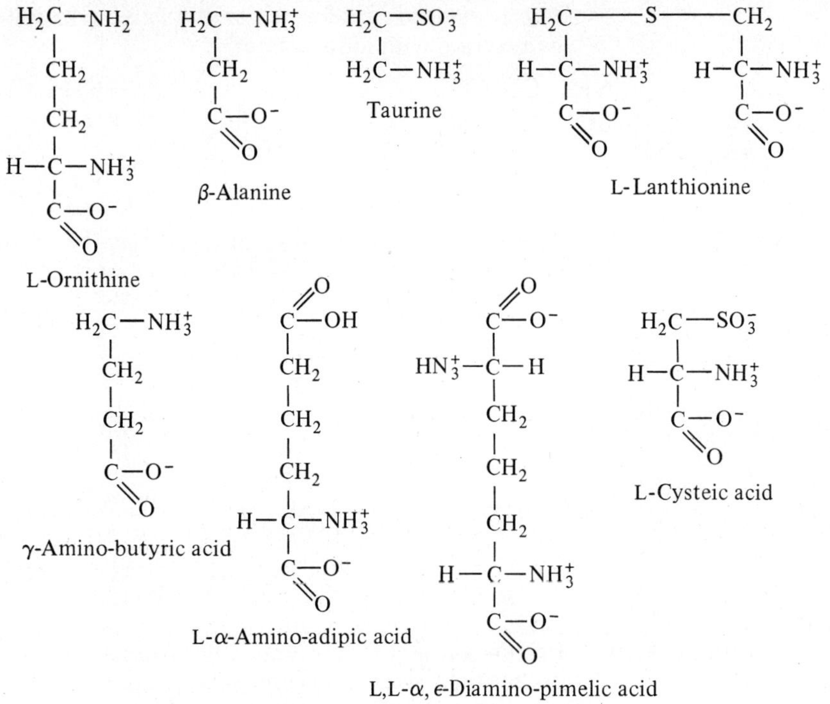

$H_2C{-}NH_2$ | CH_2 | CH_2 | $H{-}C{-}NH_3^+$ | $C(=O){-}O^-$

L-Ornithine

$H_2C{-}NH_3^+$ | CH_2 | $C(=O){-}O^-$

β-Alanine

$H_2C{-}SO_3^-$ | $H_2C{-}NH_3^+$

Taurine

$H_2C{-}S{-}CH_2$; $H{-}C{-}NH_3^+$, $C(=O){-}O^-$ (on each side)

L-Lanthionine

$H_2C{-}NH_3^+$ | CH_2 | CH_2 | $C(=O){-}O^-$

γ-Amino-butyric acid

$C(=O){-}OH$ | CH_2 | CH_2 | CH_2 | $H{-}C{-}NH_3^+$ | $C(=O){-}O^-$

L-α-Amino-adipic acid

$C(=O){-}O^-$ | $HN_3^+{-}C{-}H$ | CH_2 | CH_2 | CH_2 | $H{-}C{-}NH_3^+$ | $C(=O){-}O^-$

L,L-α, ε-Diamino-pimelic acid

$H_2C{-}SO_3^-$ | $H{-}C{-}NH_3^+$ | $C(=O){-}O^-$

L-Cysteic acid

Figure 3-3 Some examples of amino acids that do not occur in proteins.

studies of proteins (chapter 4). The reagent 1-fluoro-2,4-dinitrobenzene (FDNB) was introduced by Sanger for the quantitative labeling of amino groups in amino acids or peptides. Under weakly alkaline conditions α-amino acids are converted to yellow 2,4-dinitrophenyl derivatives, called DNP-amino acids. FDNB also reacts with the side-chain amino group (or α-amino

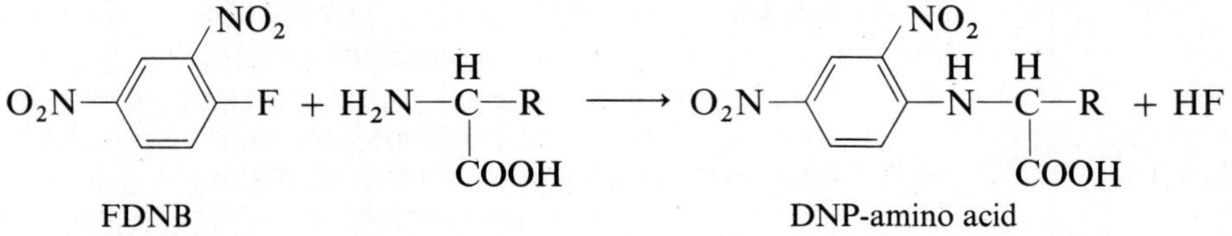

$$O_2N{-}C_6H_3(NO_2){-}F + H_2N{-}\underset{COOH}{\overset{H}{C}}{-}R \longrightarrow O_2N{-}C_6H_3(NO_2){-}\overset{H}{N}{-}\underset{COOH}{\overset{H}{C}}{-}R + HF$$

FDNB DNP-amino acid

group) of lysine, but this derivative can be readily distinguished from the DNP derivatives of the α-amino acids by the chromatographic techniques discussed in section 3-7. Thc FDNB reaction is of particular utility in the identification of the NH_2-terminal amino acid of a polypeptide chain.

Ninhydrin Reaction

A general reaction of analytical importance occurs upon heating of α-amino acids with ninhydrin (Fig. 3-4). The same blue pigment is developed in each case with an absorption maximum near 570 nm. The imino acids—proline and hydroxyproline—undergo a different reaction which yields a yellow-red color.

$$2\,\text{(ninhydrin, } \mathrm{C(OH)_2}\text{)} + \mathrm{HC(NH_3^+)(R)}-\mathrm{COO^-} \longrightarrow \text{(}\mathrm{C{-}N{=}C}\text{ blue pigment, } \mathrm{O^-}\text{)} + \mathrm{CO_2} + \mathrm{RCHO}$$

Figure 3-4
Ninhydrin reaction. The blue pigment formed is independent of the nature of the amino acid (except for proline and hydroxyproline).

3-5 OPTICAL ACTIVITY OF THE AMINO ACIDS

Ordinary unpolarized light may be regarded, for many purposes, as a collection of electromagnetic waves whose electric vectors are perpendicular to the direction of propagation but form no particular azimuthal angle with that direction. Certain optical systems, such as a Nicol prism or a Polaroid sheet, have the property of transmitting only light whose electric vectors are confined to a single plane, much as a knife blade can be inserted easily between the pages of a book only when it is aligned parallel to them. The light emerging from such an optical system has an electric vector whose periodic oscillation lies in a single plane and is said to be **plane polarized**. Unpolarized light contains all possible orientations of the electric vector (Fig. 3-5).

Unpolarized

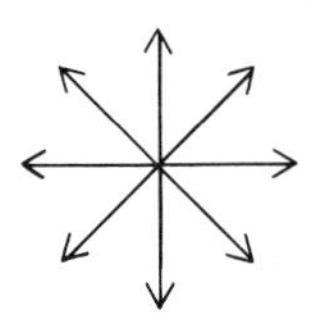

Polarized

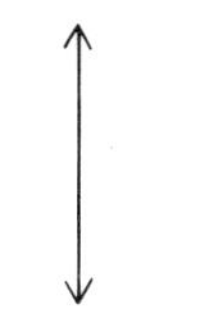

Figure 3-5
Schematic representation of unpolarized and polarized light. The arrows represent the directions of the electric vector.

Many organic molecules have the capacity, when dissolved in a solvent, to rotate the plane of polarization of a beam of plane-polarized light. The rotatory power of a substance is usually expressed as the **specific rotation** $[\alpha]$. This is defined as the rotation, in degrees, produced by the passage of a beam of polarized light through one decimeter of a solution containing 1 g/ml of optically active solute:

$$[\alpha] = \frac{\alpha}{\text{path length (dm)} \times \text{concentration (g/ml)}} \tag{3-1}$$

where α is the observed rotation in degrees.

The specific rotation is a function of the wavelength. Much data has been collected using the "D" line of sodium, whose wavelength is 589 nm. The specific rotation measured at this wavelength is often designated as $[\alpha]_D$.

Not all molecules are optically active. Optical activity is associated with a **chirality** of the molecule; that is, neither a plane nor a center of symmetry is present. This will be the case if the molecule contains a chiral carbon atom, with four different substituents:

$$\begin{array}{ccc} & a & \\ & | & \\ d- & C & -b \\ & | & \\ & c & \end{array}$$

Inspection of atomic models of molecules containing a single chiral carbon reveals that two possible configurations exist, which cannot be interconverted by rotation or translation in space. Their geometrical relationship is that of **mirror images** (Fig. 3-6). Two such optical isomers or **enantiomorphs** are equivalent in chemical and most physical properties but differ in the sign of their optical rotation. An isomer which rotates the plane of polarization to the *right* is said to be **dextrorotatory** and to have a *positive* rotation. If the direction of rotation is to the *left*, it is said to be **levorotatory** and to have a *negative* rotation. The dextro- and levorotatory forms are often designated as the (+), d, or *dextro*- and (−), l, or *levo*-isomers, respectively. A mixture containing equal quantities of two enantiomorphs is optically inactive and is termed a **racemic mixture**.

The two optical isomers of a molecule containing a center of chirality correspond to two different spatial configurations (Fig. 3-6). These are cus-

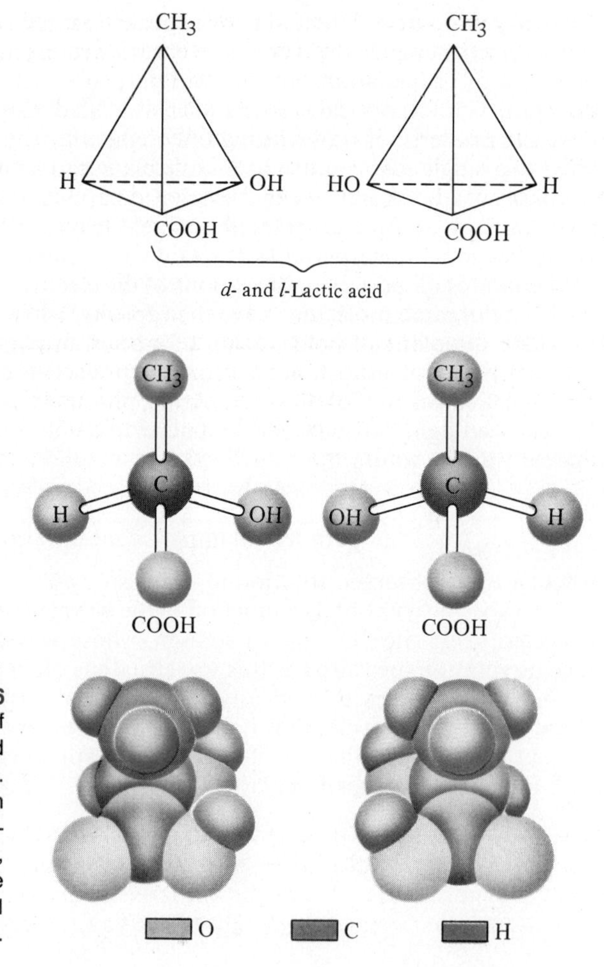

Figure 3-6
Optical isomers of lactic acid represented in three different ways. The upper portion represents the molecule as a tetrahedron, corresponding to the four bonds of the chiral carbon.

tomarily designated as D and L. The similarity in terminology to the d and l isomers is unfortunate, since the sign of rotation has no simple relationship to the configuration. The D-configuration is often levorotatory, and vice versa.

The absolute configurations of many organic molecules have been established by chemical transformations that relate them to compounds whose configurations have been determined directly by x-ray diffraction. All the amino acids, with the exception of glycine, contain a center of chirality in the α-carbon atom and can exist in two configurations. However, only amino acids of the L-configuration occur in natural proteins, although D-forms are sometimes found in the free state (Fig. 3-7). In contrast, there is no systematic pattern for the direction of the rotation, which often depends upon the state of ionization of the amino acid.

3-6 IONIZATION OF THE AMINO ACIDS

Amino acids belong to the class of weak electrolytes whose ionic state depends upon the pH and other external conditions. Each amino acid in the free state contains at least two ionizable sites, in the α-amino and α-carboxyl groups. In addition, many amino acids contain a third ionizable site in the side-chain. When amino acids are joined by peptide linkages to form a polypeptide chain, the ionization of their α-amino and α-carboxyl groups is blocked (except for the single amino group at the NH_2-terminal end and the single carboxyl group at the COOH-terminal end), so that the only other ionizable sites are those present on the side-chains (Table 3-1).

The state of charge of a free amino acid will depend primarily upon the pH. If the side-chain does not contain an ionizable group, the amino acid will exist predominantly as a zwitterion, of zero net charge, at pH's intermediate to the p*K*'s of the α-amino and α-carboxyl groups. At pH's near the α-amino p*K*, the amino acid will exist as a mixture of the neutral and alkaline forms, so that the *average* net charge will lie between 0 and −1. At pH's sufficiently alkaline to the α-amino p*K* it will exist as an anion of charge −1, while at pH's well to the acid side of the α-carboxyl p*K* it will be a cation of charge +1.

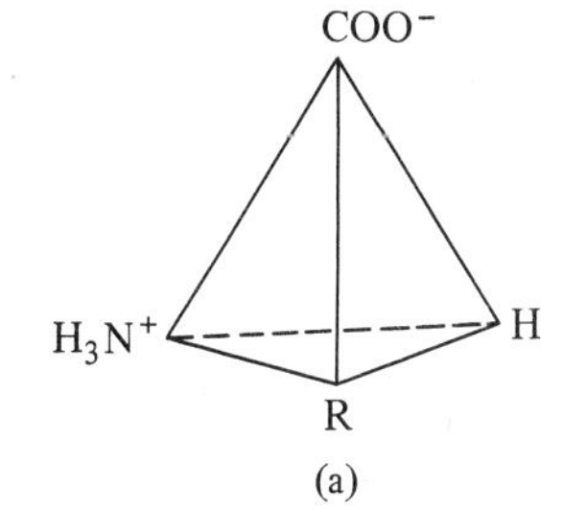

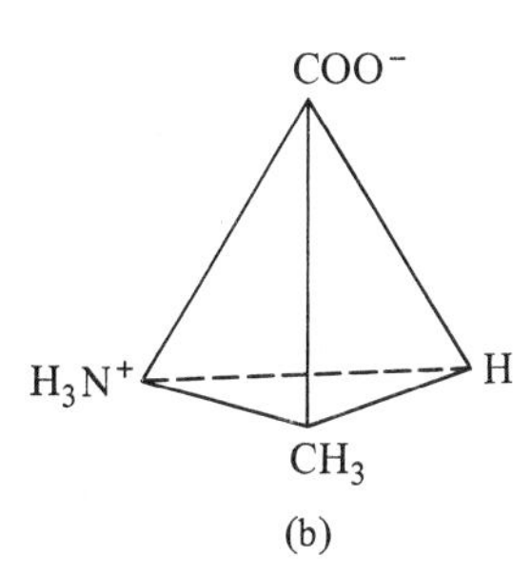

Figure 3-7
(a) General configuration of the L-amino acids. (b) Configuration of L-alanine.

$$\underset{\substack{\text{Acid form}\\ \text{Net charge} = +1}}{\text{R}-\overset{\overset{\displaystyle NH_3^+}{|}}{\underset{\underset{\displaystyle H}{|}}{C}}-\text{COOH}} \underset{OH^-}{\overset{H^+}{\rightleftharpoons}} \underset{\substack{\text{Neutral form}\\ \text{Net charge} = 0}}{\text{R}-\overset{\overset{\displaystyle NH_3^+}{|}}{\underset{\underset{\displaystyle H}{|}}{C}}-\text{COO}^-} \underset{H^+}{\overset{OH^-}{\rightleftharpoons}} \underset{\substack{\text{Alkaline form;}\\ \text{Net charge} = -1}}{\text{R}-\overset{\overset{\displaystyle NH_2}{|}}{\underset{\underset{\displaystyle H}{|}}{C}}-\text{COO}^-}$$

If the side-chain contains an ionizable group the maximum net charge is increased to +2 or −2.

An amino acid with one amino and one carboxyl group, such as glycine, exists as a diprotic acid in its fully protonated form and can donate two protons in the course of its complete titration with a base. Figure 3-8 displays the complete titration curve of glycine. The p*K* values for the amino and

Table 3-1
p*K* Values for the Ionizing Groups of Amino Acids at 25°C

Amino acid (or group)	p*K** -COOH	p*K** $-NH_2$	p*K** *side-chain*	p*K side-chain in proteins*
Glycine	2.3	9.6		
Alanine	2.3	9.7		
Leucine	2.4	9.6		
Serine	2.2	9.2		
Threonine	2.6	10.4		
Glutamine	2.2	9.1		
Aspartic acid	2.1	9.8	3.9	4.0 ± 1
Glutamic acid	2.2	9.7	4.3	4.0 ± 1
Histidine	1.8	9.2	6.0	6.3 ± 0.7
Cysteine	1.7	10.8	8.3	10.0 ± 0.5
Tyrosine	2.2	9.1	10.1	10.1 ± 0.3
Lysine	2.2	9.0	10.5	10.1 ± 0.7
Arginine	2.2	9.0	12.5	12.0 ± 0.5
NH_2-terminus (in proteins)		8		
COOH-terminus (in proteins)	3			

*These values refer to the *free* amino acid.

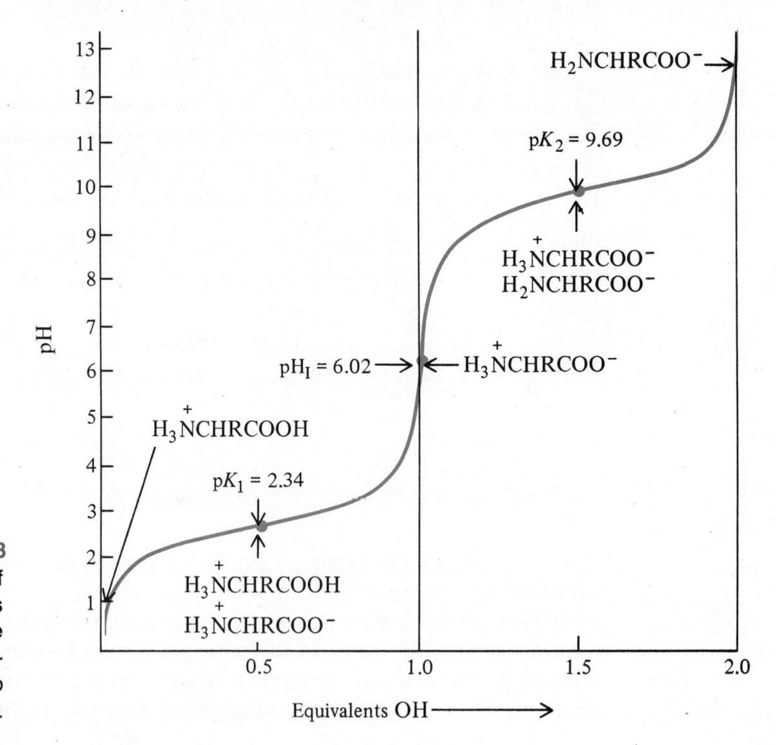

Figure 3-8
Titration curve of glycine. The curve is biphasic, reflecting the independent ionizations of the α-amino and α-carboxyl groups.

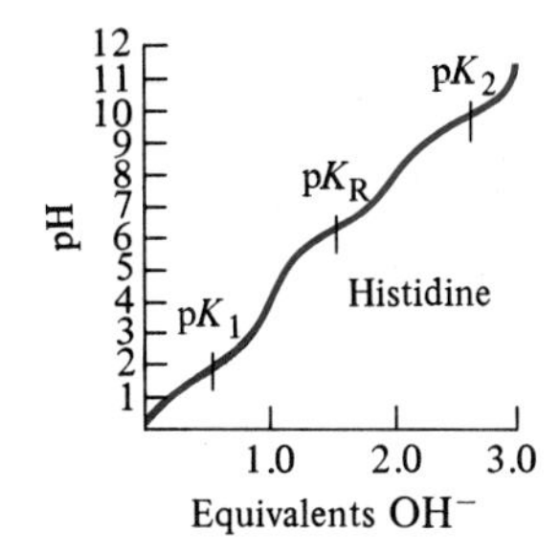

Figure 3-9
Titration curve of histidine, which contains three ionizable sites, corresponding to the α-carboxyl (COOH), imidazole (Im), and α-amino (NH_2) groups. The complete process may be represented by

$$\begin{matrix} COOH \\ ImH^+ \\ NH_3^+ \end{matrix} \underset{}{\overset{pK_1 = 1.8}{\rightleftharpoons}} \begin{matrix} COO^- \\ ImH^+ \\ NH_3^+ \end{matrix} \underset{}{\overset{pK_R = 6.0}{\rightleftharpoons}} \begin{matrix} COO^- \\ Im \\ NH_3^+ \end{matrix} \underset{}{\overset{pK_2 = 9.2}{\rightleftharpoons}} \begin{matrix} COO^- \\ Im \\ NH_2 \end{matrix}$$

carboxyl ionizations are sufficiently separated so that the curve is biphasic. At pH 2.34, the midpoint of the carboxyl ionization, equal concentrations of the species $NH_3^+—CH_2—COOH$ and $NH_3^+—CH_2—COO^-$ are present. At pH 9.6, the midpoint of the amino ionization, there exist equal concentrations of the species $NH_3^+—CH_2—COO^-$ and $NH_2—CH_2—COO^-$. Each of the two branches of the biphasic curve may be represented fairly accurately by the Henderson-Hasselbalch equation.

If the side-chain contains an ionizable group, then a third ionization zone will be superimposed upon those arising from the amino and carboxyl groups (Fig. 3-9). Table 3-1 cites values of the dissociation constants for each of the titratable sites occurring in the free amino acids. It is noteworthy that none of the amino acids, except histidine, has significant buffering capacity in the physiological pH range 6–8.

Under conditions where a net charge is present, an amino acid has the property of migrating in an electric field. The direction and velocity of movement depend upon the charge and hence upon the pH. If the pH is sufficiently alkaline so that the amino acid is entirely in the anionic form, its rate of movement or **mobility** will have its maximum negative value. (The mobility is considered *negative* if the solute migrates toward the *positive* electrode and is considered *positive* if it migrates toward the *negative* electrode.) As the pH is lowered the magnitude of the net negative charge, and hence of the mobility, will decrease until, at a pH value termed the **isoelectric point**, the mobility is zero. With a further decrease in the pH the sign of the mobility changes to positive. The magnitude of the mobility increases as the pH is further lowered. For an amino acid containing an α-carboxyl group (pK_1) and an α-amino (pK_2), the isoelectric point (pI) is equal to the *arithmetic mean* of the two pK's.

$$pI = \tfrac{1}{2}(pK_1 + pK_2)$$

3-7 SEPARATION AND ANALYSIS OF THE AMINO ACIDS AND PEPTIDES

The rapid and accurate analysis of amino acid mixtures became possible with the introduction of chromatographic techniques in the 1940's. The numerous variants of the chromatographic approach all depend upon differences in the *rates* of movement of the various components in a supporting medium, which may be a column of gel or powder or a sheet of filter paper.

Column Chromatography

The system for carrying out column chromatography is prepared by pouring the chromatographic material as a slurry into a vertical cylinder constricted at the base (Fig. 3-10). The column is retained by a porous plug of sintered glass, glass wool, or some other material permeable to solvent. The solution containing the mixture of components to be separated is placed on the top of the column and allowed to pass into it. An eluting solvent, or series of solvents, is pumped continuously through the column, or allowed to flow through it by gravity. The effluent solution is collected as a series of fractions.

The rate of movement down the column of each component is dependent upon the tenacity with which it is retained by the column material. Compounds that are weakly adsorbed move down the column relatively rapidly and appear in the earlier effluent fractions. More strongly bound substances are transported more slowly and appear in the later fractions. The order of appearance of the various components is the inverse of that of their affinity for the chromatographic medium.

Chromatographic separation is usually based primarily upon differences in molecular size or charge. **Gel filtration** relies entirely upon the former. A frequently used medium consists of granules of cross-linked preparations of the polysaccharide dextran. This material is generally known by its commercial name *Sephadex*. The porosity of the granules can be controlled by varying the degree of cross-linking. Molecules whose effective radius is large in comparison with the average pore size will be unable to penetrate the granules and will be confined to the solution exterior to the granules. The passage of such molecules down the column will be relatively unimpeded, so that they emerge in the earlier fractions. Molecules small in comparison with the average pore size can enter the granules with a consequent retarda-

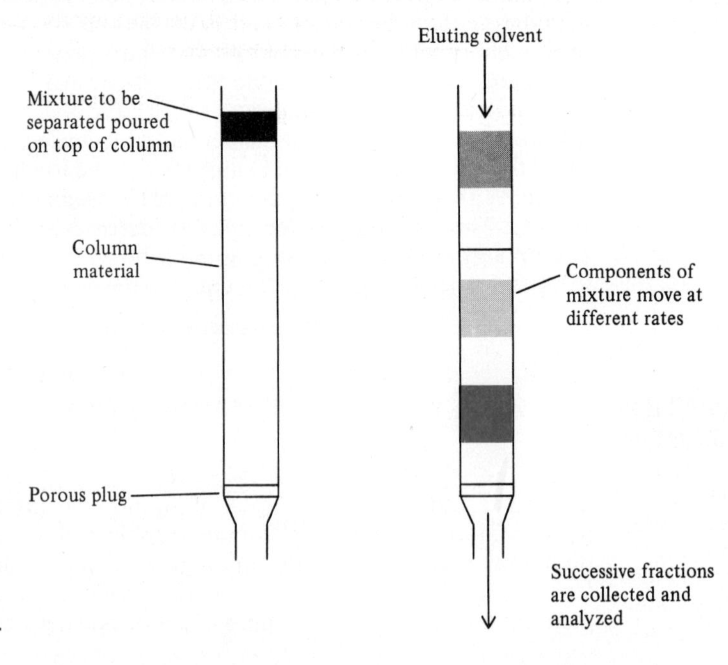

Figure 3-10 Schematic representation of chromatographic separation.

tion of their rate of movcmcnt. In general, the order of appearance in the effluent fractions of the different components will be the inverse of that of their molecular size, with the largest species appearing first. Gel filtration is not ordinarily useful for amino acid separation, but has proved to be of great utility for the fractionation of proteins and polypeptides.

The most generally useful techniques for amino acid separation depend upon differences in net charge and utilize **ion-exchange resins** as the chromatographic medium. These resins are three-dimensional cross-linked polymeric networks containing positively or negatively charged sites (Fig. 3-11). Positively charged resins are normally useful for the separation of anionic materials and are called **anion exchangers**; negatively charged resins are used to fractionate cationic substances and are termed **cation exchangers**.

The resin is supplied in the form of small beads that are sufficiently porous to permit the entry of water, electrolyte, and solute. The fulfillment of the requirement for electroneutrality within the beads demands that the fixed charges attached to the resin network be balanced by the mobile charges contributed by the electrolyte and other solutes. An uncharged solute, or one whose charge is similar to that of the resin, is not involved in the balance of charge and is not retained by the resin, so that its passage down the column is unimpeded. A solute of charge opposite to that of the resin is retained as a result of electrostatic interaction with the fixed charges, with a consequent retardation of its transport down the column.

In general, the greater the charge on the solute and the lower the electrolyte concentration, the more tenaciously will the solute be held by the resin and the slower its rate of movement will be. Since the overall process corresponds to the displacement of retained solute ions by electrolyte ions of similar charge, it is usually termed **ion exchange**. However, electrostatic forces are not the only factor, since more specific interactions of solute and resin are often superimposed upon these.

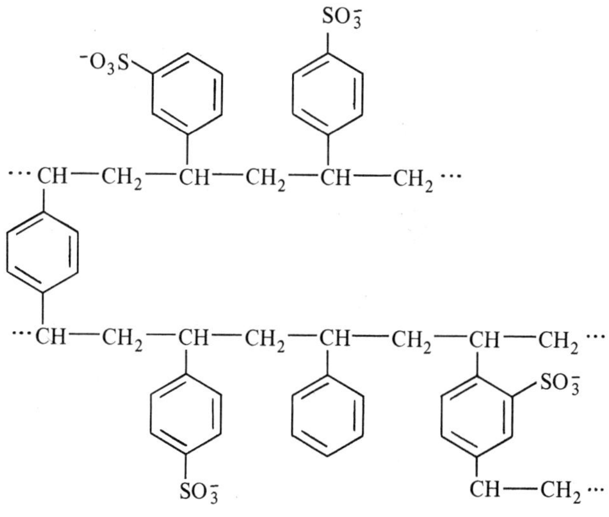

Figure 3-11
Schematic version of one type of cation-exchange resin. In this case the charged sites are sulfonate groups. These are attached to a cross-linked polystyrene network.

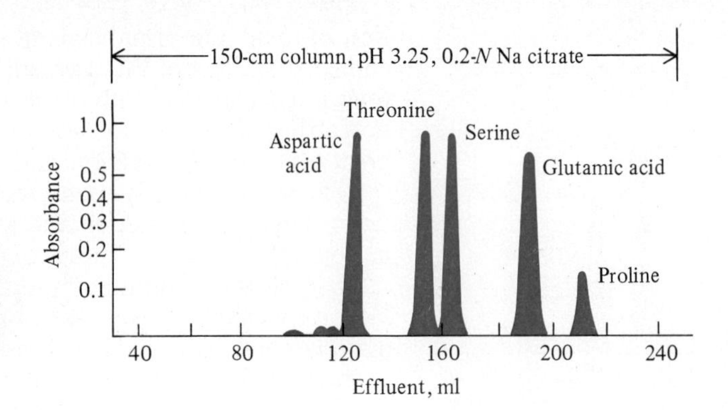

Figure 3-12
Amino acid analysis obtained with an automated instrument. The amino acids are separated by ion exchange chromatography. The ninhydrin color value is plotted.

The charge of an amino acid, and hence its affinity for the resin, is a function of the solvent pH. In practice the column is normally eluted with a solvent gradient, in which the pH and electrolyte concentration are varied continuously. By the proper choice of eluting conditions the resolving power of the system can be made very high.

The separation of amino acids is usually carried out on cation-exchange resins consisting of sulfonated polystyrene. The negative charges of the sulfonate groups are initially balanced by Na^+ ions. The mixture of amino acids at pH 3 is added to the resin. Since at this pH, all of the amino acids are cations with a net positive charge, they tend to be adsorbed by the column and to displace bound Na^+. The most basic amino acids—arginine, lysine, and histidine—are retained most strongly by electrostatic forces, while the most acidic—glutamic and aspartic acids—are held the least tenaciously. As the pH and electrolyte concentrations of the eluting solvent are progressively increased, the amino acids migrate down the column at different rates. The appearance of amino acids in the effluent solution can be monitored continuously by the ninhydrin reaction. A plot of ninhydrin color value as a function of the volume of effluent has the form of a series of discrete peaks, each of which corresponds to an amino acid, which is identifiable by its position on the chromatogram after calibration with known materials (Fig. 3-12). The quantity of each amino acid present is proportional to the area under the corresponding peak.

Today, highly automated commercial apparatuses are available for the analysis of amino acid mixtures by ion exchange chromatography, reducing the process to a routine laboratory procedure. A complete amino acid analysis for a protein may now be accomplished in a matter of hours with the requirement of only a few milligrams of material.

Paper Chromatography

In paper chromatography the separation matrix is a strip or sheet of filter paper. The method, which is a variant of **partition chromatography**, depends primarily upon the redistribution, or partition, of solute between a *stationary* aqueous phase, which is immobilized by the filter paper, and a *mobile* organic phase. A drop of solution containing the mixture of amino acids, or other materials, that are to be separated is applied near one end of a strip of paper that has been equilibrated with water. The end of the strip is dipped

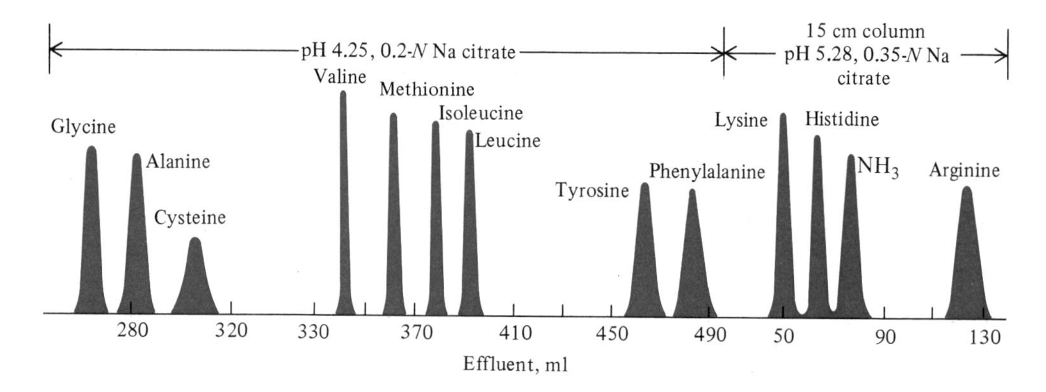

into a reservoir containing the developing solution, which is an organic solvent or mixture of solvents (Fig. 3-13).

The developing solution is drawn into the paper by capillarity. The direction of solvent movement may be upward, in *ascending* paper chromatography, or downward in *descending* paper chromatography. As the solvent front moves along the paper, each component is extracted and transported in the same direction at a rate that depends upon its relative solubility in the aqueous and organic phases, as well as upon its affinity for the filter paper matrix. At the conclusion of the separation the paper is dried and stained with ninhydrin. The amino acids appear as spots that can be eluted and estimated colorimetrically.

The ratio of the distance of movement of a solute to that of the solvent front is called the R_f value and is characteristic of the solute, the developing solvent, and other conditions. Under carefully controlled conditions it is a

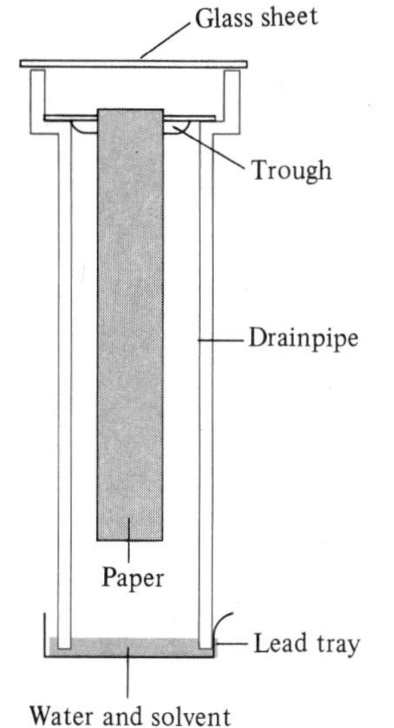

Figure 3-13
Simple device for one-dimensional paper chromatography (descending). The sample is applied as a narrow band at the top of the strip, which is placed in contact with the developing solution in the upper trough.

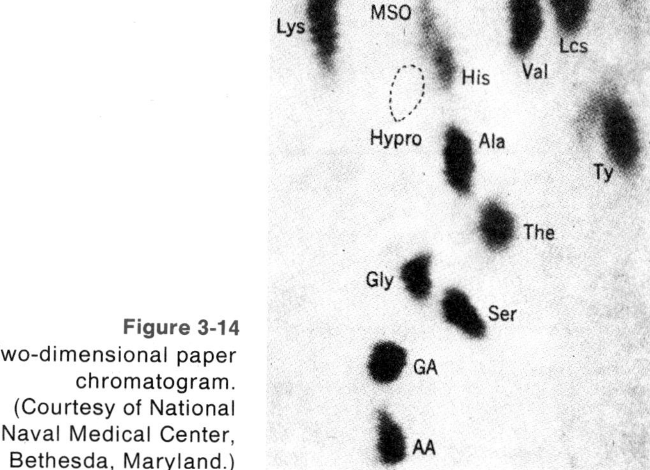

Figure 3-14
Two-dimensional paper chromatogram. (Courtesy of National Naval Medical Center, Bethesda, Maryland.)

reproducible quantity and can serve to identify an unknown amino acid.

In *two-dimensional* paper chromatography, a square sheet of filter paper is used. The solution is applied near one corner and developed in one direction with a solvent mixture. It is then dried, rotated by 90°, and exposed to a second developing solvent, whose direction of flow is at 90° to that of the first. In this manner the resolving power of the technique is greatly enhanced (Fig. 3-14).

Thin Layer Chromatography

In this form of chromatography, filter paper is replaced by a glass plate covered with a thin layer of an inert absorbent, such as cellulose or silica gel. The solution containing the solutes to be separated is applied as a series of spots at the bottom of the plate, whose lower edge is then immersed vertically into a developing solvent. The solvent rises by capillarity and the various solute components migrate at differing rates. After the solvent front reaches the upper edge, the plate is dried and the separated components rendered visible by spraying with ninhydrin or another suitable indicator. The components appear as discrete spots.

Paper Electrophoresis

An alternative approach to the analysis of amino acid or peptide mixtures is based on their varying rates of migration in an electric field. In paper electrophoresis the medium is a strip of filter paper saturated with buffer of pH such that the amino acids possess net charges (Fig. 3-15). The velocity of

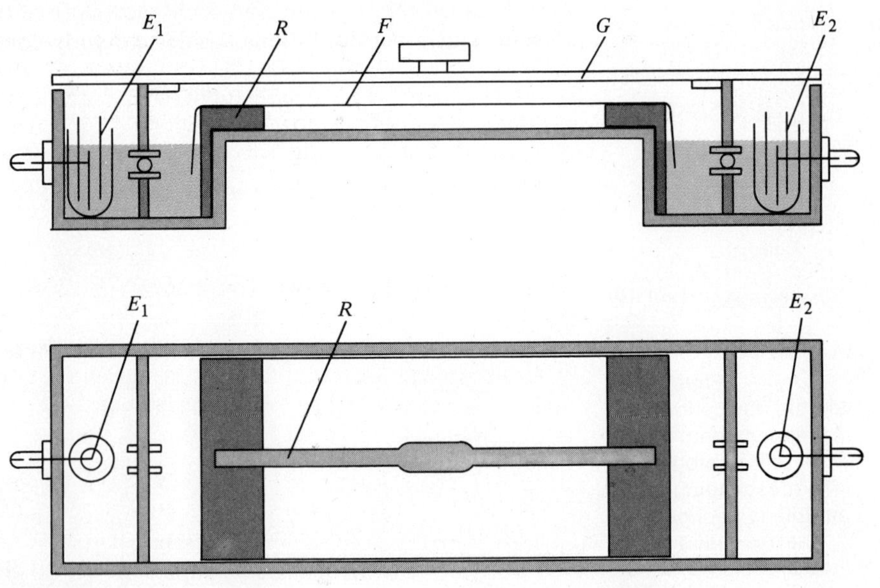

Figure 3-15
Apparatus for paper electrophoresis. E_1 and E_2 are electrodes; F is the filter paper strip, to which the mixture is applied as a narrow band; R is a support; and G is a cover to minimize evaporation. The electrode compartments are filled with buffer.

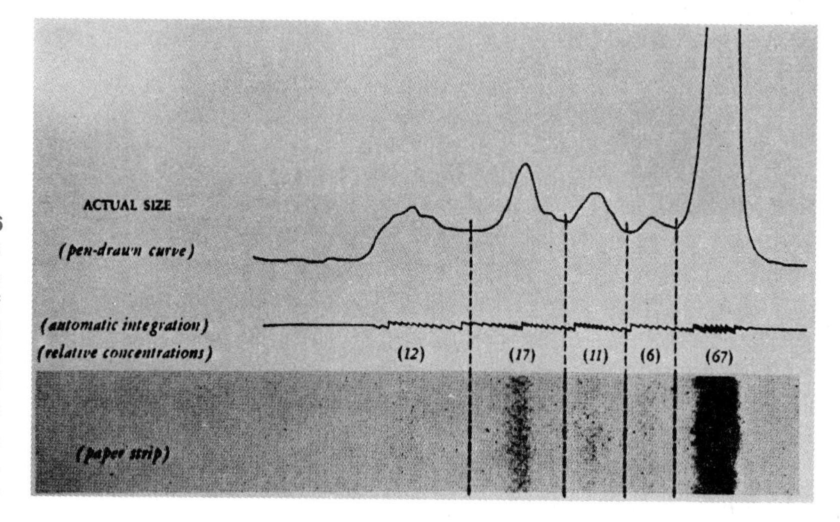

Figure 3-16 Stained paper strip showing the distribution of components. (Adapted from P. Alexander and R. Block, *Analytical Methods of Protein Chemistry*, Vol. 2, Pergamon, London, 1960.)

movement of each amino acid is a function of its net charge and hence of the pH and ionic strength of the buffer. In practice, two or more determinations with different buffers are normally required to separate all 20 amino acids.

A small volume of solution containing the mixture of amino acids, or other solutes, is applied as a thin band to one end of the strip of filter paper and an electric field is applied. Each amino acid migrates as a separate band at its characteristic velocity. In *high voltage* paper electrophoresis, several thousand volts are applied in order to achieve separation in minimal time and thereby minimize the broadening and blurring caused by diffusion. In this case special precautions must be taken to cool the strip to avoid convective disturbances. After the completion of electrophoresis the strip is dried and stained with ninhydrin (Fig. 3-16). By prior calibration with a known mixture of amino acids, each unknown amino acid may be identified from its distance of migration. The quantity of each amino acid may be estimated from the color intensity of its band as determined by scanning with a densitometer.

Fingerprinting

This term is applied to two-dimensional separations on filter paper in which electrophoresis is applied in one direction followed by chromatographic separation in the other. Very high resolution can be attained in this way. The method is often used for comparing complex mixtures of peptides that may differ in only one or two components.

Gel Electrophoresis

Although gel electrophoresis is more useful for the analysis of mixtures of proteins and polypeptides than for amino acids, it will be discussed here. In this case the electrophoresis medium is a slab or cylinder of gel (Fig. 3-17). While a number of types of gel have been successfully used, the majority of separations have employed cross-linked polyacrylamide gel. The gel matrix serves to minimize blurring of the bands of migrating solutes and to enhance resolution (Fig. 3-18). The cross-linked polyacrylamide medium preferentially retards the movement of the larger molecules present in the mixture. Thus

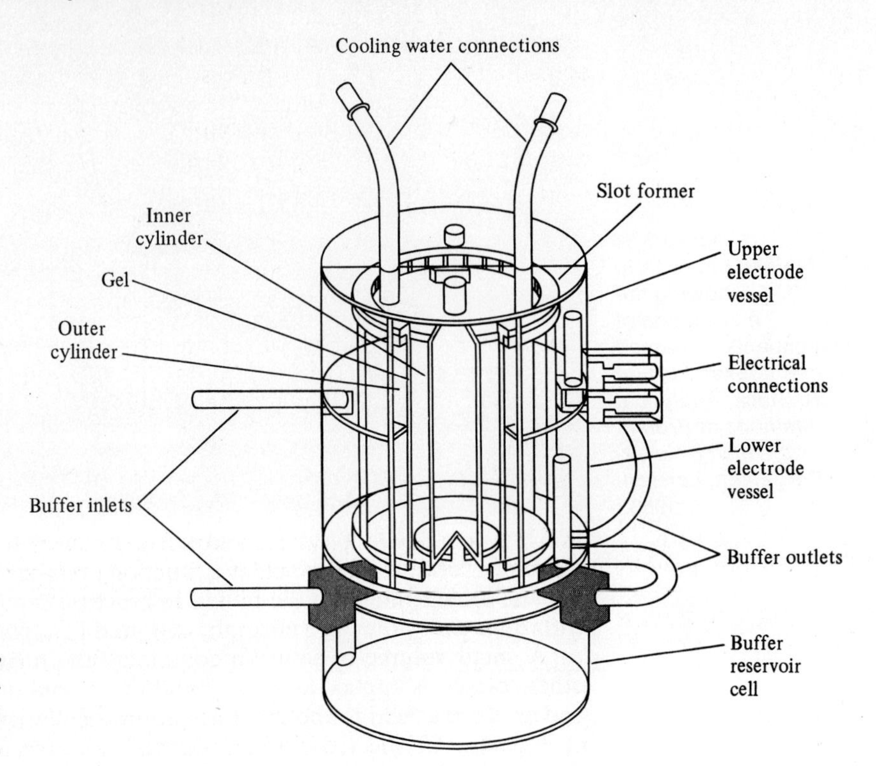

Figure 3-17 Apparatus for acrylamide gel electrophoresis.

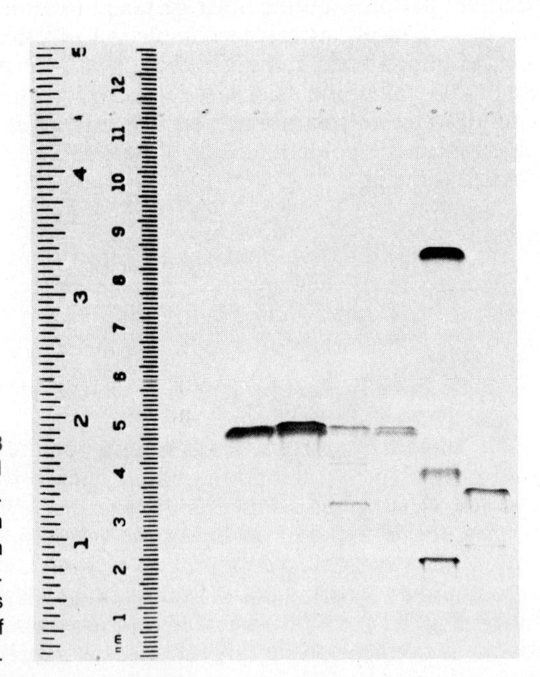

Figure 3-18 Stained gel electrophoresis slab. The two channels on the left contain homogeneous samples. The other channels contain mixtures of components.

separation on the basis of molecular size is superimposed upon separation according to molecular charge, resulting in improved resolution.

In **disc gel** electrophoresis the migrating bands are hypersharpened by the formation of a pH gradient within the gel column so that they appear as very thin bands. The very high resolution attained permits the analysis of quite small samples of complex mixtures.

SUMMARY

The proteins are polymers of fundamental subelements called **amino acids**, each of which contains an amino and a carboxyl group. The generalized structure of an amino acid may be written

$$\begin{array}{c} NH_2 \\ | \\ R\text{—}CH\text{—}COOH \end{array}$$

Here R is the **side-chain**, which may be an aliphatic hydrocarbon, an aromatic group, an acidic group, a basic group, an aliphatic hydroxyl group, or a sulfur-containing group.

Proline and **hydroxyproline** are distinctive in that their side-chains are closed rings including the nitrogen; they are thus **imino** rather than **amino** acids.

Proteins consist of one or more **polypeptide** chains, in which the amino acids are joined by **peptide** bonds arising by the condensation of amino and carboxyl groups to form an amide type linkage. Each polypeptide chain contains a free amino group at one end (the **NH_2-terminal** end) and a free carboxyl group at the other (the **COOH-terminal** end).

The side chain of **cysteine** contains a **sulfhydryl** (—SH) group. Two cysteine groups may be oxidized to form **cystine**, in which the two residues are joined by a **disulfide** (—S—S—) bridge. Cystine bridges function as **cross-links** between different polypeptide chains or different parts of the same chain.

The aromatic amino acids **tyrosine, tryptophan**, and **phenylalanine** are primarily responsible for the ultraviolet absorption spectra of proteins. Their absorption maxima are near 280 nm.

Involvement in peptide bonds suppresses the ionization of the amino and carboxyl groups of the amino acids incorporated into proteins, except for the two at the ends of each polypeptide chain.

The electrolyte behavior of proteins thus primarily reflects the side-chain ionizations of the acidic amino acids **aspartic** and **glutamic acids** and the basic amino acids **lysine, histidine**, and **arginine**.

Each amino acid undergoes a reaction with **ninhydrin** to form a blue pigment. This reaction is often used to estimate amino acids quantitatively.

Every amino acid contains at least one chiral center and may accordingly exist as **stereoisomers**. The amino acids occurring in natural proteins have the L-conformation.

Mixture of amino acids may be separated and analyzed by methods involving differential rates of transport. These include **ion-exchange chromatography, paper chromatography**, and **paper electrophoresis**.

REFERENCES

The following code is used to classify references. I: particularly useful as an introduction to the subject; R: useful primarily as a reference text; A: an advanced account of the material; H: a publication of historical importance.

General

E. J. Cohn and J. T. Edsall, *Proteins, Amino Acids, and Peptides*, Reinhold, New York (1944). (H)

A. L. Lehninger, *Biochemistry*, Worth, New York (1975). (A)

A. Meister, *Biochemistry of the Amino Acids*, Academic, New York (1968). (R)

D. E. Metzler, *Biochemistry*, Academic, New York (1979). (R)

L. Stryer, *Biochemistry*, Freeman, San Francisco (1976). (I)

Amino Acid Analysis

S. Blackburn, *Amino Acid Determination*, Dekker, New York (1968). (R)

C. H. W. Hirs, ed., "Amino Acid Analysis and Related Procedures," sec. 1 of vol. II, *Methods in Enzymology*, Academic, New York (1967). (R)

Stereochemistry

W. L. Alworth, *Stereochemistry and Its Application to Biochemistry*, Wiley-Interscience, New York (1972). (A)

REVIEW QUESTIONS

Questions marked with an asterisk are of a high level of difficulty.

*3-1 "Tyrosine hydrochloride" (Tyr·HCl) is formed by treating the "normal" pH 7 form with excess HCl and crystallizing out the resulting salt.

(a) How many moles of chloride per mole of tyrosine in the tyrosine HCl·salt? Consider the titration of 1 l of 1*M* tyrosine·hydrochloride with NaOH, assuming that *the volume does not increase* during the titration, and that the p*K*'s are 3, 9, and 11.

(b) Write the reactions and the equilibrium constant expressions with the correct *K* values for all the titration steps. Use the abbreviations TH_3^+, TH_2, TH^-, T^{2-}, but define the abbreviations in terms of complete tyrosine structures.

(c) Write all of the equations necessary to solve for the concentrations of all species after $\frac{1}{2}$*M of NaOH* has been added to the 1 l of 1-*M* Tyr·HCl, assuming that only the first titration step need be considered. Show that by neglecting small terms (1% or less), simple expressions are obtained for the concentrations of the *two* forms of *tyrosine* which predominate under these conditions. Find the *concentrations of these two forms* of tyrosine and the pH; explain why you made your assumptions (if any).

(d) Use expressions analogous to the simple ones obtained in (c) to find the concentrations of the two major forms of tyrosine (indicate what they are) and the pH under the conditions below.

Justify your assumptions.
(1) $1\frac{1}{2}M$ of NaOH have been added
(2) $2\frac{1}{2}M$ of NaOH have been added

(e) Consider the situation when just $1M$ of NaOH has been added, ignoring the third titration step. Write the complete set of equations. Make and justify the assumptions which result in a simple expression (independent of tyrosine concentration) for the pH (or $[H^+]$) under these conditions. Calculate the pH. Write an analogous expression for the pH when $2M$ of NaOH have been added, assuming that you can ignore the *first* titration step. What assumptions must be made? Calculate the pH.

(f) Consider the situation when $3M$ of NaOH have been added. Ignore the first two titration steps. Determine the concentration of $[T^{2-}]$ and $[H^+]$.

3-2 Paper electrophoresis of a mixture of Asp, Phe, Lys at pH 7.0 after 2 h at 5000 V results in three "spots" distributed as follows on the paper:

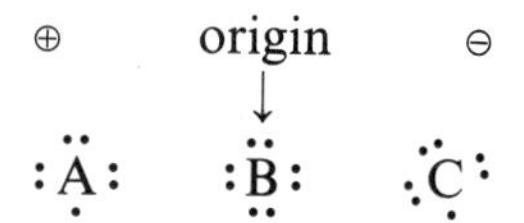

Identify the three spots, A, B, and C.
(a) A = Asp; B = Phe; C = Lys
(b) A = Lys; B = Phe; C = Asp
(c) A = Phe; B = Asp; C = Lys
(d) A = Lys; B = Asp; C = Phe
(e) none of the above

3-3 Pick the best pairs of amino acids from the list below and draw their structures. Show the interactions. Consider only forms which will be at least 5% available at pH 7.

serine, glycine, methionine, cysteine, valine, asparagine,
glutamic acid, histidine, arginine, alanine, proline

(a) Two (different) amino acids forming a hydrophobic interaction with their side chains.
(b) Two (different) amino acids forming a charge–charge attraction with their side chains.
(c) Two (different) amino acids forming an H-bond involving a side-chain *amide* nitrogen's proton and a side-chain hydroxyl oxygen.

3-4 Describe or illustrate the kinds of bonds that the side chains of the following amino acids can participate in which could be important for maintaining the conformation of proteins.
(a) Cys
(b) Lys
(c) Glu
(d) Ile

4

The Chemistry of Polypeptides and Proteins

4-1 CHEMICAL STRUCTURE AND BIOLOGICAL FUNCTION

In 1945, few events seemed less imminent than the establishment of the structure of even the simplest protein. The only aspect of the problem which appeared clear was that it presented difficulties of an altogether different order than those encountered in the case of most small molecules. Yet it was only six years before the chemical structure of insulin was essentially solved by Sanger and only 10 more before the complete three-dimensional structures of myoglobin and hemoglobin were obtained by Kendrew and by Perutz, respectively. Today, protein structural determination is rapidly becoming almost a routine procedure. Every natural protein consists of one or more polypeptide chains, which may be joined or internally cross-linked by cystine disulfide bridges (Fig. 4-1). The polypeptide chains are folded into a well-defined three-dimensional pattern, whose geometry is characteristic of the protein.

This chapter will be concerned with the purely *chemical* aspects of protein structure; that is, the number of polypeptide chains, their amino acid sequences, and the number and location of disulfide cross-links. If these structural features are known, a two-dimensional structural formula may be written for the protein, which shows the nature and relative location of all the chemical bonds but not the geometry of folding; this is termed the **primary structure**.

Why would anyone other than a protein chemist find all this interesting? There are three main reasons. The first is that primary structure determination is a very important part of solving the overall three-dimensional struc-

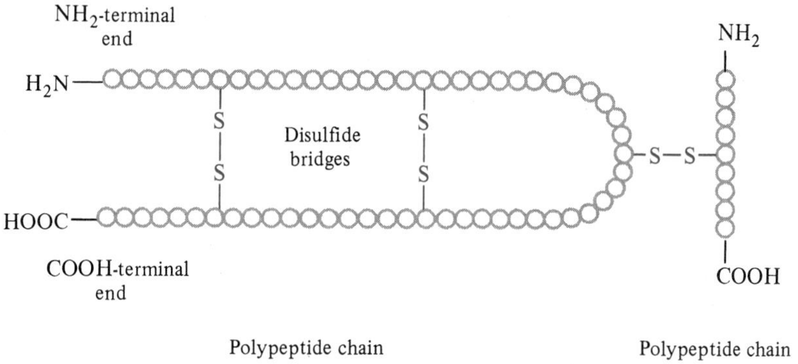

NH$_2$-terminal end

H$_2$N — HOOC — COOH-terminal end

Disulfide bridges

S—S

NH$_2$ COOH

Polypeptide chain Polypeptide chain

Figure 4-1 Schematic version of a protein consisting of two polypeptide chains joined by a cystine bridge; one chain is internally cross-linked by two cystine disulfides.

ture, which is, in turn, essential for understanding the *biological* function of the protein. We shall encounter an example of this in chapter 6, when we consider the structural basis of the activity of the enzyme chymotrypsin.

The second reason is that because not all parts of the primary structure are equally important for preserving the spatial geometry and biological functioning of the protein, *mutations* resulting in amino acid *substitutions* may often be tolerated in the less essential parts of the sequence. Such mutations tend to accumulate in the course of evolutionary time, so that the extent of the differences in primary structure between samples of the same protein isolated from different species is a measure of the time that has elapsed since the evolutionary **divergence** of the two from a common ancestor. This permitted the construction of phylogenetic "family trees," with profound consequences for our understanding of the evolutionary process and the relationships between species.

The third reason is more practical. Within a single species a mutational alteration, sometimes of only one amino acid, may sometimes result in a protein whose biological activity is impaired, but is still sufficient to permit the life of the individual to continue, at least for a time. This may be the origin of a hereditary disease, whose symptoms reflect the deficient functioning of an identifiable protein. The classical example is **sickle-cell anemia**, in which a defective hemoglobin molecule results from the substitution of a single amino acid. A knowledge of the molecular origins of this disease is proving invaluable in providing clues as to what avenues of therapy to explore.

While most of this chapter will be concerned with the details of how primary structure is determined, we shall also consider the biological implications of the observed variation with species of primary structure and the role of mutational alterations in hereditary disease.

Prosthetic Groups

Before plunging into the primary topics of this chapter, one complication should be noted. Some proteins contain a structural element that is not composed of or derived from amino acids; such proteins are said to be **conjugated**. This structural element is often essential for biological activity; in such cases it is termed a **prosthetic group**. A prosthetic group may be covalently linked to the protein or may be tightly combined via noncovalent interactions. A familiar example is provided by the **heme** groups of hemoglobin, which combine reversibly with molecular oxygen. Many enzymes also contain prosthetic groups.

4-2 PURIFICATION AND CRITERIA FOR HOMOGENEITY OF PROTEINS

Before any definitive study can be made of its structure and properties, a protein must be obtained in purified form. Generalization becomes impossible here, and each protein must be regarded as a special case. The initial step is usually the extraction of the protein, in impure form, from the tissue or biological fluid in which it naturally occurs. In the case of tissue proteins this is usually accomplished by selective extraction, followed by precipitation with a suitable agent, such as ethanol or ammonium sulfate.

The difficulty encountered in extracting a protein in native form depends both upon its intrinsic lability and upon the degree to which it is integrated into a structural matrix. In general, proteins which are present in free solution, such as the blood plasma proteins, are easier to isolate in native form than those which occur as part of a structural element, such as some mitochondrial enzymes.

The later stages of protein purification are usually carried out by chromatographic procedures. Both gel filtration and ion-exchange chromatography are widely used. Among the most generally useful ion-exchange materials are several cellulose derivatives, including the cation exchangers, carboxymethyl cellulose and phosphocellulose, as well as the anion exchanger, diethylaminoethyl cellulose. If the biological activity of the protein involves its binding of a small molecule with high affinity, then it is often possible to take advantage of this property in a purification step. In **affinity chromatography** the small molecule is linked chemically to the chromatographic material and selectively binds the protein, thereby separating it from inert contaminants, which are not retained by the column and pass through it into the effluent solution. By a subsequent change in pH, or other conditions, the bound protein is then released from the column in purified form.

Polyacrylamide gel electrophoresis is probably the most widely used of the available criteria for the homogeneity of a protein preparation. After completion of electrophoresis in the gel medium, the protein bands are rendered visible by staining with a dye, such as Amido Black or Coomassie Blue, which combines selectively with proteins.

4-3 DETERMINATION OF THE PRIMARY STRUCTURE OF PROTEINS

Amino Acid Composition

The first step in establishment of the complete primary structure of a protein is the procurement of accurate figures for the overall amino acid composition. This requires the complete hydrolysis of all peptide bonds, which is usually achieved by heating for 24 h in 6-M HCl at 100–110°C. Tryptophan is destroyed by this treatment and requires a separate alkaline hydrolysis for its determination.

$$\cdots\text{—NHCHCONHCHCO—}\cdots \xrightarrow[\text{HCl}]{\text{H}_2\text{O}} \text{R}_1\text{—}\underset{\text{H}}{\overset{\text{NH}_2}{\text{C}}}\text{—COOH} + \text{R}_2\underset{\text{H}}{\overset{\text{NH}_2}{\text{C}}}\text{—COOH} + \cdots$$

(with R_1 and R_2 as side chains on the two α-carbons of the peptide)

The resultant mixture of amino acids is analyzed quantitatively by ion-exchange chromatography as described in section 3-7. The results of such an analysis can be expressed in tabular form, giving the number of moles of each amino acid in any arbitrary weight, say 100,000 grams, of protein.

A *minimum* molecular weight, equal to the true molecular weight divided by an unknown integer, can be computed from these data as follows. If each protein molecule contains n residues of a particular amino acid, of which m moles are present per 100,000 grams, then

$$\frac{100{,}000}{m} = \frac{M}{n} \tag{4-1}$$

where M is the molecular weight of the protein and is equal to n times the minimum molecular weight, M/n. If a residue of infrequent occurrence is chosen, n will be a small integer. An approximate value of the molecular weight (to within 2%–4%) can be obtained by the use of any of several physical techniques (chapter 5). If n is small its value may often be established in this way, permitting the assignment of an accurate figure for the molecular weight. This approach is of course most readily applied to the smaller proteins, which are more likely to contain a residue which occurs only once or twice.

When the molecular weight is known, it is possible to write an empirical formula for the protein, giving the number per molecule of each kind of amino acid.

Determination of NH_2-Terminal Residues

The sequential analysis of polypeptides requires a convenient technique for identifying the terminal amino acids. The oldest procedure, which was originally developed by Sanger, utilizes FDNB (section 3-4) to label the NH_2-terminal residue. This reagent combines with the terminal α-amino group to form the DNP derivative of the peptide. The DNP derivative of the NH_2-terminal amino acid is sufficiently stable to survive complete acid hydrolysis of all peptide bonds. It may then be extracted and identified chromatographically. The bright yellow color of the label facilitates its location on a chromatogram. Lysine, whose side-chain contains an amino group, may combine with either two or one DNP groups, according to whether it is NH_2 terminal or occurs elsewhere in the polypeptide. However, in practice this causes little difficulty, as the two derivatives may be readily differentiated chromatographically.

The procedure most widely used today for the identification of NH_2-terminal groups is the **Edman degradation**, which also provides a means for the sequential removal of amino acids from the NH_2-terminal end. The peptide is initially reacted with phenylisothiocyanate at an alkaline pH (8–9).

```
R         R'          R
|         |           |                 R'
CHCONHCH-     HC——CO                   |
|          →  |    |      + NH2CHCO-
HN    S       N    S
  \  //        \\  /
   C             C
   |             |
 NHC6H5        NHC6H5

Phenylthiocarbamyl    Thiazolinone
     peptide              (II)              (III)
       (I)
                      /        \
       Aqueous acid  /          \  Anhydrous acid
                    R            R
                    |            |
                 CHCOOH          HC——CO
                    |        →   |    |
                 HN    NHC6H6    HN    NC6H5
                   \  /            \  /
                    C               C
                    ||              ||
                    S               S

              PTC amino acid     Phenylthiohydantoin
                  (IV)                 (V)
```

Figure 4-2
The reactions of the Edman degradation.

This reagent combines with the terminal α-amino group to form the phenylthiocarbamyl peptide (Fig. 4-2). This is transferred to an acid medium, which favors formation of the cyclic thiazoline derivative, with subsequent splitting off of the terminal amino acid.

In aqueous acid the terminal residue is split off largely as the phenylthiocarbamyl derivative, which slowly rearranges to form phenylthiohydantoin; in organic solvents the phenylthiohydantoin is formed directly (Fig. 4-2).

The phenylthiohydantoin derivative may be extracted and identified chromatographically. The conditions are sufficiently mild so that no peptide bonds are ruptured other than that involving the NH_2-terminal residue. The process may be repeated, permitting stepwise degradation of the polypeptide from the NH_2-terminal end. In this manner the order of the NH_2-terminal amino acids may be established. While progressive losses and other complications prevent the procedure from being continued indefinitely, it is usually possible to determine 10 or more residues. This suffices to establish the sequence of many short peptides. Commercial automated sequence analyzers or **sequenators** are available which are programmed to perform the various sequential operations of the Edman degradation.

Determination of COOH-Terminal Residues

It is also possible to degrade polypeptides sequentially from the COOH-terminal end. The enzymes carboxypeptidase A and carboxypeptidase B both catalyze the hydrolysis of peptide bonds involving the COOH-terminal residue. The *first* amino acid released is the COOH-terminal residue.

Carboxypeptidase A has a fairly broad specificity, but will not remove lysine, arginine, or proline residues. Carboxypeptidase B catalyzes the removal of lysine and arginine.

Sequence Determination of the Individual Polypeptide Chains

The identification of the NH_2-terminal groups yields the number of polypeptide chains and their NH_2-terminal residues. Further progress requires the rupture of all cystine bridges, whose persistence would hopelessly confuse the problem of sequence analysis, and the separation of the polypeptide chains. This is usually accomplished by reductive scission, followed by blocking of the free sulfhydryl groups with iodoacetate, as discussed in section 3-2. Alternatively the —S—S— bonds may be cleaved by oxidation with performic acid to cysteic acid (section 4-2). The individual polypeptide chains are separated chromatographically and analyzed separately.

The usual approach to sequence determination depends upon a comparison of the products of digestion by two different enzymes. **Proteolytic enzymes**, which catalyze the hydrolysis of peptide bonds, have a high degree of specificity, each enzyme attacking only peptide bonds joining a restricted number of amino acids (Table 4-1).

The digest formed by two enzymes of different specificity will consist of two different sets of peptide fragments. Since different bonds are attacked by the two enzymes, overlapping fragments will be produced. The two sets of peptides are separated into their components by chromatographic procedures, and the amino acid sequence of each peptide is determined by the methods described above.

Consideration of the regions of overlap permits the arrangement of the peptides in a linear sequence. For example, if a peptide of sequence

C—L—B—A—F

where the letters stand for amino acids, is isolated from the digest produced by enzyme 1 and the two peptides

F—C—L—B and A—F—D—G—L

from the digest formed by enzyme 2, it follows that the three must originate from, and form part of, the extended sequence

$$—{*}F \overset{\downarrow}{—} C—L—B{*}—A—F \overset{\downarrow}{—} D—G\quad L{*}—$$

where the arrows and asterisks indicate the bonds attacked by enzymes 1 and 2, respectively.

Table 4-1
Splitting of Peptide Bonds by Several Proteolytic Enzymes

$$—NH—\underset{R_1}{CH}—CO \vdots NH—\underset{R_2}{CH}—CO—$$

Enzyme	Specificity
Trypsin	R_1 = Lys or Arg
Chymotrypsin	R_1 = Phe, Trp, Tyr, Ile, Leu (and others)
Pepsin	R_1 = Phe, Trp, Tyr (and others)
Thermolysin	R_2 = Leu or Val

The enzymes trypsin and chymotrypsin are often used to produce the two sets of overlapping fragments. Trypsin catalyzes the hydrolysis of bonds of the type (see Fig. 3-1 for abbreviations of amino acids)

Arg—CO—NH—X
Lys—CO—NH—X

where X is some other amino acid; chymotrypsin attacks, among others, the bonds

Tyr—CO—NH—X
Trp—CO—NH—X
Phe—CO—NH—X

If a peptide is formed which is too long for its sequence to be determined by the Edman degradation already described, it may be necessary to digest it further with other enzymes to produce fragments of convenient size. Figure 4-3 illustrates this approach for a polypeptide hormone.

Hydrolytic agent	Amino acid sequence
Trypsin	Ser·Tyr·Ser·Met·Glu·His·Phe·Arg
Chymotrypsin	Arg·Try
Trypsin	Try·Gly·Lys·Pro·Val·Gly·Lys
Trypsin	Lys·Arg
Trypsin	Lys·Arg·Arg
Trypsin	Arg·Pro·Val·Lys
Acid	Pro·Val·Lys·Val·Tyr
Trypsin	Val·Tyr·Pro·Ala·Gly·Glu·Asp·Asp·Glu·Ala·Ser·Glu·Ala·Phe·Pro·Leu·Glu·Phe
Acid	Ala·Gly·Glu·Asp
Pepsin	Asp·Glu
Pepsin	Asp·Glu·Ala
Pepsin	Asp·Glu·Ala·Ser·Ser
Pepsin	Glu·Ala·Ser
Pepsin	Ser·Glu
Pepsin	Ser·Glu·Ala
Pepsin	Ser·Glu·Ala·Phe
Pepsin	Glu·Ala·Phe
Pepsin	Phe·Pro·Leu·Glu
Pepsin	Leu·Glu·Phe·Leu·Glu·Phe

Complete sequence:

Ser·Tyr·Ser·Met·Glu·His·Phe·Arg·Try·Gly·Lys·Pro·Val·Gly·Lys·Lys·Arg·Arg·Pro·Val·Lys·Val·
1 2 3 4 5 6 7 8 9 10 11 12 13 14 15 16 17 18 19 20 21 22
Tyr·Pro·Ala·Gly·Glu·Asp·Asp·Glu·Ala·Ser·Glu·Ala·Phe·Pro·Leu·Glu·Phe
23 24 25 26 27 28 29 30 31 32 33 34 35 36 37 38 39

Figure 4-3
Determination of the sequence of the pituitary hormone corticotropin or adrenocorticotropic hormone (ACTH). The peptides obtained by hydrolysis with various enzymes are shown. Consideration of the regions of overlap permits establishment of the complete linear sequence of amino acids. Sequences in parenthesis are arbitrary.

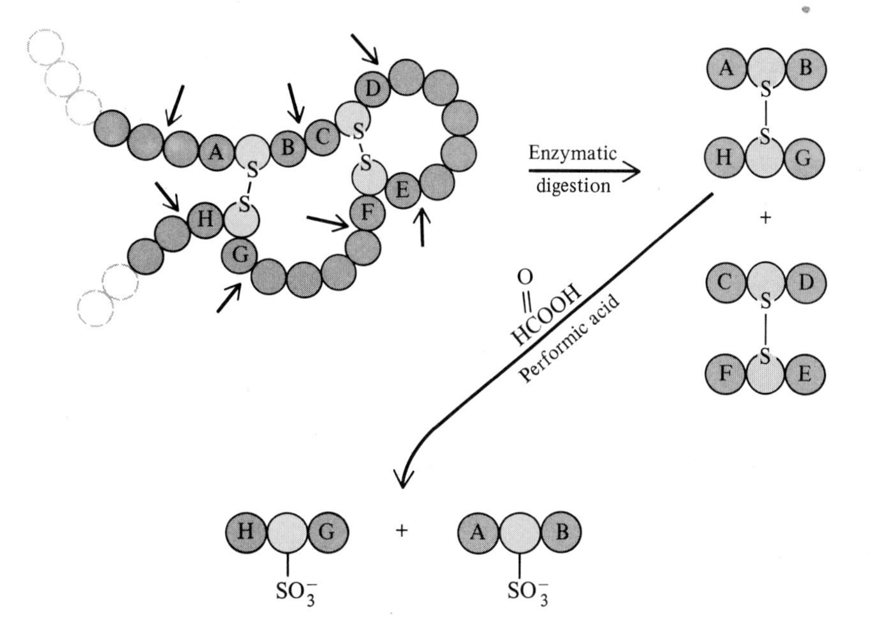

Figure 4-4 Location of disulfide bonds within a known primary sequence.

Location of Disulfide Bridges

The determination of the positions of the cystine cross-links is complicated by the possibility of disulfide interchange, which must be avoided by proper choice of conditions. By carrying out an enzymic hydrolysis of the protein *without* prior splitting of disulfide bonds, a set of cystine-containing fragments is obtained (Fig. 4-4). These fragments are separated chromatographically and cleaved by reduction or oxidation into their constituent peptides, whose sequences are determined by standard procedures. Each pair of linked peptide fragments may be located within the known amino acid sequences of the intact polypeptide chains, thereby establishing the positions of the disulfide bridges.

4-4 PROTEINS OF KNOWN PRIMARY STRUCTURE

Insulin

The successful determination of the primary structure of the protein hormone **insulin** by Sanger was a truly axial event in biochemistry. The experience acquired from this classic achievement provided much of the basis for the subsequent structural analyses of other proteins. Insulin, which is secreted by the pancreas, is important for the metabolic utilization of glucose. Its deficiency results in the symptoms of diabetes mellitus.

The basic molecular unit of insulin has a molecular weight of 5700. In aqueous solution it forms a dimer, which subsequently self-associates to yield particles of higher molecular weight.

The fundamental unit consists of two polypeptide chains joined by two cystine bridges. The smaller of the two, called the A chain, contains 20 amino acids and has glycine as the NH_2-terminal group. The A chain is internally

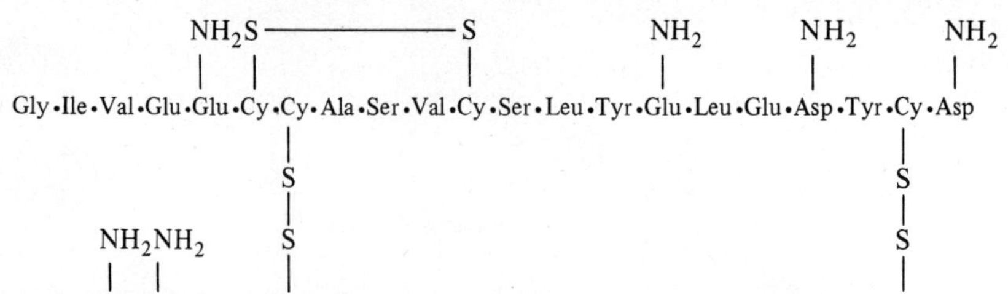

Figure 4-5
Primary structure of bovine insulin.

cross-linked by a single disulfide bridge (Fig. 4-5). The larger chain, designated the B chain, consists of 30 residues; phenylalanine is the NH_2-terminal group. It may be observed from Fig. 4-5 that several of the aspartic and glutamic acid residues occur as the amide derivatives asparagine and glutamine (Fig. 3-1).

Ribonuclease The enzyme **ribonuclease**, which catalyzes the hydrolysis of the phosphodiester bonds of RNA, consists of a single polypeptide chain internally cross-linked by four disulfide bonds (Fig. 4-6).

Lysozyme The enzyme **lysozyme** or muramidase, which catalyzes the hydrolytic rupture of certain bacterial cell walls, consists of a single polypeptide chain with four disulfide cross-links (Fig. 4-7).

4-5 GENETICALLY CONTROLLED VARIATIONS IN PRIMARY STRUCTURE

Species Variations It is a familiar feature of comparative biochemistry that parallel biological functions are performed in widely different species by proteins whose primary structures are similar, but not identical. This recurrence in altogether dissimilar organisms of proteins similar in structure and biological activity is understandable if such proteins arise from similar genes present in different species. This concept cannot however be verified by the standard method of classical genetics, since the classical tests for allelism apply only to organisms which can be crossed.

The primary structure of each protein occurring in individuals of a given species will reflect the evolutionary history of that species and, in particular, the mutational events which led to its emergence as a distinct species. In general, certain parts of the primary structure of a biologically active protein are much more essential to its function than others. An alteration in the amino acid sequence in a noncritical zone can often be tolerated without impairment of function, provided that the critical arrangement of amino acids in its center of activity remains intact.

Mutations leading to a loss or impairment of activity will normally either be directly lethal or result in a significantly reduced capacity for

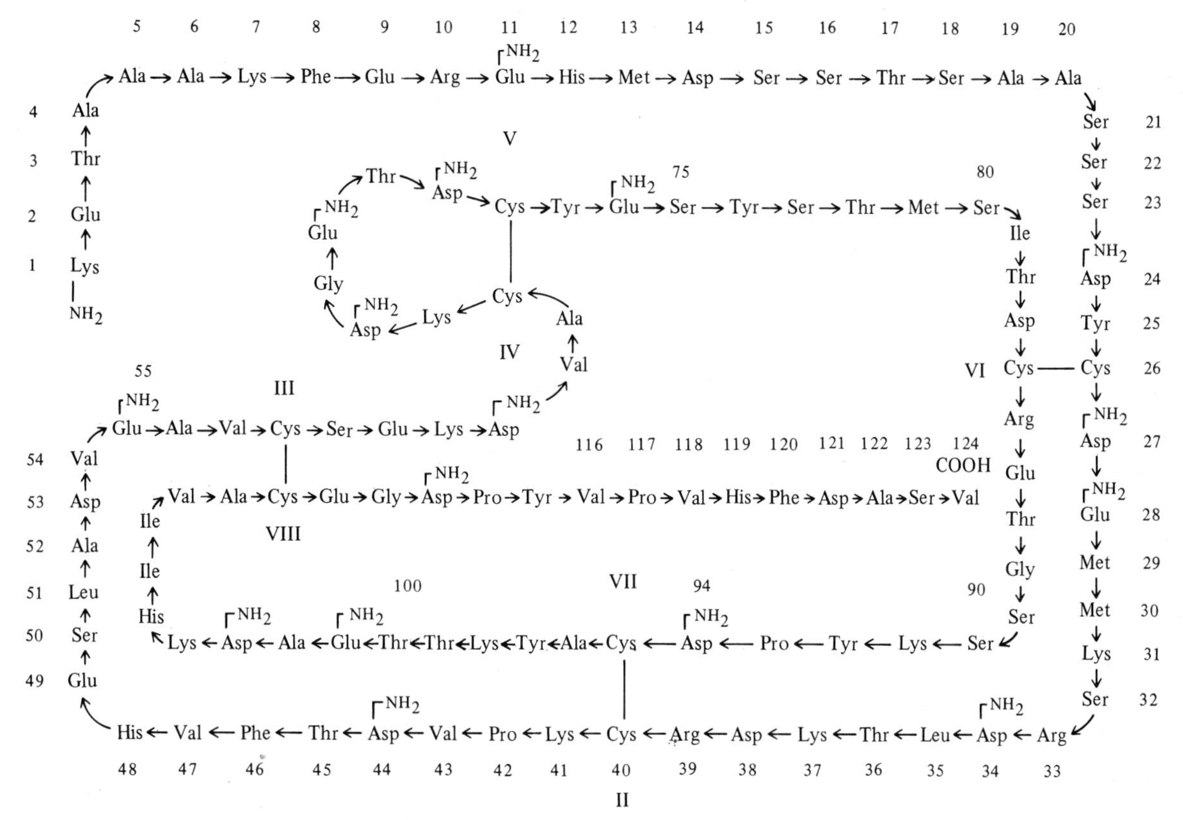

Figure 4-6
Primary structure of pancreatic ribonuclease.

survival, so that they tend to be eliminated by natural selection within a few generations. Mutations that do not interfere with function may be perpetuated to subsequent generations.

A comparative study of the primary structures of proteins of equivalent function occurring in a series of species should therefore provide an index of both the frequency of occurrence of mutations in the course of evolutionary time and the minimum primary structure that must remain intact for the persistence of activity.

As an example, consider the species variation of corticotropin (Table 4-2). The alterations in sequence encountered in four mammalian species are concentrated within a limited region of the polypeptide chain lying between residues 25 and 33 (numbering from the NH_2-terminal end). In each case the alteration has the form of a substitution, rather than an interchange or deletion.

The primary structure of insulin has also been determined for a set of mammalian species (Table 4-3). All the insulins in this group are equivalent in molecular weight and hormonal activity. The variations within the group again consist of substitutions, which are confined to a limited region within

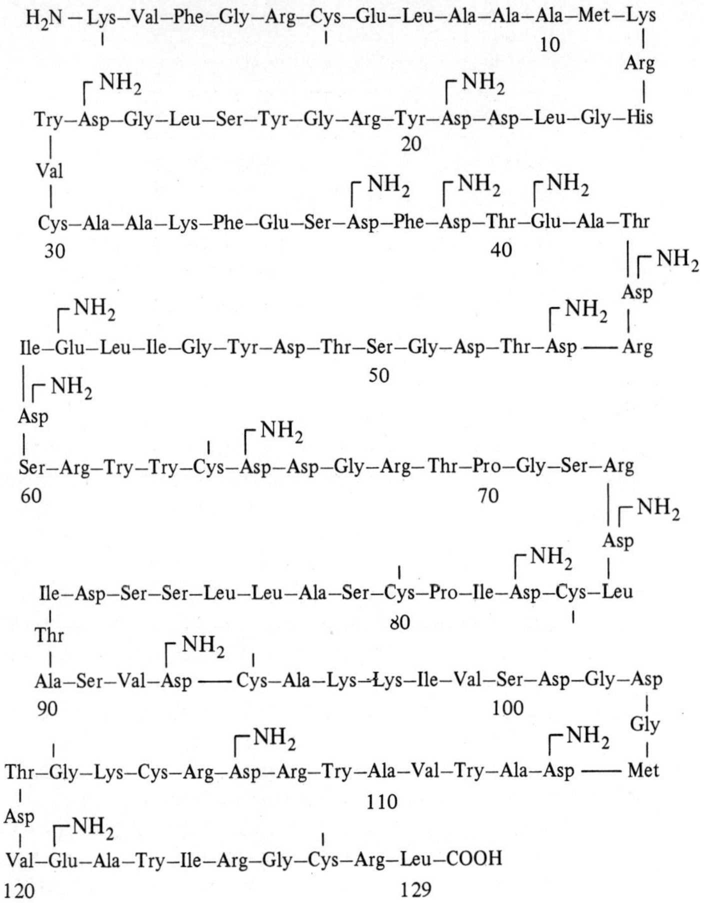

H_2N – Lys–Val–Phe–Gly–Arg–Cys–Glu–Leu–Ala–Ala–Ala–Met–Lys (10)
– Arg – His–Gly–Leu–Asp(NH_2)–Asp–Tyr–Arg–Gly–Tyr–Ser–Leu–Gly–Asp(NH_2)–Try (20)
– Val – Cys–Ala–Ala–Lys–Phe–Glu–Ser–Asp(NH_2)–Phe–Asp(NH_2)–Thr–Glu(NH_2)–Ala–Thr (30, 40)
– Asp(NH_2) – Arg – Asp(NH_2)–Thr–Asp–Gly–Ser–Thr–Asp–Tyr–Gly–Ile–Leu–Glu(NH_2)–Ile (50)
– Asp(NH_2) – Ser–Arg–Try–Try–Cys–Asp(NH_2)–Asp–Gly–Arg–Thr–Pro–Gly–Ser–Arg (60, 70)
– Asp(NH_2) – Leu–Cys–Asp(NH_2)–Ile–Pro–Cys–Ser–Ala–Leu–Leu–Ser–Ser–Asp–Ile (80)
– Thr – Ala–Ser–Val–Asp(NH_2) — Cys–Ala–Lys–Lys–Ile–Val–Ser–Asp–Gly–Asp (90, 100)
– Gly – Met — Asp(NH_2)–Ala–Try–Val–Ala–Try–Arg–Asp(NH_2)–Arg–Cys–Lys–Gly–Thr (110)
– Asp – Val–Glu(NH_2)–Ala–Try–Ile–Arg–Gly–Cys–Arg–Leu–COOH (120, 129)

Figure 4-7
Primary structure of egg white lysozyme.

Table 4-2

Species Variation of Corticotropin*

Species	*Sequence*
Pig	· · · Asp-Gly-Ala-Glu-Asp-GluNH_2-Leu-Ala-Glu · · · 25 26 27 28 29 30 31 32 33
Man	· · · Asp-*Ala*-*Gly*-Glu-Asp-GluNH_2-*Ser*-Ala-Glu · · ·
Ox	· · · Asp-Gly-*Glu*-*Ala*-*Glu*-*Asp*-*Ser*-Ala-GluNH_2
Sheep	· · · *Ala*-Gly-*Glu*-*Asp*-Asp-*Glu*-*Ala*-*Ser*-GluNH_2

*The complete structure is cited in Fig. 4-3.

Table 4-3
Species Variation in A-Chain of Insulin*

Species	Sequence
Beef	CySSCy-Ala-Ser-Val 8 9 10
Sheep	CySSCy-Ala-Gly-Val
Pig	CySSCy-Thr-Ser-Ile
Horse	CySSCy-Thr-Gly-Ile
Sperm whale	CySSCy-Thr-Ser-Ile

*The complete structure is cited in Fig. 4-5.

the disulfide loop of the A chain and to a single point within the B chain. The implication is that neither region is crucial to activity. However, if the comparison is extended to nonmammalian species, it is found that replacements may occur at almost any residues in the B chain.

An inspection of Table 4-3 reveals that a distinct pattern exists in the substitutions at positions 8, 9, and 10 of the A chain. At position A8, alanine and threonine are interchangeable; at position A9, serine and glycine; while A10 may be occupied by valine, isoleucine, or threonine. This suggests that the species variations in the gene responsible for the synthesis of this polypeptide are governed by definite chemical rules.

Unusually complete sequence information is available for the protein **cytochrome c**, which occurs in all plants, animals, and aerobic microorganisms, in which it functions in electron transport. Cytochrome c contains a single polypeptide chain of 104 residues. In the two dozen species examined thus far, only 35 residues are the same for all species. The invariant amino acids occur at irregular intervals along the chain, with the striking exception of positions 70–80, where a continuous set of 11 invariant residues is found. The number of alterations between a pair of species varies roughly as their phylogenetic separation (Fig. 4-8). Thus man and monkey differ in only one position, while mammals and tuna differ in 17–21, and mammals and yeast by up to 48 residues.

Mutational Alteration Within a Single Species

Much of what has been stated above applies as well to mutational variations within a single species. The vast majority of such mutations have a significantly deleterious effect upon the viability of the organism and are eliminated by natural selection. Occasionally a mutant form arises that, although producing an impairment of function, is preserved because of some compensating factor and transmitted to progeny, where it is associated with a hereditary disease.

The classic example of this is **sickle cell anemia**, a hereditary condition among central Africans, which derives its name from the characteristic shape of the red blood cells when deoxygenated. Sickle cells tend to clump together when deoxygenated, thereby blocking blood capillaries and interrupting the supply of oxygen to the tissues. The resulting crisis is often fatal. Individuals

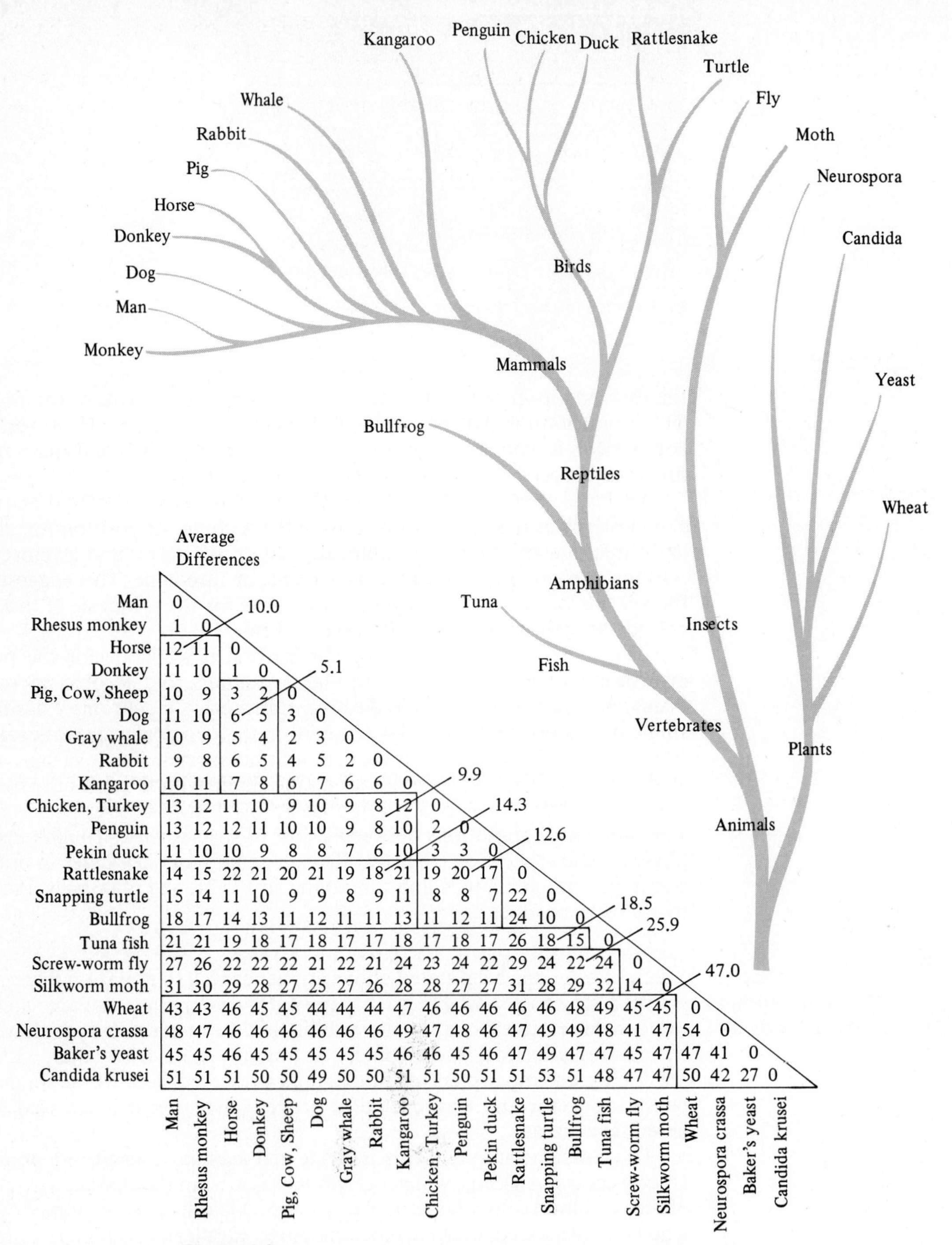

	Man	Rhesus monkey	Horse	Donkey	Pig, Cow, Sheep	Dog	Gray whale	Rabbit	Kangaroo	Chicken, Turkey	Penguin	Pekin duck	Rattlesnake	Snapping turtle	Bullfrog	Tuna fish	Screw-worm fly	Silkworm moth	Wheat	Neurospora crassa	Baker's yeast	Candida krusei
Man	0																					
Rhesus monkey	1	0																				
Horse	12	11	0																			
Donkey	11	10	1	0																		
Pig, Cow, Sheep	10	9	3	2	0																	
Dog	11	10	6	5	3	0																
Gray whale	10	9	5	4	2	3	0															
Rabbit	9	8	6	5	4	5	2	0														
Kangaroo	10	11	7	8	6	7	6	6	0													
Chicken, Turkey	13	12	11	10	9	10	9	8	12	0												
Penguin	13	12	12	11	10	10	9	8	10	2	0											
Pekin duck	11	10	10	9	8	8	7	6	10	3	3	0										
Rattlesnake	14	15	22	21	20	21	19	18	21	19	20	17	0									
Snapping turtle	15	14	11	10	9	9	8	9	11	8	8	7	22	0								
Bullfrog	18	17	14	13	11	12	11	11	13	11	12	11	24	10	0							
Tuna fish	21	21	19	18	17	18	17	17	18	17	18	17	26	18	15	0						
Screw-worm fly	27	26	22	22	22	21	22	21	24	23	24	22	29	24	22	24	0					
Silkworm moth	31	30	29	28	27	25	27	26	28	28	27	27	31	28	29	32	14	0				
Wheat	43	43	46	45	45	44	44	44	47	46	46	46	46	46	48	49	45	45	0			
Neurospora crassa	48	47	46	46	46	46	46	46	49	47	48	46	47	49	49	48	41	47	54	0		
Baker's yeast	45	45	46	45	45	45	45	45	46	46	45	46	47	49	47	47	45	47	47	41	0	
Candida krusei	51	51	51	50	50	49	50	50	51	51	50	51	51	53	51	48	47	47	50	42	27	0

Figure 4-8
Phylogenetic family tree of the cytochromes c. The table gives the average number of amino acids that are different for the indicated species.

displaying symptoms of the disease produce an aberrant form of the oxygen-carrying protein **hemoglobin** called **hemoglobin S**.

Hemoglobin S differs from the normal form, hemoglobin A, at only one point in its primary structure. Hemoglobin consists of two identical α-chains and two equivalent chains of the β-type. Hemoglobin A contains a glutamic acid residue at position 6 of each β-chain; in hemoglobin S this is replaced by a valine. This single substitution suffices to alter the properties of the molecule significantly, endowing it with a tendency to form fibrous aggregates when deoxygenated.

SUMMARY

The amino acid sequence or **primary structure** of the polypeptide chains of each protein is distinctive and characteristic of the protein. The complete determination of the chemical aspects of protein structure requires the establishment of the amino acid sequence of each polypeptide chain, as well as the number and position of the cystine cross-links. The amino acid composition may be determined by complete acid hydrolysis of all peptide bonds and analysis of the resultant amino acid mixture by ion-exchange chromatography.

The usual procedure for the determination of the primary structure involves an initial rupturing of all cystine disulfide bonds by reduction with a sulfhydryl reagent, followed by a reaction of the cysteine groups with iodoacetate. The different polypeptide chains are separated by ion-exchange chromatography and their amino acid sequences are determined independently.

Each polypeptide is cleaved into a set of nonoverlapping peptides by the hydrolysis of particular peptide bonds by a proteolytic enzyme of known specificity, such as **trypsin** or **chymotrypsin**. The short peptides are separated by ion-exchange chromatography and their individual sequences are determined by the stepwise **Edman degradation**. Repetition of the procedure, using a second proteolytic enzyme of different specificity, and consideration of the regions of overlap of the two sets of peptides permit the establishment of the complete sequence of the polypeptide. By enzymic hydrolysis of the protein without a preliminary rupture of disulfide bonds and analysis of the disulfide-linked peptides, the positions of the cystine bridges may be determined.

The primary structure of a particular protein varies with the species. The difference between two species increases with increasing phylogenetic separation of the two. The primary structure of the protein **cytochrome c** has been determined for a wide range of animal species and has permitted estimates of the points in evolutionary time at which different species diverged.

Mutational alterations may occur for a protein within a single species. If the alteration is nonlethal, but results in impaired function, a hereditary disease may arise. A well-known example is the disease **sickle-cell anemia**, which reflects the replacement of a glutamic acid with a valine in one of the polypeptide chains of hemoglobin.

REFERENCES

The following code is used to classify references. I: particularly useful as an introduction to the subject; R: useful primarily as a reference text; A: an advanced account of the material; H: a publication of historical importance.

General

M. O. Dayhoff, ed., *Atlas of Protein Sequence and Structure*, National Biomedical Research Foundation, Silver Spring, Md. (1972). (R)

R. E. Dickerson and I. Geis, *The Structure and Action of Proteins*, Benjamin, Menlo Park (1973). (I)

A. L. Lehninger, *Biochemistry*, Worth, New York (1975). (A)

D. E. Metzler, *Biochemistry*, Academic, New York (1979). (R)

L. Stryer, *Biochemistry*, Freeman, San Francisco (1976). (I)

Sequence Determination

S. B. Needleman, ed., *Protein Sequence Determination*, Springer-Verlag, Berlin (1970). (R)

R. Perham, ed., *Instrumentation in Amino Acid Sequence Analysis*, Academic, New York (1975). (R)

Insulin Sequence

F. Sanger and E. O. P. Thompson, "The Amino Acid Sequence of the Glycyl Chain of Insulin," *Biochem. J.* 53, 353 (1953). (H)

F. Sanger and H. Tuppy, "The Amino Acid Sequence in the Phenylalanyl Chain of Insulin," *Biochem. J.* 49, 463 (1951). (H)

Lysozyme

E. F. Osserman, R. E. Canfield, and S. Beychok, eds., *Lysozyme*, Academic, New York (1974). (R)

Subunit of Tobacco Mosaic Virus

C. Knight, in "Protein Structure and Function," *Brookhaven Symposia in Biology*, No. 13 (1960). (A)

Molecular Evolution

W. M. Fitch and E. Margoliash, "Construction of Phylogenetic Trees," *Science* 155, 279 (1967). (I)

REVIEW QUESTIONS

Questions marked with an asterisk are of a high level of difficulty.

4-1 The peptide shown below was treated with trypsin and "fingerprinted" by a two-dimensional technique.

(a) Indicate the breakpoints and number the spots (1, 2, and so on) beginning with the N-terminal end.

1 2 3 4 5 6 7 8 9 10 11 12 13
(N) Gly-Cys-Val-Arg-Leu-Met-Glu-Ser-Arg-Lys-Ala-Met-Asp(c)

(b) Assume that the peptide was *first* "treated" with the reagents indicated, *then* treated with trypsin and fingerprinted. In each case, indicate which old spots disappear and what the new spots (if any) will contain:

(1) iodoacetic acid ($ICH_2\overset{\overset{O}{\|}}{C}OH$)
(2) The organism which makes the peptide undergoes a mutation in which Lys is replaced by Gly.

*4-2 A collection of protein fragments was purified by chromatography, and one peak, presumably corresponding to a single fragment, was subjected to sequence analysis without further treatment. The first round on the sequenator (a device which performs succesive Edman degradations) yielded both Leu and Ala derivatives, in a 1:1 ratio. The second round did not go well, so other approaches were used. Cleavage of the untreated fragment with trypsin yielded two peptides; amino acid compositions (not sequences) were: Cys_2, Leu, Ala, Lys, Arg and Cys_2, Lys. The untreated fragment did not react with iodoacetamide $ICH_2\overset{\overset{O}{\|}}{C}NH_2$, but after treatment with mercaptoethanol, $HOCH_2CH_2SH$, each molecule of original fragment reacted with four molecules of iodoacetamide. After this reduction-alkylation treatment, the fragment consisted ot *two* peptides, one with composition Cys*, Lys, Ala, the other with Cys_3^*, Leu, Arg, Lys. (Cys* is alkylated Cys.) Trypsin treatment of the reduced-and-alkylated peptides yielded a total of four peptides with compositions [Cys*, Leu, Arg]; [Cys*, Lys]; [Cys*]; [Cys*, Ala, Lys]. Write the sequence. Indicate NH_2 terminal and COOH terminal amino acids, and any covalent bonds between amino acids.

4-3 Calculate isoelectric points for the following molecules.
(a) *Arg*: for which the pK_a's of the ionizable groups are

$-CO_2H$	2.17
$-NH_3^+$	9.04
guanidino-NH_3^+	12.48

(b) *Cys*: for which the pK_a's of the ionizable groups are

$-CO_2H$	1.71
$-NH_3^+$	10.78
-SH	8.33

(c) *Cys-Arg-Cys-Cys*: for which the pK_a's of the ionizable groups are

terminal-CO_2H	2.0
terminal-NH_3^+	10.0
-SH	8.3
guanidino-NH_3^+	12.00

4-4 Suppose that you wish to determine the sequence of a peptide whose composition is (acetyl Ala, Asp, Gly, Leu_2, Val, Pro). Which of the following treatments would be the best choice for the first step in your analysis, and why?

(a) Trypsin digestion (attacks peptide bonds with Arg or Lys on —CO— side)
(b) Chymotrypsin digestion (attacks peptide bonds with hydrophobic amino acids on —CO— side)
(c) Reduction and alkylation
(d) Edman degradation

*4-5 A polypeptide antibiotic was isolated from culture filtrates of a bacterium. Acid hydrolysis of the purified peptide yielded equimolar amounts of alanine, aspartic acid, glutamic acid, glycine, and tyrosine. When the peptide was treated first with FDNB and then acid hydrolyzed, the only DNP derivative that could be detected was that of glycine.

(a) Sketch the titration curve of the pentapeptide that you have drawn. Assume that you are titrating 100 ml of a 0.1-*M* solution of the peptide hydrochloride with 0.1-*M* KOH. Label both axes correctly. Indicate all pertinent positions and points of the curve.
(c) Indicate the net charge on the pentapeptide at each pH value listed below.

pH	Net charge
0.5	
4	
7	
13	

4-6 How many "titrable" protons are contained in the following polypeptide chain? Assume all groups are protonated at the start of the titration.

Gly-His-Asp-Lys-Phe-Ala-Val-Leu

4-7 (a) Draw the structures of as many different pentapeptides containing only lysine, aspartic acid, alanine, and tyrosine as possible.
(b) Name each pentapeptide.
(c) Draw the titration curve for one of the pentapeptides (assume you are titrating 100 ml of a 0.1-*M* solution of the peptide hydrochloride with 0.1-*M* KOH). Indicate all positions of half-neutralization and complete neutralization.
(d) Indicate the predominant chemical form that each ionizable group would have at the following pH values: 0.3, 3, 7, and 13.
(e) What would the *net* charge on the above pentapeptide be at the above pH values?
(f) In which direction would the peptide move at the above pH's in an electric field?
(g) Describe briefly what is meant by the "primary" structure of proteins.

4-8 (a) Draw the structure of the tripeptide Gly-Glu-Lys that would predominate at pH 7.0.

(b) Sketch the titration curve for this tripeptide with NaOH. Label the equivalence points and the pH values equal to the p*K*'s of the titrable groups.

(c) Calculate the isoelectric point of the tripeptide.

5

Size, Shape, and Conformation of Proteins

5-1 SPATIAL ORGANIZATION OF PROTEINS

In this chapter the focus of the discussion will shift to the molecular architecture of natural proteins, the methods for determining the three-dimensional structure, and the principles governing the geometry of folding of polypeptides. The spatial organization of proteins is crucial for their biological activity; it is at this point that the continuum between chemistry and biology becomes clear.

5-2 GENERAL ASPECTS OF PROTEIN CONFORMATION

The determination of the structure of a protein is by no means completed with the establishment of its amino acid sequence. The geometry of folding of the polypeptide chains presents an even more challenging problem, whose solution for a particular protein is even today a major achievement.

For most proteins the information obtainable by the application of physical techniques to protein solutions is sufficient to provide only a somewhat blurred image of the molecule. While highly useful for many purposes, the solution methods permit in general only indirect examination of limited aspects of the structure.

The more powerful, but also far more difficult, techniques applicable to the solid state are required to establish the details of the molecular architecture. X-ray diffraction, as applied to protein crystals, has developed into the most powerful structural probe available to the chemist, but only a few laboratories are equipped to utilize it.

It is possible to make a few generalizations about protein structure. While a few proteins, such as collagen and myosin, have elongated structures and are said to be **fibrous**, the majority are relatively compact in shape and are termed **globular**. The structures of most globular proteins are condensed and rigid with little or no internal space accessible to solvent (Fig. 5-1).

It has become customary to subdivide a discussion of protein fine structure into the one-dimensional spatial arrangement of the polypeptide backbone, or **secondary structure**, and the manner of three-dimensional folding of the polypeptide chain or **tertiary structure**. The distinction is a little artificial, as these structural aspects are interdependent to some degree. **Quaternary structure** refers to the mutual arrangement of the polypeptide chains of proteins consisting of more than one chain.

5-3 ORIGINS OF THE STABILITY OF PROTEIN STRUCTURE

The only covalent bonds occurring in proteins (apart from conjugated proteins) are peptide and disulfide linkages. The three-dimensional molecular organization is stabilized by noncovalent linkages, of which the most important are hydrogen and hydrophobic bonds, as well as ionic interactions (Table 5-1).

Although a hydrogen atom normally forms only a single true covalent bond, it may under certain conditions form a weak additional linkage to another atom. Hydrogen bonds are partially electrostatic in origin and reflect the interaction of the incompletely shielded nucleus of the hydrogen atom, which is a proton of unit positive charge, with the electronic system of another atom. In general, only electronegative atoms form hydrogen bonds. Hydrogen bonds have a definite directionality, which results from the characteristic arrangement of the bonding orbitals of the atoms involved, as well as a specific bond length, which depends upon the nature of the atoms linked. The bonds tend to attain maximum stability when the hydrogen atom lies near the vector joining the two linked atoms. The energies of formation (-1 to -5 kcal) are small in comparison with those of most covalent bonds. As a consequence, single hydrogen bonds can be broken and reformed with relative ease at ordinary temperatures. The hydrogen bonds of importance to biology are largely of the types

$$
\begin{array}{c}
-N-H\cdots O- \\
-O-H\cdots O- \\
-N-H\cdots N-
\end{array}
$$

Each $-\overset{R}{HC}-\overset{\overset{O}{\|}}{C}-\overset{H}{N}-$ repeating unit of the polypeptide backbone contains a potential hydrogen bond donor in the NH group and a potential acceptor in the C=O group. X-ray diffraction studies upon simple peptides have indicated that the C—N covalent bond distance between the CO and

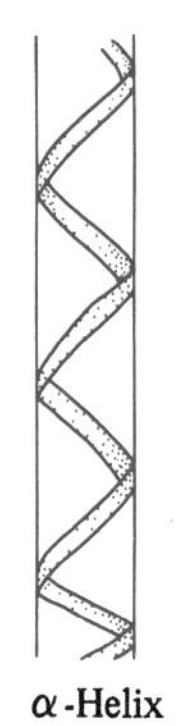

α-Helix

Random coil

Native form

Figure 5-1
Schematic versions of the structures of an extended α-helix, a structureless random coil, and a compact globular protein.

Table 5-1

Energies of Covalent and Noncovalent Bonds

Bond	*Bond Energy* (kcal/mole)
Covalent:	
C—C	83
S—S	50
Noncovalent:	
Electrostatic	20–30
Hydrogen bonds	3–7
Hydrophobic bonds	3–5
Dipole–dipole	1–2

NH groups is substantially shorter than the normal single bond length. This difference is suggestive of the presence of a major degree of double bond character (Fig. 5-2). This probably reflects an important contribution of resonance forms of the type

$$-\overset{\overset{\large O^-}{|}}{C}=\underset{\underset{\large H}{|}}{N^+}-$$

The presence of fractional negative and positive charges on the oxygen and nitrogen atoms, respectively, reinforces any hydrogen bonds involving these atoms by contributing an additional electrostatic component to the forces stabilizing the linkage.

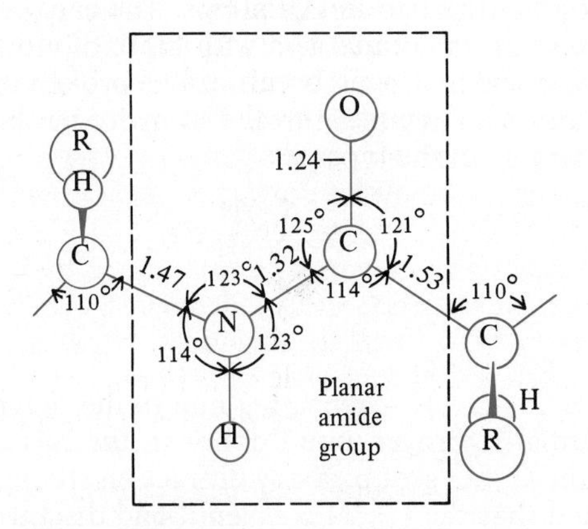

Figure 5-2
Bond angles and interatomic distances of the peptide linkage. The portion within the dashed lines is normally planar. The high energetic barrier to rotation about the C—N bond (about 21 kcal) is sufficient to keep the amide group planar. Distances are given in angstroms (1 Å = 0.1 nm).

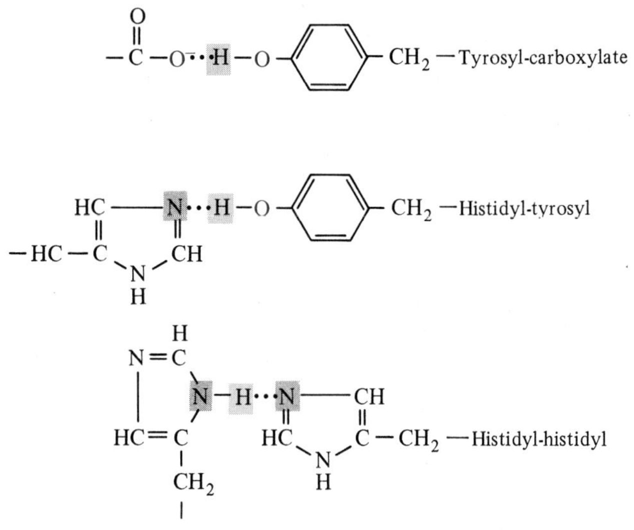

$-C(=O)-O\cdots H-O-C_6H_4-CH_2$ —Tyrosyl-carboxylate

(imidazole) $N\cdots H-O-C_6H_4-CH_2$ —Histidyl-tyrosyl

(imidazole) $N-H\cdots N$ (imidazole) $C-CH_2$ —Histidyl-histidyl

Figure 5-3 Three examples of hydrogen bonds involving side-chain groups.

Potential hydrogen bond forming groups also occur in the side-chains. The possibilities include carboxyl-tyrosine, carboxyl-imidazole, and imidazole-tyrosine bonds (Fig. 5-3). The —OH group of serine and the —SH group of cysteine may also form hydrogen bonds.

In considering the contribution of hydrogen bonds to the stabilization of protein structure, it should be recognized that water, the normal biological solvent, is itself an excellent former of hydrogen bonds both with itself and with other molecules. It can therefore rupture hydrogen bonds in proteins by competitive bond formation with the donor or acceptor groups, unless the latter are shielded from contact with solvent through envelopment in the interior of the protein. The formation by water of hydrogen bonds with polar groups on the protein surface may also help to stabilize protein structure.

Hydrophobic bonds, which are currently believed to contribute most of the structural stabilization energy for the majority of proteins, arise from the mutual cohesion of nonpolar hydrocarbon side-chains (Fig. 5-4). This does not arise from any very powerful attractive forces between such residues, but rather from their relationship to the solvent, water. Water has a high degree of short range order and may be regarded as existing in a loose and imperfect lattice, stabilized by O—H· · ·O hydrogen bonds. While charged groups can be accommodated fairly well by the water lattice, hydrocarbons, which do not form hydrogen bonds, resemble holes or imperfections in the lattice. Such a situation is energetically unfavorable, as is manifested by the low solubility of hydrocarbons in water.

When hydrocarbon groups occur in a protein as the side-chains of amino acids, the stable conformation adopted by the protein tends to be such as to minimize the area of hydrocarbon-water contact. The nonpolar amino acids are largely packed into the interior of the protein molecule, while the charged groups lie on the surface, in contact with solvent.

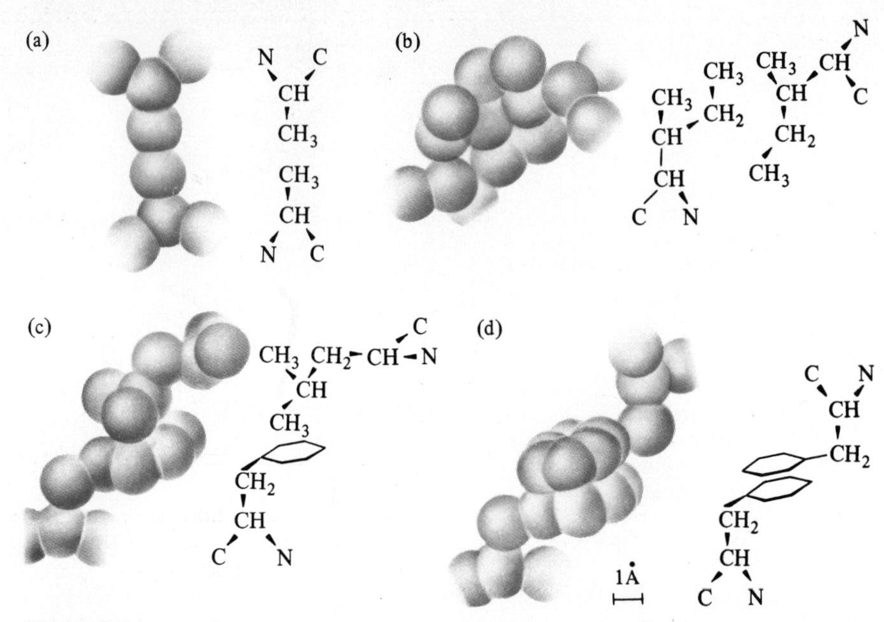

Figure 5-4
Some possible types of hydrophobic bond. (a) and (b) show bonds between two aliphatic side-chains; (c), between an aliphatic and an aromatic side-chain; and (d), between two aromatic side-chains.

The prevailing opinion today is that hydrophobic forces are primarily responsible for stabilizing the organized structure of proteins and that hydrogen bonding is of secondary importance in most cases. The extensive hydrogen bonding occurring in many proteins probably depends largely upon the integrity of the overall structure for its stability.

The importance of hydrophobic forces is not confined to stabilization of the internal structure of proteins. Hydrophobic interactions are often a dominant factor in the mutual *association* of protein molecules to form complexes in solution. In general, proteins whose *surfaces* contain a high proportion of nonpolar amino acids have a strong tendency to **self-associate** in solution.

In addition to hydrogen and hydrophobic bonding, other kinds of forces may contribute to the stabilization of protein structure in particular instances. These forces include electrostatic interaction between oppositely charged sites, dipole–dipole interactions, and ion–dipole interactions (Fig. 5-5 and Table 5-1).

The biological significance of the prominent role of noncovalent interactions lies in their selectivity, easy reversibility, and adaptability. These factors permit a rapid and graded response to changed conditions.

Other forces may contribute to the stabilization of protein structure in particular instances. These include electrostatic interaction between oppositely charged sites (Fig. 5-5), dipole–dipole interactions, and ion–dipole interactions.

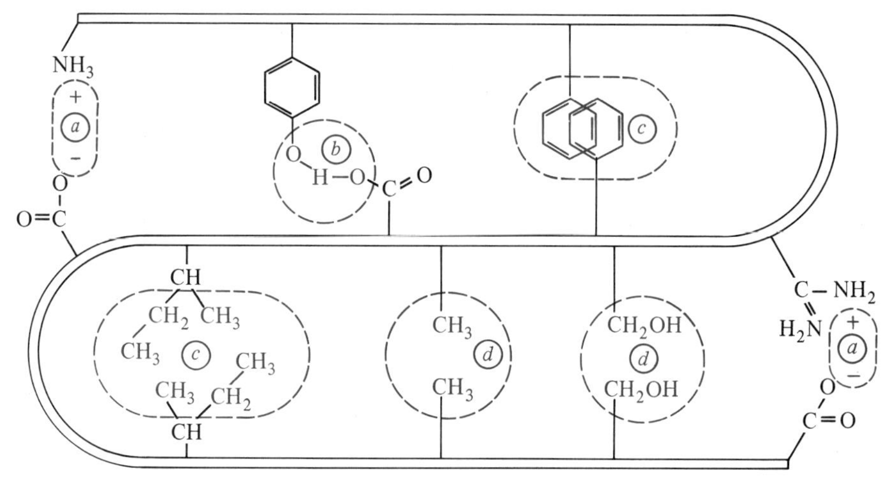

Figure 5-5
Some types of side-chain interaction that may contribute to the stabilization of protein structure: (a) electrostatic bonds (salt linkages); (b) hydrogen bonds; (c) hydrophobic bonds; and (d) dipole–dipole interactions.

5-4 HYDRODYNAMIC METHODS FOR ESTIMATING SIZE AND SHAPE

Diffusion

If the individual solute molecules of a protein solution could be observed directly, they would be seen to be in violent and erratic motion. This thermal or **Brownian** motion reflects the continual bombardment of the solute molecules by molecules of solvent.

Since the movement of each particle is, in the absence of an external force field, entirely random, an imaginary plane in a solution of uniform concentration will, in a finite time interval, be crossed by equal numbers of molecules moving in either direction. However, if the concentration of solute varies in a direction normal to the surface, the transport of solute by Brownian motion is *biased* so that more molecules cross the surface in the direction of *decreasing* concentration. The net transport of solute by thermal motion or **diffusion** always tends to erase a nonuniformity of concentration and to restore an invariant concentration. The rate of diffusion of a solute depends upon the temperature, the properties of the solvent, and, in particular, upon the size and shape of the solute molecules. In practice, the rate may be measured by observations of the blurring with time of an initially sharp boundary between solution and solvent (Fig. 5-6).

Two chambers, one containing solution and the other solvent, are initially separated by some type of movable partition. Careful removal of the partition forms a sharp boundary between solution and solvent (Fig. 5-6). As diffusion proceeds, the variation of solute concentration with distance in the direction of diffusion becomes progressively more gradual. The **gradient** of concentration, dc/dx, where x is the distance from the initial boundary position, has at all times a maximal value at $x = 0$ and decreases to zero at distances far

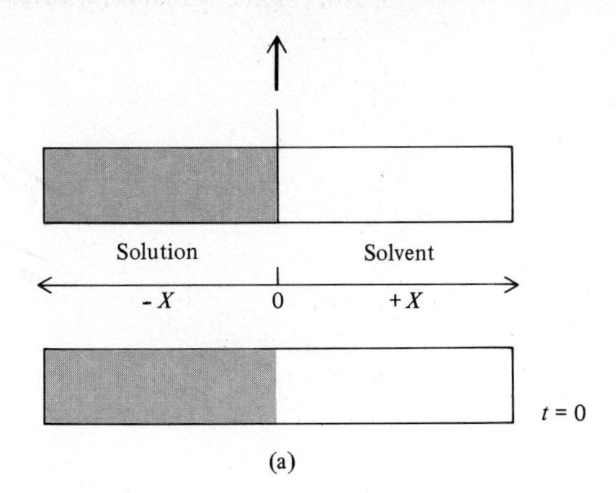

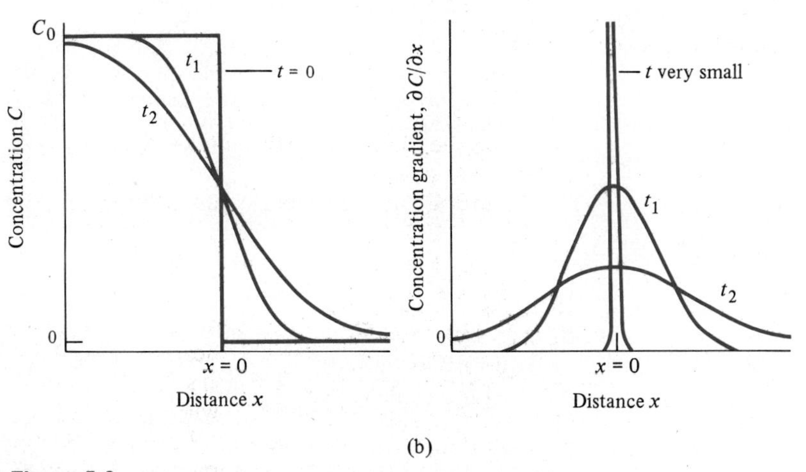

Figure 5-6
(a) Formation of a boundary between solution and solvent by withdrawal of a movable partition. (In practice the arrangement would be vertical, with the denser solution in the lower compartment.) (b) Variation with time of concentration and concentration gradient in a diffusion experiment.

from the boundary, where c is constant, being equal to the initial concentration, c_0, or to zero (Fig. 5-6).

The diffusion process can be mathematically described relatively simply, provided that diffusion occurs in only one direction and that the concentrations at the ends are unperturbed during the course of the experiment. The net mass of solute (Δw) which diffuses in the time interval Δt across a plane of area A, which is perpendicular to the direction of diffusion, is given by **Fick's first law**:

$$\Delta w = -DA\frac{dc}{dx}\Delta t \tag{5-1}$$

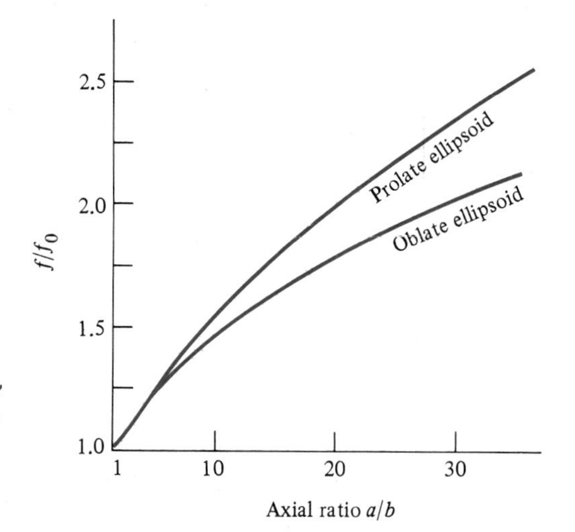

Figure 5-7
Dependence of frictional ratio upon axial ratio for a prolate (elongated) ellipsoid.

Here D is the **diffusion coefficient**, which characterizes the rate of diffusion for a particular system.

The diffusion coefficient is related to the properties of the molecule by

$$D = \frac{kT}{f} \tag{5-2}$$

Here k is the Boltzmann constant and f is the **frictional coefficient**. The value of the frictional coefficient is a measure of the frictional drag encountered by the particle in its motion through solvent. In general, the frictional coefficient of a protein molecule increases with increasing molecular size, asymmetry of shape, or permeability to solvent. For a given molecular weight the minimum value of f corresponds to a compact unsolvated sphere which is impermeable to solvent. We have in this case

$$f_o = 6\pi\eta r \tag{5-3}$$

where r is the radius of the sphere and η is the solvent viscosity. The value of the **frictional ratio**, f/f_o, is a measure of the departure of the actual protein molecule from this limiting form. If an ellipsoidal shape is assumed for the protein molecule, it is possible to compute values of f/f_o as a function of the **axial ratio**, or the ratio of the long to the short axis of the ellipsoid (Fig. 5-7). The diffusion coefficient *decreases* with *increasing* molecular weight, axial ratio, or permeability to solvent.

Diffusion is a complicating factor in all experiments involving the transport of proteins through a medium. Its effect is always to blur boundaries and reduce resolution.

Sedimentation Velocity

All terrestrial objects are subject to the force field of the earth's gravitation, which is sufficient to cause the settling out or sedimentation of coarse suspensions, such as chalk dust in water. The sedimentation of protein molecules requires much more intense fields that are most commonly obtained by centrifugal means.

A centrifuge is a device for achieving high speeds of rotation and hence high centrifugal fields. If it is capable of very high centrifugal fields, up to 300,000 times gravity, it can cause the sedimentation of proteins and is termed an **ultracentrifuge**.

The standard ultracentrifuge contains the following principal elements: (1) a duralumin rotor, capable of withstanding high centrifugal fields; (2) an electrical or gas turbine drive unit, which causes rotation of the rotor; (3) a collimated light source for illuminating the ultracentrifuge cell; and (4) a Schlieren or interference optical system for observing sedimentation.

The ultracentrifuge cell, which consists of a hollow centerpiece placed between two quartz windows and inserted in a cylindrical case, is positioned within a cylindrical well in the rotor (Fig. 5-8). The rotor is irradiated from below by a collimated light beam that intercepts the cell once for each revolution (Fig. 5-8). The speed of the rotor is sufficiently high so that there is no perceptible flicker and the illumination of the cell appears to be constant.

(a)

(b)

Figure 5-8
(a) An ultracentrifuge cell, assembled (left) and disassembled (center), and an ultracentrifuge rotor (right). The cell fits into one of the cylindrical wells in the rotor. The centerpiece, which contains a sector-shaped cavity, is the third object from the top in the disassembled cell. (b) A portion of an analytical ultracentrifuge, showing the suspended rotor. The light beam intercepts the rotor from below and then enters the optical system. (Courtesy of National Naval Medical Center, Bethesda, Maryland.)

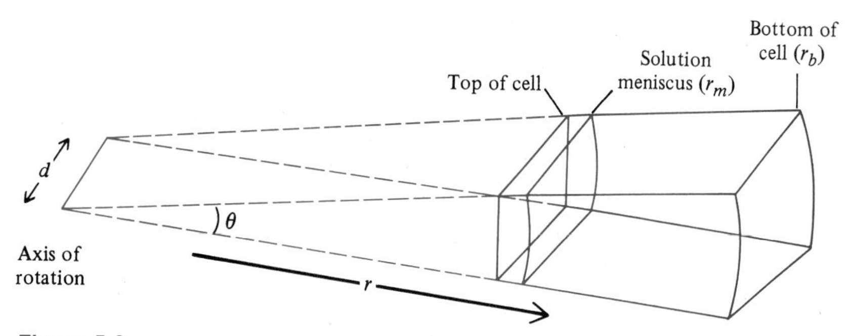

Figure 5-9
Direction of the centrifugal force within the ultracentrifuge cell. The volume occupied by the solution is shown. This corresponds to the sector-shaped cavity within the **centerpiece** of the cell (Fig. 5-8). The cavity has a width equal to d and the sector angle is θ.

The solution to be observed is placed in a sector-shaped cavity within the cell centerpiece. The cell is positioned so that the wide end of the sector points outward from the axis of rotation. Sedimentation occurs radially outward (Fig. 5-9).

After passing through the cell the light beam enters an optical system which permits direct observation of the sedimentation process. If a Schlieren optical system is used, a cell image is obtained which shows the refractive index gradient as a function of radial distance from the axis of rotation (Fig. 5-10). This may be photographed at a series of time intervals.

In **sedimentation velocity** the rotor is driven at a speed sufficiently high so that sedimentation proceeds essentially to completion, the solute being ultimately packed into a thin layer at the outer periphery of the cell. As sedimentation progresses a boundary between protein solution and pure solvent emerges from the meniscus and moves with time toward the outer edge of the cell. The boundary appears as a peak in the curve of refractive index gradient versus distance (Figs. 5-11 and 5-12). If two solutes of different sedimentation rate are present, two peaks will be observed (Fig. 5-13).

The centrifugal force per unit mass is equal to

$$\omega^2 r(1 - \bar{V}\rho_o)$$

where ω is the angular velocity of rotation (equals 2π times the revolutions per second) and r is the radial distance from the axis of rotation. $\bar{V}$ is the **partial specific volume** (ml/g), which is equal to the volume increment per gram of protein; ρ_o is the solvent density. The term $(1 - \bar{V}\rho_o)$ is the buoyancy factor, which follows from Archimedes' law.

The centrifugal force per mole is $\omega^2 r(1 - \bar{V}\rho_o)M$, where M is molecular weight. If the velocity of sedimentation is constant, this must be balanced by an equal and opposite force arising from the frictional resistance of the

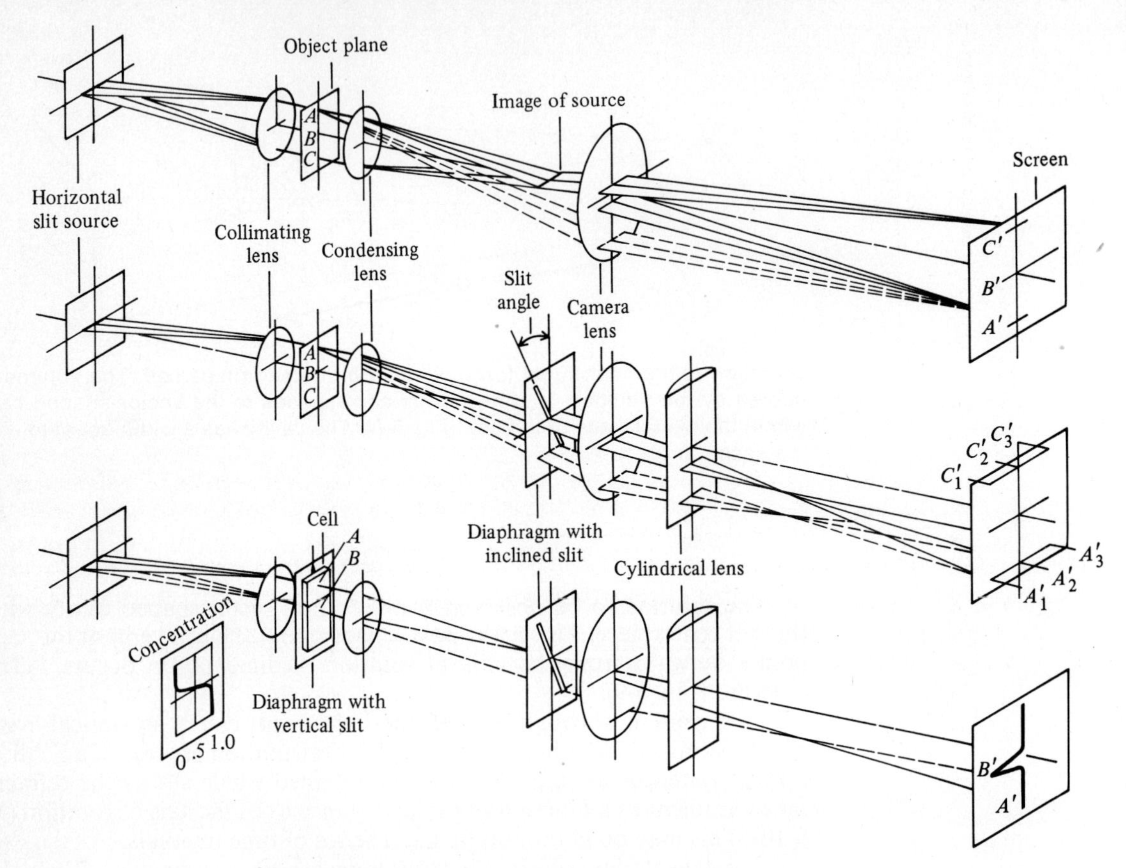

Figure 5-10

Diagram of a schlieren optical system. A horizontal slit serves as a source. A collimating lens produces a parallel bundle of rays that pass through the cell. In the absence of a refractive index gradient these are brought to a common focus by a condensing lens to form an image of the slit. The presence of a refractive index gradient in the vertical direction results in a displacement of the rays, which no longer form a common focus. The interception of the rays by an inclined slit yields, with the aid of further optical elements, an image of the refractive index gradient.

medium. This is equal to $N_o f(dr/dt)$, where dr/dt is the velocity of movement of the boundary and N_o is Avogadro's number.

Equating the two balanced forces,

$$\omega^2 r(1 - \bar{V}\rho_o)M = N_o f \frac{dr}{dt} \tag{5-4}$$

or

$$\frac{1}{\omega^2 r}\frac{dr}{dt} = \frac{(1 - \bar{V}\rho_o)M}{N_o f}$$

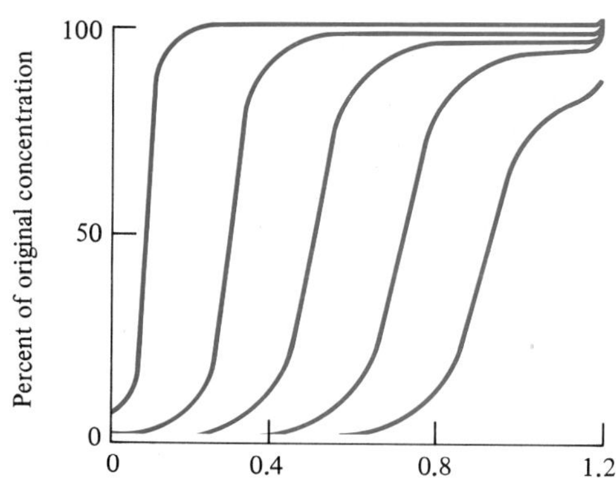

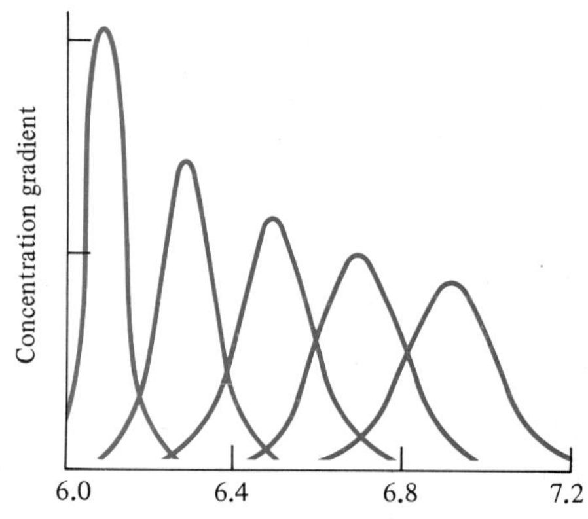

Figure 5-11

Variation of concentration and of concentration gradient with time in a sedimentation velocity measurement for a single solute. From left to right corresponds to increasing time.

The quantity $(1/\omega^2 r)(dr/dt)$ is designated as the **sedimentation** coefficient, S. It should be independent of rotor speed and instrumental characteristics. Introducing this parameter into Equation (5-4),

$$M = \frac{SN_o f}{(1 - \bar{V}\rho_o)} \tag{5-5}$$

and from Equation (5-2).

$$M = \frac{S}{(1 - \bar{V}\rho_o)} \frac{RT}{D} \tag{5-6}$$

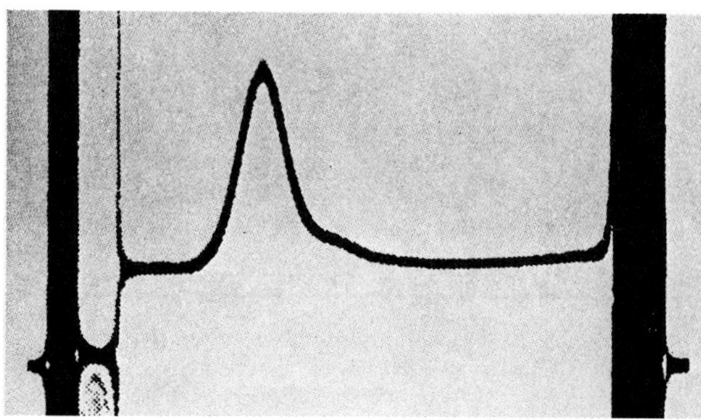

Figure 5-12

Schlieren diagram for the sedimentation of a protein, depicting the variation of refractive index gradient (or concentration gradient) with distance. Sedimentation proceeds from left to right. The solution contains a major component corresponding to the principal peak, plus a minor component corresponding to the small "shoulder" to the right of the peak. The thin vertical line at the left is the meniscus.

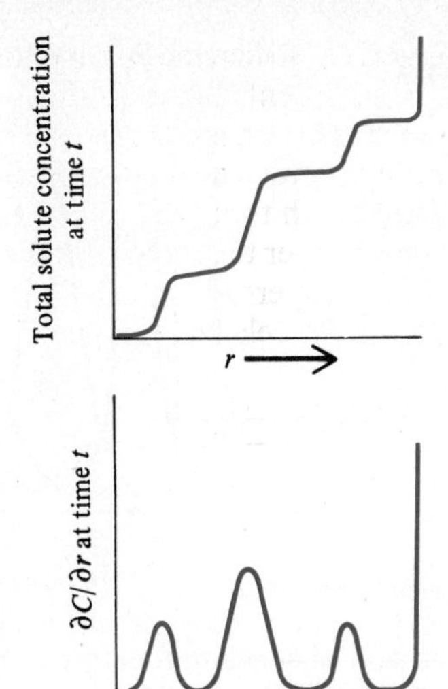

Figure 5-13 Concentration and concentration gradient as a function of distance for the sedimentation of a system containing three components.

where R = gas const = $N_o k$.

The molecular weight of a homogeneous protein system may in this way be computed from combined measurements of the sedimentation and diffusion coefficients.

The sedimentation coefficient is itself a useful parameter for protein characterization. The value of S measured at a given temperature, T_1, may be converted to a second temperature, T_2, by the equation

$$S_2 = S_1 \frac{\eta_1}{\eta_2} \frac{(1 - \bar{V}\rho_o)_2}{(1 - \bar{V}\rho_o)_1} \tag{5-7}$$

where the subscripts refer to the two temperatures.

If the molecular weight is known, the frictional coefficient and the frictional ratio may be computed from Equation (5-5) (Table 5-2).

Table 5-2
Frictional Ratios of Some Proteins

Protein	Molecular Weight	f/f_o
Human plasma albumin	68,500	1.29
Bovine cytochrome c	13,400	1.19
Human hemoglobin A	66,000	1.11
Horse myoglobin	17,600	1.10
Insulin	12,650	1.07
Enolase	63,300	1.00
Human fibrinogen	340,000	2.34
Cod myosin	500,000	3.63

Sedimentation Equilibrium

Unlike sedimentation velocity, **sedimentation equilibrium** is not a dynamic method. In this variant of ultracentrifugation the rotor is driven at a speed sufficiently low so that sedimentation is effectively countered by diffusion and does not proceed to completion. Instead an equilibrium distribution of solute within the cell is approached with time (Fig. 5-14). This distribution depends upon molecular weight. The higher the molecular weight, the more the solute is concentrated in the bottom layers of the cell.

At equilibrium the following relation holds at all points within the cell:

$$M = \frac{RT}{(1 - \bar{V}\rho_o)\omega^2 r} \frac{1}{c} \frac{dc}{dr} \tag{5-8}$$

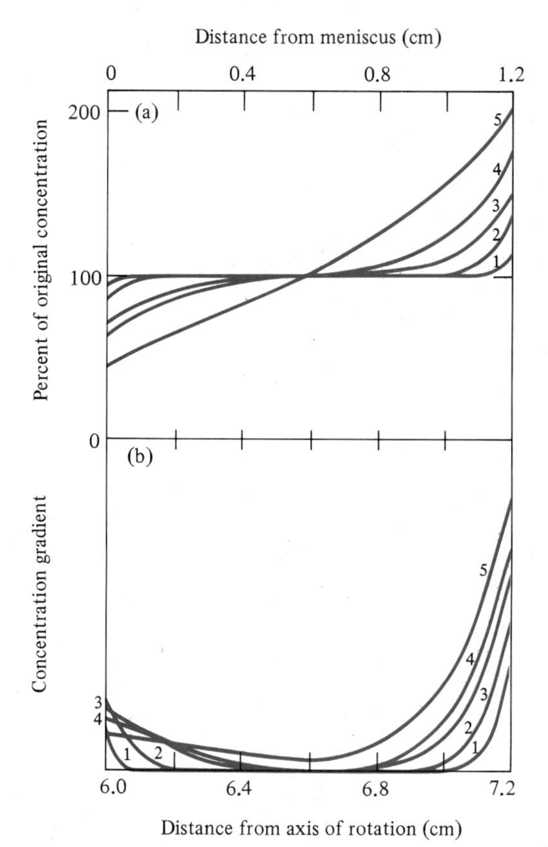

Figure 5-14
The approach with time to an equilibrium distribution of solute in a sedimentation equilibrium measurement. Curves 1–5 correspond to increasing times. In (a) the *concentration* relative to the original (uniform) concentration is plotted. For example, in curve 5 the concentration increases from 50% of the original at the meniscus, to a value equal to (100% of) the original at 0.5 cm from the meniscus, and to twice (200% of) the original at 1.2 cm from the meniscus. In (b) the concentration gradients corresponding to curves 1–5 are plotted.

Equation (5-8) may be integrated to yield

$$\ln c = \frac{(1 - \bar{V}\rho_o)\omega^2 r^2}{2RT} M + \text{const} \qquad \textbf{(5-9)}$$

The molecular weight may be obtained from the slope of a linear plot of ln c versus r^2.

Sedimentation equilibrium is probably the most precise of the various physical techniques available for determining the molecular weights of proteins. It has however the disadvantage of being relatively laborious and time consuming, as well as requiring elaborate equipment.

Sucrose Gradient Centrifugation

This variant of ultracentrifugal analysis has proved to be of great utility for the characterization of mixtures, particularly when the component of interest is present in low concentration. **Sucrose gradient centrifugation** dispenses with optical measurements during the course of centrifugation. The analytical cell described earlier is replaced by an ordinary plastic centrifuge tube, which is shaped like a test tube.

Basically, the technique depends upon the stabilization of concentration gradients by a density gradient formed by sucrose. The density gradient allows removal of the centrifuge tube at the conclusion of centrifugation and the direct sampling of successive layers of solution without disruption of the concentration distribution of the solute species.

A rotor of the "swinging bucket" type is used, with hinged tube holders which swing out at right angles to the axis of rotation when the rotor is in motion. The centrifuge tube is thereby oriented in the direction of the centrifugal field, so that sedimentation proceeds radially outward, from top to bottom of the tube.

A linear sucrose gradient is formed within the tube by filling it carefully from a device in which an initially concentrated sucrose solution draining into the centrifuge tube is continuously replenished by a more dilute solution, so that the concentration of sucrose decreases progressively from the bottom of the tube, which is filled first, to the top, which is filled last. A small volume of protein solution is then layered on top of the sucrose gradient. The tube is then placed in the "swinging bucket" rotor and spun at 20,000–50,000 rpm for several hours.

The rotor is then stopped, the tube removed, and a puncture carefully made in its bottom with a needle. The solution is then allowed to drain dropwise into a series of test tubes, so that a set of fractions is collected. The earliest fractions come from the bottom of the tube and the subsequent fractions from progressively higher layers. The fractions are then analyzed for protein content from the absorbance at 280 nm, or for enzymic activity (Figs. 5-15 and 5-16).

If the partial specific volume of each component is less than about 0.8 ml/g, the distance sedimented, and hence the numerical order of the fraction in which it attains its highest concentration, is linear with respect to time. The ratio of the distances traveled for two different proteins is equal to the ratio of their sedimentation coefficients. By including one or more *marker* proteins of known sedimentation coefficient, the value for an un-

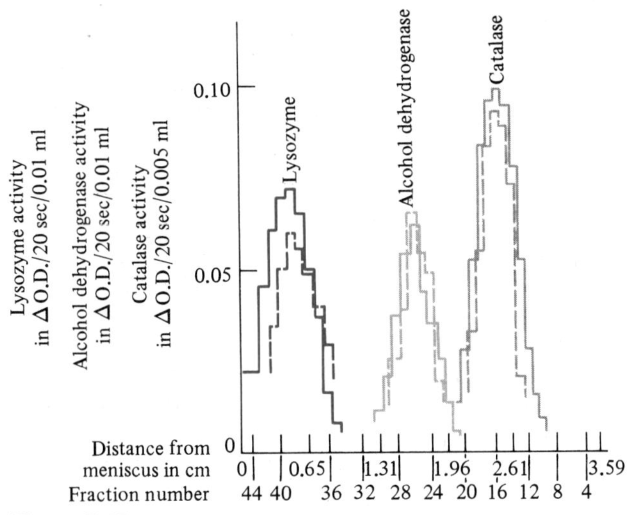

Figure 5-15
Results of sucrose gradient sedimentation for three enzymes as monitored by enzymic activity. The ordinate denotes the activity in arbitrary assay units. The molecular weight of the three proteins increases from lysozome to catalase.

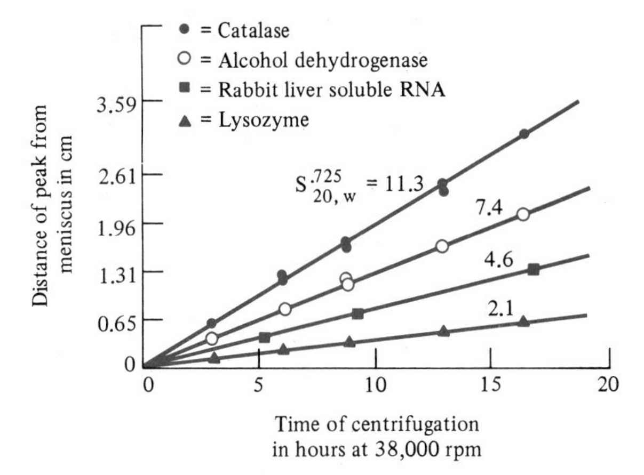

Figure 5-16
Demonstration of linearity of sedimentation in a sucrose gradient for four proteins.

known material may be computed by interpolation. From a knowledge of the volume of each fraction and the dimensions of the tube, the average distance traveled may be computed from the distribution among the fractions.

In contrast to standard sedimentation velocity in which the distributions of different sedimenting species overlap (Fig. 5-13), two components of differing sedimentation rate can be resolved completely by sucrose gradient centrifugation into separate sedimenting zones. Sucrose gradient centrifugation is accordingly termed a **zonal** technique. Since enzymic activity, or some other biological activity, may be used to monitor concentration, the method is applicable to impure enzymes in which the enzyme of interest is contaminated with inactive material.

5-5 ALTERNATIVE TECHNIQUES FOR MOLECULAR WEIGHT DETERMINATION

Gel Filtration

The technique of gel filtration has already been discussed as a method for the separation of different species by the criterion of molecular size (section 3-7). The rate of transport of a protein through a column containing Sephadex or an equivalent material will in general be dependent upon both its size and its shape. However, apart from a few fibrous proteins of exceptionally high axial ratio, the effects of shape are relatively minor and the rate of movement, for practical purposes, may be regarded as a function of the molecular weight alone.

The elution of a protein from a gel filtration column is usually monitored by its absorption of ultraviolet light at 270–280 nm. A plot of the optical density of the eluent solution versus the volume of eluting solvent has the form of a roughly symmetrical peak whose midpoint corresponds to the **elution volume** of the protein (Fig. 5-17). If several proteins of sufficiently different molecular weight are present, they will be eluted as separate peaks with different elution volumes. For a particular column under specified conditions, the elution volume will be characteristic of the molecular weight of the protein.

By calibration with a series of proteins of known molecular weight, a curve of molecular weight as a function of elution volume may be constructed (Fig. 5-18). If the elution volume of an unknown protein is determined, its molecular weight may be found from the standard curve by interpolation.

If the unknown protein has a biological activity which can be assayed, then its elution volume may be determined by monitoring this activity in-

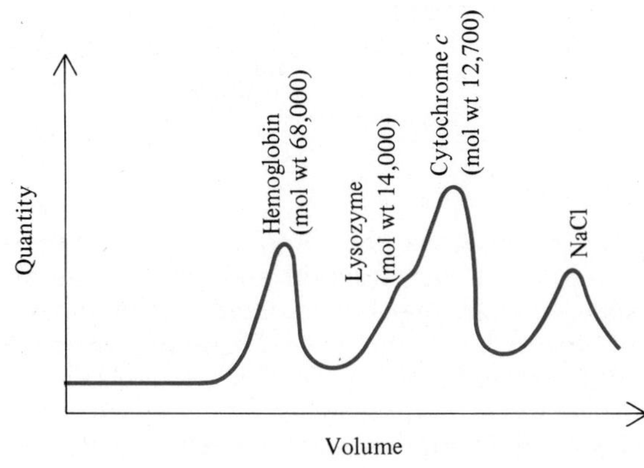

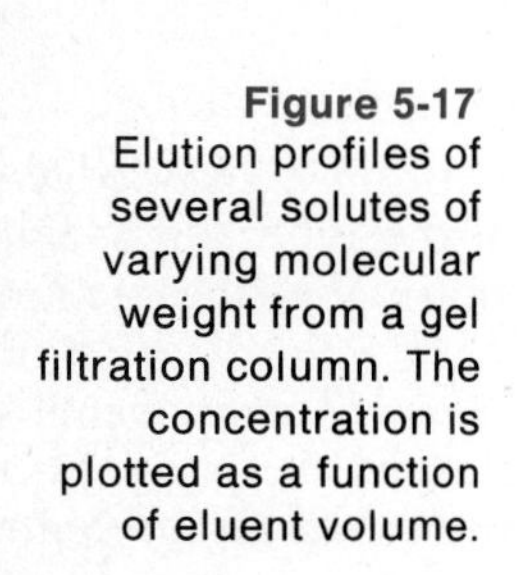
Figure 5-17 Elution profiles of several solutes of varying molecular weight from a gel filtration column. The concentration is plotted as a function of eluent volume.

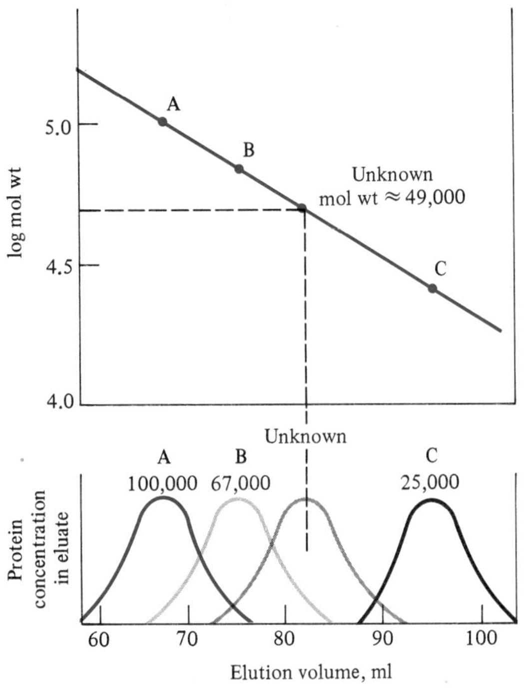

Figure 5-18
Determination of the molecular weight of an unknown protein by gel filtration. The column is calibrated by determining the elution profiles of a series of proteins of known molecular weight (A, B, and C). The elution volume is taken as the volume corresponding to the midpoint of the elution profile. If the elution volume of the unknown protein is measured, its molecular weight may be located by interpolation.

stead of the optical density. In this way it is possible to determine the elution volume, and hence the molecular weight, of a protein which is present as a component of an impure mixture. This is responsible for much of the usefulness of the method, as it permits characterization of a protein prior to rigorous purification.

While gel filtration does not equal sedimentation equilibrium in precision, it is usually possible to obtain molecular weights to within 5%. This is adequate for many purposes.

SDS Gel Electrophoresis

While gel electrophoresis (section 3-7) normally separates proteins by the criteria of both size and electric charge, it is possible to choose conditions such that the rate of movement of a protein is a function of molecular weight alone. In the presence of sufficient levels of the anionic detergent sodium dodecyl sulfate (SDS), the charged groups present on the protein are swamped by the extensive binding of detergent anions, so that differences in charge are effectively suppressed. If the disulfide bridges present in the native protein are ruptured by prior treatment with a reducing agent (section 3-2), the detergent moreover appears to convert proteins to similar molecular shapes. Under these conditions each protein migrates as a distinct band at a velocity characteristic of its molecular weight. By calibration with a set of known proteins, the molecular weight of an unknown protein may be determined from its distance of migration in a given time interval.

Like gel filtration, **SDS gel electrophoresis** does not provide the highest precision. However, because of its speed, ease, and requirement for only minimal quantities of protein, it has become standard procedure for the routine characterization of proteins.

5-6 SECONDARY STRUCTURE OF PROTEINS

α-Helix

The first major success in the protracted struggles of chemists and crystallographers to gain some insight into the spatial geometry of proteins came only in 1951, after decades of unrewarding effort and speculation, with the proposal of the α-helical conformation by Pauling, Corey, and their coworkers. Remarkably enough, this achievement almost coincided with the determination of the primary structure of insulin by Sanger.

The **α-helix** was originally proposed by Pauling and Corey on largely theoretical grounds. In searching for the intrinsically most stable configuration of a polypeptide chain they were guided by the following tenets:

(1) The maximum possible number of hydrogen bonds between CO and NH groups is formed.
(2) The peptide (—CONH—) group is planar, as is the case in oligopeptides.
(3) The N—H bond for each N—H· · ·O bond does not deviate by more than 30° from the vector joining the N and O atoms.
(4) The orientations about C—N and C—C single bonds are near the potential energy minima for rotation about these bonds.
(5) The spatial progression from one residue to the next is constant.

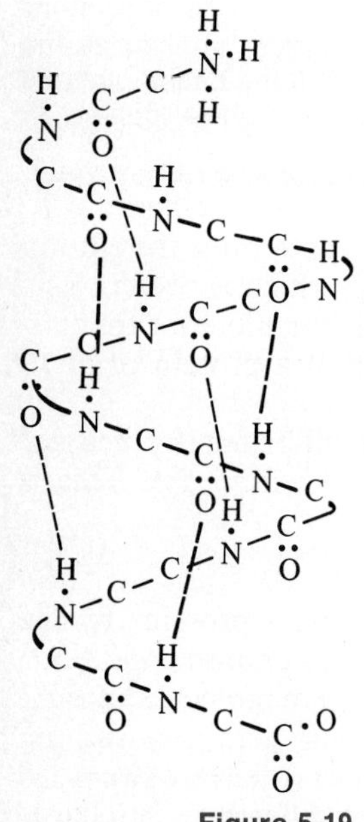

Figure 5-19 Hydrogen bonding occuring in the α-helix.

Only one helical structure involving a single polypeptide chain was found to satisfy these requirements. The α-helix consists of a single strand twisted about a helical axis (Fig. 5-19). The structure is stabilized by hydrogen bonds between C=O and NH groups which are roughly parallel to the helical axis. The helix is *nonintegral*, with 3.6 amino acid residues per turn. The **pitch** or spacing between successive turns is 0.54 nm. The distance between consecutive atoms in the direction of the helical axis is 0.15 nm (Fig. 5-20).

The α-helix is a highly rigid and asymmetric structure. A polypeptide existing as a perfect α-helix would have a very extended and rodlike configuration. The dimensions of most proteins are much too compact to be consistent with this model. However, it is possible to reconcile a compact overall shape with a high helical content, provided that the latter is distributed among a set of *short* helical segments, separated by nonhelical zones that permit their folding into a compact shape. This appears indeed to be the case for the handful of globular proteins whose structures are known in detail.

X-ray diffraction and other data early provided evidence for the forma-

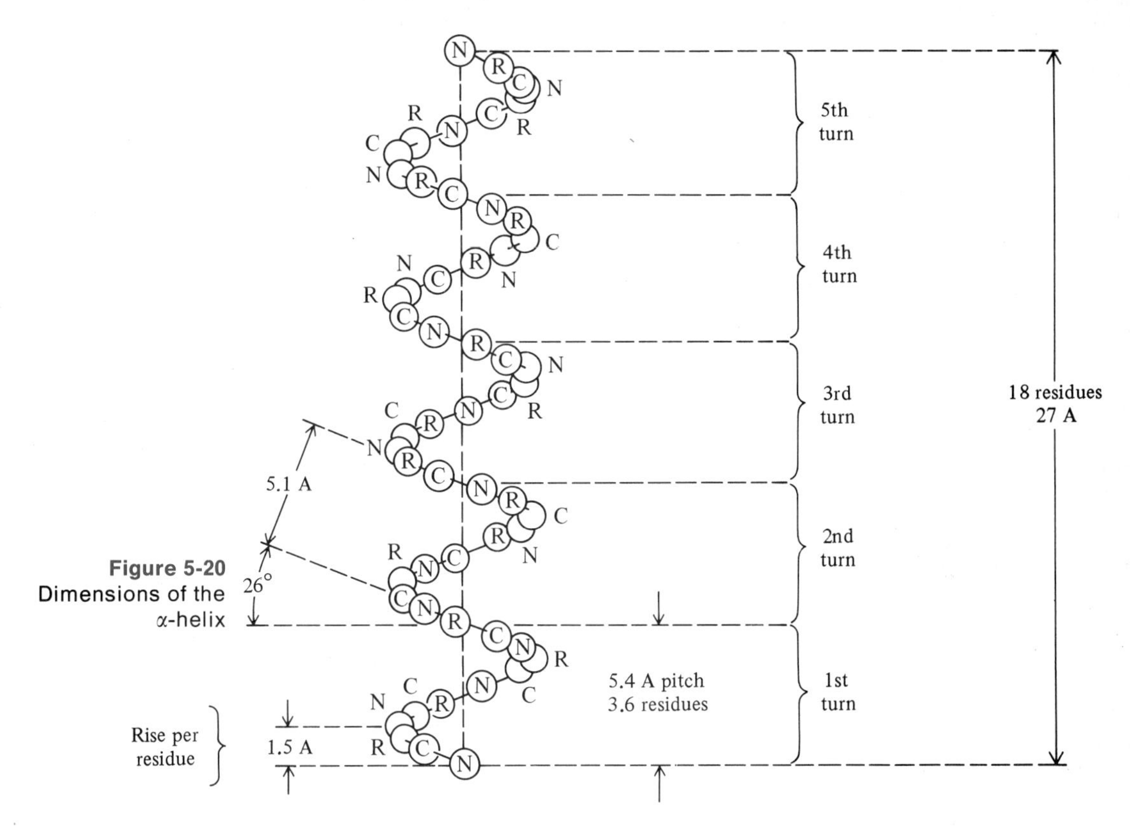

Figure 5-20 Dimensions of the α-helix

tion of the α-helix in synthetic polypeptides containing only one kind of amino acid. Its presence as an important structural element in proteins is now well established, as a result of more recent x-ray diffraction studies upon protein crystals.

Other Polypeptide Conformations

The ***β*-pleated sheet** structure (or *β*-structure) (Figs. 5-21 and 5-22), which may occur in proteins of the silk fibroin-*β*-keratin group, consists of a set of side-by-side polypeptide chains which are stretched to their maximum extension. The chains may run either all in the same direction (parallel pleated sheet) or in alternate directions (antiparallel pleated sheet). The CO and NH groups of neighboring chains are hydrogen bonded.

The structure of collagen, which differs from the models described above, will be discussed in chapter 24.

5-7 X-RAY CRYSTALLOGRAPHY

X rays are a form of electromagnetic radiation resulting from the bombardment of a metallic target by a stream of electrons. While x rays vary in wavelengths from 0.001 to 100 nm, most crystallographic work has been done with wavelengths of 0.15 nm.

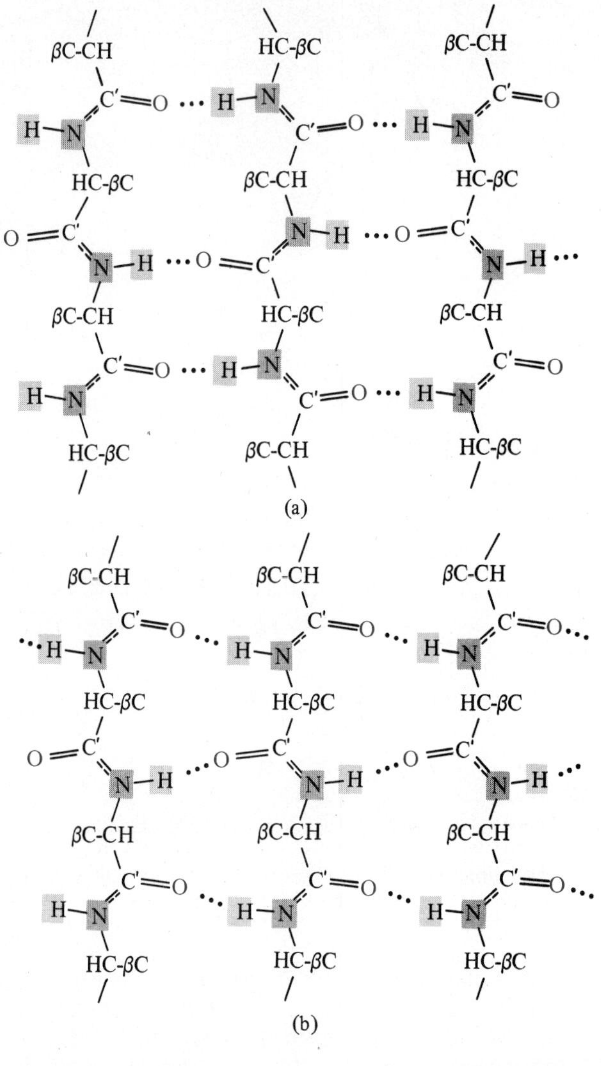

Figure 5-21
(a) Antiparallel and (b) parallel β-pleated sheet structures. The chains are linked by interchain N—H · · · O hydrogen bonds.

X-ray diffraction is only useful for investigating solid-state systems with a high degree of regularity in their spatial geometry. Macroscopic systems whose three-dimensional organization is completely ordered are termed **crystals**. Crystals are characterized by a definite periodicity of structure and may be regarded as generated by the indefinite repetition of a basic three-dimensional unit, called the unit cell (Fig. 5-23). This contains a small integral number of molecules.

Every atom in the molecule forms part of a space lattice consisting of the collection of the equivalent atoms in all molecules of the crystal. If the molecule contains *s* atoms, then the crystal is made of *s* overlapping space lattices, of identical geometry, each of which contains only one kind of atom.

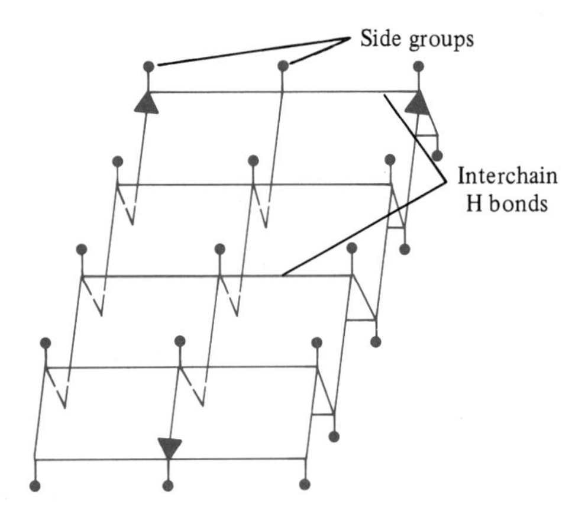

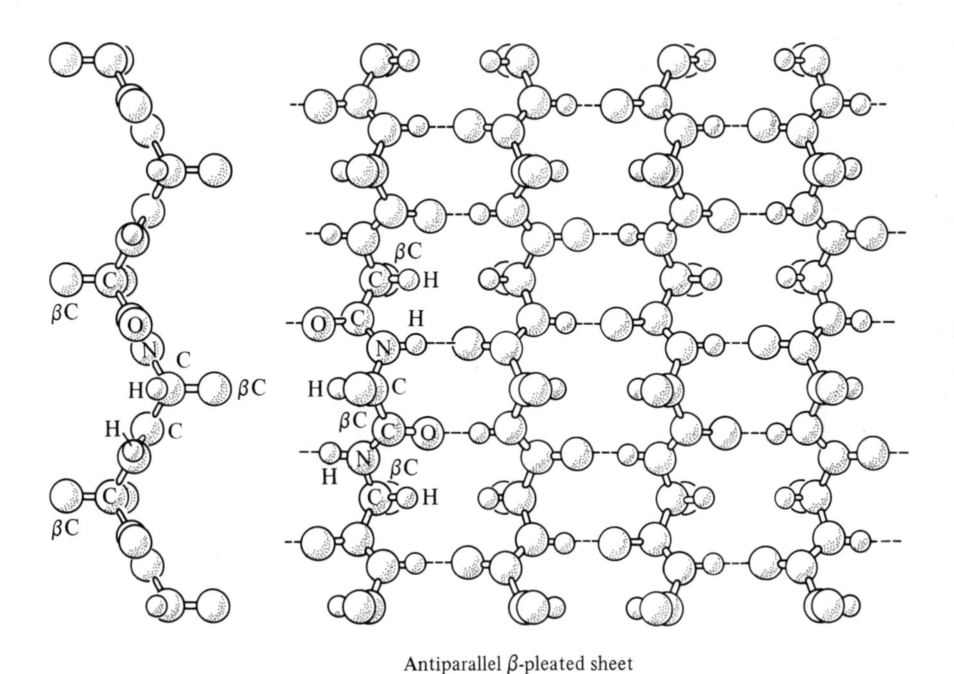

Figure 5-22
Another representation of the β-pleated sheet structure. The side-chains lie alternately above and below the pleated sheet.

If a protein crystal is placed in the path of a thin beam of monochromatic x rays and rotated about an axis perpendicular to the direction of the beam, then a photographic plate placed behind the crystal will register (in addition to the transmitted beam) a set of beams arising from scattering, or

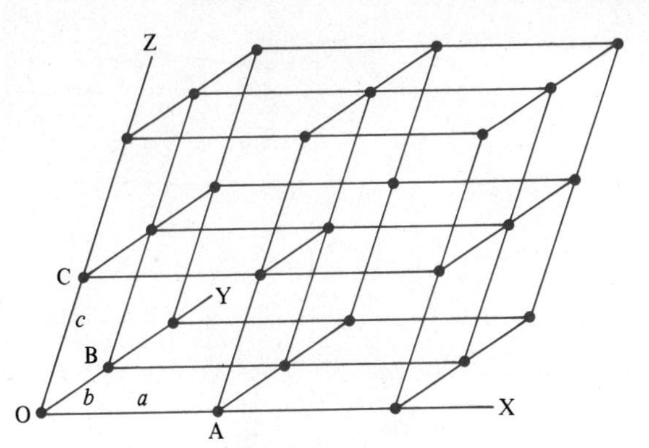

Figure 5-23
A simple crystal lattice, showing the basic repeating unit or unit cell.

diffraction, by the atoms of the crystal (Fig. 5-24). The pattern consists of a set of discrete spots of varying spacing and intensity.

Inspection of a representative x-ray diffraction photograph of a protein crystal (Fig. 5-25) reveals a number of regularities. The spots are not distributed at random, but fall on a series of horizontal lines. The intensities of the spots on each line vary in a regular manner.

Any crystal, because of its generation by repetition of a basic unit, will contain a large number of planar layers of like atoms. The characteristic diffraction pattern of the crystal arises from the mutual interference or reinforcement of beams reflected by different layers (Fig. 5-26). The occurrence of reinforcement requires that the beams reflected from a set of parallel planes be in phase. This will be the case if the difference in path length to the point of observation of rays reflected from successive layers is an integral multiple of the wavelength λ. This holds true only for particular angles θ

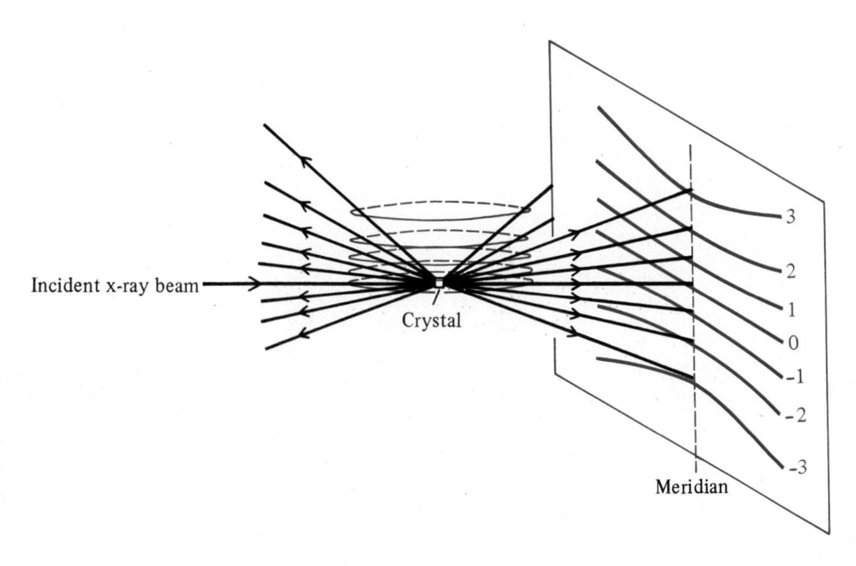

Figure 5-24
Diffraction of x rays by a crystal. The horizontal lines are the **layer lines** or loci of spots.

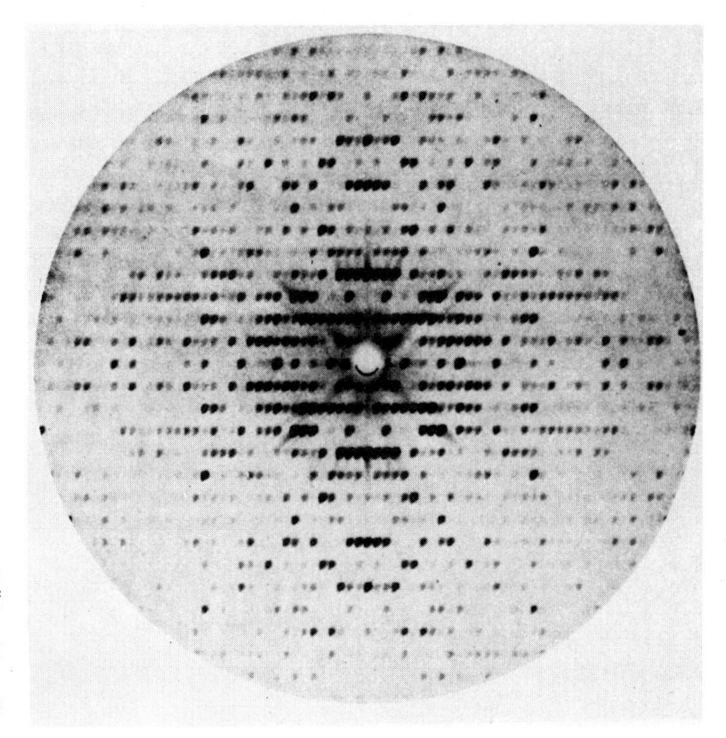

Figure 5-25
X-ray diffraction pattern of a crystal of the protein myoglobin. (Courtesy of National Naval Medical Center, Bethesda, Maryland.)

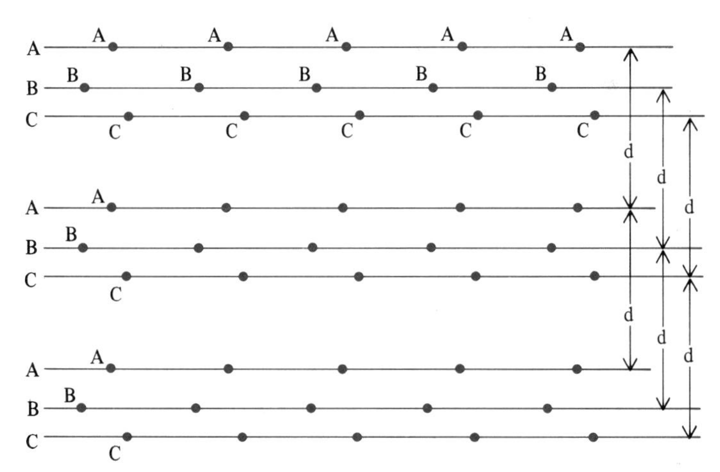

Figure 5-26
Layers of like atoms within a crystal.

of incidence (Fig. 5-27). The conditions for reinforcement are concisely stated by the **Bragg equation**:

$$n\lambda = 2d \sin \theta \qquad \textbf{(5-10)}$$

where d is the spacing of the planes.

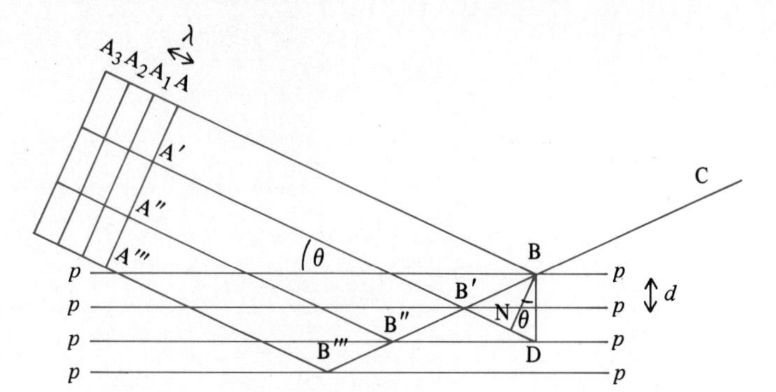

Figure 5-27 Relationships involved in the Bragg equation.

Intense diffracted rays will occur only in certain directions for which the Bragg condition is satisfied. In other directions the beams will not be in phase, so that destructive interference will greatly reduce their intensity. Each spot in the pattern corresponds to the fulfillment of the Bragg condition for a particular set of planes.

The protein crystallographer has essentially two kinds of information to work with. These are the *positions* of the spots, which are determined by the shape and dimensions of the unit cell, and their *intensities*, which are determined by the mutual positions of the atoms of the unit cell and hence by the structure of the molecule.

It is possible to predict the diffraction pattern of a molecule of known structure by tedious but basically straightforward calculations. Working backwards to obtain the structure from the pattern is far more difficult. This is because an essential piece of information, the *phases* of the diffracted rays, cannot be determined from the photograph. In favorable cases they can be determined by examining the effects of selective chemical modification of the protein upon the diffraction pattern. While the labor involved has been greatly reduced by the availability of modern computers, the process remains exceedingly complex and demanding and has been carried to completion for only a limited number of proteins.

An alternative approach is to propose a reasonable model structure, compute a diffraction pattern, and compare the actual and predicted patterns. Deviations are progressively reduced by systematically refining the original model. This procedure has been successfully applied to DNA.

5-8 AN EXAMPLE OF A PROTEIN OF KNOWN STRUCTURE

Lysozyme

The enzyme **lysozyme** causes the lysis and disruption of certain bacterial cells by catalyzing the hydrolytic cleavage of the polysaccharide components of their cell walls. Lysozyme has the further distinction of being among the first proteins whose three-dimensional structures were successfully determined by x-ray diffraction.

Lysozyme consists of a single polypeptide chain of 129 amino acids, with a molecular weight of 14,600. The chain is cross-linked by four disulfide bridges. Its amino acid sequence, which shows no indication of any obvious regularity or periodicity, is shown in Fig. 4-8.

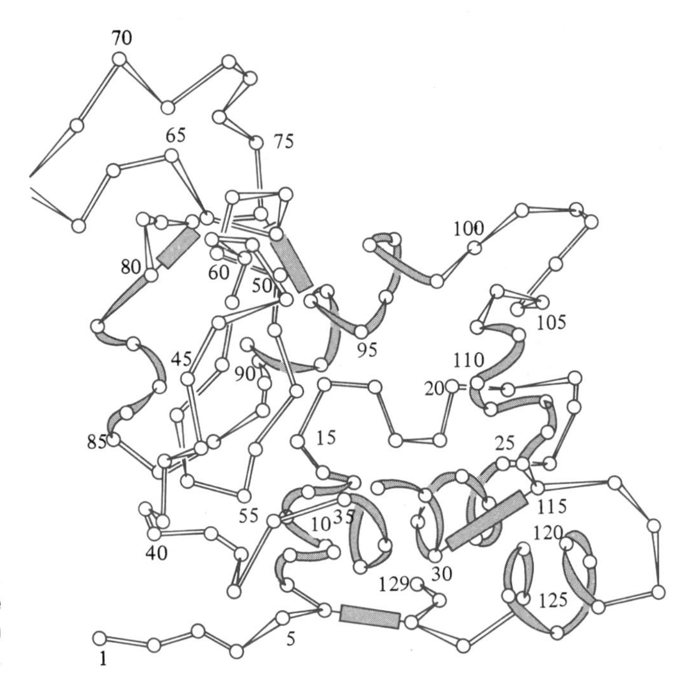

Figure 5-28 Folding of the polypeptide chain of lysozyme.

The molecular architecture of lysozyme was established by the crystallographic work of D. C. Phillips and co-workers. The molecule has approximately the shape of a compact spheroid of dimensions 4.5 × 3.0 × 3.0 nm. A prominent feature of its structure is the presence of an antiparallel pleated sheat (Fig. 5-28) involving residues 41–45 and 50–54 (Fig. 5-29), with the connecting residues 46–49 folded into a hairpin loop. The pleated sheet is stabilized by N—H· · ·O═C hydrogen bonds involving the peptide groups of the above chain segments. Three short α-helical regions are present corresponding to residues 5–15, 24–34, and 88–96. The interior of the molecule consists almost entirely of nonpolar amino acids, underlining the importance of hydrophobic contacts in stabilizing the molecular structure.

5-9 GENERAL CONCLUSIONS ABOUT PROTEIN STRUCTURE

The list of proteins whose three-dimensional structures are known is now substantial and is growing steadily. A number of generalizations have emerged, the most important of which include:

(1) The α-helical content of globular proteins generally has the form of a set of short helical segments that may collectively comprise a major fraction of the molecule.
(2) Proline residues, which cannot be incorporated into an α-helix because of the absence of a hydrogen atom on the α-nitrogen, are confined to the nonhelical regions, being often found in the bends between helical segments.

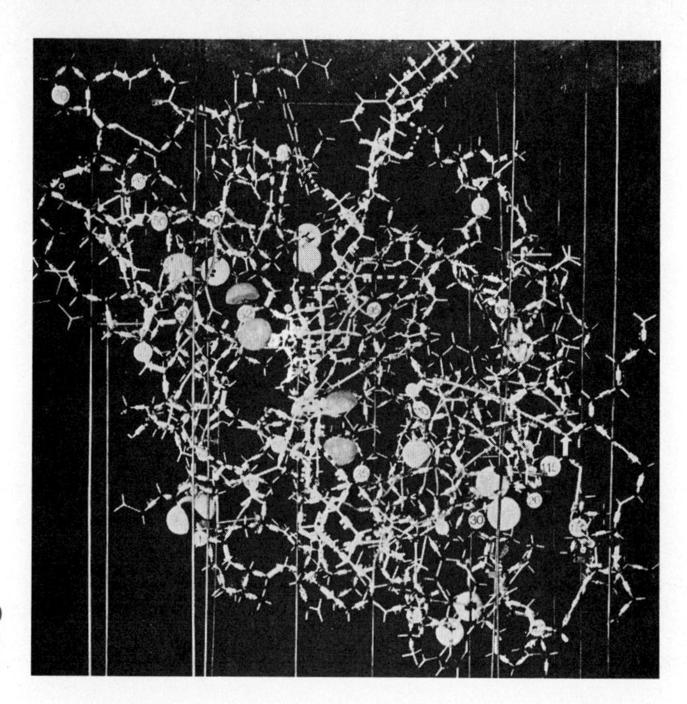

Figure 5-29
Model of the structure of lysozyme

(3) Charged groups are located on the periphery of the molecule, while nonpolar hydrocarbon side-chains tend to be largely packed into the interior.
(4) A comparison of proteins of different species has shown that considerable latitude in sequence is compatible with essentially the same protein conformation. In particular, many substitutions of nonpolar groups are without important structural consequences. However, substitutions which would place a charged group in the interior do not generally occur.

It has already been stated that hydrophobic bonds provide most of the stabilization energy for globular proteins. Chemical agents which disrupt hydrophobic bonds, such as detergents, urea, and guanidine hydrochloride, produce a general loss of organized structure or **denaturation**. In a strongly denaturing solvent, such as 10-M urea or 6-M guanidine hydrochloride, proteins tend to revert to the state of unorganized and randomly coiled polypeptides, which correspond to collections of different conformations governed by purely statistical factors. The optical rotatory dispersion and circular dichroism spectra of such denatured systems show an absence of contributions from organized structural elements such as the α-helix and the β-pleated sheet. The frictional ratio increases dramatically, suggesting that the molecule exists in a more open form with internal space accessible to solvent. This and other physical evidence has led to the conclusion that, in these strongly denaturing solvents, most globular proteins are essentially devoid of organized structure and that any noncovalent contacts are temporary and transient. Denaturation may also be caused by alteration of pH or by heating to above a critical temperature range.

5-10 THE AMINO ACID SEQUENCE OF A PROTEIN DICTATES ITS THREE-DIMENSIONAL STRUCTURE

Of the vast number of possible conformations that the polypeptide chain of a typical globular protein is capable of adopting, only one corresponds to the biologically active native form. The question of the mechanism whereby a freshly synthesized polypeptide assumes its correct conformation attracted much speculation but proved to have a refreshingly simple answer. The correct conformation is assumed spontaneously and automatically without the intervention of an external agent.

The experimental evidence for this far-reaching conclusion came largely from the work of Anfinsen and co-workers. A series of proteins, including the enzymes lysozyme and ribonuclease, were converted to a structureless state, devoid of cystine cross-links, by reduction of all disulfide bonds with mercaptoethanol in a strongly denaturing solvent, such as 9-*M* urea or 6-*M* guanidine hydrochloride. If the denaturing agent is removed and the disulfide bonds allowed to reform by air oxidation, all or most of the protein may be reconverted to the native form (Fig. 5-30).

The implication of this result is that the native structures of most proteins are overwhelmingly favored thermodynamically and thus tend to be assumed spontaneously. The structureless polypeptide is presumably sufficiently mobile to explore a range of conformations and select that of greatest stability. This automatically leads to the correct pairing of cystine half-residues.

When the reformation of the native structure is carried out *in vitro*, a significant fraction of incorrect disulfide pairing usually occurs. It is uncertain to what extent this also occurs *in vivo*. An enzyme present in the cytoplasm has been isolated which catalyzes the interchange of disulfide bonds and increases the yield and rate of formation of native protein. It is possible that the function of this **disulfide-interchange catalyzing enzyme** is to “reshuffle” the disulfide bonds, eliminating any incorrect pairings and facilitating the attainment of the most stable arrangement.

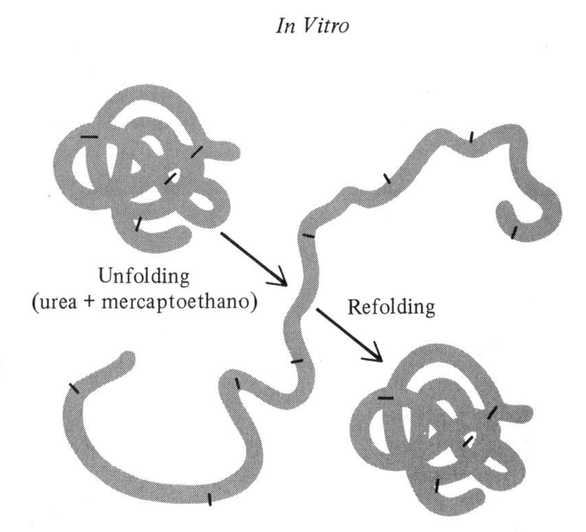

Figure 5-30 Conversion of a protein to a structureless state, followed by reformation of the native structure.

SUMMARY

The three-dimensional structures of proteins are highly organized and specific. In some cases the size and shape of a protein may be estimated by direct **electron microscopic** observation. The most accurate physical technique for obtaining the molecular weight is by **sedimentation equilibrium** measurements using an **ultracentrifuge**. Two faster, but more approximate methods are by **gel filtration** or **SDS gel electrophoresis**.

A recurring structural feature in many proteins is the **α-helix**, which is stabilized by CO···HN hydrogen bonds involving peptide groups of the polypeptide backbone. The α-helical content of most globular proteins consists of short helical segments separated by **randomly coiled** regions. Proline residues do not fit into an α-helix and generally occur in the bends between helices. Another structural element which is often observed is the **pleated sheet** or **β-structure** which is stabilized by CO···HN hydrogen bonds between extended and parallel polypeptide chains.

The balance of the protein, which is not incorporated into structures of either of the above types, is nevertheless also highly organized. Interactions of the side-chains are very important in stabilizing protein structure; **hydrophobic bonds** are probably the single most important stabilizing factor.

The complete three-dimensional structure of a protein may be determined by **x-ray crystallography**. The diffraction of a beam of x rays by a single protein crystal produces a characteristic pattern of spots on a photographic film, whose positions and intensities reflect the spatial arrangement of the atoms within the crystal and may be analyzed to compute their positional coordinates.

A remarkable conclusion of crystallographic studies of the same protein obtained from a series of different species is that many amino acids may be substituted without altering the overall molecular organization. Another generalization is that charged groups normally are confined to the surface of the protein in contact with solvent; the interior is largely composed of nonpolar amino acids.

A protein may be converted into a randomly coiled state devoid of organized structure by reduction of all disulfide bonds under strongly denaturing conditions. Upon removal of the denaturant and oxidative reformation of the disulfide bridges, the native structure is spontaneously reformed. The important conclusion follows that the three-dimensional structure is essentially dictated by the amino acid sequence.

REFERENCES

The following code is used to classify references. I: particularly useful as an introduction to the subject; R: useful primarily as a reference text; A: an advanced account of the material; H: a publication of historical importance.

General

V. A. Bloomfield and R. E. Harrington (eds.), *Biophysical Chemistry*, Freeman, San Francisco (1975). (A)

R. H. Haschemeyer and A. E. V. Haschemeyer, *Proteins: A Guide to Study by Physical and Chemical Methods*, Wiley, New York (1973). (I)

H. Neurath and R. L. Hill, eds., *The Proteins*, 3rd. ed., vols. 1–4, Academic, New York (1976). (R)

C. Tanford, *Physical Chemistry of Macromolecules*, Wiley-Interscience, New York (1961). (A)

K. E. Van Holde, *Physical Biochemistry*, Prentice-Hall, Englewood Cliffs, N.J. (1970). (I)

Electron Microscopy

E. M. Slayter, in *Physical Principles and Techniques of Protein Chemistry*, S. J. Leach, ed., Academic, New York (1969). (R)

Ultracentrifugation

J. H. Coates, in *Physical Principles and Techniques of Protein Chemistry*, S. J. Leach, ed., Academic, New York (1969). (I)

H. Schachman, *Ultracentrifugation in Biology*, Academic, New York (1958). (I)

J. W. Williams, *Ultracentrifugation of Macromolecules: Modern Topics*, Academic, New York (1972). (A)

Gel Filtration

G. K. Ackers, "Analytical Gel Chromatography of Proteins," *Adv. Prot. Chem.* 24, 343 (1970). (A)

H. Determann, *Gel Chromatography*, Springer-Verlag, Berlin (1968). (R)

Optical Rotation and Circular Dichroism

P. Crabbe, *ORD and CD in Chemistry and Biochemistry: An Introduction*, Academic, New York (1972). (I)

J. T. Yang, in *A Laboratory Manual of Analytical Methods in Protein Chemistry*, P. Alexander and H. P. Lundgren, eds., vol. 5, Pergamon, Oxford (1968). (R)

The α-Helix

L. Pauling, R. B. Corey, and H. R. Branson, "The Structure of Proteins: Two Hydrogen-Bonded Configurations of the Polypeptide Chain," *Proc. Natl. Acad. Sci. U.S.* 37, 205 (1951). (*H*)

L. Pauling and R. B. Corey, "Helical Configurations of Polypeptide Chains: Structure of Proteins of the α-Keratin Type," *Nature* 171, 59 (1953). (H)

X-Ray Diffraction

T. L. Blundell and L. Johnson, *Protein Crystallography*, Academic, New York (1976). (A)

H. R. Wilson, *Diffraction of X-Rays by Proteins, Nucleic Acids, and Viruses*, St. Martin's, New York (1968). (I)

Folding of Protein Chains

C. B. Anfinsen, "The Principles that Govern the Folding of Polypeptide Chains," *Science* 181, 223 (1973). (H)

REVIEW QUESTIONS

Questions marked with an asterisk are of a high level of difficulty.

*5-1 Below is a cross section of the structure of an important enzyme in a creature living on a planet where the ocean (and its body fluid) is a methanol–water mixture (at pH 7). You can assume that all possible covalent bonds between adjacent residues (horizontally, vertically, *or* diagonally) are formed.

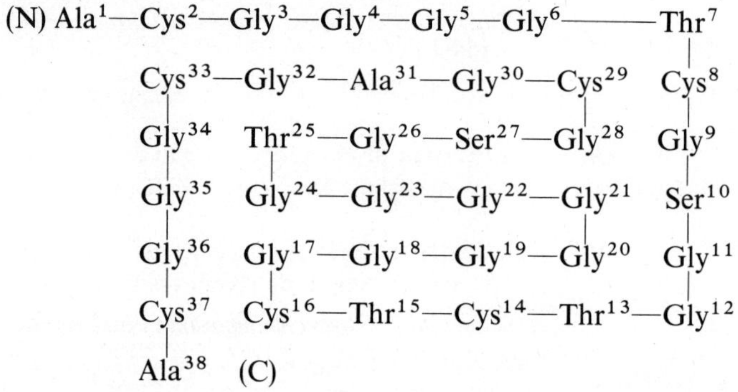

(N) Ala^1—Cys^2—Gly^3—Gly^4—Gly^5—Gly^6———Thr^7

Cys^{33}—Gly^{32}—Ala^{31}—Gly^{30}—Cys^{29} Cys^8

Gly^{34} Thr^{25}—Gly^{26}—Ser^{27}—Gly^{28} Gly^9

Gly^{35} Gly^{24}—Gly^{23}—Gly^{22}—Gly^{21} Ser^{10}

Gly^{36} Gly^{17}—Gly^{18}—Gly^{19}—Gly^{20} Gly^{11}

Cys^{37} Cys^{16}—Thr^{15}—Cys^{14}—Thr^{13}—Gly^{12}

Ala^{38} (C)

If the planet's ocean and the creature's fluids gradually become 100% water during the next million years, indicate whether the (independent) mutations listed below would be expected to be favorable (+), unfavorable (−), or of marginal importance (0) in preserving the *original* conformation.

(a) Gly^4 to Asp
(b) Gly^{23} to Met
(c) Gly^{36} to Val
(d) Gly^{22} to Leu
(e) Gly^{12} to Lys
(f) Gly^{19} to Glu
(g) Ser^{27} to Cys

5-2 Which of the following groups of amino acids would *most* likely be found on the inside of a protein?
(a) Asn,Gln,Lys
(b) Phe,Ile,Val
(c) Asp,Glu,His
(d) Tyr,Lys,Gln
(e) Gly,Lys,Asn

5-3 Indicate which of the statements below are *true* about α-helical conformations or β-sheet conformations (or both).
(a) They are found in globular proteins, mostly on the inside.
(b) Most peptide groups are not involved in H-bonding to water.
(c) Some rotation is required about the amide bond N-(carbonyl-carbon).
(d) Each turn of the helix contains exactly four amino acid residues.
(e) The H of one amide nitrogen is H-bonded to the carbonyl O of the next amino acid residue in the polypeptide chain.

5-4 Consider all of the factors in the equations for sedimentation and diffusion. The speed at which any macromolecule moves in a centrifugal force field must (yes or no):
(a) vary as the first power of the molecular weight, that is, (molecular weight)$^{+1.0}$

(b) vary as (diffusion coefficient)$^{-1.0}$
(c) vary as (frictional coefficient)$^{+1.0}$
(d) be dependent on the net charge on the macromolecule
(e) decrease with the density of the macromolecule
(f) depend on the density of the solvent

*5-5 A mixture of the following four proteins was purified:

Protein	*Isoelectric Point (pI)*	*Molecular Weight*
A	6.0	30,000
B	7.2	50,000
C	7.2	100,000
D	8.6	30,000

Step I was used to separate A from B + C + D; Step II to separate C from B + D; Step III to separate B from D. For each of the techniques listed below, indicate the step (or *none*) for which it was most likely to be used, and answer the accompanying question:

Technique	*Step*	*Question*
Ion-exchange chromatography on diethylaminoethyl cellulose at pH 8	?	Which protein(s) is (are) expected to stick most tightly to the column?
Ion-exchange chromatography on phosphocellulose at pH 9.0	?	(Same as above)
Adjustment of pH to 5.9; brief centrifugation at medium speed	?	Which protein(s) is (are) in the precipitate?
Molecular exclusion chromatography (gel filtration)	?	Which protein(s) elute(s) first? Last?

5-6 Polyarginine (Arg-Arg-Arg · · ·) in aqueous solution at pH 7 exists exclusively as a random coil (no secondary structure).
(a) Why is this so?
(b) How might you modify conditions so that polyarginine would assume an ordered secondary structure? What secondary structure is most likely to form?

5-7 Discuss and criticize the kind of approximation involved in assuming an ellipsoidal model for the shape of an actual protein.

5-8 Compute the length of an α-helical polypeptide 1000 amino acid units long.

5-9 What is the axial ratio of the polypeptide of problem 5-8?

The Enzymes and Biocatalysis

6-1 GENERAL PROPERTIES OF ENZYMES

An enzyme is a protein with **catalytic** properties; that is, it increases the velocity of a chemical reaction without being itself consumed or irreversibly altered by the process. The great majority of the more than 1000 different proteins, which have been proven to exist in living systems, are enzymes; it would be rash to predict the number which will ultimately be identified. Apart from catalytic activity, there is no general property which sets enzymes as a class apart from other proteins; they may be purified and characterized by the usual procedures.

Within the cell enzymes occur at locations appropriate to their function. Some enzymes, like those involved in **glycolysis** (chapter 10), occur in free solution in the cytoplasm. Other enzymes, such as those concerned with the **TCA cycle** (chapter 12), occur in **mitochondria** or other **supramolecular** structures. Enzymes may be integrated into **membranes**. In the higher species, it is common for enzymes to be synthesized in a tissue different than that in which they function; an example is the digestive enzymes **trypsin** and **chymotrypsin**, which are formed (as inactive precursors) in the pancreas, but are active in the intestine. While some enzymes, such as the two cited above, act independently, other enzymes, such as those that mediate the synthesis of **fatty acids** (chapter 13), are organized into complex structures, in which the mutual location of the different components is important for their collective activity.

The catalytic properties of an enzyme are localized in a restricted region termed the **active site**. Only a small fraction of the amino acid residues of the protein are usually involved in the active center. Indeed, the balance of the protein may function primarily to maintain the elements of the catalytic center in the correct juxtaposition.

It is of course impossible to generalize as to the overall structure of enzymes. They are quite as diverse in structural organization and properties as other proteins. Any generalization about the nature of the active center is equally unjustified, although important correlations have emerged for certain classes of enzyme.

An important general characteristic of enzymes is their **specificity**. Enzymes can catalyze only a very restricted set of reactions involving closely related molecules as reactants or **substrates**. The specificity often extends to the spatial configuration of the substrate. In many cases, a change from the D- to the L-configuration, or vice versa, is sufficient to abolish activity.

For any chemical reaction, the three questions of interest to the scientist concern the **mechanism** of the reaction, or the nature of the intermediates involved in the transition from reactants to products; the position of **chemical equilibrium** and the factors governing it; and the reaction **kinetics** or the factors controlling the *velocity* of the transformation of reactants. Enzymic reactions are no exception to this rule.

For obvious reasons, the mechanism of enzymic reactions does not lend itself to discussion in general terms; the mechanisms of individual reactions will be discussed as they are encountered. In this chapter we shall consider the latter two topics in inverted order, dealing with kinetics prior to equilibria; this is not illogical since the two subjects are approached by distinct and independent avenues.

It should be acknowledged at the onset that the observed net velocity of any reaction generally corresponds to a *balance* between *forward* and *reverse* reactions and will depend therefore necessarily upon where the composition of the mixture of reactants and products is with respect to the composition characteristic of chemical equilibrium; if the mixture is at chemical equilibrium, the net velocity will of course be zero, as the forward and reverse reactions balance each other. It is possible to describe the kinetics of an enzymic reaction in simple terms only for the case where the system is far from equilibrium, so that the reverse reaction is negligible in comparison with the forward reaction. This will be the case at the *onset* of the reaction before products have accumulated to a significant extent. The discussion of enzyme kinetics presented in section 6-3 will apply to *initial* velocities.

Why should enzyme kinetics be worthy of the attention of the biologist or biochemist who is not a specialist in this area? There are two principal reasons. The first is that, as we shall see, kinetics is often a primary source of information as to the mechanism of the reaction. The second is that biological systems are generally far removed from equilibrium so that the direction a series of biochemical transformations takes may be governed by kinetic rather than equilibrium considerations.

The question of the position of chemical equilibrium requires the altogether different approach of **thermodynamics**, which will be considered in section 6-4. The importance of this approach lies in its unique ability to predict the direction which a reaction will take (that is, forward or reverse) for a given set of concentrations of reactants and products.

We will turn next to a detailed consideration of the enzyme **chymotrypsin**, for which there is unusually complete information relating structure and mechanism. The chapter will conclude with a discussion of the **regulation** of enzyme activity within living cells.

6-2 TYPES OF ENZYMIC REACTIONS

It should be emphasized that enzymes cannot alter the position of chemical equilibrium or force an intrinsically improbable reaction to occur; rather, they overcome *kinetic* barriers so as to accelerate the approach to equilibrium. A consequence of enzymic specificity is that one of a set of possible reactions may be selectively accelerated. The degree of acceleration is typically enormous, factors of 10^6 or more being commonplace. The majority of biological reactions would not occur at significant rates in the absence of enzymic catalysis.

A partial list of the kinds of chemical reaction catalyzed by enzymes is given in Table 6-1. A large fraction (but not all) of enzymic reactions fall into three categories:

(1) **Isomerization**—a molecule of substrate is transformed by internal rearrangement, without net consumption of the solvent or a second substrate:

$$A \rightarrow B$$

(2) **Hydrolysis**—the substrate is decomposed by reaction with water:

$$AB + H_2O \rightarrow AH + BOH$$

(3) **Transfer**—a group is transferred between two different substrates:

$$AB + C \rightarrow A + BC$$

The catalytic action of all enzymes appears to involve the formation of a transient complex with the substrate. Since the enzyme may be recovered in unchanged form at the completion of the process, it is clear that complex formation does not involve any *permanent* chemical bonds. The complex may be stabilized by secondary forces, including hydrogen, hydrophobic, and electrostatic bonds, as well as weak covalent bonds. Only in a fraction of the known enzymic reactions is there sufficient information to justify the proposal of explicit models.

The combination of enzyme with substrate is reversible. An enzyme-substrate complex may **dissociate** to reform the reactants or **decompose** to form the product and free enzyme. For a reaction involving only a single substrate,

$$\underset{\text{Enzyme}}{E} + \underset{\text{Substrate}}{S} \rightarrow \underset{\text{Complex}}{ES} \rightarrow \underset{\text{Products}}{P} + E \qquad \textbf{(6-1)}$$

The decomposition of the complex may itself be stepwise:

$$E + S \rightleftarrows ES \rightarrow ES' \rightarrow P + E \qquad \textbf{(6-2)}$$

The activity of many enzymes requires the participation of an additional molecule or ion termed a **cofactor**. These may be grouped into two categories:

Table 6-1
The Principal Classes of Enzymes*

Class	Group Affected	Example
(1) Oxidoreductases	CHOH	Lactate dehydrogenase
	CH—CH	Acyl CoA dehydrogenase
	$CH—NH_2$	Amino acid oxidase
(2) Transferases	Methyl	Guanidinoacetate methyltransferase
	Hydroxymethyl	Serine hydroxymethyl-transferase
	Acyl	Choline acetyltransferase
	Amino	Transaminase
(3) Hydrolases	Carboxylic ester	Esterase, lipase
	Phosphoric monoester	Phosphatase
	Phosphoric diester	Ribonuclease
	Glycosidic bond	Amylase
	Peptide	Trypsin
(4) Lyases	C—C	Aldolase
	C—O	Fumarate hydratase
(5) Isomerases	Aldose–ketose	Phosphoglucoisomerase
(6) Ligases	Forms C—O bonds	Amino acid activating enzyme
	Forms C—N bonds	Glutamine synthetase
	Forms C—C bonds	Acetyl CoA carboxylase

*Only a few of the many subclasses are cited here. For a complete list see *Report of the Commission on Enzymes*, International Union of Biochemistry Symposium,.Pergamon, New York, 1961, vol. 20; M. Florkin and E. H. Stotz, *Comprehensive Biochemistry*, Elsevier, Amsterdam, 1965, vol. 13.

(1) **Metal ions**. The active centers of many enzymes can only function if the enzyme is complexed with a particular metallic cation. In some cases this is so strongly bound as to be virtually an integral part of the enzyme structure; in others, the association with enzyme is transient and reversible. The list of metals effective in particular instances includes Co^{2+}, Zn^{2+}, Mg^{2+}, Ca^{2+}, and Mn^{2+}.

(2) **Coenzymes**. Many enzymic reactions involve the transfer of a group between a substrate and an organic molecule called a coenzyme. This is the case, in particular, for numerous oxidation-reduction processes. Oxidation of the substrate is accompanied by reduction of the coenzyme, or vice versa. The coenzyme is often reoxidized by a second reaction. The coenzyme may have only a transient association with the enzyme, being indistinguishable from a second substrate, or it may be tightly and permanently bound. In the latter case it is often termed a *prosthetic* group.

Primers differ from coenzymes in being chemically incorporated into, or complexed with, the product. Primers invariably resemble the product chemically and may be considered to be a form of substrate.

Enzymic Assay Procedures

The technique chosen to measure the velocity of an enzymic reaction must of course depend upon the nature of the reaction. It is usually desirable to measure the *initial* velocity. Either the rate of disappearance of the substrate or the rate of appearance of the product may be monitored, the latter being usually more convenient.

A change in a physical property, such as absorbance for visible or ultraviolet light or intensity of fluorescence, often accompanies the conversion of substrate into product and provides an index of the rate of reaction. If a release or an uptake of hydrogen ions is associated with the reaction, this may be monitored by continuous titration with standardized acid or base at constant pH ("pH-stat titration"). An evolution of gas may be followed by manometric procedures. Alternatively, the reaction may be followed by direct chemical analysis.

6-3 ENZYME KINETICS

Michaelis-Menten Model

The account of enzyme kinetics to be presented here will be concerned with reactions involving a single substrate, according to the mechanism presented in Equation (6-1). Of the three general types of reaction cited in section 6-2, types (1) and (2) fall into this category. This is the case for hydrolytic reactions since, under the usual conditions, water is normally present in sufficient excess so that its concentration may be regarded as effectively invariant and need not enter the kinetic scheme explicitly.

The classical model developed by Michaelis and Menten and by Haldane permits a simple quantitative description of systems of this kind, provided that the following assumptions are valid:

(1) The total concentration of substrate is not significantly depleted in the course of the reaction.
(2) The concentration of the enzyme-substrate complex rapidly attains a constant **steady-state** value, so that the formation of complex is balanced by **dissociation** into reactants and **decomposition** into products. In practice the steady-state concentration is generally attained very rapidly, the initial transient phase lasting for only a fraction of a second.

The rate of formation of the enzyme-substrate complex is given by

$$\left(\frac{d[\mathrm{ES}]}{dt}\right)_f = k_1[\mathrm{E}][\mathrm{S}] \qquad \textbf{(6-3)}$$

where [E], [S], and [ES] are the concentrations of *free* enzyme, substrate, and complex, respectively, and k_1 is the second order rate constant for the forward reaction.

The rate of **dissociation** of the complex into reactants is given by

$$\left(\frac{d[\mathrm{ES}]}{dt}\right)_{\mathrm{dis}} = -k_2[\mathrm{ES}] \qquad \textbf{(6-4)}$$

The rate of **decomposition** into products is given by

$$\left(\frac{d[\text{ES}]}{dt}\right)_{\text{dec}} = -k_3[\text{ES}] \tag{6-5}$$

In Equations (6-4) and (6-5), k_2 and k_3 are the first order rate constants for the corresponding processes.

In the *steady state* the rate of the forward reaction is equal to the *sum* of those which remove complex, so that the concentration of complex does not change with time:

$$\frac{d[\text{ES}]}{dt} = 0 = k_1[\text{E}][\text{S}] - k_2[\text{ES}] - k_3[\text{ES}] \tag{6-6}$$

The concentration of free enzyme is equal to the total enzyme concentration, $[\text{E}_o]$, minus the concentration of complex:

$$[\text{E}] = [\text{E}_o] - [\text{ES}] \tag{6-7}$$

Upon substitution and rearrangement

$$[\text{ES}] = k_1[\text{E}_o][\text{S}]/(k_1[\text{S}] + k_2 + k_3) \tag{6-8}$$

The overall velocity, v, of the enzymic reaction is given by

$$\begin{aligned} v &= k_3[\text{ES}] \\ &= \frac{k_1k_3[\text{E}_o][\text{S}]}{k_1[\text{S}] + k_2 + k_3} \\ &= \frac{k_3[\text{E}_o][\text{S}]}{[\text{S}] + (k_2 + k_3)/k_1} \end{aligned} \tag{6-9}$$

The ratio $(k_2 + k_3)/k_1$ has been designated as the **Michaelis constant.** If $k_2 \gg k_3$, it is equivalent to a true equilibrium constant for dissociation of the complex species. Equation (6-9) may be rewritten in terms of the Michaelis constant (K_m) as

$$v = \frac{k_3[\text{E}_o]}{1 + K_m/[\text{S}]} \tag{6-10}$$

or, in reciprocal form,

$$\frac{1}{v} = \frac{1}{k_3[\text{E}_o]} + \frac{1}{[\text{S}]}\frac{K_m}{k_3[\text{E}_o]} \tag{6-11}$$

In the absence of complicating factors a plot of $1/v$ versus $1/[\text{S}]$ should be linear, with an intercept equal to $1/k_3[\text{E}_o]$ and a slope equal to $K_m/k_3[\text{E}_o]$ (Figs. 6-1 and 6-2). Since $k_3[\text{E}_o]$ is the **maximum velocity** (V) approached at very high substrate concentrations, Equation (6-10) may be rewritten

$$\frac{V}{v} = 1 + \frac{K_m}{[\text{S}]} \tag{6-12}$$

The decomposition into products may occur by a *stepwise* mechanism, as

$$\text{ES} \xrightarrow{k_3'} \text{ES}' \xrightarrow{k_3''} \text{products} \tag{6-13}$$

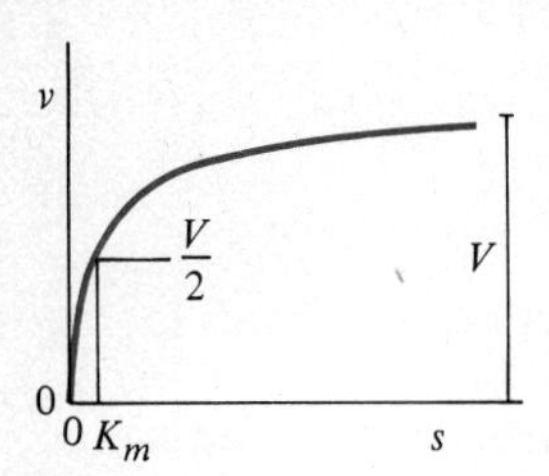

Figure 6-1
Dependence of initial reaction velocity upon substrate concentration for an enzymic reaction obeying the Michaelis–Menten mechanism. The substrate concentration at which the reaction velocity is equal to one-half the maximum velocity (V) is numerically equal to the Michaelis constant (K_m). Note the *hyperbolic* shape of the curve.

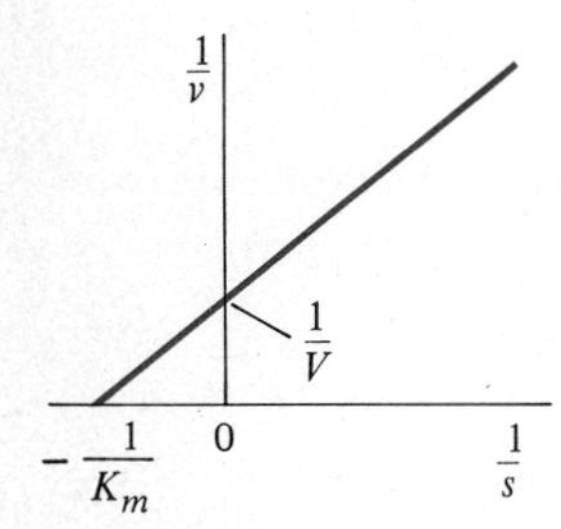

Figure 6-2
Determination of the maximum velocity and the Michaelis constant of an enzymic reaction by a double reciprocal plot. This kind of plot is often designated as a **Burk–Lineweaver** plot. The linear form of this plot makes extrapolation easier. The *ordinate* intercept is equal to the *reciprocal* of the maximum velocity ($1/V$). The *abscissa* intercept is equal to the *reciprocal* of the Michaelis constant ($1/K_m$).

The kinetics are still described by an equation of the form of Equation (6-10). The observed parameter $k_{3,\text{obs}}$ is related to k_3' and k_3'' by

$$\frac{1}{k_{3,\text{obs}}} = \frac{1}{k_3'} + \frac{1}{k_3''} \tag{6-14}$$

Enzyme Inhibition

Many enzymes may be inhibited by combination with a small molecule or ion. **Irreversible** inhibition usually involves the formation of a covalent bond. The degree of inhibition is in this case equal to the mole ratio of inhibitor to enzyme.

Reversible inhibitors may be classified as **competitive** or **noncompetitive**, although intermediate cases are known. In competitive inhibition, the inhibitor interferes with the binding of the substrate, presumably by occupying the same site, or elements of the site. The degree of inhibition for a fixed concentration of competitive inhibitor depends upon the substrate concentration. With increasing level of substrate the inhibitor is crowded competitively off the active site, so that the extent of inhibition is reduced. Competitive inhibitors usually resemble structurally all or part of the substrate.

The basic Michaelis-Menten model may be readily modified to account for the influence of a competitive or noncompetitive inhibitor, provided that it is valid to assume that a true equilibrium exists between bound and free inhibitor:

$$\mathrm{E} + \mathrm{I} \underset{k_5}{\overset{k_4}{\rightleftarrows}} \mathrm{EI} \tag{6-15}$$

$$\mathrm{E} + \mathrm{S} \underset{k_2}{\overset{k_1}{\rightleftarrows}} \mathrm{ES} \overset{k_3}{\rightarrow} \text{products}$$

For fully competitive inhibition, the enzyme can bind either inhibitor or substrate but not both. An equilibrium constant, K_i, may be defined for the dissociation of the enzyme-inhibitor complex:

$$K_i = \frac{k_5}{k_4} = \frac{[\mathrm{E}][\mathrm{I}]}{[\mathrm{EI}]} \tag{6-16}$$

and $[\mathrm{EI}] = [\mathrm{E}][\mathrm{I}]/K_i$.
Using the relationship

$$[\mathrm{E}_o] = [\mathrm{E}] + [\mathrm{ES}] + [\mathrm{EI}] \tag{6-17}$$

and making the same steady-state assumption as before, one obtains finally

$$v = \frac{k_3[\mathrm{E}_o]}{1 + \dfrac{K_m}{[\mathrm{S}]}\left(1 + \dfrac{[\mathrm{I}]}{K_i}\right)} \tag{6-18}$$

or

$$\frac{V}{v} = 1 + \frac{K_m}{[\mathrm{S}]}\left(1 + \frac{[\mathrm{I}]}{K_i}\right)$$

For fully competitive inhibition, the values of V and k_3, as obtained from the intercept of the (linear) plot of $1/v$ versus $1/[\mathrm{S}]$, are independent of the concentration of inhibitor. The slope of V/v as a function of $1/[\mathrm{S}]$ is equal to $K_m(1 + [\mathrm{I}]/K_i)$. The *apparent* value of the Michaelis constant thus increases with inhibitor concentration.

In the case of completely noncompetitive inhibition the combination of enzyme with inhibitor interferes with the conversion, but not the *binding*, of substrate. In this case the inhibitor is bound at a *different* site from that at which the substrate is attached and influences the catalytic site *indirectly*. The general equations are as follows:

$$\mathrm{E} + \mathrm{S} \underset{k_2}{\overset{k_1}{\rightleftarrows}} \mathrm{ES} \overset{k_3}{\rightarrow} \text{products} \tag{6-19}$$

$$\mathrm{E} + \mathrm{I} \underset{k_5}{\overset{k_4}{\rightleftarrows}} \mathrm{EI}; \quad K_i = \frac{k_5}{k_4}$$

$$\mathrm{EI} + \mathrm{S} \underset{k_2}{\overset{k_1}{\rightleftarrows}} \mathrm{EIS} \overset{k_6}{\rightarrow} \text{products} \tag{6-20}$$

$$\mathrm{ES} + \mathrm{I} \underset{k_5}{\overset{k_4}{\rightleftarrows}} \mathrm{EIS} \overset{k_6}{\rightarrow} \text{products}$$

where $k_6 \ll k_3$.

Completely noncompetitive inhibition is only possible if the Michaelis constant is a true equilibrium constant ($k_3 \ll k_2$). Otherwise an alteration in k_3 must inevitably change K_m, in view of the definition of the latter quantity. Upon setting K_m equal to k_2/k_1, introducing the steady-state assumption, and solving the kinetic equations, one obtains finally

$$\frac{v}{[\mathrm{E}_o]} = \frac{k_3 + k_6\dfrac{[\mathrm{I}]}{K_i}}{\left(1 + \dfrac{K_m}{[\mathrm{S}]}\right)\left(1 + \dfrac{[\mathrm{I}]}{K_i}\right)} \tag{6-21}$$

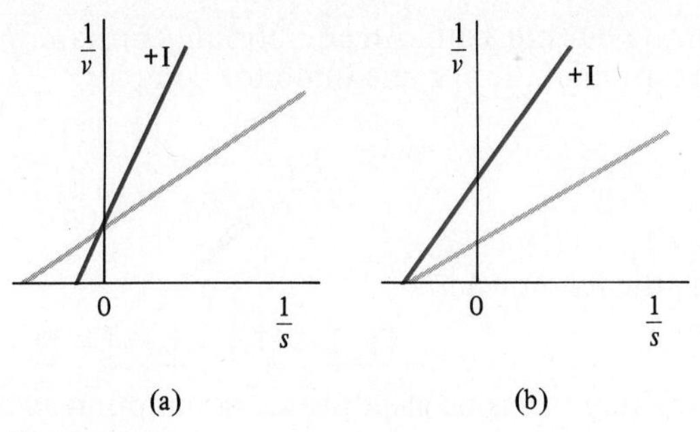

Figure 6-3
Kinetic behavior predicted for (a) competitive and (b) noncompetitive inhibition. In the former case the apparent value of maximum velocity, as reflected by the intercept on the vertical axis, is unchanged in the presence of inhibitor, while the apparent value of the Michaelis constant increases; in the latter case the reverse is true.

The Michaelis constant is in this case *independent* of the inhibitor concentration. However, the *apparent* value of k_3 computed by formal application of the Michaelis-Menten equation is now equal to

$$\frac{k_3 + k_6 \dfrac{[\mathrm{I}]}{K_i}}{1 + \dfrac{[\mathrm{I}]}{K_i}}$$

or, if $k_6 = 0$, to

$$\frac{k_3}{1 + \dfrac{[\mathrm{I}]}{K_i}}$$

Figure 6-3 illustrates the kinetic behavior predicted for competitive and noncompetitive inhibition. In both cases the intercept on the vertical axis of a plot of $1/v$ versus $1/[\mathrm{S}]$ is equal to the reciprocal of the *apparent* value of the maximum velocity $(1/V)$, as already shown for the case where inhibition is absent [Equation (6-11)], while the slope, or the intercept on the horizontal axis, yields the *apparent* value of K_m, or $-1/K_m$, respectively.

Dependence of Rate Upon pH

The velocity of an enzymic reaction is, in general, dependent upon the pH. If the active center contains one or more ionizable sites, it may exist in several states of ionization, not all of which are active. For example, in the case where the active center contains two ionizable groups, one of which must be protonated and the other unprotonated for activity to be present, there will be an optimum pH range, on either side of which activity is lost.

$$\underset{\text{Inactive}}{\mathrm{E(AH^+, BH^+)}} \overset{\mathrm{H^+}}{\leftrightarrows} \underset{\text{Active}}{\mathrm{E(AH^+, B)}} \overset{\mathrm{OH^-}}{\rightleftarrows} \underset{\text{Inactive}}{\mathrm{E(A, B)}} \qquad \textbf{(6-22)}$$

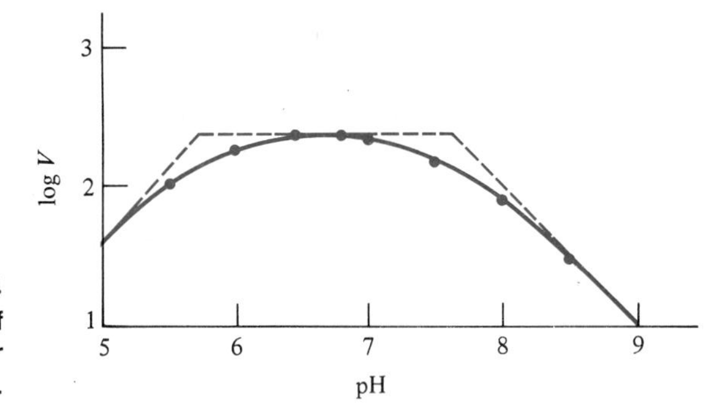

Figure 6-4
pH dependence of reaction velocity for the enzyme fumarase.

In the above case the optimal zone will lie at pH's intermediate to the pK's of groups A and B. Figure 6-4 illustrates the effect.

6-4 THERMODYNAMICS OF BIOCHEMICAL REACTIONS

Free Energy

In the language of thermodynamics, biological systems may be said to be "open"; that is, they exchange mass and energy with their surroundings. The maintenance of life in any biological organism is generally dependent upon a continuous supply of chemical energy from the external environment. This is the case even for mature organisms which perform no external work and whose state seemingly does not change with time. This requirement is a consequence of the intrinsic instability of biological systems, which are always far removed from a condition of thermodynamic equilibrium. In order to survive, living systems must continually synthesize proteins, nucleic acids, and other biopolymers from smaller constituents. The concentrations within the living cell of all these substances are far in excess of those predicted for a thermodynamically stable system. Indeed the death of an organism may, in a sense, be regarded as a reassertion of the claims of thermodynamics, corresponding to a transition from a thermodynamically unfavored to a favored state.

While the *rate* at which a chemical reaction proceeds may be influenced by the presence of suitable catalysts, including enzymes, the *direction* of the reaction is controlled entirely by thermodynamic principles. The most useful of the thermodynamic parameters for predicting the course of biochemical processes is **free energy**.

The free energy, G, has the same dimensions as energy, heat, or work, and is expressed in the same units, usually as calories or joules. For a pure substance under specified conditions, the free energy per mole is a well-defined quantity which has been measured and tabulated for many compounds of biological interest. The usefulness of free energy stems partly from the fact that it may be identified with that fraction of the total energy which is potentially available for external work.

For any chemical process the change in free energy, ΔG, is given by:

$$\Delta G = \Delta H - T\,\Delta S \tag{6-23}$$

Here ΔH is the change in **enthalpy** or heat content, which, at constant pressure, is equal to the energy change plus the (usually wasted) work done on or by the system against atmospheric pressure if a volume change occurs. The parameter ΔS or **entropy** is a measure of the change in randomness or disorder of the system, a positive value of ΔS corresponding to an increase in randomness. The term $T\,\Delta S$ loosely represents that part of the total energy change which is dissipated in the form of increased random molecular motions and hence is not available for useful work. For any *spontaneous* process, the entropy of the *universe* (system plus surroundings) increases.

The following principle is of central importance to chemical thermodynamics. *Any spontaneously occurring process is accompanied by a decrease in free energy of the system* ($\Delta G < 0$). A process, chemical or otherwise, which is characterized by a positive value of ΔG, cannot occur spontaneously. If it is to occur at all, energy must be made available from an external source.

For a system at thermodynamic equilibrium, no net change in free energy occurs with time ($\Delta G = 0$), irrespective of the *rates* of any chemical processes occurring. The free energy of such a system is at a *minimum*. The approach to equilibrium reflects the tendency of free energy toward a minimal value. Natural processes can only run "downhill."

The free energy change is directly related to the equilibrium constant governing a chemical reaction and to the direction of chemical change. For a reversible chemical reaction of the general type

$$\mathrm{A} + \mathrm{B} \rightleftarrows \mathrm{C} + \mathrm{D} \tag{6-24}$$

an equilibrium constant may be defined by

$$K = \frac{[\mathrm{C}][\mathrm{D}]}{[\mathrm{A}][\mathrm{B}]} \tag{6-25}$$

where [A] and [B] are the molar concentrations of reactants, while [C] and [D] are the molar concentrations of products.

If the concentrations of all four species are such that the ratio on the right-hand side of Equation (6-25) is equal to K, no net reaction will occur and the composition of the mixture is unchanged with time. If the ratio is less than K then the reaction will occur in the forward direction, with a net formation of C and D. If the ratio is greater than K, the reaction will occur in the reverse direction, with a net conversion of C and D to A and B.

For *any* composition of the system, the free energy change accompanying the reaction of one mole each of A and B to form one mole each of C and D is given by*

$$\begin{aligned} \Delta G &= \Delta G^\circ + \mathrm{RT} \ln \frac{[\mathrm{C}][\mathrm{D}]}{[\mathrm{A}][\mathrm{B}]} \\ &= \Delta G^\circ + \mathrm{RT} \ln \sigma \end{aligned} \tag{6-26}$$

*The total quantities of A, B, C, and D present are supposed to be sufficiently large so that no significant change in relative composition occurs and σ is effectively constant. Or, in other words, ΔG is the free energy change per mole for infinitesimal extents of reaction.

Here ΔG° is the **standard free energy change** and is equal to the free energy change per mole when all reactants are present at a concentration of one mole per liter; that is, $[A] = [B] = [C] = [D] = 1$.

When reactants and products are at chemical equilibrium, $\Delta G = 0$ and

$$\begin{aligned}\Delta G^\circ &= -\mathrm{RT}\ln \sigma_{\text{equilibrium}} \\ &= -\mathrm{RT}\ln\left\{\frac{[C][D]}{[A][B]}\right\}_{\text{equilibrium}} \\ &= -\mathrm{RT}\ln K\end{aligned} \tag{6-27}$$

The standard free energy change is thus directly related to the equilibrium constant. The more negative the value of ΔG°, the further is the equilibrium displaced toward the products C and D.

If ΔG° is known, it is possible to predict the direction in which a chemical reaction proceeds for a particular set of concentrations of reactants and products. If [A], [B], [C], and [D] are such that

$$\Delta G^\circ + \mathrm{RT}\ln \sigma = \Delta G < 0 \tag{6-28}$$

then the reaction will proceed in the *forward* direction (as written) with a net formation of C and D. If

$$\Delta G^\circ + \mathrm{RT}\ln \sigma = \Delta G > 0 \tag{6-29}$$

then the reaction will go in the *reverse* direction with a net formation of A and B. If

$$\Delta G^\circ + \mathrm{RT}\ln \sigma = 0 \tag{6-30}$$

then the system is at equilibrium and no net reaction can occur in either direction.

The preceding discussion says nothing as to the *rate* of reaction. Often a reaction with a large and negative value of ΔG will proceed extremely slowly because of kinetic barriers. This is the case for a mixture of the gases hydrogen and oxygen, whose reaction to form water is immeasurably slow in the absence of a catalyst, but occurs with extreme rapidity if platinum black is present.

Figure 6-5 illustrates the energy profile for a system of this kind. Although the free energy of the final equilibrium mixture, in which the reactants are almost completely converted to water, is much less than that for the initial mixture of H_2 and O_2, the system must surmount the barrier of a large and positive **energy of activation**. The catalyst does not alter the values of the initial and final free energies; its action is confined to lowering the activation energy.

The same principles apply to biochemical reactions, most of which become possible only with the aid of enzyme catalysts.

Coupled Reactions

Many biochemical systems superficially appear to defy the preceding generalities. Reactions which seem to proceed despite an unfavorable value

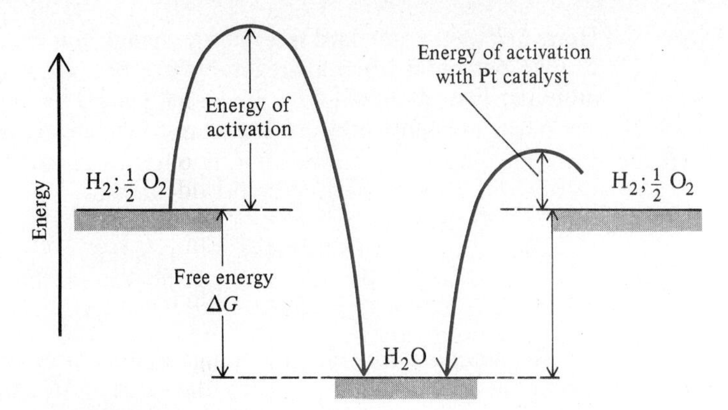

Figure 6-5
Free energy profile for the reaction between H_2 and O_2. The curves show the change in free energy of the system as reactants are converted to products in the absence and presence of a catalyst. In the latter case the activation energy is reduced.

of ΔG are very common. A dramatic example is the previously mentioned case of the biosynthesis of proteins and nucleic acids, which proceeds far beyond the extent predicted for an equilibrium system.

The inconsistency is of course only apparent. An **endergonic** reaction, with a positive value of ΔG, can occur only if it is *coupled* with an **exergonic** process, for which ΔG is negative. This is only possible if the two reactions have a common intermediate, one of the products of the first being consumed by the second.

$$\begin{aligned} A + B \rightarrow C + D \quad & \Delta G_1 > 0 \\ C + E \rightarrow F + K \quad & \Delta G_2 < 0 \end{aligned} \tag{6-31}$$

If the *total* free energy change for the sum of the two reactions is negative, then the coupled pair will proceed in the forward direction.

$$A + B + E \rightarrow D + F + K \quad \Delta G = \Delta G_1 + \Delta G_2 < 0 \tag{6-32}$$

By this means the compound D may be formed to an extent much greater than would be possible if the first reaction were isolated. The second reaction is said to *drive* the first. This behavior may of course also be explained on a kinetic basis. If product C is removed by the second reaction as fast as it is formed, then the first reaction will never be able to approach equilibrium. The existence of such coupled reactions explains the paradox whereby many biochemical syntheses appear superficially to proceed "uphill" in the direction of increasing free energy.

The above is a consequence of the fact that biological organisms are not in a state of chemical equilibrium. A closed system at equilibrium does not require energy for its maintenance; nor can energy, or work, be obtained from it. Biological systems are rather in a *dynamic* steady state, which tends toward equilibrium, but does not attain it. Their persistence in the steady state requires a supply of energy from an external source.

6-5 PROTEOLYTIC ENZYMES

Zymogen Activation

Many enzymes which catalyze the hydrolysis of peptide bonds are synthesized in the form of inactive precursors or **zymogens**. The conversion of the zymogen to the active form is often catalyzed by another enzyme and usually involves a splitting of peptide bonds.

Two particularly well-understood examples of zymogen activation are the formation of trypsin and chymotrypsin from their respective precursors trypsinogen and chymotrypsinogen. Both zymogens are synthesized in specialized cells of the pancreas and subsequently excreted to the small intestine, where activation occurs.

Trypsinogen consists of a single polypeptide chain of molecular weight 24,000 which is cross-linked by six disulfide bridges. It may be activated by any of several peptidases, including trypsin itself and enterokinase, which occurs in intestinal secretions and is probably the physiological agent.

The activation of trypsinogen is accomplished by the scission of a single Lys-Ile peptide bond and the subsequent removal of the NH_2-terminal sequence Val-$(Asp)_4$-Lys to form a new NH_2-terminal isoleucine. An important change in optical rotation accompanies the process, suggesting that a conformational change occurs. It has been proposed that removal of the four negatively charged aspartate groups relieves electrostatic stress sufficiently to permit a folding of the NH_2-terminal end of the polypeptide so as to form the active conformation.

Chymotrypsinogen is a single polypeptide chain of molecular weight 25,000 cross-linked by five cystine bridges. It may be activated by trypsin to form a series of active derivatives (Fig. 6-6).

If activation is carried out rapidly at high trypsin levels, self-digestion of the products is minimized. Under these conditions the initial active product, π-chymotrypsin, may be identified. This differs from the zymogen solely by the hydrolysis of a single Arg-Ile bond, which is part of a closed

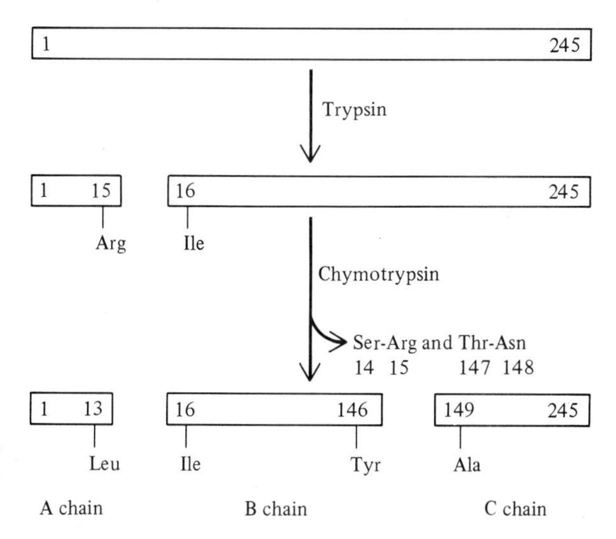

Figure 6-6 Activation of chymotrypsinogen by trypsin.

loop formed by cystine. No peptide is split off at this stage. Thus a seemingly minor covalent modification suffices to convert this protein from an inactive to an essentially completely active form.

The hydrolysis of a single Leu-Ser bond converts π to δ-chymotrypsin, with the loss of one residue each of serine and arginine and the formation of a new NH_2-terminal isoleucine. Rapid activation terminates here, with the formation of the relatively stable derivative δ-chymotrypsin.

In summary, a sequence Leu-Ser-Arg-Ile exists in chymotrypsinogen. During rapid activation the Arg-Ile bond is first split to form π-chymotrypsin. Removal of the Ser-Arg dipeptide yields δ-chymotrypsin.

If activation is carried out by prolonged ($\sim$48 h) incubation with low levels of trypsin, slow secondary reactions involving the action of chymotrypsin itself have time to occur. The principal product of this complex process is α-chymotrypsin, which is the classical "chymotrypsin." Its formation involves, in addition to the above changes in the Leu-Ser-Arg-Ile sequence, a rupture of peptide bonds in a second Tyr-Thr-Asn sequence. The Thr-Asn dipeptide is split off to form a new NH_2-terminal alanine and a new COOH-terminal tyrosine (Fig. 6-6).

As in the case of trypsinogen, the activation of chymotrypsinogen is accompanied by a major conformational change, as reflected by alterations in optical activity and other physical properties.

Trypsin and chymotrypsin, together with pepsin, carboxypeptidase, and elastase, comprise the group of digestive enzymes (Table 6-2) all of which are originally synthesized as zymogens and are subsequently converted to the active form. In each case activation occurs by peptide bond scission.

The four pancreatic zymogens—trypsinogen, chymotrypsinogen, proelastase, and procarboxypeptidase—appear to be activated by trypsin following an initial formation of catalytic amounts of trypsin by the action of enterokinase upon trypsinogen. Enterokinase is produced by cells lining the duodenum. The activation of trypsinogen by enterokinase is thus the controlling step for the activation of this set of enzymes. In this way their digestive action upon proteins is synchronized. The concerted activity of this group of proteolytic enzymes is essential, since each is specific for peptide bonds involving a limited number of side-chains (Table 6-2).

Table 6-2

The Major Digestive Enzymes

Site of Zymogen Synthesis	*Zymogen*	*Enzyme*	*Specificity*
Pancreas	Chymotrypsinogen	Chymotrypsin	Trp, Tyr, Phe, Ile
Pancreas	Trypsinogen	Trypsin	Lys, Arg
Pancreas	Proelastase	Elastase	Smaller, neutral side-chains
Pancreas	Procarboxypeptidase	Carboxypeptidase	COOH-terminal residues
Stomach	Pepsinogen	Pepsin	Broad

6-6 THE STRUCTURAL BASIS FOR THE ACTIVITY OF CHYMOTRYPSIN

Chymotrypsin catalyzes the hydrolysis of peptide, ester, and amide bonds involving the **α-carboxyl** group of an **aromatic** amino acid (tyrosine, phenylalanine, or tryptophan). The specificity is somewhat blurred since bonds involving the α-carboxyls of asparagine, glutamine, isoleucine, leucine, and methionine are also attacked at a finite rate. The rate is depressed if a free α-amino group is adjacent to the bond hydrolyzed; activity is restored if the amino group is acylated.

Substrates of chymotrypsin are known which are not derivatives of amino acids. One of these, *p*-nitrophenyl acetate (NPA), has been extensively used in studies of the mechanism of action of chymotrypsin.

The Active Center

Chymotrypsin and many other proteolytic enzymes, including trypsin, are completely and stoichiometrically inhibited by one mole of diisopropylfluorophosphate (DFP). The single DFP molecule that combines with the enzyme is attached to a reactive serine group in each case.

Partial hydrolysis of the DFP-derivatives and subsequent sequence analysis has permitted the identification of the amino acid sequences adjacent to the active serine, Ser-195. For both chymotrypsin and trypsin a Gly-Asp-Ser*-Gly- sequence occurs. Similar sequences have been found in other proteolytic enzymes.

From the pH dependence of activity and other evidence it has been concluded that a histidine group is also involved in the active center (Fig. 6-7).

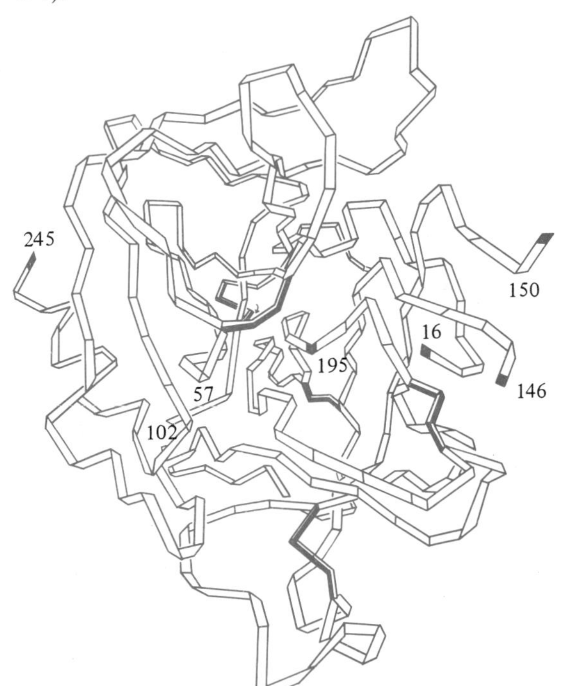

Figure 6-7 Location of the active center of chymotrypsin with respect to its three-dimensional structure.

Much of the available information as to the catalytic site of chymotrypsin has been obtained from kinetic studies with NPA, which is hydrolyzed to *p*-nitrophenol and acetate. This substrate has the advantage of permitting observation of the reaction by spectrophotometric techniques, because of the major change in absorbance accompanying the hydrolysis.

By the use of a stopped-flow device for rapid mixing of enzyme and substrate and special spectrophotometric techniques for observation, it has been possible to extend measurements to very short reaction times, of the order of 10^{-2} sec. This has permitted examination of the initial **transient phase** which precedes attainment of the steady state and lasts only a few seconds.

This approach has revealed that the reaction takes place in at least three distinguishable steps:

(1) $E + S \rightarrow ES_1$
(2) $ES_1 \rightarrow ES_2 + P_1$
(3) $ES_2 \rightarrow E + P_2$

Step (1), which is rapid, represents the initial formation of the enzyme-substrate complex. The remaining steps correspond to the conversion of substrate. Step (2) results in the formation of an acylated derivative of the enzyme and the liberation of one of the products, *p*-nitrophenol. The probable site of attachment of the acetyl group is the active serine. In the final step the acetyl group is released and transferred to a water molecule to form the other product, acetate (Fig. 6-8).

The catalytic activity of chymotrypsin depends upon the unusual properties of Ser-195. As the—CH_2OH group of serine in proteins is usually quite inert, it is clear that the microenvironment of this group in chymotrypsin must have distinctive features.

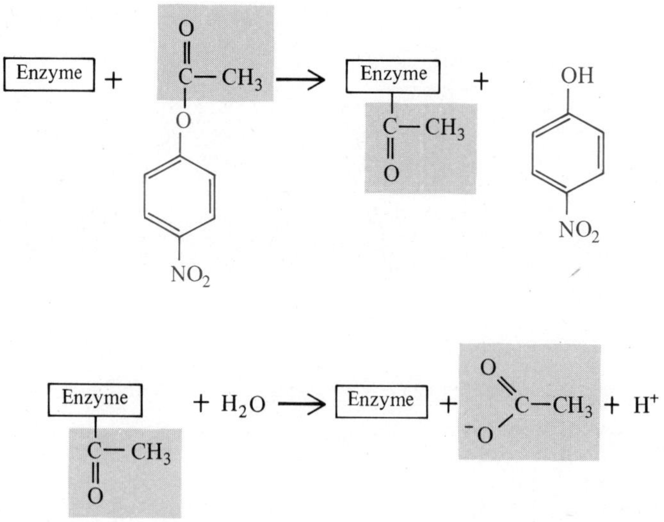

Enzyme + O=C(—CH$_3$)—O—C$_6$H$_4$—NO$_2$ ⟶ Enzyme—C(=O)—CH$_3$ + HO—C$_6$H$_4$—NO$_2$

Enzyme—C(=O)—CH$_3$ + H_2O ⟶ Enzyme + O=C(—CH$_3$)—O$^-$ + H^+

Figure 6-8
Proposed mechanism for the hydrolysis of *p*-nitrophenyl acetate by chymotrypsin.

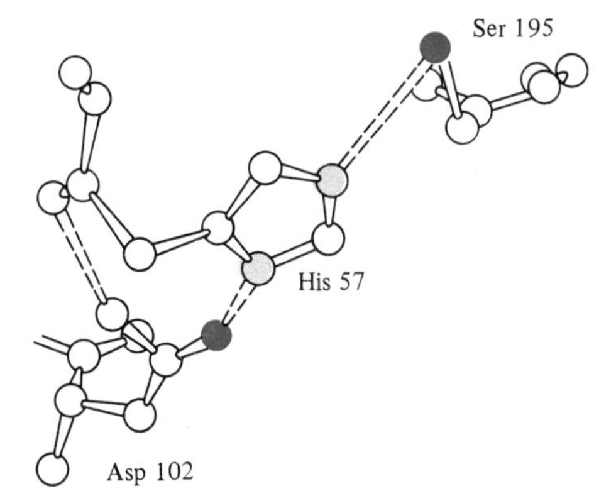

Figure 6-9
Elements of the active site of chymotrypsin.

X-ray crystallographic studies by Blow and others have revealed that this is indeed the case and have provided a plausible explanation. In chymotrypsin His-57 is adjacent to Ser-195, while the carboxyl-containing side-chain of Asp-102 is nearby (Fig. 6-9). It is the interaction of these three residues which is responsible for the catalytic activity of chymotrypsin. His-57 is hydrogen-bonded to both Asp-102 and to Ser-195 (Fig. 6-10). The carboxyl group of Asp-102 will tend to attract a proton from the histidine which in turn pulls a proton away from the hydroxyl group of serine. As a consequence of the operation of this **charge relay network** a small but significant fraction of the active center exists in the form shown in the lower half of Fig. 6-10, with a negative charge upon the oxygen atom of Ser-195.

The presence of this negative charge completely alters the properties of Ser-195, converting it from a relatively inert group to a potent nucleophile. This can attack the carbonyl carbon of either the acetyl group of NPA or of a peptide linkage (Fig. 6-11).

The catalyzed hydrolysis of a peptide bond begins with the reaction of the serine oxygen with the peptide carbonyl carbon to form a transient

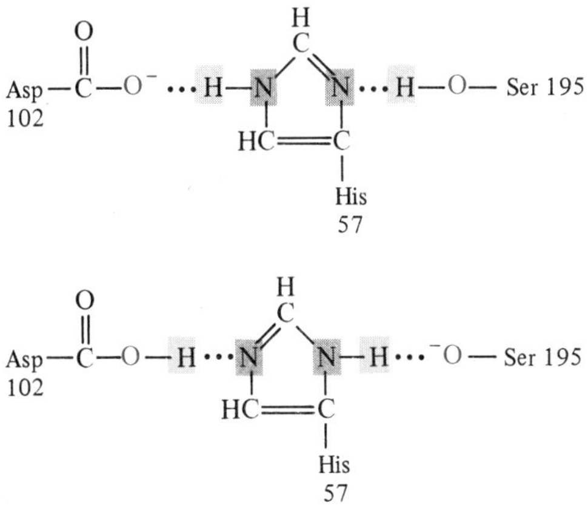

Asp 102 —C(=O)—O⁻ ··· H—N(His 57 imidazole: HC, C–H, HC=C)N ··· H—O— Ser 195

Asp 102 —C(=O)—O—H ··· N(His 57 imidazole: HC, C–H, HC=C)N—H ··· ⁻O— Ser 195

Figure 6-10
Charge relay system of chymotrypsin.

tetrahedral intermediate. Cleavage of the peptide bond occurs by donation of a proton by His-57 to the peptide nitrogen (Fig. 6-11). The amino component of the former peptide bond in now hydrogen bonded to His-57, while the acyl component is covalently attached to Ser-195.

In the final stage of the process the amino component diffuses away and is replaced by a water molecule at the active site. Hydrolytic deacylation occurs in much the same way as the initial peptide cleavage. In this case the charge relay network draws a proton away from water to form an OH^- ion that subsequently attacks the carbonyl carbon of the acyl group linked to Ser-195. As in the case of acylation a tetrahedral intermediate is transiently formed. A proton is then donated by His-57 to the oxygen of Ser-195, resulting in the release of the acyl group. Upon diffusion of the latter, the enzyme reverts to its initial state and is ready for another catalytic cycle.

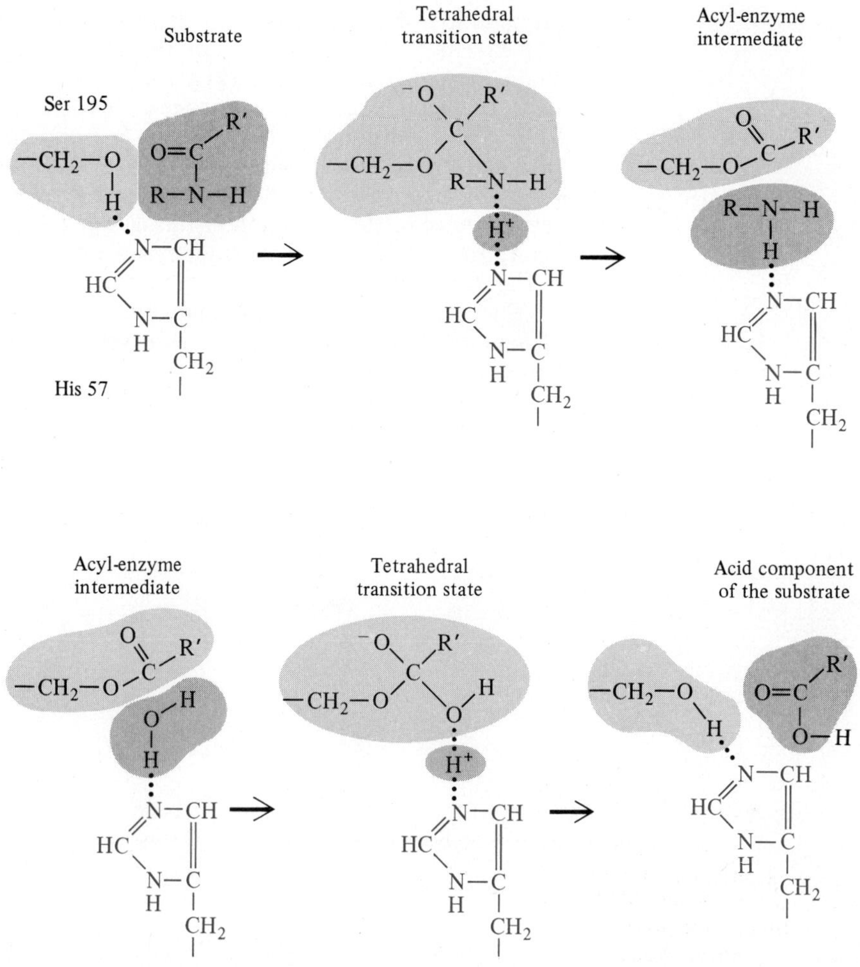

Figure 6-11
Hydrolysis of a peptide bond by chymotrypsin.

Activation of Chymotrypsinogen

The establishment of the three-dimensional structures of chymotrypsin and chymotrypsinogen by x-ray crystallography has permitted a plausible explanation of the proteolytic activation of chymotrypsinogen. Chymotrypsinogen has the shape of a compact spheroid. The α-helical content is very low. While the charged groups lie generally on the surface, the critical group Asp-194 is anomalously withdrawn to some degree into the interior.

The hydrolytic scission of the peptide bond between Arg-15 and Ile-16 gives rise to a new α-carboxyl and a new α-amino group. The protonated α-amino group of Ile-16 interacts electrostatically with the negatively charged α-carboxylate group of Asp-194 (Fig. 6-12). The electrostatic force between the two oppositely charged groups pulls them into proximity, dragging portions of the polypeptide chain with them. The resultant minor conformational change brings the elements of the active site into the correct juxtaposition. The structural rearrangement also produces a nonpolar cavity whose dimensions are appropriate for the binding of nonpolar aromatic side-chains, such as tryptophan, tyrosine, and phenylalanine. This may be responsible for the specificity characteristics of chymotrypsin.

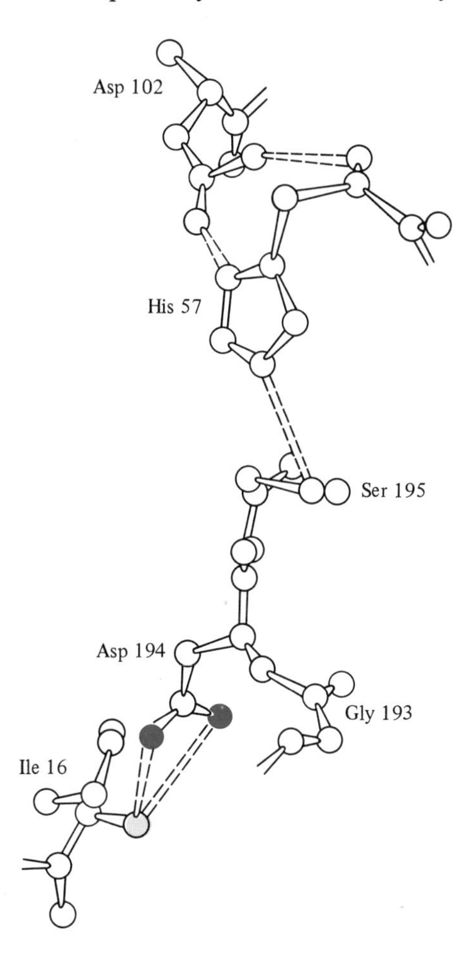

Figure 6-12 Relationship of the amino acids involved in the activation of chymotrypsin.

6-7 THE CHYMOTRYPSIN FAMILY OF ENZYMES

The pancreatic enzymes trypsin and elastase are remarkably similar to chymotrypsin in molecular architecture. Almost half of the amino acid sequences of the three are equivalent and the modes of folding of their polypeptide chains, as deduced from x-ray diffraction, are quite similar.

All three possess active serines and are inhibited stoichiometrically by DFP. In addition, all three contain the same amino acid sequence, Gly-Asp-Ser*-Gly-Gly-Pro about the active serine. All three are activated proteolytically. It is probable that their catalytic mechanisms are analogous and depend upon a charge relay mechanism.

The only dramatic differences in properties of the three lie in their substrate specificities, particularly with respect to the nature of the amino acid contributing the peptide carbonyl. While chymotrypsin prefers aromatic side-chains, the specificity of trypsin is confined absolutely to the basic amino acids lysine and arginine, and that of elastase favors the smaller uncharged side-chains.

The explanation probably resides in the differing nature of the substrate binding site in each case. The nonpolar cavity in proximity to the active center of chymotrypsin can accommodate a bulky nonpolar side-chain. Trypsin contains a similar cavity, but with a negatively charged aspartic group replacing an uncharged group. Electrostatic interaction between the negatively charged aspartic and the positively charged side-chains of lysine or arginine probably accounts for the substrate preference of trypsin. The dimensions of the corresponding pocket of elastase are such as to exclude the larger side-chains.

It is probable that trypsin, chymotrypsin, and elastase arose from the action of the evolutionary process upon a common ancestor. The differences between the three reflect the summation of the mutational events occurring in the course of evolutionary time.

6-8 OTHER PROTEOLYTIC ENZYMES

Pepsin

Pepsin, the principal protein-hydrolyzing enzyme of the stomach, is formed by the activation of the inactive precursor pepsinogen, which is secreted by the gastric mucosa. Pepsinogen consists of a single polypeptide chain of molecular weight 42,000 which is cross-linked by three cystine bridges.

The conversion of pepsinogen to pepsin occurs spontaneously in acid solution, or by the action of pepsin itself. The reaction is autocatalytic, since pepsin is a product, although it passes through an intermediate stage in which pepsin is complexed with a peptide inhibitor. During the activation process pepsinogen is cleaved to form the pepsin–inhibitor complex plus five small peptides of total molecular weight 4,000.

$$\underset{42{,}000}{\text{pepsinogen}} \xrightarrow[\text{pepsin}]{\text{acid or}} \underset{38{,}000}{\text{pepsin–inhibitor complex}} + \underset{4{,}000}{\text{5 peptides}}$$

$$\underset{38{,}000}{\text{pepsin–inhibitor complex}} \longrightarrow \underset{35{,}000}{\text{pepsin}} + \underset{3{,}000}{\text{inhibitor}}$$

The substrate specificity of pepsin is less restrictive than those of trypsin or chymotrypsin. Many peptide bonds are attacked at varying rates. Optimal activity is attained at pH's below 3.

Subtilisin

This bacterial enzyme, whose specificity is broad, resembles trypsin and chymotrypsin in possessing a reactive serine, whose combination with DFP blocks enzymic activity. The prevailing evidence indicates that the proteolytic activity of **subtilisin** arises from a charge relay system similar to that occurring in chymotrypsin.

However, there is little structural similarity between the two enzymes, which differ widely in amino acid sequence and three-dimensional molecular organization. Subtilisin is not cross-linked, while chymotrypsin contains five disulfide bridges. The amino acid sequence Gly-Asp-Ser*-Gly-Gly-Pro about the active serine of chymotrypsin is replaced by Gly-Thr-Ser*-Met-Ala-Ser in the case of subtilisin.

It is thus most unlikely that subtilisin and chymotrypsin have a common ancestor. It appears rather that they represent, at the molecular level, an example of the **evolutionary convergence** so familiar in living systems.

Papain

Papain, an enzyme of molecular weight 23,000 that occurs in papaya fruit, is a **sulfhydryl** rather than a serine enzyme. A cysteine group occurs in its active center and is essential for activity. A striking feature of the three-dimensional structure of papain is the presence of a deep crevice containing the active site. Like subtilisin, papain has a broad specificity, attacking a variety of peptide bonds.

The Carboxypeptidases

Carboxypeptidase A and **carboxypeptidase B** are pancreatic enzymes which are synthesized as the inactive precursors procarboxypeptidase A and B. Procarboxypeptidase A is a complex of three subunits with a molecular weight near 88,000. It is activated by cleavage by trypsin to form ultimately carboxypeptidase A of molecular weight 35,000.

Carboxypeptidase A is an example of a **metallo-enzyme**; Zn is an essential component of its active site. The Zn atom lies in a broad depression in the surface of the roughly spherical molecule and is tetrahedrally coordinated, with three of its four ligands coming from the oxygens or nitrogens of Glu, His, and Lys and the fourth contributed by the substrate.

The carboxypeptidases, whose specificities have been discussed in chapter 4, differ from the other proteolytic enzymes described above in that their hydrolytic action is confined to the carboxyl terminus of a polypeptide or oligopeptide, so that they catalyze the stepwise removal of amino acids from the COOH-terminal end.

6-9 ALLOSTERIC ENZYMES

Feedback Control of Enzymes

Living systems have a number of mechanisms for controlling the level of a biochemical intermediate available to them. Among the most important

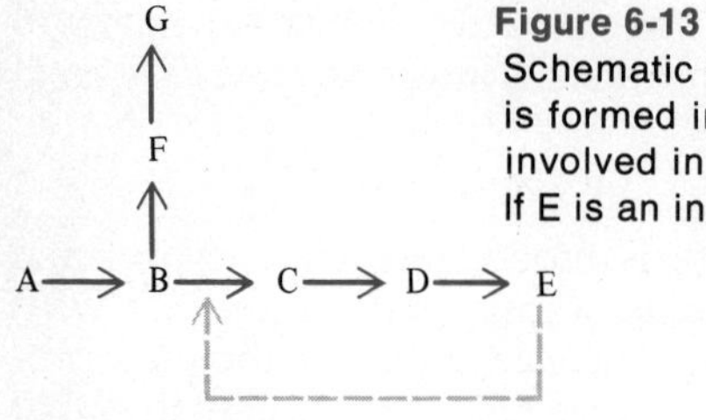

Figure 6-13
Schematic representation of the feedback principle. A product, E, which is formed in a metabolic pathway, influences the activity of an enzyme involved in an earlier stage of the pathway in the conversion of B to C. If E is an inhibitor, the pathway may be diverted to the formation of G.

of these is the self-regulating process whereby an intermediate controls its own biosynthesis by modifying the activity of an enzyme involved in its formation (Fig. 6-13).

This is a biological version of the familiar principle of **feedback** control, whereby a system is furnished information as to the results of its operation and modifies its activity accordingly. Without control mechanisms of some kind, biosynthetic processes would operate in a wasteful manner, with intermediates accumulating in excess of need.

In particular cases, the modification of enzymic activity may take the form of either inhibition or activation. In either case the enzyme must be able to *recognize* the regulatory intermediate so as to display an inherent specificity for binding the substance.

There are endless variations of the basic scheme. For example, in a branched metabolic pathway, two intermediates lying on different branches may both act as modifiers of a single enzyme in such a way that maximal inhibition is achieved only by the simultaneous action of both intermediates. In this way an excessive concentration of one intermediate will not block the biosynthesis of the other.

Allosterism

The general designation **allosteric** has been applied to enzymes whose activity is influenced by the binding of a modifier at a **regulatory** site and that exhibit **cooperativity** in the binding of substrate or modifier. The sensitivity of the response of allosteric enzymes to variations in the level of substrates or modifiers is generally enhanced by the existence of pronounced *cooperativity* of binding, that is, the binding of one molecule of modifier increases the affinity of the enzyme for binding additional molecules. The binding of modifier may enhance the binding of more modifier (**homotropic cooperativity**) or of substrate (**heterotropic cooperativity**). The presence of cooperativity implies the existence of *multiple* binding sites with strong mutual interactions. It results in a sharpening of the response to modifier, so that the effects of a change in modifier level are amplified.

All known allosteric enzymes have been found to consist of combinations of two or more subunits known as **protomers**. Catalytic sites for the conversion of substrate are present, as well as sites for the binding of modifier. The latter are usually distinct from the catalytic sites. The modifier may be the substrate itself, a product of the enzymic reaction, an intermediate lying elsewhere on the metabolic pathway, or an unrelated molecule. The protomers may be identical or different.

In the enzymes of this class which possess separate allosteric and catalytic sites, the regulatory modifiers are bound by the former and the influence of the bound modifier upon the catalytic site is indirect. In most cases this is believed to occur via a conformational change induced by the modifier which alters the catalytic site.

It has become customary to employ the term allosterism in a somewhat wider sense so as to encompass the cooperative interactions involved in the binding of a single ligand molecule to a protein containing multiple sites, even if the protein is not an enzyme. Thus, the oxygen-transporting protein hemoglobin is often referred to as an allosteric protein, since the cooperative binding of oxygen by its four subunits obeys principles similar to those involved in the binding of substrate or modifier by an allosteric enzyme.

Models of Allosteric Behavior

We shall consider here only the simplest case where the enzyme consists of two or more equivalent subunits, the substrate is itself a modifier, and the allosteric regulation occurs via modification of the *binding* of substrate. Each protomer contains a single site for the binding of substrate and may exist in either of two conformations, called R and T (Fig. 6-14). Form R has a high affinity for substrate, while form T has low or zero affinity. Forms R and T are interconvertible. In the absence of substrate, an equilibrium constant, L, may be defined by

$$\mathrm{L} = \frac{[\mathrm{T}]}{[\mathrm{R}]} \tag{6-33}$$

where T and R are the respective concentrations of the two forms in the absence of bound substrate.

We may also define an equilibrium constant (K_{R}) for the binding of a molecule of substrate by a (hypothetical) isolated R subunit. For simplicity, we shall assume that form T does not bind substrate at all.

$$K_{\mathrm{R}} = \frac{[\mathrm{RS}]}{[\mathrm{R}][\mathrm{S}]} \tag{6-34}$$

where [RS] is the concentration of the complex species.

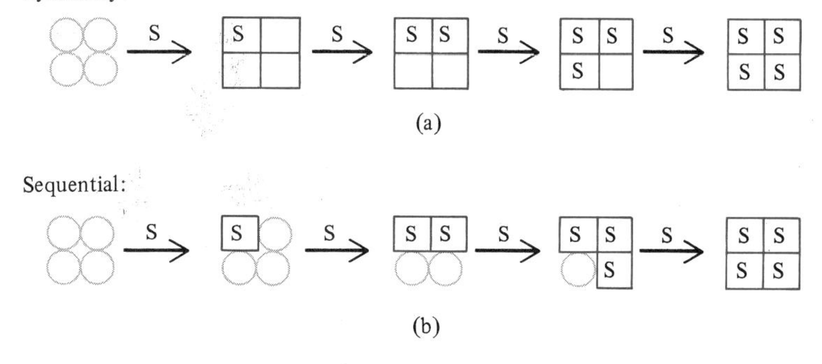

Figure 6-14 Schematic depiction of the conformational changes involved in the the binding of modifier by a tetrameric allosteric enzyme. (a) Monod–Wyman–Changeux model; (b) Koshland–Nemethy–Filmer model.

At this point a basic distinction arises between the principal theoretical models. In the **symmetry** model of Monod, Wyman, and Changeux *all* the protomers of an enzyme molecule are postulated to exist either in the R or the T conformation with no mixed species allowed. Thus the binding of a single substrate molecule by an enzyme molecule, all of whose subunits are in the T form, requires the *simultaneous* transition of all subunits to the R form (Fig. 6-14). For an enzyme consisting of q subunits:

$$R_q \rightleftarrows T_q \tag{6-35}$$

$$R_q + S \rightleftarrows R_qS$$

The fraction of occupied sites, θ, is then given by

$$\theta = \frac{K_R[S](1 + K_R[S])^{q-1}}{L + (1 + K_R[S])^q} \tag{6-36}$$

Equation (6-36) predicts a *sigmoidal*, rather than hyperbolic, shape for the variation of θ with [S] (Fig. 6-15). The fraction of enzyme sites which are occupied increases more rapidly with increasing substrate level (after a critical level of substrate is reached) than would be predicted for simple Michaelis-Menten behavior; the system is cooperative. The binding of substrate shifts the conformational equilibrium in the direction of the R form. When the sites are completely saturated, all of the subunits are in the R conformation.

The binding of an allosteric activator other than the substrate can be described in an equivalent manner. The above model can also account for the action of an allosteric inhibitor. In this case the inhibitor is preferentially bound by the T form and shifts the conformational equilibrium in the direction of the inactive conformation. An enzyme that is subject to both allosteric

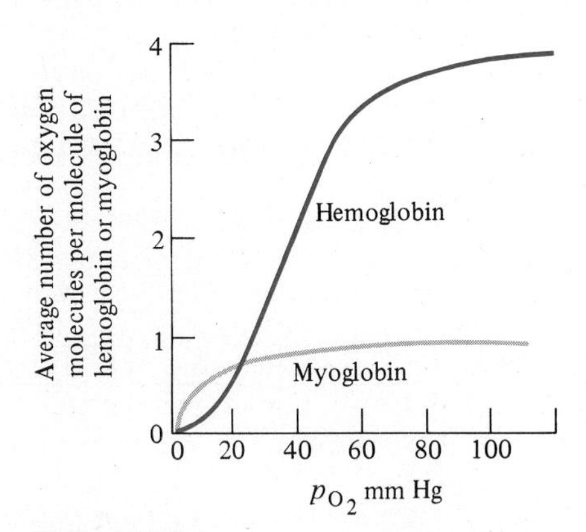

Figure 6-15
An illustration of the contrast between the cooperative and noncooperative binding of a ligand by a protein. The examples shown are the noncooperative binding of oxygen by myoglobin, which has only one subunit, and the cooperative binding of oxygen by hemoglobin, which consists of four subunits. Note the sigmoidal (S-shaped) character of the latter curve.

activation and allosteric inhibition may be highly responsive to a change in conditions.

The alternative **sequential** model developed by Koshland, Nemethy, and Filmer differs from that of Monod, Wyman, and Changeux in allowing mixed enzyme states in which R and T forms coexist (Fig. 6-14). The binding of substrate by a subunit and its consequent confinement to the R state does not force the other subunits into the R conformation. Cooperativity arises from the modification of intersubunit interactions as a consequence of the binding of substrate.

In practice it is difficult to differentiate between the two models from binding data alone. While the application of physical techniques, in principle, may permit a choice between the two, the question remains unsettled in most cases. There is of course no guarantee that the central assumption of only two conformational states may not fail in some instances.

SUMMARY

Enzymes are proteins with **catalytic** properties, which are able to increase the rates of specific chemical reactions. The catalytic activity is localized in a restricted area of the protein surface termed the **active center**. The action of an enzyme proceeds via an initial reversible binding of the **substrate** to form an **enzyme-substrate** complex; this may either **dissociate** to reform the reactants or **decompose** to yield the products.

The kinetics of an enzymic reaction may be interpreted on a simple basis for the case of a single substrate, as for an isomerization of the type $A \rightarrow P$, or for the case where the second substrate is present in overwhelming excess, as for a hydrolytic reaction of the type $A + H_2O \rightarrow P$. The **Michaelis-Menten** model assumes that the concentration of the enzyme-substrate complex is constant during the **steady-state** phase and derives an equation relating the observed velocity as a function of substrate level to the **Michaelis constant**, which determines the degree of saturation of the enzyme with substrate, and the **limiting velocity** when the enzyme is completely saturated with substrate.

Many enzymes require for their activity the interaction with a small molecule or ion. This may be reversibly bound or may be effectively integrated into the enzyme structure. The examples include **metal ions; coenzymes**, which are organic molecules participating in the reaction; and **primers**, which resemble the product and may be incorporated into it.

While enzymes accelerate the **rate** of a biochemical reaction, they cannot influence the position of **equilibrium**, which is governed by **thermodynamic** considerations. A chemical reaction always proceeds in the direction corresponding to a *decrease* in **free energy**. The predicted free energy change per mole may be related to the **standard free energy change** for unit concentrations of reactants and products and the set of actual concentrations. A thermodynamically unfavored reaction may be *driven* by a second favorable reaction with which it is *coupled.*

Among the best understood enzymic mechanisms is that for the hydrolysis of a peptide bond or of the synthetic substrate *p*-nitrophenol acetate by the proteolytic enzyme **chymotrypsin**. In the active center of chymotrypsin, **histidine** 57 and **serine** 195 are bonded in such a way as to promote a **nucleophilic** attack of the serine oxygen upon the carboxyl carbon of the peptide or ester bond. A transient **acyl** derivative of the serine hydroxyl is formed. The histidine group subsequently promotes the transfer of the acyl group to the second substrate, water.

Chymotrypsin is synthesized as the inactive precursor **chymotrypsinogen**. Full activation occurs via proteolytic scission of a single peptide bond. Subsequent proteolysis by chymotrypsin itself yields a series of active enzymes.

In biological systems the regulation of the rate of a stepwise metabolic pathway is often achieved by a **feedback** mechanism, whereby a product formed by a particular reaction step modifies the activity of an **allosteric** enzyme that catalyzes a reaction occurring earlier in the pathway. Allosteric enzymes always contain two or more **subunits** and generally contain an allosteric site that is distinct from the catalytic site. The binding of a **modifier** by the allosteric site induces a **conformational change** in the subunit which alters the activity of the catalytic site. The conformational change is transmitted by a cooperative mechanism to other subunits, thereby sharpening the response of the enzyme to modifier.

REFERENCES

The following code is used to classify references. I: particularly useful as an introduction to the subject; R: useful primarily as a reference text; A: an advanced account of the material; H: a publication of historical importance.

General

S. A. Bernhard, *The Structure and Function of Enzymes*, Benjamin, New York (1968). (A)

P. D. Boyer, ed., *The Enzymes*, Academic, New York (1970). (R)

S. P. Colowick and N. O. Kaplan, *Methods in Enzymology*, Academic, New York (1954). (R)

M. Dixon and E. C. Webb, *Enzymes*, 2nd ed., Academic, New York (1963). (A)

H. Gutfreund, *Enzymes: Physical Principles*, Wiley-Interscience, New York (1972). (A)

Chymotrypsin

D. M. Blow, in *The Enzymes*, P. D. Boyer, ed., vol. 3, p. 185, Academic, New York (1971). (A)

R. Henderson, C. S. Wright, G. P. Hess, and D. M. Blow, *Cold Spring Harbor Symp. Quant. Biol.* 36, 63 (1971). (A)

Allosterism

D. E. Koshland, G. Nemethy, and D. Filmer, *Biochemistry* 5, 365 (1966). (H)

J. Monod, J. P. Changeux, and F. Jacob, *J. Mol. Biol.* 6, 306 (1963). (H)

J. Monod, J. Wyman, and J. P. Changeux, *J. Mol. Biol.* 12, 88 (1965). (H)

REVIEW QUESTIONS

Questions marked with an asterisk are of a high level of difficulty.

*6-1 Suggest an amino acid side-chain that might be in the active sites of enzymes that employ the kinds of catalysis listed below; draw the side-chain structure in the active form and estimate the amount of the active form which will be present at pH 7.2 (10%, 1%, 0.1%, and so on), assuming normal pK's.
- (a) catalysis by nucleophilic attack
- (b) general base catalysis (accept a proton from the substrate)
- (c) nonpolar environment in active site

6-2 An enzyme catalyzes the phosphorylation of α-D-glucose only to produce 6-phosphoglucose as product. If the enzyme is allowed to operate on 50 ml of a 0.1-M solution of glucose in aqueous solution, how many moles of final product can theoretically be produced? Assume the phosphorylation reaction is essentially irreversible.

6-3 What will the velocity (v) of an enzyme-catalyzed reaction be equal to when $[S] = K_m$, in the presence of a competitive inhibitor for which $[I] = K_i$?

6-4 Answer the following questions regarding the kinetics of an enzyme-catalyzed reaction.
- (a) For an enzyme that obeys simple Michaelis-Menten kinetics, what is the V_{max} in μmole/min if $v = 35$ μmole/min when $[S] = K_m$?
- (b) What is the K_m of this enzyme if $v = 40$ μmole/min when $[S] = 2 \times 10^{-5} M$?
- (c) If I is a competitive inhibitor of the enzyme with a K_i of $4 \times 10^{-5} M$, what will be the value of v when $[S] = 3 \times 10^{-2} M$ and $[I] = 3 \times 10^{-5} M$?

6-5 Which of the following statements are correct?
- (a) Equilibrium thermodynamics can be useful in biochemistry for
  - (1) understanding the forces responsible for maintaining macromolecular conformations.
  - (2) predicting the rates of biochemical reactions.
  - (3) determining the energy requirements of living cells at thermal equilibrium.
  - (4) understanding the mechanisms of enzyme catalysis of biochemical reactions.
  - (5) none of the above
- (b) In general, systems at constant temperature and pressure will always change spontaneously in the direction of
  - (1) lower energy.
  - (2) higher energy.
  - (3) lower entropy.
  - (4) higher entropy.
  - (5) none of the above

6-6 List at least two reasons why almost all enzyme-catalyzed reactions show a pH optimum.

6-7 How would you determine whether a specific inhibitor of an enzyme-catalyzed reaction was a competitive or noncompetitive inhibitor?

6-8 Define the following terms:

(a) coenzyme
(b) activator
(c) prosthetic group
(d) specific activity
(e) enzyme denaturation
(f) zymogen
(g) active site
(h) K_m
(i) zero order kinetics
(j) first order kinetics

6-9 A certain enzyme-catalyzed reaction has a velocity of 65 μmole/l min at a pH of 7.2. When the reaction was buffered at pH 8.2, the velocity was 125 μmole/l min. What reasons can you suggest for the observed effect of the pH on the reaction velocity?

6-10 Draw curves showing how the velocity of an enzyme-catalyzed reaction varies with the following variables: (label axes)

(a) time (at a substrate concentration very low relative to the K_m)
(b) enzyme concentration
(c) substrate concentration, in the presence of a competitive inhibitor

6-11 The Michaelis-Menten equation, derived to describe the rate of an enzyme-catalyzed reaction S → P, has the form $v = A[S]/(B + [S])$, where A and B are constants.

(a) What is the simplest reaction mechanism which yields this expression?
(b) Which of the following *must be assumed* to obtain this expression?
  (1) Binding of substrate to enzyme is at thermodynamic equilibrium.
  (2) There is only one intermediate enzyme form besides free enzyme.
  (3) The amount of substrate converted into product during the measurement is much less than S.

6-12 Fill in the blanks with the *most correct* thermodynamic terms.

(a) All chemical and physical changes are accompanied by an overall increase in the ______ of the universe.
(b) The activities of living cells always cause a decrease in the ______ of the environment.

6-13 Allosteric control of enzymic activity involves which of the following?

(a) binding to the active site by the allosteric modifier
(b) a conformational change in protein structure
(c) binding of the allosteric modifier to a site on the enzyme separate from the active site
(d) both (b) and (c)
(e) none of the above

7

Coenzymes and Biological Oxidations

7-1 THE FUNCTION OF COENZYMES

The activity of many enzymes is dependent upon the participation of a small organic molecule termed a **coenzyme**. If this is tightly combined with the enzyme, it is often referred to as a **prosthetic group**. A loosely bound coenzyme, whose association with the enzyme is only transient, is a **cosubstrate**.

Coenzymes of the latter class usually function as donors or acceptors of a particular group that is transferred to, or from, a substrate molecule. Their reaction with each substrate obeys a definite stoichiometry and is noncatalytic in nature. However, such coenzymes often act in conjunction with two different enzymes, acting as an acceptor in one reaction and as a donor in the other. In this way the coenzyme serves to *couple* the two enzymic reactions and functions, in a sense, as a catalyst.

When the coenzyme is tightly combined with the enzyme it generally acts to transfer a group between two different substrates, functioning as an acceptor for the first substrate and as a donor for the second. After the second transfer the coenzyme returns to its initial state and may repeat the process.

Many coenzymes have a close structural relationship to **vitamins**. The latter are essential nutritional components that must be supplied in the diet since the organism is incapable of synthesizing them. In general, a vitamin is the chief, or sole, constituent of a coenzyme. The deficiency of a particular vitamin in the diet results in characteristic clinical manifesta-

Table 7-1
The Principal Coenzymes and the Groups That They Transfer

Coenzyme	*Group Transferred*
Coenzyme A	Acyl group
Tetrahydrofolic acid	Formyl group and other one-carbon groups
Biotin	CO_2
Pyridoxal phosphate	Amino group of an amino acid
Thiamine pyrophosphate	Aldehyde
Nicotinamide adenine dinucleotide (NAD^+)	Hydrogen (hydride ion)
Nicotinamide adenine dinucleotide phosphate ($NADP^+$)	Hydrogen (hydride ion)
Flavin mononucleotide (FMN)	Hydrogen
Flavin-adenine dinucleotide (FAD)	Hydrogen
Lipoic acid	Hydrogen; acetyl group
Coenzyme Q	Hydrogen

tions. Some of the most devastating diseases known to history, including scurvy, pellagra, and beriberi, fall in this category.

Table 7-1 lists the principal coenzymes, together with the nature of the groups transferred.

7-2 COENZYMES THAT TRANSFER GROUPS OTHER THAN HYDROGEN

Coenzyme A

Coenzyme A transfers carboxylic acids. The structure of coenzyme A (CoA) is relatively complex, consisting of one residue each of pantetheine phosphate and adenosine-3′,5′-diphosphate (Fig. 7-1). The latter is an example of a **nucleotide** (chapter 18). Pantothenic acid, the major constituent of pantetheine (Fig. 7-1), is one of the B vitamins.

The part of coenzyme A which is active in transferring acyl groups is the end bearing the sulfhydryl group. The most important of its derivatives is **acetyl CoA**. As the removal of the acetyl group is highly favored thermodynamically, this is a potent acetylating agent.

The reactions of acetyl CoA include the following:

(1) **Ester or amide formation** A hydrogen atom of an alcohol or an amine may be replaced by the acyl group:

$$H_3C{-}\overset{\overset{\displaystyle O}{\|}}{C}{\sim}SCoA + HO{-}R \longrightarrow H_3C{-}\overset{\overset{\displaystyle O}{\|}}{C}{-}O{-}R + HSCoA$$

Here, HSCoA represents the uncombined form of coenzyme A, in which the sulfhydryl group is unsubstituted. The wavy line ($\sim$) in the structure of acetyl CoA denotes an unstable **high energy bond**.

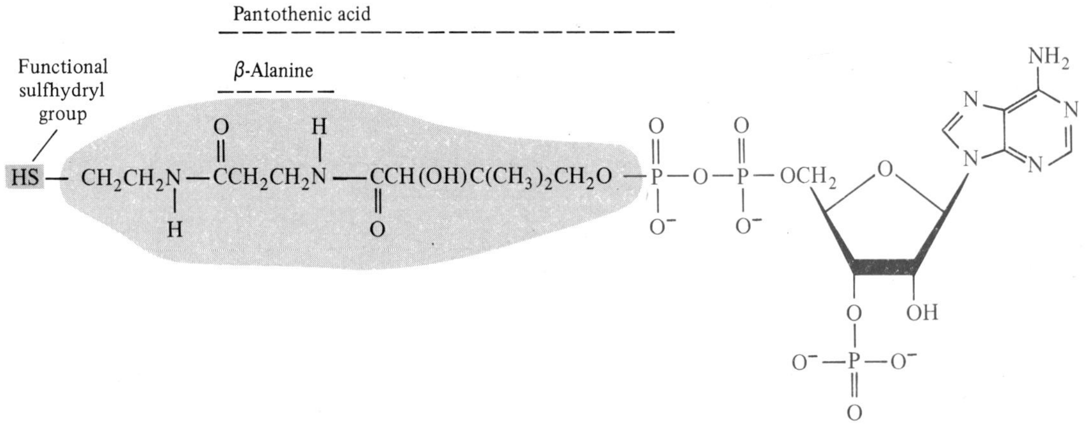

$$\mathrm{HS{-}CH_2CH_2\overset{H}{N}{-}\overset{O}{\overset{\|}{C}}CH_2CH_2\overset{H}{N}{-}\underset{O}{\underset{\|}{C}}CH(OH)C(CH_3)_2CH_2O{-}\overset{O}{\overset{\|}{\underset{O^-}{P}}}{-}O{-}\overset{O}{\overset{\|}{\underset{O^-}{P}}}{-}OCH_2}\text{-ribose(3'-}OPO_3^{2-}\text{)-adenine}$$

Figure 7-1
Structure of coenzyme A. Here (and in Figs. 7-13 and 7-14) a ribose group is shown in the **Haworth** form. For a discussion of this form see section 9-2. The portion of the molecule to the left of the diphosphate group is called pantetheine.

(2) **Condensation** Acetyl CoA may transfer an acetyl group to a second molecule:

$$H_3C{-}\overset{O}{\overset{\|}{C}}{\sim}SCoA + H_3C{-}\overset{O}{\overset{\|}{C}}{\sim}SCoA \longrightarrow H_3C{-}\overset{O}{\overset{\|}{C}}{-}CH_2{-}\overset{O}{\overset{\|}{C}}{\sim}SCoA + HSCoA$$

The unusual reactivity of the methyl group of acetyl CoA may stem from a *fractional negative charge* arising from its polarization via the thioester bond. The methyl carbon thereby becomes a good nucleophile.

(3) **Methyl group reactions** An important example is the formation of citric acid:

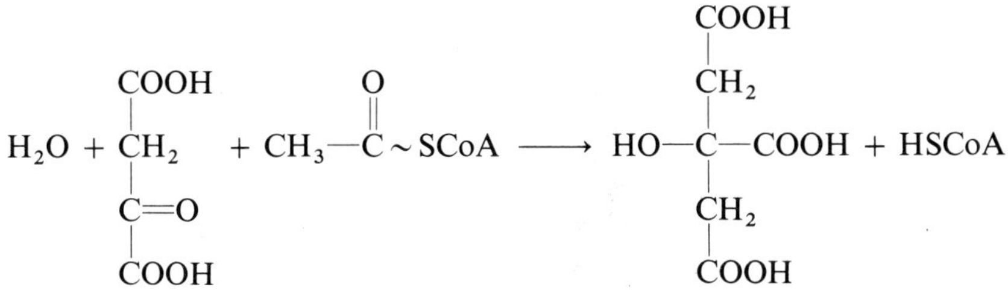

$$H_2O + \underset{\text{Oxaloacetic acid}}{\begin{array}{c}COOH\\|\\CH_2\\|\\C{=}O\\|\\COOH\end{array}} + CH_3{-}\overset{O}{\overset{\|}{C}}{\sim}SCoA \longrightarrow \underset{\text{Citric acid}}{\begin{array}{c}COOH\\|\\CH_2\\|\\HO{-}C{-}COOH\\|\\CH_2\\|\\COOH\end{array}} + HSCoA$$

As in case (2) the activation of the methyl group permits acylation.

Tetrahydrofolic Acid Folic acid (Fig. 7-2) is a vitamin composed of **pteroic** acid plus glutamic acid. The former component contains a **pteridine** ring, a fused heterocyclic

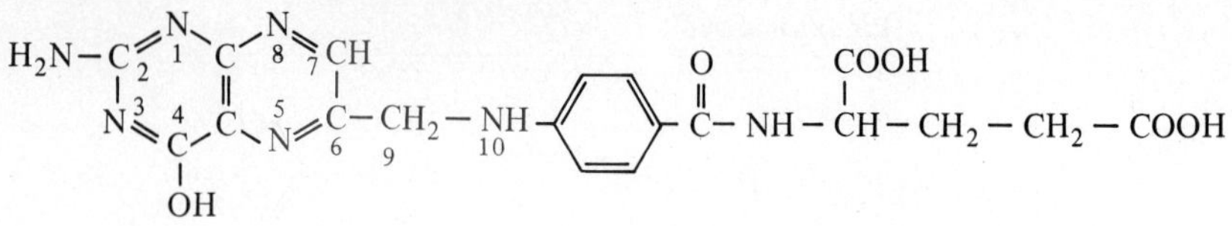

Figure 7-2
Structure of folic acid.

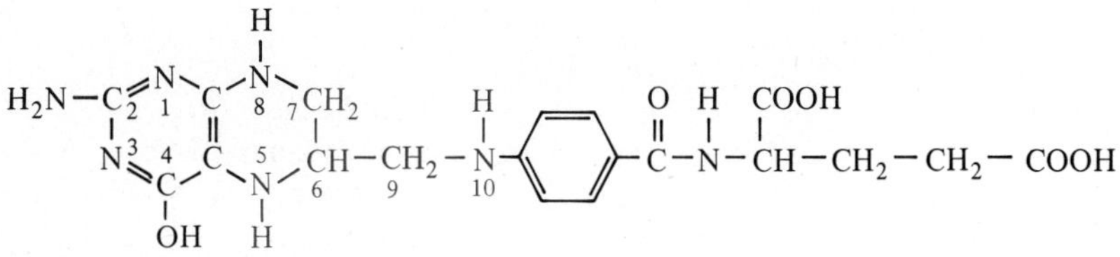

Figure 7-3
Structure of tetrahydrofolic acid.

ring system. In tetrahydrofolic acid the pteridine ring is hydrogenated (Fig. 7-3).

Folic acid derivatives figure in the transfer of *single carbon* units, which may be in the form of formyl, formaldehyde, hydroxymethyl, methyl, or formimino groups. It is the reduced form, **tetrahydrofolic acid** (FH_4), which is active in the one-carbon transfers.

The transfer function is localized in the part of the molecule containing the N_5 and N_{10} atoms. Two derivatives that are active in the transfer of formyl groups are N_5- or N_{10}-formyltetrahydrofolic acid or **active formate**. These may be represented in condensed form by

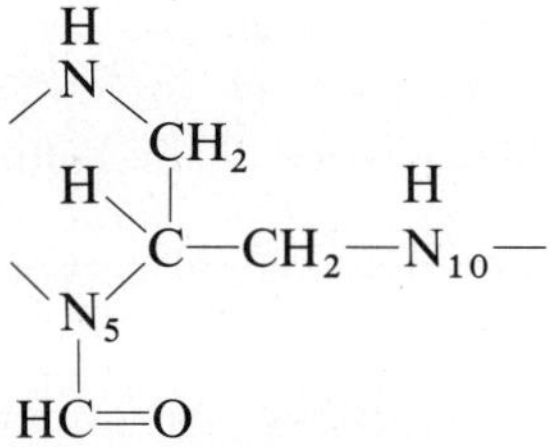

N_5-formyl FH_4 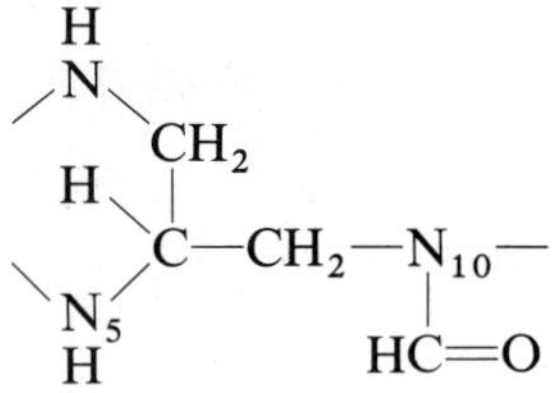 $\qquad$ N_{10}-formyl FH_4

The N_5-formyl and N_{10}-formyl derivatives may be interconverted by an enzymic reaction.

Two other derivatives of importance are N_5,N_{10}-methylene-FH_4 (Fig. 7-4), in which N_5 and N_{10} are joined by a methylene group, and N_5,N_{10}-methenyl-FH_4 (or N_5,N_{10}-anhydroformyl-FH_4) (Fig. 7-5), in which the

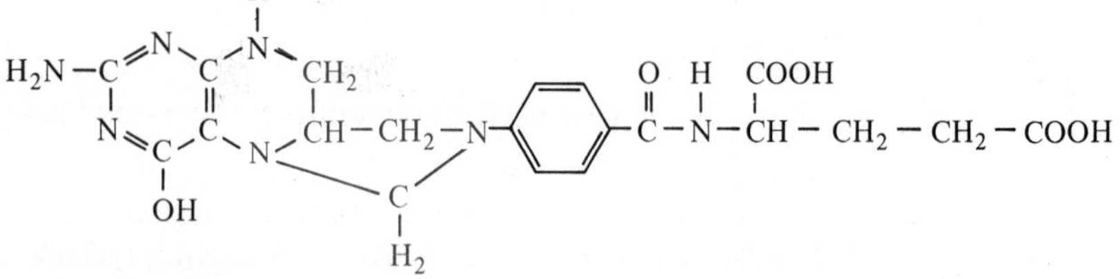

Figure 7-4
Structure of N_5,N_{10}-methylenetetrahydrofolic acid.

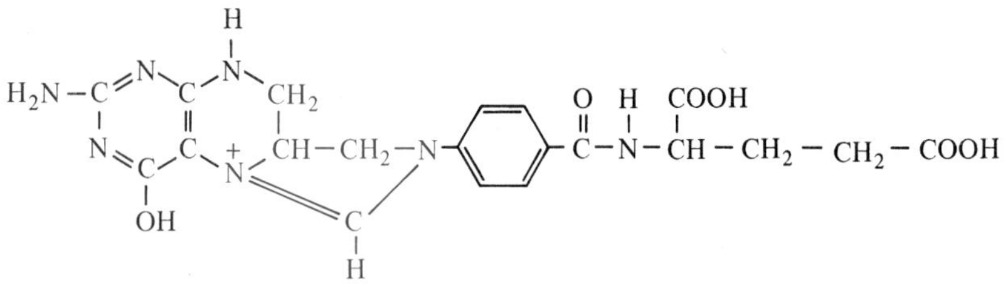

Figure 7-5 Structure of N_5,N_{10}-anhydroformyltetrahydrofolic acid (N_5,N_{10}-methenyltetrahydrofolic acid).

methylene group is partially dehydrogenated. The two are enzymically interconvertible by a dehydrogenase. N_5,N_{10}-methenyl-FH_4 may revert to N_{10}-formyl-FH_4 in alkaline solution by the addition of OH^-. The critical structural regions are

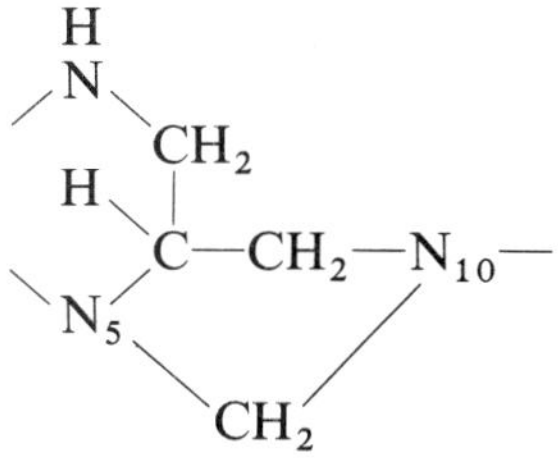

N_5,N_{10}-methylene-FH_4 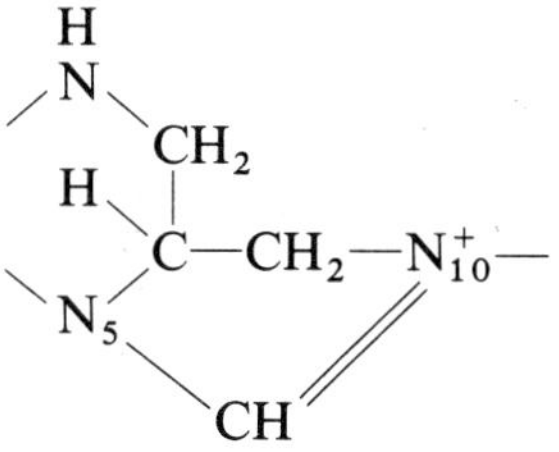 N_5,N_{10}-methenyl-FH_4

Formaldehyde reacts directly with FH_4 to yield N_5,N_{10}-methylene-FH_4, which is involved in several enzymic processes.

Other forms frequently encountered in metabolic pathways are N_{10}-hydroxymethyl-FH_4 and N_5-formimino-FH_4.

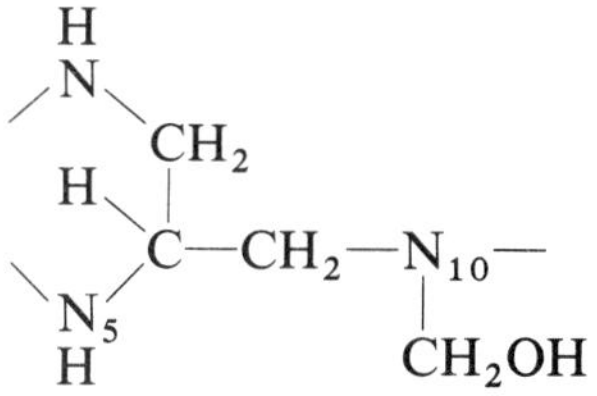

N_{10}-hydroxymethyl-FH_4 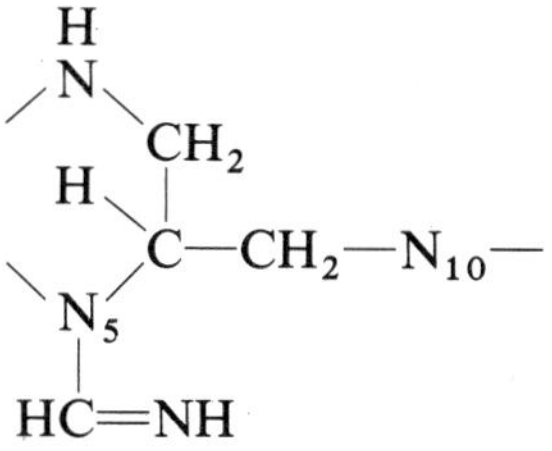 N_5-formimino-FH_4

Biotin

Biotin, which has been isolated from liver and egg yolk, is a cyclic derivative of urea, with an attached thiophane ring (Fig. 7-6). Biotin is normally linked

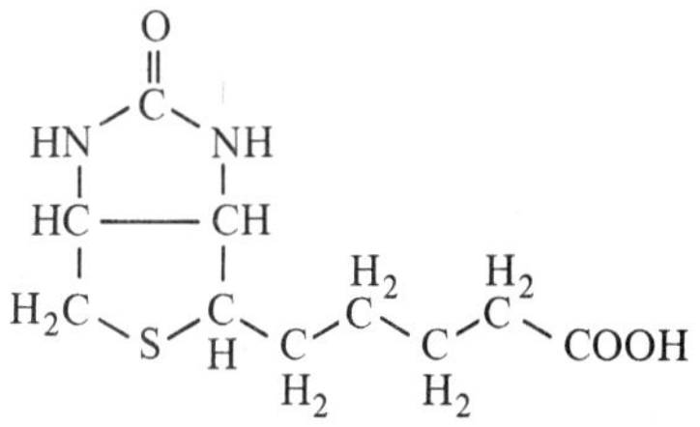

Figure 7-6 Structure of biotin.

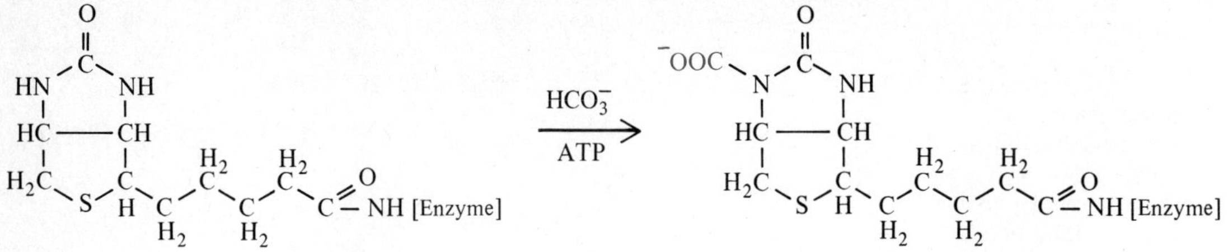

Figure 7-7
Activation of CO_2 by biotin.

to its enzyme by a peptide bond and is regarded as a prosthetic group (Fig. 7-7).

The biological function of biotin is to bind and activate carbon dioxide (Fig. 7-7). **Active carbon dioxide** figures in many carboxylation reactions.

Pyridoxal Phosphate

This coenzyme is important in amino acid metabolism. It is related to pyridoxine, or vitamin B_6, and to pyridoxamine phosphate (Fig. 7-8).

Pyridoxal phosphate acts as a cofactor for several different classes of enzyme, including decarboxylases and aminotransferases. It is believed that many transamination reactions involving amino acids proceed via formation of a Schiff's base (Fig. 7-9).

Thiamine Pyrophosphate

Thiamine (vitamin B_1), the chief component of this coenzyme, was among the first vitamins to be recognized. Its deficiency in the human diet produces the characteristic symptoms of beriberi.

Thiamine (Fig. 7-10) contains two separate heterocyclic rings: one pyrimidine and one thiazol. The coenzyme itself is the pyrophosphate of thiamine.

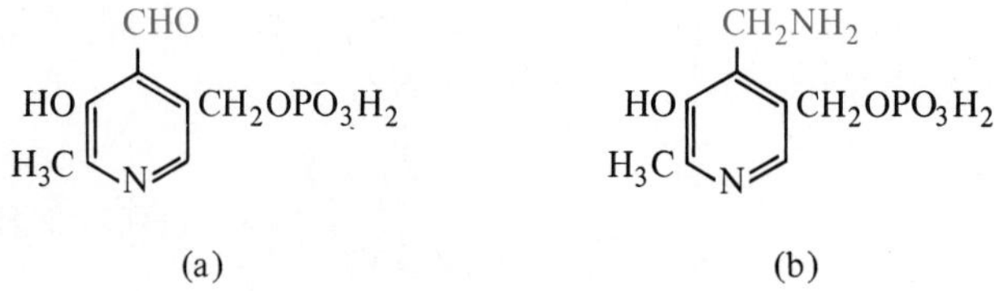

Figure 7-8
Structures of (a) pyridoxal phosphate and (b) pyridoxamine phosphate.

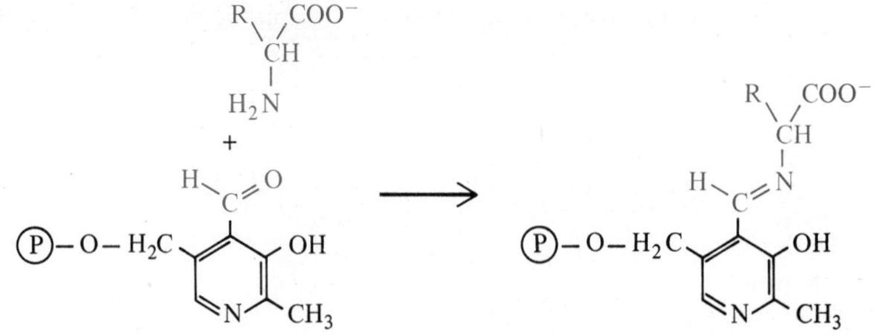

Figure 7-9
Formation of a Schiff's base by pyridoxal phosphate with an amino acid.

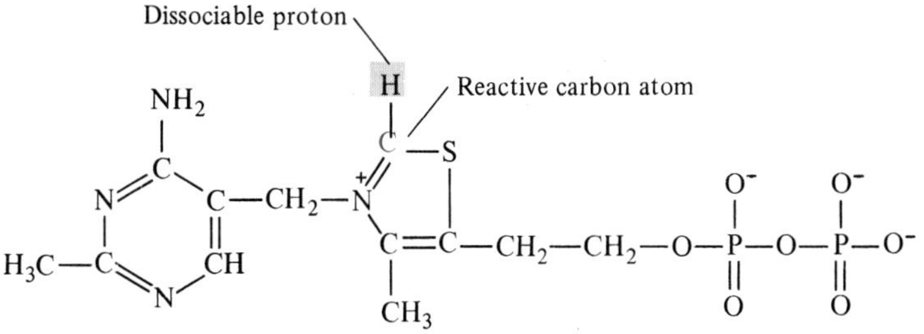

Figure 7-10
Strucutre of thiamine pyrophosphate.

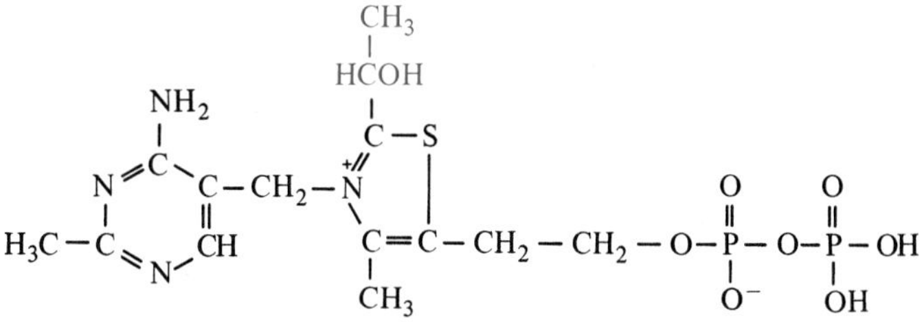

Figure 7-11
Active acetaldehyde.

Since it is normally tightly bound to the enzyme, it must be considered a prosthetic group.

Thiamine pyrophosphate acts as a carrier for **active acetaldehyde** and **active glycoaldehyde**. In these derivatives the aldehyde group is attached to the thiazol ring as indicated (Fig. 7-11).

Thiamine pyrophosphate functions in the oxidative decarboxylation of α-keto acids. CO_2 is split off and the aldehyde residue is transferred by thiamine pyrophosphate to lipoic acid (section 7-4).

S-Adenosylmethionine

This coenzyme acts as a carrier of methyl groups and may be regarded as an active form of methionine in transmethylation reactions. The molecule contains one residue each of methionine and the nucleoside adenosine (Fig. 7-12). The sulfur atom forms part of a "sulfonium" group and is positively charged (Fig. 7-12).

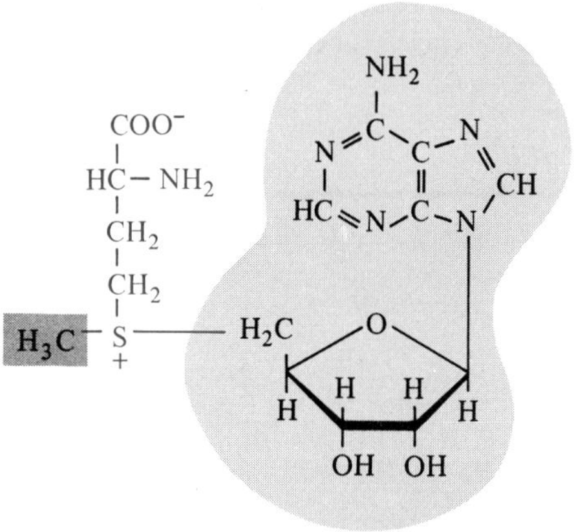

Figure 7-12
Structure of *S*-adenosylmethionine.

7-3 COENZYMES THAT TRANSFER HYDROGEN

The Nicotinamide Coenzymes

Nicotinamide is a derivative of nicotinic acid or niacin, another B vitamin, which serves to prevent the nutritional disease pellagra. **Nicotinamide adenine dinucleotide** (NAD^+) contains one nicotinamide group attached by an N-glycosidic linkage to ribose, which is joined by a double ester of pyrophosphoric acid to the ribose portion of the nucleoside adenosine (Fig. 7-13).

Nicotinamide adenine dinucleotide phosphate ($NADP^+$) differs from NAD^+ in that the adenosine component is esterified by an additional phosphate group in the 2′-position of its ribose moiety. The function of both coenzymes is the reversible uptake and subsequent transfer of hydrogen. In the oxidized, or nonhydrogenated, form the ring nitrogen of the nicotinamide portion is quaternary and carries a single positive charge (Figs. 7-13 and 7-14). In the reduced, or hydrogenated, form the nitrogen loses its charge and the nicotinamide ring consequently loses its aromatic character, retaining only two double bonds (Fig. 7-15). The hydrogenated coenzymes are designated as NADH and NADPH.

The hydrogenation of NAD^+ and $NADP^+$ produces a major change in absorption spectrum, which may be utilized to monitor enzymic reactions involving these coenzymes (Fig. 7-16). An absorption peak which is not present in the oxidized forms arises at 340 nm.

The nicotinamide coenzymes figure in many dehydrogenation reactions, including numerous dehydrogenations of primary and secondary alcohol groups. The reactions are usually reversible. Much of the importance of these coenzymes lies in their capability of reversibly transferring hydrogen (and electrons) between two different substrates, thereby coupling the action of two different enzymes.

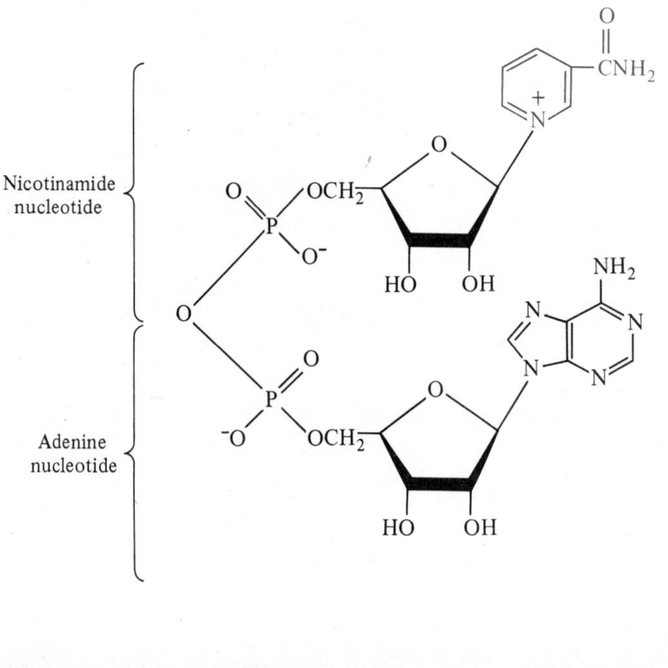

Figure 7-13
Structure of NAD^+.

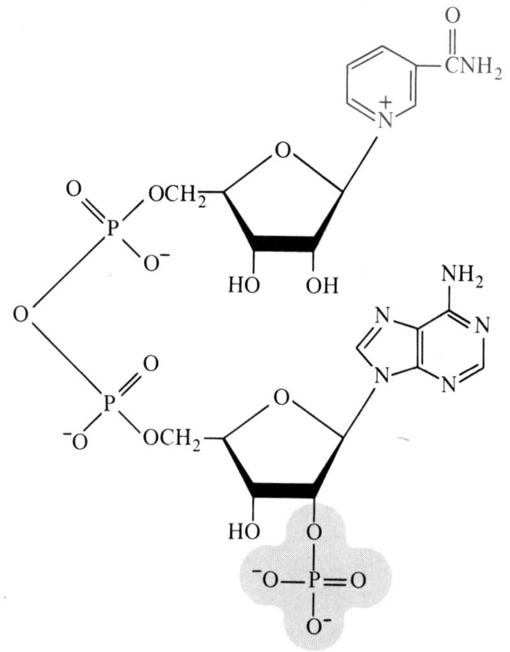

Figure 7-14
Structure of $NADP^+$.

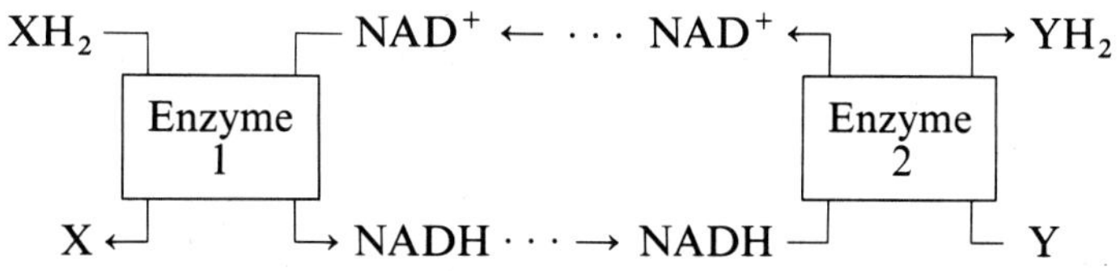

$$XH_2 + NAD^+ \rightarrow X + NADH + H^+$$

$$H^+ + Y + NADH \rightarrow YH_2 + NAD^+$$

In the conversion of NAD^+ to NADH the latter acquires two electrons and the half-reaction may be written

$$NAD^+ + H^+ + 2e^- \rightarrow NADH$$

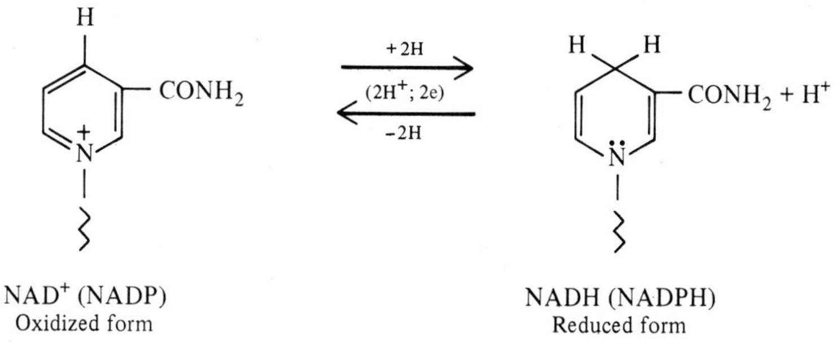

NAD^+ (NADP)
Oxidized form

NADH (NADPH)
Reduced form

Figure 7-15
Conversion of NAD^+ (or $NADP^+$) to NADH (or NADPH). The changes occurring in the active nicotinamide portion are shown.

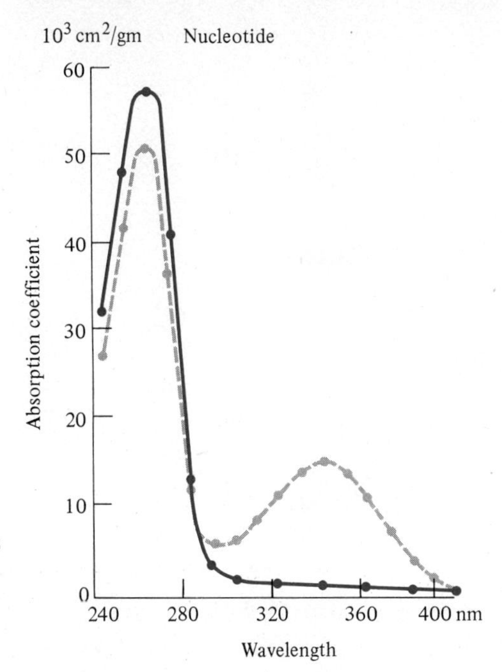

Figure 7-16
Ultraviolet absorption spectra of NADH (–○–) and NAD^+ (–●–).

The Flavin Coenzymes

These contain **riboflavin** or vitamin B_2, which is a derivative of isoalloxazine (Fig. 7-17). The side-chain is ribitol, a pentahydroxy compound.

The two flavin coenzymes are **flavin mononucleotide** (FMN), or riboflavin-5′ phosphate, and **flavin adenine dinucleotide** (FAD). The latter contains, in addition to riboflavin, a residue of adenosine-5′-diphosphate (Fig. 7-18).

Like the nicotinamide coenzymes, FMN and FAD are concerned with hydrogen transfer. The isoalloxazine ring of riboflavin can add two hydro-

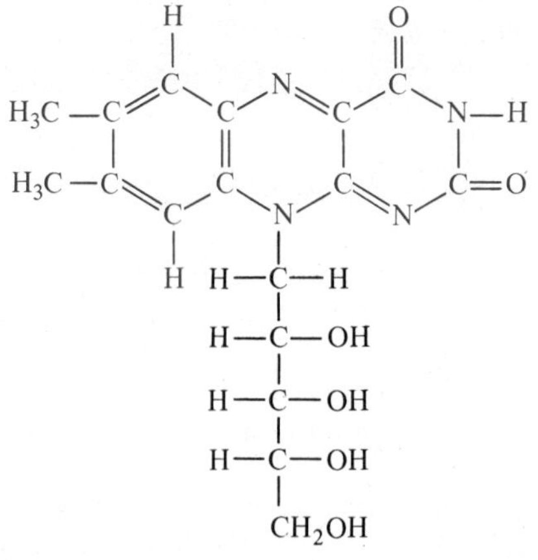

Figure 7-17
Structure of riboflavin.

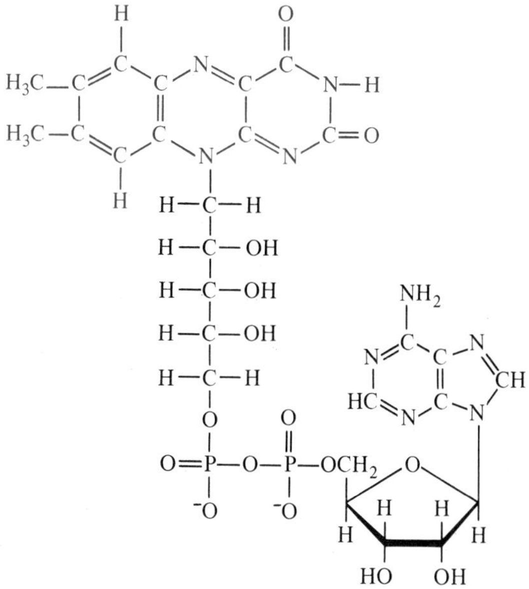

Figure 7-18
Structure of FAD.

gens at the N_1 and N_{10} positions (Fig. 7-19). The flavin coenzymes are generally tightly bound to their enzymes and are considered to be prosthetic groups.

Hydrogen is transferred to the prosthetic group by the action of the enzyme. The flavin coenzyme is usually reoxidized by another enzyme system, permitting repetition of the process. The formation of the reduced flavin coenzymes involves the transfer of two electrons.

Lipoic Acid

Lipoic acid, or thioctic acid, is a cyclic disulfide containing a side-chain terminating in a carboxyl group (Fig. 7-20). The latter is usually attached to the enzyme protein by an amide linkage.

Lipoic acid is of central importance in oxidative decarboxylation, where it functions for hydrogen transfer and as a factor in the formation of active acetate. The reduced form of lipoic acid contains two sulfhydryl groups (Fig.

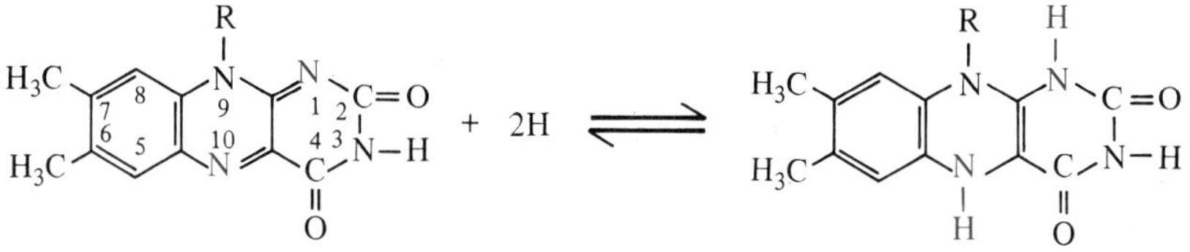

$+ \; 2H \; \rightleftharpoons$

Figure 7-19
Oxidized and reduced forms of riboflavin. R represents the ribitol group. The oxidized form of the isoalloxazine group is on the left and the reduced form is on the right. Hydrogen atoms are added to the N atoms at the 1 and 10 positions as shown.

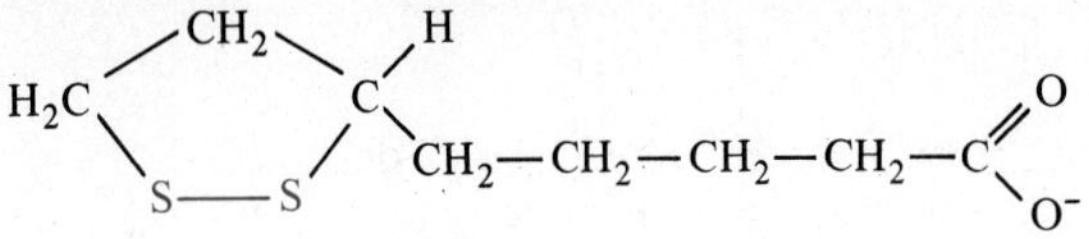

Figure 7-20
Oxidized form of lipoic acid (1,2-dithiolane-3-valeric acid).

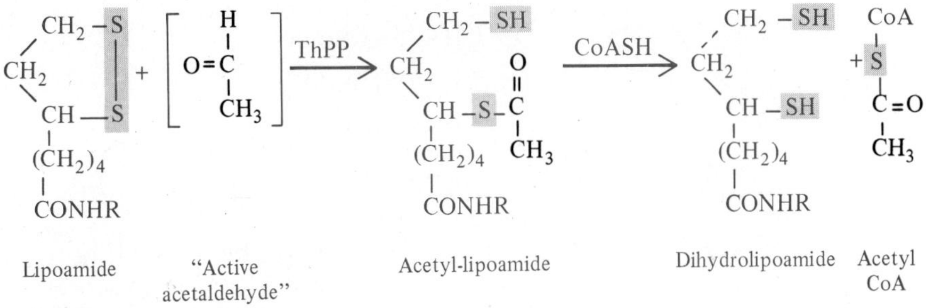

CH_2 H
H_2C C $CH_2—CH_2—CH_2—CH_2—C$ O O^-
S—S

CH_2—S, CH_2, CH—S, $(CH_2)_4$, CONHR + [H, O=C, CH_3] $\xrightarrow{ThPP}$ CH_2—SH, CH_2, CH—S—C(=O)CH_3, $(CH_2)_4$, CONHR $\xrightarrow{CoASH}$ CH_2—SH, CH_2, CH—SH, $(CH_2)_4$, CONHR + CoA—S—C=O—CH_3

Lipoamide "Active acetaldehyde" Acetyl-lipoamide Dihydrolipoamide Acetyl CoA

Figure 7-21
Reduced lipoic acid as a carrier of an acyl group.

7-21) and can accept an acetyl group from active acetaldehyde. The acetyl group is then transferred to HSCoA and the disulfide form of lipoic acid is reformed by a dehydrogenase.

Coenzyme Q

This coenzyme, sometimes called **ubiquinone**, is a quinone derivative linked to a long unsaturated hydrocarbon side-chain. The latter is a polymer of isoprene units ($—CH_2—CH{=}\overset{CH_3}{\overset{|}{C}}—CH_2—$). In mammals the most abundant form contains 10 isoprene units and is accordingly designated CoQ_{10} (Fig. 7-22).

Coenzyme Q is prominent in the terminal phases of oxidative metabolism, being reduced by $FMNH_2$ and $FADH_2$ and in turn reducing the first of a series of cytochromes that comprise the electron transport chain (chapter 12).

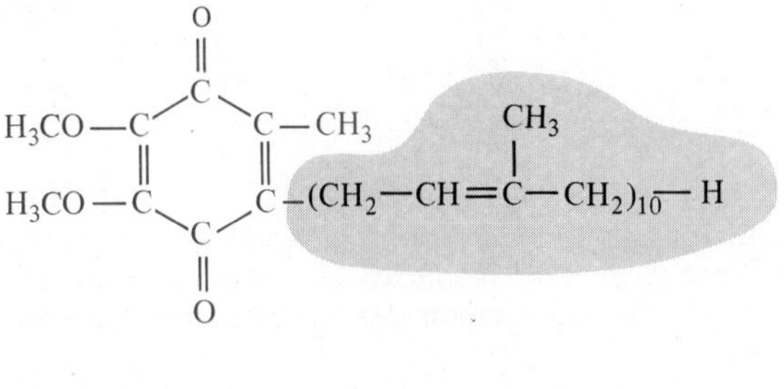

O
$H_3CO—C$ C $C—CH_3$ CH_3
$H_3CO—C$ C $C—(CH_2—CH{=}C—CH_2)_{10}—H$
O

Figure 7-22
Structure of coenzyme Q_{10}.

7-4 OXIDATION AND REDUCTION

The term **oxidation** originally designated chemical processes in which oxygen was directly involved; it is used today in a broader sense to encompass all reactions involving a loss of one or more electrons. Conversely, the acquisition of one or more electrons is termed **reduction**. Oxidation and reduction are most readily visualized for inorganic ions.

$$\begin{aligned} &\text{Oxidation:} \quad Fe^{2+} \rightarrow Fe^{3+} + e^- \\ &\text{Reduction:} \quad Fe^{3+} + e^- \rightarrow Fe^{2+} \end{aligned} \tag{7-1}$$

The oxidation of molecular hydrogen may be written

$$H_2 \rightarrow 2H^+ + 2e^- \tag{7-2}$$

The electrons must be accepted by some oxidizing agent, which is reduced. If ferric ion functions as the oxidizing agent,

$$\begin{aligned} H_2 &\rightarrow 2H^+ + 2e^- \\ 2Fe^{3+} + 2e^- &\rightarrow 2Fe^{2+} \\ \text{Sum:} \quad H_2 + 2Fe^{3+} &\rightarrow 2H^+ + 2Fe^{2+} \end{aligned} \tag{7-3}$$

When molecular oxygen functions as an oxidizing agent, it may acquire either two or four electrons:

$$\begin{aligned} O_2 + 2e^- &\rightarrow O_2^{2-} \xrightarrow{2H^+} H_2O_2 \\ O_2 + 4e^- &\rightarrow 2O^{2-} \xrightarrow{4H^+} 2H_2O \end{aligned} \tag{7-4}$$

The combination of the above reactions results in the formation of water:

$$\begin{aligned} 2H_2 &\rightarrow 4H^+ + 4e^- \\ O_2 + 4e^- &\rightarrow 2O^{2-} \\ 4H^+ + 2O^{2-} &\rightarrow 2H_2O \\ \text{Sum:} \quad 2H_2 + O_2 &\rightarrow 2H_2O \end{aligned} \tag{7-5}$$

All oxidation-reduction processes can be formally expressed in terms of the transfer of electrons. For example, the overall reaction for the oxidation of hydroquinone may be written

$$\text{Hydroquinone (OH, OH)} + \tfrac{1}{2}O_2 \longrightarrow \text{Quinone (O, O)} + H_2O \tag{7-6}$$

However, it may be alternatively written in terms of electron transfer, as a dissociation of hydrogen ions being followed by a transfer of electrons.

If ferric ion is the oxidizing agent,

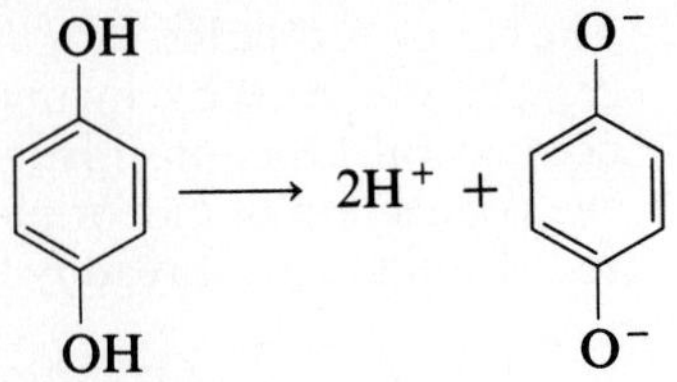

$$\text{HO-C}_6\text{H}_4\text{-OH} \longrightarrow 2H^+ + {}^-\text{O-C}_6\text{H}_4\text{-O}^- \quad (7\text{-}7)$$

$$^-\text{O-C}_6\text{H}_4\text{-O}^- \longrightarrow \text{O=C}_6\text{H}_4\text{=O} + 2e^-$$

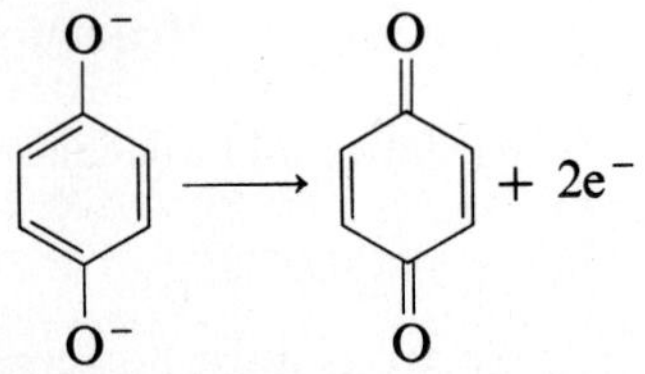

$$2Fe^{3+} + 2e^- \longrightarrow 2Fe^{2+}$$

$$\text{Sum: } \text{HO-C}_6\text{H}_4\text{-OH} + 2Fe^{3+} \longrightarrow \text{O=C}_6\text{H}_4\text{=O} + 2Fe^{2+} + 2H^+$$

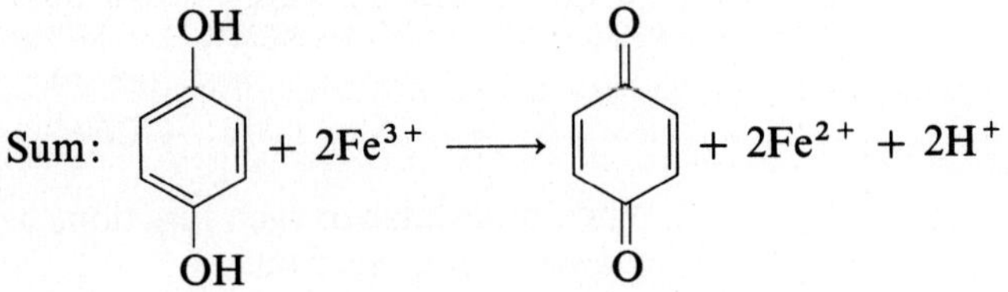

In actuality, the oxidation of hydroquinone is stepwise, proceeding by way of **semiquinone**, which is a free radical containing an unpaired electron:

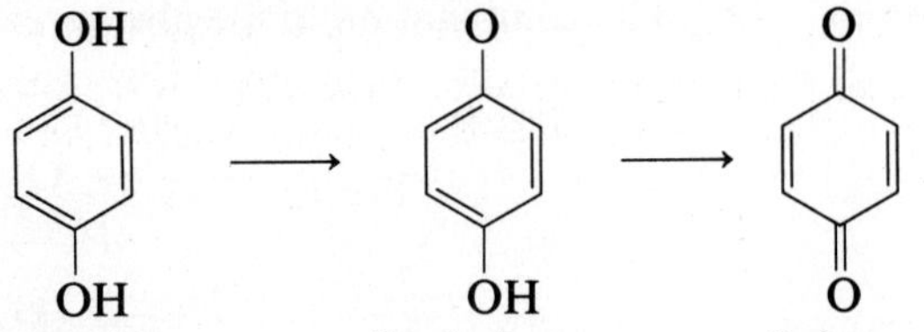

$$\underset{\text{Hydroquinone}}{\text{HO-C}_6\text{H}_4\text{-OH}} \longrightarrow \underset{\text{Semiquinone}}{\text{O-C}_6\text{H}_4\text{-OH}} \longrightarrow \underset{\text{Quinone}}{\text{O=C}_6\text{H}_4\text{=O}} \quad (7\text{-}8)$$

The dehydrogenation of a primary alcohol may also be written in terms of electron transfer:

$$CH_3\text{—}CH_2\text{—}OH \longrightarrow CH_3\text{—}CH_2\text{—}O^- + H^+ \quad (7\text{-}9)$$

$$CH_3\text{—}CH_2\text{—}O^- \longrightarrow CH_3\text{—}CH\text{=}O + H^+ + 2e^-$$

$$\text{Sum:}\quad CH_3-\underset{\underset{H}{|}}{\overset{\overset{H}{|}}{C}}-OH \longrightarrow CH_3-\overset{\overset{H}{|}}{C}=O + 2H$$

Many biological oxidations are equivalent to the withdrawal of hydrogen.

7-5 THE REDOX POTENTIAL

In a nonenzymic oxidation-reduction process it is not essential that the reactants be in contact. Instead of the electrons being transferred by direct interaction, they may also be transported over a conducting wire. The passage of electrons through the wire constitutes an electric current (Fig. 7-23).

In the already mentioned case of the oxidation of molecular hydrogen by Fe^{3+} ion, it is possible to construct two half-cells (Fig. 7-23), one of which contains a solution with Fe^{3+} and Fe^{2+} ions, while the other contains H^+ ion and molecular hydrogen. The electrodes are usually platinum, which permits a rapid equilibration between H^+ ion and H_2 gas. The two half-cells are connected by an external wire joining the two electrodes and by a salt bridge between the two solutions (Fig. 7-23).

In one half-cell the reaction is

$$Fe^{3+} + e^- \rightarrow Fe^{2+} \qquad \textbf{(7-10)}$$

and in the other

$$\tfrac{1}{2}H_2 \rightarrow H^+ + e^- \qquad \textbf{(7-11)}$$

The reaction proceeds in the same way as if the reactants were mixed, with the sole difference that the electrons are transferred via the external circuit. A potential difference is developed between the electrodes immersed in the two half-cells, whose magnitude, in volts, may be measured by standard methods. This electrical potential is effectively a measure of the relative

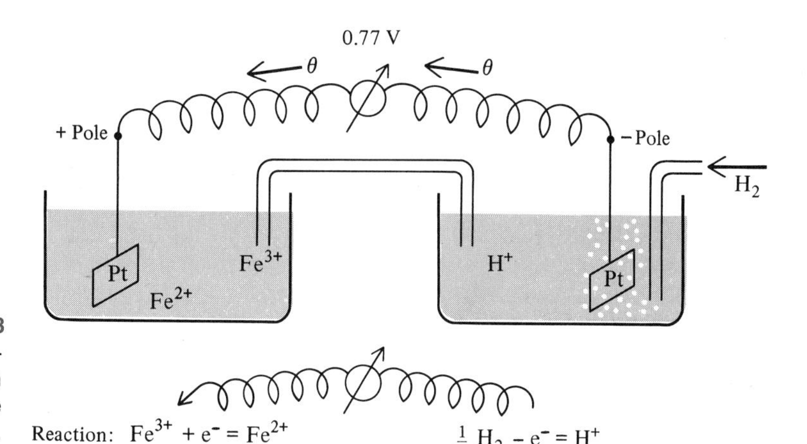

Figure 7-23
A simple electrochemical cell, in which the reaction is the reduction of Fe^{3+} by H_2.

affinities for electrons of Fe^{3+} and H^+ and is termed the **redox potential** of the Fe^{2+}/Fe^{3+} system.

The hydrogen half-cell (H_2 pressure = 1 atmosphere; molarity of H^+ = 1.0) has arbitrarily been selected as the standard reference cell for all systems. A positive potential by this convention indicates that hydrogen is oxidized by the second system; that is, the reaction in the hydrogen half-cell is $\frac{1}{2} H_2 \rightarrow H^+ + e^-$. Redox potentials are tabulated for the standard concentration of 1*M* for the oxidized and reduced forms.

The redox potentials of metallic cations are very important in inorganic chemistry. The well-known electromotive series is a list of the metals in the order of increasing positive potential.

In a broader sense, the redox potential is a measure of the affinity of the substance for electrons. A negative redox potential implies that the substance has a lower affinity for electrons than H_2, while a positive potential implies the reverse. Thus, a strong reducing agent has a negative redox potential, while a powerful oxidizing agent has a positive potential.

Relation of Redox Potential to Free Energy

In order to compare the redox potentials of different metallic half-cells, it is necessary to choose a set of standard conditions. The convention is to take as the standard half-cell a metal electrode immersed in a solution of its ions whose effective molarity (activity) is unity. The potential developed by this system with respect to a standard hydrogen half-cell is defined as the **normal electrode potential**, E_o.

If two normal electrodes are connected, as described earlier, the potential difference developed between them is equal to the arithmetic difference between the two normal electrode potentials. For example, the normal electrode potential of the Zn^{2+}/Zn system is -0.76 V. That of the Fe^{2+}/Fe system is -0.44 V. If the two are connected, the iron electrode is positive with respect to the zinc by $0.76 - 0.44 = 0.32$ V.

Some metals, such as platinum, are essentially inert when immersed in a solution. These are suitable for use as electrodes in registering potentials developed by solutions containing redox systems. An example is the already mentioned use of the platinum electrode to measure the potential of the Fe^{3+}/Fe^{2+} system. In this case the electrode is not involved in any chemical reaction, but acts as a passive conducting agent for the transfer of electrons.

If a platinum electrode is immersed in a solution containing the oxidized and reduced forms of a particular substance, such as Fe^{3+} and Fe^{2+}, then the potential, *E*, developed with respect to the normal hydrogen electrode is given by

$$E = E_o + \frac{RT}{nF} \ln \frac{[\text{oxidized form}]}{[\text{reduced form}]} \tag{7-12}$$

Here E_o, a constant, is the electrode potential when the concentration ratio of oxidized and reduced forms is unity; n is the number of electrons transferred; and F is the Faraday, or the charge per mole (= 96,500 Coulombs).

For the Fe^{3+}/Fe^{2+} system, $n = 1$, and Equation (7-12) becomes

$$E = E_o + \frac{RT}{F} \ln \frac{[Fe^{3+}]}{[Fe^{2+}]} \tag{7-13}$$

When $[Fe^{3+}] = [Fe^{2+}]$, $E = E_0 = 0.747$. [This system should, of course, be differentiated from the Fe (metal)/Fe^{2+} system mentioned earlier.]

In many cases hydrogen ions enter directly into the oxidation/reduction reaction. For the quinone/hydroquinone (Q/HQ) system:

$$\mathrm{HQ} \leftrightarrows \mathrm{Q} + 2\mathrm{H}^+ + 2\mathrm{e}^- \tag{7-14}$$

so that $n = 2$.

For a system of this kind, Equation (7-12) is replaced by

$$E = E_o + \frac{RT}{nF} \ln \frac{[\text{oxidized form}]}{[\text{reduced form}]} + \frac{RT}{F} \ln [\mathrm{H}^+] \tag{7-15}$$

The term involving $[H^+]$ becomes equal to zero when $H^+ = 1$. However, the use of $[H^+] = 1$ as the standard state is unrealistic for most biochemical systems in which the process is catalyzed by an enzyme, since few enzymes are active at so low a pH. It is therefore convenient to define a new parameter, E_o', by

$$\begin{aligned} E_o' &= E_o + \frac{RT}{F} \ln [\mathrm{H}^+] \\ &= E_o - 0.06 \text{ pH} \end{aligned} \tag{7-16}$$

so that Equation (7-15) becomes

$$E = E_o' + \frac{RT}{nF} \ln \frac{[\text{oxidized form}]}{[\text{reduced form}]} \tag{7-17}$$

For the quinone/hydroquinone system, $n = 2$, and we have

$$E = E_o' + \frac{RT}{2F} \ln \frac{\mathrm{Q}}{\mathrm{HQ}} \tag{7-18}$$

Usually, E_o' is chosen to correspond to the value at pH 7. The use of Equation (7-18) is formally equivalent to introducing a new standard hydrogen half-cell whose potential with respect to the normal hydrogen electrode is -0.42 $(= 0.06 \times 7.0)$

Since the reaction in a galvanic cell is reversible and the electrical energy thereby generated is available for external work, it is to be expected that the redox potential is related to the free energy change for the overall process.

If two half-cells are connected, the difference in redox potentials is a measure of the free energy change for the corresponding complete reaction. To interconvert the two, we have

$$\Delta G = -nF\,\Delta E \tag{7-19}$$

If the concentrations of oxidized and reduced forms in the two half-cells correspond to the equilibrium values for the overall reaction, then ΔG and ΔE are both equal to zero and there is no net reaction. The magnitude of ΔE is thus a measure of the departure of the complete system from a state of chemical equilibrium.

More explicitly, if A and B, AH_2 and BH_2 represent the oxidized and reduced forms of the two systems, the complete reaction is

$$\mathrm{A} + \mathrm{BH}_2 \rightleftarrows \mathrm{AH}_2 + \mathrm{B} \tag{7-20}$$

and the reactions in the two half-cells are

$$2e^- + A + 2H^+ \rightleftarrows AH_2 \qquad (7\text{-}21)$$
$$BH_2 \rightleftarrows B + 2H^+ + 2e^-$$

The corresponding electrical potentials of the two half-cells are given by

$$E_A = E'_{o,A} + \frac{RT}{nF}\ln\frac{[A]}{[AH_2]} \qquad (7\text{-}22)$$
$$E_B = E'_{o,B} + \frac{RT}{nF}\ln\frac{[B]}{[BH_2]}$$

For the potential difference we have

$$\begin{aligned}\Delta E &= E_A - E_B \qquad (7\text{-}23)\\ &= \Delta E'_o + \frac{RT}{nF}\ln\frac{[A]}{[AH_2]} - \frac{RT}{nF}\ln\frac{[B]}{[BH_2]}\\ &= \Delta E'_o - \frac{RT}{nF}\ln\frac{[B][AH_2]}{[BH_2][A]}\end{aligned}$$

If E_A is more positive than E_B, so that $\Delta E > 0$, then A oxidizes BH_2 and the reaction proceeds in the *forward* direction:

$$A + BH_2 \rightarrow AH_2 + B \qquad (7\text{-}24)$$

If $\Delta E < 0$, then AH_2 is oxidized by B and the reaction proceeds in the *reverse* direction:

$$AH_2 + B \rightarrow A + BH_2 \qquad (7\text{-}25)$$

In either case the reaction proceeds until equilibrium is attained, when $\Delta E = 0$ and

$$\begin{aligned}\Delta E'_o &= \frac{RT}{nF}\ln\left\{\frac{[B][AH_2]}{[A][BH_2]}\right\}_{\text{equilibrium}} \qquad (7\text{-}26)\\ &= \frac{RT}{nF}\ln K\end{aligned}$$

where K is the equilibrium constant. Also, since $\Delta G^\circ = -RT \ln K$,

$$\Delta G^\circ = -nF\,\Delta E'_o \qquad (7\text{-}27)$$

The magnitude of $\Delta E'_o$ is an index of the position of the equilibrium and of the relative oxidizing power of the reactants. If $E'_{o,A}$ is more positive than $E'_{o,B}$, then A will oxidize BH_2 in a mixed solution of the two compounds. The relative oxidizing power of a series of compounds is that of their values of E'_o. The more positive the value of E'_o, the greater is the oxidizing power.

Redox Potentials of Biological Systems

Any reversible chemical reaction involving the oxidation of one reactant and the reduction of a second, or the transfer of electrons from the first to the second, comprises a redox system characterized by an electrochemical potential. Oxidation-reduction processes in which one reactant is a coenzyme are of central importance in metabolic conversions.

For example, the oxidation of a metabolite YH_2 by NAD^+ may be represented as the sum of the half-reactions:

$$YH_2 \rightarrow Y + 2H^+ + 2e^-; \quad E_Y = E'_{o,Y} + \frac{RT}{2F}\ln\frac{[Y]}{[YH_2]} \tag{7-28}$$

$$NAD^+ + H^+ + 2e^- \rightarrow NADH; \quad E_{NAD^+} = E'_{o,NAD^+} + \frac{RT}{2F}\ln\frac{[NAD^+]}{[NADH]}$$

$$\text{Sum:} \quad YH_2 + NAD^+ \rightarrow Y + NADH + H^+$$

$$\Delta E = E'_{o,NAD^+} - E'_{o,Y} + \frac{RT}{2F}\ln\frac{[NAD^+][YH_2]}{[Y][NADH]}$$

If $\Delta E > 0$, then $\Delta G(= -2F\,\Delta E)$ is negative and the reaction proceeds in the forward direction as written so that NAD^+ oxidizes YH_2. If $\Delta E < 0$ then $\Delta G > 0$ and the reaction goes in reverse; NADH is oxidized.

An important biological example of the latter case is the reduction of pyruvate to lactate by NADH.

$$\underset{\text{Pyruvate}}{CH_3COCOO^-} + NADH + H^+ \rightarrow \underset{\text{Lactate}}{CH_3CHOHCOO^-} + NAD^+ \tag{7-29}$$

The two half-reactions are

$$CH_3COCOO^- + 2H^+ + 2e^- \rightarrow CH_3CHOHCOO^-; \quad E'_o = -0.19\text{ V} \tag{7-30}$$

$$NAD^+ + H^+ + 2e^- \rightarrow NADH; \quad E'_o = -0.32\text{ V}$$

The difference of the two yields the desired reaction, for which $\Delta E'_o = 0.13$ V. From this we also have

$$\begin{aligned}\Delta G^{\circ\prime} &= -nF\,\Delta E'_o \\ &= -2 \times 23 \times 0.13 \\ &= -6\text{ kcal/mole}\end{aligned} \tag{7-31}$$

(Here the faraday is expressed in kilocalories per volt and is equal to 23 kcal/V.)

If all reactants and products were $1M$ in concentration, then $\Delta G'$ would equal $\Delta G^{\circ\prime}$ and the reaction would proceed in the forward direction as

Table 7-2
Redox Potentials for Some Coenzymes

Coenzyme	E'_o (volts)
NAD^+/NADH	−0.32
$NADP^+$/NADPH	−0.32
Riboflavin	−0.185
Cytochrome b	0.07
Ubiquinone	0.10

written; NADH would reduce pyruvate (or pyruvate would oxidize NADH).

In general, the magnitude of E'_o is an index of the oxidizing power of a coenzyme. The more positive the value of E'_o, the greater is the oxidizing power. Table 7-2 lists the values of E'_o for a series of coenzymes.

SUMMARY

A **coenzyme** is an organic molecule required for the activity of an enzyme. A tightly bound coenzyme is often termed a **prosthetic group**. A loosely bound coenzyme may equally well be regarded as a **cosubstrate.**

Coenzymes usually function as **donors** or **acceptors** of a particular group, which is transferred to, or from, a substrate. Coenzymes often act in conjunction with two different enzymes, acting as an acceptor in one reaction and as a donor in the other, thereby coupling the two enzymic reactions. However a coenzyme differs from a true substrate in being ultimately returned to its original state.

Many coenzymes are closely related to **vitamins**, which are essential dietary components. In general, a vitamin is the chief, or sole, constituent of a coenzyme.

The coenzymes may be grouped into those which transfer hydrogen and figure in oxidation-reduction processes and those which transfer groups other than hydrogen.

The following coenzymes fall into the latter category. **Coenzyme A** (CoA), which contains a **sulfhydryl** group, is active in transferring acyl groups. The most important CoA derivative is acetyl CoA, which figures prominently in both oxidative metabolism and biosynthetic mechanisms. **Tetrahydrofolic acid** (FH_4) is an agent for the transfer of **single carbon units**, including formyl, formaldehyde, hydroxymethyl, methyl, or formimino groups. **Biotin**, which is usually present as a prosthetic group, binds and activates CO_2; it participates in many carboxylation reactions. **Pyridoxal phosphate** is a cofactor for several classes of enzymes, including **decarboxylases** and aminotransferases. It is important in amino transfer reactions involving amino acids. **Thiamine pyrophosphate** is a carrier for **active acetaldehyde** and active **glycoaldehyde**, functioning in the oxidative decarboxylation of α-keto acids. *S*-adenosyl methionine is a carrier of methyl groups and is important in transmethylation reactions.

The coenzymes which transfer hydrogen and occur in oxidized and reduced forms include the following. **Nicotinamide adenine dinucleotide** (NAD^+) and **nicotinamide adenine dinucleotide phosphate** ($NADP^+$) function in the reversible uptake of hydrogen in forming the reduced forms NADH and NADPH. The nicotinamide coenzymes participate in numerous dehydrogenation reactions, including many dehydrogenations of primary and secondary alcohol groups. The **flavin** coenzymes, **flavin mononucleotide** (FMN) and **flavin-adenine dinucleotide** (FAD) likewise exist in oxidized and reduced forms and figure in many reactions involving oxidation or reduction; the reduced forms, $FMNH_2$ and $FADH_2$, contain two added hydrogen atoms. The flavin coenzymes usually occur as prosthetic groups.

Lipoic acid, which may exist in an oxidized **disulfide** or a reduced **sulfhydryl** form, can transfer both a hydrogen and an acyl group.

Many biochemical processes involve **oxidation** or **reduction**, which correspond respectively to the withdrawal or addition of electrons.

A **galvanic cell** is a **redox** system in which the solution containing the substance oxidized is separated from that containing the substance reduced and the transfer of electrons occurs through an external circuit. A galvanic cell is characterized by an **electrode potential**, E. A particular biological oxidizing agent, such as NAD^+ or FAD, may be characterized by its **normal electrode potential**, E'_o, which is developed under standard conditions with respect to a standard hydrogen half-cell. The more positive the value of E'_o, the greater the oxidizing power is.

REFERENCES

The following code is used to classify references. I: particularly useful as an introduction to the subject; R: useful primarily as a reference text; A: an advanced account of the material; H: a publication of historical importance.

General

M. Dixon and E. C. Webb, *Enzymes*, Academic, New York (1964). (R)

A. L. Lehninger, *Biochemistry*, Worth, New York (1975). (A)

L. Stryer, *Biochemistry*, Freeman, San Francisco (1976). (I)

Coenzymes

R. L. Blakley, *The Biochemistry of Folic Acid and Related Pteridines*, North-Holland, Amsterdam (1969). (R)

M. Florkin and E. H. Stotz (eds.), *Metabolism of Vitamins and Trace Elements*, in *Comprehensive Biochemistry*, vol. 2, Elsevier, Amsterdam (1970). (R)

D. W. Hutchinson, *Nucleotides and Coenzymes*, Wiley, New York (1964). (A)

J. Knappe, *Mechanism of Biotin Action*, Ann. Rev. Biochem. 39, 757 (1970). (A)

A. F. Wagner and K. Folkers, *Vitamins and Coenzymes*, Interscience, New York (1964). (R)

Redox Potentials

H. G. Bray and K. White, *Kinetics and Thermodynamics in Biochemistry*, Academic, New York (1957). (I)

T. E. King and M. Klingenberg, eds., *Electron and Coupled Energy Transfer in Biological Systems*, vols. 1 and 2, Dekker, New York (1971). (A)

REVIEW QUESTIONS

Questions marked with an asterisk are of a high level of difficulty.

7-1 What is the electromotive force in volts developed by an electrode immersed in the following solutions at pH 7.0 and 25°C?

(a) 1.0-mM NAD^+ and 5-mM NADH
(b) 0.1-mM NAD^+ and 0.01-mM NADH
(c) 0.1-mM NAD^+ and 0.1-mM NADH

7-2 List the following in the order of increasing tendency to remove electrons:
(a) NAD^+
(b) cytochrome c
(c) riboflavin
(d) cytochrome b

7-3 What is the standard free energy change ($\Delta G^{\circ\prime}$) for the reaction

$$NAD^+ + H_2 \rightarrow NADH + H^+?$$

*7-4 Suggest a reason why FMN and FAD are usually prosthetic groups while NAD^+ is not.

7-5 If the standard pH were chosen as 8.0, rather than 7.0, how would E_o' change?

7-6 For the redox reaction $AH_2 + B \rightleftharpoons BH_2 + A$, $\Delta G^{\circ\prime} = -10$ kcal. If a suitable electrode exists for the two half-cells, what potential is developed when $[A] = [AH_2] = [B] = [BH_2] = 3$ mM? What is the equilibrium constant?

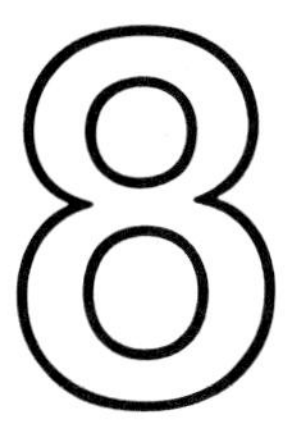

Nutrition

8-1 REQUIREMENTS OF LIVING SYSTEMS

Biological systems are set apart from the inanimate world by the activites they engage in: they grow, replicate, and act upon their environment and each other in various ways. In order to continue these activities they require a continuous supply of mass and energy from their surroundings; if the supply is interrupted, they speedily rejoin the inanimate world.

In this chapter we will depart from the specific chemical considerations of the last few chapters to adopt a point of view which is both more biological and more general. We now ask the question: what do biological organisms need to obtain from their surroundings in order to survive and why is this so? The answers are different for different organisms and, after an initial survey, we shall concentrate on the nutritional requirements of mammals. This will lead us logically into the next group of chapters, which will deal with the biological fuels and their utilization by living cells.

8-2 TRANSFER OF MASS AND ENERGY BETWEEN LIVING CELLS AND THEIR ENVIRONMENT

In order to survive, grow, and reproduce, living systems require a supply of both mass and energy from their surroundings. In the case of **photosynthetic** organisms, which include green plants and algae, as well as some species of bacteria, a major fraction of the energy requirement is derived from the radiant energy of sunlight. For all other living systems, the energy requirement must be satisfied by the metabolism of ingested nutritional material.

The reaction schemes that comprise metabolism may be grouped into biosynthetic and degradative pathways. The two pathways are nearly always distinct and differentiated and, in eukaryotic cells, generally occur in different cellular compartments. Both processes are mediated by enzymes and are subject to reversible allosteric control. The biosynthetic pathways require a source of energy, which is supplied in the form of **high energy compounds** generated by the oxidative metabolism of foodstuffs.

With respect to the source of carbon needed for the biosynthesis of cellular constituents, most organisms fall into two categories. The **chemoorganotrophs**, which include all higher animals, the nonphotosynthetic plants, and the majority of bacteria, derive both their carbon and their energy from the metabolism of consumed organic compounds. The **photolithotrophs**, which include the photosynthetic green plants, blue-green algae, and photosynthetic bacteria, obtain their carbon from atmospheric CO_2 and their energy from sunlight.

Two less numerous categories are the **chemolithotrophs** whose carbon source is CO_2, but which derive energy from oxidation-reduction reactions involving inorganic compounds, and the **photoorganotrophs** whose carbon comes from organic compounds, but which utilize sunlight as their energy source. Both of the above are confined to certain species of bacteria.

Living cells whose carbon source is organic compounds (**heterotrophs**), which include the chemoorganotrophs and the photoorganotrophs, may be further grouped into two classes by a different criterion. The metabolism of both involves the stepwise oxidation or transfer of electrons from the original organic nutrients and their metabolic products and requires an ultimate acceptor of electrons or **oxidizing agent**. For **aerobic** cells this electron acceptor is molecular oxygen; **anaerobic** cells employ some other molecule. Many cells can function in either way, but generally use oxygen preferentially when it is availablc; most cells of higher organisms fall into this category. Not all of the cells of a particular organism are necessarily of the same class and many instances are known of considerable metabolic adaptability.

A considerable variation also exists in the source of nitrogen, a constituent of many biomolecules. Atmospheric nitrogen is too inert chemically to be used directly by most organisms, the great majority of which must procure their nitrogen as a constituent of some compound. This may, depending upon the species, be a simple inorganic molecule, such as nitrate, or a more complex organic molecule, such as an amino acid.

A few living systems, such as the nitrogen-fixing bacteria, have the capability of reducing nitrogen and incorporating it into compounds that can be utilized by heterotrophs. The combined action of a set of soil microorganisms converts atmospheric nitrogen into nitrate. The majority of plants obtain their nitrogen from the soil in the form of nitrate, which they convert into amino acids, nucleotides, ammonia, and other products. Plant proteins are consumed by heterotrophs that ultimately return nitrogen to the soil by excretion or by the death and decay of the organism. A continuous exchange of nitrogen occurs between the soil and living systems.

8-3 ENERGY SOURCES FOR BIOLOGICAL ORGANISMS

Metabolism of Foodstuffs as an Energy Source

The principal biological fuels, whose oxidation provides the primary energy source for aerobic organisms are carbohydrates, fats, and proteins. The degradative processing of foodstuffs may be regarded as occurring in three phases. In the first of these, the large molecules of food are converted to smaller molecules that can more readily enter the metabolic pathways. Proteins are hydrolyzed to amino acids, while polymeric carbohydrates, such as starch, are hydrolyzed to simple sugars, such as glucose, and fats are hydrolyzed to glycerol and fatty acids.

In the next phase these small molecules are degraded to a few common products that form the points of departure for the further stages of oxidative metabolism. The most important intermediate is the **acetyl group** of acetyl CoA, to which fats, sugars, and several amino acids are ultimately converted.

The final phase consists of the linked processes of the **tricarboxylic acid cycle, respiration**, and **oxidative phosphorylation** (chapter 12). Acetyl groups enter this pathway in the form of acetyl CoA and are completely oxidized to CO_2 and H_2O. The oxidation of acetyl units drives the formation of the high-energy molecule **adenosine triphosphate** (ATP) (Fig. 8-1) by means of the coupled reactions of oxidative phosphorylation. It is in this form that the energy released in aerobic organisms by the oxidative metabolism of foodstuffs becomes available to the cells.

ATP as an Energy Carrier

ATP is a **nucleotide** consisting of the purine adenine, the sugar ribose, and an esterified triphosphate group. In the related molecule **adenosine diphosphate** (ADP), the triphosphate group is replaced by a diphosphate.

At neutral pH the triphosphate portion of ATP possesses 3.5 negative charges (Fig. 8-1). As a consequence of the mutual electrostatic repulsion of these, as well as reduced resonance stabilization energy, ATP is thermodynamically unstable with respect to hydrolysis to ADP. The reaction

$$\text{ATP} + \text{H}_2\text{O} \rightarrow \text{ADP} + \text{P}_i \quad (\Delta G^{\circ\prime} = -7.3 \text{ kcal})$$

is thermodynamically favored. For the levels of ATP, ADP, and P_i typical of most cells, the value of $\Delta G'$ is about -12 kcal.

ATP may also be hydrolyzed to **adenosine monophosphate** (Fig. 8-1) and pyrophosphate (PP_i). The decrease in free energy is similar in magnitude to that for the hydrolysis to ADP:

$$\text{ATP} + \text{H}_2\text{O} \rightarrow \text{AMP} + \text{PP}_i \quad (\Delta G^{\circ\prime} = -7.3 \text{ kcal})$$

The free energy released by the hydrolysis of ATP serves to drive **endergonic** reactions that require a source of free energy, such as muscle contraction and the active transport of ions across membranes. Conversely, ATP is generated from ADP and P_i by the harnessing of radiant energy in photosynthetic systems or by the oxidation of biological fuels in other organisms. The ATP $\rightleftarrows$ ADP conversion is the basic means of free energy transfer in biological systems. ATP is sometimes compared to the charged state of a storage battery and ADP to its discharged form.

Biological reactions exist that are driven by other nucleotides which are structurally analogous to ATP. These include cytidine triphosphate (CTP), uridine triphosphate (UTP), and guanosine triphosphate (GTP). The corresponding diphosphate species are CDP, UDP, and GDP (Fig. 8-2). We

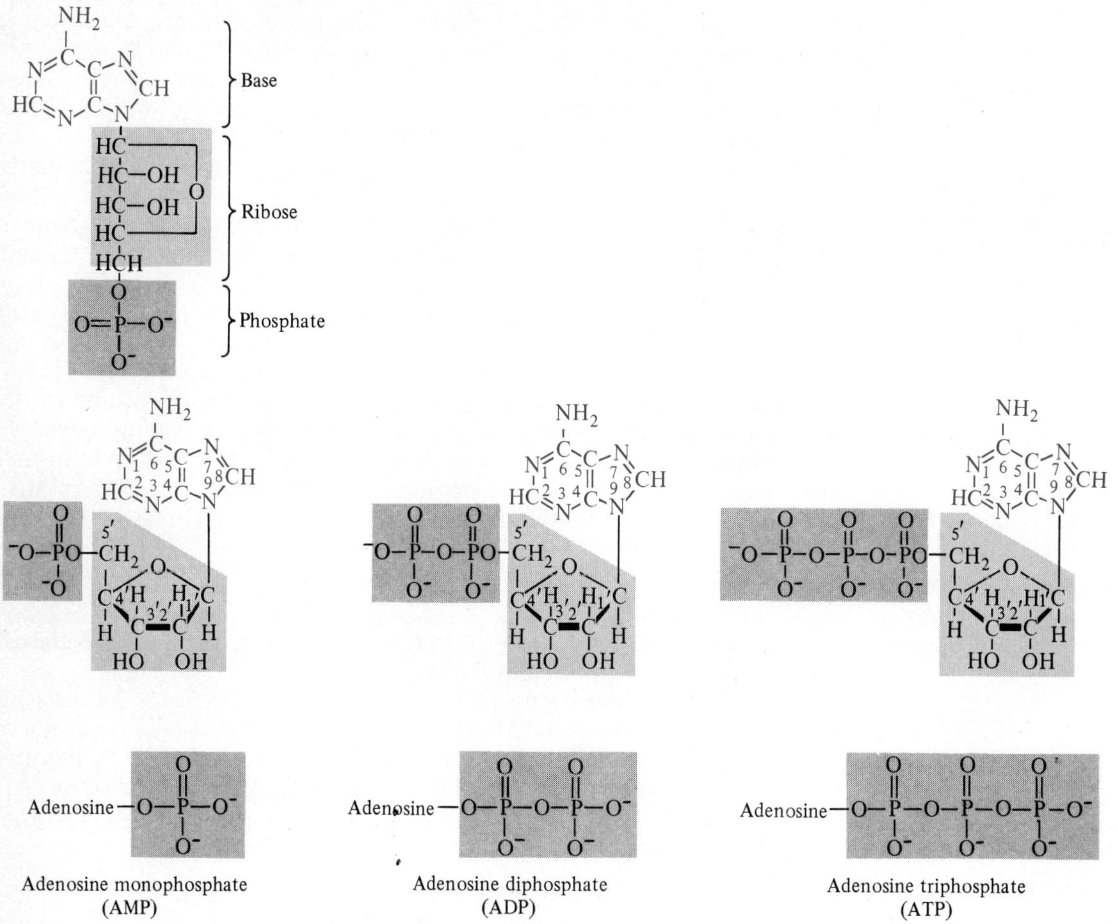

Base

Ribose

Phosphate

Adenosine monophosphate (AMP)

Adenosine diphosphate (ADP)

Adenosine triphosphate (ATP)

Figure 8-1
Structures of (left to right) adenosine monophosphate (AMP), adenosine diphosphate (ADP), and adenosine triphosphate (ATP). The structure of AMP is shown in two ways : the upper is a conventional chemical structure, while the lower is a stereochemical formula giving some idea of the spatial arrangement of the atoms (section 9-1). The ring of the adenine base is planar. Note the numbering systems used for the ring positions of adenine and ribose.

shall also encounter the cyclic phosphoesterified nucleotide **cyclic AMP** (cAMP) (Fig. 8-3).

Electron Carriers

The ultimate products of the oxidative metabolism of foodstuffs by aerobic organisms are CO_2 and H_2O. The overall process could equally well be regarded as a reduction of, or transfer of electrons to, molecular O_2. The transfer of electrons does not occur directly. Instead, the fuel molecules, or

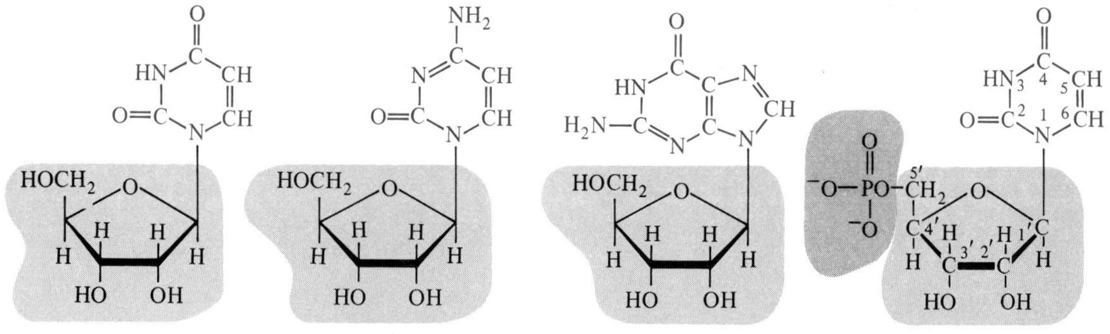

Uridine Cytidine Guanosine UMP

Figure 8-2
Stereochemical structural formulas for the **nucleosides** uridine, cytidine, and guanosine. The mono-, di-, and tri-phosphate esters CMP, CDP, and CTP have the same relation to these as AMP, ADP, and ATP have to adenosine. The structure of UMP is also shown.

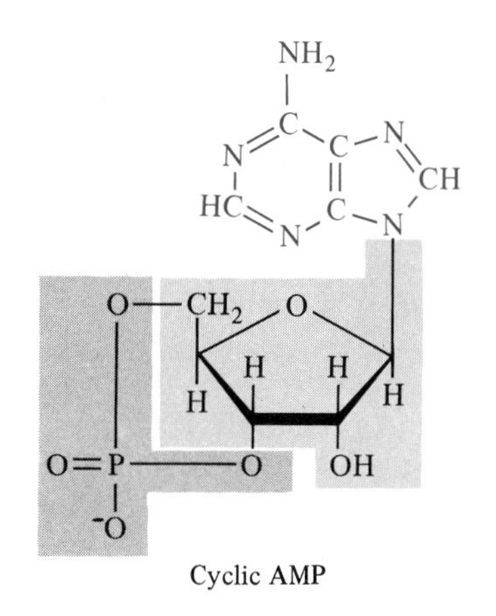

Figure 8-3
Structure of cyclic AMP (cAMP).

their degradation products, transfer electrons to, or are oxidized by, specific electron carriers. The most important of these are the nicotinamide derivatives NAD^+ and $NADP^+$ and the flavin derivative FAD (section 7-8). The reduced forms of all three of these react very slowly with O_2 in the absence of an enzyme catalyst. Analogously, ATP is hydrolyzed only at a slow rate in the absence of a catalyst. The **kinetic stability** of these compounds enables enzymes to regulate their action closely and is indispensable to their biological function.

There is a basic difference in the function of NADH and NADPH. While NADH is important in the formation of ATP by oxidative metabolism, NADPH figures in biosynthetic reactions.

Table 8-1
Essential and Nonessential Amino Acids in Mammals*

Essential	*Nonessential*
Arginine	Alanine
Histidine	Asparagine
Isoleucine	Aspartic acid
Leucine	Cysteine
Lysine	Glutamic acid
Methionine	Glutamine
Phenylalanine	Glycine
Threonine	Proline
Tryptophan	Serine
Valine	Tyrosine

*This table is derived from experiments using white rats.

8-4 AMINO ACIDS REQUIREMENTS

The ability of biological organisms to synthesize the organic molecules required for their functioning is quite variable. While some organisms can generate all of their necessary nitrogenous compounds from simple molecules, such as ammonia, most species require an external source of some of their essential nitrogen-containing organic molecules.

Many of the higher animals, including man, are unable to synthesize a number of amino acids. These amino acids, which are termed **essential amino acids**, must be supplied preformed in the diet. For example, albino rats can synthesize from simpler molecules only 10 of the 20 amino acids needed for the formation of proteins (Table 8-1); the others must be present as dietary components. Higher animals are able to synthesize the remaining, **nonessential** amino acids from ammonia, but cannot utilize nitrate or atmospheric nitrogen for this purpose. In contrast, the higher plants are able to manufacture all of the amino acids needed for protein synthesis from nitrate, nitrite, or ammonia.

Microorganisms show a wide variance in their ability to manufacture amino acids. In this respect they range from species for which nearly all amino acids must be supplied to those for which none are essential.

8-5 REQUIREMENTS FOR SPECIALIZED COMPOUNDS

Vitamins

Vitamins are organic molecules required by all living systems that cannot be synthesized by higher animals and must therefore be supplied in their diet. While the biochemical function of vitamins is basically the same in all biological organisms, the higher animals have, in the course of evolutionary time, lost the ability to synthesize them.

Table 8-2

The Water-Soluble Vitamins

Vitamin	*Coenzyme*	*Deficiency Disease*
Thiamine (vitamin B_1)	Thiamine pyrophosphate	Beriberi
Riboflavin (vitamin B_2)	FMN; FAD	
Nicotinate (niacin)	NAD^+; $NADP^+$	Pellagra
Pyridoxine, pyridoxal, and pyridoxamine (vitamin B_6)	Pyridoxal phosphate	
Pantothenate	Coenzyme A	
Biotin		
Folate	Tetrahydrofolate	
Cobalamin (vitamin B_{12})	Cobamide coenzymes	Pernicious anemia
Ascorbic acid (vitamin C)		Scurvy

The vitamins are often loosely classified into two groups. These are the **fat-soluble** vitamins A, D, E, and K and the **water-soluble** vitamins B_1, B_2, B_{12}, C, nicotinic acid, pantothenic acid, pyridoxine, biotin, folic acid, and lipoic acid.

The chemically dissimilar class of compounds termed the **B vitamins** are either coenzymes or constituents of coenzymes (Table 8-2 and Fig. 8-4). Thiamine (vitamin B_1), riboflavin (vitamin B_2), and nicotinate (niacin) have already been encountered as the electron-carrying portions of thiamine Thiamine (vitamin B_1), riboflavin (vitamin B_2), and nicotinate (niacin) have already been encountered as the electron-carrying portions of thiamine pyridoxal, and pyridoxamine (vitamin B_6) are precursors of pyridoxal phosphate. The reduced forms of folic acid and biotin are themselves coenzymes; biotin normally occurs in covalent attachment to carboxylase enzymes. Cobalamin (vitamin B_{12}) (Fig. 8-5) is essential for the normal development of red blood cells. The active coenzyme is a derivative of cobalamin containing a **5-deoxyadenosyl** group (Fig. 8-5).

The biological function of ascorbic acid (vitamin C) is still poorly understood. It is an essential dietary component in only a few of the higher animals, including man and the monkeys. Ascorbic acid is a strong reducing agent and readily loses its hydrogens to form dehydroascorbic acid.

While the water-soluble vitamins are present in trace quantities in a wide variety of foods, their level is highly variable and deficiencies may occur under conditions of restricted diet. Deficiencies of several of the vitamins produce characteristic clinical manifestations (Table 8-2), which have been designated as diseases.

The fat-soluble vitamins belong chemically to the class of compounds called **terpenes**, and discussion of these will be deferred to section 16-1.

Other Essential Organic Compounds

Mammals have a requirement for two unsaturated fatty acids, linoleic and γ-linolenic acids (section 13-2). Plants are a good source of fatty acids.

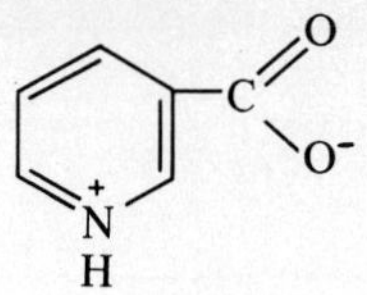

Nicotinate
(Niacin)

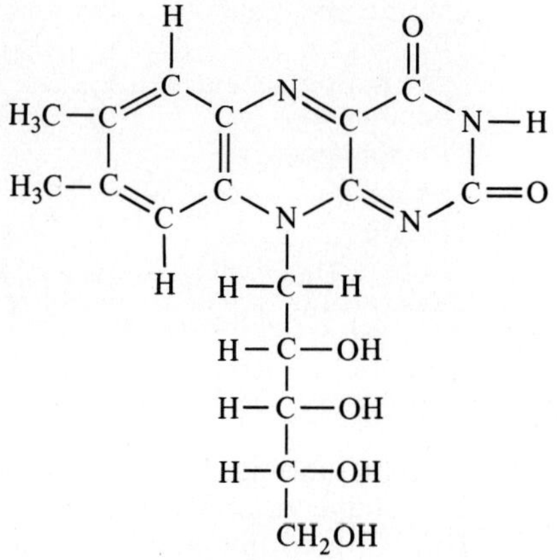

Riboflavin
(Vitamin B_2)

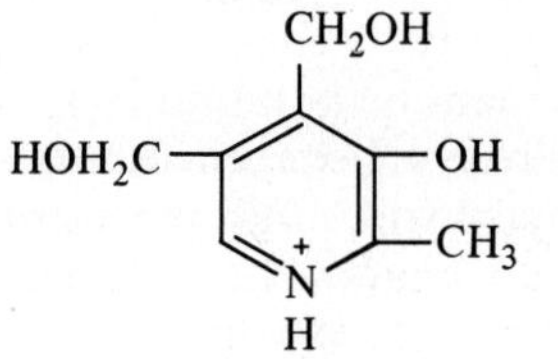

Pyridoxine
(A form of vitamin B_6)

Figure 8-4
Structures of several vitamins.

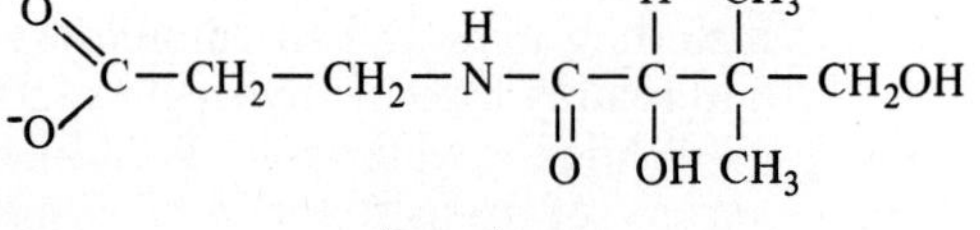

Pantothenate

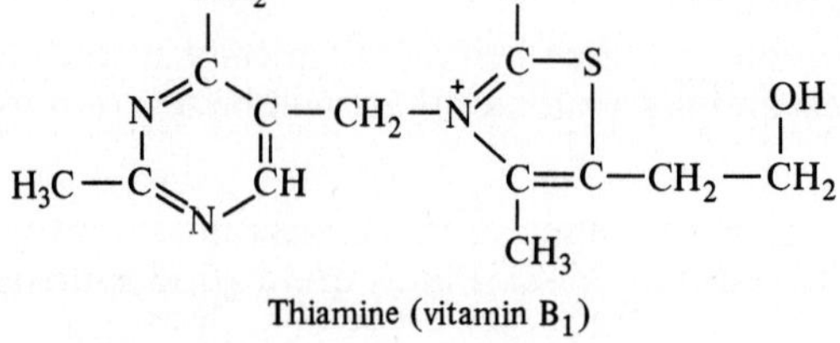

Thiamine (vitamin B_1)

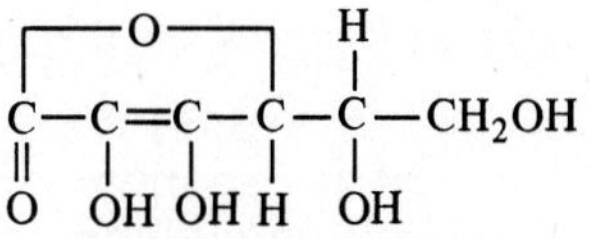

Ascorbic acid

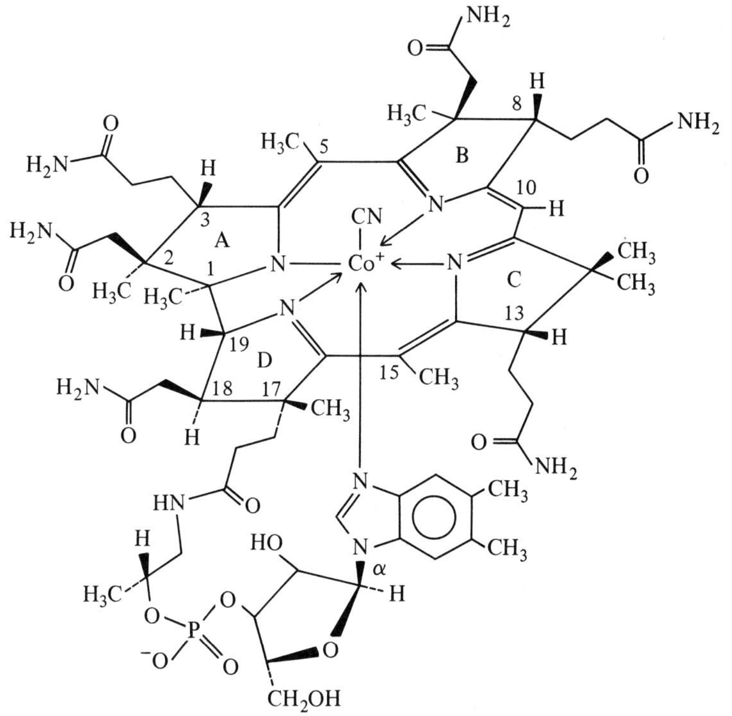

Figure 8-5
Structure of vitamin B_{12}.

SUMMARY

The purposes of cellular metabolism are to extract chemical energy from sunlight or from nutritional fuels, to form the biopolymers essential for the functioning of the cells, and to provide the biological molecules required for specialized cellular functions.

Phototrophic organisms, which include plants and some bacteria, can utilize sunlight directly as an energy source; **chemotrophic** organisms, which include most microorganisms and all higher animals, must obtain their chemical energy by some form of oxidative metabolism of nutrients.

Auxotrophic organisms, such as plants and a few bacteria, utilize CO_2 as a carbon source; **heterotrophic** organisms, which include animals and most microorganisms, require complex organic molecules and must subsist on the products of other cells. Only a few organisms, such as the nitrogen-fixing bacteria, can use atmospheric nitrogen as a nitrogen source; most species require nitrogen supplied in a chemically combined form. Plants generally obtain nitrogen from the soil as **nitrate**; animals procure it as **amino acids**.

The chemical energy derived by cells from the metabolic processing of nutrients is conserved in the form of **adenosine triphosphate** (ATP), which is a kind of bioenergetic "traveler's check" whose stored energy can be utilized for a wide variety of processes. A major purpose of the **oxidative metabolism** occurring in animal and bacterial cells is the harnessing of the energy released by the oxidation of fats, carbohydrates, and proteins for the synthesis of ATP.

While most bacteria can synthesize all of the necessary amino acids from simpler nitrogenous compounds, such as ammonia, the higher animals have partially lost this ability and require half of the amino acids to be supplied in the diet. Animals also require a dietary source of **vitamins**, which are components or precursors of coenzymes.

REFERENCES

The following code is used to classify references. I: particularly useful as an introduction to the subject; R: useful primarily as a reference text; A: an advanced account of the material; H: a publication of historical importance.

General

E. Baldwin, *Dynamic Aspects of Biochemistry*, Cambridge University, London (1967). (I)

S. Dagley and D. E. Nicholson, *An Introduction to Metabolic Pathways*, Wiley, New York (1970). (I)

L. L. Ingraham and A. B. Pardee, *Free Energy and Entropy in Metabolism*, D. M. Greenberg, ed., in *Metabolic Pathways*, 3rd ed., vol. 1, p. 2, Academic, New York (1967). (A)

A. L. Lehninger, *Bioenergetics*, 2nd ed., Benjamin, Menlo Park, California (1972). (A)

W. B. Wood, *The Molecular Basis of Metabolism*, McGraw-Hill, New York (1974). (A)

ATP

R. A. Alberty, "Effect of pH and Metal Ion Concentration on the Equilibrium Hydrolysis of Adenosine Triphosphate to Adenosine Diphosphate," *J. Biol. Chem.* 243, 1337 (1968). (A)

J. Rosing and E. C. Slater, "The Value of $\Delta G^{\circ\prime}$ for the Hydrolysis of ATP," *Biochim. Biophys. Acta* 267, 275 (1972). (A)

H. M. Kalckar, *Biological Phosphorylations: Development of Concepts*, Prentice-Hall, Englewood Cliffs, New Jersey (1969). (A)

Inorganic Compounds in Metabolism

H. J. M. Bowen, *Trace Elements in Biochemistry*, Academic, New York (1966). (R)

W. G. Hoekstra, J. W. Suttie, H. E. Ganther, and W. Mertz, eds., *Trace Element Metabolism in Animals*, Univ. Park Press, Baltimore (1974). (R)

Vitamins

H. F. Deluca and J. W. Suttie, *The Fat-soluble Vitamins*, Wisconsin, Madison (1970). (A)

S. F. Dyke, *The Chemistry of the Vitamins*, Interscience, New York (1965). (A)

W. H. Sebrell, Jr., and R. S. Harris, eds., *The Vitamins*, 2nd ed., Academic, New York (1967–1972). (A)

REVIEW QUESTIONS

8-1 Indicate which of the following statements are true or false.
(a) Biosynthetic pathways generate energy; breakdown pathways consume energy.
(b) Biosynthetic pathways often use NADPH as a reducing cofactor; breakdown pathways use NADH instead of NADPH.
(c) Biosynthetic and breakdown pathways are controlled independently of one another.
(d) All of the above are false.
(e) None of the above is false.

8-2 The energy generated by metabolism of fuel molecules is used for:
(a) synthesis of macromolecules from smaller "building blocks."
(b) transport of certain ions out of the cell.
(c) muscle contraction and movement.
(d) synthesis of storage fat.
(e) all of the above

8-3 The free energy change for the complete oxidation of 1 mole of glucose to CO_2 and H_2O is 686 kcal. What is the maximum number of moles of ATP that could be formed under physiological conditions from ADP by the oxidation of one mole of glucose if the efficiency were 100%?

8-4 What is the $\Delta G'$ for the hydrolysis of ATP to ADP at pH 7.0 and 25°C when

$$[\text{ATP}] = 0.01\ \text{m}M \text{ and } [\text{ADP}] = [\text{P}_i] = 1\ \text{m}M?$$

What is $\Delta G^{\circ\prime}$?

The Carbohydrates

9-1 THE BIOLOGICAL ROLE OF THE CARBOHYDRATES

In the animal world the **carbohydrates** are of the very highest nutritional importance, serving both as major biological fuels, whose degradative metabolism generates ATP, and as sources of carbon for biosynthetic processes. Certain carbohydrate derivatives, the **mucopolysaccharides**, function as important structural components of animal tissues. In plants the carbohydrate **cellulose** is the principal structural biopolymer.

Chapters 9–11 will be concerned with the chemical nature, the metabolic conversions, and the biosynthesis of carbohydrates. The present chapter will be concerned with background information about the covalent structures, stereochemistry, and reactions of the carbohydrates, without which their biological role is unintelligible. We shall begin with the **sugars**, which are the basic monomeric units, and then proceed to consider the polymeric carbohydrates, including both those which serve as biological fuels and those that have a purely structural function.

Chapter 10 deals with the metabolic conversions of carbohydrates, including those that yield ATP and those that are important in biosynthetic pathways. Chapter 11 will be concerned with the light-dependent biosynthesis of carbohydrates by plants, which is the primary source of carbohydrate and upon which other forms of life ultimately depend; it is here that biochemistry impacts upon agronomy and agriculture.

9-2 MONOSACCHARIDES

The group of substances traditionally misnamed the **carbohydrates** collectively comprise by far the most abundant class of organic compounds. Although largely of plant origin, they account for a major portion of the diet of most animals, including man. They are probably the most important biological fuel for the majority of organisms, outranking fats and proteins as suppliers of energy requirements.

The carbohydrates include the simple sugars or **monosaccharides**, their derivatives, and their polymers or **polysaccharides**. Two structurally related polymers of the monosaccharide D-glucose, **starch** and **glycogen**, function as storage fuels in plants and animals, respectively, while a structurally different polymer of D-glucose, **cellulose**, is the component of plant tissues largely responsible for their strength and rigidity.

The monosaccharides are polyhydroxy aldehydes or ketones. The two basic monosaccharide series, which are termed the **aldoses** and **ketoses**, have as their simplest structural prototypes glyceraldehyde and dihydroxyacetone, respectively,

$$\begin{array}{c} HC{=}O \\ | \\ HC{-}OH \\ | \\ H_2C{-}OH \\ \text{D-Glyceraldehyde} \end{array} \qquad \begin{array}{c} H_2C{-}OH \\ | \\ C{=}O \\ | \\ H_2C{-}OH \\ \text{Dihydroxyacetone} \end{array}$$

The higher monosaccharides of both series may be visualized as being derived from the 3-carbon prototypes by the insertion of one or more HCOH groups between the carbonyl and primary alcohol groups. The generalized structures of the two series may be written

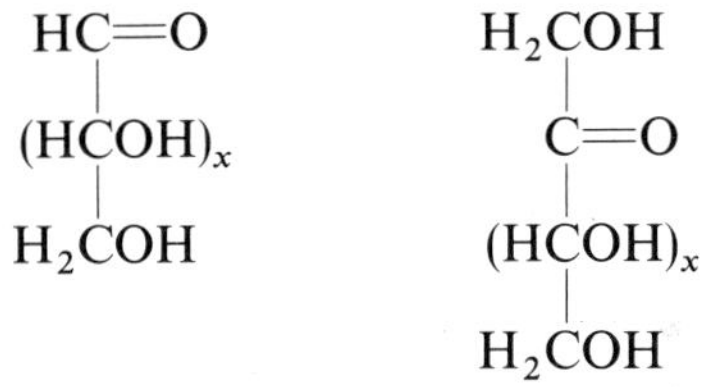

$$\begin{array}{c} HC{=}O \\ | \\ (HCOH)_x \\ | \\ H_2COH \\ \\ \\ \text{Aldose series} \end{array} \qquad \begin{array}{c} H_2COH \\ | \\ C{=}O \\ | \\ (HCOH)_x \\ | \\ H_2COH \\ \text{Ketose series} \end{array}$$

The principal members of the aldose and ketose series are shown in Figs. 9-1 and 9-2.

Stereochemistry of the Monosaccharides

With the exception of dihydroxyacetone, all of the monosaccharides contain at least one chiral, or asymmetric, carbon atom and hence can exist as stereoisomers. An aldose with N carbon atoms has N − 2 chiral carbon atoms and a ketose has N − 3. Glyceraldehyde, which has one chiral center, can exist in D- or L-forms (Fig. 9-3). The configurations of other monosaccharides are assigned on the basis of their relationship to glyceraldehyde. If the configuration of the chiral carbon most removed from the carbonyl group is like that of D-glyceraldehyde, the sugar is arbitrarily assigned a D-configuration; if its configuration resembles that of L-glyceraldehyde, the sugar is assigned an L-configuration (section 3-5). Monosaccharides derived from D-glyceraldehyde by the addition of HCOH groups are regarded as being of the D-configuration; the converse holds for derivatives of L-glyceraldehyde.

If two or more chiral carbon atoms are present, stereoisomers are possible which are not enantiomorphs (section 3-5; Figs. 9-1 and 9-2). These generally differ in physical and chemical properties and are regarded as distinct and

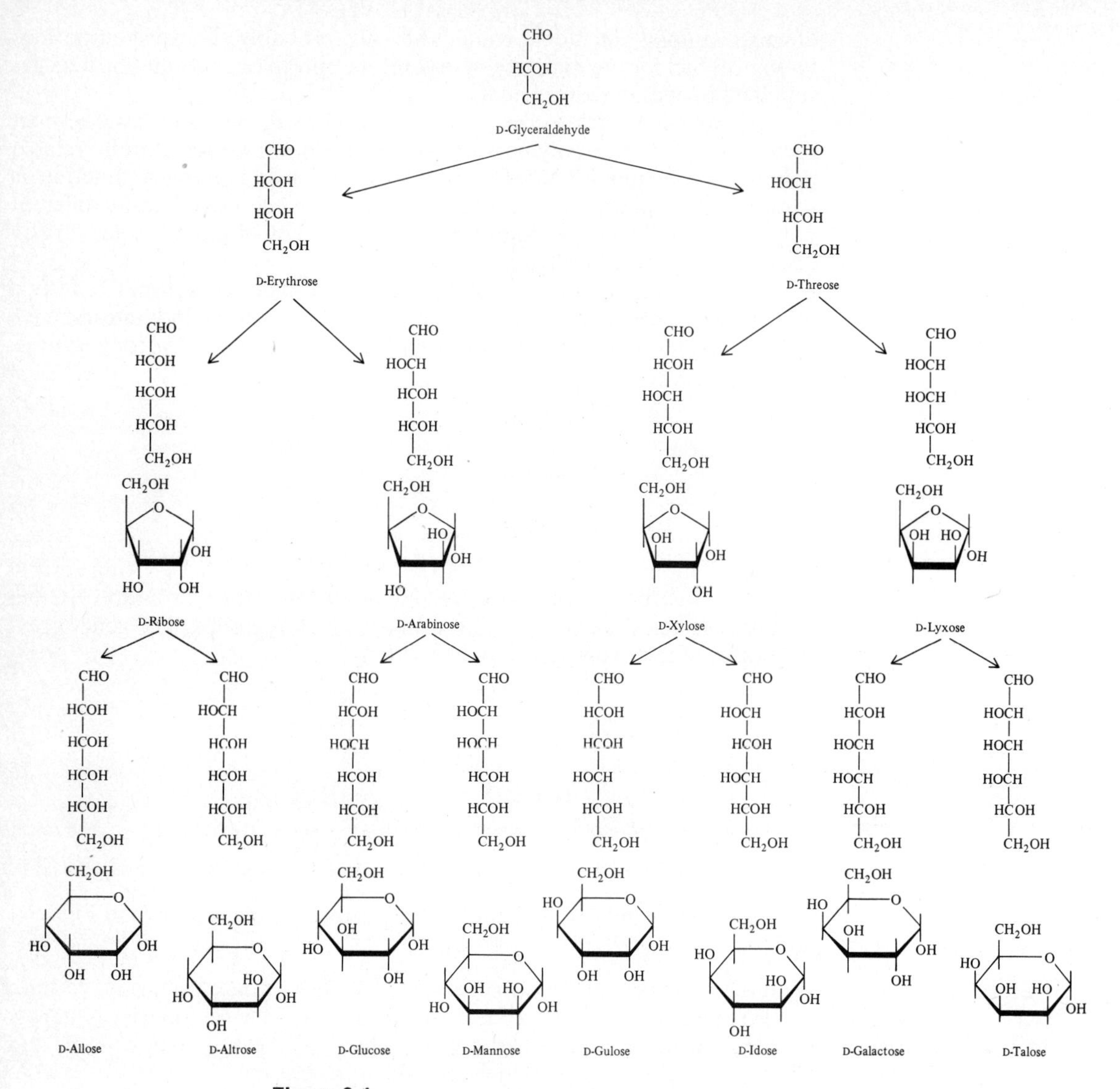

Figure 9-1
Aldose series of monosaccharides, represented as linear and ring structures.

different sugars, as threose and erythrose (Fig. 9-1) or xylulose and ribulose (Fig. 9-2). In general, if x chiral centers are present, there are 2^{x-1} different sugars, each of which can exist as a D- or L-form.

The great majority of the sugars of biological importance are of the D-configuration, although a few L-sugars, such as L-sorbose, occur in nature.

Two sugars which differ only with respect to the configuration of a single carbon atom are said to be **epimers**. D-glucose and D-mannose are epimers with respect to carbon atom 2.

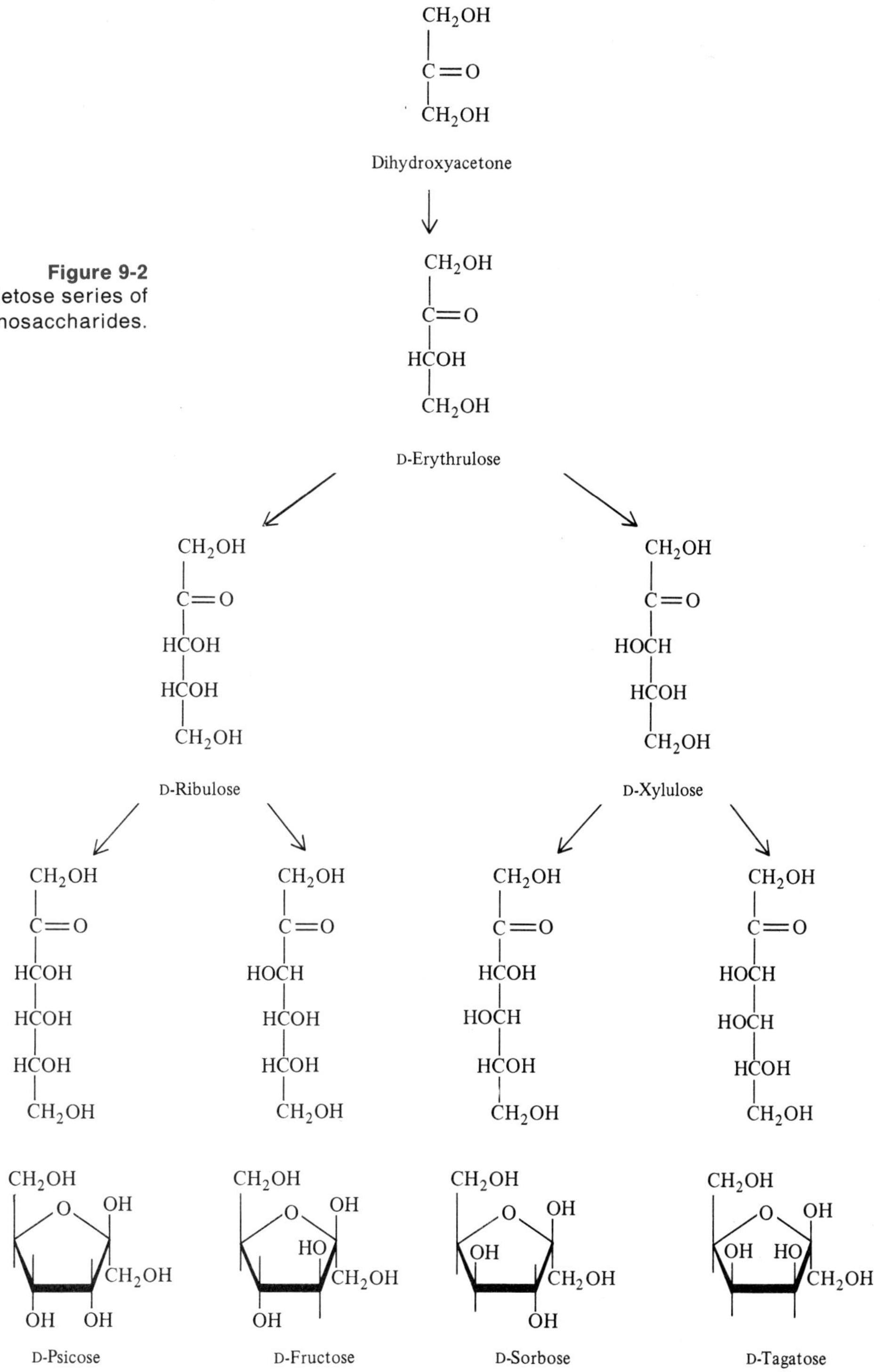

Figure 9-2
Ketose series of monosaccharides.

CH_2OH
$C{=}O$
CH_2OH

Dihydroxyacetone

CH_2OH
$C{=}O$
HCOH
CH_2OH

D-Erythrulose

CH_2OH
$C{=}O$
HCOH
HCOH
CH_2OH

D-Ribulose

CH_2OH
$C{=}O$
HOCH
HCOH
CH_2OH

D-Xylulose

CH_2OH
$C{=}O$
HCOH
HCOH
HCOH
CH_2OH

D-Psicose

CH_2OH
$C{=}O$
HOCH
HCOH
HCOH
CH_2OH

D-Fructose

CH_2OH
$C{=}O$
HCOH
HOCH
HCOH
CH_2OH

D-Sorbose

CH_2OH
$C{=}O$
HOCH
HOCH
HCOH
CH_2OH

D-Tagatose

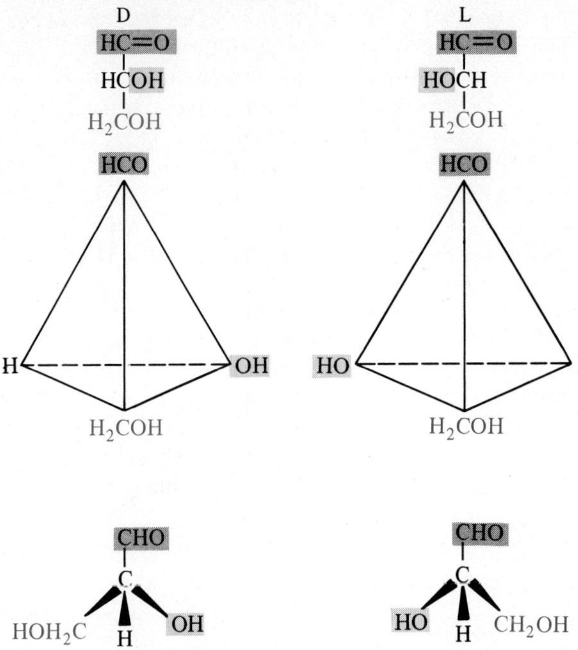

Figure 9-3
D- and L-forms of glyceraldehyde.

Ring Structures

The linear (Fischer) structural formulas of Figs. 9-1 and 9-2 actually provide an incomplete picture of the structures of the higher monosaccharides, many of which behave as if they possessed one additional chiral center not apparent in the linear formulas. For monosaccharides with five or more carbon atoms, the linear structures exist to only a minor extent in solution, in a dynamic equilibrium with closed ring hemiacetal forms (Fig. 9-4).

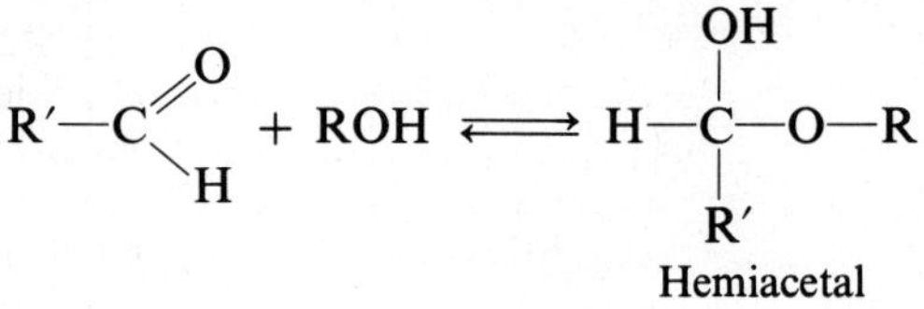

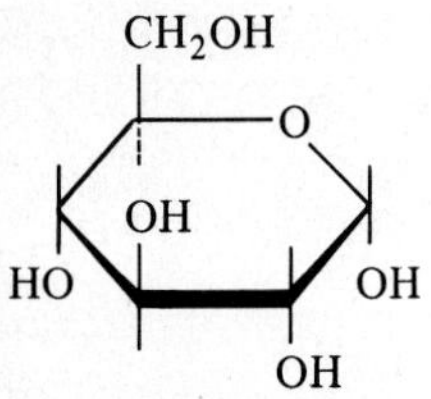

α-D-Glucose

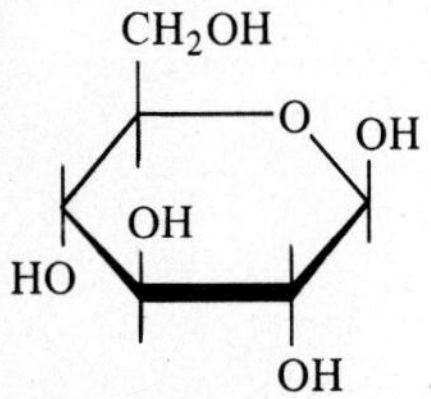

β-D-Glucose

Figure 9-5
Haworth ring formulas for α-D-glucose and β-D-glucose.

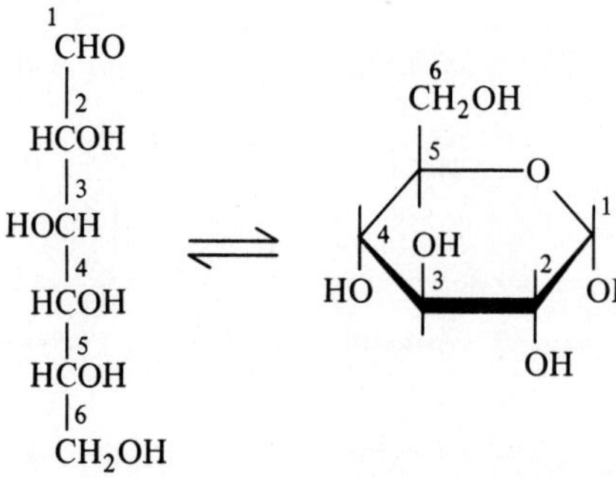

Figure 9-4
Linear and ring forms of D-glucose. Note the numbering system for the carbon atoms, which designates the aldehyde carbon C_1.

β-D-Fructofuranose

β-D-Arabinofuranose

Figure 9-6
Ring structures of β-D-fructofuranose and β-D-arabinofuranose.

Hemiacetal formation in the monosaccharides results in a five-membered **furanose** or six-membered **pyranose** ring containing one oxygen atom. The majority of the aldoses of biochemical interest exist in the pyranose form.

The Fischer-type formulas give a somewhat misleading concept of the actual spatial arrangement of the ring structures. The alternative Haworth formulas show the ring in perspective, with thickened lines indicating the edge nearest the viewer (Figs. 9-5 and 9-6). When, as in the usual representation, the oxygen is on the side away from the viewer and the C_1 carbon (numbering from the aldehyde end) is on his right, then the terminal CH_2OH group points upward for D-sugars (Fig. 9-5).

In formation of the hemiacetal ring the C_1 carbon thereby gains a new substituent, creating a new center of chirality. Two additional configurational isomers thereby become possible: these are termed the α- and β-forms. They are readily interconverted by way of the open chain form, which however exists only in trace quantities in the equilibrium mixture. The α- and β-forms are not enantiomorphs and generally differ in physical properties, including specific rotation. In the case of D-glucose

$$\begin{array}{ccccc} \alpha\text{-D-glucose} & \rightleftharpoons & \text{aldehyde form} & \rightleftharpoons & \beta\text{-D-glucose} \\ [\alpha]_D = 112^\circ & & & & [\alpha]_D = 19^\circ \end{array}$$

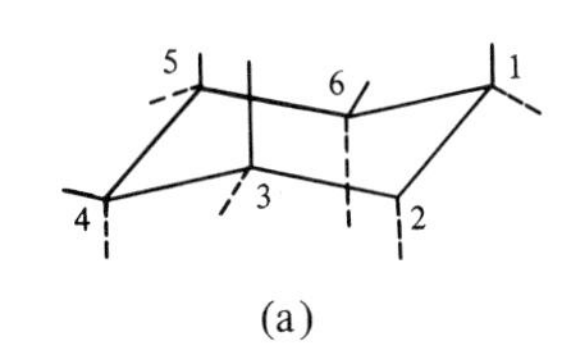

(a)

When pure α-D-glucose is dissolved in water, the specific rotation ($[\alpha]_D$) slowly decreases from the initial value of 112° to 52°, which corresponds to the equilibrium mixture of α- and β-forms. The same final value is approached with time by a solution of β-D-glucose. This phenomenon is called *muta-rotation.*

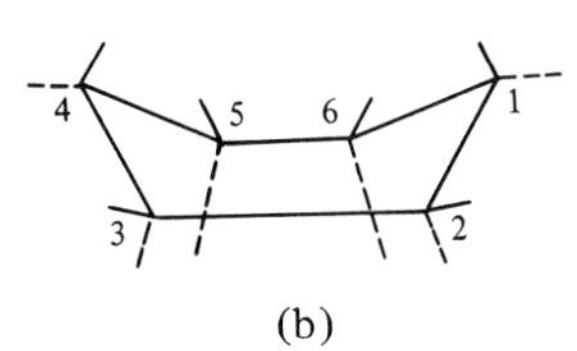

(b)

Figure 9-7
(a) Chair and (b) boat conformations of cyclohexane.

While glucose, both in the free state and in polymeric carbohydrates, exists as a 6-membered *pyranose* ring, several other sugars, including fructose and arabinose (Fig. 9-6) form 5-membered *furanose* rings. In the conventional nomenclature for the ring forms, the complete name includes the name of the sugar, its configuration, the type of ring, and the configuration of the hydroxyl group attached to the hemiacetal carbon. The latter configuration is assigned on the basis of the orientation of the hydroxyl group attached to the hemiacetal carbon. If the hydroxyl is on the same side of the structure as the ring oxygen in the Fischer formulas, the configuration is α; if it is on the opposite side, the conformation is β. Thus the β-isomer of the furanose form of D-fructose is designated as *β-D-fructofuranose.*

"Chair" and "boat" forms

The pyranose ring is not in reality the planar hexagon depicted by the Haworth formulas, which would be inconsistent with the valence angle of carbon. The 6-membered ring of cyclohexane can exist in puckered "chair" or "boat" forms (Fig. 9-7). The chair conformation is the more stable of the two and corresponds to that adopted by glucose (Fig. 9-8). In this conformation the secondary hydroxyls are not all equivalent, as is reflected by their varying chemical reactivities and physical properties.

For most purposes the Fischer and Haworth formulas are adequate and will accordingly be retained in the text.

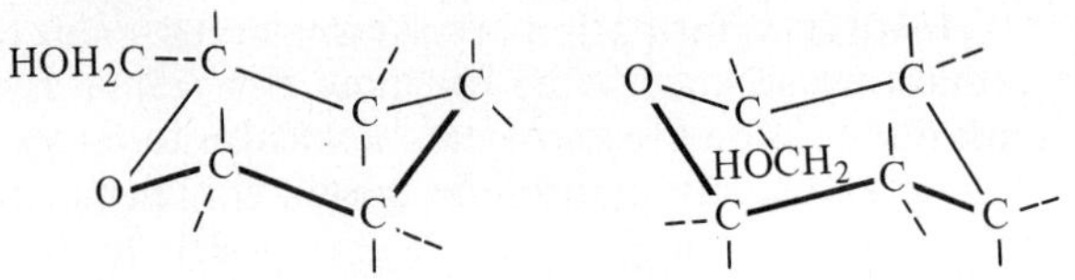

Figure 9-8
Chair conformation of α-D-glucose.

9-3 CHEMICAL PROPERTIES AND DERIVATIVES OF THE MONOSACCHARIDES

Glycosides

The hemiacetal hydroxyl reacts readily with an alcohol in the presence of dilute acid to form a mixed acetal, or glycoside:

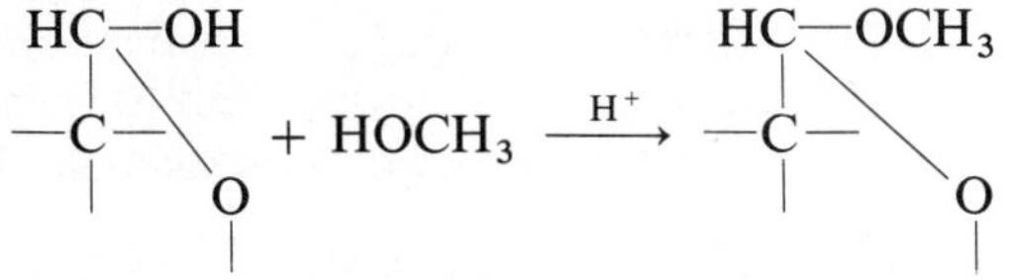

$$\text{HC—OH} + HOCH_3 \xrightarrow{H^+} \text{HC—OCH}_3$$

If the sugar is glucose, the corresponding acetal is called a **glucoside**.

Since glycoside formation may occur with either the α- or β-isomer of the hemiacetal form of a monosaccharide, there exist corresponding α- and β-glycosides. Conversion to the glycoside abolishes the dynamic equilibrium between the α- and β-configurations and immobilizes the sugar derivative as an α- or β-form.

The monosaccharides occur less abundantly in the free state than as derivatives or polymers. The monosaccharide subunits of the polymeric carbohydrates are generally joined by **glycosidic** bonds.

Methylation of Alcoholic Hydroxyls

Methylation of the hydroxyls other than the hemiacetals, which are generally less reactive, requires relatively drastic treatment, such as methyl iodide plus silver oxide or dimethyl sulfate plus alkali. This reaction has often been used to establish which of the hydroxyls are available for substitution and which are blocked through involvement in a hemiacetal or glycosidic linkage. For example, the C_5 hydroxyl of methyl **glucoside** (methyl glucopyranoside) is not methylated, since it is involved in the hemiacetal ring, while the C_2, C_3, C_4, and C_6 hydroxyls are converted to methoxy ($—OCH_3$) groups (Fig. 9-9).

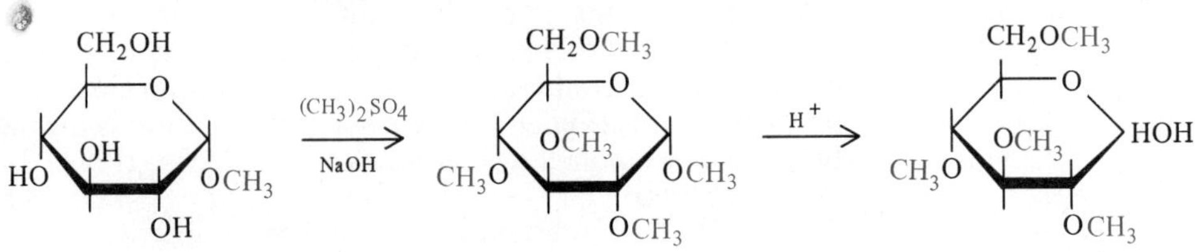

Figure 9-9
Methylation of the secondary hydroxyls of glucose by dimethyl sulfate followed by hydrolytic demethylation of the hemiacetal group.

Phosphoric Acid Esters

The hydroxyl groups of monosaccharides can be esterified. The esters formed with phosphoric acid are of particular biological importance, since it is in this form that the monosaccharides enter the pathways of oxidative metabolism. Two examples are **glucose-6-phosphate** and **glucose-1-phosphate** (Fig. 9-10).

Amino Derivatives

The amino sugars arise formally by the replacement of a hydroxyl by an amino group. The amino derivative of glucose, **glucosamine** (2-amino-2-deoxy-D-glucose), and that of galactose, **galactosamine** (2-amino-2-deoxy-D-galactose), are of considerable biochemical interest. Both are substituted in the 2-position (Fig. 9-11). Glucosamine is a component of many polysaccharides of vertebrate tissues, while galactosamine is found in glycolipids and the cartilage polysaccharide chondroitin sulfate.

Deoxysugars

The most important of the deoxysugars, **2-deoxy-D-ribose**, is a component of deoxyribonucleic acid (DNA), the basic carrier of genetic information. The other deoxysugars of biochemical importance include rhamnose and fucose.

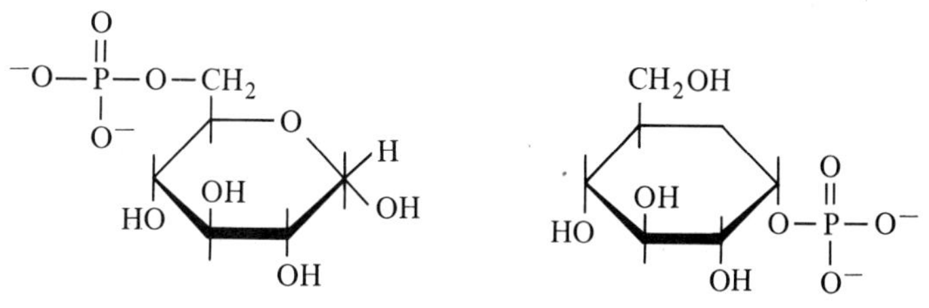

Glucose-6-phosphate

α-D-Glucose-1-phosphate

Figure 9-10 Two biologically important phosphoric acid esters of glucose

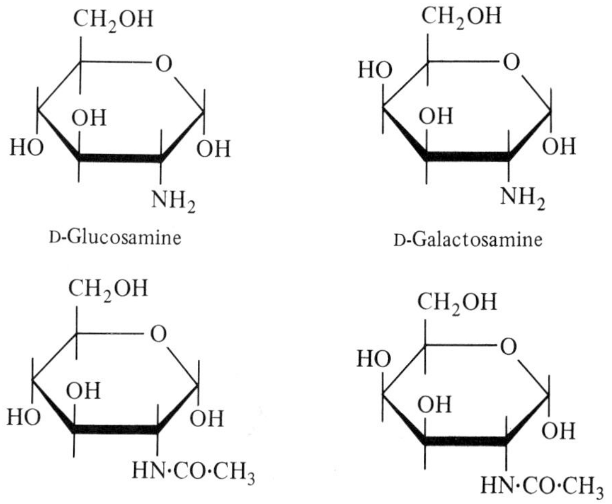

D-Glucosamine

D-Galactosamine

N-Acetyl-D-glucosamine

N-Acetyl-D-galactosamine

Figure 9-11 Structures of two important amino sugars and their N-acetyl derivatives.

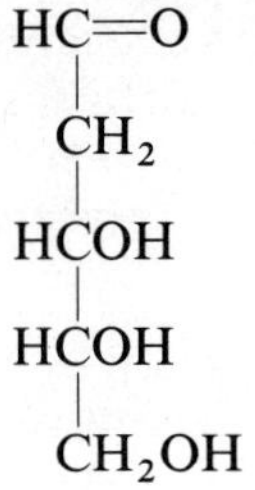

$$\begin{array}{ccc}
\begin{array}{c} HC{=}O \\ | \\ CH_2 \\ | \\ HCOH \\ | \\ HCOH \\ | \\ CH_2OH \end{array} &
\begin{array}{c} HC{=}O \\ | \\ HOCH \\ | \\ HCOH \\ | \\ HCOH \\ | \\ CH_3 \end{array} &
\begin{array}{c} HC{=}O \\ | \\ HCOH \\ | \\ HCOH \\ | \\ HOCH \\ | \\ HOCH \\ | \\ CH_3 \end{array} \\
\text{2-Deoxy-D-ribose} & \text{L-Fucose} & \text{L-Rhamnose}
\end{array}$$

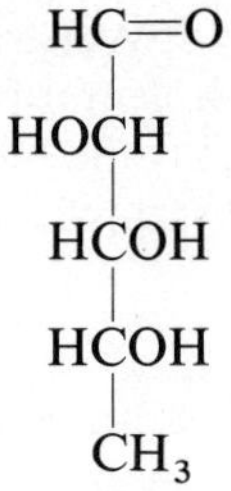

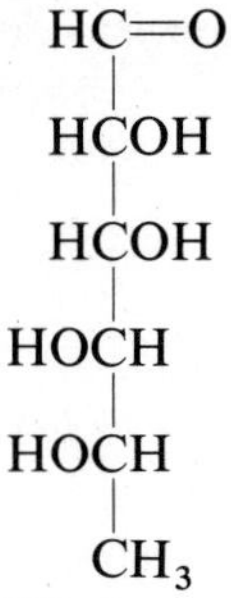

Reactions in Alkaline Medium

In alkaline solution the monosaccharides undergo a complex set of interconversions leading, in the case of glucose, to over 100 derivatives. Among the more interesting of the possible transformations is a rearrangement that takes place at low temperatures in dilute alkali.

The mechanism involves **enolization**, or the migration of a hydrogen from a carbon atom to an adjacent carbonyl, to form an unsaturated alcohol:

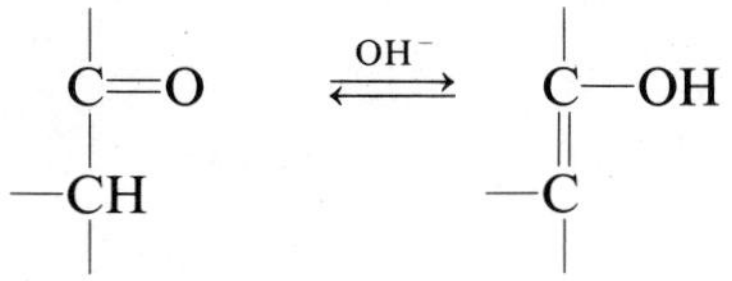

$$\begin{array}{c} | \\ C{=}O \\ | \\ -CH \\ | \end{array} \quad \overset{OH^-}{\rightleftharpoons} \quad \begin{array}{c} | \\ C{-}OH \\ \| \\ -C \\ | \end{array}$$

Aldehyde or ketone — Enol

In this way glucose is converted to an equilibrium mixture containing mannose and fructose:

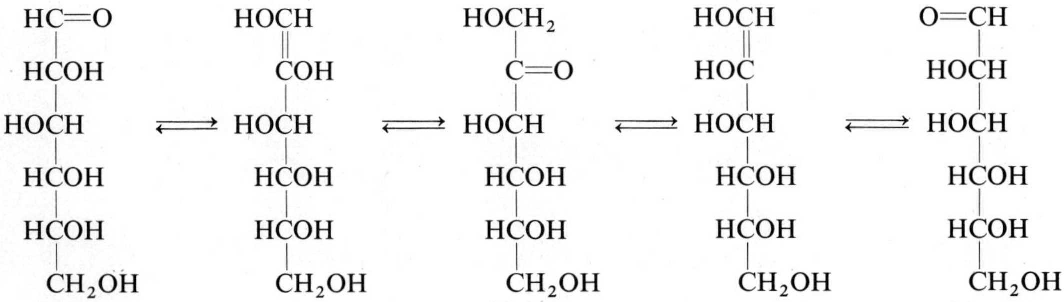

$$\begin{array}{c} HC{=}O \\ | \\ HCOH \\ | \\ HOCH \\ | \\ HCOH \\ | \\ HCOH \\ | \\ CH_2OH \\ \text{Glucose} \end{array} \rightleftharpoons
\begin{array}{c} HOCH \\ \| \\ COH \\ | \\ HOCH \\ | \\ HCOH \\ | \\ HCOH \\ | \\ CH_2OH \\ \end{array} \rightleftharpoons
\begin{array}{c} HOCH_2 \\ | \\ C{=}O \\ | \\ HOCH \\ | \\ HCOH \\ | \\ HCOH \\ | \\ CH_2OH \\ \text{Fructose} \end{array} \rightleftharpoons
\begin{array}{c} HOCH \\ \| \\ HOC \\ | \\ HOCH \\ | \\ HCOH \\ | \\ HCOH \\ | \\ CH_2OH \\ \end{array} \rightleftharpoons
\begin{array}{c} O{=}CH \\ | \\ HOCH \\ | \\ HOCH \\ | \\ HCOH \\ | \\ HCOH \\ | \\ CH_2OH \\ \text{Mannose} \end{array}$$

Muramic And Neuraminic Acids

These derivatives are important components of the structural polysaccharides occurring in the cell walls of bacteria and higher organisms. Both are derivatives of amino sugars (Fig. 9-12).

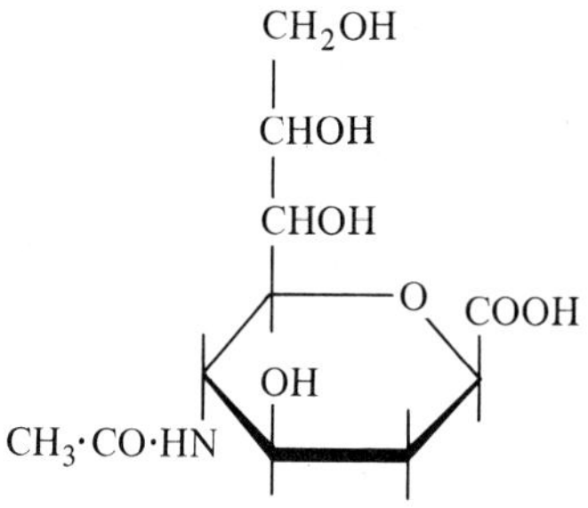

N-Acetylneuraminic acid

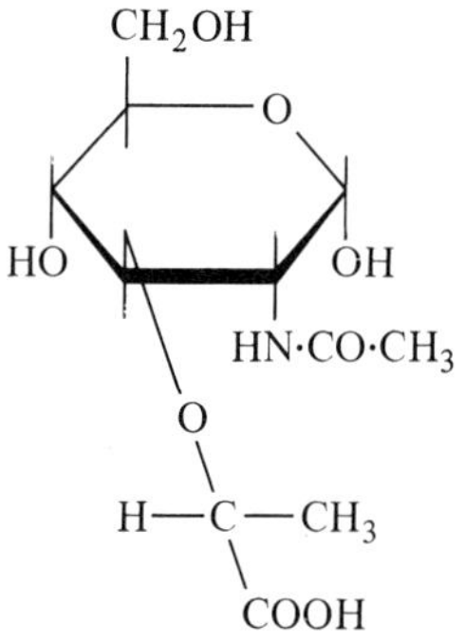

N-Acetylmuramic acid

Figure 9-12
Structures of N-acetyl derivatives of neuraminic and muramic acids. The former derivative is called **sialic acid**.

9-4 OLIGOSACCHARIDES

The name oligosaccharide is loosely applied to molecules that may be converted to monosaccharides by complete hydrolysis and contain from two to about ten monosaccharide units. The demarcation between oligosaccharides and polysaccharides of low molecular weight is arbitrary.

Oligosaccharides are combinations of monosaccharides united by glycosidic bonds, with the loss of one molecule of water for each bond formed. While each glycosidic bond must involve at least one hemiacetal carbon, the position of the second carbon is variable. A complete structural determination for an oligosaccharide requires the identification of the monosaccharide units, the positions joined by glycosidic bonds, and the configuration (α or β) of the linkages.

A commonly used approach for structural determination is exhaustive methylation, usually with alkaline dimethyl sulfate, followed by hydrolysis to methylated monosaccharides. These may be isolated and identified by chromatographic procedures. This treatment replaces all hydroxyls by methoxyl ($—OCH_3$) groups, except for those blocked through involvement in glycosidic bonds or hemiacetal rings. The methoxyl groups are sufficiently stable

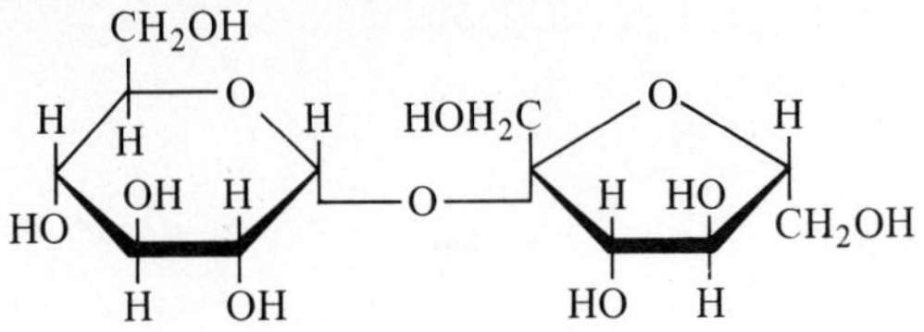

Figure 9-13
Structure of sucrose.

to survive complete acid hydrolysis of the glycosidic bonds, except for those formed by the hemiacetal hydroxyls, which are converted to free hemiacetal. Consideration of the distribution of the methoxyl groups among the various possible locations on the monosaccharides places definite limits upon the possible positions of the glycosidic linkage.

The assignment of the configuration of the glycosidic bond (α versus β) may often be made from optical rotatory or enzymatic evidence. If the oligosaccharide contains only D-glucose units, a *decrease* in $[\alpha]_D$ during the mutarotation accompanying acid hydrolysis suggests that the configuration is α, since the α-D-glucose formed rapidly by hydrolysis has a higher specific rotation (112.2°) than the β-D-glucose (18.7°) into which it is slowly converted. Conversely, an increase in $[\alpha]_D$ suggests that the configuration is β.

Enzymes exist which exhibit selectivity in catalyzing the hydrolysis of α- and β-glycosidic bonds. The α-D-glucosidase of yeast splits glycosidic bonds formed by α-D-glucose, but not those formed by β-D-glucose. The β-D-glucosidase of almond emulsin has an opposite specificity. The susceptibility of a particular glycosidic bond to hydrolysis by enzymes of known specificity often provides evidence as to its configuration.

Sucrose

Sucrose, O-β-fructofuranosyl-(2,1)-α-D-glucopyranoside, is the familiar "sugar" of everyday life. While it is of widespread occurrence in plants, its common commercial sources are sugar cane or beet.

An unusual feature of the structure of sucrose is the absence of a potentially free carbonyl group. The oxygen bridge linking the two monosaccharides joins the C_1 carbon of glucose and the C_2 carbon of fructose and formally arises by the elimination of water between the hemiacetal and hemiketal hydroxyls (Fig. 9-13). Consequently, sucrose does not react with reagents specific for carbonyl groups and is termed a **nonreducing** sugar.

The complete hydrolysis of sucrose produces a 1:1 mixture of glucose and fructose. The process is accompanied by a change in optical rotation, whose sign changes from positive to negative. For this reason the process is often called **inversion** of sucrose. The reaction is catalyzed by hydrogen ion, as well as by a class of enzymes called **invertases**.

Maltose

Maltose, O-α-D-glucopyranosyl-(1,4)-β-D-glucopyranose, consists of two glucose units joined by an α glycosidic bridge between the C_1 and C_4 carbons (designated as α, 1, 4, or α, $1 \rightarrow 4$). In contrast to sucrose, one of the hemiacetal hydroxyls remains unsubstituted (Fig. 9-14), so that the disaccharide retains a potentially free and reactive carbonyl. While the glycosidic linkage of

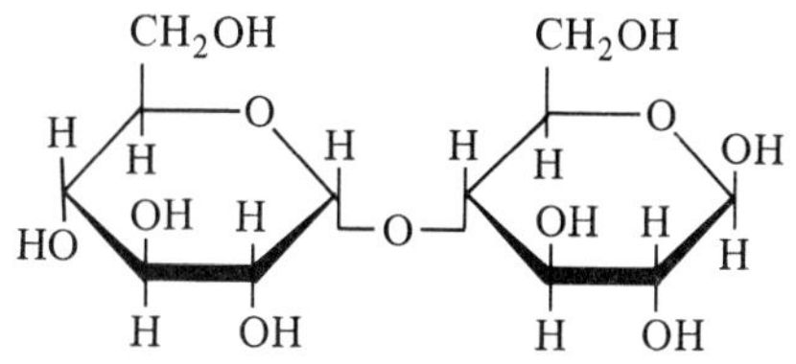

Figure 9-14
Structure of maltose.

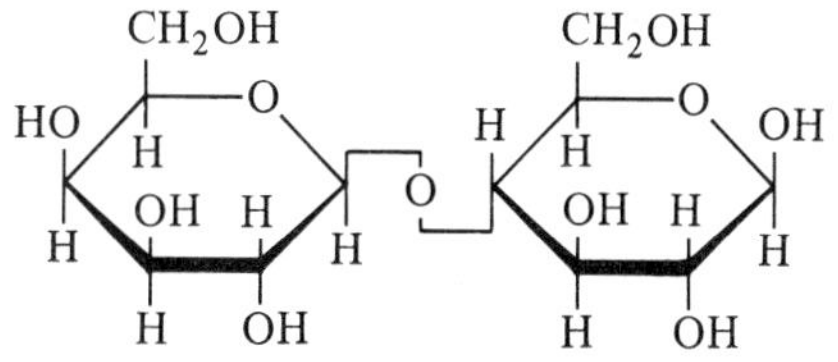

Figure 9-15
Structure of lactose.

maltose is immobilized in the α-configuration, the free hemiacetal exists as an equilibrium mixture of the α and β forms.

Maltose is the chief product of the hydrolysis of starch by certain enzymes including salivary and pancreatic amylase. It is not of frequent occurrence in biological systems in the free state.

Lactose

Lactose, O-β-galactopyranosyl-(1,4)-β-D-glucopyranose, which occurs in milk, contains one mole each of galactose and glucose, joined by a glycosidic bond between the C_1 and C_4 carbons, respectively (Fig. 9-15).

9-5 POLYSACCHARIDES

Most of the carbohydrate content of living tissues consists of polysaccharides, which are polymers of the simple sugars. These usually function either as structural elements, which confer rigidity upon soft tissues, or as nutritional reservoirs, which may be converted as needed to metabolically utilizable sugar derivatives. Polysaccharides account for a major part of the dry weight of higher plants and seaweeds.

The glycosidic bond is the universal mode of linkage of the monosaccharide units occurring in polysaccharides. Homopolymers of a single monosaccharide and copolymers of two or more different monosaccharides are both of wide occurrence. The chains may be linear or branched. The monosaccharide units are usually joined according to a regular repeating scheme.

As usually prepared, polysaccharides are very heterogeneous with respect to molecular size. In view of the extensive degradation that may accompany their isolation, the measured average molecular weight, as determined by physicochemical techniques, may be very different from that in the natural state.

Amylose

Starch, the principal reserve carbohydrate of plants, is a mixture of two different polysaccharides, amylose and amylopectin. Starch is deposited in

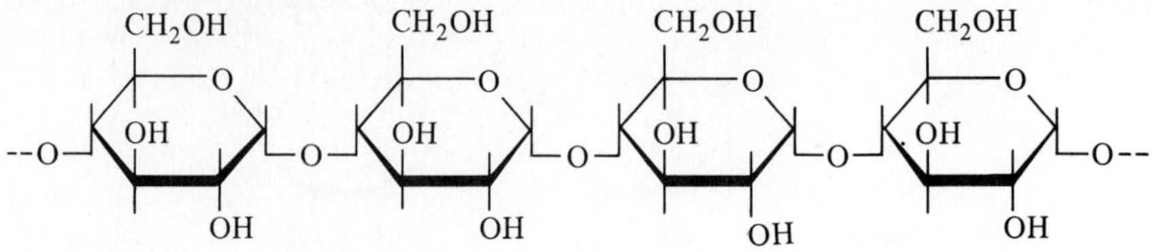

Figure 9-16
Structure of amylose.

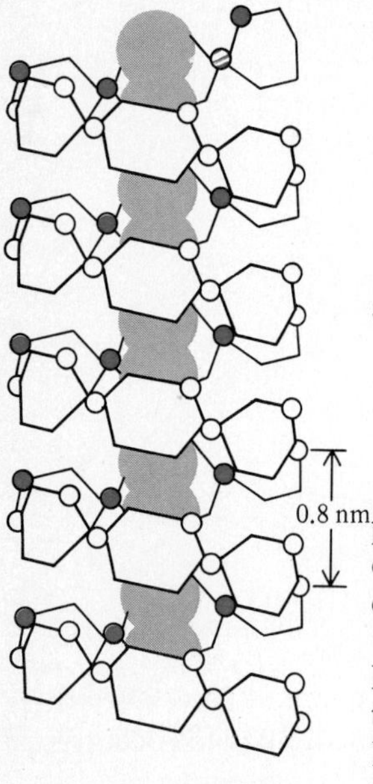

Figure 9-17
Structure of the helical complex formed by amylose with I_2.

plants as granules, 3–100 μm in diameter, from which the two components can be extracted with water and separated by selective precipitation with organic solvents.

Amylose is converted quantitatively to D-glucose by complete acid hydrolysis, indicating that it is a polymer of this sugar alone. The nature of the glycosidic linkages was established by methylation studies (Fig. 9-16). Exhaustive methylation, followed by complete acid hydrolysis, yields 2,3,6-tri-O-methylglucose, with a minor quantity of 2,3,4,6-tetra-O-methylglucose. The latter is derived from the terminus of the amylose chain that lacks a free hemiacetal group. (The other terminus possesses a free hemiacetal and is converted by methylation and subsequent hydrolysis to 2,3,6-tri-O-methylglucose.) From evidence of this kind it has been concluded that the glucose units of amylose are joined by α,1,4 glycosidic bonds. Optical rotatory studies have established that the glycosidic linkages are in the α configuration. In summary, amylose consists of linear chains of D-glucose united by α,1,4 glycosidic bonds. There is no evidence for any significant degree of branching. The molecular weight ranges from 10^4 to 10^6.

Physical studies have shown that amylose in aqueous solution has a helical conformation. A helical molecular complex of a deep blue color is formed with I_2 (Fig. 9-17).

Amylopectin

Amylopectin, the other constituent of starch, also consists of polymers of D-glucose. However, exhaustive methylation, followed by complete acid hydrolysis, produces, in addition to the derivatives obtained from amylose, a significant quantity of 2,3-di-O-methylglucose (Fig. 9-18). The blocking of the C_6 hydroxyl to methylation reflects the presence of numerous branch points, the mode of attachment of the branches being by α,1,6 linkages. Except for the branch points, the glucose units are joined by α,1,4 glycosidic bonds, as in amylose.

Amylopectin consists of chains of glucose residues joined by α,1,4 glycosidic linkages, with branches connected by α,1,6 linkages occurring at a frequency of roughly one for every 20–30 glucose units. Physical measurements indicate that the amylopectin molecule is compact and roughly symmetrical in shape, rather like a dense shrub. The molecular weights are quite high—up to 10^8.

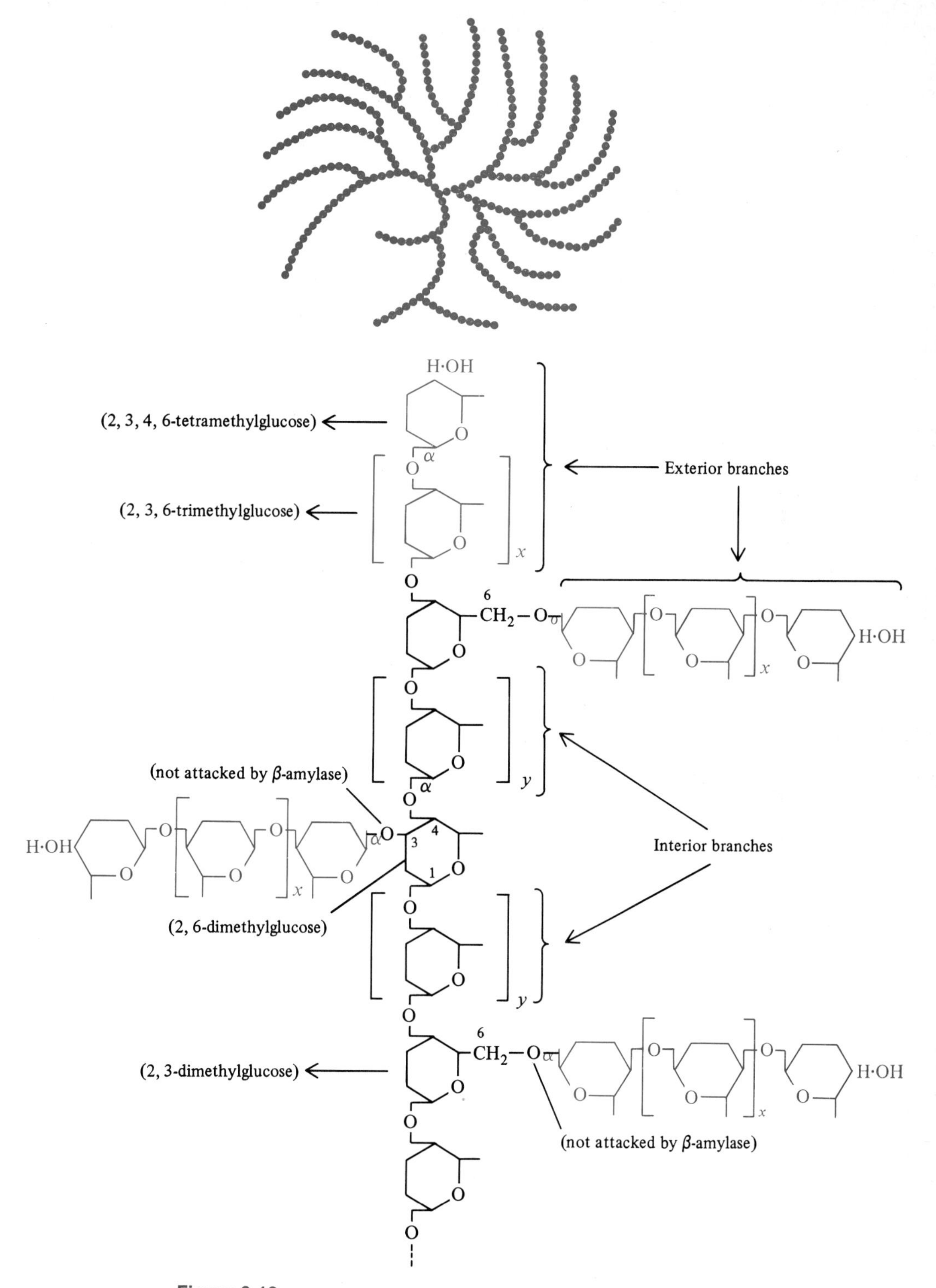

Figure 9-18

Structure of amylopectin as established by methylation studies. The upper diagram is a macroscopic view.

Glycogen

Glycogen, the principal reserve carbohydrate of animals, has a chemical structure resembling that of amylopectin, likewise consisting of α,1,4 linked glucose chains, with branches joined by α,1,6 bonds. However, the extent of branching is greater than for amylopectin, about one branch for every 10–12 glucose units being present.

Cellulose

Cellulose, the chief structural carbohydrate of plants, is probably the most abundant organic compound found in nature. While it occurs most extensively in the higher plants, it is also present in the lower plants, as well as in some bacteria. The 98% cellulose seed hairs of the cotton plant are among the purest natural sources. The principal commercial source is wood, where it occurs with lignin, xylans, pectic substances, and a multitude of minor components.

Cellulose consists of unbranched chains of glucose linked by β,1,4 glycosidic bonds (Fig. 9-19). While it is chemically similar to amylose, except for the configuration at the C_1 position, its physical properties are totally different. The geometry of the β-glycosidic linkages is such that close lateral alignment of the strands can occur, permitting extensive interchain hydrogen bonding. The resulting ordered bundles (Fig. 9-20) resist solution in water, so that, in contrast to amylose, cellulose must be chemically modified to permit solution.

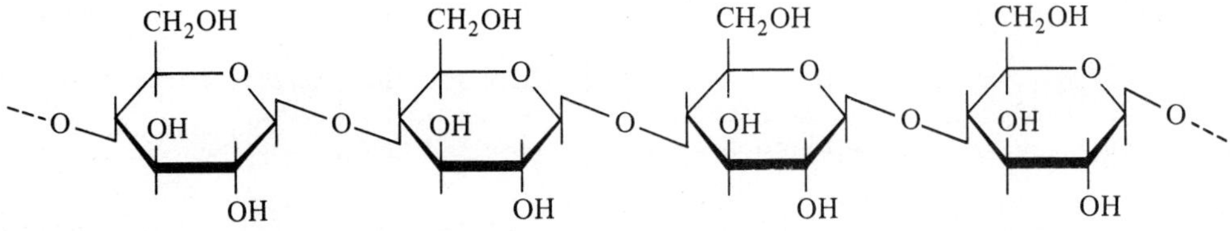

Figure 9-19
Structure of cellulose.

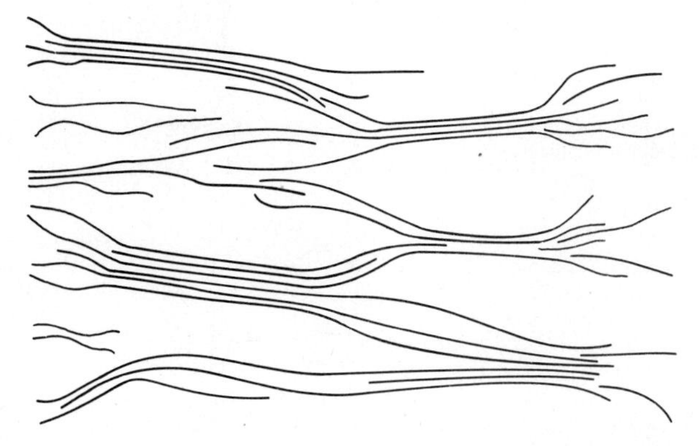

Figure 9-20
Ordered and amorphous regions in cellulose fibers.

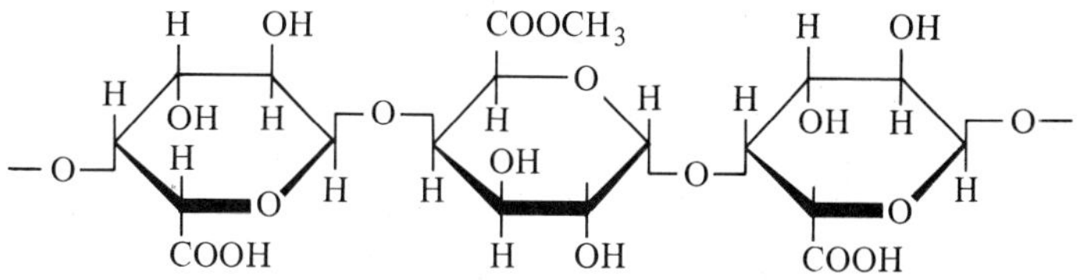

Figure 9-21
Structure of pectic acid.

The combined application of electron microscopy and x-ray diffraction has permitted some generalizations as to the organization of cellulose strands in the cell wall of the plant. Cellulose molecules tend to coalesce into fibrils which contain both crystalline and amorphous regions. The former are generally about 5 nm in diameter and 60 nm or more in length. Estimates of the molecular weight of native cellulose vary widely. It is likely that the molecular weight is at least 10^5.

Pectic Polyuronides (Pectins)

The pectins are of universal occurrence in plant tissues, accounting for a significant fraction of the dry weight of many fruits. The distribution of pectins within the plant suggests that their function may be primarily structural.

The structural unit of pectic acid, the chief constituent of the pectins, is D-galacturonic acid (Fig. 9-21). The units are linked by α,1,4 glycosidic bonds. The carboxyl groups exist in part as methyl esters, the extent of methylation varying widely with the source. The reported molecular weights range from 10^4 to 10^5.

Dextrans

The dextrans, which are formed by yeasts and bacteria, are highly branched polymers of glucose. The linkages other than the branch points are α,1,6 bonds. The mode of linkage at the branch points (1,2; 1,3; or 1,4) and the distances between branches are characteristic of the species.

The dextrans have found extensive practical use both clinically as a plasma extender, and, in cross-linked form, as a gel filtration medium for biochemical separations.

Hyaluronic Acid

Hyaluronic acid belongs to the class of **mucopolysaccharides**, each of which has a disaccharide repeating unit containing an amino sugar. Hyaluronic acid consists of alternating residues of D-glucuronic acid and *N*-acetyl-D-glucosamine. Unlike the polysaccharides described earlier, two kinds of glycosidic linkage are present (Fig. 9-22), the bonds alternating between β,1,3 and β,1,4.

Hyaluronic acid is found in many animal tissues, often as a complex with protein. It occurs in highest concentration in the umbilical cord, joint fluids, and vitreous humor. Solutions of hyaluronic acid are viscous and

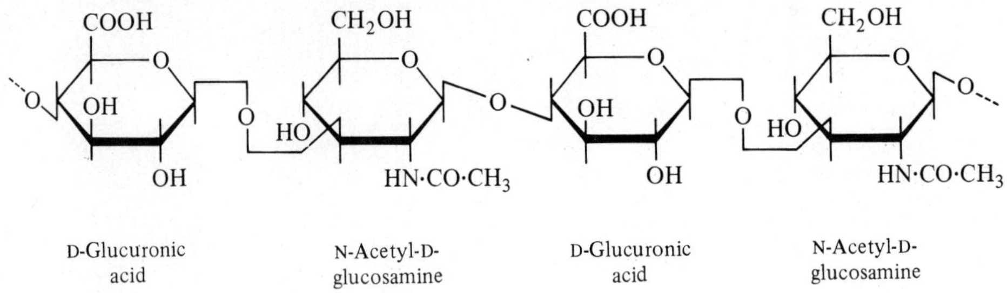

D-Glucuronic acid — N-Acetyl-D-glucosamine — D-Glucuronic acid — N-Acetyl-D-glucosamine

Figure 9-22
Structure of hyaluronic acid.

contribute to the lubricant properties of synovial and other body fluids. In tissues it acts as a cementing substance and functions in tissue barriers which oppose penetration by bacteria.

Chondroitin Sulfates These are among the most important mucopolysaccharides, occurring in the ground substances of mammalian tissues and in cartilage, where they exist as complexes with protein.

Chondroitin sulfate A has a structure similar to that of hyaluronic acid, except that: (1) *N*-acetyl-D-*galactos*amine replaces *N*-acetyl-D-*glucos*amine; (2) position 4 of galactosamine is esterified with sulfate. Chondroitin sulfate C is equivalent to the A form, except that the sulfate is at position 6 of galactosamine instead of position 4 (Fig. 9-23).

Heparin This mucopolysaccharide is found in liver and in arterial walls. It is also a well-known anticoagulant. The repeating unit is a disaccharide consisting

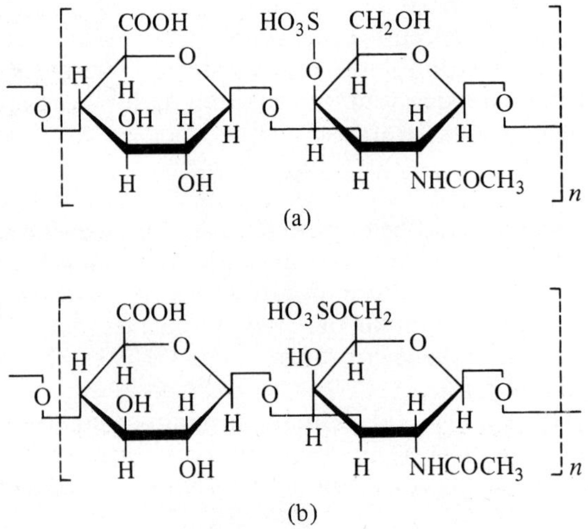

(a)

(b)

Figure 9-23
Structure of chondroitin sulfate. (a) The repeating unit of chondroitin sulfate A. (b) The repeating unit of chondroitin sulfate C.

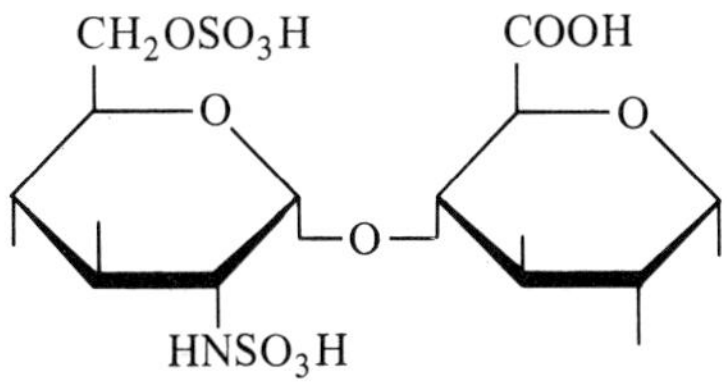

Figure 9-24 Repeating unit of heparin.

of D-glucuronic acid plus a residue of D-glucosamine-*N*-sulfate with a second sulfate esterified at the C_6 position. The linkages are α,1,4 (Fig. 9-24).

Xylans These polysaccharides occur in lignified tissues of higher plants and in many algae. The somewhat complex chemical structure of the xylans is based upon linear chains of D-xylopyranose units, joined by α,1,4 linkages (Fig. 9-25). A number of short branches may be attached to the xylose groups at the C_2 or C_3 positions. Among the common substituents are L-arabofuranose, D-glucopyranuronic acid, 4-O-methyl-D-glucopyranuronic acid, and the disaccharide D-xylopyranose-1,2-L-arabofuranose.

The xylans are quite variable with respect to the number and composition of the side-chains. The xylans of wheat flour are heavily substituted with arabinose. Those of many hard woods contain little arabinose, but are rich in 4-O-methyl-D-glucopyranuronic acid.

Muropeptide The high internal osmotic pressure of bacteria would lead in some environments to swelling and rupture of the cellular membrane if protection were not conferred by a rigid cell wall. This generally consists of polysaccharide chains covalently cross-linked by short peptide chains (Fig. 9-26) to form a network enclosing the cell. This is termed a **peptidoglycan**. The repeating unit of the polysaccharide chains is a disaccharide of *N*-acetyl-D-glucosamine and *N*-acetylmuramic acid joined by a β,1,4 glycosidic linkage. The peptide chains are attached to the carboxyl groups of the *N*-acetylmuramic acid residues. While the composition of the peptide depends upon the bacterial species, L-alanine, D-alanine, and D-glutamic acid are generally present. The presence of D-amino acids confers resistance to ordinary proteolytic enzymes. However, the β,1,4 glycosidic bonds of the polysaccharide backbone are susceptible to hydrolysis by lysozyme. Bacterial cell wall biosynthesis is inhibited by the antibiotic penicillin.

Figure 9-25 Structure of a xylan. Here the substituent X may be L-arabinofuranose, D-glucopyranuronic acid, 4-O-methyl-D-glucopyranuronic acid, or D-xylopyranose-1,2-L-arabinofuranose.

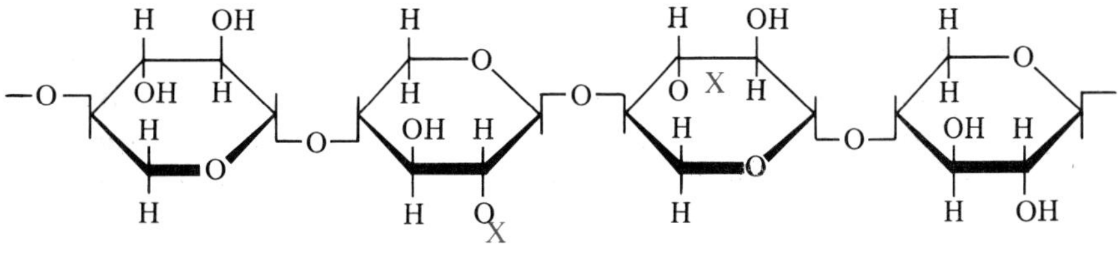

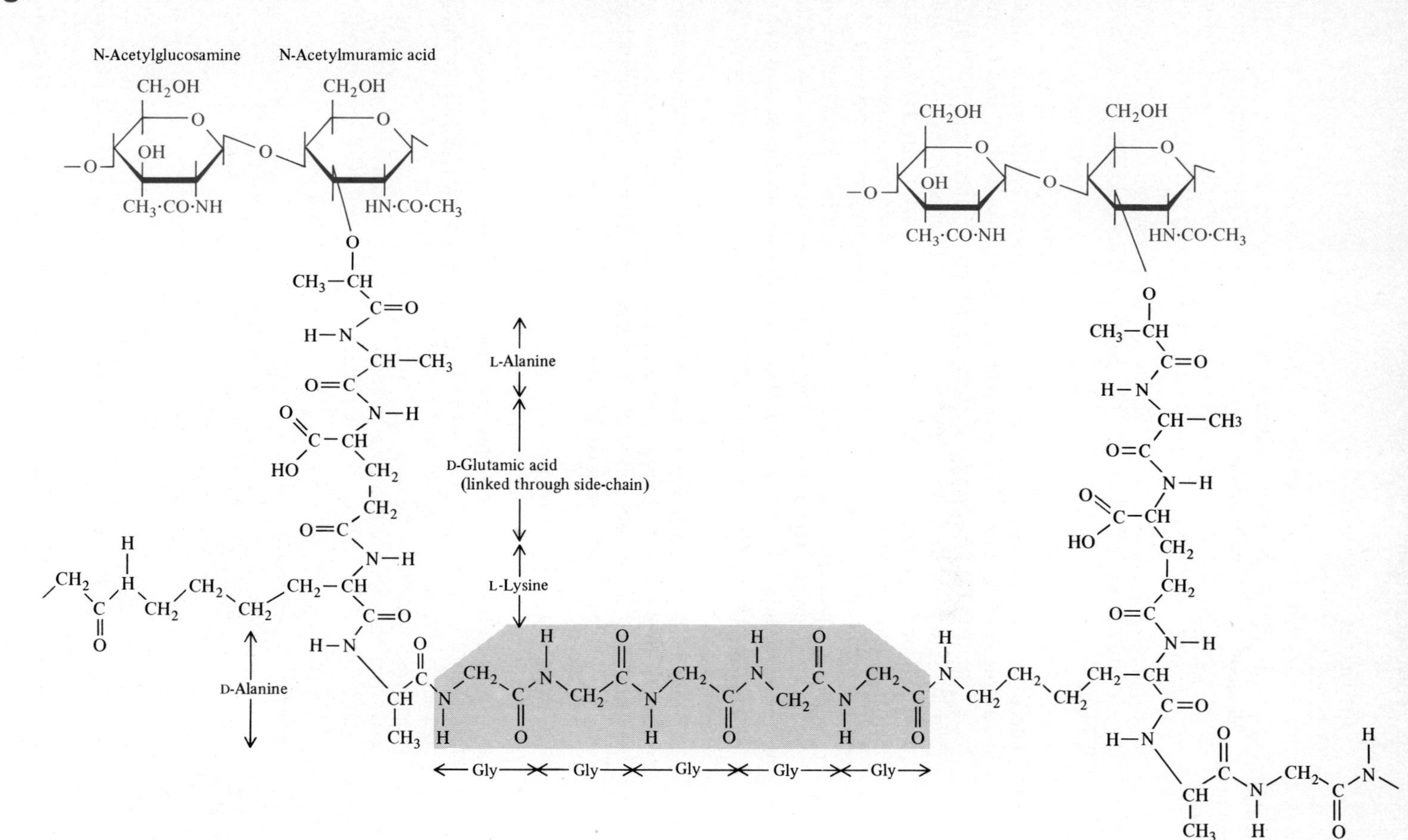

Figure 9-26
Repeating unit of the muropeptide occurring in bacterial cell walls.

9-6 GLYCOPROTEINS

Many proteins contain short oligosaccharide chains that are covalently attached; these are termed **glycoproteins**. Examples of this class are very numerous and include enzymes, protein hormones, and structural proteins, as well as proteins occurring in cell surfaces. The normal mode of attachment is by an O-glycosidic bond to the hydroxyl group of serine or threonine, or by an *N*-glycosidic linkage to the amide nitrogen of asparagine. While a wide range of monosaccharides occurs in glycoproteins, *N*-acetylneuraminic acid and *N*-acetylglucosamine are particularly common.

In many glycoproteins, such as ribonuclease and ovalbumin, the carbohydrate content consists of only one, or a few, short polysaccharide chains, accounting for only a few percent of the total mass. In most cases the function of the carbohydrate moiety is unknown. However, in the **mucins**, which function as lubricants, the carbohydrate content may be 50% or more. The presence in the mucins of many negatively charged *N*-acetylneuraminic acid groups has a dominant influence upon their physical properties, causing them to assume the shape of expanded coils of high viscosity. The mucin occurring in the salivary gland of sheep contains about 10^3 carbohydrate chains per protein molecule; these consist of α-D-*N*-acetylneuraminic acid and α-D-*N*-acetylglucosamine.

Cell Coats

Unlike plant cells, the cells in the tissues of higher animals are not enclosed by rigid walls. However, a gelatinous outer layer of complex composition is present outside the cell membrane. The principal components of the cell coats are mucopolysaccharides, lipids, and glycoproteins. These outer layers are responsible for the mutual recognition and adhesion of cells from the same tissues and dominate the antigenic properties of cells.

9-7 ENZYMIC HYDROLYSIS OF POLYSACCHARIDES

Amylases

These enzymes catalyze the hydrolytic degradation of starch to maltose units. Starch may be digested in two ways. **α-Amylase**, α,1,4-glucan 4-glucanohydrolase, occurs in saliva and pancreatic juice and contributes to the digestion of starch in the gastrointestinal tract. This enzyme attacks amylose and amylopectin at interior points of the chains in a random manner to produce ultimately a mixture of glucose and maltose units; the latter are resistant to further hydrolysis (Fig. 9-27).

β-Amylase (α,1,4-glucan maltohydrolase), which occurs in plants, likewise hydrolyzes the α,1,4 glycosidic bonds of amylose and amylopectin. However, the hydrolysis begins at the chain termini, removing consecutive maltose units, and ultimately yields maltose quantitatively (Fig. 9-28).

Neither enzyme attacks the α,1,6 glycosidic bonds of amylopectin, so that the end product of the action of either amylase is a highly branched dextrin. A debranching enzyme, **α,1,6-glucosidase**, is capable of hydrolyzing the α,1,6 linkages at the branch points, so that the combined action of this enzyme and α-amylase converts amylopectin to glucose and maltose.

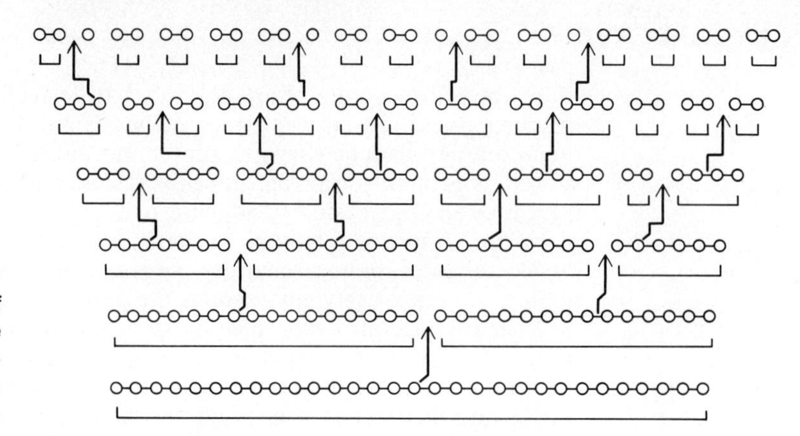

Figure 9-27
Schematic version of the action of α-amylase (α,1,4-glucan-4-glucanohydrolase) on amylose.

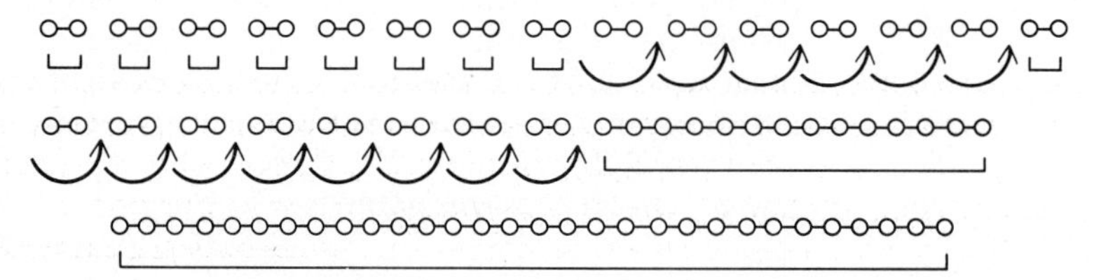

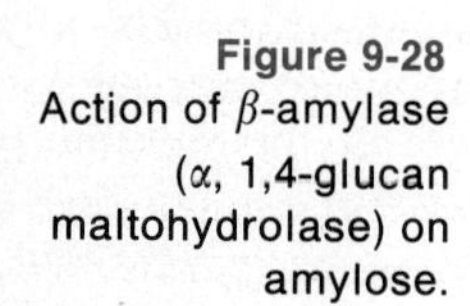

Figure 9-28
Action of β-amylase (α, 1,4-glucan maltohydrolase) on amylose.

Cellulases

Hydrolytic enzymes that attack the β,1,4 bonds of cellulose are not ordinarily found in animals, but are of common occurrence in microorganisms. The latter are generally essential for the digestion of cellulose by ruminants, such as cows.

SUMMARY

The **carbohydrates** include the **simple sugars** or **monosaccharides** and their polymers and derivatives. The monosaccharides are **polyhydroxy alcohols** or **polyhydroxy ketones**, whose overall composition is $(CH_2O)_x$; the two series are termed **aldose** and **ketose**, respectively. Because of the presence of multiple chiral centers, the monosaccharides exist as stereoisomers. Those related to **D-glyceraldehyde** are assigned the **D-conformation**; those derived from **L-glyceraldehyde** are said to have the **L-conformation**.

For the higher monosaccharides, the aldehyde or keto forms exist only in trace quantities in tautomeric equilibrium with closed ring **hemiacetal** forms. Hemiacetal formation generates a new chiral center and an additional –OH group of enhanced reactivity, which may have an α- or β-configuration.

The more important monosaccharides include the six-carbon aldoses **D-glucose** and **D-galactose**, the 6-carbon ketose **fructose**, and the 5-carbon aldose **ribose**. 2-Deoxy-D-ribose is also important as a constituent of DNA. (The carbon atoms of monosaccharides are numbered sequentially from the end closest to the carbonyl.)

Oligomers and polymers of monosaccharides arise via the formation of **glycosidic bonds** by the elimination of H_2O between a hemiacetal hydroxyl and another hydroxyl of a second monosaccharide. The important oligosaccharides include the disaccharides **lactose, maltose**, and **sucrose**, as well as the trisaccharide **raffinose**.

The primary reserve carbohydrates of plants are **amylose**, which is a linear polymer of glucose units joined by glycosidic bonds, and **amylopectin**. Amylopectin differs from amylose in containing numerous branches joined by glycosidic bonds between positions 1 and 6 (α,1,6). **Cellulose**, the principal structural carbohydrate of plants, is, like amylose, an unbranched polymer of glucose units joined by 1,4 glycosidic bonds. However, the two differ in the configuration about the hemiacetal carbon; the glycosidic bonds of amylose are α,1,4 while those of cellulose are β,1,4. The physical properties of the two are very different; amylose is water soluble while cellulose is insoluble and partially crystalline. The chief reserve carbohydrate of animals is **glycogen**, which resembles amylopectin in structure, but is more highly branched.

Polymers of the amino sugars **D-glucosamine, D-galactosamine**, and ***N*-acetyl-D-glucosamine**, as well as the sugar derivatives ***N*-acetylmuramic acid** and ***N*-acetylneuraminic acid**, are important components of **structural polysaccharides**. The cell walls of many bacteria contain **murein** or **peptidoglycan**, which consists of linear β,1,4 polymers of *N*-acetylglucosamine and *N*-acetylmuramic acid joined by polypeptide bridges to form a saclike network enclosing the cell.

Several important structural polysaccharides are polymers of *N*-acetyl-D-galactosamine and **glucuronic acid**, a glucose derivative containing a carboxyl group at the 6-position. These include **hyaluronic acid**, **chondroitin**, and **chondroitin sulfate**, in which *N*-acetyl-D-galactosamine is esterified with H_2SO_4 at the 6-position. These polymers belong to the class of **mucopolysaccharides**. Mucopolysaccharides have a slippery or sticky consistency. They function as biological lubricants or intercellular cements.

Many proteins contain covalently linked carbohydrate chains and are termed **glycoproteins**. The carbohydrate content may have a significant influence upon their properties.

REFERENCES

The following code is used to classify references. I: particularly useful as an introduction to the subject; R: useful primarily as a reference text; A: an advanced account of the material; H: a publication of historical importance.

General

E. A. Davidson, *Carbohydrate Chemistry*, Holt, New York (1967). (R)

A. L. Lehninger, *Biochemistry*, Worth, New York (1975). (A)

D. Metzler, *Biochemistry*, Academic, New York (1979). (R)

W. W. Pigman and D. Horton, eds., *The Carbohydrates*, vols. 1A and 1B, Academic, New York (1972). (R)

Oligosaccharides

R. W. Bailey, *Oligosaccharides*, Macmillan, New York (1965). (R)

Glycoproteins

A Gottschalk, ed., *Glycoproteins: Composition, Structure, and Function*, 2nd ed., Elsevier, Amsterdam (1972). (A)

R. D. Marshall, "Glycoproteins," *Ann. Rev. Biochem.* 41, 673 (1972). (A)

Mucopolysaccharides

E. C. Heath, "Complex Polysaccharides," *Ann. Rev. Biochem.* 40, 29 (1971). (A)

G. Quintirelli, ed., *The Chemical Physiology of Mucopolysaccharides*, Little, Brown, Boston (1967). (A)

REVIEW QUESTIONS

Questions marked with an asterisk are of a high level of difficulty.

9-1 Answer the following.
- (a) What is a chiral (asymmetric) carbon atom?
- (b) Draw the structure of any β-D-aldoheptose in the pyranose ring form. (Use Fischer projection or Haworth ring structure.)
- (c) How many asymmetric carbon atoms does the above sugar have?
- (d) How many stereoisomers of the above sugars are theoretically possible?
- (e) Draw the structure of the **anomer** (α-form) of the above β-D-aldoheptose.
- (f) Draw the structure of the enantiomorph of the above β-D-aldoheptose.
- (g) Draw the structure of an **epimer** (other than the anomer) of the above sugar.
- (h) Draw the structure of a diastereoisomer of the above β-D-aldoheptose.
- (i) Draw the structure of a structural isomer of the above β-D-aldoheptose.
- (j) Draw the structure of the same β-D-aldoheptose shown in (b) in the furanose ring form.
- (k) If the sugar drawn in (j) were subjected to exhaustive methylation, and then mild acid hydrolysis, what would the final product be (formula or name)?

9-2 Draw the structures of the repeating basic unit of the following polysaccharides:
- (a) amylose
- (b) amylopectin
- (c) cellulose

9-3 A disaccharide composed of glucose and mannose was isolated from a plant. Preliminary experiments established that the two monosaccharides were connected by a β-glycosidic linkage. Exhaustive methylation of the disaccharide, followed by mild acid hydrolysis yielded two products: 2,3,4,6-tetramethylglucose, and 2,3,4,-trimethyl-mannose. Draw the structure of the original disaccharide using the Haworth convention. Indicate clearly the manner in which the two monosaccharides are linked, and the correct ring structure of each.

*9-4 A new oligosaccharide was analyzed by enzymatic and chemical techniques. Treatment with an enzyme which cleaves maltose yielded a trisaccharide and no other products. Exhaustive methylation, followed by acid hydrolysis, gave a mixture of 2,3-O-methyl galactose and 2,3,6-O-methyl glucose in the ratio 1:2, and no other products. Draw a structure or structures for this oligosaccharide. You can use abbreviations for portions of the structure if you define them with a drawing.

9-5 Treatment of glycogen with an α-1,6-glucosidase *only* would yield:
(a) maltose
(b) glucose
(c) unbranched polysaccharide chains
(d) both (a) and (b)
(e) none of the above

9-6 Show the *mechanism* by which α- and β-D-glucose are interconverted in aqueous solution at pH 7.0.

9-7 Draw the structure of any disaccharide which is a reducing sugar. Number the carbon atoms. What kind of bond connects the two monosaccharide units?

9-8 Maltose is exhaustively methylated, hydrolyzed, and then treated with $NaBH_4$. Draw the structure of the products. Show the pathway from maltose to these products.

10

Carbohydrate Metabolism

10-1 CARBOHYDRATES AS BIOLOGICAL FUELS

The biochemists of the 19th century recognized that the carbohydrates consumed by animals and bacteria are largely oxidized to CO_2 and H_2O; during the same period it became clear that the reverse process occurred in green plants.

Respiration:

$$C_6H_{12}O_6\,(\text{glucose}) + 6O_2 \rightarrow 6CO_2 + 6H_2O + \text{energy}$$

Photosynthesis:

$$6CO_2 + 6H_2O + \text{solar energy} \rightarrow C_6H_{12}O_6 + 6O_2$$

The ultimate source of chemical energy for all living systems is the radiant energy of sunlight, which is trapped in plants by the reactions of photosynthesis and utilized for the synthesis of carbohydrates and other organic molecules. The oxidative processing of carbohydrates by animals releases **chemical energy**, which is diverted to the synthesis of the high energy compounds, such as ATP, that are needed to drive endergonic processes essential for the growth and functioning of cells and tissues.

The free energy released by the complete biological oxidation of a mole of glucose to CO_2 and H_2O is exactly the same (686 kcal) as if the same quantity of glucose were incinerated in a crucible. However, while the latter process would simply dissipate the released energy as heat, the *controlled* and *stepwise* biological reaction permits the trapping and storage in high energy compounds of much of this energy.

Most of the metabolic conversions of carbohydrates actually occur in the absence of molecular oxygen, which is the ultimate electron acceptor or oxidizing agent. The terminal phases of oxidative metabolism, in which molecular oxygen is directly involved and in which most of the ATP synthesis occurs, will be discussed in chapter 12. In the present chapter we shall be concerned with the initial stages of the metabolic transformation of carbohydrates, in which molecular oxygen does not figure. These initial stages, which include the important pathway termed **glycolysis**, take place in cell cytoplasms, while the later phases occur in the mitochondria. We shall see that glycolysis can generate ATP even in the absence of an oxygen supply.

Only a few carbohydrates are of major nutritional significance for animals. The only monosaccharides present in significant quantities in the normal diet are glucose and fructose, neither of which is usually present to more than a minor extent. Among the disaccharides, lactose is important in infant nutrition, while sucrose is present in highly variable quantities. By far the most important carbohydrates in human nutrition are the starches, amylose and amylopectin, both of which are synthesized in plants, with a lesser contribution from glycogen. Amylose and amylopectin occur in flour, potatoes, and vegetables, while glycogen is found in meat.

The metabolism of carbohydrates by animals is largely channeled through glucose, to which the above carbohydrates are hydrolyzed by digestive enzymes.

10-2 DIGESTION OF CARBOHYDRATES

Polysaccharides of high molecular weight cannot readily penetrate the mucosa of the gastrointestinal tract. An essential preliminary step to the absorption and metabolic processing of the polymeric carbohydrates is their hydrolytic degradation to small molecules by the digestive system.

The digestion of the starches begins with their exposure to saliva. The α-amylase present in saliva initiates the hydrolytic degradation of amylose and amylopectin. The importance of salivary digestion is variable, depending on the time elapsing before the amylase is inactivated by exposure to the gastric juices.

Little digestion of starch occurs during its passage through the stomach. The most important digestive site is the small intestine, where starch is exposed to the action of pancreatic α-amylase, which resembles the digestive action of saliva.

The **oligosaccharases** present in the small intestine include maltase, sucrase, and lactase, which hydrolyze their respective disaccharide substrates to monosaccharides, as well as an α-1,6-glucosidase, which hydrolyzes oligosaccharides containing an α,1,6 glucosidic bond. Unlike α-amylase, enzymes of this group are not secreted into the intestinal juice, but act within the mucosal cells lining the intestine.

While polysaccharides containing α,1,4 or α,1,6 glycosidic bonds are digested by the enzymes of the mammalian gastrointestinal tract, those containing β,1,4 bonds are resistant. Cellulose is not digested directly by most species, although some mammals harbor bacteria which produce **cellulases**, thereby allowing cellulose to make an important contribution to their diet.

Carbohydrates are normally transported across the intestinal mucosa and absorbed in the form of monosaccharides, although a limited absorption of disaccharides may occur. The monosaccharides, especially glucose, are thus the point of departure for the transformations of oxidative metabolism.

The absorption of sugars from the intestine does not appear to occur solely by ordinary diffusion, in view of the lack of critical dependence upon the concentration gradient and the varying absorption rates for different hexoses. It is probable that the transport of monosaccharides from the intestine proceeds by **active transport**, which requires metabolically derived energy.

10-3 BIOCHEMICAL CONVERSIONS OF GLUCOSE

While the carbon atoms of glucose may ultimately enter any of several metabolic pathways, glucosc itself is almost invariably first phosphorylated to glucose-6-phosphate (Fig. 10-1). The reaction, which is catalyzed by the widely distributed **hexokinases**, consists of the phosphorylation of glucose by adenosine triphosphate (ATP) and the conversion of the latter to adenosine diphosphate (ADP).

$$\text{glucose} + \text{ATP} \xrightarrow[\text{hexokinase}]{Mg^{2+}} \text{glucose-6-phosphate} + \text{ADP}$$

This reaction is so strongly exergonic as to be effectively irreversible under normal conditions, with a value of $\Delta G^{\circ\prime}$ close to 5 kcal.

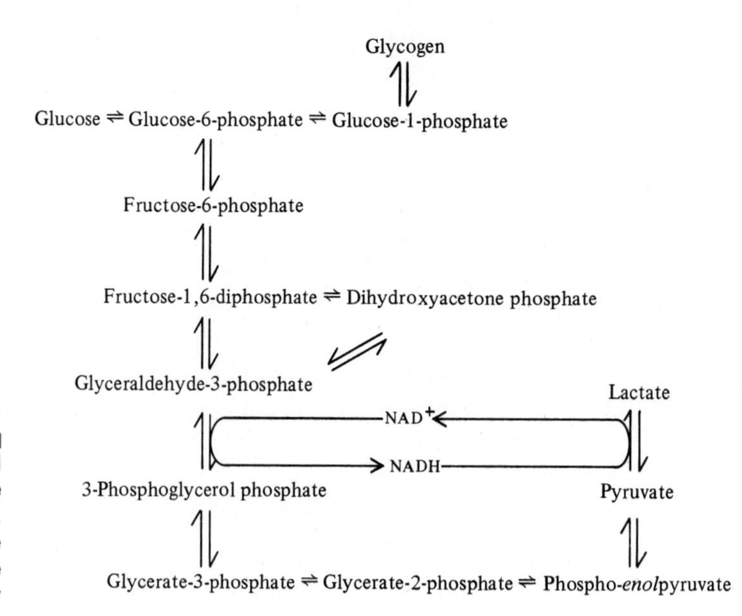

Figure 10-1 (a) Chemical transformations of the glycolytic pathway. (b) A more complete depiction of the glycolytic pathway showing the structures of intermediates.

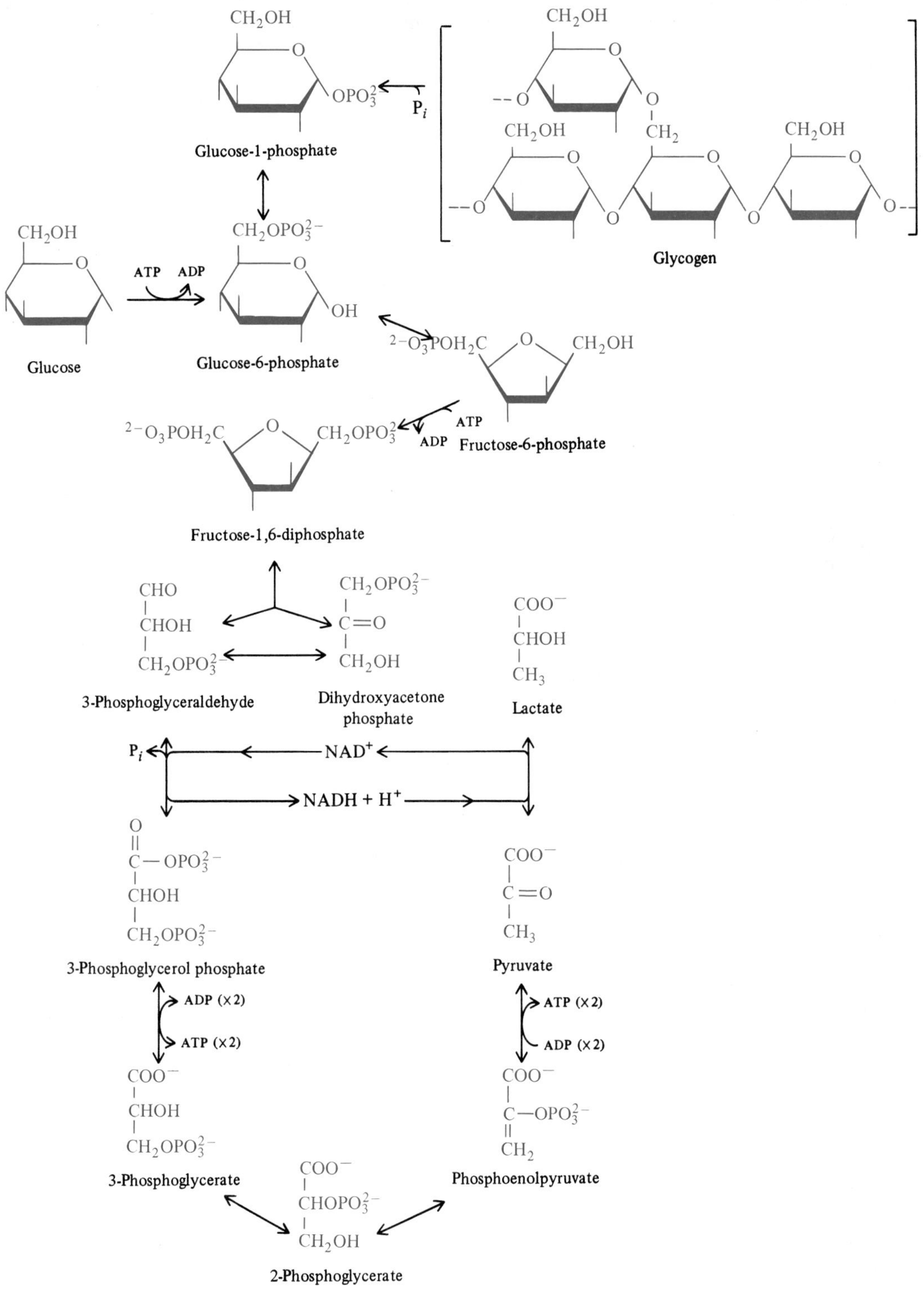

Glycogen

P_i

Glucose-1-phosphate

Glucose

ATP ADP

Glucose-6-phosphate

Fructose-6-phosphate

ATP ADP

Fructose-1,6-diphosphate

3-Phosphoglyceraldehyde

Dihydroxyacetone phosphate

Lactate

P_i

NAD^+

$NADH + H^+$

3-Phosphoglycerol phosphate

Pyruvate

ADP (×2)

ATP (×2)

ATP (×2)

ADP (×2)

3-Phosphoglycerate

Phosphoenolpyruvate

2-Phosphoglycerate

(b)

Figure 10-1 *(continued)*

The phosphate ester of glucose has much less capacity than glucose itself to penetrate cell membranes and, as a result, tends to be confined to the intracellular space for subsequent transformations.

The hexokinases of brain and of yeast are relatively nonspecific and catalyze the conversion of glucose, mannose, and fructose to the corresponding 6-phosphates. Muscle contains two distinct enzymes for the phosphorylation of glucose and fructose, called **glucokinase** and **fructokinase**, respectively. While muscle glucokinase catalyzes the formation of glucose-6-phosphate, the fructokinase synthesizes fructose-1-phosphate.

Carbohydrate metabolism largely begins with **glucose-6-phosphate**. Its principal metabolic destinies include the following:

(1) The C_1 carbonyl may be oxidized to form **6-phosphogluconic** acid. This is the initial reaction of a specialized pathway whereby pentose is formed and NADPH generated (section 10-6).
(2) Conversion to **glucose-1-phosphate** is the initial step in the biosynthesis of various nucleoside diphosphate esters of glucose, which are the starting materials for a variety of biosynthetic reactions (sections 10-8 and 10-9).
(3) In liver, kidney, and intestine, glucose-6-phosphate may be hydrolyzed to **glucose** and **inorganic phosphate**. This reaction is not of course the reverse of the hexokinase reaction, the sum of the two being equivalent to the hydrolysis of ATP to ADP plus phosphate.
(4) The most important of the transformations of glucose-6-phosphate is its conversion to **fructose-6-phosphate**, which subsequently enters the reactions of **glycolysis** (Fig. 10-1).

10-4 GLYCOLYSIS

Glycolysis is one of several metabolic pathways whereby organisms extract chemical energy from organic nutrients. It is often defined as the anaerobic breakdown of carbohydrates by the organism. In actuality the aerobic and anaerobic pathways differ solely in the fate of the product, pyruvic acid. If an adequate oxygen supply is available, this intermediate may enter a subsequent pathway, termed the **Krebs** or **tricarboxylic acid (TCA) cycle**, where it is burned to CO_2 and H_2O. If the oxygen supply is inadequate, the pathway terminates with the conversion of pyruvic to lactic acid. This reaction is important in muscle, which subsequently exports lactic acid to the liver, a highly vascular organ well supplied with oxygen. The ultimate result of the combined operation of glycolysis and the subsequent phases of oxidative metabolism is the combustion of glucose to CO_2 and H_2O and the associated synthesis of ATP.

The reactions of glycolysis, unlike those of the TCA cycle, occur in the cytoplasm; the corresponding enzymes are not associated with any subcellular particles.

The conversion of glucose to glucose-6-phosphate, which precedes its entry into the glycolytic pathway, consumes a molecule of ATP. This loss must be made good if there is to be a net gain of ATP.

Interconversion of Hexose Phosphates Glucose-6-phosphate is reversibly transformed to the ketose, fructose-6-phosphate, by the action of **glucosephosphate isomerase** (Fig. 10-1). The equilibrium ratio of the phosphate ester of glucose to that of fructose is about 2.3 ($\Delta G^{\circ\prime} = 0.4$ kcal).

Fructose-6-phosphate is further phosphorylated by ATP to form fructose-1,6-diphosphate, in a reaction catalyzed by **phosphofructokinase**:

$$\text{fructose-6-phosphate} + \text{ATP} \rightarrow \text{fructose-1,6-diphosphate} + \text{ADP} \qquad (\Delta G^{\circ\prime} = -3.4 \text{ kcal})$$

This reaction is strongly exergonic and practically irreversible.

Splitting of Glucose The next reaction in the glycolytic sequence represents the first actual cleavage of the hexose molecule. **Fructose diphosphate aldolase** catalyzes the reversible cleavage of fructose-1,6-diphosphate between the C_3 and C_4 positions to yield dihydroxyacetone phosphate and D-glyceraldehyde-3-phosphate (Fig. 10-2). This reaction is endergonic and is enabled to proceed only by being driven by subsequent, energetically more favorable, steps ($\Delta G^{\circ\prime} = 5.7$ kcal).

Fructose diphosphate aldolase also catalyzes a number of condensations of dihydroxyacetone phosphate with aldehydes to yield sugar phosphates. One example is the combination of dihydroxyacetone phosphate with L-glyceraldehyde to form L-sorbose-1-phosphate.

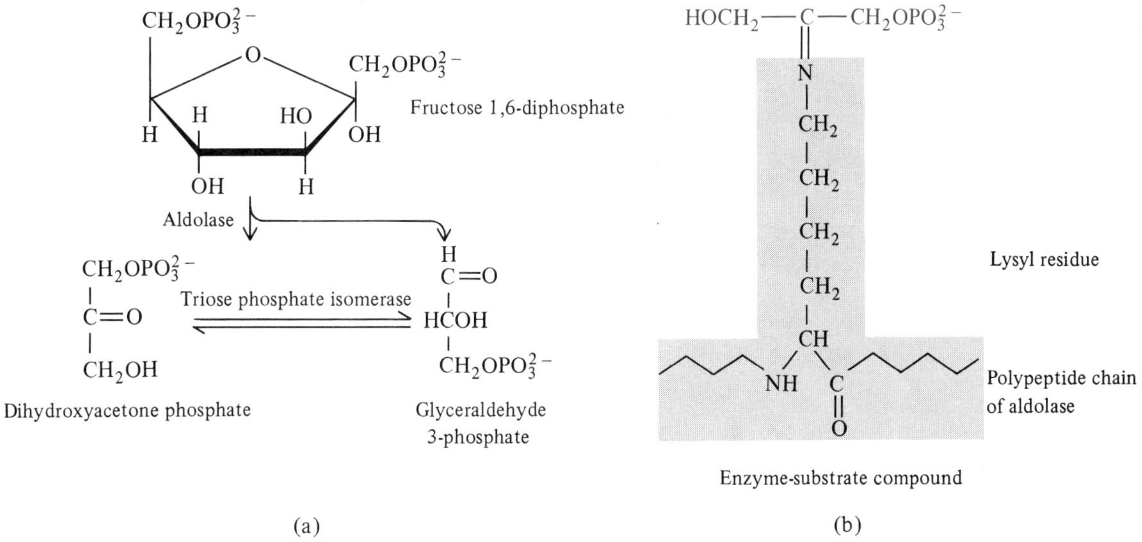

Fructose 1,6-diphosphate

Aldolase

Triose phosphate isomerase

Dihydroxyacetone phosphate

Glyceraldehyde 3-phosphate

(a)

$HOCH_2—C—CH_2OPO_3^{2-}$

Lysyl residue

Polypeptide chain of aldolase

Enzyme-substrate compound

(b)

Figure 10-2
(a) Splitting of fructose-1,6-diphosphate by fructosediphosphate aldolase. (b) The postulated structure of the enzyme-substrate complex, which involves formation of a Schiff's base between the ε-amino group of a lysine residue and the carbonyl group of dihydroxyacetone phosphate (section 10-5).

The two products of the splitting of the fructose diphosphate are reversibly interconverted by **triose phosphate isomerase**:

$$\begin{array}{c} CH_2OH \\ | \\ C{=}O \\ | \\ H_2COPO_3^{2-} \\ \text{Dihydroxyacetone phosphate} \end{array} \rightleftharpoons \begin{array}{c} HC{=}O \\ | \\ HCOH \\ | \\ H_2COPO_3^{2-} \\ \text{Glyceraldehyde-3-phosphate} \end{array} \quad (\Delta G^{\circ\prime} = 1.8 \text{ kcal})$$

Oxidation of Glyceraldehyde-3-Phosphate

Glyceraldehyde-3-phosphate is next oxidized and phosphorylated to the acid anhydride 3-phosphoglyceroyl phosphate (1,3-diphosphoglycerate) by the NAD^+-requiring enzyme **glyceraldehydephosphate dehydrogenase** in the presence of inorganic phosphate:

$$\begin{array}{c} HC{=}O \\ | \\ HCOH \\ | \\ H_2COPO_3^{2-} \\ \text{Glyceraldehyde-3-phosphate} \end{array} + NAD^+ + P_i \rightleftharpoons \begin{array}{c} O{-}PO_3^{2-} \\ | \\ C{=}O \\ | \\ HCOH \\ | \\ H_2COPO_3^{2-} \\ \text{3-Phosphoglyceroyl phosphate} \end{array} + NADH + H^+ \quad (\Delta G^{\circ\prime} = 1.5 \text{ kcal})$$

The acid anhydride, a high energy compound, next serves to phosphorylate ADP to ATP in a reaction catalyzed by **3-phosphoglycerate kinase**; the strongly exergonic nature of the hydrolysis of the mixed anhydride, for which $\Delta G^{\circ\prime}$ is about -10 kcal, suffices to render the entire reaction energetically feasible:

$$\begin{array}{c} O{-}PO_3^{2-} \\ | \\ C{=}O \\ | \\ HCOH \\ | \\ H_2COPO_3^{2-} \\ \text{3-Phosphoglyceroyl phosphate} \end{array} + ADP \xrightleftharpoons{Mg^{2+}} \begin{array}{c} COOH \\ | \\ HCOH \\ | \\ H_2COPO_3^{2-} \\ \text{3-Phosphoglyceric acid} \end{array} + ATP \quad (\Delta G^{\circ\prime} = -4.5 \text{ kcal})$$

The net result of the pair of coupled reactions is the oxidation of glyceraldehyde-3-phosphate to 3-phosphoglyceric acid, with the concomitant reduction of NAD^+ and the phosphorylation of ADP ($\Delta G^{\circ\prime} = -3.0$ kcal):

$$\text{glyceraldehyde-3-phosphate} + P_i + NAD^+ + ADP \rightleftharpoons \text{3-phosphoglyceric acid} + NADH + ATP + H^+$$

This is a particularly striking illustration of the coupling of an exergonic oxidation with the synthesis of an energy carrier, ATP. The chemical energy released by oxidation of the aldehyde group of glyceraldehyde-3-phosphate is stored in ATP, rather then degraded as heat.

Conversions of 3-Phosphoglyceric Acid

The 3-phosphoglyceric acid arising from the above reactions is isomerized to 2-phosphoglyceric acid by the action of **phosphoglyceromutase**. The reaction appears to proceed via a 2,3-diphosphoglyceric acid intermediate.

$$P_i + \begin{array}{c} COOH \\ | \\ HCOH \\ | \\ H_2COPO_3^{2-} \end{array} \rightleftharpoons \begin{array}{c} COOH \\ | \\ HCOPO_3^{2-} \\ | \\ H_2COPO_3^{2-} \end{array} \rightleftharpoons \begin{array}{c} COOH \\ | \\ HCOPO_3^{2-} \\ | \\ H_2COH \end{array} + P_i \quad (\Delta G^{\circ\prime} = 1.1 \text{ kcal})$$

3-Phosphoglyceric acid — 2,3-Diphosphoglyceric acid — 2-Phosphoglyceric acid

A molecule of water is removed from 2-phosphoglyceric acid by the action of **phosphopyruvate hydratase (enolase)** to yield phosphoenolpyruvic acid:

$$\begin{array}{c} COOH \\ | \\ HCOPO_3^{2-} \\ | \\ H_2COH \end{array} \overset{Mg^{2+}}{\rightleftharpoons} \begin{array}{c} COOH \\ | \\ COPO_3^{2-} \\ \| \\ CH_2 \end{array} + H_2O \quad (\Delta G^{\circ\prime} = 0.4 \text{ kcal})$$

2-Phosphoglyceric acid — Phosphoenolypyruvic acid

Formation of Pyruvic Acid

Finally, ADP is phosphorylated by phosphoenolpyruvic acid to form ATP and pyruvic acid, in a reaction catalyzed by **pyruvate kinase**:

$$\begin{array}{c} COOH \\ | \\ COPO_3^{2-} \\ \| \\ CH_2 \end{array} + ADP \rightleftharpoons \begin{array}{c} COOH \\ | \\ C{=}O \\ | \\ CH_3 \end{array} + ATP \quad (\Delta G^{\circ\prime} = -7.5 \text{ kcal})$$

Phosphoenolpyruvic acid — Pyruvic acid

The high negative free energy ($\Delta G^{\circ\prime} = -12$ kcal) associated with the hydrolysis of the phosphate ester of the enol form of pyruvic acid renders the phosphate transfer exergonic. The enol to keto conversion provides the basic thermodynamic driving force for this reaction. Phosphoenolpyruvate resembles a cocked spring ready to go off.

The reactions catalyzed by pyruvate kinase and 3-phosphoglycerate kinase together produce four molecules of ATP for each molecule of *original* glucose, since they involve two 3-carbon fragments of glucose. If the two molecules of ATP consumed in the initial reactions of glycolysis are subtracted, there is a net gain of two molecules.

Formation of Lactic Acid

In the presence of an adequate oxygen supply, pyruvic acid is converted by oxidative decarboxylation to acetyl CoA, which subsequently enters the TCA cycle (section 12-3). However, an alternative mode of reaction exists for pyruvic acid, which does not accumulate under anaerobic conditions.

One of the steps of glycolysis, the oxidation of 3-phosphoglyceraldehyde to 3-phosphoglyceric acid, requires the reduction of NAD^+ to NADH. In an aerobic environment NADH is reoxidized by the mitochondrial electron transport system. Since only a limited quantity of NAD^+ is present within the cell, anaerobic glycolysis would soon be forced to a halt if there existed no mechanism for the reoxidation of NADH. This is the function of **lactate dehydrogenase**, which catalyzes the reaction:

$$\underset{\text{Pyruvic acid}}{\begin{array}{c} COOH \\ | \\ C{=}O \\ | \\ CH_3 \end{array}} + NADH + H^+ \rightleftharpoons \underset{\text{Lactic acid}}{\begin{array}{c} COOH \\ | \\ HCOH \\ | \\ CH_3 \end{array}} + NAD^+ \quad (\Delta G^{\circ\prime} = -6.0 \text{ kcal})$$

In skeletal muscle the concentrations of reactants and products are such that they favor the formation of lactic acid, which tends to accumulate if pyruvic acid is not continuously removed. Lactic acid is a metabolic dead end. It can enter the mainstream of oxidative metabolism only after reconversion to pyruvate.

The net result of *anaerobic* glycolysis may be written:

$$\text{glucose} + 2\text{ ADP} + 2P_i \rightleftharpoons 2\text{ lactate} + 2\text{ ATP} + 2\text{ }H_2O \quad (\Delta G^{\circ\prime} = -32.4 \text{ kcal})$$

Glycolysis provides a means for the rapid biosynthesis of ATP in relatively oxygen-poor tissues, including skeletal muscle. Many bacteria, including those responsible for the souring of milk, rely upon anaerobic glycolysis for their energy requirements. The evolutionary origins of this pathway probably lie in the development of the earliest living systems in an oxygen-poor environment.

Unlike the phosphorylated glucose derivatives, lactate can readily diffuse through cellular membranes and is therefore easily removed from the cell. In this way it may be transported from skeletal muscle by the blood to the *liver* and other *oxygen-rich* organs for further oxidative processing.

An Overview of Glycolysis

If one imagines that glucose represents a peak of stored free energy and that its ultimate products, CO_2 and H_2O, correspond to a valley far below, then glycolysis may be thought of as a relatively gradual uppermost slope leading from the summit and to the more precipitous descent of the later stages of oxidative metabolism.

The standard free energy change ($\Delta G^{\circ\prime}$) for the complete oxidation of glucose to CO_2 and H_2O is -686 kcal/mole. The conversion of glucose to lactate by anaerobic glycolysis releases about 8% of this, or 56 kcal/mole. The synthesis of two moles of ATP from ADP and inorganic phosphate accounts for 16 kcal, or about 30% of the 56 kcal. The latter percentage is an index of the efficiency of glycolysis in storing chemical energy.

In higher animals the *ultimate* product of glycolysis is pyruvate, which is subsequently converted to acetyl CoA. In the case of oxygen-poor tissues, such as muscle, the lactic acid formed is exported to organs with a more ample oxygen supply, where it is oxidized to pyruvate, that is converted to acetyl CoA. In many anaerobic microorganisms lactate is the primary end product of glycolysis, In yeast and a few other microorganisms pyruvate is converted to ethanol.

The Pasteur Effect

Although molecular oxygen does not figure in the reactions of glycolysis, the rate of glycolysis is nevertheless linked to oxygen consumption, as was first recognized by Pasteur. The rates of formation of lactic acid and of consumption of glucose in tissue preparations and bacteria are generally lower under aerobic than under anaerobic conditions.

The biological utility of the Pasteur effect is clear. In the presence of an adequate oxygen supply many cells oxidize pyruvic acid completely to CO_2 and H_2O. The synthesis of ATP by the more efficient reactions of the TCA cycle and the associated processes of oxidative phosphorylation permits the energy requirements of the cell to be fulfilled with the consumption of less glucose. The suppression of glycolysis by oxygen thereby enables the cell to conserve glucose. The probable origin of the effect is the inhibition of phosphofructokinase by the high cellular ATP levels arising in the presence of oxygen (section 10-5).

Alcoholic Fermentation

In many microorganisms, including yeasts, the reaction sequence leading to pyruvic acid is the same as for animal tissues. However, in yeasts the reversible conversion of pyruvic to lactic acid is replaced by a decarboxylation to acetaldehyde; the reaction is catalyzed by **pyruvate decarboxylase**, with thiamine pyrophosphate (TPP) as a cofactor:

$$CH_3COCOOH \underset{TPP}{\overset{Mg^{2+}}{\rightleftarrows}} CH_3CHO + CO_2$$

The acetaldehyde is reduced by NADH to ethanol, with far-reaching consequences for human history, in a reaction catalyzed by **alcohol dehydrogenase**:

$$CH_3CHO + NADH + H^+ \rightleftarrows CH_3CH_2OH + NAD^+$$

As in the case of the conversion of pyruvic to lactic acid, the reduction of acetaldehyde is accompanied by the regeneration of NAD^+.

Metabolism of Alcohol

Ethanol is a dietary component for only one mammalian species, for which its significance fluctuates wildly among different individuals. While some uncertainty remains as to the details of its metabolism, there is little doubt as to the course of the main pathway.

The liver is the principal organ involved in the oxidation of ethanol. The only enzyme known to act directly upon ethanol is **alcohol dehydrogenase**, which converts it to acetaldehyde in the presence of NAD^+:

$$C_2H_5OH + NAD^+ \rightleftarrows NADH + H^+ + CH_3CHO$$

The acetaldehyde is subsequently largely oxidized further to acetyl CoA, which enters the mainstream of oxidative metabolism.

$$CH_3CHO + NAD^+ + HSCoA \rightleftarrows CH_3CO{-}SCoA + NADH + H^+$$

Alternative Metabolic Fates of Pyruvic Acid

The most important of the metabolic schemes entered by pyruvic acid is conversion to acetyl CoA and entry into the TCA cycle for ultimate conversion to CO_2 and H_2O (chapter 12). In addition to the already discussed

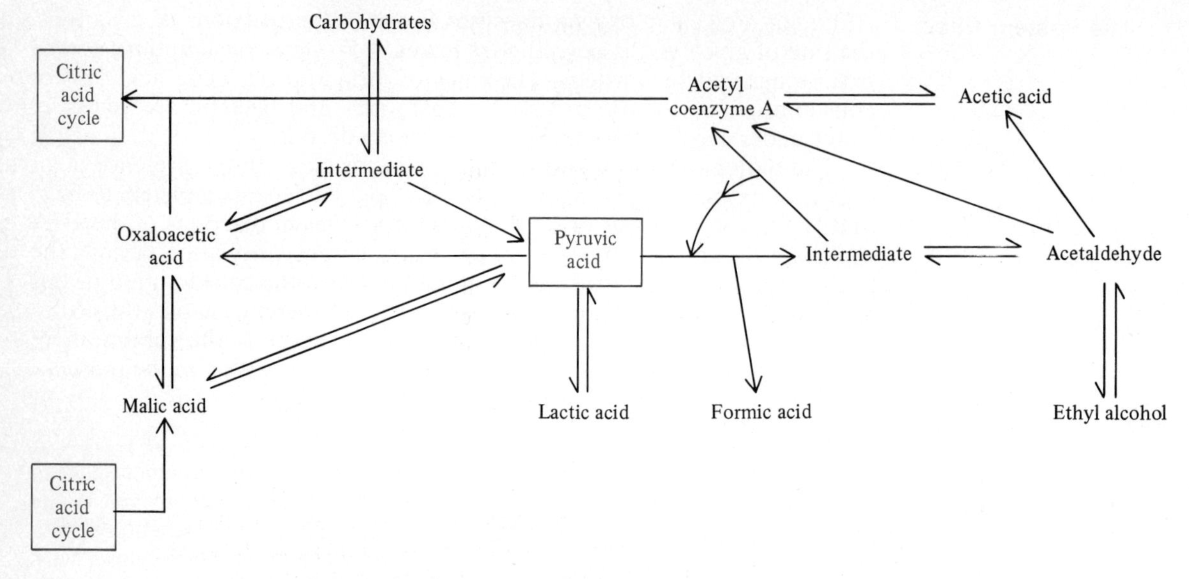

Figure 10-3
Alternative metabolic pathways of pyruvic acid.

reduction to lactic acid, it may also be transformed to oxaloacetic acid, reconverted to carbohydrate, or transaminated to alanine (Fig. 10-3).

The conversion to oxaloacetic acid, which occurs in both bacteria and animal tissues, may proceed via CO_2 fixation, as catalyzed by **malic enzyme** (officially designated malate dehydrogenase, decarboxylating: NADP):

$$CH_3COCOOH + CO_2 + NADPH + H^+ \rightleftarrows \underset{\text{Malic acid}}{HOOCCH_2CHOHCOOH} + NADP$$

Malic acid is an intermediate of the TCA cycle and may be oxidized to oxaloacetic acid by NAD^+ in the presence of **malate dehydrogenase**, a different enzyme from that cited above (section 12-3). In the liver, pyruvate may also be converted *directly* to oxaloacetate by **pyruvate carboxylase**, a biotin-containing mitochondrial enzyme:

$$CH_3COCOOH + CO_2 + ATP + H_2O \rightleftarrows \text{oxaloacetate} + ADP + P_i$$

10-5 ENZYMES OF GLYCOLYSIS

Hexokinase and Glucokinase

The widely distributed enzyme hexokinase has a rather blurred specificity, catalyzing the phosphorylation of many hexoses in addition to D-glucose, including D-mannose and D-fructose. The molecular weight of the yeast enzyme is 96,000. The hexokinase of animal tissues is an allosteric enzyme with regulatory properties; it is inhibited by its own product, glucose-6-phosphate. In this way the synthesis of glucose-6-phosphate is blocked when an adequate supply is present.

Glucokinase, which is present in liver, is specific for D-glucose. Since its Michaelis constant, K_m, is quite high (1 mM), its action is important only at high levels of glucose. It is not inhibited by glucose-6-phosphate.

Phosphofructokinase

This enzyme, which catalyzes the phosphorylation of fructose-6-phosphate to fructose-1,6-diphosphate, has a molecular weight of 380,000 and consists of multiple subunits. It has allosteric properties and is regulated by a number of allosteric modifiers, including the inhibitors ATP and citrate and the activators ADP and AMP. When the cellular level of ATP is high or the energy-yielding compound citrate is present, the enzyme is inhibited, thereby bringing glycolysis to a halt. If the cellular balance between ATP and ADP or AMP favors the latter, phosphofructokinase and glycolysis are stimulated. This enzyme is thus of central importance in the regulation of the glycolytic pathway.

Fructose Diphosphate Aldolase

The aldolase occurring in the skeletal muscle of higher animals consists of four subunits with a combined molecular weight of 160,000. Variants of aldolase are found in other tissues which differ in amino acid composition from the muscle variety and each other; however, the four-subunit structure appears to be general. The reaction catalyzed by aldolase is reversible; the value of $\Delta G^{\circ\prime}$ for the splitting of fructose-1,6-diphosphate is +5.7 kcal. Despite the positive value of $\Delta G^{\circ\prime}$, the reaction proceeds in the forward direction under intracellular conditions.

The action of aldolase in either direction appears to involve the formation of a Schiff's base between the ε-amino group of a lysine residue and the carbonyl group of dihydroxyacetone phosphate (Fig. 10-2). The formation of fructose-1,6-diphosphate follows the transient loss of a proton by the Schiff's base to form a carbanion. This attacks the aldehyde carbon atom of glyceraldehyde-3-phosphate to yield the hexose diphosphate, which leaves the enzyme after acquiring a proton. In the splitting of fructose-1, 6-diphosphate, these steps occur in reverse sequence.

Glyceraldehyde-phosphate Dehydrogenase

This enzyme consists of four identical subunits of total molecular weight 140,000; each subunit contains a binding site for NAD^+. This coenzyme oxidizes, that is, accepts electrons from, the aldehyde group of glyceraldehye-3-phosphate.

Glyceraldehydephosphate dehydrogenase is a sulfhydryl enzyme, whose active —SH group must remain in the reduced state for enzymic function. Reagents, such as iodoacetate, which combine with sulfhydryl groups, inactivate the enzyme.

After an initial binding of NAD^+ by the enzyme, the aldehyde group of the substrate forms a **thiohemiacetal** linkage with the active sulfhydryl. This is followed by the oxidation of the covalently bound glyceraldehyde-3-phosphate and the transfer of a hydrogen to bound NAD^+, which is then dissociated. The substrate is now attached by a **thioester** bond to the enzyme. In the final step of the reaction the acyl group is transferred from the sulfhydryl group to inorganic phosphate to form the product, 3-phosphoglyceroyl phosphate.

$$\begin{array}{c} \text{enzyme} \\ | \\ \text{SH} \\ + \\ \text{HC}{=}\text{O} \\ | \\ \text{HCOH} \\ | \\ \text{H}_2\text{COPO}_3^{2-} \end{array} \longrightarrow \begin{array}{c} \text{enzyme} \\ | \\ \text{S} \\ | \\ \text{HCOH} \\ | \\ \text{HCOH} \\ | \\ \text{H}_2\text{COPO}_3^{2-} \end{array} \xrightarrow{\text{NAD}^+} \begin{array}{c} \text{enzyme} \\ | \\ \text{S} \\ | \\ \text{C}{=}\text{O} \\ | \\ \text{HCOH} \\ | \\ \text{H}_2\text{COPO}_3^{2-} \end{array} \xrightarrow{P_i} \begin{array}{c} \text{enzyme} \\ | \\ \text{SH} \\ + \\ \text{O}{=}\text{C}{-}\text{OPO}_3^{2-} \\ | \\ \text{HCOH} \\ | \\ \text{H}_2\text{COPO}_3^{2-} \end{array}$$

The energetics of the reaction are most readily understood if it is formally written as the sum of two equivalent half-reactions, the first of which is strongly exergonic and the second strongly endergonic.

$$\text{RCHO} + \text{NAD}^+ + \text{H}_2\text{O} \rightarrow \text{RCOOH} + \text{NADH} + \text{H}^+ \quad (\Delta G^{\circ\prime} = -10.3 \text{ kcal})$$

$$\text{RCOOH} + \text{HPO}_4^{2-} \rightarrow \text{RCOOPO}_3^{2-} + \text{H}_2\text{O} \quad (\Delta G^{\circ\prime} = +11.8 \text{ kcal})$$

Sum:

$$\text{RCHO} + \text{HPO}_4^{2-} + \text{NAD}^+ \rightarrow \text{RCOOPO}_3^{2-} + \text{NADH} + \text{H}^+ \ (\Delta G^{\circ\prime} = +1.5 \text{ kcal})$$

The importance of the role of NAD^+ now becomes evident. The oxidation of the aldehyde by NAD^+ is sufficiently exergonic to drive the thermodynamically unfavored formation of the acid anhydride, to which it is coupled in the enzymic reaction.

Glyceraldehydephosphate dehydrogenase is an allosteric enzyme. Its principal modifier is NAD^+, whose binding by one subunit decreases the affinity of the other subunits for NAD^+.

Enolase

Enolase has a molecular weight of 85,000. The enzyme requires Mg^{2+} or Mn^{2+} for activity; the divalent cation is bound by the enzyme prior to the substrate. Fluoride is a strong inhibitor.

Pyruvate Kinase

The activity of this allosteric enzyme, whose molecular weight is 250,000, is likewise dependent upon Mg^{2+} or Mn^{2+}. There is an additional requirement for one of the alkali metal cations K^+, Rb^+, or Cs^+; K^+ is the *in vivo* activator. Different variants of pyruvate kinase have been isolated from liver and from muscle. The liver form is inhibited by ATP and activated by fructose-1,6-diphosphate and high levels of phosphoenolpyruvate. An intracellular excess of ATP depresses the activity of pyruvate kinase and thereby slows glycolysis. The metabolic fuel citrate, which is processed by the TCA cycle, is also an inhibitor. Pyruvate kinase functions as one of the regulatory elements of glycolysis, whereby the activity of this pathway is adapted to the changing requirements of the cell.

Lactate Dehydrogenase

This enzyme, whose molecular weight is close to 134,000, consists of four subunits of molecular weight 33,500. Lactate dehydrogenase can exist as varying combinations of two kinds of subunit. These combinations yield at least five electrophoretically distinct species or **isozymes**, all of which possess enzymic activity.

These subunits have been designated as H (heart) and M (muscle) forms, according to the tissue in which they predominate. The five isozymes have the compositions H_4, H_3M, H_2M_2, HM_3, and M_4. The distribution of isozymes varies with the tissue and changes during embryonic development. While all of the above species have the same basic catalytic activity, there are significant differences between the H and M forms with respect to kinetic constants. The H_4 isozyme occurring in heart has a higher value of K_M for pyruvate and a lower value of V_{max} than the M_4 form present in skeletal muscle. This is consistent with the requirements of the two tissues. Heart muscle does not normally form lactate except in situations of deficient oxygen supply; its pyruvate is instead diverted to other pathways for ultimate oxidation to CO_2 and H_2O. It therefore has less need for the rapid conversion of pyruvate to lactate than does skeletal muscle, which tends to form lactate. The existence of isozymes permits a graded adaptation of the enzyme to the needs of different tissues.

Regulation of Glycolysis

The operation of the glycolytic pathway must be controlled and adapted to the needs of the organism, if it is not to function wastefully. The prevailing evidence is that the step catalyzed by phosphofructokinase is normally rate limiting and that it is at this point in the pathway that feedback control is exercised. The pacemaker role of the phosphofructokinase reaction is a consequence of its essentially irreversible character and of the fact that it is very far from equilibrium in tissues.

Phosphofructokinase is an allosteric enzyme; it is inhibited by high ATP levels and stimulated by AMP. When the ATP concentration becomes excessively high the activity of this enzyme and the rate of the glycolytic pathway are reduced so as to depress the rate of synthesis of ATP; as the ATP level falls, the activity of phosphofructokinase increases again.

Phosphofructokinase is also inhibited by citrate, an intermediate of the TCA cycle, which occurs in the mitochondria (chapter 12). The export of accumulated citrate from the mitochondria into the cytoplasm provides a message to the cell that degradative metabolism is proceeding more rapidly than necessary. The inhibition of phosphofructokinase by citrate depresses the rate of glycolysis and, indirectly, that of the TCA cycle.

While the phosphofructokinase reaction is the primary control point of glycolysis, additional regulatory mechanisms are available. Hexokinase and pyruvate kinase are allosteric enzymes and subject to feedback control. Hexokinase is inhibited by high levels of its product, glucose-6-phosphate, thereby avoiding excessive accumulation of the latter. Pyruvate kinase, like phosphofructokinase, is inhibited by ATP, so that its activity is switched off if the concentration of ATP rises to an excessive value.

10-6 THE PHOSPHOGLUCONATE PATHWAY

The most important of the alternative pathways for glucose metabolism has been named the **phosphogluconate oxidative pathway** or the **hexose monophosphate shunt.** Its particular interest stems from its provision of a mechanism for the complete oxidation of glucose which is independent of the TCA cycle, as well as a means for the synthesis of D-ribose.

As in the case of glycolysis, the participating enzymes occur in the cytoplasm. Four of the glycolytic enzymes are also involved in the phosphogluconate pathway. The operation of the pathway is cyclic (Fig. 10-4).

The initial reaction is the oxidation of glucose-6-phosphate by $NADP^+$, catalyzed by **glucose-6-phosphate dehydrogenase**:

$$\text{Glucose-6-phosphate (HCOH–HCOH–HOCH–HCOH–HC(–O–)–H}_2\text{COPO}_3^{2-}) + NADP^+ \longrightarrow NADPH + H^+ + \text{6-Phosphoglucono-}\delta\text{-lactone (C=O–HCOH–HOCH–HCOH–HC(–O–)–H}_2\text{COPO}_3^{2-}) \longrightarrow \text{6-Phosphogluconic acid (COOH–HCOH–HOCH–HCOH–HCOH–H}_2\text{COPO}_3^{2-})$$

Glucose-6-phosphate; 6-Phosphoglucono-δ-lactone; 6-Phosphogluconic acid

The δ-lactone is hydrolyzed to the acid by a gluconolactonase. The equilibrium strongly favors the oxidized forms.

The 6-phosphogluconic acid is further oxidized by **6-phosphogluconate dehydrogenase**, with $NADP^+$ and Mg^{2+} as cofactors; the hypothetical β-keto intermediate decomposes to form D-ribulose-5-phosphate:

$$NADP^+ + \text{6-Phosphogluconic acid (COOH–HCOH–HOCH–HCOH–HCOH–H}_2\text{COPO}_3^{2-}) \xrightarrow{Mg^{2+}} \text{COOH–HCOH–C=O–HCOH–HCOH–H}_2\text{COPO}_3^{2-} + NADPH + H^+ \longrightarrow \text{D-Ribulose-5-phosphate (CH}_2\text{OH–C=O–HCOH–HCOH–H}_2\text{COPO}_3^{2-}) + CO_2$$

6-Phosphogluconic acid; D-Ribulose-5-phosphate

The ribulose-5-phosphate undergoes subsequent isomerization to xylulose-5-phosphate, in a reaction catalyzed by **ribulosephosphate-3-epimerase**, and to

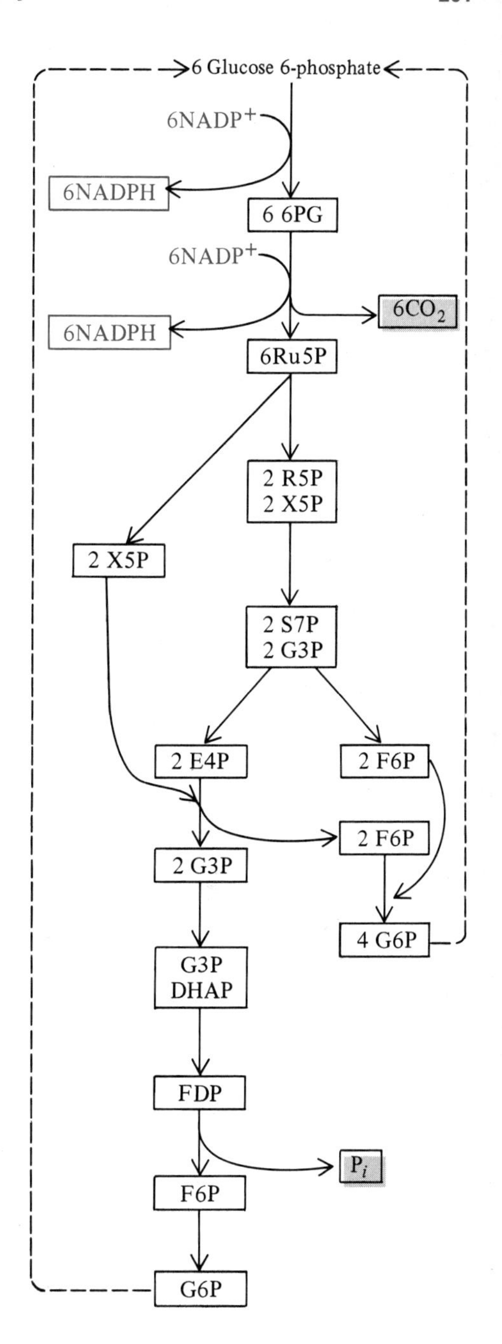

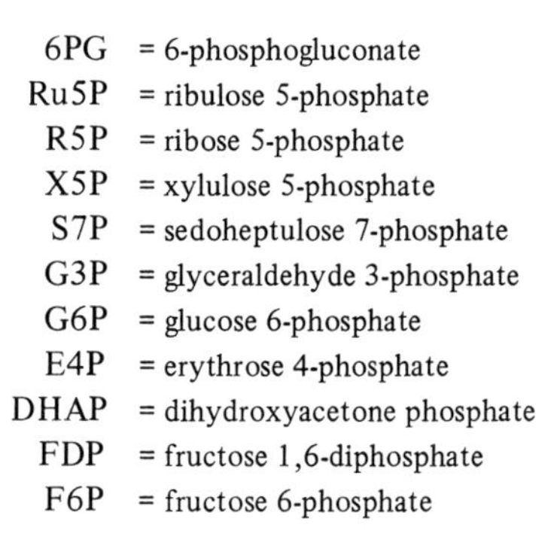

Figure 10-4
Phosphogluconate oxidative pathway.

ribose-5-phosphate, catalyzed by **ribosephosphate isomerase**:

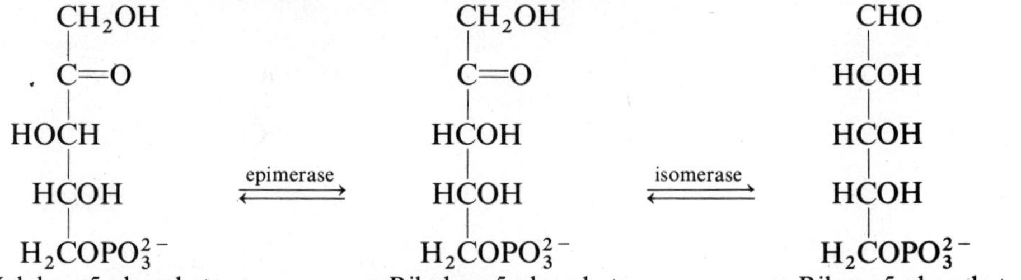

$$
\begin{array}{ccccc}
CH_2OH & & CH_2OH & & CHO \\
| & & | & & | \\
C{=}O & & C{=}O & & HCOH \\
| & & | & & | \\
HOCH & & HCOH & & HCOH \\
| & \xrightleftharpoons{\text{epimerase}} & | & \xrightleftharpoons{\text{isomerase}} & | \\
HCOH & & HCOH & & HCOH \\
| & & | & & | \\
H_2COPO_3^{2-} & & H_2COPO_3^{2-} & & H_2COPO_3^{2-} \\
\text{D-Xylulose-5-phosphate} & & \text{D-Ribulose-5-phosphate} & & \text{D-Ribose-5-phosphate}
\end{array}
$$

Two carbons are transferred from xylulose-5-phosphate to ribose-5-phosphate to form sedoheptulose-7-phosphate, in a reaction catalyzed by **transketolase**:

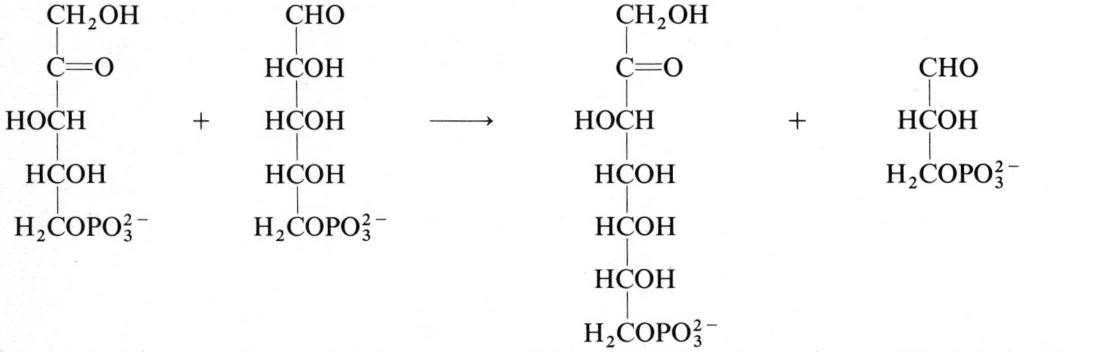

$$
\begin{array}{ccccccc}
CH_2OH & & CHO & & CH_2OH & & \\
| & & | & & | & & \\
C{=}O & & HCOH & & C{=}O & & CHO \\
| & & | & & | & & | \\
HOCH & + & HCOH & \longrightarrow & HOCH & + & HCOH \\
| & & | & & | & & | \\
HCOH & & HCOH & & HCOH & & H_2COPO_3^{2-} \\
| & & | & & | & & \\
H_2COPO_3^{2-} & & H_2COPO_3^{2-} & & HCOH & & \\
 & & & & | & & \\
 & & & & HCOH & & \\
 & & & & | & & \\
 & & & & H_2COPO_3^{2-} & & \\
\text{Xylulose-5-phosphate} & & \text{Ribose-5-phosphate} & & \text{Sedoheptulose-7-phosphate} & & \text{Glyceraldehyde-3-phosphate}
\end{array}
$$

The reaction mechanism is analogous to that for the oxidative decarboxylation of pyruvic acid (chapter 12). Transketolase utilizes thiamine pyrophosphate as a cofactor and requires Mg^{2+}.

The first three carbon atoms of sedoheptulose-7-phosphate are transferred to glyceraldehyde-3-phosphate by the action of **transaldolase**:

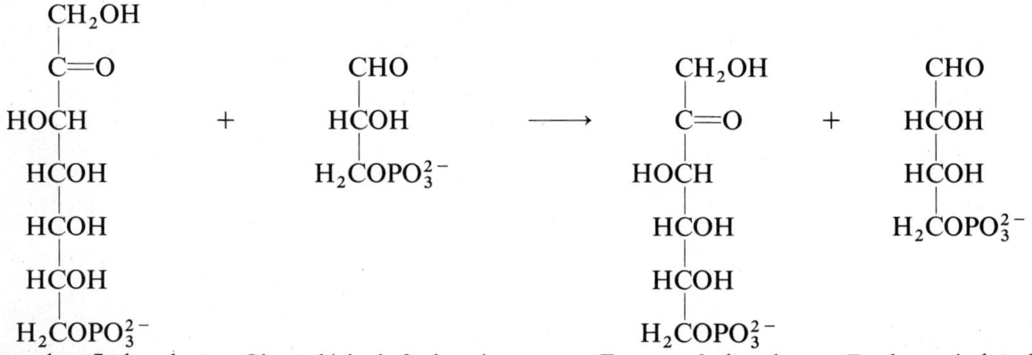

$$
\begin{array}{ccccccc}
CH_2OH & & & & & & \\
| & & & & & & \\
C{=}O & & CHO & & CH_2OH & & CHO \\
| & & | & & | & & | \\
HOCH & + & HCOH & \longrightarrow & C{=}O & + & HCOH \\
| & & | & & | & & | \\
HCOH & & H_2COPO_3^{2-} & & HOCH & & HCOH \\
| & & & & | & & | \\
HCOH & & & & HCOH & & H_2COPO_3^{2-} \\
| & & & & | & & \\
HCOH & & & & HCOH & & \\
| & & & & | & & \\
H_2COPO_3^{2-} & & & & H_2COPO_3^{2-} & & \\
\text{Sedoheptulose-7-phosphate} & & \text{Glyceraldehyde-3-phosphate} & & \text{Fructose-6-phosphate} & & \text{Erythrose-4-phosphate}
\end{array}
$$

The erythrose-4-phosphate arising from the transaldolase reaction accepts a 2-carbon unit from xylulose-5-phosphate to yield fructose-6-phosphate:

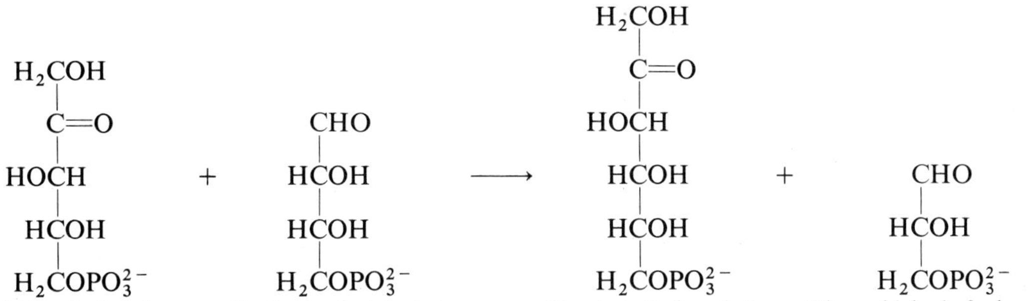

$$\begin{array}{c} H_2COH \\ | \\ C{=}O \\ | \\ HOCH \\ | \\ HCOH \\ | \\ H_2COPO_3^{2-} \end{array} + \begin{array}{c} CHO \\ | \\ HCOH \\ | \\ HCOH \\ | \\ H_2COPO_3^{2-} \end{array} \longrightarrow \begin{array}{c} H_2COH \\ | \\ C{=}O \\ | \\ HOCH \\ | \\ HCOH \\ | \\ HCOH \\ | \\ H_2COPO_3^{2-} \end{array} + \begin{array}{c} CHO \\ | \\ HCOH \\ | \\ H_2COPO_3^{2-} \end{array}$$

Xylulose-5-phosphate Erythrose-4-phosphate Fructose-6-phosphate Glyceraldehyde-3-phosphate

The reaction is catalyzed by transketolase.

The remaining reactions of the cycle are already familiar from the discussion of glycolysis. Glyceraldehyde-3-phosphate is partially converted by triose-phosphate isomerase to dihydroxyacetone phosphate. The two are then condensed by the action of fructosediphosphate aldolase to form fructose-1,6-diphosphate. Fructose-1,6-diphosphate is partially dephosphorylated by hexose diphosphatase to form fructose-6-phosphate:

$$\text{fructose-1,6-diphosphate} + H_2O \rightarrow \text{fructose-6-phosphate} + P_i$$

The Function of the Phosphogluconate Pathway

While the phosphogluconate pathway is not the primary mechanism for extracting energy from the oxidation of glucose, it nevertheless is of primary biological importance as it carries out several indispensable functions. Perhaps the most important function in most cells is the formation of NADPH, a crucial reducing agent involved in many biosynthetic reactions, including especially the formation of fats from acetyl CoA.

A second major function is the formation of pentoses from hexoses. D-Ribose-5-phosphate, which is essential for the biosynthesis of nucleic acids, arises in this way. An additional function is the conversion of pentoses to hexoses, which can enter the glycolytic pathway.

Animal cells differ widely in the prominence of the phosphogluconate pathway. It is virtually absent in skeletal muscle, but very active in tissues, such as liver, mammary gland, and adipose tissue, that are extensively engaged in fatty acid biosynthesis and require an ample supply of NADPH.

In green plants a modified version of this pathway has a somewhat different function, figuring in the incorporation of CO_2 into glucose in the dark reactions of photosynthesis (chapter 11).

10-7 HEXOSE TRANSFORMATIONS

The hexose sugars present in living tissues are all derivable from glucose-6-phosphate by a limited number of general reactions. Only an outline of these will be presented here.

Aldose-Ketose Conversions

The 6-phosphates of glucose, mannose, and fructose are interconvertible by a set of glucosephosphate isomerases, which have already been mentioned in the discussion of glycolysis:

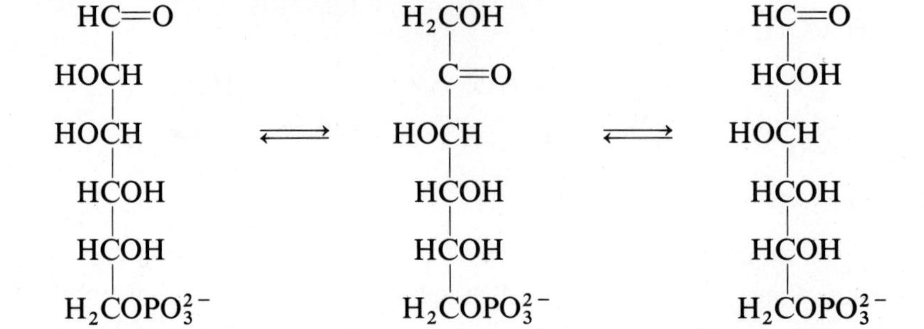

$$\begin{array}{ccccc}
\text{HC=O} & & \text{H}_2\text{COH} & & \text{HC=O} \\
| & & | & & | \\
\text{HOCH} & & \text{C=O} & & \text{HCOH} \\
| & & | & & | \\
\text{HOCH} & \rightleftarrows & \text{HOCH} & \rightleftarrows & \text{HOCH} \\
| & & | & & | \\
\text{HCOH} & & \text{HCOH} & & \text{HCOH} \\
| & & | & & | \\
\text{HCOH} & & \text{HCOH} & & \text{HCOH} \\
| & & | & & | \\
\text{H}_2\text{COPO}_3^{2-} & & \text{H}_2\text{COPO}_3^{2-} & & \text{H}_2\text{COPO}_3^{2-} \\
\text{Mannose-6-phosphate} & & \text{Fructose-6-phosphate} & & \text{Glucose-6-phosphate}
\end{array}$$

The three hexose phosphates thus form a common metabolic pool.

Mutases and Kinases

A **mutase** catalyzes the transfer of a phosphate group between two positions in the *same* molecule. An example is the interconversion of the 6- and 1-phosphates of glucose by **phosphoglucomutase**. This reaction requires the presence of glucose-1,6-diphosphate and Mg^{2+}, suggesting that an interchange of phosphate between the enzyme and glucose-1,6-diphosphate may be involved:

enzyme + glucose-1,6-diphosphate ⇄
enzyme-O-PO_3^{2-} + glucose-1-phosphate or glucose-6-phosphate

In support of this model, the radioisotope ^{32}P is redistributed among all three sugar esters after addition of any single labeled ester.

The kinases catalyze the transfer of phosphate from ATP to *another* molecule. Phosphoglucokinase converts glucose-1-phosphate to glucose-1, 6-diphosphate.

10-8 SYNTHESIS OF OLIGO- AND POLYSACCHARIDES

Sugar Esters of Nucleoside Diphosphates

Many biochemical conversions of carbohydrates proceed via an intermediate consisting of a sugar ester of the terminal phosphate of a nucleoside diphosphate. One of the most important of these is **uridine diphosphate glucose** (UDP-glucose), whose structure is shown in Fig. 10-5. The formation of this intermediate is catalyzed by an enzyme of the group called **glycosyl-1-phosphate nucleotidyltransferases** or in the older nomenclature **pyrophosphorylases.** In the present case, uridine triphosphate, a **nucleoside triphosphate** (section 8-2) combines with glucose-1-phosphate, with the splitting out of pyrophosphate. This reaction is representative of the mechanism of synthesis of an entire class of nucleoside diphosphate sugar esters:

nucleoside triphosphate + sugar-1-phosphate →
PP_i + nucleoside diphosphate sugar ester

Epimerization

An inversion of configuration about a single carbon atom, or epimerization, recurs frequently in carbohydrate metabolism. One example is the epi-

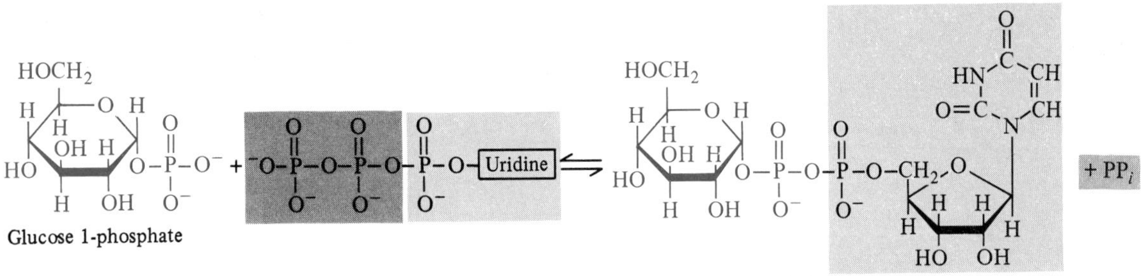

Glucose 1-phosphate

Uridine diphosphate glucose (UDP-glucose)

Figure 10-5
Biosynthesis of uridine diphosphate glucose.

merization of UDP-glucose to form UDP-galactose. In liver this reaction is catalyzed by **UDP-glucose 4-epimerase**, an NAD^+-requiring enzyme.

In liver, galactose is linked metabolically with glucose by the above epimerase and by **hexose-1-phosphate uridylyltransferase**, which catalyzes the interchange of the two sugars between galactose-1-phosphate and UDP-glucose:

$$\text{UDP-glucose} + \text{galactose-1-phosphate} \leftrightharpoons \text{UDP-galactose} + \text{glucose-1-phosphate}$$

The epimerization of UDP-galactose converts it to UDP-glucose.

$$\text{UDP-galactose} \leftrightharpoons \text{UDP-glucose}$$

In liver, galactose is phosphorylated by ATP at the C-1 position by the action of **galactokinase** to form galactose-1-phosphate. The above set of reactions enables galactose to form part of a common metabolic pool with glucose. In this way the galactose derived from the lactose of milk can enter the main pathways of carbohydrate metabolism. In human beings a genetic impairment of the enzymes catalyzing the interconversion of glucose and galactose results in the hereditary disease **galactosemia**, which causes blindness and mental disorders.

Biosynthesis of Glycosides

The formation of glycosides is strongly endergonic, so that glycosidic bonds are intrinsically unstable with respect to hydrolysis. In living systems, glycoside formation never occurs directly through the combination of two monosaccharides, but always proceeds by way of a phosphorylated intermediate.

An important general route for disaccharide synthesis involves UDP derivatives, as is illustrated by the following mechanism for the formation of sucrose in many plants, as catalyzed by **sucrose phosphate synthase**:

$$\text{UDP-glucose} + \text{fructose-6-phosphate} \rightarrow \text{UDP} + \text{sucrose-6}'\text{-phosphate}$$

The sucrose-6′-phosphate is subsequently hydrolyzed to sucrose in a reaction catalyzed by **sucrose phosphatase**:

$$\text{sucrose-6}'\text{-phosphate} \xrightarrow{H_2O} \text{sucrose} + P_i$$

A second mechanism for glycoside formation is **transglycosylation**, whereby a monosaccharide is exchanged between two glycosides. For example:

$$\text{sucrose} + \text{L-sorbose} \rightarrow \text{D-glucosido-L-sorbose} + \text{fructose}$$

The disaccharide lactose is synthesized in the mammary gland from glucose and UDP-galactose in a reaction catalyzed by the **lactose synthase** system. The latter consists of an enzyme, **galactosyl transferase**, plus a protein modifier, α-lactalbumin. In the absence of α-lactalbumin, galactosyl transferase catalyzes the transfer of galactose from UDP-galactose to the acceptor *N*-acetyl-D-glucosamine:

$$\text{UDP-galactose} + N\text{-acetyl-D-glucosamine} \rightarrow \text{UDP} + N\text{-acetyllactosamine}$$

In the absence of α-lactalbumin glucose does not act as an acceptor to a significant extent, since the K_m for glucose is high. A high value of K_m implies inefficient binding of the substrate by the enzyme (chapter 6). Complex formation with α-lactalbumin reduces the K_m of galactosyl transferase for glucose sufficiently to permit glucose to replace *N*-acetyl-D-glucosamine as acceptor, so that lactose is synthesized:

$$\text{UDP-galactose} + \text{D-glucose} \rightarrow \text{UDP} + \text{lactose}$$

Biosynthesis of Polysaccharides

In plants and animals, polysaccharides are usually synthesized from monosaccharide units, the immediate precursor being a nucleoside diphosphate sugar:

$$n\ \text{UDP—sugar} \rightarrow (\text{sugar})_n + n\ \text{UDP}$$

It is by this kind of mechanism that the storage carbohydrates glycogen and starch are formed in animals and higher plants, respectively.

The starting material for the pathway is glucose-1-phosphate, which is derived from glucose-6-phosphate by the action of phosphoglucomutase. Glucose-1-phosphate is combined with UTP to form UDP-glucose, in a reaction catalyzed by **glucose-1-phosphate uridyltransferase**:

$$\alpha\text{-D-glucose-1-phosphate} + \text{UTP} \rightarrow \text{UDP-D-glucose} + \text{PP}_i$$

The UDP-glucose is subsequently added to the nonreducing end of an amylose or glycogen chain by formation of an α,1,4 glycosidic bond.

10-9 SYNTHESIS AND DEGRADATION OF GLYCOGEN

Glycogen, the principal storage carbohydrate of animals, is mobilized by cells through phosphorolysis to form glucose-1-phosphate, which is subsequently converted to glucose-6-phosphate by the action of phosphoglucomutase:

$$(\text{glucose})_n + \text{P}_i \rightarrow (\text{glucose})_{n-1} + \text{glucose-1-phosphate}$$

$$\text{glucose-1-phosphate} \rightarrow \text{glucose-6-phosphate}$$

Glucose-6-phosphate may enter the glycolysis pathway.

The above reactions enable the organism to draw upon the reserve carbohydrate when an energy need arises which cannot be otherwise satisfied. The reversible phosphorolysis of glycogen is catalyzed by **glycogen phosphorylase**; only the forward reaction occurs for the levels of reactants and products found in living cells.

Glycogen phosphorylase, whose molecular weight is 195,000, is a dimer consisting of two identical subunits. Its inactive form, **phosphorylase b**, may be activated noncovalently by the binding of AMP, or by covalent modification through phosphorylation by ATP of a serine residue (Ser-14); the latter mechanism, whose product is termed **phosphorylase a**, is the preferred biological route. The covalent activation of phosphorylase b is catalyzed by **phosphorylase kinase** and reversed by **phosphorylase phosphatase**:

$$\text{4ATP} + \text{2 phosphorylase b} \xrightarrow[\text{phosphorylase kinase}]{} \text{phosphorylase a} + \text{4ADP}$$

$$\text{phosphorylase a} + 4H_2O \xrightarrow[\text{phosphorylase phosphatase}]{} \text{2 phosphorylase b} + 2P_i$$

Each subunit of glycogen phosphorylase contains a residue of pyridoxal-5′-phosphate (PLP) (Fig. 10-6), which is attached to the protein by an aldimine linkage to a lysine group. X-ray crystallographic studies have located the PLP group, which is in proximity to the catalytic site, where the substrates glucose-1-phosphate or P_i are bound (Fig. 10-6). This is in harmony with the finding that PLP is essential for catalytic activity; its removal results in an inactive enzyme.

Glycogen phosphorylase is an allosteric enzyme, whose activity is modulated by the activator AMP and the inhibitors glucose and ATP. The specific binding sites for modifiers which have been identified by x-ray crystallography include, in addition to the catalytic and PLP sites mentioned earlier, an AMP site, a glycogen attachment site, and a nucleoside (section 8-2) site (Fig. 10-6).

The AMP binding sites are located at the interface between the two subunits and involve elements of both polypeptide chains. Chemical or physical treatment that produces dissociation of the dimer invariably causes a loss of the catalytic activity of phosphorylase b. The binding of AMP is essential for the activity of phosphorylase b and substantially enhances that of phosphorylase a. Since the AMP site is located 3 nm from the catalytic site (Fig. 10-6), this indicates that the conformational change accompanying the binding of AMP is transmitted over a considerable distance.

The glycogen attachment site, which is distinct and well separated from the catalytic site (Fig. 10-6), provides the means whereby phosphorylase is associated with glycogen in a supramolecular structure. Glycogen is of course also transiently bound at the catalytic site.

The nucleoside site, which binds the allosteric inhibitors adenosine and caffeine, is located about 1 nm from the catalytic site (Fig. 10-6). Both adenosine and caffeine appear to act in a synergistic manner with glucose, which is bound at the catalytic site, so that their combined effect is much greater than their predicted individual effects.

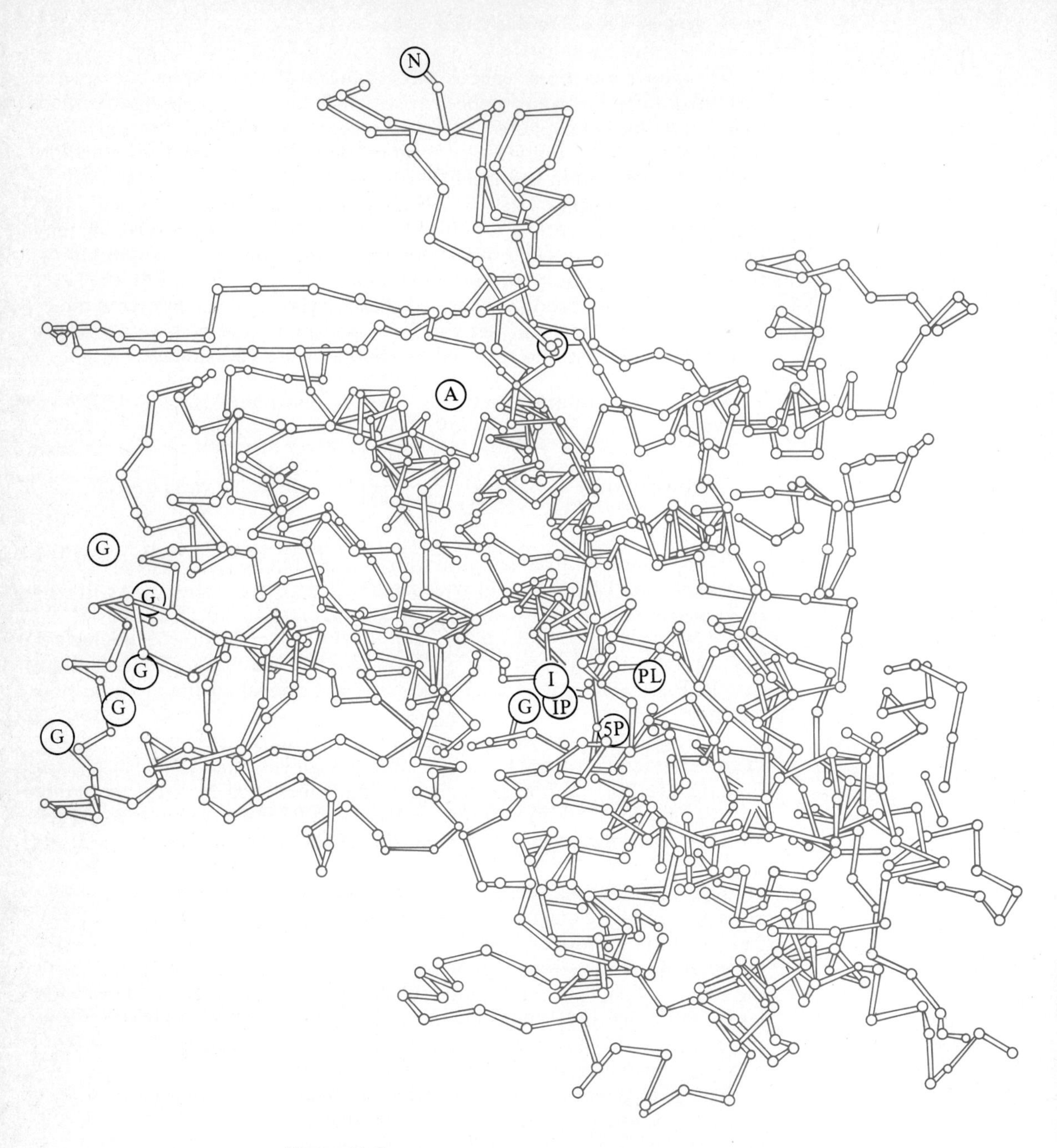

Figure 10-6
Structure of the monomer unit of glycogen phosphorylase a, as determined by the x-ray crystallographic studies of R. J. Fletterick and co-workers [P.J. Kasvinsky, *et al.*, J. Biol. Chem. 253, 3343 (1978)]. Here N indicates the NH_2-terminus-; A, the binding site of AMP; GGGGG, a binding site for glycogen; G1P, the binding site for the substrate, glucose-1-phosphate; and PL, the pyridoxal-5′-phosphate group. The allosteric modifier glucose is bound at the G1P site.

10-10 CONTROL OF GLYCOGEN DEGRADATION

In skeletal muscle, glycogen phosphorylase occurs, together with glycogen, in supramolecular structures called **glycogen particles**. These structures also contain a set of enzymes responsible for the controlled activation and deactivation of phosphorylase and for the synthesis of glycogen.

The synthesis and degradation of glycogen depend on the opposed activities of the enzymes **glycogen phosphorylase** and **glycogen synthase**. The latter catalyzes the transfer of glucose units from UDP-glucose to glycogen:

$$(\text{glucose})_{n-1} + \text{UDP—glucose} \underset{\substack{\text{glycogen}\\\text{synthase}}}{\rightleftharpoons} (\text{glucose})_n + \text{UDP}$$

The activity of glycogen synthase, like that of glycogen phosphorylase, is regulated by phosphorylation. However, in this case it is the phosphorylated form that is inactive and the dephosphorylated form which has catalytic activity. The same enzyme, phosphorylase phosphatase, which deactivates phosphorylase a, also dephosphorylates and activates glycogen synthase (Fig. 10-7).

Phosphorylase kinase, the enzyme responsible for the phosphorylation and activation of glycogen phosphorylase, contains three kinds of subunits, which have been designated α, β, and γ, and whose molecular weights are 145,000, 130,000, and 45,000, respectively. The structure of the intact enzyme corresponds to $(\alpha\beta\gamma)_4$. Phosphorylase kinase is itself activated by phosphorylation of a site on the β-subunit. The phosphate donor is ATP

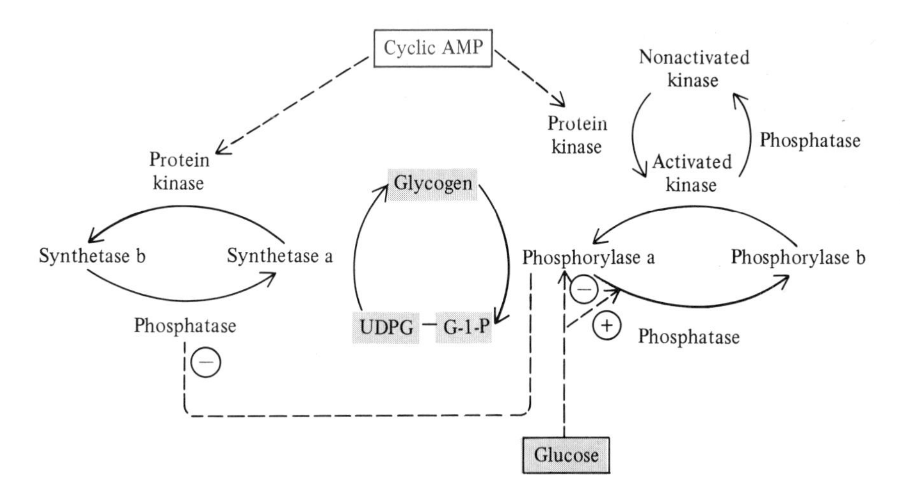

Figure 10-7
Simultaneous control of glycogen synthesis and degradation. The cAMP generated by the action of adenyl cyclase (Fig. 8-3) causes phosphorylation and activation of phosphorylase kinase and phosphorylation and deactivation of glycogen synthase. Both reactions are catalyzed by cAMP-dependent protein kinase and reversed by the same phosphatase. In liver glycogen phosphorylase activity is also regulated by the allosteric inhibitor glucose.

(Fig. 10-7); the reaction is catalyzed by **cyclic AMP-dependent protein kinase**. Cyclic AMP (cAMP) is an allosteric activator.

$$\underset{\text{(inactive)}}{\text{phosphorylase kinase}} + \text{ATP} \xrightarrow[\substack{\text{cAMP-}\\ \text{dependent}\\ \text{kinase}}]{} \underset{\text{(active)}}{\text{phosphorylase kinase}} + \text{ADP}$$

The activity of phosphorylase kinase requires Ca^{2+}.

Cyclic AMP is generated by the action of **adenylate cyclase** upon ATP. The inactive form of adenylate cyclase is, in turn, activated by binding the hormone epinephrine, which is secreted by the adrenal medulla upon stress-induced stimulation (section 22-3).

The complete scheme (Fig. 10-8) is a particularly compelling illustration of biological control and of the amplified response of the organism to external stimulus. The existence of a cascade of enzymes, each of which activates the next member of the series allows the release of minute quantities of epinephrine to be ultimately expressed in the mobilization of significant amounts of storage carbohydrate.

In skeletal muscle a second regulatory mechanism depends upon the Ca^{2+} requirement for phosphorylase kinase activity. The reception of a nerve signal at the neuromuscular junction (section 24-2) results in the release of free Ca^{2+}, which is directly involved in the initiation of contraction. It is likely that the increase in the Ca^{2+} level also stimulates the action of phosphorylase kinase, so that the mobilization of reserve carbohydrate is synchronized with muscle activity.

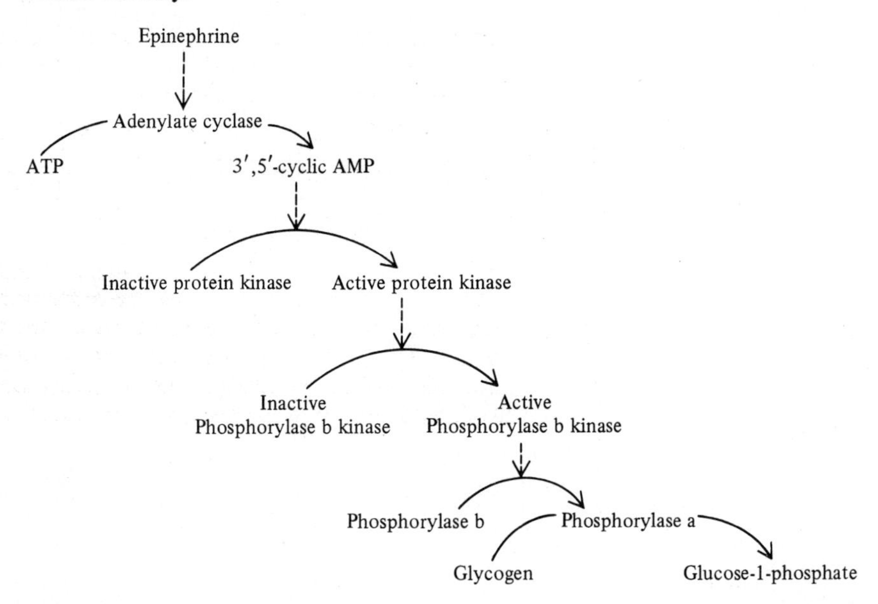

Figure 10-8
Set of sequential reactions responsible for the activation of glycogen phosphorylase. The cAMP-dependent protein kinase which mediates the activation of phosphorylase kinase is also responsible for the phosphorylation of glycogen synthase (Fig. 10-7).

The Balance Between Glycogen Synthesis and Degradation

Cyclic AMP is of central importance for the control of both the formation of glycogen and its mobilization as glucose-1-phosphate. A hormone-stimulated rise in the level of cAMP results in the phosphorylation and activation of phosphorylase kinase and in the phosphorylation and deactivation of glycogen synthase (Fig. 10-7). Both reactions are catalyzed by cAMP-dependent protein kinase and reversed by phosphorylase phosphatase. The phosphorylation and activation of glycogen phosphorylase b are catalyzed by the activated form of phosphorylase kinase and reversed by phosphorylase phosphatase (Fig. 10-7).

In this way, the hormonal response to external conditions of stress requiring the mobilization of reserve carbohydrate is such that it suppresses glycogen synthesis and stimulates glycogen phosphorylation. It should be noted that the basic control mechanisms in both cases involve *covalent* modification rather than the noncovalent binding of an allosteric modifier, and that the covalent modification in each case is a phosphorylation. We shall encounter other examples of this mode of regulation.

In liver, which, unlike muscle, contains significant quantities of glucose, a second scheme of allosteric control is superimposed upon the above. Glucose is an allosteric inhibitor of phosphorylase a. An important and poorly understood synergistic effect of insulin exists, so that a rise in insulin or glucose level results in a major reduction of phosphorylase activity.

SUMMARY

Living cells derive much of their chemical energy from the controlled oxidative degradation of carbohydrates, such as glucose. The energy which is thereby released in a stepwise manner is conserved by the synthesis of ATP.

The initial stages of the metabolic processing of glucose do not involve oxygen and may proceed under anaerobic conditions. A major anaerobic pathway is **glycolysis**, which is important in muscle, an oxygen-poor tissue. The overall reaction of glycolysis is glucose + 2ADP + $2P_i \rightarrow$ 2 lactic acid + 2ATP + $2H_2O$. The **lactic acid** formed may be transported to the liver and enter other pathways which reform blood glucose.

In the initial phase of glycolysis, D-glucose is first enzymically phosphorylated by ATP and then split into two molecules of **glyceraldehyde-3-phosphate**. Other sugars are also converted into this intermediate. In the subsequent phase of glycolysis, glyceraldehyde-3-phosphate is oxidized enzymically by NAD^+, with uptake of P_i, to yield **3-phosphoglyceroyl phosphate**. The acyl phosphate group of the latter is used to phosphorylate ADP to form ATP and **3-phosphoglycerate**, which is isomerized to **2-phosphoglycerate**. The latter compound is dehydrated by enolase to form **phosphoenolpyruvate**, whose phosphate group is transferred to ADP to yield ATP and pyruvic acid. In the presence of an adequate oxygen supply pyruvic acid is converted to acetyl CoA, which enters other pathways; under anaerobic conditions pyruvate is reduced to lactate by NADH in a reaction catalyzed by **lactate dehydrogenase**. In this way the NAD^+ consumed earlier is replaced. The efficiency of energy recovery by glycolysis is close to 30%.

In some microorganisms, such as yeast, glycolysis is replaced under anaerobic conditions by **alcoholic fermentation**. This is equivalent to glycolysis for the steps leading to pyruvate formation; however, the pyruvate is decarboxylated to form CO_2 and **acetaldehyde**, which is reduced to **ethanol**.

Many cells, especially those of tissues active in fatty acid or steroid synthesis, such as liver, possess an alternative pathway for glucose degradation, termed the **phosphogluconate pathway**. The initial reaction is the dehydrogenation of glucose-6-phosphate to 6-phosphogluconate. This is a multifunctional pathway which interconverts hexoses and pentoses, provides a source of **D-ribose-5-phosphate,** and generates NADPH needed in other pathways.

The key precursors in the biosynthesis of oligo- and polysaccharides are the **nucleoside diphosphate** sugars (NDP-sugars). An important example is **uridine diphosphate glucose** (UDP-glucose), which is enzymically interconvertible with **UDP-galactose**. The NDP-sugars are formed by reaction of the corresponding nucleoside triphosphate with the sugar-1-phosphate. In plants sucrose or sucrose-6′-phosphate may be formed by the enzymic condensation of UDP-glucose with fructose or fructose-6-phosphate, respectively; hydrolysis of sucrose-6′-phosphate yields sucrose.

Glycogen, the primary storage carbohydrate of animals, is formed by the addition of glucose units from UDP-glucose to preexisting glycogen, in a reaction catalyzed by **glycogen synthase.**

The glycogen reserve may be drawn upon by the action of **glycogen phosphorylase**, which catalyzes a **phosphorolysis** of glycogen to **glucose-l-phosphate**. The latter may be converted enzymically to glucose-6-phosphate and may be hydrolyzed to glucose. Glycogen phosphorylase is activated through phosphorylation by ATP, in a reaction catalyzed by **phosphorylase kinase**. The latter enzyme is itself activated via phosphorylation by a **cyclic AMP-dependent kinase**. The level of cyclic AMP is subject to hormonal control. The mobilization of reserve carbohydrate is thus subject to an elaborate system of regulation.

REFERENCES

The following code is used to classify references. I: particularly useful as an introduction to the subject; R: useful primarily as a reference text; A: an advanced account of the material; H: a publication of historical importance.

General

M. Florkin and E. H. Stotz, eds., *Carbohydrate Metabolism*, vol. 17 of *Comprehensive Biochemistry*, Elsevier, New York (1967). (R)

A. L. Lehninger, *Biochemistry*, Worth, New York (1975). (A)

L. Stryer, *Biochemistry*, Freeman, San Francisco (1975). (I)

A. White, P. Handler, and E. Smith, *Principles of Biochemistry*, McGraw-Hill, New York (1964). (R)

Glycolysis

B. Axelrod, in *Glycolysis*, in D. M. Greenberg, ed., *Metabolic Pathways*, vol. 1, Academic, New York (1967). (R)

F. S. Rolleston and E. A. Newsholme, "Control of Glycolysis in Cerebral Cortex Slices," *Biochem. J.* 104, 524 (1967). (A)

Glycolytic Enzymes

S. P. Colowick, *The Hexokinases*, in P. D. Boyer, ed., *The Enzymes*, vol. 9, Academic, New York (1973). (R)

R. Everse and N. O. Kaplan, "Lactate Dehydrogenases," *Adv. Enzymol.* 37, 61 (1973). (R)

C. Y. Lai and B. L. Horecker, "Aldolase," *Essays Biochem.* 8, 149 (1972). (A)

C. Villar-Palasi and J. Larner, "Glycogen Metabolism and Glycolytic Enzymes," *Ann. Rev. Biochem.* 39, 639 (1970). (R)

Glycogen Phosphorylase

E. H. Fischer, A. Pocker, and J. C. Saari, "The Structure, Function, and Control of Glycogen Phosphorylase," *Essays Biochem.* 6, 23 (1970). (I)

R. J. Fletterick, J. Sygusch, H. Semple, and N. B. Madsen, "The Structure of Phosphorylase a at 3.0 Å Resolution and its Ligand Binding Sites at 6 Å," *J. Biol. Chem.* 251, 1642 (1976). (H)

N. B. Madsen and R. J. Fletterick, "X-Rays Reveal Phosphorylase Architecture," *Trends Biol. Sci.* 2, 145 (1977). (I)

Regulation

E. R. Stadtman, "Allosteric Regulation of Enzyme Activity," *Adv. Enzymol.* 28, 42 (1966). (R)

REVIEW QUESTIONS

Questions marked with an asterisk are of a high level of difficulty.

10-1 What compounds are present in significant quantities at equilibrium when pure aldolase acts on a mixture of fructose-1,6-diphosphate (1 m*M*) and D-glyceraldehyde-3-phosphate (10 m*M*)?

*10-2 Write a balanced equation for the conversion of D-fructose to lactic acid if the fructose is phosphorylated by hexokinase.

10-3 Compute the free energy change at pH 7.0 and 25°C for the formation of lactate from glucose when [glucose] = 10 m*M*; [phosphate] = 1 m*M*; [ADP] = 1 m*M*; [ATP] = 5 m*M*; [lactate] = 4 m*M*.

10-4 For an isolated system, what initial concentration of pure fructose-1,6-diphosphate would be converted to products by 50% at equilibrium by pure aldolase?

10-5 For the reaction glucose + phosphate ⇄ glucose-6-phosphate + H_2O, $\Delta G^{\circ\prime} = 3.3$ kcal at 25°C and pH 7. What is the equilibrium constant?

10-6 What is the lowest concentration at which malate must be present to make the reaction malate ⇄ fumarate + H_2O ($\Delta G^{\circ\prime} = 0.75$ kcal) proceed to the right at pH 7.0 and 25°C, if [fumarate] = 3 m*M*?

10-7 Would it be feasible to form glucose from lactate by a simple reversal of the glycolytic pathway? Why or why not?

11

Photosynthesis

11-1 THE PHOTOSYNTHETIC APPARATUS OF PLANTS

Occurrence of Photosynthesis

It is possible to group living cells into two categories with respect to their mechanisms for extracting energy from their environment. **Heterotrophic** cells, which include all cells of higher animals, require a supply of chemical energy from an external source, which is usually supplied in the form of organic molecules of complex structure, such as fats, carbohydrates, and proteins. **Autotrophic** cells largely satisfy their energy requirements by direct utilization of the radiant energy of sunlight. Cells of this kind occur in the leaves of green plants and in algae, and include those of certain bacteria. The complex set of reactions whereby radiant energy is trapped and harnessed for the synthesis of carbohydrates and other materials is collectively known as **photosynthesis**.

The net result of photosynthesis may be loosely regarded as the reverse of the respiration of heterotrophic organisms. Photosynthetic systems incorporate H_2O and atmospheric CO_2 into carbohydrates and release excess oxygen as O_2. Since this process is thermodynamically unfavored, an external energy source is required. This is supplied by solar radiant energy.

The assimilation of CO_2 is not of course confined to plants. There are numerous instances of CO_2 incorporation by mammalian systems. It is the *utilization* of *light* for biosynthetic reactions which is characteristic of plants. The overall stoichiometry may be represented by

$$6\ CO_2 + 6\ H_2O + h\nu \rightarrow 6\ O_2 + C_6H_{12}O_6$$

The above equation is actually somewhat misleading in that it fails to convey that *all* of the oxygen atoms of the evolved O_2 come from H_2O, as is currently

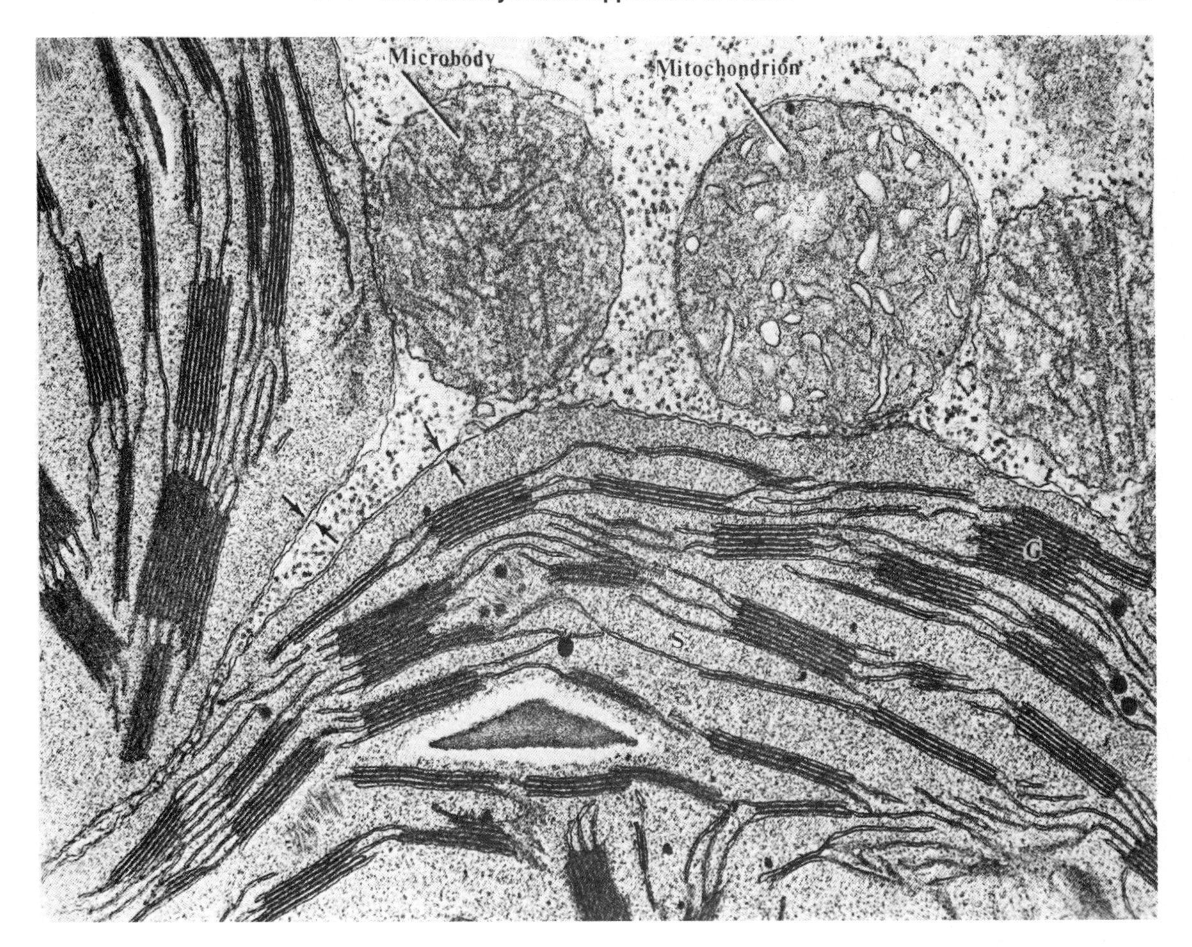

Figure 11-1
Electron microscope photograph of a chloroplast showing grana. (Courtesy of E. H. Newcomb and P. J. Gruber.)

believed. A better representation might be provided by

$$6\ CO_2 + 12\ H_2O^* \rightarrow 6\ O_2^* + C_6H_{12}O_6 + 6\ H_2O$$

In green plants photosynthesis takes place in specialized organelles called **chloroplasts** (Fig. 11-1). If a suitable electron acceptor, such as ferricyanide, is present, isolated chloroplasts are capable of evolving O_2 upon irradiation with visible light, even if CO_2 is absent. This indicates that CO_2 reduction is not obligatory for the evolution of O_2 and that such unnatural electron acceptors as ferricyanide can substitute for CO_2. Since the ferricyanide is simultaneously reduced to ferrocyanide, it further suggests that the basic photochemical event involves the light-induced transfer of an electron to an acceptor.

In eukaryotic green plants H_2O is split to form O_2, thereby providing hydrogen atoms for the reduction of CO_2. The photosynthetic system is the only biological oxidizing agent powerful enough to dehydrogenate water.

11-2 PHOTOSYNTHETIC PIGMENTS

Chloroplasts

The chloroplasts of green plants are membrane-enclosed subcellular particles about 1–10 μm in diameter. Like mitochondria, they contain DNA and are self-replicating. By careful physical treatment they may be isolated in intact form from some plant tissues, such as spinach leaves.

Again in parallel to mitochondria, chloroplasts possess both a continuous outer membrane and a continuous inner membrane which is arranged into parallel folds called **lamellae** (Fig. 11-1). In well-defined regions the lamellae enlarge to form flat sacs which are stacked in parallel arrays called **grana** (Fig. 11-1). The lamellae membranes contain the photosynthetic pigments involved in the light-dependent phase of photosynthesis.

Chlorophylls and Accessory Pigments

The chloroplasts of the higher plants contain two forms of chlorophyll, a and b (Fig. 11-2). Both are magnesium-porphyrin complexes possessing a hydrocarbon side-chain which arises from the esterification of the alcohol **phytol** with a propionic acid ring substituent. The two forms differ solely in one ring substituent (Fig. 11-2).

The absorption spectra of chlorophyll a in intact cells of green plants contain at least four major bands at 662, 670, 677, and 683 nm, although a solution of chlorophyll a in acetone shows only a single band at 678 nm (Fig. 11-3). This suggests that the chlorophyll of green plant cells exists in a multiplicity of environments and that its spectrum is dependent upon the environment. The resultant broadening of the absorption band yields a more efficient absorption of light.

If the efficiency of visible light in producing oxygen evolution by green plant cells is measured as a function of wavelength, an **action spectrum** of photosynthesis may be obtained (Fig. 11-4). Over much of the visible range its shape corresponds closely to the summed absorption spectra of chlorophylls a and b, suggesting that the absorption of light by these molecules is the dominant initial event in photosynthesis. The action spectrum shows an abrupt drop at wavelengths above 680 nm, falling more rapidly than the absorption spectrum. This observation has proved to be of importance in the development of current theories of photosynthesis.

The chloroplasts of green plants contain, in addition to chlorophylls a and b, a set of **accessory pigments** which may be grouped into the two major classes of **carotenes** and **xanthophylls** (Fig. 11-5). Both are polyisoprenoid molecules (section 16-1), whose major difference is the presence of oxygen-containing terminal groups of the xanthophylls.

The accessory pigments functions as supplementary light traps for regions of the visible spectrum which are incompletely covered by chlorophyll. However, the radiant energy absorbed by the accessory pigments is effective in photosynthesis only via transfer to chlorophyll, which is the indispensable photosynthetic pigment.

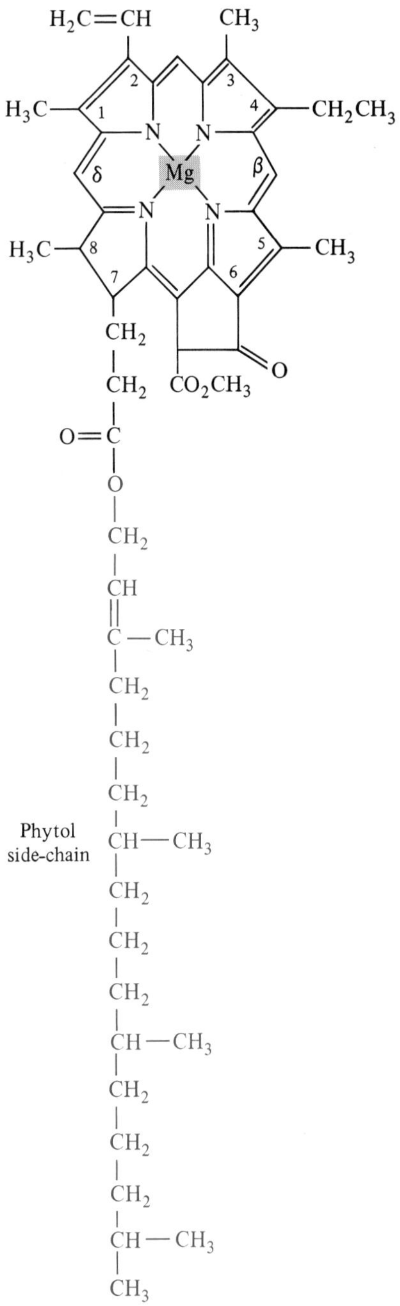

Figure 11-2
Structure of chlorophyll a. In chlorophyll b, the CH_3-group at position 3 is replaced by a —CHO group.

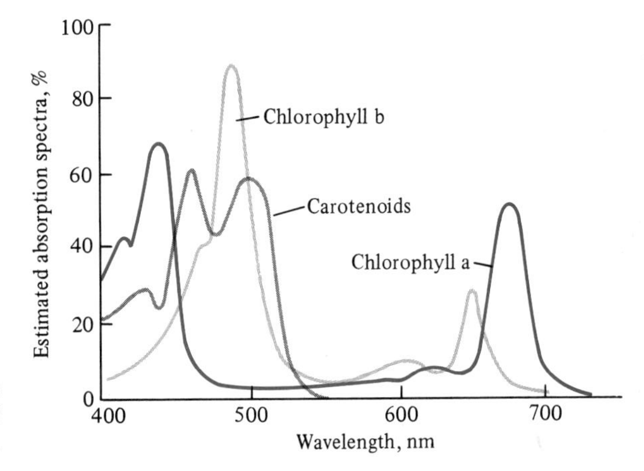

Figure 11-3
Absorption spectra of chlorophyll a and b.

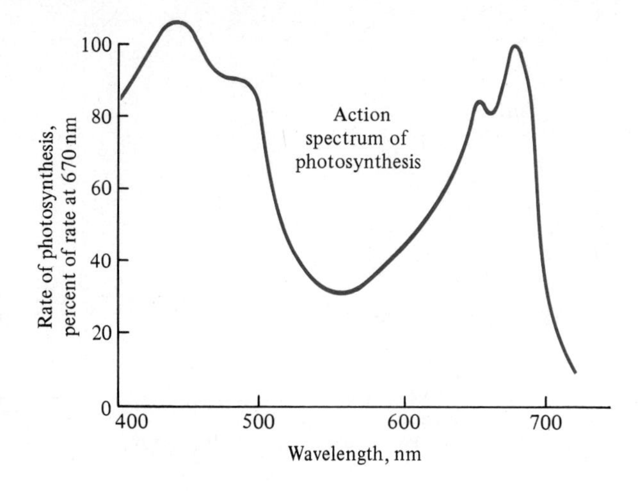

Figure 11-4
General appearance of the photochemical action spectrum of green plants. Note the sharp drop at wavelengths above 680 nm.

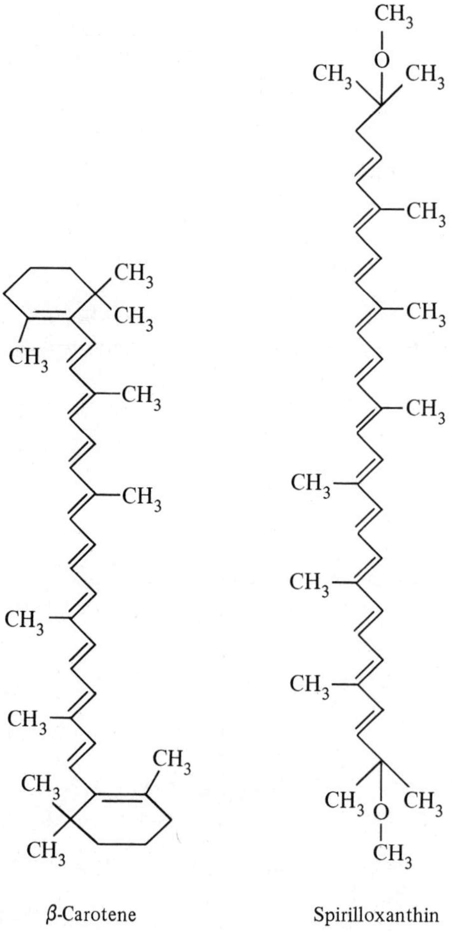

β-Carotene

Spirilloxanthin

Figure 11-5
Structure of a carotene, β-carotene, and of a xanthophyll, spirilloxanthin.

11-3 THE PRIMARY PHOTOSYNTHETIC EVENTS

The Two Photosystems

The rate of oxygen evolution by chloroplasts upon irradiation by light of a particular wavelength, divided by the number of quanta of radiant energy absorbed at that wavelength, yields the **relative quantum efficiency** of photosynthesis. If only a single type of photoreceptor were present, the quantum efficiency should be independent of wavelength over the entire absorption band. In actuality, the quantum efficiency falls abruptly at wavelengths above about 680 nm. However, if the system is simultaneously irradiated with light of wavelength greater than 680 nm and with light of shorter wavelength, it is found that the rate of photosynthesis exceeds the sum of the rates observed for separate irradiation at the two wavelengths.

These findings have led to the conclusion that photosynthesis in chloroplasts involves the operation of two distinct photosystems, both of which can utilize light of wavelength less than 680 nm, but only one of which can be driven by light of longer wavelength. The system activated by long wavelengths is called **photosystem I**, while the short wavelength system is termed **photosystem II**.

The two photosystems are structurally distinct. By careful fractionation of chloroplasts it is possible to isolate particles with the characteristics of photosystem I or of photosystem II. While both photosystems contain chlorophyll a and chlorophyll b, the proportion of chlorophyll a is higher in photosystem I.

The photosynthetic units comprising photosystem I each contain a molecule of chlorophyll with exceptional properties, which has an absorption maximum at 700 nm and which is bleached upon illumination of the cell. This pigment, which has been named P700, probably consists of a molecule of chlorophyll a combined with a specific protein. An analogous pigment, P680, is associated with photosystem II.

The pigments P700 and P680 both function as collection centers for **excitons**, or quanta of excitation energy, from the other chlorophyll molecules in the photosynthetic unit. While light is absorbed by most of the chlorophyll molecules in each photosynthetic unit, only the P700 and P680 pigments actually mediate the conversion of radiant energy into chemical energy. The radiant energy absorbed by the other chlorophyll molecules migrates through the photosynthetic unit until it reaches, and is trapped by, the P700 or P680 pigment. Migration occurs through the array of chlorophyll and accessory pigment molecules by a process termed **exciton transfer**, which is the radiationless transfer of excitation energy between molecules. The process terminates with the acceptance of an exciton by a single molecule of P700 or P680, which subsequently loses an electron. Both the initial absorption of a photon of radiant energy and its ultimate capture by a P700 or P680 center are very rapid processes, occurring in less than a nanosecond.

Photosynthesis requires the action of both photosystem I and photosystem II. Photosystem I generates NADPH, while photosystem II splits water, thereby forming O_2.

After capture of an exciton by P700, an electron is ejected and transferred to a primary electron acceptor, leaving the P700 chlorophyll in an oxidized and electron-deficient form (Chl^+). Electrons generated in this way flow

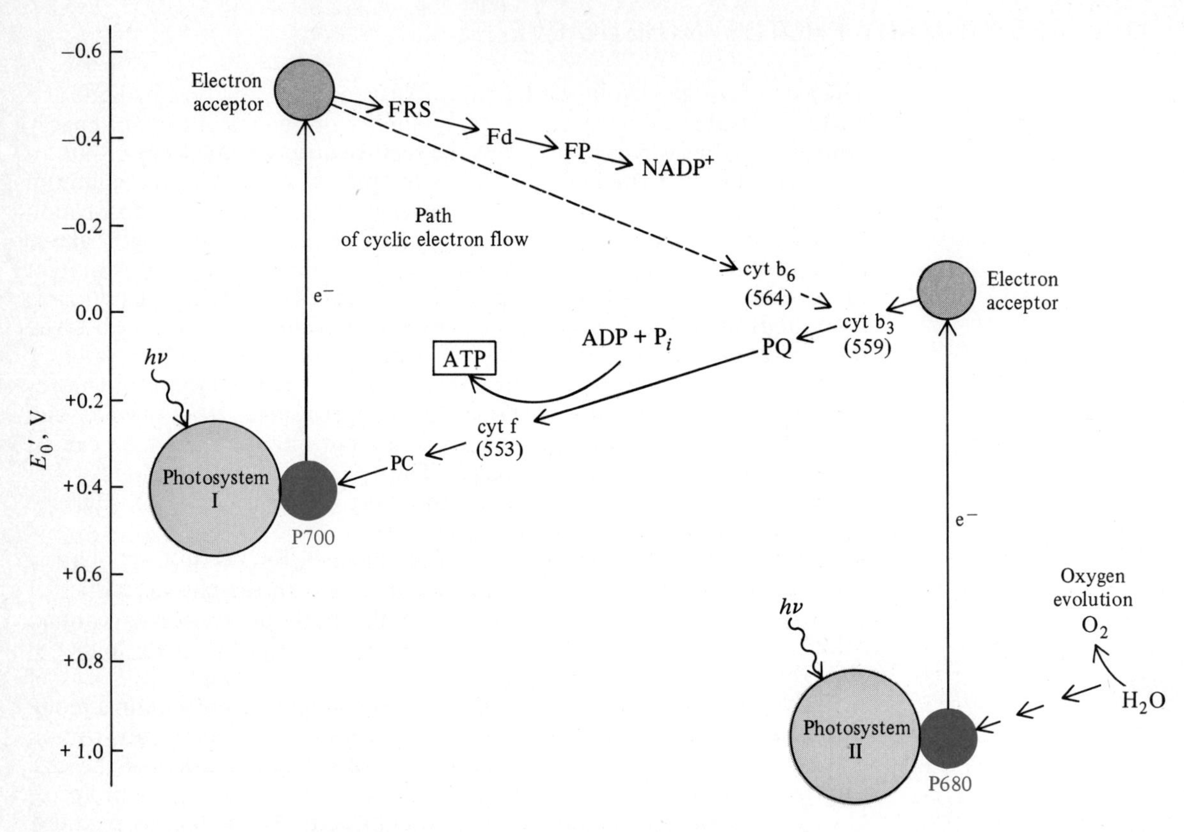

Figure 11-6
Schematic version of the mutual relationship of photosystem I and II, showing the approximate positions of their components on the electrode potential scale.

through a chain of electron carriers to $NADP^+$, causing its reduction to NADPH. Two electrons are required to reduce each molecule of $NADP^+$.

In detail, the light-dependent phase of the process terminates with the transfer of an exciton to P700. The remainder of the reactions mediated by photosystem I are redox processes occurring by electron transfer (Fig. 11-6). While P700 in the ground state has a positive redox potential (about +0.5 V), upon excitation it develops a strong negative potential which is sufficient to cause the reduction of $NADP^+$ ($E_o' = -0.32$ V). After transfer of an electron to, and reduction of, a primary electron acceptor, the latter in turn reduces an iron- and sulfur-containing protein called **ferredoxin reducing substance** (FRS).

The reduced form of FRS transfers its electron to **ferredoxin** (Fd), an iron-sulfur protein of molecular weight 11,600. The active site of ferredoxin contains an iron atom which can exist in oxidized or reduced states. The reduced form of ferredoxin transfers its electron to $NADP^+$ in a reaction catalyzed by the flavoprotein (FP) enzyme **ferredoxin-NADP reductase,** which contains

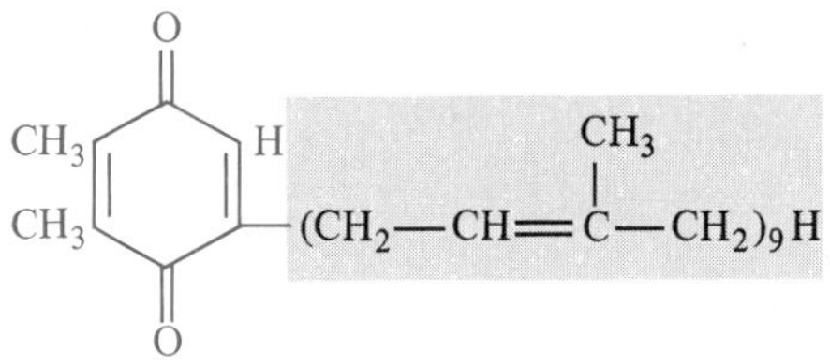

Figure 11-7
Structure of plastoquinone A, which is abundant in algae and green plants. Other plastoquinones differ from this structure in the length of the side-chain and in the substituents in the quinone ring.

FAD as its prosthetic group. The NADPH formed by photosystem I ultimately reduces CO_2 to synthesize carbohydrate.

The initial photochemical events involved in the operation of photosystem I leave P700 in an electron-deficient state. The completion of the cycle requires the restoration of an electron to P700. This is provided by photosystem II. The pigment P680 captures absorbed radiant energy via exciton transfer, in parallel to the behavior of P700. Upon excitation it likewise loses an electron, which is transferred to P700 by way of a set of electron carriers. The electron flow from photosystem II to photosystem I is coupled to, and drives, the phosphorylation of ADP to form ATP (Fig. 11-6).

The initial acceptor of the electron lost upon photoexcitation of P680 has been designated Q. It is probably a plastoquinone (Fig. 11-7). The reduced form of Q subsequently transfers an electron to an electron carrier chain that includes a second plastoquinone acceptor PQ. The remainder of the electron carrier chain includes two heme-containing **cytochromes**, cytochrome b_{559} and cytochrome f (Fig. 11-6) (section 17-5). Finally, the electron is passed to the copper-containing protein **plastocyanin** (PC), which serves as the immediate electron donor to P700.

The sequence cyt $b_{559} \rightarrow$ PQ $\rightarrow$ cyt f is coupled to the synthesis of ATP. This process has many points of similarity to the formation of ATP in mitochondria (section 12-6).

The restoration of an electron to the electron-deficient form of P680 occurs by electron transfer from water, which serves as the ultimate electron donor (Fig. 11-6). In order to form one molecule of oxygen (O_2) it is necessary to remove four electrons from two water molecules. The electron-deficient form of P680 is a sufficiently strong oxidant to oxidize (extract electrons from) water. The detailed nature of the process is still uncertain. One or more Mn^{2+} ions are believed to be somehow involved.

The transfer of electrons from photosystem II to photosystem I is not the only process in chloroplasts which can drive a linked phosphorylation of ADP. Additional ATP may be formed by a **cyclic photophosphorylation** associated with photosystem I (Fig. 11-6). This cyclic electron flow can be detected only indirectly by an effect produced by the flow, namely, the synthesis of ATP. Cyclic electron flow is believed to be a type of bypass which provides an alternative to the transfer of electrons from P700 to $NADP^+$. The electrons lost by P700, instead of passing to $NADP^+$, are transferred through a set of electron carriers back to P700. The phosphorylation of ADP to ATP is coupled to this process. In contrast to the noncyclic flow described earlier, there is no net reduction of an electron acceptor, but only the coupled formation of ATP.

In summary, the net results of the operation of photosystems I and II are as follows:

(1) $NADP^+$ is reduced to NADPH.
(2) Water is split to form molecular oxygen (O_2).
(3) ADP is phosphorylated to ATP.

11-4 BIOSYNTHESIS OF CARBOHYDRATES

The Secondary Phase of Photosynthesis

The generation of NADPH and ATP make possible the incorporation of CO_2 into carbohydrate. This process is only indirectly dependent upon the primary photochemical event and may occur in the dark, provided that the above cofactors are made available.

The nature of the cyclic pathway of CO_2 fixation was established largely through the radioisotope tracer studies of Calvin and his associates, using the green algae *Chlorella pyrenoidosa.* Many of the reactions and enzymes of the Calvin cycle recur in other pathways.

The experimental approach depended upon the differential labeling of intermediates with radioactive ^{14}C. Growing suspensions of algae were supplied with $^{14}CO_2$ by adding labeled HCO_3^- to the medium. After varying periods of incubation, samples of algae were withdrawn and killed by suspension in methanol. In this way the period of CO_2 fixation could be sharply defined.

An extract of the killed algae was next analyzed by two-dimensional paper chromatography. The distribution of radioactivity on the chromatogram was determined by placing it in contact with an x-ray film. Corresponding to each radioactive spot, a darkened area appeared on the film. The spots were identified by comparison with known compounds.

The earliest *stable* compound to be identified was 3-phosphoglyceric acid (PGA). Since the incorporated ^{14}C was found to occur in the carboxyl group, there was a definite implication that $^{14}CO_2$ incorporation proceeded via a carboxylation.

If illumination was switched off *simultaneously* with the introduction of labeled CO_2, PGA formation continued for some time with little further conversion to other intermediates. If illumination was continued, labeled PGA did not accumulate and radioactivity appeared in compounds identified as the triose phosphates glyceraldehyde-3-phosphate and dihydroxyacetone phosphate. This suggested that the latter arise from PGA by reduction and isomerization. Since the reduction of PGA is NADPH-dependent, this reaction proceeds in the dark only until the limited supply of this cofactor is exhausted, as the production of NADPH by the photochemical reaction ceases in the absence of light.

If photosynthesis was allowed to proceed until uniform labeling of all intermediates was achieved and the illumination was then halted, their relative concentrations changed rapidly with time. A labeled pentose derivative, ribulose-1,5-diphosphate, disappeared rapidly, while labeled PGA continued to accumulate. The levels of the other intermediates remained relatively constant.

This result suggested that PGA was derived from ribulose-1,5-diphosphate. If the synthesis, but not the conversion, of the latter depended upon a substance produced by the light reaction, it would be expected that PGA would accumulate at its expense in the dark. The substance was subsequently identified as ATP.

The simultaneous decay in ribulose-1,5-diphosphate and accumulation of PGA, together with the location of PGA at the earliest stage of CO_2 incorporation, further suggested that CO_2 fixation occurred via a carboxylation of ribulose-1,5-diphosphate to yield an unstable intermediate which decomposed to form PGA (Fig. 11-8).

This model was confirmed by the complementary experiment of suddenly withdrawing the CO_2 while leaving the light on. In this case the PGA level declined rapidly, while ribulose-1,5-diphosphate accumulated.

From these and other experiments it was possible to put together the cyclic mechanism of CO_2 assimilation. The principal steps are as follows, choosing an arbitrary point of departure:

(1) Ribulose-5-phosphate is phosphorylated by ATP to form ribulose-1,5-diphosphate in a reaction catalyzed by **phosphopentokinase**.

ribulose-5-phosphate + ATP → ribulose-1,5-diphosphate + ADP

(2) The next, and crucial, step is the addition of CO_2 to ribulose-1,5-diphosphate to form two molecules of PGA, perhaps by way of an unstable 6-carbon intermediate, which is split hydrolytically. The carboxylation reaction is catalyzed by **ribulosediphosphate carboxylase**.

CO_2 + ribulose-1,5-diphosphate → 2,3-phosphoglycerate

(3) PGA is phosphorylated by ATP, in a reaction catalyzed by **phosphoglycerate kinase**, to form 3-phosphoglyceroyl phosphate and then reduced by NADPH to yield glyceraldehyde-3-phosphate, in a reaction catalyzed by **triose-phosphate dehydrogenase ($NADP^+$)**. Part of the glyceraldehyde-3-phosphate is subsequently converted to dihydroxyacetone phosphate by **triosephosphate isomerase**.

3-phosphoglycerate + ATP → 3-phosphoglyceroyl phosphate + ADP

3-phosphoglyceroyl phosphate + NADPH + H^+ → $NADP^+$ + P_i + glyceraldehyde-3-phosphate

glyceraldehyde-3-phosphate → dihydroxyacetone phosphate

These are the last steps in which the cofactors formed in the photochemical reaction are involved directly.

(4) The remaining steps result in the synthesis of three molecules of ribulose-5-phosphate from five molecules of triose phosphate (Fig. 11-8). The reaction sequence represents a combination of steps from the glycolytic and phosphogluconate pathways (sections 10-4 and 10-6). Dihydroxyacetone phosphate is condensed with glyceraldehyde-3-phosphate to

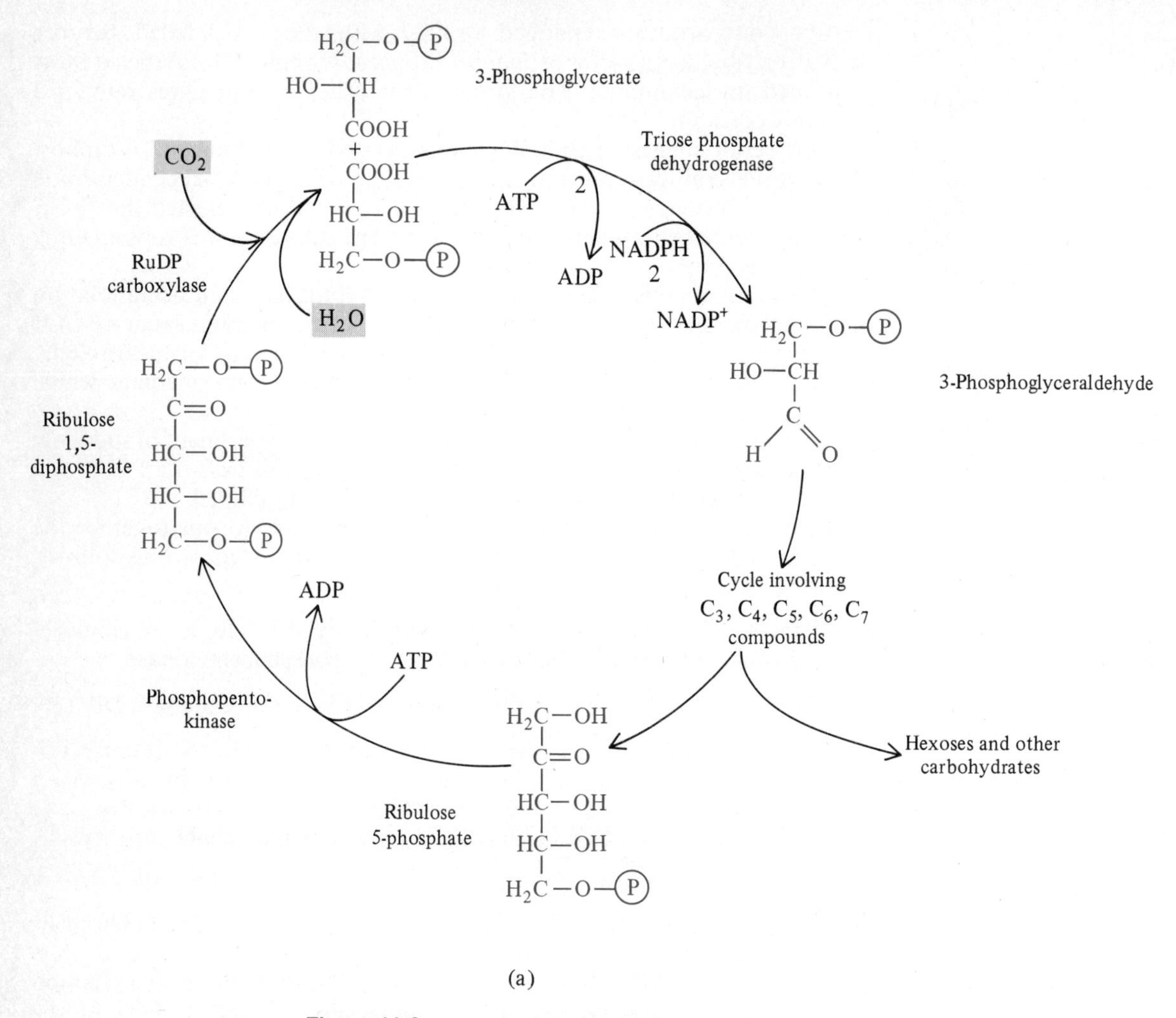

Figure 11-8
(a) The Calvin cycle, in which CO_2 is incorporated into carbohydrate. (b) A more complete version of the Calvin cycle, showing all of the conversions.

yield fructose-1,6-diphosphate, which is dephosphorylated to fructose-6-phosphate. Part of the latter is isomerized to glucose-6-phosphate, which is dephosphorylated to glucose (Fig. 11-8).

Two carbon atoms of fructose-6-phosphate are transferred to glyceraldehyde-3-phosphate to form xylulose-5-phosphate (five carbons) and erythrose-4-phosphate (four carbons). The latter condenses with dihydroxyacetone phosphate to produce sedoheptulose-1,7-diphosphate (seven carbons), which is dephosphorylated to sedoheptulose-7-phosphate.

By transfer of two carbons from sedoheptulose-7-phosphate to glyceraldehyde-3-phosphate, one molecule each of ribose-5-phosphate and

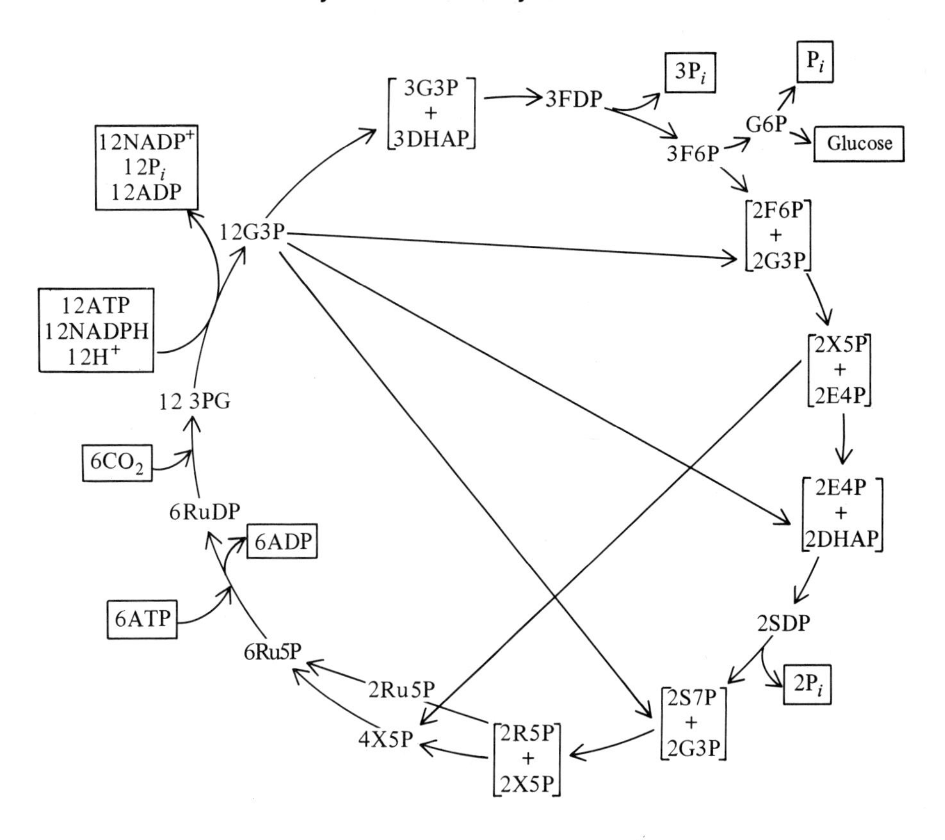

3PG = 3-phosphoglyceric acid
G3P = glyceraldehyde 3-phosphate
DHAP = dihydroxyacetone phosphate
FDP = fructose 1,6-diphosphate
F6P = fructose 6-phosphate
G6P = glucose 6-phosphate
E4P = erythrose 4-phosphate
X5P = xylulose 5-phosphate
SDP = sedoheptulose 1,7-diphosphate
S7P = sedoheptulose 7-phosphate
R5P = ribose 5-phosphate
Ru5P = ribulose 5-phosphate
RuDP = ribulose 1,5-diphosphate

(b)

Figure 11-8 *(continued)*

xylulose-5-phosphate is formed. The latter are both isomerized to ribulose-5-phosphate, thereby completing the cycle (Fig. 11-8).

Energetics of Photosynthesis

The overall equation for the biosynthesis of hexose in green plants by the mechanism summarized above may be written

$$6CO_2 + 18ATP + 12NADPH + 12H^+ + 12H_2O \rightarrow \text{hexose} + 18P_i + 18ADP + 12NADP^+$$

To reduce one molecule of CO_2, three molecules of ATP and two of NADPH are required. The NADPH is formed by the reduction of $NADP^+$ following the ejection of electrons by photosystem I (section 11-3). The equation for the reduction of $NADP^+$ by ferredoxin is

$$2Fd(II) + 2H^+ + NADP^+ \rightarrow 2Fd(III) + NADPH + H^+$$

where Fd represents ferredoxin.

Two electrons derived from photosystem I are thus required for each molecule of NADPH and four for two molecules of NADPH. Four electrons are concurrently ejected by photosystem II, yielding one molecule of O_2 from the splitting of water.

It is uncertain how many molecules of ATP are produced by the photophosphorylation of ADP that is coupled to the flow of electrons from photosystem II to photosystem I. There is some evidence that one ATP is formed for each electron transported. If this is correct, then four molecules of ATP will be produced by the transport of four electrons. Four molecules of ATP would be more than what is required to reduce one molecule of CO_2.

One quantum of absorbed light is required to eject each electron. According to the model a total of eight quanta would be absorbed for each molecule of CO_2 reduced. This is consistent with the experimental value of the quantum requirement obtained by direct measurement.

The standard free energy change per mole, $\Delta G^{\circ\prime}$, for the total combustion of one molecule of glucose to CO_2 and H_2O is -686 kcal. Conversely, the value for the synthesis of one molecule of glucose from six molecules each of CO_2 and H_2O is $+686$ kcal, or 114 kcal for each molecule of CO_2.

The energy of a quantum of radiant energy of wavelength 700 nm is about 41 kcal/mole. (This value is equal to Avogadro's number times the energy of one photon of light of this wavelength.) If the figure of eight quanta is accepted as the requirement for the reduction of one molecule of CO_2, the thermodynamic efficiency of photosynthesis is about 30%.

Alternative Pathways

In some green plants, including sugar cane, sorghum, and maize, the incorporation of CO_2 occurs by a different pathway. In this group of plants, CO_2 is initially incorporated into phosphoenolpyruvate by **phosphoenolpyruvate carboxylase**:

$$\text{phosphoenolpyruvate} + CO_2 \rightarrow \text{oxaloacetate} + P_i$$

The oxaloacetate formed in this way may be converted to aspartate by **transamination**:

$$\text{oxaloacetate} + \text{glutamate} \rightarrow \text{aspartate} + \alpha\text{-ketoglutarate}$$

Alternatively, it may be reduced to malate by NADPH, in a reaction catalyzed by **malate dehydrogenase.**

$$\text{oxaloacetate} + \text{NADPH} + \text{H}^+ \rightarrow \text{L-malate} + \text{NADP}^+$$

Plants whose CO_2 fixation occurs by this mechanism, which are termed **C_4 plants**, possess two distinct kinds of photosynthetic cells. These are the **bundle-sheath** cells, which surround the veins of the leaves, and the **mesophyll** cells, which are located in proximity to the **bundle-sheath** cells and in which the CO_2 incorporation described above occurs.

In C_4 plants the malate formed in the mesophyll cells by the reduction of oxaloacetate is transported to the bundle-sheath cells where it is decarboxylated by **malic enzyme**:

$$\text{malate} + \text{NADP}^+ \rightarrow \text{pyruvate} + CO_2 + \text{NADPH} + \text{H}^+$$

The CO_2 released in this way is incorporated into ribulose-1,5-diphosphate as described in the preceding section. The remaining steps of the pathway leading to the synthesis of hexose are the same as for other green plants and collectively comprise the Calvin cycle. The pyruvate arising from the above reaction is transported back to the mesophyll cells and reconverted to phosphoenolpyruvate.

The biological utility of the C_4 pathway is that it takes advantage of the very high affinity for CO_2 of phosphoenolpyruvate carboxylase, which enables mesophyll cells to collect CO_2 with high efficiency.

Photorespiration

The cells of green plants, like those of other eukaryotic organisms, contain mitochondria and are engaged in respiration and oxidative phosphorylation in the absence of illumination. The dark reaction consumes substrates formed by photosynthesis.

Respiration also occurs during active photosynthesis in the presence of light. However, this is not the same process as mitochondrial respiration; it is not linked to the phosphorylation of ADP and is not influenced by the characteristic inhibitors of the mitochondrial process.

The principal substrate for photorespiration is glycolic acid, which is oxidized by molecular oxygen to glyoxylic acid and hydrogen peroxide in a reaction catalyzed by **glycolate oxidase**, a flavin enzyme:

$$\underset{\text{Glycolic acid}}{\begin{array}{c}CH_2OH\\|\\COOH\end{array}} + O_2 \longrightarrow \underset{\text{Glyoxylic acid}}{\begin{array}{c}CHO\\|\\COOH\end{array}} + H_2O_2$$

The hydrogen peroxide formed in this reaction is decomposed by **catalase**. The glyoxylic acid formed may enter any of several alternative pathways to be converted to such products as formate, oxalate, or glycine, depending upon the species.

The glycolic acid involved in photorespiration is derived from ribulose diphosphate. Molecular oxygen can replace CO_2 in the reaction catalyzed by ribulose-diphosphate carboxylase. The resultant oxygenation of ribulose diphosphate produces phosphoglycolic acid and 3-phosphoglyceric acid:

$$
\begin{array}{l}
\text{CH}_2\text{OPO}_3\text{H}_2 \\
| \\
\text{C=O} \\
| \\
\text{HCOH} \\
| \\
\text{HCOH} \\
| \\
\text{CH}_2\text{OPO}_3\text{H}_2 \\
\text{Ribulose diphosphate}
\end{array}
\xrightarrow{O_2}
\begin{array}{l}
\text{CH}_2\text{OPO}_3\text{H}_2 \\
| \\
\text{COOH} \\
\\
\text{Phosphoglycolic acid}
\end{array}
+
\begin{array}{l}
\text{COOH} \\
| \\
\text{HCOH} \\
| \\
\text{CH}_2\text{OPO}_3\text{H}_2 \\
\text{3-Phosphoglyceric acid}
\end{array}
+ 2H^+
$$

The phosphoglycolic acid undergoes enzymic hydrolysis to yield glycolic acid.

Photorespiration, which is not important for C_4 plants, detracts from the efficiency of photosynthesis since the synthesis of ribulose diphosphate consumes both ATP and NADPH. In plants other than the C_4 group the rate of photorespiration is sufficiently high to reduce the net rate of photosynthesis to much less than the maximal value. The efficiency of photosynthesis is much higher in C_4 plants.

SUMMARY

In **photosynthetic** organisms, which include plants and some bacteria, the radiant energy of sunlight is used to drive the biosynthesis of carbohydrate; the overall equation for green plants is $6CO_2 + 6H_2O \rightarrow \text{glucose} + 6O_2$. The initial phase or **light reaction** involves the absorption of quanta of radiant energy by the pigment **chlorophyll**, which occurs in membrane-bounded organelles called **chloroplasts. Accessory pigments** are also present and their absorbed radiant energy is transferred to chlorophyll. Two distinct **photosystems**, I and II, are present. A migration occurs of absorbed radiant energy until it is captured by a specialized chlorophyll molecule, termed P700 for photosystem I and P680 for photosystem II.

The excitation of P700 results in the expulsion of an **electron**, which is transferred via a series of carriers, including **ferredoxin**, to lead ultimately to the reduction of $NADP^+$ to NADPH. The electron lost by photosystem I is replaced by an electron expelled by the excited P680 molecule of photosystem II, which has its own set of pigments. The electron is transferred via a chain of **electron carriers**, which include **plastoquinone, cytochromes**, and **plastocyanin**. The transfer of electrons down the chain is coupled with, and drives, reactions that form ATP from ADP. The electrons lost by photosystem II are replaced by the decomposition of, and withdrawal of electrons from, H_2O to yield O_2.

A cyclic **photophosphorylation** process is also present in which electrons expelled from P700 are not used to reduce $NADP^+$, but are instead shunted to the main electron carrier chain and returned to P700. This process is also coupled to the phosphorylation of ADP.

The light reaction of photosynthesis concludes with the generation of ATP and NADPH. The subsequent reactions are not radiation dependent. In most nontropical plants, CO_2 is incorporated into a precursor of glucose by a **dark reaction** with **ribulose-1, 5-diphosphate** to yield **3-phosphoglycerate**. Six molecules of CO_2 are ultimately converted into glucose by the **Calvin cycle**, whose reactions utilize the ATP and NADPH formed by the light reaction. Some of the reactions of the Calvin cycle are also part of the phosphogluconate and glycolytic pathways.

In some tropical plants, CO_2 is incorporated by the alternative **C_4 pathway**, in which CO_2 is first incorporated into 4-carbon acids in **mesophyll** cells. The **oxaloacetate** initially formed is converted to **malate** or **aspartate**, which are transported to **bundle-sheath** cells and decarboxylated. The CO_2 which is thereby released enters the Calvin cycle.

REFERENCES

The following code is used to classify references. I: particularly useful as an introduction to the subject; R: useful primarily as a reference text; A: an advanced account of the material; H: a publication of historical importance.

General

R. P. F. Gregory, *Biochemistry of Photosynthesis*, Wiley, New York (1971). (A)

D. W. Krogmann, *The Biochemistry of Green Plants*, Prentice-Hall, Englewood Cliffs, New Jersey, (1973). (I)

E. Rabinowitch and Govindjee, *Photosynthesis*, Academic, New York (1969). (A)

D. A. Walker and A. R. Crofts, "Photosynthesis," *Ann. Rev. Biochem.* 89, 389 (1970). (R)

The Primary Phase of Photosynthesis

D. I. Arnon, "The Light Reactions of Photosynthesis," *Proc. Natl. Acad. Sci. U.S.* 68, 2883 (1971). (H)

M. Dienes, *Energy Conversion by the Photosynthetic Apparatus*, Brookhaven Symposia in Biology, vol. 19 (1967). (A)

B. Ke, "The Primary Electron Acceptor of Photosystem I," *Biochim. Biophys. Acta* 301, 1 (1973). (H)

A. Trebet, "Energy Conservation in Photosynthetic Electron Transport of Chloroplasts," *Ann. Rev. Plant Physiol.* 25, 423 (1974). (R)

M. Tribe and P. Whittaker, *Chloroplasts and Mitochondria*, Arnold, London (1972). (A)

The Biosynthesis of Carbohydrates

J. A. Bassham, "The Control of Photosynthetic Carbon Metabolism," *Science* 172, 526 (1971). (A)

O. Bjorkman and J. Berry, "High-Efficiency Photosynthesis," *Sci. Am.* 229, 80 (1973). (I)

M. Calvin and J. A. Bassham, *The Photosynthesis of Carbon Compounds*, W. A. Benjamin, New York (1971). (I)

L. Zelitch, *Photosynthesis, Photorespiration, and Plant Productivity*, Academic, New York (1971). (R)

REVIEW QUESTIONS

11-1 Assuming 100% efficiency, what potential difference is required for the synthesis of one molecule of ATP at pH 7.0 and 25°C?

11-2 P700 in the ground state has a standard potential (E'_o) of about 0.4 V. Could it reduce $NADP^+$ under these conditions? What value of E'_o must the excited P700 pigment attain in order to become a reducing agent for $NADP^+$?

11-3 What are the only truly light-dependent processes in photosystems I and II?

11-4 Much effort has been devoted to developing hybrids between temperate plants, which incorporate CO_2 by the Calvin cycle, and tropical C_4 plants. What is the purpose of these projects?

11-5 Why does the photoionization of P700 increase its effectiveness as a reducing agent?

11-6 Why is photorespiration undesirable from the point of view of the agronomist?

12

The Tricarboxylic Acid Cycle and Oxidative Phosphorylation

12-1 BIOLOGICAL SIGNIFICANCE

We arrive now at a crucial metabolic pathway that recurs in all cells and generates the bulk of the high energy phosphate compounds upon which life depends. This pathway represents a point of **metabolic convergence** into which are fed the products of the initial processing of carbohydrates, fats, and amino acids. These enter the **tricarboxylic acid** (TCA) cycle, whose operation generates the *reduced* forms of the nicotinamide and flavin coenzymes, NADH, $FMNH_2$, and $FADH_2$. The **reoxidation** of these by molecular oxygen is highly exergonic and serves to *drive* the *coupled* reactions that actually synthesize ATP. The reoxidation occurs by an indirect **electron transfer** mechanism.

This will be our first encounter with a pathway that occurs within **organized supramolecular** structures called **mitochondria**. As we shall see, this structural organization is essential for the operation of the pathway.

The TCA cycle also has an important **biosynthetic** function, which includes the formation of intermediates required for the biosynthesis of glucose, in the process termed **gluconeogenesis**.

12-2 OXIDATIVE METABOLISM

The general recognition that biological systems are engaged in processes that have a superficial resemblance to combustion dates from the 18th century. Indeed, the *net* result of the oxidative metabolism by the animal body of such foodstuffs as glucose or a fatty acid is formally equivalent to their complete combustion to carbon dioxide and water. The total energies released are equal. In the case of glucose

$$C_6H_{12}O_6 + 6O_2 \rightarrow 6CO_2 + 6H_2O \quad (\Delta G^{\circ\prime} = -686 \text{ kcal})$$

It is not possible, however, to extend the analogy between combustion and biological oxidation much further. When glucose is burned completely by direct reaction with molecular oxygen, according to the above equation, only a small fraction of the energy released can be diverted to any useful purpose, as most of it is lost in an uncontrolled evolution of heat. In contrast, biological oxidations proceed in a series of steps, each of which releases only a small part of the total energy. The controlled and stepwise nature of the process allows the released energy to be trapped by the synthesis of high energy compounds. These high energy compounds are used to drive the multitude of endergonic reactions whereby living systems can, for a time, seemingly defy the laws of thermodynamics.

The most important of these high energy compounds is ATP, the universal carrier of chemical energy in *usable form*. The maintenance of biological functions is dependent upon the continuous supply of energy in the form of ATP or related compounds.

The synthesis of ATP must itself overcome severe thermodynamic barriers and can only be accomplished by coupling with exergonic reactions. In the cells of organisms that lack the capacity for photosynthesis, this synthesis is achieved by the processes of oxidative metabolism.

The most important source of metabolic energy for animals is the stepwise oxidation of fats and carbohydrates. The oxidative pathways of both nutrients converge in a common terminal phase. It is at this stage that most of the energy made available by biological oxidation is stored as high energy compounds.

Prior to their entry into the terminal pathways of oxidative metabolism, carbohydrate molecules are broken down into 2-carbon fragments, which are supplied as **active acetate (acetyl CoA)**. This is converted into CO_2 and H_2O by a series of stepwise reactions catalyzed by a set of mitochondrial enzymes.

This reaction sequence is variously known as the **Krebs cycle**, the **citric acid cycle**, or the **tricarboxylic acid (TCA) cycle**. The actual generation of ATP is achieved by another set of reactions that are coupled with the TCA cycle and are collectively designated as **oxidative phosphorylation**.

The net result of the TCA cycle is the supply of electrons to the **respiratory chain**, an array of reversibly oxidizable components, which in turn reduce molecular oxygen and drive the reactions of oxidative phosphorylation.

12-3 MITOCHONDRIA

Both the enzymes of the TCA cycle and the oxidative phosphorylation system are localized within the mitochondria (section 1-2). The former do not appear to be incorporated into any supramolecular structure and are released into solution by any physical treatment that disrupts the mitochondrial membrane. In contrast, the enzymes and other proteins involved in oxidative phosphorylation are integrated into the inner mitochondrial membrane.

The mitochondria are ellipsoidal organelles, whose dimensions are typically $0.5 \times 2\ \mu m$. They have the unusual structural feature of possessing a double membrane system (Fig. 12-1). The inner membrane is folded into a set of internal ridges, termed **cristae**. The inner membrane thus divides the interior of the mitochondrion into two spaces. The **intermembrane space** is bounded by the inner and outer membranes, while the **matrix** is enclosed by the inner membrane. The enzymes of the TCA cycle occur in the matrix.

The two membranes differ substantially in properties. While the outer membrane is readily penetrated by most small molecules and ions, the inner membrane has a highly selective permeability. Many molecules traverse it only with the aid of specific protein carriers. This difference in properties reflects a difference in composition. While the outer membrane is about 50% protein, almost 80% of the inner membrane is protein and over 50 different species are present. The balance of both membranes is lipid (section 15-1). The protein constituents of the mitochondrial membranes are generally quite difficult to isolate from the membranes in native form.

The matrix itself is about 50% protein and has a gel-like consistency. The physical state of the matrix is dependent upon the external environment of the mitochondria. The addition of ADP to a suspension of mitochondria causes a pronounced shrinkage of the matrix space and a change in the appearance of the cristae, which become more tightly folded.

The inner surface of the inner membrane is studded with spherical particles, about 9 nm in diameter, which are linked to the membrane surface by a narrow stalk. These particles have proved to be of central importance in the process of oxidative phosphorylation (section 12-7).

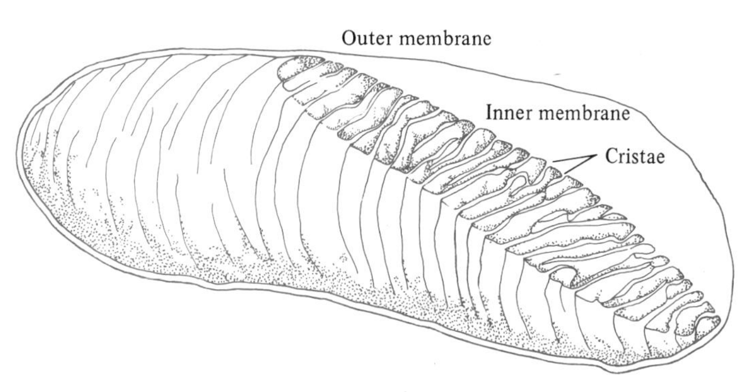

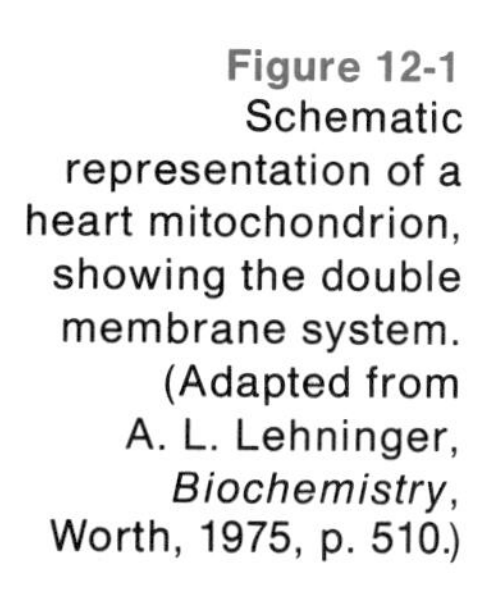
Figure 12-1 Schematic representation of a heart mitochondrion, showing the double membrane system. (Adapted from A. L. Lehninger, *Biochemistry*, Worth, 1975, p. 510.)

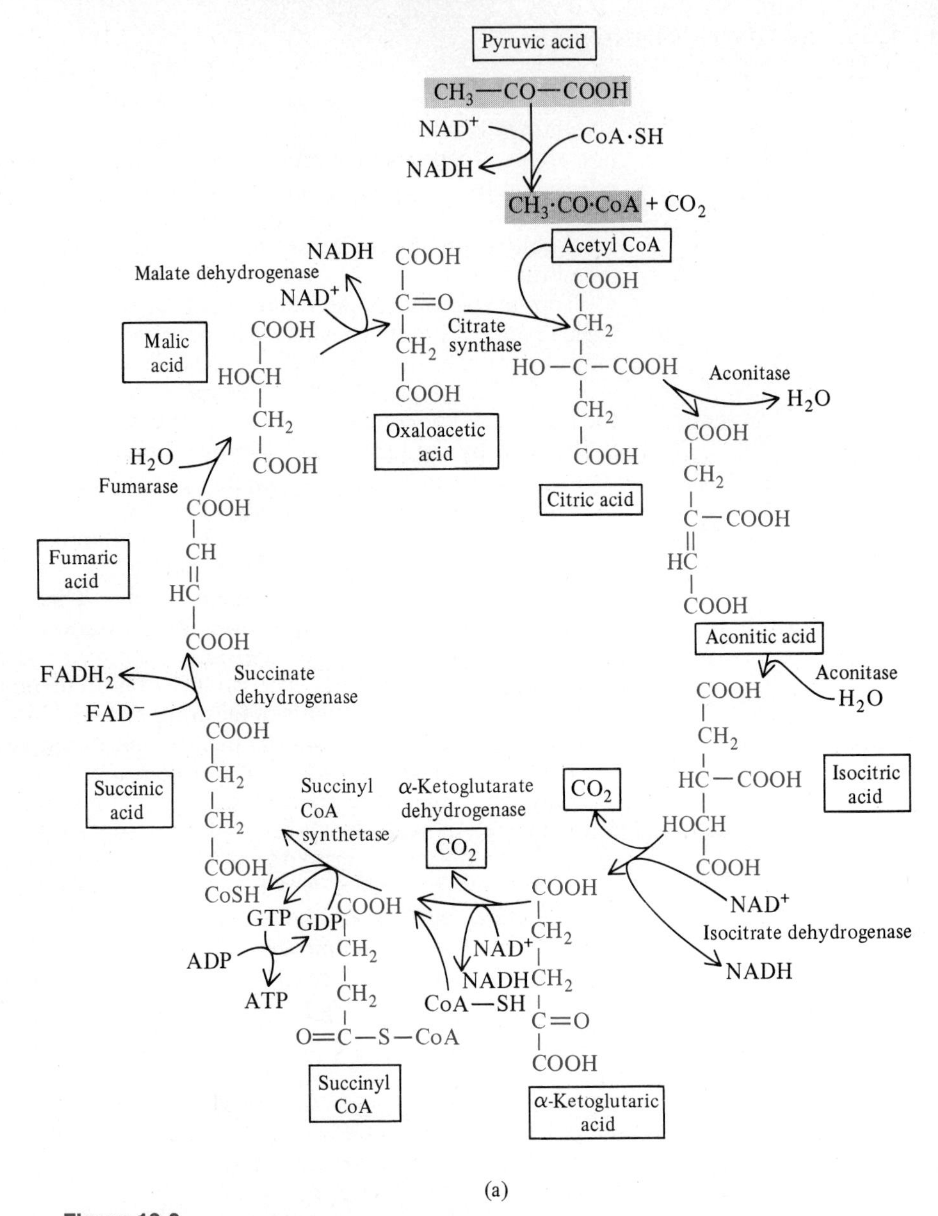

Figure 12-2
The reactions of the TCA cycle and its relation to the respiratory chain. The carboxyl groups of the compounds shown in (a) are ionized under physiological conditions. Not all intermediates are shown.

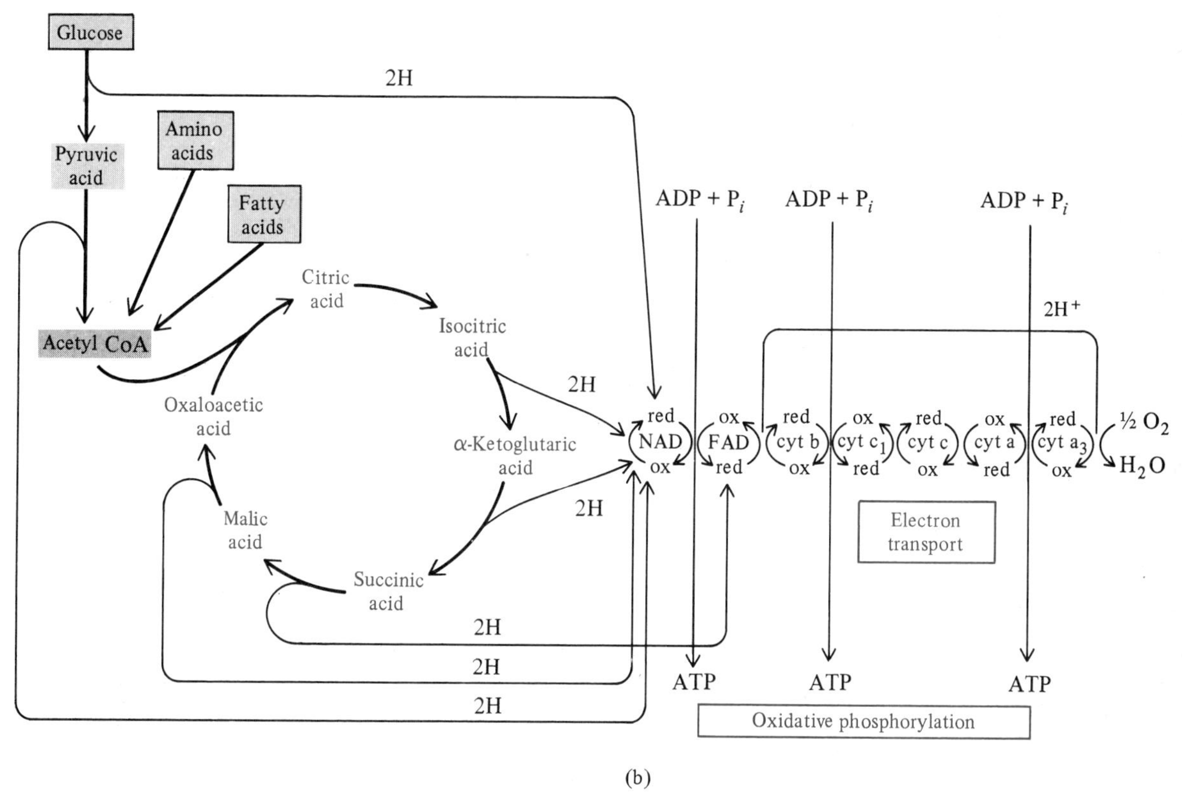

Figure 12.2 *(continued)*

12-4 THE TCA CYCLE

Molecular oxygen does not figure directly in the operation of the TCA cycle. The direct involvement of oxygen occurs only after the transfer of electrons by the complicated oxidation-reduction processes of the respiratory chain. In the formation of CO_2 from acetyl CoA by the TCA cycle, the coenzymes NAD^+ and $NADP^+$ are converted to their reduced forms. It is the *re-oxidation* of these coenzymes by the respiratory chain that yields most of the energy used for the synthesis of ATP.

The initial metabolic conversions of carbohydrates and fats lead ultimately to the formation of acetyl CoA. The TCA cycle (Fig. 12-2) may be said to begin with the condensation of the acetyl group of acetyl CoA with the 4-carbon compound oxaloacetic acid to form the 6-carbon molecule citric acid. By the stepwise removal and recovery of water, citric acid is isomerized to isocitric acid, which undergoes oxidation to oxalosuccinic acid. Oxalosuccinic acid is subsequently decarboxylated to α-ketoglutaric acid. The latter intermediate is converted by oxidative decarboxylation to succinic acid. Succinic acid is then oxidized to fumaric acid, which is converted to malic acid by the addition of water. Finally, malic acid is oxidized to oxaloacetic acid, thereby completing the cycle.

For each turn of the cycle one molecule of active acetate (acetyl CoA) is consumed and two molecules of CO_2 are evolved. The regeneration of oxaloacetic acid in the final step of the cycle permits the process to continue indefinitely, as long as acetyl CoA is supplied and the products are removed.

The TCA cycle generates a pool of common intermediates that relate it to other pathways. Oxaloacetic acid and succinyl CoA are also products of amino acid metabolism (chapter 20). The TCA cycle should not be thought of as set apart from the multitude of other metabolic processes occurring in the mitochondria, but rather as closely integrated with them.

12-5 REACTIONS OF THE TCA CYCLE

Origins of Acetyl CoA

Acetyl CoA or active acetate arises from the metabolism of a variety of foodstuffs. One important source is the metabolism of carbohydrates (chapter 10), which ultimately converts each molecule of glucose into two molecules of pyruvic acid. The pyruvic acid can enter the mitochondria, where it undergoes oxidation, with the simultaneous evolution of CO_2, reduction of NAD^+, and formation of acetyl CoA. The *net* reaction is as follows:

$$\begin{array}{l} CH_3 \\ | \\ C{=}O \\ | \\ COOH \end{array} + \text{CoA—SH} + NAD^+ \longrightarrow CH_3CO\text{—SCoA} + CO_2 + NADH + H^+$$

The complete process is quite complex (Fig. 12-3), involving the action of three different enzymes organized into the **pyruvate dehydrogenase complex**. Pyruvic acid first reacts with thiamine pyrophosphate (TPP), becoming decarboxylated in the process, to yield active acetaldehyde (α-hydroxyethyl thiamine pyrophosphate) in a reaction catalyzed by pyruvate dehydrogenase. The reaction is similar to that occurring in the decarboxylation of pyruvic acid during alcoholic fermentation.

Without leaving the complex, the hydroxyethyl group of active acetaldehyde is next dehydrogenated to form an acetyl group, which is transferred to a sulfhydryl group of the *reduced form* of lipoic acid (Fig. 12-3). This step is catalyzed by **lipoate acetyltransferase**, which has lipoic acid as a covalently attached prosthetic group. The reduction of lipoic acid to dihydrolipoic acid occurs by the transfer of hydrogen atoms from the hydroxyethyl group to the disulfide group of lipoic acid.

In the third step the acetyl group is enzymically transferred to the sulfhydryl group of coenzyme A to form acetyl CoA, which leaves the enzyme complex. Finally, dihydrolipoic acid is reoxidized to lipoic acid via the transfer of hydrogen atoms to the FAD prosthetic group of **lipoamide dehydrogenase**, thereby forming $FADH_2$. The latter is subsequently reoxidized to FAD by NAD^+, while remaining bound by the enzyme.

The pyruvate dehydrogenase complex occurring in heart mitochondria, which has a molecular weight of 7×10^6, consists of a core of 60 lipoate

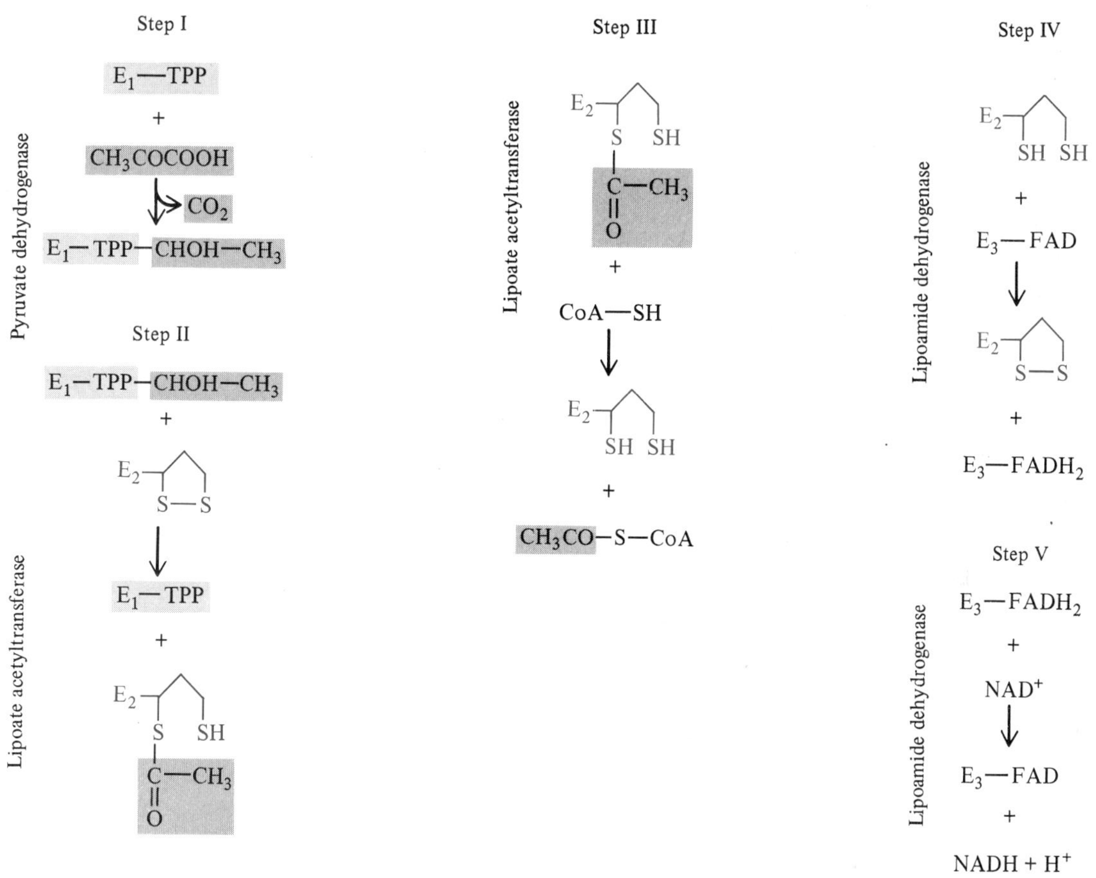

Figure 12-3
Conversion of pyruvic acid to acetyl CoA by the pyruvate dehydrogenase system.

acetyltransferase molecules, to which are attached 20 molecules of pyruvate dehydrogenase and six of lipoamide dehydrogenase, plus two other enzymes which have a regulatory function.

The latter include **pyruvate dehydrogenase kinase**, which catalyzes the phosphorylation by ATP of pyruvate dehydrogenase. The phosphorylated form, which is catalytically inactive, is dephosphorylated and reactivated by **pyruvate dehydrogenase phosphatase**, a Mg^{2+}-dependent enzyme. In this manner the activity of the pyruvate dehydrogenase complex is regulated by the level of ATP. Since the phosphatase is stimulated by Ca^{2+}, the activity of the complex is also sensitive to the concentration of this ion.

The product of the combined operation of the TCA cycle and the oxidative phosphorylation system is ATP. When the level of ATP is high, the pyruvate dehydrogenase is phosphorylated and deactivated, thereby reducing the rate of synthesis of ATP. As the ATP level falls, the pyruvate dehydrogenase is competitively reactivated by dephosphorylation and the rate of formation of ATP rises again.

Formation of Citric Acid

The first reaction of the TCA cycle proper is the combination of acetyl CoA with oxaloacetic acid, which is catalyzed by **citrate synthase**:

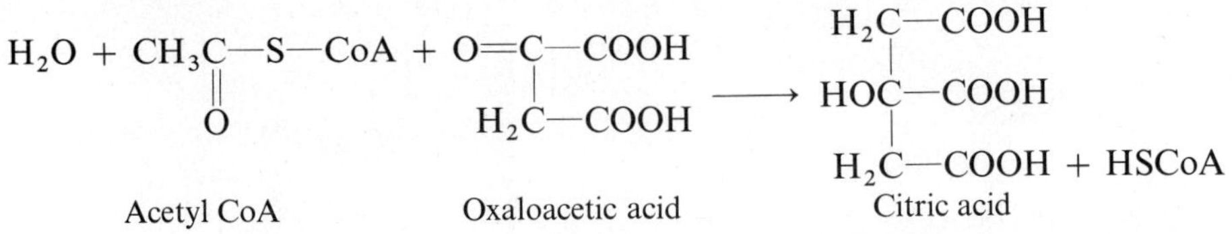

$$H_2O + \underset{\text{Acetyl CoA}}{CH_3\overset{}{\underset{\underset{O}{\|}}{C}}—S—CoA} + \underset{\text{Oxaloacetic acid}}{\begin{array}{l} O{=}C—COOH \\ \quad | \\ H_2C—COOH \end{array}} \longrightarrow \underset{\text{Citric acid}}{\begin{array}{l} H_2C—COOH \\ \quad | \\ HOC—COOH \\ \quad | \\ H_2C—COOH \end{array}} + HSCoA$$

This reaction is strongly exergonic; under physiological conditions $\Delta G^{\circ\prime} = -7.8$ kcal.

Conversion of Citric to Isocitric Acid

The reversible transformation of citric acid to *cis*-aconitic acid, and of the latter to isocitric acid, is catalyzed by a single enzyme, **aconitase**; the enzyme requires as cofactors Fe^{2+} and cysteine or glutathione.

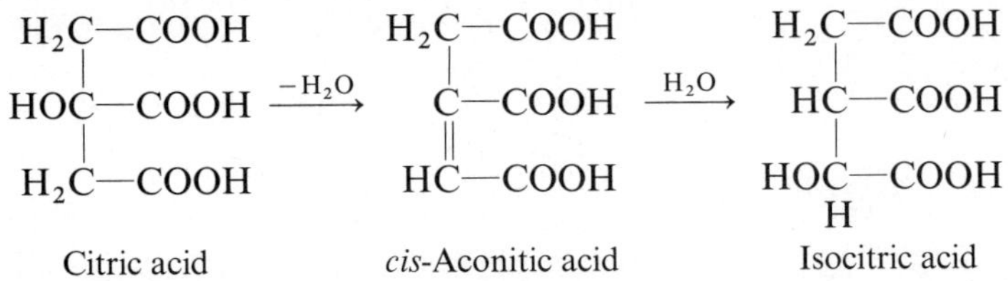

$$\underset{\text{Citric acid}}{\begin{array}{l} H_2C—COOH \\ \quad | \\ HOC—COOH \\ \quad | \\ H_2C—COOH \end{array}} \xrightarrow{-H_2O} \underset{\textit{cis}\text{-Aconitic acid}}{\begin{array}{l} H_2C—COOH \\ \quad | \\ \;\; C—COOH \\ \quad \| \\ \;HC—COOH \end{array}} \xrightarrow{H_2O} \underset{\text{Isocitric acid}}{\begin{array}{l} H_2C—COOH \\ \quad | \\ \;HC—COOH \\ \quad | \\ HOC—COOH \\ \quad H \end{array}}$$

While the mixture at thermodynamic equilibrium would contain a preponderance of citric acid, under physiological conditions isocitric acid is continually removed, so that the reaction proceeds to the right. *cis*-Aconitate normally remains combined with the enzyme during the reaction.

Formation of α-Ketoglutaric Acid

Isocitric acid is next converted to α-ketoglutaric acid by a reaction involving the reduction of NAD^+ or $NADP^+$. Oxalosuccinic acid is formed as an intermediate and is subsequently decarboxylated to yield α-ketoglutaric acid. The entire reaction is catalyzed by **isocitrate dehydrogenase**:

$$\underset{\text{Isocitric acid}}{\begin{array}{l} H_2C—COOH \\ \quad | \\ \;HC—COOH \\ \quad | \\ HOC—COOH \\ \quad H \end{array}} + NAD^+ \longrightarrow NADH + \underset{\text{Oxalosuccinic acid}}{\begin{array}{l} H_2C—COOH \\ \quad | \\ \;HC—COOH \\ \quad | \\ O{=}C—COOH \end{array}} \longrightarrow \underset{\alpha\text{-Ketoglutaric acid}}{\begin{array}{l} H_2C—COOH \\ \quad | \\ H_2C \\ \quad | \\ O{=}C—COOH \end{array}} + CO_2$$

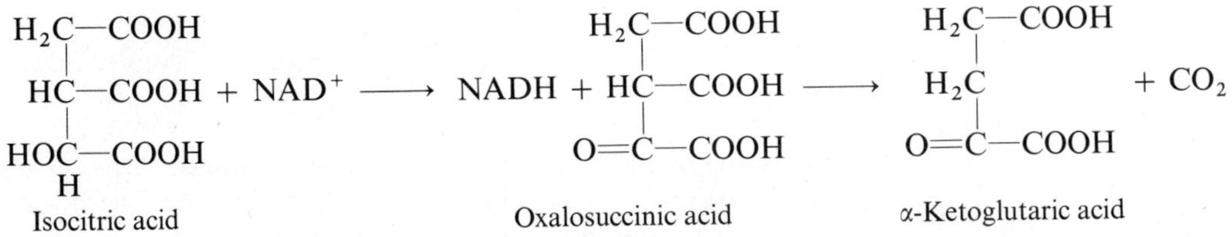

The complete reaction is strongly exergonic. Under physiological conditions $\Delta G^{\circ\prime}$ is about -5 kcal. It should be noted that the CO_2 released in this reaction (and in the subsequent conversion of α-ketoglutaric acid to succinyl CoA) is derived from **oxaloacetate**.

Both an NAD^+- and an $NADP^+$-dependent isocitrate dehydrogenase occur in mitochondria; the former is probably the enzyme of primary importance in the TCA cycle. The NAD^+-dependent isocitrate dehydrogenase consists of eight identical subunits of total molecular weight 380,000. The enzyme is allosterically stimulated by ADP and requires Mg^{2+} for activity.

The allosteric enhancement of activity by ADP is important for the regulation of the enzyme. The depletion of the ATP supply of the cell as a result of the demands of energy-requiring processes is accompanied by a rise in the ADP level. This, in turn, stimulates the action of isocitrate dehydrogenase and the overall activity of the TCA cycle, thereby tending to restore the ATP level. ATP and NADH are allosteric inhibitors.

Conversion of α-Ketoglutaric Acid to Succinic Acid

α-Ketoglutaric acid represents an important junction between the metabolic pathways of carbohydrates and amino acids. Glutamic acid may be converted to α-ketoglutaric acid by transamination or oxidation (chapter 20). Several other amino acids are interconvertible with glutamic acid and hence are potential precursors of α-ketoglutaric acid.

The formation of succinic acid from α-ketoglutaric acid proceeds by a rather complicated set of reactions (Fig. 12-4) that are quite analogous to those involved in the oxidation of pyruvic acid. In the initial phase α-ketoglutaric acid is oxidized and transferred as a succinyl group to coenzyme A. The

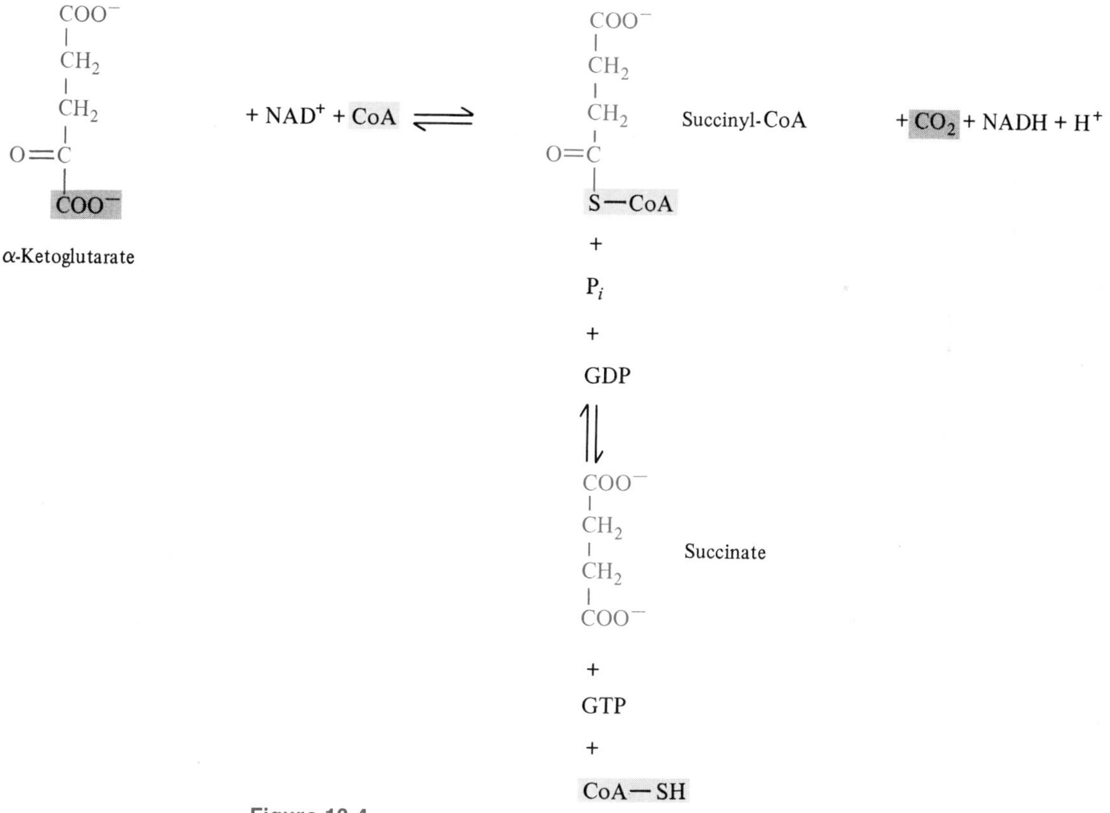

$$\underset{\alpha\text{-Ketoglutarate}}{^{-}OOC{-}CH_2{-}CH_2{-}C({=}O){-}COO^{-}} + NAD^{+} + CoA \rightleftharpoons \underset{\text{Succinyl-CoA}}{^{-}OOC{-}CH_2{-}CH_2{-}C({=}O){-}S{-}CoA} + CO_2 + NADH + H^{+}$$

$$\text{Succinyl-CoA} + P_i + GDP \rightleftharpoons \underset{\text{Succinate}}{^{-}OOC{-}CH_2{-}CH_2{-}COO^{-}} + GTP + CoA{-}SH$$

Figure 12-4
Formation of succinic acid from α-ketoglutaric acid.

reaction is catalyzed by the **α-ketoglutarate dehydrogenase** system.

$$\begin{array}{l} \text{COOH} \\ | \\ \text{CH}_2 \\ | \\ \text{CH}_2 \\ | \\ \text{C=O} \\ | \\ \text{COOH} \\ \alpha\text{-Ketoglutaric acid} \end{array} + \text{HS—CoA} + \text{NAD}^+ \longrightarrow \begin{array}{l} \text{COOH} \\ | \\ \text{CH}_2 \\ | \\ \text{CH}_2 \\ | \\ \text{C=O} \\ | \\ \text{S—CoA} \\ \text{Succinyl CoA} \end{array} + \text{NADH} + \text{H}^+ + \text{CO}_2$$

The reaction is strongly exergonic, with a value of $\Delta G^{\circ\prime}$ equal to -8 kcal. As in the pyruvate dehydrogenase case, thiamine pyrophosphate, lipoic acid, FAD, and NAD^+ participate as coenzymes. Without leaving the surface of the enzyme complex, the succinyl group is transferred to lipoic acid and then to coenzyme A.

Succinyl CoA is next converted to succinic acid in a reaction with guanosine diphosphate (GDP) and inorganic phosphate.

$$\text{succinyl CoA} + \text{GDP} + \text{P}_i \rightarrow \text{succinic acid} + \text{GTP} + \text{CoA} \quad (\Delta G^{0\prime} = -0.7 \text{ kcal})$$

The reaction is catalyzed by **succinyl CoA synthetase**. A covalent phosphoenzyme is formed as an intermediate:

$$\begin{aligned} \text{enzyme} + \text{P}_i + \text{succinyl CoA} &\rightarrow \text{enzyme-succinyl phosphate} + \text{CoA} \\ \text{enzyme-succinyl phosphate} &\rightarrow \text{enzyme-phosphate} + \text{succinic acid} \\ \text{enzyme-phosphate} + \text{GDP} &\rightarrow \text{enzyme} + \text{GTP} \end{aligned}$$

It is believed that succinyl phosphate, a mixed anhydride of the two acids, is formed on the active site and that its phosphate group is subsequently transferred to the imidazole ring of a histidine group of the enzyme.

The terminal phosphate of GTP is transferred to ADP in a reaction catalyzed by a *nucleoside diphosphate kinase.*

$$\text{GTP} + \text{ADP} \rightarrow \text{GDP} + \text{ATP}$$

Formation of Fumaric Acid

Succinic acid is oxidized to fumaric acid by the action of *succinate dehydrogenase.* This enzyme is a flavoprotein containing covalently bound flavin adenine dinucleotide (FAD). Unlike the other enzymes involved in the TCA cycle, it is tightly combined with the inner mitochondrial membrane and is difficult to isolate in soluble form. Succinate dehydrogenase obtained from beefheart mitochondria has a molecular weight close to 100,000 and contains one FAD molecule and eight iron atoms. Its FAD unit is reduced to $FADH_2$ (section 7-3) in the reaction:

$$\text{enzyme—FAD} + \begin{array}{l} \text{COOH} \\ | \\ \text{CH}_2 \\ | \\ \text{CH}_2 \\ | \\ \text{COOH} \\ \text{Succinic acid} \end{array} \longrightarrow \begin{array}{c} \text{HC—COOH} \\ \| \\ \text{HOOC—CH} \\ \text{Fumaric acid} \end{array} + \text{enzyme—FADH}_2$$

Succinate dehydrogenase is a member of the respiratory chain, having a close relationship with its other components. The reduced enzyme can donate electrons to an acceptor within the system.

Formation of Malic and Oxaloacetic Acids

Fumaric acid is converted to malic acid by the addition of water. The reaction is catalyzed by **fumarase**:

$$\begin{array}{c} HC-COOH \\ \| \\ HOOC-CH \end{array} + H_2O \rightleftharpoons \begin{array}{c} H_2C-COOH \\ | \\ HOC-COOH \\ H \end{array}$$

Malic acid

The reaction is freely reversible, with a value of $\Delta G^{\circ\prime}$ close to zero.

Fumarase consists of four polypeptide chains, with a total molecular weight of 200,000. Its action is stereospecific, only the L-isomer of malic acid being formed.

In the final reaction of the TCA cycle, malic acid is oxidized to oxaloacetic acid by malate dehydrogenase and NAD^+.

$$\begin{array}{c} H_2C-COOH \\ | \\ HOC-COOH \\ H \end{array} + NAD^+ \longrightarrow \begin{array}{c} H_2C-COOH \\ | \\ O=C-COOH \end{array} + NADH + H^+ \quad (\Delta G^{\circ\prime} = 7.1 \text{ kcal})$$

The oxaloacetic acid formed can combine with acetyl CoA for another turn of the cycle.

Net Result of the TCA Cycle

Four of the reactions of the TCA cycle are dehydrogenations. In three of these NAD^+ is converted to NADH; there are isocitrate → oxalosuccinate; α-ketoglutarate → succinyl CoA; and malate → oxaloacetate.

For the cycle to turn indefinitely in the absence of an oxygen supply, NAD^+ would have to be made continuously available. In actuality the coenzyme is only present in catalytic amounts and is constantly regenerated by reoxidation. This is achieved by the respiratory chain, whose action results in the reconversion of NADH to NAD^+ and in the formation of H_2O through the transfer of hydrogen to molecular oxygen.

One revolution of the TCA cycle releases little energy in the form of ATP. Almost 90% of the total chemical energy released is due to the subsequent operations of the respiratory chain and the formation of water.

The TCA cycle is by no means merely a degradative pathway. In addition to accepting the products of fat, carbohydrate, and protein metabolism, it is involved in the synthesis of glucose and several amino acids and provides a general pool of intermediate products.

Regulation of the TCA Cycle

The operation of the TCA cycle is subject to control at several points. The activity of the pyruvate dehydrogenase complex, which provides acetyl CoA for the cycle, is abolished by ATP-dependent phosphorylation of the dehydrogenase and restored by dephosphorylation. An excessively high level of ATP thus depresses pyruvate dehydrogenase activity and reduces the supply of acetyl CoA for the TCA cycle, thereby slowing the operation of the cycle. As the ATP concentration falls, the dehydrogenase is competitively dephosphorylated and activity is recovered.

However, it is the citrate synthase-catalyzed reaction that is the chief pacemaker of the TCA cycle. The velocity of this reaction is primarily governed by the concentrations of its substrates, acetyl CoA and oxaloacetate, and by the level of succinyl CoA, which is a competitive inhibitor. An undue accumulation of succinyl CoA, a later intermediate of the cycle, reduces citrate synthase activity and decreases the overall rate of the cycle. This is another example of feedback control. Citrate synthase is an allosteric enzyme and is inhibited by ATP and NADH.

The NAD^+-dependent isocitrate dehydrogenase reaction is also subject to regulation. This enzyme has a positive requirement for ADP as an allosteric activator; its activity is stimulated as the balance between ATP and ADP shifts in favor of the latter. Finally, the succinate dehydrogenase reaction is sensitive to a number of modifiers, including the promoters ATP and succinate and especially the powerful inhibitor oxaloacetate.

The existence of this complex set of controls insures that the TCA cycle will be sensitively responsive to the needs of the organism and that it will avoid the wasteful accumulation of intermediates in excess of need.

12-6 METABOLIC PATHWAYS LINKED TO THE TCA CYCLE

Synthesis of Glucose from Noncarbohydrate Precursors

In animals the primary site for **gluconeogenesis** or the formation of glucose from noncarbohydrate precursors is the liver. The process does not occur to an important extent in muscle or brain, which have a requirement for glucose. Gluconeogenesis helps to maintain a glucose level in the blood that is adequate for the needs of the latter two organs.

The initial reaction is the stepwise formation of phosphoenolpyruvate from pyruvate generated by glycolysis. Pyruvate is first carboxylated to oxaloacetate by the action of **pyruvate carboxylase**, a biotin enzyme:

$$\text{pyruvic acid} + CO_2 + \text{ATP} + H_2O \rightarrow \text{oxaloacetate} + \text{ADP} + P_i + 2H^+$$

Oxaloacetate is next decarboxylated and phosphorylated to form phosphoenolpyruvate in a reaction catalyzed by **phosphoenolpyruvate carboxykinase**:

$$\text{oxaloacetate} + \text{GTP} \rightarrow \text{phosphoenolpyruvate} + \text{GDP} + CO_2$$

The sum of these reactions is

$$\text{pyruvate} + \text{ATP} + \text{GTP} + H_2O \rightleftarrows \text{phosphoenolpyruvate} + \text{ADP} + \text{GDP} + P_i + 2H^+ \qquad (\Delta G^{\circ\prime} = 0.2 \text{ kcal})$$

The net reaction is endergonic, but only barely so, and is thermodynamically feasible. Both stepwise reactions are driven at the expense of a high energy phosphate bond. It should be noted that the formation of phosphoenolpyruvate does not occur by reversal of the pyruvate kinase-catalyzed reaction of the glycolytic pathway, which would be very unfavorable thermodynamically ($\Delta G^{\circ\prime} = 7.5$ kcal).

The phosphoenolpyruvate that is formed from pyruvate by the above reactions is converted to fructose-1,6-diphosphate by retracing the intervening steps of the glycolytic pathway, each of which is adequately reversible.

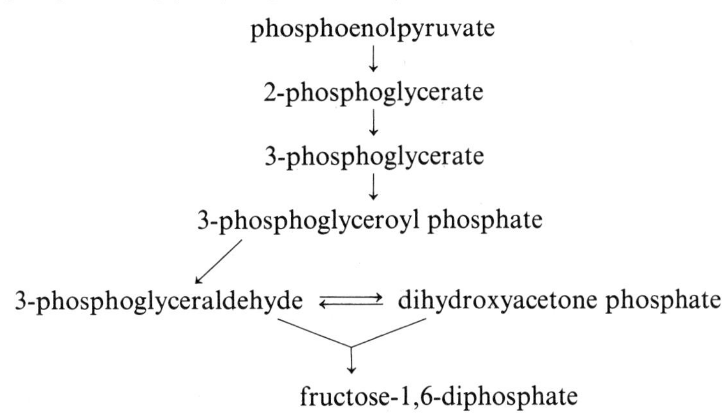

Fructose-1,6-diphosphate is hydrolyzed to fructose-6-phosphate by **fructose diphosphatase**:

$$\text{fructose-1,6-diphosphate} + H_2O \rightarrow \text{fructose-6-phosphate} + P_i$$

This reaction is virtually irreversible; the value of $\Delta G^{\circ\prime}$ is -4.0 kcal. This hydrolytic step bypasses the thermodynamically unfavored reversal of the corresponding glycolytic step, namely, the phosphorylation of fructose-6-phosphate by phosphofructokinase.

Fructose-6-phosphate is subsequently reversibly converted to glucose-6-phosphate by a reversal of the glycolytic reaction, which is catalyzed by glucosephosphate isomerase.

$$\text{fructose-6-phosphate} \rightleftarrows \text{glucose-6-phosphate}$$

Finally, glucose-6-phosphate is hydrolyzed to glucose by glucose-6-phosphatase:

$$\text{glucose-6-phosphate} + H_2O \rightarrow \text{glucose} + P_i \quad (\Delta G^{\circ\prime} = -3.3 \text{ kcal})$$

This reaction is another instance of the bypassing of an endergonic reversal of a glycolytic reaction. In this case the formation of glucose does not occur by a reversal of the hexokinase-catalyzed reaction, but rather by an alternative hydrolytic detour, which is thermodynamically more feasible. Gluconeogenesis retraces the readily reversible steps of the glycolytic sequence and circumvents those that are essentially irreversible.

Relationship of Gluconeogenesis to Other Pathways

The involvement of oxaloacetate in the initial steps of gluconeogenesis provides a link between this process and the TCA cycle, since oxaloacetate lies on the latter pathway. All of the intermediates of the TCA cycle may thus function as precursors of glucose.

Although oxaloacetate is formed within the mitochondria by the action of pyruvate carboxylase, the other reactions of gluconeogenesis occur in the cytoplasm. Oxaloacetate itself does not cross the mitochondrial membrane.

Its transport to the cytoplasm requires its reduction to malate by NADH in a reaction catalyzed by the *mitochondrial* form of malate dehydrogenase:

$$\text{oxaloacetate} + \text{NADH} \rightarrow \text{malate} + \text{NAD}^+$$

The malate thereby formed can cross the membrane into the cytoplasm, where its reoxidation to oxaloacetate is catalyzed by the *cytoplasmic* form of malate dehydrogenase

$$\text{malate} + \text{NAD}^+ \rightarrow \text{oxaloacetate} + \text{NADH}$$

Since the metabolism in higher animals of many amino acids converts them to either pyruvate or an intermediate of the TCA cycle, these amino acids are also precursors of glucose.

Gluconeogenesis and Glycolysis

Gluconeogenesis requires a supply of pyruvate. In animals the primary source is the lactate and pyruvate produced by the glycolysis occurring in active skeletal muscle. Both lactate and pyruvate can readily penetrate cell membranes and enter the blood for transport to the liver. While in skeletal muscle the ratio of NADH to NAD^+ is such so as to favor the formation of lactate at the expense of pyruvate by lactate dehydrogenase, in liver the reverse is true; lactate entering the liver is oxidized to pyruvate. Pyruvate may then enter the pathway of gluconeogenesis in liver. The glucose thereby produced returns to the skeletal muscle via the blood and re-enters the glycolytic pathway, which yields ATP.

In this way the lactate generated by skeletal muscle, which would otherwise be a metabolic dead end, is diverted to a useful purpose. This is another example of how the metabolism occurring in liver complements that of skeletal muscle. The complete process may be visualized as a cycle of the form:

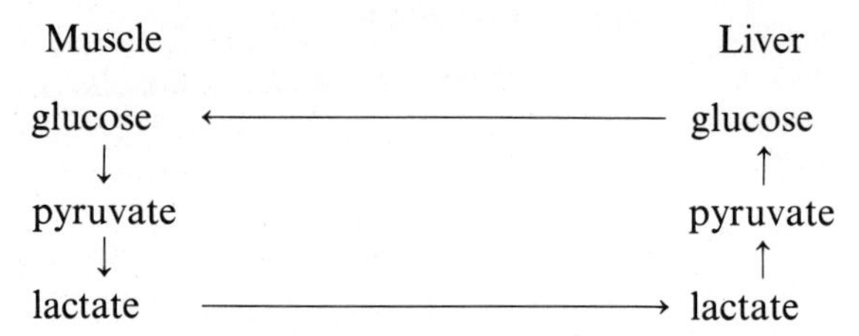

The Glyoxylate Cycle

In many plants and microorganisms a modified version of the TCA cycle occurs, whose function is to enable these organisms to carry out the biosynthesis of carbohydrate from acetyl CoA derived from fatty acids. In the **glyoxylate** cycle, those steps of the TCA cycle that involve the formation of CO_2 are absent. The two carbon atoms of the acetyl group of acetyl CoA are thereby conserved for incorporation into carbohydrate.

The initial stages of the glyoxylate cycle are equivalent to those of the TCA cycle, as far as the formation of isocitrate. However, at this point the reactions that, in the TCA cycle, lead to the formation of α-ketoglutarate are replaced by the cleavage of isocitrate by **isocitrate lyase** to yield succinate and glyoxylate ($\text{H—}\overset{\overset{\text{O}}{\|}}{\text{C}}\text{—COO}^-$). Glyoxylate then combines with a second molecule of acetyl CoA to form malate, in a reaction catalyzed by **malate synthase**. Malate is then oxidized by malate dehydrogenase to oxaloacetate, which may condense with acetyl CoA to initiate another turn of the cycle:

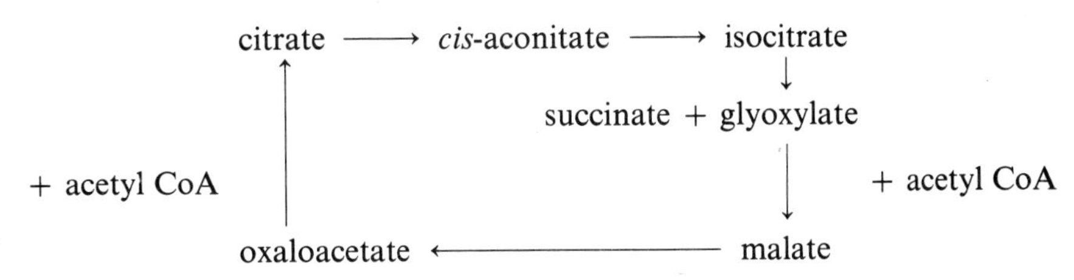

Each turn of the glyoxylate cycle consumes two molecules of acetyl CoA and produces one molecule of succinate. Succinate is an intermediate of the TCA cycle and may serve as a precursor in gluconeogenesis.

In plants and microorganisms, the glyoxylate and TCA cycles coexist and share the same enzymes with two exceptions. Isocitrate lyase and malate synthase are located not in the mitochondria, but in specialized cytoplasmic particles termed **glyoxysomes**.

The glyoxylate cycle, which is not present in higher animals, is particularly important in plant seeds, which are thereby enabled to utilize acetyl groups obtained from the oxidation of fats for the synthesis of carbohydrate via succinate and gluconeogenesis.

12-7 THE RESPIRATORY CHAIN AND OXIDATIVE PHOSPHORYLATION

The Respiratory Chain

The factors involved in the respiratory chain and oxidative phosphorylation occur in the mitochondria and appear to be part of a highly organized system, which is integrated into the mitochondrial membrane. The reoxidation of NADH does not occur by direct reaction of the reduced coenzyme with molecular oxygen, but rather through a sequence of intermediate steps. The release of chemical energy is also stepwise. A portion of the chemical energy released is used to synthesize ATP from ADP and inorganic phosphate. The stepwise nature of the process gives the cell better control over the stepwise release of energy and permits more efficient capture of released energy in the form of ATP. The overall results of the operation of the respiratory chain and the associated processes of oxidative phosphorylation may be summarized as follows:

(1) NADH is reoxidized to NAD^+.
(2) The hydrogen atoms removed from NADH are combined with oxygen to form water.
(3) ATP is formed from ADP and P_i.

The respiratory chain may be visualized of as a series of reversibly oxidizable components of graded redox potentials. The NADH produced by the TCA cycle is reoxidized by the FMN-containing flavoprotein **NADH dehydrogenase**, which is the initial element of the respiratory chain (Fig. 12-2).

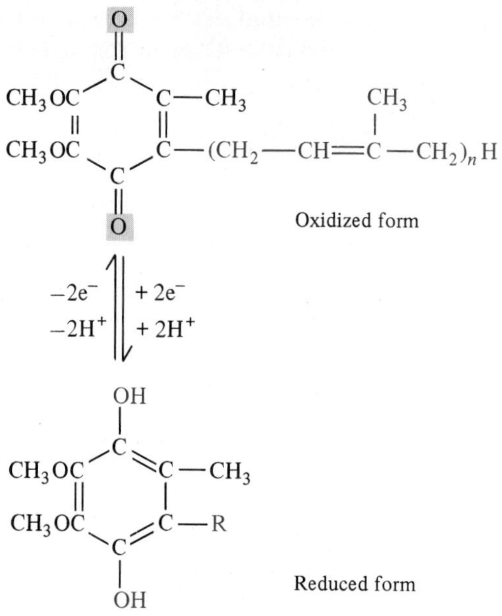

CH_3OC $C-CH_3$ CH_3

CH_3OC $C-(CH_2-CH=C-CH_2)_nH$

Oxidized form

$-2e^-$ $-2H^+$ $+2e^-$ $+2H^+$

OH

CH_3OC $C-CH_3$

CH_3OC $C-R$

OH

Reduced form

Figure 12-5 Oxidized and reduced forms of coenzyme Q. In mammals, *n* is usually equal to 10.

An iron- and sulfur-containing protein, followed by coenzyme Q (Fig. 12-5), occur next in the sequence.

Other flavoproteins, including succinate dehydrogenase and **acyl-CoA dehydrogenase** (not a component of the TCA cycle), contribute reducing equivalents to the respiratory chain via ubiquinone (coenzyme Q). Coenzyme Q is the last element of the respiratory chain whose reduction and reoxidation involve the transfer of hydrogen (Fig. 12-5). Thereafter, only electrons are transferred down the remainder of the chain.

The remaining components of the respiratory chain consist of a series of **cytochromes**, which are **heme**-containing proteins, plus two or more iron-sulfer proteins. The heme prosthetic groups of the cytochromes (section 17-2) each contain an iron atom, which may exist in an oxidized Fe(III) or reduced Fe(II) state. The passage of electrons down the respiratory chain is accompanied by the reversible transition of the cytochromes between the two states. The spatial organization of the respiratory chain components is important for this process.

The respiratory chain terminates with the transfer of electrons from cytochromes aa_3 to molecular oxygen and the combination of the latter with hydrogen ions to form water (Fig. 12-6 and Table 12-1).

Several inhibitors have been found to block electron transport in the respiratory chain. Rotenone, a plant toxin, inhibits NADH dehydrogenase. The antibiotic antimycin A blocks electron transport from cytochrome b to cytochrome c_1. Hydrogen cyanide and carbon monoxide block the final step of electron transfer from cytochrome aa_3 to molecular oxygen.

Several of the stages of the respiratory chain involve the reaction or liberation of hydrogen ions (Table 12-1). It is now believed that this may be an important factor in the related processes of oxidative phosphorylation.

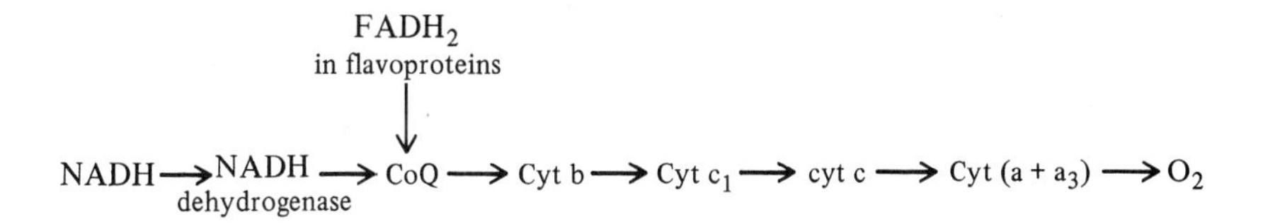

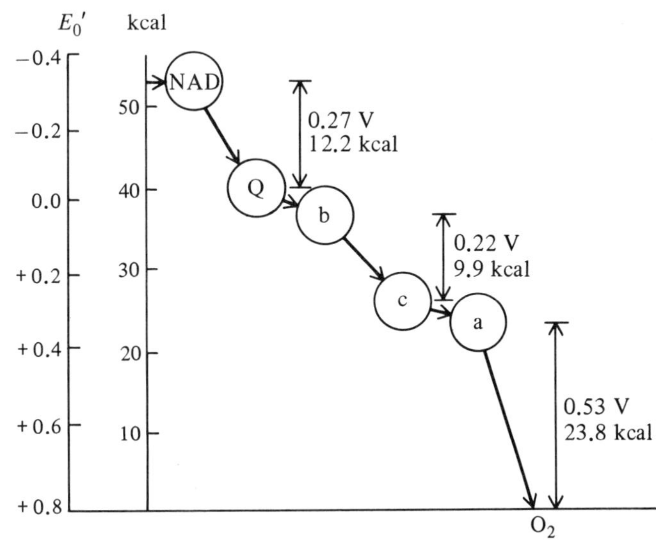

Figure 12-6
Constituents of the respiratory chain and their approximate electrode potentials. Not all components are shown here. The incompletely characterized iron-sulfur proteins are not included.

Table 12-1

Reactions of the Respiratory Chain

$FMN + NADH + H^+ \rightleftharpoons FMNH_2 + NAD^+$
$FMNH_2 + 2Fe(III)\cdot protein \rightleftharpoons FMN + 2Fe(II)\cdot protein + 2H^+$
$2Fe(II)\cdot protein + 2H^+ + Q \rightleftharpoons 2Fe(III)\cdot protein + QH_2$
$QH_2 + 2$ cytochrome b · Fe(III) $\rightleftharpoons$ Q + 2 cytochrome b · Fe(II) + $2H^+$
2 cytochrome b · Fe(II) + 2 cytochrome c · Fe(III) $\rightleftharpoons$ 2 cytochrome b · Fe(III) + 2 cytochrome c · Fe(II)
2 cytochrome c · Fe(II) + 2 cytochrome a · Fe(III) $\rightleftharpoons$ 2 cytochrome c · Fe(III) + 2 cytochrome a · Fe(II)
2 cytochrome a · Fe(II) + 2 cytochrome a_3 · Fe(III) $\rightleftharpoons$ 2 cytochrome a · Fe(III) + 2 cytochrome a_3 · Fe(II)
2 cytochrome a_3 · Fe(II) + $\frac{1}{2}O_2$ + $2H^+$ $\rightleftharpoons$ H_2O + 2 cytochrome a_3 · Fe(II)

In this table **FMN** represents the prosthetic group of the flavoprotein NADH dehydrogenase; **Fe(III)·protein** and **Fe(II)·protein** stand for the oxidized and reduced forms of the prosthetic group of an iron-sulfur protein, respectively; and **Q** and **QH_2** represent the oxidized and reduced forms of coenzyme Q, respectively.

Oxidative Phosphorylation

The transfer of electrons down the respiratory chain from NADH to molecular oxygen is the primary source of the energy utilized for the formation of ATP by the coupled phosphorylation of ADP. For each molecule of NADH reoxidized by the respiratory chain, three molecules of ATP are formed from ADP and inorganic phosphate. This suggests that, at three loci in the sequence of electron carriers comprising the respiratory chain, a portion of the energy liberated in oxidation-reduction reactions is diverted to drive the endergonic phosphorylation of ADP.

The net result of the linked processes of electron transport and oxidative phosphorylation may be written

$$\text{NADH} + \text{H}^+ + 3\text{ADP} + 3\text{P}_i + \tfrac{1}{2}\text{O}_2 \rightarrow \text{NAD}^+ + 3\text{ATP} + 4\text{H}_2\text{O}$$

While the complete reaction is exergonic ($\Delta G^{\circ\prime} = -30.8$ kcal), the phosphorylation of ADP is endergonic:

$$3\text{ADP} + 3\text{P}_i \rightarrow 3\text{ATP} + 3\text{H}_2\text{O} \qquad (\Delta G^{\circ\prime} = 21.9 \text{ kcal})$$

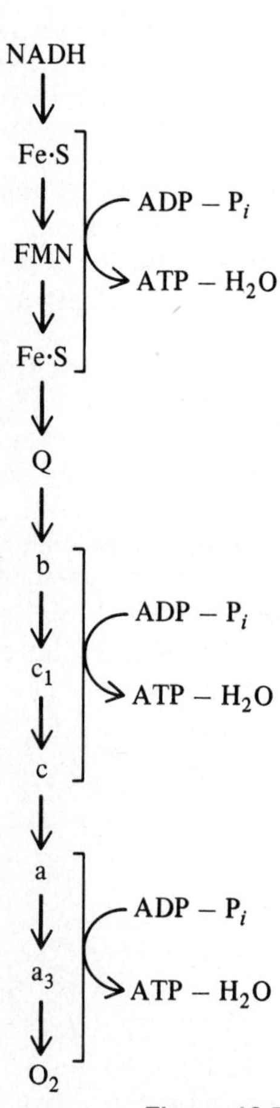

Figure 12-7 Probable location of the sites of oxidative phosphorylation with respect to the elements of the respiratory chain.

The location of the three segments of the respiratory chain associated with ADP phosphorylation (Fig. 12-7) has been suggested by consideration of the magnitudes of the decrease in free energy for the various steps of the sequence of oxidation-reduction reactions. At three points in this series of reactions the stepwise decrease in free energy is large enough theoretically to furnish the energy required to drive the phosphorylation of a single molecule of ADP to ATP. These three sites (Fig. 12-7) occur between NADH and ubiquinone, between cytochrome b and cytochrome c, and between cytochrome a and molecular oxygen. This tentative identification has been confirmed by direct observation of fragments of the respiratory chain system.

Studies of the molecular mechanism of oxidative phosphorylation have been hampered by the existence of the various protein components as a highly organized system that is integrated into the inner mitochondrial membrane. Attempts to examine the individual steps have generally employed indirect methods. One valuable experimental approach has involved the use of specific inhibitors to separate the individual reactions. These include **uncoupling reagents**, which block the phosphorylation of ADP but do not interfere with electron transport. They thus *uncouple* the two processes. This class of inhibitors, of which 2,4-dinitrophenol is an example, typically stimulate the rate of oxygen consumption by intact mitochondria in the absence of ADP. Uncoupling agents are believed to function by causing the deactivation of some form of high energy intermediate or state generated by electron transport.

A group of antibiotics called **ionophores**, block oxidative phosphorylation in the presence of certain monovalent cations. An example is **valinomycin**, which requires K^+ for inhibitory activity. The mechanism of inhibition is different from that of the uncoupling agents and stems from the ability of the ionophores to form lipid-soluble complexes with monovalent cations and thereby facilitate their transit across the inner mitochondrial membrane. In the case of valinomycin, the complex formed with K^+ readily passes through the membrane out of the matrix, thereby compelling the diversion of the energy generated by electron transport to pump the K^+ cations back into the matrix.

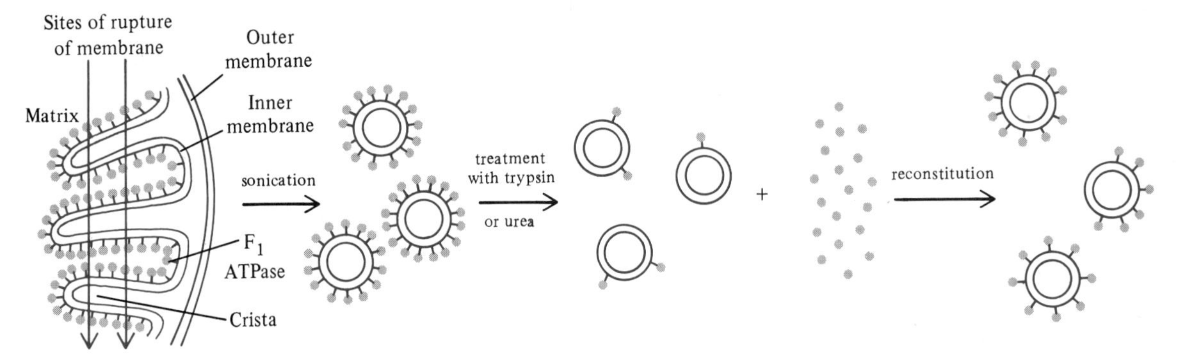

Figure 12-8
Formation of reconstituted vesicles from the mitochondrial inner membrane. These retain the inner membrane spheres and can carry out oxidative phosphorylation. (Adapted from A. L. Lehninger, *Biochemistry,* Worth, 1975).

Intact mitochondria have the capacity to hydrolyze ATP. This **ATPase** activity is normally quite low, but is greatly increased in the presence of uncoupling agents, such as 2,4-dinitrophenol. The ATPase reaction represents the reverse of the process whereby ATP is formed by the phosphorylation of ADP.

It is possible by various chemical and physical treatments, including exposure to **digitonin** and sonic irradiation, to obtain mitochondrial fragments that can carry out both electron transport and oxidative phosphorylation. These sealed **vesicles** probably arise from the pinching off and resealing of cristae during the disruption of mitochondria (Fig. 12-8). Their outer surface retains the spherical globules observed in intact mitochondria (section 12-2) on the inner membrane surface.

By mechanical agitation of the vesicle suspensions, or by trypsin-urea treatment, it is possible to separate the vesicles from their attached spheres. The stripped vesicles retain electron transport properties, but cannot carry out oxidative phosphorylation. The spheres are not active in electron transport, but possess ATPase activity. From experiments of this kind it has been concluded that the mitochondrial inner membrane contains the electron transport system, while the inner membrane spheres are responsible for the coupling of ATP synthesis to electron transport.

Models for Oxidative Phosphorylation

The oldest plausible mechanism that was advanced to account for oxidative phosphorylation was the **chemical-coupling theory**. This postulated that the electron transport process generates a compound of high energy, which in turn serves as a reactant in a linked reaction that phosphorylates ADP. Coupled processes that utilize a common intermediate recur frequently in metabolic pathways; several examples have been encountered in the descriptions of glycolysis and the TCA cycle. In the present case the greatest obstacle to the acceptance of this model has been the failure of investigators, despite intensive and protracted efforts, to detect the postulated high energy intermediate. However, it probably cannot be ruled out altogether.

A second mechanism that has attracted adherents suggests that electron transport induces a conformational change in a protein of the respiratory chain and that the energy thereby stored in a high energy state of the protein is utilized to drive the endergonic phosphorylation of ADP. However, the direct evidence for this model is slender.

The currently favored mechanism of those proposed to account for the driving of oxidative phosphorylation by electron transport is the **chemiosmotic** hypothesis. The concept of a high energy intermediate or conformation is replaced by that of a high energy state of the *membrane*. The chemiosmotic hypothesis postulates that the transport of electrons down the respiratory chain serves to produce a gradient of $[H^+]$ concentration across the inner mitochondrial membrane and it is this electrochemical gradient that couples the processes of electron flow and phosphorylation.

Some of the reactions of the respiratory chain *release* hydrogen ions, while others *consume* them. The chemiosmotic model requires that the proteins of the respiratory chain be so organized spatially that the H^+-producing reactions occur on the outer surface of the inner membrane and the H^+-consuming reactions occur on the inner surface (Fig. 12-9). In this way electron transport tends to produce a *deficiency* of H^+ ions within the matrix and an excess outside the mitochondrial membrane in the intermembrane space. This is a thermodynamically unstable situation, which cannot be relieved by diffusion of H^+ ions back into the matrix, since the inner mitochondrial membrane is impermeable to H^+ ions.

The chemiosmotic hypothesis proposes that the imbalance of H^+ ions is countered by a reaction which phosphorylates ADP in such a way as to tend to restore a uniform concentration of H^+ ions. This reaction is normally written

$$ADP + P_i \rightarrow ATP + H_2O$$

However, under certain conditions it might take the form

$$ADP + P_i \rightarrow ATP + H^+ + OH^-$$

If the geometrical arrangement within the membrane of the enzyme catalyzing this reaction were appropriate, it is plausible that H^+ ions might be delivered to the H^+-deficient matrix, where they combine with excess OH^- ions to form H_2O, while the OH^- ions formed by the reaction are dispatched to the external solution where they combine with excess H^+ ions.

The chemiosmotic model is consistent with the observations that the mitochondrial membrane must be physically intact for oxidative phosphorylation to occur and that uncoupling agents, such as 2,4-dinitrophenol, cause the mitochondrial membrane to become permeable to H^+ ions.

In summary, the chemiosmotic model postulates that the reactions of the respiratory chain tend to produce a thermodynamically unstable or *high energy* state characterized by an imbalance of H^+ ion concentration. Because of the impermeability of the mitochondrial membrane to H^+ ions, this condition cannot be removed by diffusion. Instead it is corrected by a coupled chemical reaction which simultaneously synthesizes ATP and delivers H^+ and OH^- ions to the inner and outer solutions in such a way as to eliminate the imbalance of H^+ ions. In actuality a steady state is attained in which the

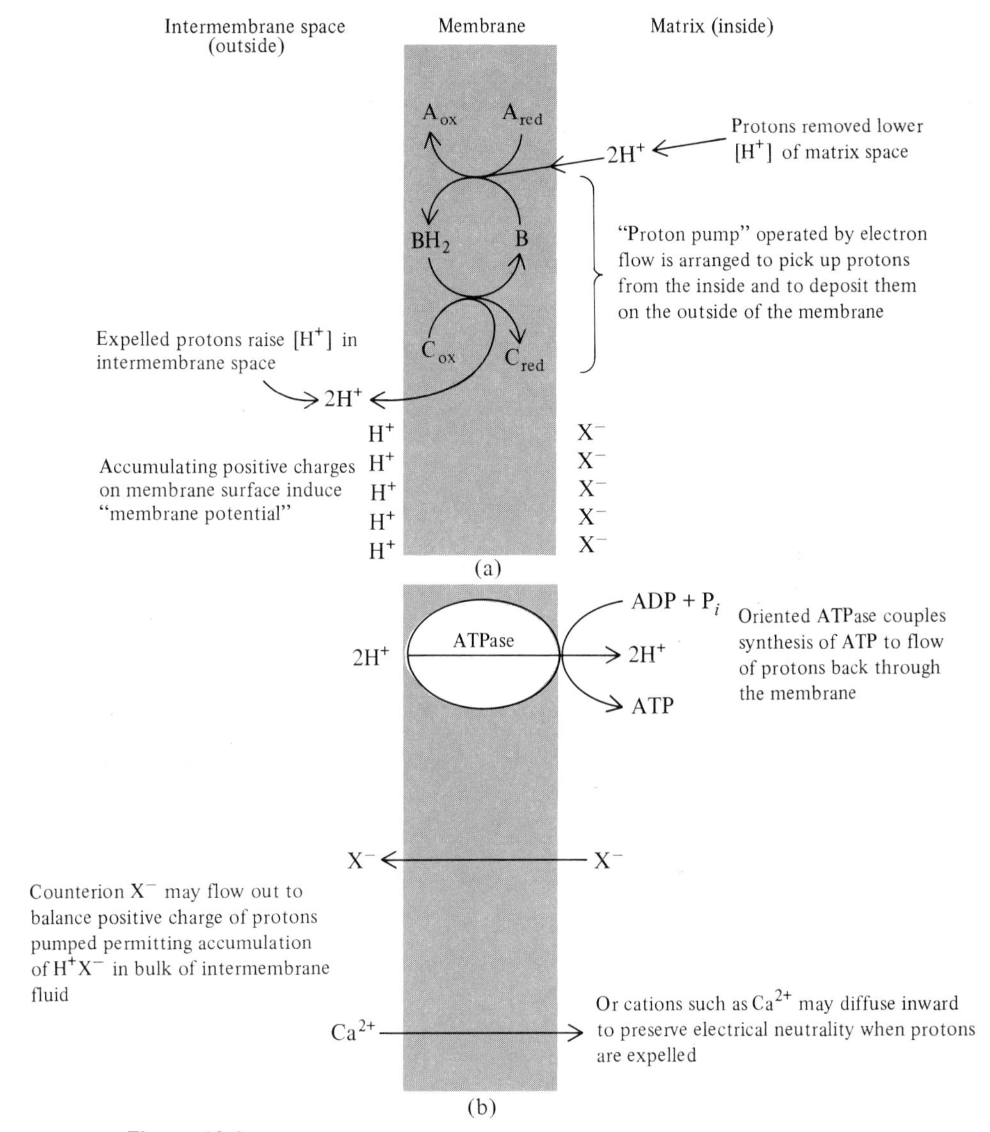

Figure 12-9
Schematic depiction of the chemiosmotic model. (a) shows the transport of H^+ ions across the inner membrane to yield a gradient of H^+ (b) shows the relief of this gradient by the geometrically specific release of H^+ and OH^- ions by the coupled phosphorylation of ADP. (Adapted from D. Metzler, *Biochemistry*, Academic, 1977.)

two processes are balanced, so that the net gradient of H^+ ions is small. In thermodynamic terms the transport of electrons down the respiratory chain may be said to *drive* the reaction synthesizing ATP.

It should be stressed that, while the chemiosmotic hypothesis is the most convincing of the models that have been invoked to account for oxidative phosphorylation, it has not been proven beyond all doubt and many details of its operation remain obscure.

SUMMARY

The degradative metabolic pathways of carbohydrates, fatty acids, and amino acids converge in the formation of **acetyl CoA**. Under aerobic conditions the degradation of glucose does not terminate with the formation of lactic acid but proceeds via pyruvate to yield acetyl CoA by the action of **pyruvate dehydrogenase**. The **tricarboxylic acid cycle** or **TCA cycle**, which occurs in the **mitochondria**, begins with the **citrate synthase**-catalyzed combination of acetyl CoA with oxaloacetate to yield **citric acid**. Citrate is next converted to **isocitrate**, which is in turn oxidized to α-ketoglutarate by NAD^+ in a reaction catalyzed by isocitrate dehydrogenase. The α-ketoglutarate is next oxidized to **succinyl CoA**, which reacts with GDP and P_i to yield **succinate** and GTP. Succinate undergoes oxidation to **fumarate** by **succinate dehydrogenase**. Fumerate is converted to **malate** by **fumarase**; malate is oxidized by NAD^+ and **malate dehydrogenase** to oxaloacetate, thereby completing the cycle.

Several intermediates of the TCA cycle are utilized in biosynthetic reactions. A synthetic pathway leads from oxaloacetate to glucose. This pathway begins with the conversion of oxaloacetate to **phosphoenolpyruvate**, whose subsequent conversion to glucose occurs via a pathway that utilizes the reversible steps of glycolysis and bypasses the irreversible steps in **gluconeogenesis**. Since the oxidative degradation of many amino acids yields pyruvate or a TCA cycle intermediate, these amino acids are also precursors of glucose.

In the final phase of the metabolic degradation of fats, carbohydrates, and amino acids, electrons derived from intermediates of the TCA cycle are transported down a series of **electron carriers**, known as the **respiratory chain**, to molecular oxygen. The passage of electrons down the respiratory chain is coupled with the synthesis of ATP from ADP in a process called **oxidative phosphorylation**. Electrons enter the respiratory chain from NADH, succinate dehydrogenase, and several flavoproteins. The respiratory chain contains coenzyme Q, one or more iron-sulfur proteins, and a series of cytochromes. Whereas the enzymes of the TCA cycle occur in the mitochondrial **matrix**, the respiratory chain is located in the mitochondrial **membrane**.

Most of the ATP generated by oxidative metabolism is formed by the reactions of oxidative phosphorylation which are linked with the respiratory chain; three sites within the latter have been identified as providing energy for phosphorylation. According to the **chemiosmotic hypothesis**, the coupling mechanism involves an electrochemical gradient of H^+ ions across the **inner mitochondrial membrane**. The transport of electrons down the chain generates a gradient of H^+ ions across the membrane. The thermodynamically favored tendency to eliminate this gradient effectively drives the synthesis of ATP.

REFERENCES

The following code is used to classify references. I: particularly useful as an introduction to the subject; R: useful primarily as a reference text; A: an advanced account of the material; H: a publication of historical importance.

The TCA Cycle

P. D. Boyer, *The Enzymes*, vols. 5–7, Academic, New York (1971). (R)

T. W. Goodwin, ed., *The Metabolic Roles of Citrate*, Academic, New York (1968). (A)

H. A. Krebs, "The History of the Tricarboxylic Acid Cycle," *Perspect. Biol. Med.* 14, 154 (1970). (I)

J. M. Lowenstein, ed., *The Citric Acid Cycle*, vol. 13, *Methods in Enzymology*, Academic, New York (1969). (R)

L. U. Reed, "Multienzyme Complexes," *Acc. Chem. Res.* 7, 40 (1974). (I)

T. P. Singer, E. B. Kearny, and W. C. Kenney, "Succinate Dehydrogenase," *Adv. Enzymol.* 37, 189 (1973). (R)

The Respiratory Chain

R. Lemberg and J. Barrett, *Cytochromes*, Academic, New York (1973). (A)

W. W. Wainio, *The Mammalian Respiratory Chain*, Academic, New York (1970). (A)

Oxidative Phosphorylation

H. Kalchar, *Biological Phosphorylation: Development of Concepts*, Prentice-Hall, Englewood Cliffs, New Jersey (1969). (I)

E. Racker, ed., *Membranes of Mitochondria and Chloroplasts*, American Chemical Society Monograph 165, Van Nostrand Reinhold, New York (1970). (R)

E. Racker, "The Two Faces of the Inner Mitochondrial Membrane," *Essays in Biochemistry* 6, 1 (1970). (I)

V. Skulachev, "Energy Transformations in the Respiratory Chain," *Curr. Top. Bioenerg.* 4, 127 (1972). (A)

Chemiosmotic Hypothesis

G. D. Greville, "A Scrutiny of Mitchell's Chemiosmotic Hypothesis," *Curr. Top. Bioenerg.* 3, 1 (1968). (I)

P. Mitchell, *Chemiosmotic Coupling and Energy Transduction*, Glynn Research, Bodmin, England (1968). (I)

REVIEW QUESTIONS

Questions marked with an asterisk are of a high level of difficulty.

12-1 Which of the following statements is false?

(a) Under anaerobic (low O_2) conditions, 2ATP and 0NADH are produced from 1 glucose in the glycolytic pathway.

(b) The key enzymes which regulate the *rate* of the TCA cycle are *isocitrate dehydrogenase* and *citrate synthetase*.

(c) Each glucose molecule degraded by glycolysis produces one molecule of pyruvate, which can feed into the TCA cycle if O_2 is present.

(d) Monosaccharides other than glucose can be converted to glucose-1-P or to fructose-6-P and then broken down by glycolysis.
(e) The first stage of glycolysis uses up energy (is endergonic).

12-2 For one "turn" of the TCA cycle (starting with acetyl CoA) which of the following statements is false?
(a) Two molecules of pyruvate are oxidized completely to CO_2.
(b) One GTP is produced.
(c) Two CO_2 are produced.
(d) Three NADH are produced.
(e) One $FADH_2$ is produced.

12-3 The compounds below are all intermediates in glycolysis or the TCA cycle:

(a) O^- — $C{=}O$ — $C{=}O$ — CH_3

(b) (furanose ring, O) with OPO_3^{2-} and OPO_3^{2-}

(c) (furanose ring, O) with OPO_3^{2-}

(d) O^- — $C{=}O$ — $C{=}O$ — CH_2 — CH_2 — $C{=}O$ — ^-O

(e) OPO_3^{2-} — $C{=}O$ — $HCOH$ — $CH_2OPO_3^{2-}$

(1) Name the intermediate into which the sugar fructose can be converted for oxidation by the cell.
(2) Name the compound which lies in the TCA cycle.
(3) Name the intermediate which is from the second stage of glycolysis.
(4) Name the intermediate which is the *substrate* for the enzyme *pyruvate dehydrogenase.*

12-4 How many molecules of CO_2 result from the complete oxidation of lactose?
(a) 6 (b) 3 (c) 12 (d) 14 (e) 13

12-5 Starting from pyruvate, how many potential ATP's (plus GTP's) can be synthesized using the TCA cycle and oxidative phosphorylation?

12-6 Which of the following reactions is catalyzed by a "kinase"?

(a) $H_2C(OH)—CH(OPO_3^-)—COO^- \rightleftharpoons H_2O + H_2C{=}C(OPO_3^-)(CO_2^-)$

(b) $H^+ + CH_3—\overset{O}{\overset{\|}{C}}—COO^- + NADH \rightleftharpoons$

$CH_3C(OH)(H)—COO^- + NAD^+$

(c) 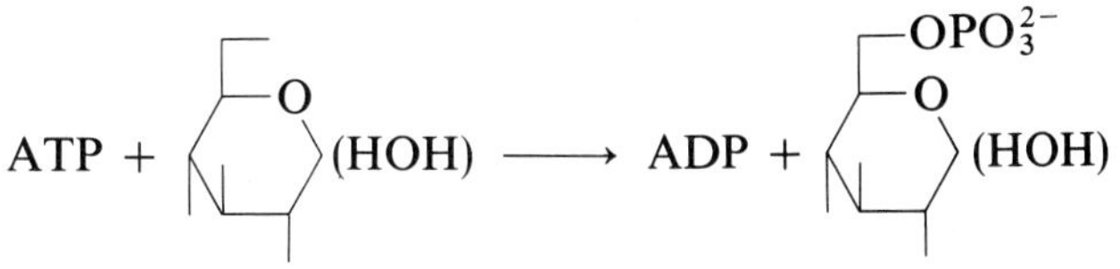 ATP + [ring—O, (HOH)] $\longrightarrow$ ADP + [ring—OPO_3^{2-}, O, (HOH)]

(d) $H_2C{=}C(OPO_3^-)(COO^-)$ + ADP $\longrightarrow$ ATP + $H_3C—\overset{O}{\overset{\|}{C}}—COO^-$

12-7 Which of the following statements is false for oxidative phosphorylation?
(a) An oxidized cytochrome has a heme ring containing Fe^{2+}
(b) The sequence of reactions in the respiratory chain is

$$H^+ + NADH—CoQ—cytb—cytc_1—cytc—cyta,a_3—O_2 \rightarrow H_2O$$

(c) One ATP is generated when CoQ is *reduced.*
(d) The components of the respiratory chain are on the inside mitochondrial membrane.
(e) Two ATP are produced for every $FADH_2$ from the TCA cycle.

12-8 The number of ATP molecules produced by complete oxidation of one glucose molecule by way of glycolysis, the TCA cycle, and oxidative phosphorylation is
(a) 2 (b) 36 (c) 38 (d) 18 (e) 33

12-9 Which of the following statements is *true* about the *synthesis* of glucose-6-P from *pyruvate*?
(a) All steps of glycolysis are by-passed, so a completely different set of enzymes is used.
(b) Pyruvate is the regulator molecule for the control enzymes.
(c) Amino acids are converted into phosphoenolpyruvate before being made into glucose.

(d) Steps in glycolysis which have large, negative $\Delta G^{\circ\prime}$'s are bypassed; different reactions and enzymes are used at these steps in synthesis.
(e) none of these

*12-10 Write a balanced equation for the complete combustion of citric acid to CO_2 and H_2O by the combined operation of the TCA cycle and oxidative phosphorylation.

*12-11 Write a balanced equation for the above process if oxidative phosphorylation, but not respiration, is inhibited.

13

The Triacylglycerol Fats

13-1 BIOLOGICAL ROLES

Chapters 13 and 14 will deal with two different varieties of the somewhat blurred biochemical category termed the **lipids**. Both varieties share the structural feature of having much of their mass contributed by the **hydrocarbon chains of fatty acids**. However, their biological functions are very different. The **triacylglycerols** serve primarily as very efficient **biological fuels**. In animals they have a secondary function in thermal insulation and shock absorption. The **phospholipids**, which we shall consider in chapter 14, have a *structural* function, acting as essential components of *membranes*.

This chapter describes the metabolic breakdown and biosynthesis of the **triacylglycerols**. The degradation and formation of their fatty acid components occur in different cellular locations. Fatty acid degradation takes place within the mitochondria, while synthesis is extramitochondrial. The biosynthesis of fatty acids involves the action of an interesting organized multienzyme system.

13-2 LIPIDS

The term **lipid** has been somewhat loosely applied to those constituents of animal or plant tissues that are insoluble in water, but can be dissolved and extracted by organic solvents, such as chloroform, ether, benzene, or hexane. This operationally defined group of compounds is chemically diverse. The more familiar examples include triacylglycerol fats, phosphatides, glycolipids, waxes, terpenes, and steroids.

The various classes of lipid have little in common except solubility and will accordingly be discussed separately. The **triacylglycerol** fats, sometimes called *triglycerides*, are the fat depots in animals. They are the principal storage reservoirs for lipids. The triacylglycerol fats have a highly important function in human and animal nutrition as biochemical fuels and nutritional reserves. In contrast to the case of carbohydrates, the animal body is able to deposit substantial quantities of these fats in depots.

In mammals the major site of storage of triacylglycerols is the cytoplasm of **adipose** or **fat cells**. In these specialized cells the triacylglycerols form large globules, which occupy a major part of the cell volume. Carbohydrates consumed in excess of nutritional requirements tend to be largely converted to triacylglycerols and stored in the appropriate tissues. In times of need this reserve is drawn upon. Quite apart from their nutritional significance, fat deposits have the additional function of providing thermal and physical insulation for the organs of the animal body.

Like the carbohydrates, fats may enter the pathways of oxidative metabolism to be converted ultimately to CO_2 and H_2O. The terminal stages of oxidative metabolism are the same for both classes of compounds; the oxidation of fats thus provides an important alternative source of ATP. The caloric value of fats is high—about twice that of carbohydrates on a weight basis (9 kcal/g versus 4 kcal/g).

13-3 FATTY ACIDS

The triacylglycerol fats are esters of glycerol with any of a number of linear monocarboxylic acids termed **fatty acids**. Since glycerol contains three hydroxyl groups, it can form mono-, di-, or triesters, which are designated as mono-, di-, or triacylglycerols:

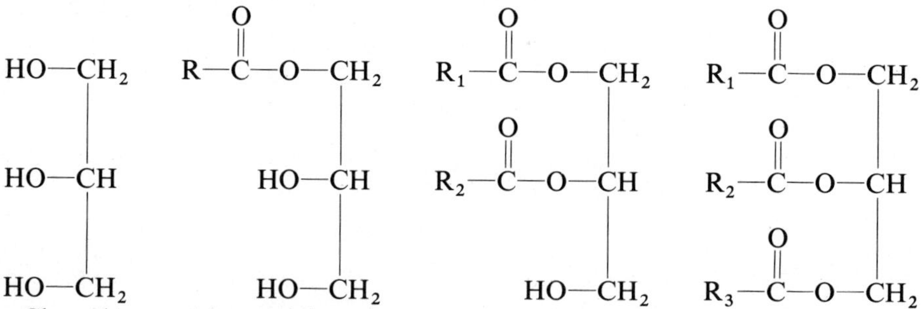

$$\begin{array}{l}
\text{HO—CH}_2 \\
\quad\;\; | \\
\text{HO—CH} \\
\quad\;\; | \\
\text{HO—CH}_2
\end{array}
\qquad
\begin{array}{r}
\text{R—}\overset{\text{O}}{\overset{\|}{\text{C}}}\text{—O—CH}_2 \\
| \\
\text{HO—CH} \\
| \\
\text{HO—CH}_2
\end{array}
\qquad
\begin{array}{r}
\text{R}_1\text{—}\overset{\text{O}}{\overset{\|}{\text{C}}}\text{—O—CH}_2 \\
| \\
\text{R}_2\text{—}\overset{\text{O}}{\overset{\|}{\text{C}}}\text{—O—CH} \\
| \\
\text{HO—CH}_2
\end{array}
\qquad
\begin{array}{r}
\text{R}_1\text{—}\overset{\text{O}}{\overset{\|}{\text{C}}}\text{—O—CH}_2 \\
| \\
\text{R}_2\text{—}\overset{\text{O}}{\overset{\|}{\text{C}}}\text{—O—CH} \\
| \\
\text{R}_3\text{—}\overset{\text{O}}{\overset{\|}{\text{C}}}\text{—O—CH}_2
\end{array}$$

Glycerol — Monoacylglycerol — Diacylglycerol — Triacylglycerol

Natural fats are always complex mixtures of different triacylglycerols, most of which contain two or three different fatty acids.

The ester linkages of the triacylglycerols may be hydrolyzed by heating in alkaline solution, thereby releasing the fatty acids in ionized form. This reaction, called **saponification**, is the classical method for preparing soaps:

$$\begin{array}{l} R-\overset{\overset{O}{\|}}{C}-O-CH_2 \\ \qquad\qquad\quad | \\ R-\overset{\overset{O}{\|}}{C}-O-CH \\ \qquad\qquad\quad | \\ R-\overset{\overset{O}{\|}}{C}-O-CH_2 \end{array} \xrightarrow[NaOH]{H_2O} 3R-\overset{\overset{O}{\|}}{C}-O^- + \begin{array}{l} HO-CH_2 \\ \qquad\quad | \\ HO-CH \\ \qquad\quad | \\ HO-CH_2 \end{array}$$

Lipases

The lipases are enzymes which catalyze the hydrolysis of fatty acid esters. The fatty acid components of neutral fats are split off in a stepwise manner to yield ultimately free glycerol.

The lipases are widely distributed in tissues with high concentrations in intestinal walls, liver, and pancreas. The specificities of the pancreatic and intestinal-wall enzymes are somewhat different; the former hydrolyzes only the two terminal (α) ester groups of a triacylglycerol, while the latter attacks all three.

The products of hydrolysis enter different metabolic pathways. Glycerol behaves as a carbohydrate, while the fatty acids are oxidized by a different mechanism.

Saturated Fatty Acids

The free fatty acids are unbranched hydrocarbon chains terminating in a carboxyl group, which is ionized under physiological conditions. The hydrocarbon may be completely saturated (Table 13-1) or it may contain one or more double bonds (Table 13-2). The number of carbon atoms in naturally occurring fatty acids is usually even.

The solubility in polar solvents of the un-ionized fatty acids is governed by the chain length. Acetic and propionic acids are completely miscible in water. With increasing chain length, the solubility in water rapidly declines; *n*-octanoic acid and its higher homologs are virtually insoluble.

The melting and boiling points of the fatty acids rise with chain length, although not in a curvilinear fashion. Saturated fatty acids of chain length below 10 carbon atoms are oils at room temperature; those of higher chain length are solids.

The hydrocarbon chains of the saturated fatty acids are flexible and can adopt a wide range of conformations because of the freedom of rotation about each C—C single bond. The most probable conformation is the fully extended form (Fig. 13-1), which has the lowest energy.

Unsaturated Fatty Acids

The presence of one or more double bonds has a profound influence upon the physical properties of a fatty acid, lowering the melting point and enhancing the solubility in nonpolar solvents. The common unsaturated fatty acids are liquids at room temperature. In the singly unsaturated fatty

Table 13-1

The Saturated Fatty Acids

Empirical Formula	*Common Name*	*Systematic Name*	*Structure*	*Melting Point* (°C)
$C_2H_4O_2$	Acetic		CH_3COOH	16.6
$C_3H_6O_2$	Propionic		CH_3CH_2COOH	−22.0
$C_4H_8O_2$	*n*-Butyric		$CH_3(CH_2)_2COOH$	− 7.9
$C_6H_{12}O_2$	Caproic	*n*-Hexanoic	$CH_3(CH_2)_4COOH$	− 2.0
$C_8H_{16}O_2$	Caprylic	*n*-Octanoic	$CH_3(CH_2)_6COOH$	16.0
$C_9H_{18}O_2$	Pelargonic	*n*-Nonanoic	$CH_3(CH_2)_7COOH$	—
$C_{10}H_{20}O_2$	Capric	*n*-Decanoic	$CH_3(CH_2)_8COOH$	31.5
$C_{12}H_{24}O_2$	Lauric	*n*-Dodecanoic	$CH_3(CH_2)_{10}COOH$	44.2
$C_{14}H_{28}O_2$	Myristic	*n*-Tetradecanoic	$CH_3(CH_2)_{12}COOH$	53.9
$C_{16}H_{32}O_2$	Palmitic*	*n*-Hexadecanoic	$CH_3(CH_2)_{14}COOH$	63.1
$C_{18}H_{36}O_2$	Stearic*	*n*-Octadecanoic	$CH_3(CH_2)_{16}COOH$	69.6
$C_{20}H_{40}O_2$	Arachidic	*n*-Eicosanoic	$CH_3(CH_2)_{18}COOH$	75.6
$C_{22}H_{44}O_2$	Behenic	*n*-Docosanoic	$CH_3(CH_2)_{20}COOH$	—
$C_{24}H_{48}O_2$	Lignoceric	*n*-Tetracosanoic	$CH_3(CH_2)_{22}COOH$	—
$C_{26}H_{52}O_2$	Cerotic	*n*-Hexacosanoic	$CH_3(CH_2)_{24}COOH$	—

* The most common fatty acids in animal fats.

Table 13-2

The Unsaturated Fatty Acids

Empirical Formula	*Trivial Name*	*Systematic Name*	*Structure*	*Melting Point* (°C)
$C_{16}H_{30}O_2$	Palmitoleic*	*cis*-$\Delta^{9:10}$-Hexadecenoic	$CH_3(CH_2)_5CH{=}CH(CH_2)_7COOH$	− 0.5
$C_{18}H_{34}O_2$	Oleic*	*cis*-$\Delta^{9:10}$-Octadecenoic	$CH_3(CH_2)_7CH{=}CH(CH_2)_7COOH$	13.4
$C_{18}H_{34}O_2$	Elaidic	*trans*-$\Delta^{9:10}$-Octadecenoic	$CH_3(CH_2)_7CH{=}CH(CH_2)_7COOH$	51.5
$C_{18}H_{34}O_2$	Vaccenic	$\Delta^{11:12}$-Octadecenoic	$CH_3(CH_2)_5CH{=}CH(CH_2)_9COOH$	—
$C_{18}H_{32}O_2$	Linoleic*	*cis,cis*-$\Delta^{9:10:12:13}$-Octadecadienoic	$CH_3(CH_2)_4CH{=}CHCH_2CH{=}$ $CH(CH_2)_7COOH$	− 5.0
$C_{18}H_{30}O_2$	Linolenic*	$\Delta^{9:10,12:13,15:16}$-Octadecatrienoic	$CH_3CH_2CH{=}CHCH_2CH{=}CHCH_2CH{=}$ $CH(CH_2)_7COOH$	−11.0
$C_{18}H_{30}O_2$	Eleostearic	$\Delta^{9:10,11:12,13:14}$-Octadecatrienoic	$CH_3(CH_2)_4CH{=}CH-CH{=}CH-CH{=}$ $CH(CH_2)_7COOH$	—
$C_{20}H_{32}O_2$	Arachidonic	$\Delta^{5:6,8:9,11:12,14:15}$-Eicosatetraenoic	$CH_3(CH_2)_4CH{=}CHCH_2CH{=}CHCH_2{-}$ $CH{=}CHCH_2{-}CH{=}CH(CH_2)_3COOH$	49.5
$C_{22}H_{34}O_2$	Clupanodonic	Docosapentaenoic		—

* The most common unsaturated fatty acids in animal fats.

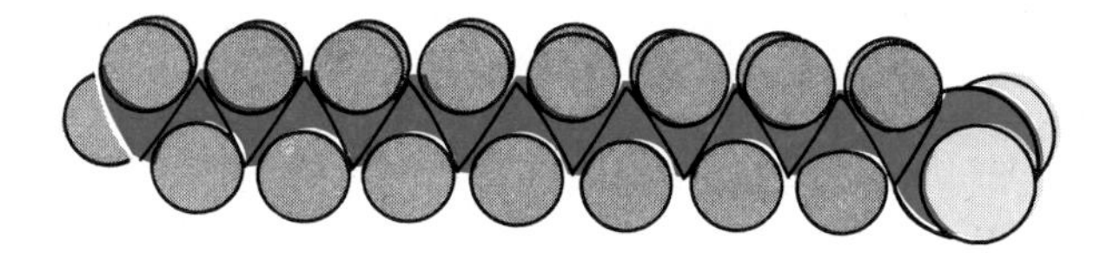

$$\begin{array}{ccccccccccccccccc} & H & H & H & H & H & H & H & H & H & H & H & H & H & H & & \\ & | & | & | & | & | & | & | & | & | & | & | & | & | & | & & O \\ H_3C- & C- & C- & C- & C- & C- & C- & C- & C- & C- & C- & C- & C- & C- & C- & C & \nearrow\!\!\!\!\!\!= \\ & | & | & | & | & | & | & | & | & | & | & | & | & | & | & & \searrow O^- \\ & H & H & H & H & H & H & H & H & H & H & H & H & H & H & & \end{array}$$

Palmitate

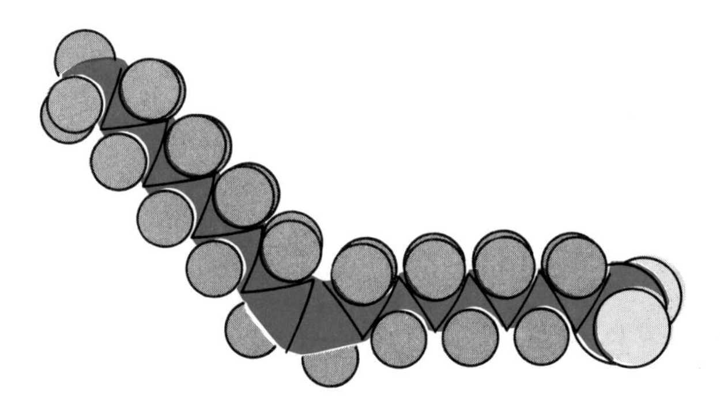

$$\begin{array}{cccccccccccccccccccc} & H & H & H & H & H & H & H & H & H & H & H & H & H & H & H & H & & \\ & | & | & | & | & | & | & | & | & | & | & | & | & | & | & | & | & & O \\ H_3C- & C- & C- & C- & C- & C- & C- & C- & C= & C- & C- & C- & C- & C- & C- & C- & C- & C & \nearrow\!\!\!\!\!\!= \\ & | & | & | & | & | & | & | & & & | & | & | & | & | & | & | & & \searrow O^- \\ & H & H & H & H & H & H & H & & & H & H & H & H & H & H & H & & \end{array}$$

Oleate

Figure 13-1 Conformations of a saturated and an unsaturated fatty acid.

acids of animal fats, the double bond is usually in the 9,10 position. This is, in particular, the case for oleic and palmitoleic acids, the most abundant of the unsaturated fatty acids of animal origin.

The absence of rotational freedom about the axis of a C=C double bond makes possible the type of spatial isomerism termed *cis-trans* isomerism (Fig. 13-1):

$$\begin{array}{l} H{-}C{-}(CH_2)_7{-}CH_3 \\ \quad\;\; \| \\ H{-}C{-}(CH_2)_7{-}COOH \end{array} \qquad \begin{array}{l} CH_3{-}(CH_2)_7{-}C{-}H \\ \qquad\qquad\quad\;\; \| \\ \qquad\qquad H{-}C{-}(CH_2)_7{-}COOH \end{array}$$

Oleic acid (*cis*) Elaidic acid (*trans*)

The naturally occurring unsaturated fatty acids are generally in the *cis* conformation.

Multiply unsaturated fatty acids also occur in animal tissues. The distribution of double bonds is commonly of the form

$$—CH{=}CH—CH_2—CH{=}CH—$$

rather than of the *conjugated* type

$$—CH{=}CH—CH{=}CH—$$

Linoleic, linolenic, and arachidonic acids (Table 13-2) are the most abundant unsaturated fatty acids found in animals. Linoleic and linolenic acids are *essential* dietary components in mammals; that is, they cannot be synthesized by the body and must be supplied from external sources, especially plants, in which they are abundant. The essential fatty acids are precursors of the hormonelike compounds called **prostaglandins**.

Hardening of Fats

The unsaturated bonds of fatty acids can be hydrogenated by reaction with molecular hydrogen in the presence of a catalyst. Conversion to the corresponding saturated form raises the melting point of oily fats, so that they become solids at room temperature. This reaction is important in the manufacture of *shortening*, which has a significant, although geographically variable, role in human nutrition.

Soaps

The name **soaps** is loosely applied to the alkali metal salts (usually Na^+ or K^+) of the fatty acids. The common soaps, such as sodium stearate, form molecular clusters or micelles in aqueous solution at concentrations above a **critical micelle concentration**. The stabilizing factor for soap micelles is the cohesive hydrophobic interaction of the hydrocarbon chains (chapter 5).

The detergent activity of soaps is related to their ability to envelope an insoluble particle in a cluster, thereby endowing it with a net charge and enabling it to remain in suspension.

13-4 DIGESTION OF THE TRIACYLGLYCEROL FATS

The mammalian body is able to synthesize from other nutrients, including carbohydrates, the saturated fatty acids and triacylglycerol fats required for normal growth. Despite the nonessential character of dietary fats, they are an important nutrient material, particularly as a concentrated source of energy.

The triacylglycerol fats undergo partial or complete hydrolysis in the gastrointestinal tract to form glycerol and free fatty acids:

$$\begin{array}{l} H_2COOCR \\ \;| \\ HCOOCR' \\ \;| \\ H_2COOCR'' \end{array} + 3H_2O \longrightarrow \begin{array}{l} H_2COH \\ \;| \\ HCOH \\ \;| \\ H_2COH \end{array} + RCOOH + R'COOH + R''COOH$$

The small intestine is the principal site of triacylglycerol digestion. Here the ingested fats encounter the **bile** and the pancreatic secretions. Bile itself does not hydrolyze fats. However, the bile salts appear to function as emulsifying agents which solubilize fats and render them more susceptible to the action of lipases.

The hydrolysis of triacylglycerol fats by pancreatic lipase is stepwise and occurs most rapidly at the α positions:

$$\begin{array}{ll} H_2COH & \alpha \\ | & \\ HCOH & \beta \\ | & \\ H_2COH & \alpha \end{array}$$

In view of the possibility of acyl migration, the occurrence of hydrolysis at the β position is uncertain. Complete hydrolysis is not essential for absorption and certain fatty acid esters that are resistant to hydrolysis are nevertheless absorbed in intact form.

The action of bile salts plus the pancreatic and other lipases encountered in the small intestine converts dietary fat to a complex mixture of free fatty acids, as well as mono-, di-, and triacylglycerols in emulsified form. Most of this material is transported across the wall of the small intestine.

Fats are made available to the mammalian body by a complicated mechanism. The free fatty acids of shorter chain length (less than about 10 carbon atoms) are absorbed directly in nonesterified form by the **portal vein** route and are supplied directly to the liver. This mechanism is particularly important for milk, which is rich in the short-chain fatty acids.

The longer chain fatty acids, whether absorbed in the free state or as triacylglycerols, enter the circulation via the **thoracic duct**, a lymphatic vessel draining the intestine. They appear here almost entirely in the esterified form.

13-5 DISTRIBUTION OF TRIACYLGLYCEROL FATS

Plasma Lipid

Immediately after a meal, normal human plasma may contain a milligram or so per milliliter of triacylglycerol fat. The fat content of the plasma exists primarily as protein complexes, or **lipoproteins**, and as **chylomicrons**, which are particles consisting of lipid plus 1%–2% protein. Both particles contain other lipids as well, including phosphatides (chapter 14) and cholesterol (chapter 16). The administration of the mucopolysaccharide **heparin** results in a rapid diminution of plasma lipid content, accompanied by a significant reduction in turbidity.

Triacylglycerols do not persist in plasma for long periods. Excess triacylglycerols are deposited in specialized **adipose cells** within several hours after a meal. These cells contain large globules of triacylglycerols, which comprise a major fraction of their mass. The incorporation of triacylglycerols into adipose cells involves their partial or complete hydrolysis by **lipoprotein lipase** to yield fatty acids and glycerol, which are recombined within the adipose cells to regenerate triacylglycerols.

Depot Fat

In mammals fat is by far the most important nutritional reservoir. This arrangement contrasts with that of plants, in which carbohydrate is the chief nutritional reserve. Adipose cells are of wide distribution, occurring in muscle, mammary glands, the abdominal cavity, and under the skin. They are not merely passive storage repositories, but have a high metabolic activity and are rapidly adaptable to the requirements of the organism.

The release of triacylglycerols from adipose tissue involves their hydrolysis by lipases, whose action is subject to hormonal control. The free fatty acids produced in this way are transferred to the blood, where they are reversibly bound by the protein **serum albumin** and transported to various tissues.

13-6 DEGRADATION OF FATTY ACIDS

The triacylglycerols released from adipose tissue are hydrolyzed by lipases to glycerol and free fatty acids. The glycerol is subsequently converted to L-glycerol-3-phosphate by **glycerol kinase** plus ATP:

$$\text{glycerol} + \text{ATP} \rightarrow \text{ADP} + \text{L-glycerol-3-phosphate}$$

L-glycerol-3-phosphate is next oxidized to dihydroxyacetone phosphate by the NAD^+-dependent enzyme **glycerol-3-phosphate dehydrogenase**:

$$\text{L-glycerol-3-phosphate} + NAD^+ \rightarrow \text{dihydroxyacetone phosphate} + NADH + H^+$$

The above reactions occur in the cytoplasm. The dihydroxyacetone phosphate formed lies on the glycolytic pathway.

The free fatty acids derived from the hydrolysis of triacylglycerols undergo oxidative degradation to CO_2 and H_2O by a process called β-oxidation. This involves an initial enzymic activation in the cytoplasm. The activated fatty acids subsequently enter the mitochondria for oxidation.

Activation of Fatty Acids

The cytoplasmic activation of the fatty acids involves their conversion to fatty acyl CoA in a reaction catalyzed by an **acyl CoA synthetase**:

$$\underset{\text{Fatty acid}}{\text{RCOOH}} + \text{ATP} + \text{HSCoA} \longrightarrow \underset{\text{Fatty acyl CoA}}{\text{RCO—S—CoA}} + \text{AMP} + PP_i$$

The reaction forms a thioester bond between the carboxyl group of the fatty acid and the sulfhydryl group of CoA. The reaction is driven by the exergonic cleavage of ATP to form AMP and inorganic pyrophosphate (PP_i). At least three different acyl CoA synthetases have been identified, each of which is specific for a particular range of fatty acid chain lengths.

Since the pyrophosphate is subsequently hydrolyzed to inorganic phosphate by a phosphodiesterase, this reaction is essentially irreversible. It should be noted that *two* high-energy bonds of ATP are utilized. AMP is phosphorylated by ATP to form ADP in a reaction catalyzed by *adenylate kinase*:

$$\text{ATP} + \text{AMP} \rightarrow 2\text{ADP}$$

There is thus a net consumption of *two* molecules of ATP associated with the formation of fatty acyl CoA.

Transport Across the Mitochondrial Membrane

The inner mitochondrial membrane presents a barrier to the entry of fatty acyl CoA. Its entry is facilitated by the transfer of the acyl group to **carnitine** in a reversible reaction catalyzed by **carnitine acyltransferase**:

$$\underset{\text{Fatty acyl CoA}}{R{-}\underset{\underset{O}{\|}}{C}{-}S{-}CoA} + \underset{\text{Carnitine}}{(CH_3)_3N^+{-}CH_2{-}CHOH{-}CH_2{-}COOH} \longrightarrow$$

$$\underset{\text{Acyl carnitine}}{(CH_3)_3N^+{-}CH_2{-}\underset{\underset{\underset{\underset{R}{|}}{C=O}}{|}}{\underset{|}{\underset{O}{}}}{CH}{-}CH_2{-}COOH} + HSCoA$$

The acyl carnitine readily crosses the inner mitochondrial membrane into the matrix, where the fatty acyl group is transferred back to CoA by a second carnitine acyltransferase located in the inner mitochondrial membrane. Since the value of $\Delta G^{\circ\prime}$ for the above reaction is quite small, the reaction may readily proceed in either direction, depending on the levels of reactants and products:

$$\text{acyl carnitine} + \text{CoA} \rightleftarrows \text{acyl CoA} + \text{carnitine}$$

The intramitochondrial fatty acyl CoA now enters the pathway of oxidative degradation, catalyzed by enzymes located within the mitochondrial matrix.

Oxidative Degradation

The degradation of fatty acids by β-oxidation occurs by the sequential removal of successive 2-carbon fragments from the carboxyl terminus.* The acyl CoA is first dehydrogenated to form an α,β-unsaturated derivative in a reaction catalyzed by an **acyl CoA dehydrogenase**; enzymes of this group contain FAD as the active group:

$$R{-}CH_2CH_2CH_2COSCoA + FAD \rightarrow FADH_2 + RCH_2CH{=}CHCOSCoA$$

Four different acyl CoA dehydrogenases have been identified, each of which is specific for a particular range of fatty acid chain lengths.

The $FADH_2$ group of the reduced acyl CoA dehydrogenase is reoxidized by electron transfer to the respiratory chain by way of another flavoprotein, which functions as an intermediate electron carrier.

---

* The earliest evidence as to the mechanism of fatty acid degradation was obtained by the German biochemist F. Knoop in 1904, who fed rabbits fatty acids substituted at the terminal methyl group with a phenyl group. When the number of carbons in the fatty acid was *even*, phenylacetic acid appeared in the urine; when the number of carbons was *odd*, benzoic acid was excreted. This pattern was observed irrespective of the length of the fatty acid chain. Knoop proposed that oxidation occurred at the β-carbon to form a β-keto acid, which was subsequently cleaved to yield acetic acid and a fatty acid shortened by two carbon atoms.

The unsaturated fatty acyl CoA derivative is next converted by **enoyl CoA hydratase** to a β-hydroxy compound by addition of a molecule of water:

$$R—CH_2CH=CHCOSCoA + H_2O \rightarrow R—CH_2CHOHCH_2COSCoA$$

The β-hydroxy compound is subsequently dehydrogenated by **3-hydroxyacyl CoA dehydrogenase** to form the β-keto derivative, with concomitant reduction of a molecule of NAD^+:

$$R—CH_2CHOHCH_2COSCoA + NAD^+ \rightarrow R—CH_2COCH_2COSCoA + NADH + H^+$$

Finally, the β-keto compound reacts further with CoA to form acetyl CoA and the acyl CoA derivative of the fatty acid with *two* fewer carbon atoms. The reaction is catalyzed by **thiolase**:

$$R—CH_2COCH_2COSCoA + HSCoA \rightarrow R—CH_2COSCoA + CH_3COSCoA$$

Several different forms of thiolase exist, whose specificities are directed to different fatty acid chain lengths.

The overall process whereby two carbon atoms are removed from the acyl CoA derivative of a fatty acid may be summarized by

$$R—CH_2CH_2CH_2COSCoA + FAD + NAD^+ + HSCoA + H_2O \rightarrow$$
$$R—CH_2COSCoA + CH_3COSCoA + FADH_2 + NADH + H^+$$

The acetyl CoA thereby generated shares the usual metabolic destiny of this intermediate, being for the most part burned to CO_2 and H_2O by the combined operation of the TCA cycle and the respiratory chain.

The $FADH_2$ and NADH generated by the above reactions are reoxidized by the respiratory chain. The linked processes of oxidative phosphorylation yield five molecules of ATP for each molecule of acetyl CoA formed. The oxidation of acetyl CoA ultimately produces an additional 12 molecules of ATP.

The process may be repeated for the shortened fatty acyl CoA (Fig. 13-2). The complete reaction for the oxidation of palmitoyl CoA may be summarized by

$$CH_3(CH_2)_{14}COSCoA + 23O_2 + 131P_i + 131\ ADP \rightarrow$$
$$CoA + 16CO_2 + 146H_2O + 131\ ATP$$

Since two molecules of ATP are consumed in the conversion of palmitic acid to palmitoyl CoA, the net yield of ATP from the complete oxidation of one molecule of palmitic acid is 129. The overall reaction may be written

$$CH_3(CH_2)_{14}COO^- + 23O_2 + 129P_i + 129\ ADP \rightarrow$$
$$16CO_2 + 145H_2O + 129\ ATP$$

Degradation of Fatty Acids with Odd-Numbered Chains

A fatty acid containing an even number of carbon atoms is degraded entirely to acetyl CoA by repetition of the stepwise mechanism described above. Fatty acids with an odd number of carbon atoms yield, in addition to acetyl CoA, a molecule of propionyl CoA.

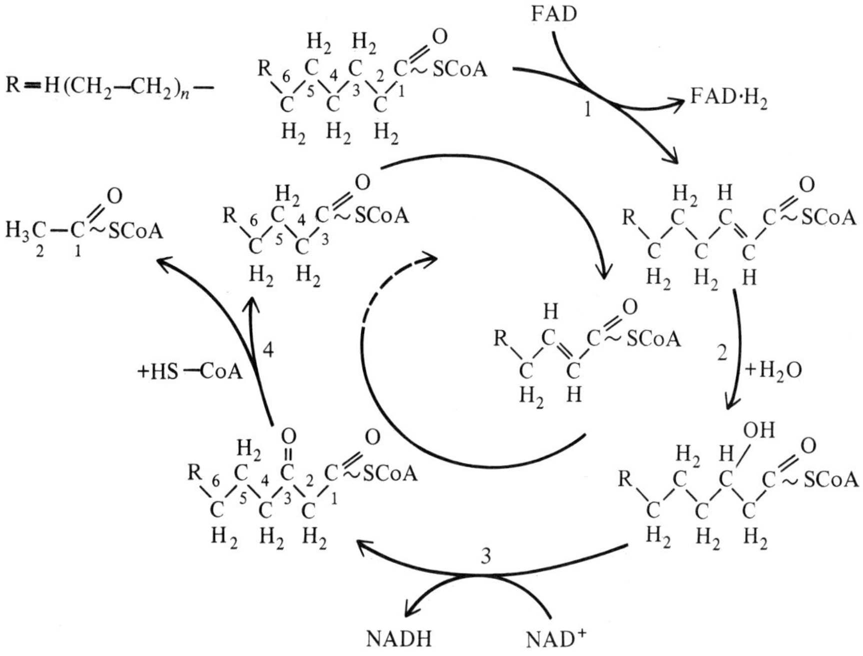

Figure 13-2 Stepwise oxidative degradation of fatty acids.

In animal tissues propionyl CoA is converted to succinyl CoA, an intermediate of the TCA cycle. The initial step is the conversion of propionyl CoA to methylmalonyl CoA, by fixation of CO_2 with consumption of one molecule of ATP. The reaction is catalyzed by **propionyl CoA carboxylase**, with Mg^{2+} and biotin as cofactors:

$$CH_3CH_2COSCoA + ATP + CO_2 + H_2O \longrightarrow ADP + P_i + \underset{\underset{\underset{\text{O}}{\|}}{\text{C—SCoA}}}{\overset{\text{COOH}}{\overset{|}{\text{HCCH}_3}}}$$

Propionyl CoA — Methylmalonyl CoA

Methylmalonyl CoA contains a chiral center and may exist as two stereoisomers, only one of which is synthesized in the above reaction. The next step is its isomerization by **methylmalonyl CoA racemase** to a mixture of both stereoisomers, one of which is subsequently converted to succinyl CoA by **methylmalonyl CoA mutase**:

methylmalonyl CoA → methylmalonyl* CoA
methylmalonyl* CoA → succinyl CoA

13-7 BIOSYNTHESIS OF FATTY ACIDS

The mammalian body is not altogether dependent upon external sources for a supply of the fatty acids. Both saturated and unsaturated fatty acids (except for linoleic and linolenic acids) may be synthesized from acetyl CoA. This is the basis for the all too familiar observation that a portion of dietary carbohydrate may be converted to fat and a high carbohydrate diet may lead to obesity.

The biosynthesis of fatty acids from acetyl CoA may occur by at least two distinct mechanisms; one occurs in the mitochondria; the main pathway is extramitochondrial and occurs in the cytoplasm.

Extramitochondrial Biosynthesis

The initial step in fatty acid synthesis by this mechanism is the carboxylation of acetyl CoA to form malonyl CoA in a reaction catalyzed by **acetyl CoA carboxylase**:

$$CH_3COSCoA + CO_2 + ATP \longrightarrow ADP + P_i + \begin{array}{c} COOH \\ | \\ CH_2 \\ | \\ O{=}C{-}SCoA \end{array}$$

Malonyl CoA

Acetyl CoA carboxylase contains biotin as its prosthetic group.

Acetyl CoA is formed in the mitochondria by the oxidative decarboxylation of pyruvate. While the mitochondrial membrane is impermeable to acetyl CoA itself, the acetyl group may be transferred to carnitine to form acetyl carnitine (section 13-5), which passes from the mitochondrial matrix to the cytoplasm, where its acetyl group is transferred to cytoplasmic CoA.

In addition, citrate, which is formed in mitochondria from acetyl CoA and oxaloacetate, may be transported across the mitochondrial membrane into the cytoplasm, where it is converted to acetyl CoA by the action of **ATP-citrate lyase**:

$$\text{citrate} + \text{ATP} + \text{CoA} \rightarrow \text{acetyl CoA} + \text{ADP} + P_i + \text{oxaloacetate}$$

Malonyl CoA is the starting point for the subsequent reactions of fatty acid biosynthesis, which are catalyzed by a set of enzymes organized into the **fatty acid synthetase** system. In mammalian cells the fatty acid synthetase system consists of seven proteins that collectively form an organized multienzyme complex, which, in the case of yeast, has a particle weight of 2.3 million. An important nonenzymic component of the system is **acyl carrier protein** (ACP), to which the elongating fatty acid chain is attached.

Acyl Carrier Protein

Acyl carrier protein, which has been isolated from a variety of sources, contains a covalently attached prosthetic group, **4-phosphopantetheine** (Fig. 13-3). The sulfhydryl group contributed by the latter serves as the point of attachment of the fatty acyl derivatives, in analogy to the function of CoA in fatty acid oxidation. The polypeptide portion is of relatively low molecular weight; *E. coli* has a polypeptide molecular weight of 10,000.

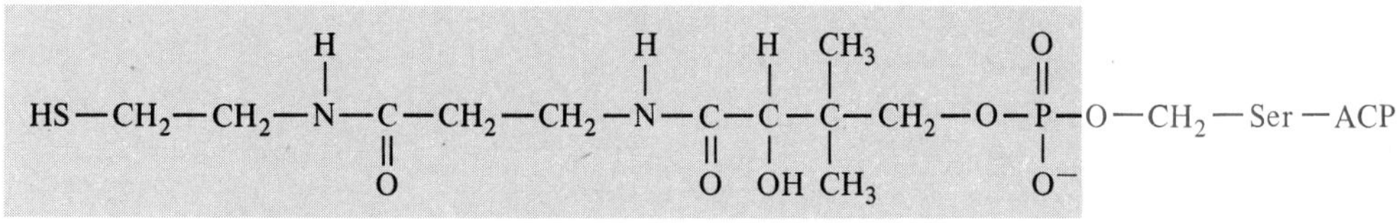

$$\mathrm{HS{-}CH_2{-}CH_2{-}\underset{\displaystyle H}{\overset{|}{N}}{-}\underset{\displaystyle O}{\overset{\|}{C}}{-}CH_2{-}CH_2{-}\underset{\displaystyle H}{\overset{|}{N}}{-}\underset{\displaystyle O}{\overset{\|}{C}}{-}\overset{H}{\underset{OH}{C}}{-}\overset{CH_3}{\underset{CH_3}{C}}{-}CH_2{-}O{-}\overset{O}{\underset{O^-}{P}}{-}O{-}CH_2{-}Ser{-}ACP}$$

Figure 13-3
Structure of the active group of acyl carrier protein (ACP). The sulfhydryl group of phosphopantetheine forms a thioester with the fatty acid.

Prior to the initiation of fatty acid synthesis, acetyl CoA combines with the sulfhydryl group of ACP in a reaction catalyzed by **ACP-acyltransferase**, a component of the fatty acid synthetase system.

$$CH_3CO{-}S{-}CoA + HS{-}ACP \rightarrow CH_3CO{-}S{-}ACP + HS{-}CoA$$

The acetyl group is next transferred from ACP to the cysteine sulfhydryl group of the enzyme **β-ketoacyl-ACP synthase**, a component of the fatty acid synthetase complex:

$$CH_3CO{-}S{-}ACP + HS\text{-}enzyme \rightarrow ACP{-}SH + CH_3CO{-}S{-}enzyme$$

Formation of Fatty Acids

The malonyl—S—CoA formed in the acetyl CoA carboxylase reaction next reacts with ACP to form malonyl—S—ACP in a reaction catalyzed by **malonyl transferase** (Fig. 13-4):

$$\text{malonyl}{-}S{-}CoA + HS{-}ACP \rightarrow \text{malonyl}{-}S{-}ACP + HSCoA$$

At this stage of the process a malonyl group is combined with ACP and an acetyl group is attached to a sulfhydryl group of β-ketoacyl-ACP synthase.

The acetyl group is next transferred to malonyl-S-ACP, with simultaneous removal of the malonyl carboxyl as CO_2 (Fig. 13-4), thereby forming acetoacetyl-S-ACP:

$$CH_3CO{-}S{-}\text{enzyme} + \text{malonyl}{-}S{-}ACP \rightarrow \text{acetoacetyl}{-}S{-}ACP + CO_2 + HS{-}\text{enzyme}$$

The acetoacetyl-S-ACP is now reduced to form D-β-hydroxybutyryl-S-ACP in a reaction catalyzed by **β-ketoacyl-ACP reductase**, an NADPH-dependent enzyme.

$$CH_3COCH_2CO{-}S{-}ACP + NADPH + H^+ \rightarrow CH_3CHOHCH_2CO{-}S{-}ACP + NADP^+$$

The D-β-hydroxybutyryl-S-ACP is subsequently dehydrated to the *trans*-α,β form of crotonyl-S-ACP by the action of **enoyl-ACP hydratase**:

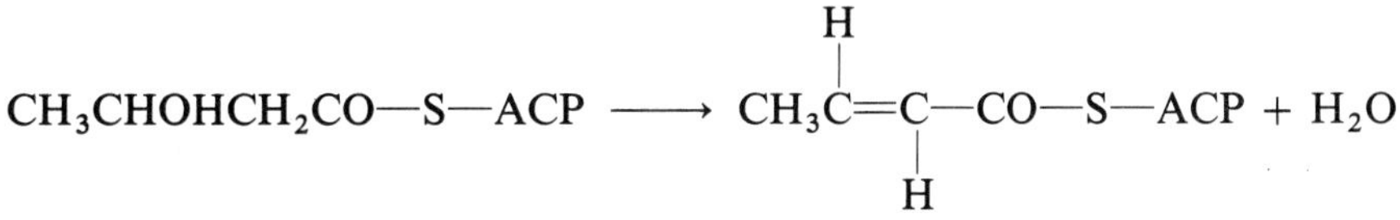

$$CH_3CHOHCH_2CO{-}S{-}ACP \longrightarrow CH_3\overset{\displaystyle H}{\overset{|}{C}}{=}\underset{\displaystyle H}{\underset{|}{C}}{-}CO{-}S{-}ACP + H_2O$$

Crotonyl-S-ACP

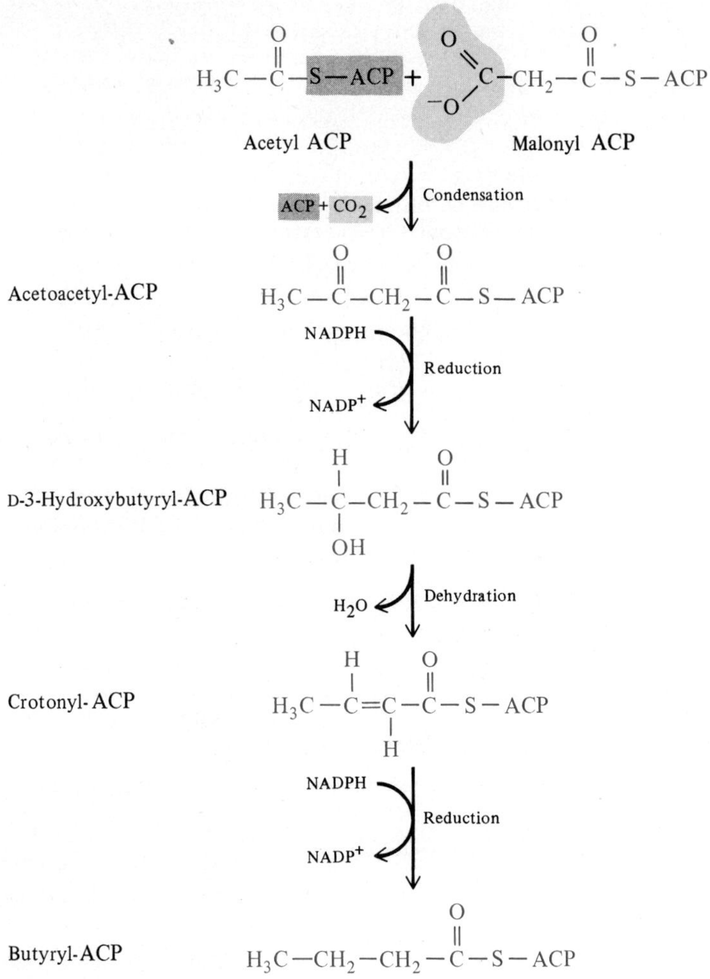

$$H_3C-\overset{O}{\overset{\|}{C}}-S-ACP + {}^{-}O-\overset{O}{\overset{\|}{C}}-CH_2-\overset{O}{\overset{\|}{C}}-S-ACP$$

Acetyl ACP　　Malonyl ACP

Condensation ($ACP + CO_2$ released)

Acetoacetyl-ACP

$$H_3C-\overset{O}{\overset{\|}{C}}-CH_2-\overset{O}{\overset{\|}{C}}-S-ACP$$

Reduction ($NADPH \rightarrow NADP^+$)

D-3-Hydroxybutyryl-ACP

$$H_3C-\underset{OH}{\overset{H}{C}}-CH_2-\overset{O}{\overset{\|}{C}}-S-ACP$$

Dehydration (H_2O released)

Crotonyl-ACP

$$H_3C-\overset{H}{C}=\underset{H}{C}-\overset{O}{\overset{\|}{C}}-S-ACP$$

Reduction ($NADPH \rightarrow NADP^+$)

Butyryl-ACP

$$H_3C-CH_2-CH_2-\overset{O}{\overset{\|}{C}}-S-ACP$$

Figure 13-4 Condensation of acetyl-S-ACP with malonyl-S-ACP.

Crotonyl-S-ACP is in turn reduced to butyryl-S-ACP by NADPH in a reaction catalyzed by **enoyl-ACP reductase**:

$$CH_3\overset{H}{C}=\underset{H}{C}-CO-S-ACP + NADPH + H^+ \longrightarrow CH_3CH_2CH_2CO-S-ACP + NADP^+$$

The subsequent steps of fatty acid biosynthesis are analogous to the reactions summarized above. The butyryl group is transferred to the sulfhydryl group of β-ketoacyl-ACP synthase, while the ACP molecule acquires a malonyl group from malonyl CoA. There follows the condensation of malonyl-S-ACP with butyryl-S-β-ketoacyl-ACP synthase to form β-ketohexanoyl-S-ACP plus CO_2, as before. The remaining steps of the cycle

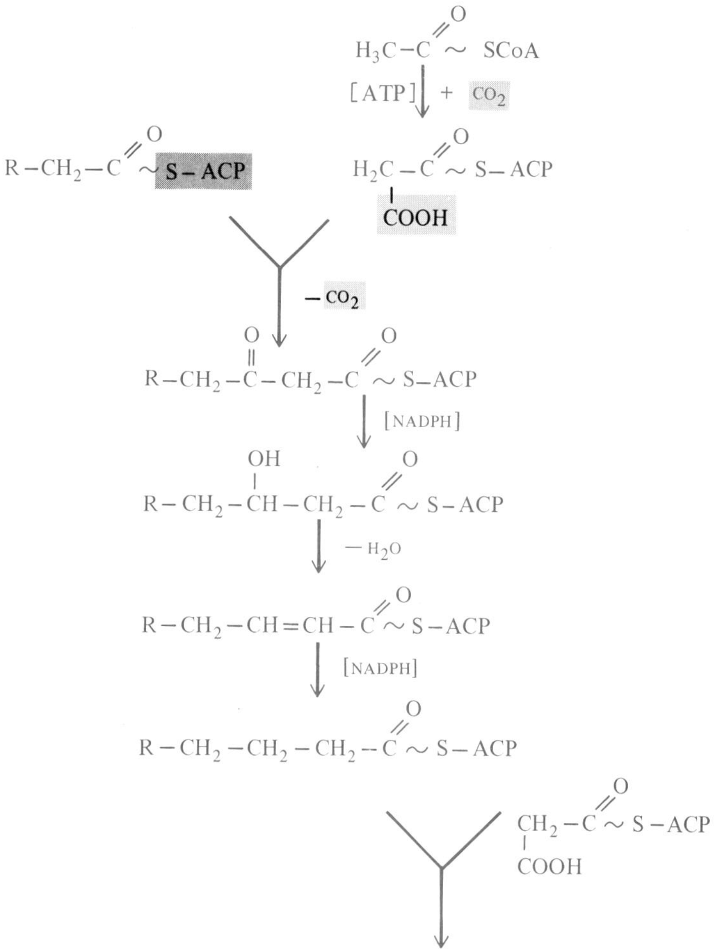

$$H_3C-C(=O) \sim SCoA \xrightarrow{[ATP] \; + \; CO_2} H_2C(COOH)-C(=O) \sim S-ACP$$

$$R-CH_2-C(=O) \sim S-ACP \; + \; H_2C(COOH)-C(=O) \sim S-ACP \xrightarrow{-CO_2} R-CH_2-C(=O)-CH_2-C(=O) \sim S-ACP$$

$$\xrightarrow{[NADPH]} R-CH_2-CH(OH)-CH_2-C(=O) \sim S-ACP$$

$$\xrightarrow{-H_2O} R-CH_2-CH=CH-C(=O) \sim S-ACP$$

$$\xrightarrow{[NADPH]} R-CH_2-CH_2-CH_2-C(=O) \sim S-ACP$$

$$+ \; CH_2(COOH)-C(=O) \sim S-ACP \longrightarrow$$

Figure 13-5
Stepwise synthesis of a fatty acid chain.

are now repeated. By repetitions of the cycle the fatty acid chain is progressively elongated (Fig. 13-5).

The enlargement of the fatty acid chain does not continue indefinitely. In most organisms the longest fatty acid synthesized is palmitic acid. This limitation is a consequence of the chain-length specificity of β-ketoacyl-ACP synthase, whose ability to accept fatty acyl groups from ACP ceases abruptly for chain lengths greater than 14 carbon atoms. The end product of the synthetic process is, after seven cycles, palmitoyl ACP. This may be released by the action of a **thioesterase** to form free palmitic acid, or it may alternatively enter other synthetic pathways leading to phospholipids and triacylglycerols. The complete reaction for the biosynthesis of palmitic acid from acetyl CoA may be written

$$8CH_3CO-S-CoA + 14NADPH + 14H^+ + 7ATP + H_2O \rightarrow$$
$$CH_3(CH_2)_{14}COOH + 8CoA + 14NADP^+ + 7ADP + 7P_i$$

Formation of Longer Fatty Acids

While the action of the extramitochondrial fatty acid synthetase system in most organisms ceases with the formation of palmitic acid, fatty acids of longer chain lengths may be produced by a separate mechanism. Mitochondria contain an enzyme system capable of elongating palmitic and other saturated fatty acids by adding acetyl groups from acetyl CoA to the carboxyl terminal end. Palmitoyl CoA is condensed with acetyl CoA to yield β-ketostearyl CoA. This is reduced by NADPH to β-hydroxystearyl-CoA. The latter is in turn dehydrated to yield an unsaturated derivative of stearyl-CoA, which is subsequently reduced to stearyl-CoA by NADPH.

An independent system for the elongation of fatty acyl CoA is present in the endoplasmic reticulum.

Biosynthesis of Unsaturated Fatty Acids

The unsaturated fatty acids palmitoleic and oleic acids differ from their respective precursors, palmitic and stearic acids, in possessing a *cis* double bond at the Δ^9 position (Table 13-2). In most aerobic organisms, this double bond is introduced by a complex enzyme system occurring in the endoplasmic reticulum of liver and adipose tissue cells. The reaction involves

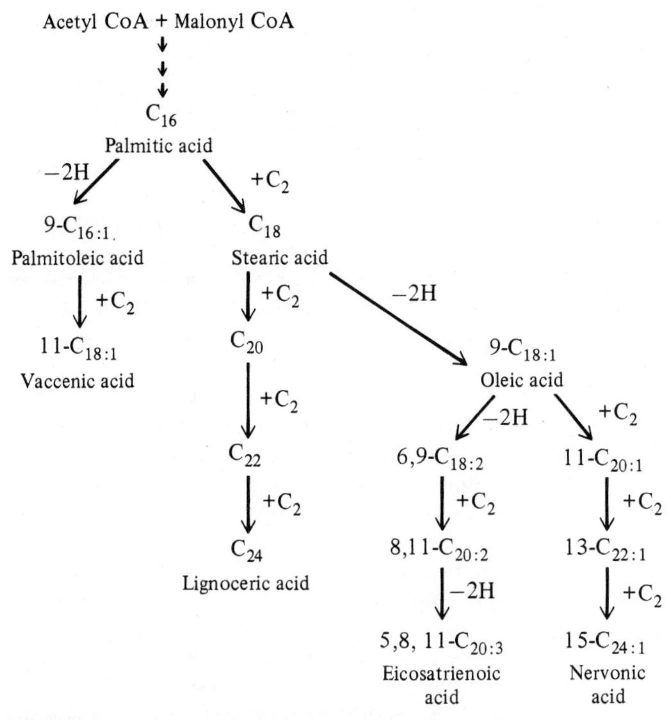

Figure 13-6
Outline of the biosynthetic pathway for the formation of unsaturated fatty acids. Here the numerical prefix designates the double bond position, while the first subscript indicates the number of carbon atoms and the second the number of double bonds.

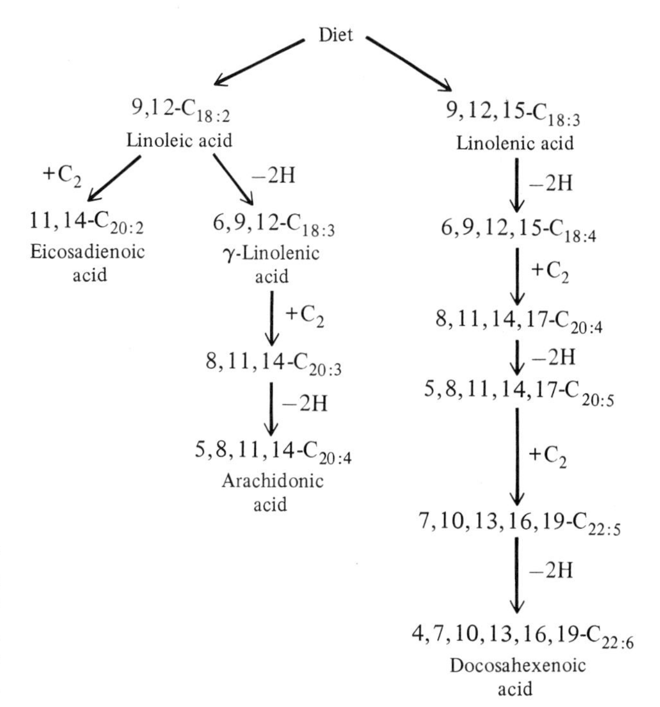

Figure 13-7 Outline of the scheme of biosynthesis of unsaturated fatty acids from the essential fatty acids.

molecular oxygen and NADPH:

$$\text{palmitoyl—CoA} + \text{NADPH} + H^+ + O_2 \rightarrow \text{palmitoleyl—CoA} + \text{NADP}^+ + 2H_2O$$

The multiply unsaturated or **polyenoic** fatty acids occurring in mammals are all derived from a group of four precursors: palmitoleic, oleic, linoleic, and linolenic acids. The latter two of these are not synthesized by mammals and must be obtained from plant sources.

Figures 13-6 and 13-7 outline the pathways whereby the higher polyenoic acids are derived from the above precursors in mammalian systems. The elongation of the fatty acid chains occurs at the carboxyl end. Desaturation occurs by a process analogous to that cited above.

13-8 SYNTHESIS OF TRIACYLGLYCEROL FATS

In mammalian tissues fatty acids occur in the free state to only a limited extent, being stored primarily as triacylglycerols. The biosynthesis of fats takes place in many tissues, with the liver the most important organ. The cytoplasm is the site of synthesis.

Formation of the triacylglycerol fats occurs via the acyl CoA derivatives of the fatty acids, which are incorporated into **phosphatidic acid** (section 14-1), a triacylglycerol precursor, by the reaction of fatty acyl CoA with

L-glycerol-3-phosphate, in a stepwise reaction catalyzed by **glycerolphosphate acyltransferase**:

$$R_1COS\text{—}CoA + \begin{array}{c} H_2COH \\ | \\ HOCH \\ | \\ H_2COPO_3H_2 \end{array} \longrightarrow \begin{array}{c} H_2COOCR_1 \\ | \\ HOCH \\ | \\ H_2COPO_3H_2 \end{array} + CoA$$

L-Glycerol-3-phosphate

$$\begin{array}{c} H_2COOCR_1 \\ | \\ HOCH \\ | \\ H_2COPO_3H_2 \end{array} + R_2COS\text{—}CoA \longrightarrow \begin{array}{c} H_2COOCR_1 \\ | \\ R_2COOCH \\ | \\ H_2COPO_3H_2 \end{array} + CoA$$

Phosphatidic acid

In the next step of the pathway, phosphatidic acid is hydrolyzed to form a diacylglycerol by the action of **phosphatidate phosphatase**:

$$\begin{array}{c} H_2COOCR_1 \\ | \\ R_2COOCH \\ | \\ H_2COPO_3H_2 \end{array} + H_2O \longrightarrow \begin{array}{c} H_2COOCR_1 \\ | \\ R_2COOCH \\ | \\ H_2COH \end{array} + P_i$$

The diacyglycerol subsequently reacts with a third molecule of fatty acyl CoA to form a triacylglycerol in a reaction catalyzed by **diacylglycerol acyltransferase**:

$$R_3CO\text{—}S\text{—}CoA + \begin{array}{c} H_2COOCR_1 \\ | \\ R_2COOCH \\ | \\ H_2COH \end{array} \longrightarrow \begin{array}{c} H_2COOCR_1 \\ | \\ R_2COOCH \\ | \\ H_2COOCR_3 \end{array} + CoA$$

The triacylglycerols present in adipose tissue generally contain two or more different fatty acids.

The L-glycerol-3-phosphate, which is a precursor of the triacylglycerols, is synthesized in animals and plants by two mechanisms. Dihydroxyacetone phosphate, which is a product of the aldolase-catalyzed reaction of glycolysis, is converted to L-glycerol-3-phosphate by the NAD^+-dependent enzyme **glycerol-3-phosphate dehydrogenase**:

$$\begin{array}{l} CH_2OH \\ | \\ C{=}O \\ | \\ CH_2OPO_3H_2 \end{array} + NADH + H^+ \longrightarrow \begin{array}{c} H_2COH \\ | \\ HOCH \\ | \\ H_2COPO_3H_2 \end{array} + NAD^+$$

Alternatively, free glycerol may be phsophorylated by ATP in a reaction catalyzed by **glycerol kinase**:

$$ATP + glycerol \xrightarrow{Mg^{2+}} \text{L-glycerol-3-phosphate} + ADP$$

13-9 KETONE BODIES

Fatty acid synthesis and breakdown normally proceed to completion without an appreciable accumulation of intermediates. Under some conditions by-products appear in the blood, which are traditionally misnamed **ketone bodies**. These include acetone, D-β-hydroxybutyric acid, and acetoacetic acid. All are derived from acetoacetyl CoA, a normal intermediate in fatty acid metabolism, which is also formed from acetyl CoA by a thiolase-catalyzed reaction:

$$2CH_3CO{-}S{-}CoA \rightleftharpoons CH_3COCH_2CO{-}S{-}CoA + CoA$$

In liver, acetoacetyl CoA is largely converted to β-hydroxy-β-methyl-glutaryl CoA:

$$CH_3COCH_2CO{-}S{-}CoA + CH_3CO{-}S{-}CoA + H_2O \longrightarrow CoA + HOOCCH_2{-}\overset{\overset{\large OH}{|}}{\underset{\underset{\large CH_3}{|}}{C}}{-}CH_2CO{-}S{-}CoA$$

β-Hydroxy-β-methylglutaryl CoA

Free acetoacetate arises by the splitting of β-hydroxy-β-methylglutaryl CoA:

$$HOOCCH_2{-}\overset{\overset{\large OH}{|}}{\underset{\underset{\large CH_3}{|}}{C}}{-}CH_2CO{-}S{-}CoA \longrightarrow CH_3COCH_2COOH + CH_3CO{-}S{-}CoA$$

Acetone is produced by the spontaneous decarboxylation of acetoacetate. Another product arising from acetoacetate is D-β-hydroxybutyric acid, which is formed by the reduction of acetoacetate by NADH in the mitochondria. The reaction is catalyzed by **D-β-hydroxybutyric acid dehydrogenase**:

$$CH_3COCH_2COOH + NADH + H^+ \rightarrow CH_3CHOHCH_2COOH + NAD^+$$

In normal metabolism the acetyl CoA formed by fatty acid oxidation enters the TCA cycle via condensation with oxaloacetate. In fasting or diabetes there is a deficiency in the oxaloacetate supply and acetyl CoA is diverted to the formation of ketone bodies. The level of oxaloacetate is lowered if carbohydrate is not available or is not metabolized at a normal rate.

The principal site of formation of ketone bodies is the liver. They are transported from the liver by the blood to peripheral tissues. The presence of important quantities of ketone bodies in the blood is an abnormal, disease-related condition, whose most common clinical origin is diabetes.

SUMMARY

The class of biomolecules termed **lipids** is defined operationally by their solubility in nonpolar solvents. The **fatty acids** consist of linear hydrocarbon chains terminating in a carboxyl group; the number of carbon atoms is usually even. The chains may be saturated or unsaturated; the most important members of the two groups have 12–20 and 16–20 carbon atoms, respectively.

The **triacylglycerols** consist of esters of glycerol with fatty acids; they are the **fat depots** in animals, the principal storage reservoirs for lipids. In mammals the major storage site of triacylglycerols is the cytoplasm of **adipose** cells.

The triacylglycerol fats are partially hydrolyzed in the intestine to form a mixture of glycerol and free fatty acids, as well as mono-, di-, and triacylglycerols. The mixture crosses the intestinal wall; the shorter fatty acids are supplied directly to the liver, while the longer fatty acids, in free or combined form, enter the circulation via the **thoracic duct** and are transported to tissues. The incorporation of triacylglycerols into adipose cells involves their partial or complete hydrolysis by **lipoprotein lipase** to glycerol and fatty acids, followed by their recombination within the adipose cells to reform triacylglycerols. The release of triacylglycercols from adipose tissue involves their hydrolysis by lipases. The free fatty acids are transported to various tissues by the blood as complexes with serum albumin.

The glycerol formed by the hydrolysis of triacylglycerols released from adipose tissue is converted to **glycerol-3-phosphate**, which is subsequently converted to **dihydroxyacetone phosphate**. The latter compound enters the glycolytic pathway. The fatty acids are converted to **fatty acyl CoA** in the cytoplasm. The acyl group is transferred to **carnitine** and crosses into the mitochondria, where it is transferred back to CoA and enters a degradative pathway.

The degradation of fatty acyl CoA occurs by the sequential removal of successive 2-carbon fragments from the carboxyl terminus of the fatty acid to form the fatty acyl CoA with two less carbon atoms. The process involves a dehydrogenation to yield an **α, β-unsaturated derivative**, followed by the addition of H_2O to form a **β-hydroxy derivative**, which is oxidized to a **β-keto** derivative. This reacts further with CoA to yield acetyl CoA and the fatty acyl CoA with two fewer carbon atoms. The process may be repeated. The acetyl CoA formed is largely burned to CO_2 and H_2O by the combined operation of the TCA cycle and the respiratory chain.

Mammals can synthesize fatty acids from acetyl CoA. Since the latter is a product of carbohydrate metabolism, carbohydrates may be metabolic precursors of fatty acids. The biosynthesis of fatty acids occurs predominantly in the cytoplasm.

The initial step in fatty acid biosynthesis is the carboxylation of acetyl CoA to form **malonyl CoA**. The subsequent reactions are catalyzed by an organized set of enzymes termed the **fatty acid synthetase** system; this includes **acyl carrier protein** (ACP), which contains a reactive —SH group, to which the elongating fatty acyl chain is attached. The malonyl CoA is transferred to ACP. The malonyl-S-ACP next reacts with an acyl group attached to the —SH group of **β-keto-acyl-ACP synthase**; the product is **acetoacetyl-S-ACP**. This is reduced by NADPH to form **β-hydroxybutyryl-S-ACP**, which is dehydrated to **crotonyl-S-ACP**. Hydrogenation of the latter by NADPH forms **butyryl-S-ACP**. Repetition of the above process results in progressive elongation of the fatty acid chain.

The incorporation of fatty acids into triacylglycerols occurs via their acyl CoA derivatives, which are incorporated into **phosphatidic acid**. This is hydrolyzed to form a diacylglycerol to which a third fatty acid is added to form the triacylglycerol.

REFERENCES

The following code is used to classify references. I: particularly useful as an introduction to the subject; R: useful primarily as a reference text; A: an advanced account of the material; H: a publication of historical importance.

General

M. Florkin and E. H. Stotz, eds., *Lipid Metabolism*, vol. 18 of *Comprehensive Biochemistry*, American Elsevier, New York (1967). (R)

M. L. Gurr and A. T. James, *Lipid Biochemistry: An Introduction*, Cornell, Ithaca (1971). (I)

F. Lynen, "The Pathway from 'Activated Acetic Acid' to the Terpenes and Fatty Acids" in *Nobel Lectures: Physiology or Medicine (1963–1970)*, p. 103, American Elsevier, New York (1972). (I)

Fatty Acids

K. S. Markley, *Fatty Acids*, Interscience, New York (1960). (R)

Oxidation

A. J. Fulco, "Metabolic Alterations of Fatty Acids," *Ann. Rev. Biochem.* 43, 147 (1974). (R)

Carnitine

I. B. Fritz, "Carnitine and Its Role in Fatty Acid Metabolism," *Adv. Lipid Res.* 1, 285 (1963). (R)

Biosynthesis of Fatty Acids

J. J. Volpe and P. R. Vagelos, "Saturated Fatty Acid Biosynthesis and Its Regulation," *Ann. Rev. Biochem.* 42, 21 (1973). (R)

REVIEW QUESTIONS

Questions marked with an asterisk are of a high level of difficulty.

*13-1 An unknown fatty acid reacts with one mole of H_2 to yield stearic acid. Heating the unknown fatty acid with a catalyst converts some of it into oleic acid; no atoms are lost or gained in the process. The unknown has a higher melting temperature than oleic but a lower melting temperature than stearic. Write the most likely structure and explain the melting behavior.

*13-2 As a general rule, the double bonds in fatty acid side chains are *cis* not *trans*. What is the likely biological significance of this observation from the fact that the degree of saturation of fatty acids in the membranes of some cells will change as a function of the temperature of the environment?

13-3 Synthesis of one palmitic acid molecule [$CH_3(CH_2)_7COOH$] involves
(a) loss of $7CO_2$'s.
(b) loss of 7 acyl carrier proteins.
(c) use of 7 NADH's.
(d) all of the above
(e) none of the above

13-4 Which of the following statement(s) is (are) true?
(a) In fatty acid synthesis the chain is carried by ACP (acyl carrier protein), whereas in fatty acid breakdown the chain is carried by CoA.
(b) Synthesis involves malonyl CoA.
(c) NADH carries the —H for fatty acid oxidation.
(d) All of the above are true.
(e) None of the above is true.

13-5 Most of fatty acid oxidation takes place
(a) in the cytoplasm.
(b) in the nucleus.
(c) in the mitochondria.
(d) in adipose tissue cells.
(e) on the outside of the mitochondrial outer membrane.

13-6 The hydrolysis of a mixture of triacylglycerols produces free oleic, stearic, and palmitic acids. What are the structures of all possible molecular species present in the original sample?

13-7 A 10-g sample of triacylglycerols obtained from an animal source required 57.6 ml of 0.40-*M* NaOH for complete hydrolysis and conversion of its fatty acids to soaps. What are the average molecular weight and chain length of the fatty acids present?

*13-8 Write a balanced equation for the biosynthesis of myristic acid from acetyl CoA, NADPH, and ATP.

13-9 Why is the phosphogluconate pathway important for fatty acid biosynthesis?

14

Phosphatides and Related Compounds

14-1 PHOSPHOLIPIDS

Like the triacylglycerol fats, the phospholipids are fatty acid derivatives. Their fatty acid component likewise accounts for most of their mass and has an important influence upon their physical properties. However, the biological roles of the two classes of lipids are altogether different. Whereas the triacylglycerols function primarily as biological *fuels*, the phospholipids are important *structural* components of biological *membranes*.

The term **phosphatide** or **phospholipid** has traditionally been applied to a rather heterogeneous set of diesters of phosphoric acid:

$$Y{-}O{-}\overset{\overset{\displaystyle O}{\|}}{\underset{\underset{\displaystyle O^-}{|}}{P}}{-}O{-}X$$

In the **glycerophosphatide** series Y is a diacylglycerol; in the **sphingolipids** it is **sphingosine** or a derivative (Fig. 14-1). X is one of the nitrogenous alcohols choline, ethanolamine, or L-serine, or the polyhydroxy compound inositol (Fig. 14-1).

The parent compound of the glycerophosphatide series contains a chiral carbon atom in its glycerol moiety and could equally well be termed D-glycerol-1-phosphate or L-glycerol-3-phosphate. This ambiguity has led to

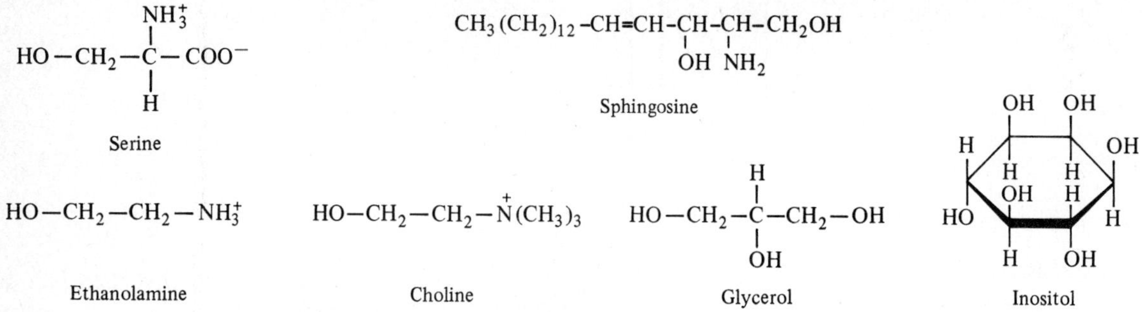

$$HO-CH_2-\underset{H}{\overset{NH_3^+}{C}}-COO^-$$

Serine

$$CH_3(CH_2)_{12}-CH{=}CH-\underset{OH}{CH}-\underset{NH_2}{CH}-CH_2OH$$

Sphingosine

$$HO-CH_2-CH_2-NH_3^+$$

Ethanolamine

$$HO-CH_2-CH_2-\overset{+}{N}(CH_3)_3$$

Choline

$$HO-CH_2-\underset{OH}{\overset{H}{C}}-CH_2-OH$$

Glycerol

Inositol

Figure 14-1
Structures of some common components of phospholipids.

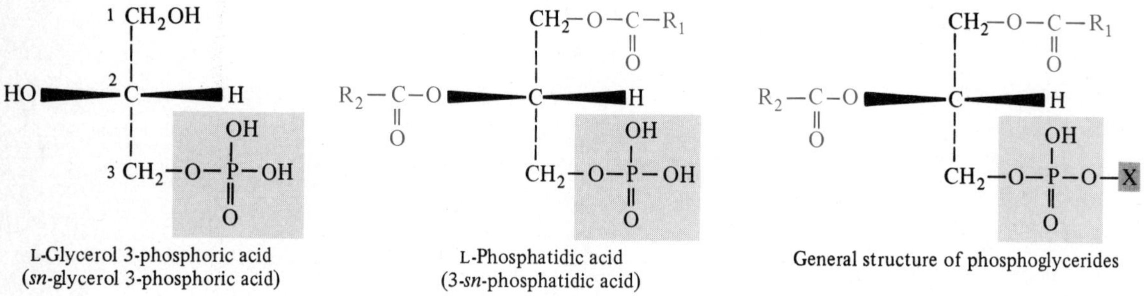

L-Glycerol 3-phosphoric acid
(*sn*-glycerol 3-phosphoric acid)

L-Phosphatidic acid
(3-*sn*-phosphatidic acid)

General structure of phosphoglycerides

Figure 14-2
Stereochemical configuration of the glycerophosphatides. The wedge-shaped bonds project out of the plane of the paper toward the viewer; the dotted lines project away from the viewer.

the introduction of the convention of **stereospecific numbering** (sn) of the carbon atoms (Fig. 14-2), which designates the parent compound as sn-glycerol-3-phosphate, which belongs to the L-stereochemical configuration and is related to L-glyceraldehyde (Fig. 9-3).

The distribution of the phosphatides is very different from that of the triacylglycerol fats. They are almost totally absent from depot fat. Their biological significance arises largely from their role as important structural components of natural membranes. As such, they occur in the cells of all tissues.

The phosphatides differ chemically from the triacylglycerols in possessing a charged site (Fig. 14-3), which modifies their properties substantially. Since they contain both a polar head group and nonpolar hydrocarbon tails, in aqueous solution they tend to form aggregates in which the polar heads are exposed to solvent and the hydrophobic tails are sequestered inside so as to be in mutual contact. The preferred structure is a **bilayer** (Fig. 14-4).

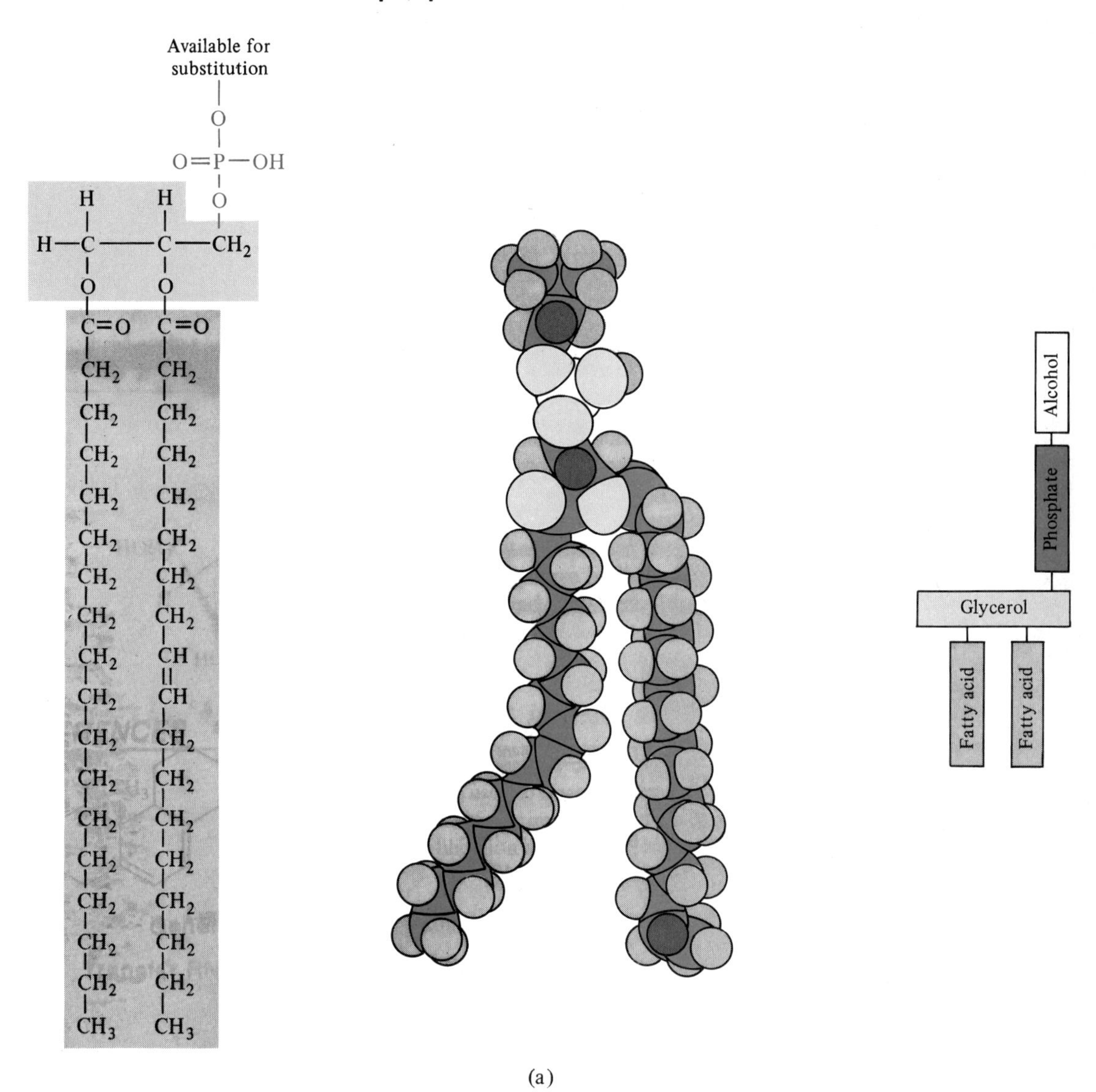

Figure 14-3
(a) Generalized structure of a glycerophosphatide and a space-filling model of phosphatidyl choline.

Phospholipid bilayers may be prepared as spherical systems or **liposomes**, or as planar bilayer membranes. Both have the bilayer structure described above and are stabilized primarily by hydrophobic interactions of the hydrocarbon tails, with a contribution from electrostatic and hydrogen-bonding interactions between the polar head and water molecules.

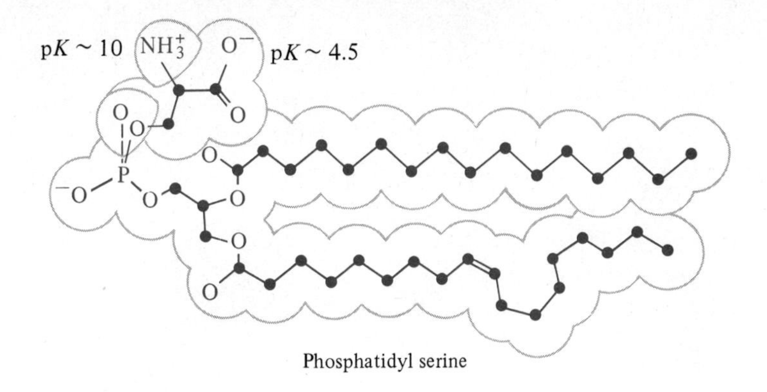

Phosphatidyl serine

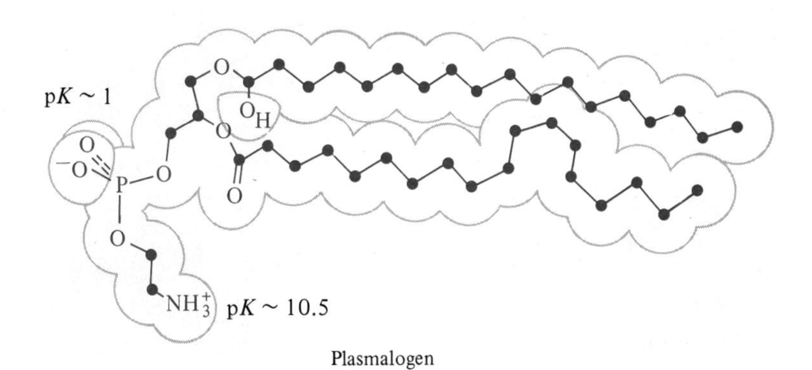

Plasmalogen

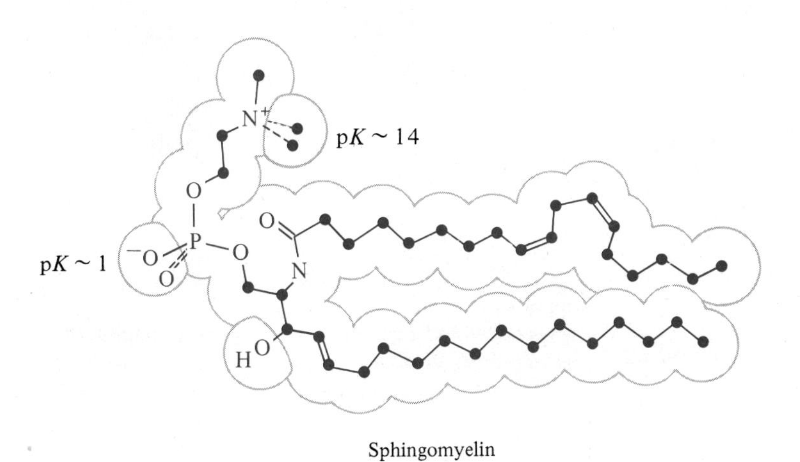

Sphingomyelin

(b)

Figure 14-3
(b) Comparison of the sizes and shapes of several phospholipids.

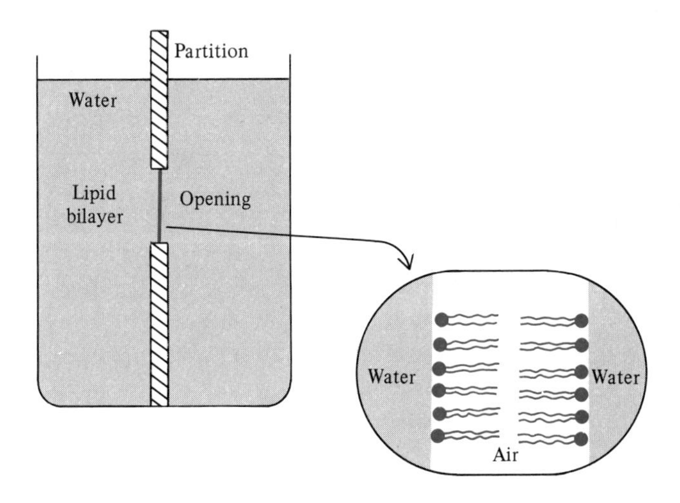

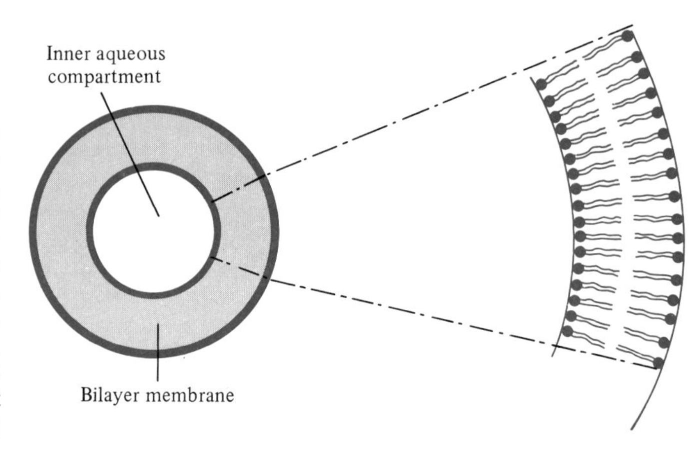

Figure 14-4 Planar bilayer membrane and a spherical liposome formed from phospholipid molecules. (From *Biochemistry* by Lubert Stryer. W. H. Freeman and Company. Copyright © 1975.)

Liposomes may be formed by sonicating a suspension of phospholipids in an aqueous medium. They consist of spherical bilayers, about 5 nm thick, which enclose an aqueous compartment (Fig. 14-4). The diameters are 10–100 nm. Planar bilayer membranes with widths of a millimeter or more can be formed across a hole in a partition separating two aqueous chambers.

14-2 GLYCEROPHOSPHATIDES

The glycerophosphatides are derivatives of L-α-phosphatidic acid, a glycerol derivative in which one hydroxyl is esterified with phosphoric acid and the other two with fatty acids. This compound has already been encountered as an intermediate in the biosynthesis of triacylglycerols (section 13-7):

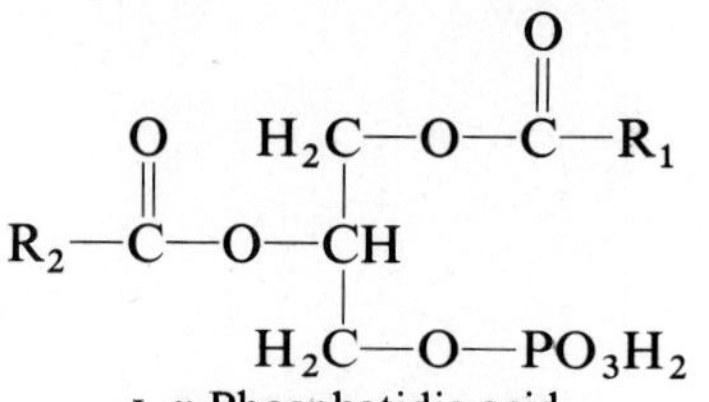

$$\begin{array}{l} \qquad\qquad\quad H_2C{-}O{-}\overset{\displaystyle O}{\overset{\|}{C}}{-}R_1 \\ R_2{-}\overset{\displaystyle O}{\overset{\|}{C}}{-}O{-}CH \\ \qquad\qquad\quad H_2C{-}O{-}PO_3H_2 \end{array}$$

L-α-Phosphatidic acid

Lecithin or **phosphatidyl choline** is a phosphatide in which phosphoric acid is esterified by choline:

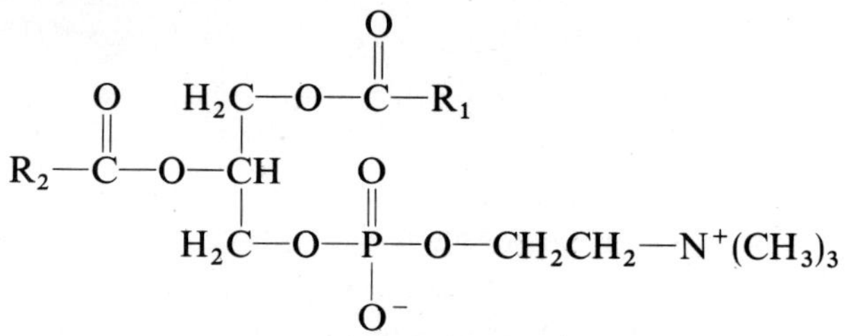

$$\begin{array}{l} \qquad\qquad\quad H_2C{-}O{-}\overset{\displaystyle O}{\overset{\|}{C}}{-}R_1 \\ R_2{-}\overset{\displaystyle O}{\overset{\|}{C}}{-}O{-}CH \\ \qquad\qquad\quad H_2C{-}O{-}\underset{\displaystyle O^-}{\underset{|}{\overset{\displaystyle O}{\overset{\|}{P}}}}{-}O{-}CH_2CH_2{-}N^+(CH_3)_3 \end{array}$$

Phosphatidyl choline

Since choline contains a quaternary nitrogen, it is a strong base that is completely ionized. Lecithin is hence a **zwitterion** over a wide pH range, with both positive and negative charges.

The lecithins contain the same fatty acids that occur in the triacylglycerols. The fatty acid esterified at the terminal (α) position of glycerol is usually saturated (stearic or palmitic), while that at the central (β) position is generally unsaturated (linolenic or oleic). Other fatty acids are sometimes found.

The other important phosphatidic acid derivatives are **phosphatidyl ethanolamine** and **phosphatidyl serine:**

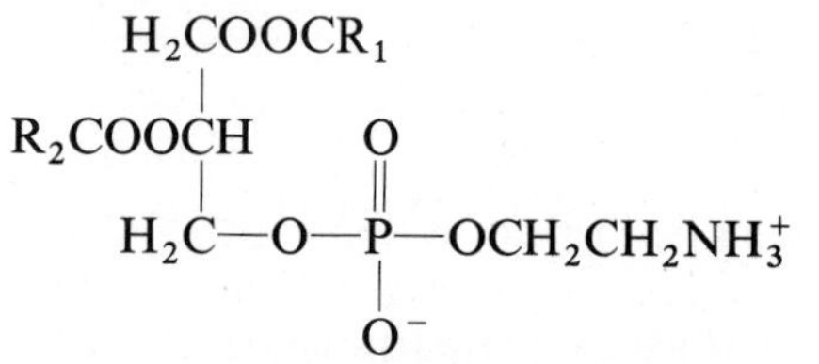

$$\begin{array}{l} \quad H_2COOCR_1 \\ R_2COOCH \\ \quad H_2C{-}O{-}\underset{\displaystyle O^-}{\underset{|}{\overset{\displaystyle O}{\overset{\|}{P}}}}{-}OCH_2CH_2NH_3^+ \end{array}$$

Phosphatidyl ethanolamine (cephalin)

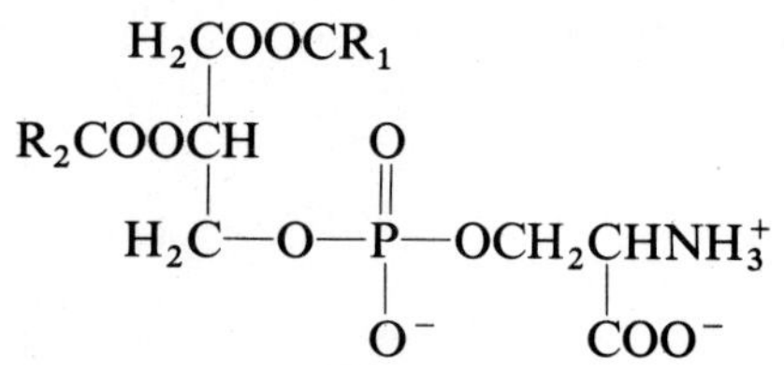

$$\begin{array}{l} \quad H_2COOCR_1 \\ R_2COOCH \\ \quad H_2C{-}O{-}\underset{\displaystyle O^-}{\underset{|}{\overset{\displaystyle O}{\overset{\|}{P}}}}{-}OCH_2\underset{\displaystyle COO^-}{\underset{|}{C}}HNH_3^+ \end{array}$$

Phosphatidyl serine

While both are zwitterions at neutral pH, the primary amino group is a weak electrolyte and is protonated only over a restricted range of pH.

Serine is a biochemical precursor of ethanolamine. Since phosphatidyl ethanolamine can be methylated to form the corresponding lecithins, the three phosphatides are biochemically related.

In the **plasmalogens** the fatty acid at the terminal position of glycerol is replaced by a *cis* α-β unsaturated ether; the central position is esterified by an

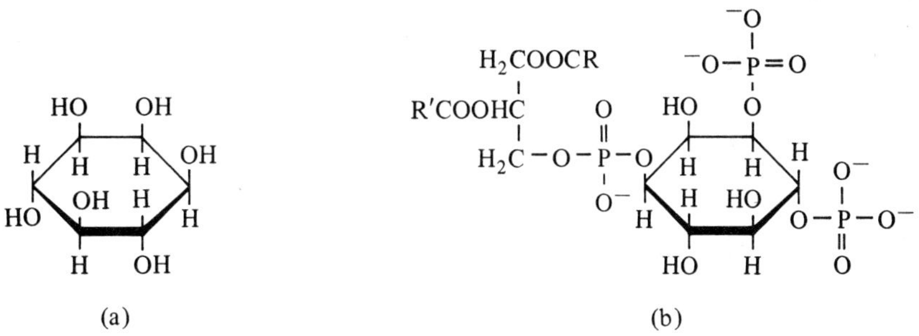

Figure 14-5 Structure of (a) myo-inositol and of (b) 1-phosphatidyl-L-myo-inositol-4,5-diphosphate.

(a) (b)

unsaturated fatty acid, while the base is ethanolamine or serine:

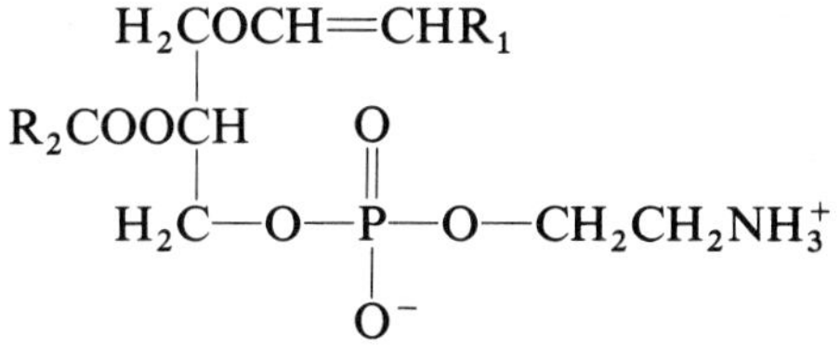

$$
\begin{array}{l}
H_2COCH{=}CHR_1 \\
R_2COOCH \\
H_2C{-}O{-}P(=O)(O^-){-}O{-}CH_2CH_2NH_3^+
\end{array}
$$

Ethanolamine plasmalogen
(phosphatidal ethanolamine)

The plasmalogens are particularly abundant in the membranes of muscle and nerve tissue.

The **inositol phosphatides** or **inositides** contain the cyclic hexahydroxy alcohol inositol (Fig. 14-5). Nine different stereoisomers are possible for inositol. In some inositol phosphatides the inositol group is esterified with one or more residues of phosphoric acid (Fig. 14-5). These compounds occur in the brain.

14-3 SPHINGOLIPIDS

The phosphatides of the sphingosine series consist of diesters of phosphoric acid with choline and sphingosine, or a derivative. In the **sphingomyelins**, which occur in brain and nervous tissue, a fatty acid residue is in amide linkage to the amino group of sphingosine. There is some structural similarity to the lecithins:

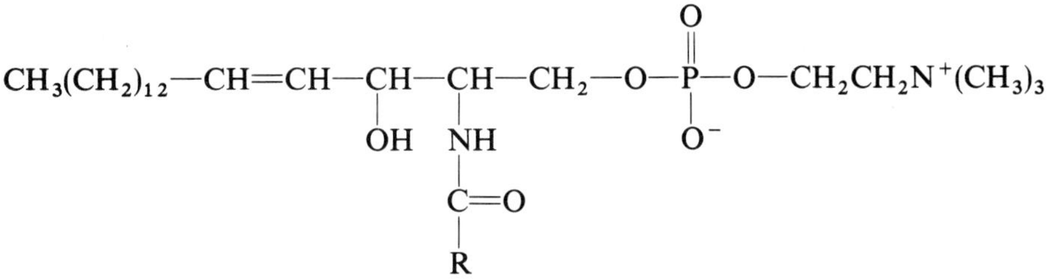

$$
CH_3(CH_2)_{12}{-}CH{=}CH{-}CH(OH){-}CH(NH{-}C(=O){-}R){-}CH_2{-}O{-}P(=O)(O^-){-}O{-}CH_2CH_2N^+(CH_3)_3
$$

Sphingomyelin
(R is a fatty acid group)

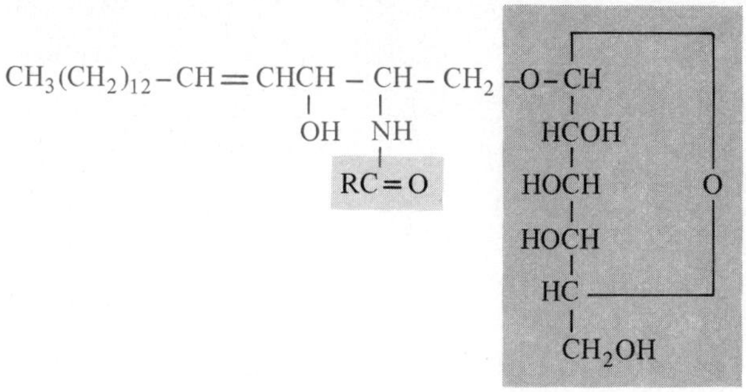

$$CH_3(CH_2)_{12}-CH=CHCH(OH)-CH(NH-C(=O)R)-CH_2-O-CH\ [HCOH-HOCH-HOCH-HC(-O-)]-CH_2OH$$

Figure 14-6 Structure of a cerebroside. RC=O is the fatty acyl component. The sphingosine-fatty acid portion (without the sugar) is a **ceramide**.

Sphingosine may be partially replaced by dihydrosphingosine:

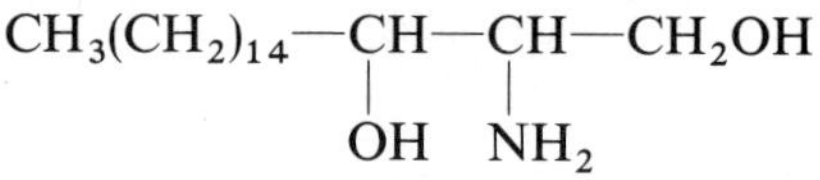

$$CH_3(CH_2)_{14}-\underset{OH}{CH}-\underset{NH_2}{CH}-CH_2OH$$

Dihydrosphingosine

The sphingomyelins are mixtures of species with different fatty acid contents. Two 24-carbon acids, lignoceric (saturated) and nervonic (unsaturated) are particularly prominent.

Cerebrosides

The cerebrosides are structurally related to the sphingomyelins, but contain no phosphorus and are uncharged. The phosphorylcholine group of the sphingomyelins is replaced by a monosaccharide unit, usually D-galactose, in glycosidic linkage (Fig. 14-6). The fatty acid has 22–26 carbon atoms and often has a double bond or a hydroxyl group in the α-position. One class of cerebrosides, **sulfatides**, contains a sulfate ester of galactose.

Gangliosides

These are carbohydrate-containing lipids of complex structure, which, like the cerebrosides, contain no phosphorus. Their components include fatty acids, glucose, galactose, hexoseamine, sphingosine, and *N*-acetylneuraminic acid (Fig. 14-7). Gangliosides are particularly abundant in the brain, but have also been identified in nonneural tissue.

Hereditary Diseases Involving Complex Lipids

A number of serious genetically controlled diseases are known that reflect the excessive accumulation of particular lipids in specific tissues. In all of these the accumulation results from a deficiency of one or more of the enzymes controlling degradation of the lipid; mental retardation and neurological disorders are common.

Two of these diseases, **Tay-Sachs disease**, which produces an abnormally high level of a **ganglioside** in brain, and **Gaucher's disease**, which results in extensive deposition of cerebrosides, are particularly common among European Jews. **Fabry's disease** and **Niemann-Pick** disease produce high levels of **sphingolipid** and **sphingomyelin**, respectively.

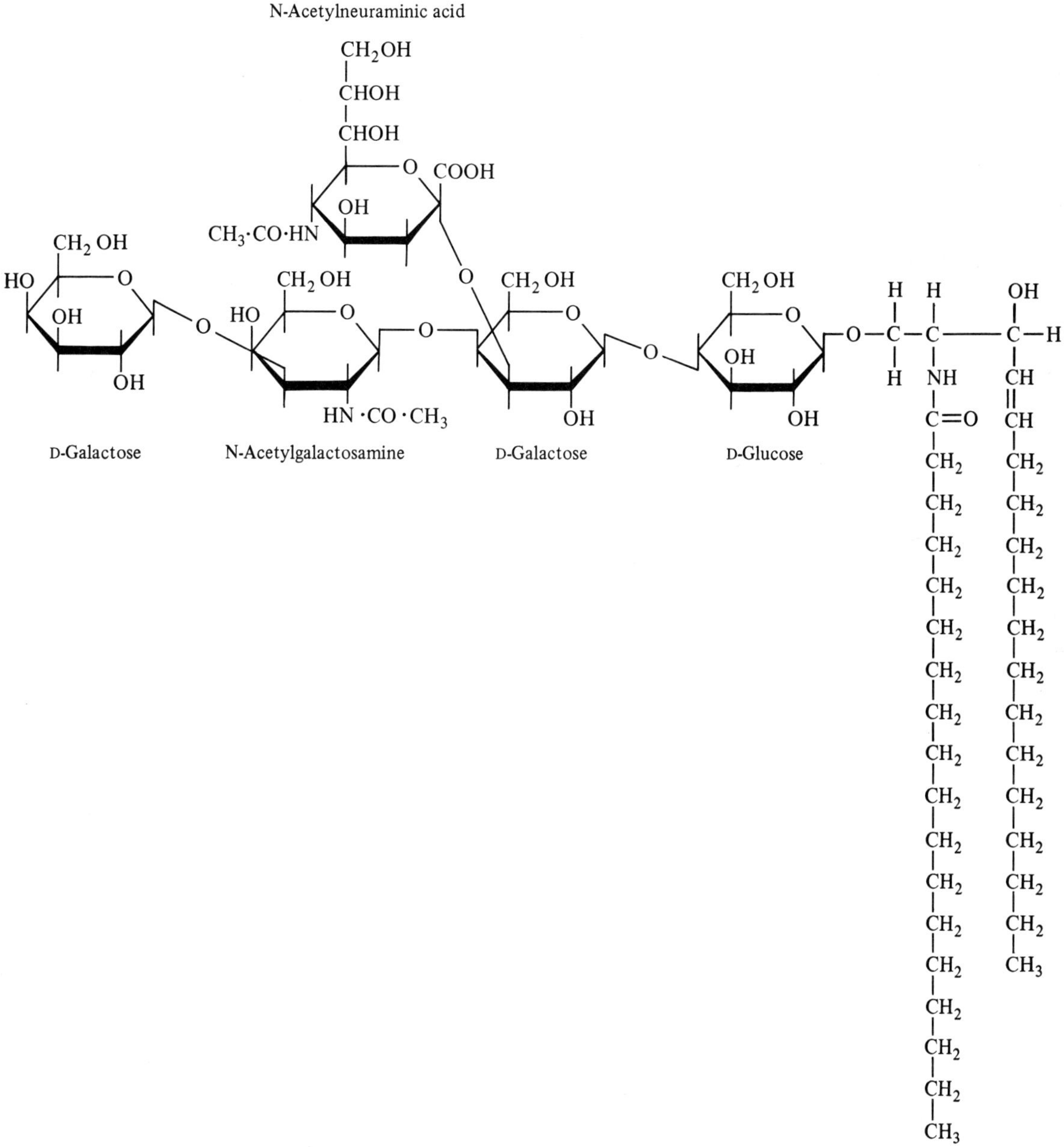
N-Acetylneuraminic acid
CH2OH
CHOH
CHOH
COOH
OH
CH3·CO·HN
CH2 OH
HO
O
OH
OH
CH2 OH
HO
HN·CO·CH3
CH2 OH
OH
CH2 OH
OH
OH
H H
C C
H NH
C=O
CH2
CH3
OH
C—H
CH
CH
CH2
CH3
D-Galactose
N-Acetylgalactosamine
D-Galactose
D-Glucose

Figure 14-7
Structure of a ganglioside.

14-4 METABOLISM OF THE PHOSPHATIDES

Biosynthesis of Phosphatidic Acid

Phosphatidic acid itself may arise by several biosynthetic routes. One of these, its formation from L-glycerol-3-phosphate and acyl CoA, has already been discussed (section 13-7). Phosphatidic acid may also be synthesized by esterification of the fatty acyl group with dihydroxyacetone phosphate. The pathway is as follows:

$$R_1\text{—}\overset{\text{O}}{\overset{\|}{\text{C}}}\text{—S—CoA} + \begin{array}{c} H_2C\text{—OH} \\ | \\ C{=}O \\ | \\ H_2C\text{—O—}PO_3H_2 \end{array} \longrightarrow \begin{array}{c} H_2C\text{—O—CO—}R_1 \\ | \\ C{=}O \\ | \\ H_2C\text{—O—}PO_3H_2 \end{array} + \text{CoA}$$

$$\begin{array}{c} H_2C\text{—O—CO—}R_1 \\ | \\ C{=}O \\ | \\ H_2C\text{—O—}PO_3H_2 \end{array} + \text{NADPH} + H^+ \longrightarrow \begin{array}{c} H_2C\text{—O—CO—}R_1 \\ | \\ HO\text{—CH} \\ | \\ H_2C\text{—O—}PO_3\text{—}H_2 \\ \text{Lysophosphatidic acid} \end{array}$$

$$\begin{array}{c} H_2C\text{—O—CO—}R_1 \\ | \\ HO\text{—CH} \\ | \\ H_2C\text{—O—}PO_3H_2 \end{array} + R_2\text{—}\overset{\text{O}}{\overset{\|}{\text{C}}}\text{—S—CoA} \longrightarrow \begin{array}{c} H_2C\text{—O—CO—}R_1 \\ | \\ R_2\text{—CO—O—CH} \\ | \\ H_2C\text{—O—}PO_3H_2 \\ \text{L-}\alpha\text{-Phosphatidic acid} \end{array}$$

The initial reaction of the above is catalyzed by **dihydroxyacetone phosphate acyltransferase**.

Biosynthesis of Phosphatidyl Ethanolamine

The formation of phosphatidyl ethanolamine in animal cells has as its initial step the phosphorylation of ethanolamine by ATP in a reaction catalyzed by **ethanolamine kinase**:

$$H_2N\text{—}CH_2CH_2\text{—OH} + \text{ATP} \rightarrow \underset{\text{Phosphoethanolamine}}{H_2N\text{—}CH_2CH_2\text{—}OPO_3^{2-}} + \text{ADP}$$

Phosphoethanolamine subsequently reacts with cytidine triphosphate (CTP) to form cytidine diphosphoethanolamine (CDP-ethanolamine) (Fig. 14-8):

$$\text{CTP} + \text{phosphoethanolamine} \rightarrow \text{CDP-ethanolamine} + PP_i$$

The reaction is catalyzed by **phosphoethanolamine cytidylyltransferase**.

Finally, the phosphoethanol portion of CDP-ethanolamine is transferred to diacylglycerol to form phosphatidyl ethanolamine in a reaction catalyzed by **phosphoethanolamine transferase**.

$$\text{CDP-ethanolamine} + \text{diacylglycerol} \rightarrow \text{phosphatidyl ethanolamine} + \text{CMP}$$

The enzymes involved in the above reactions are incorporated into the endoplasmic reticulum.

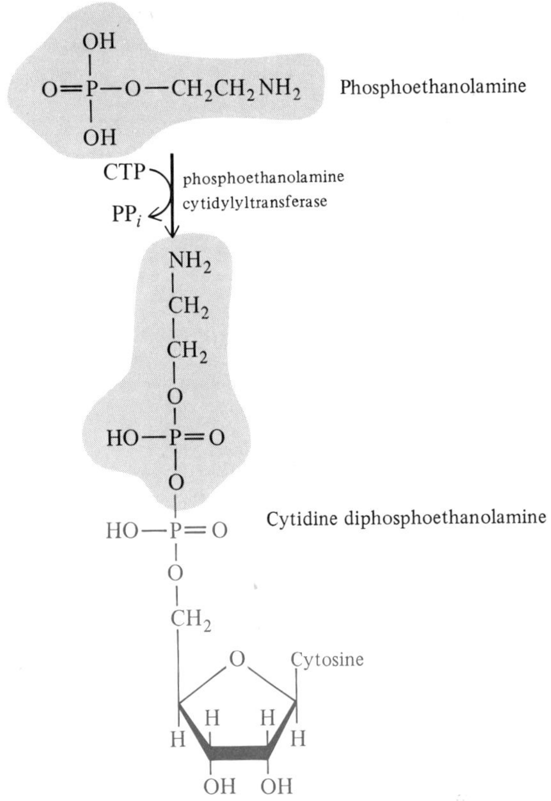

$O{=}P(OH)_2{-}O{-}CH_2CH_2NH_2$ Phosphoethanolamine

CTP → PP_i phosphoethanolamine cytidylyltransferase

$NH_2{-}CH_2{-}CH_2{-}O{-}P(OH)(=O){-}O{-}P(OH)(=O){-}O{-}CH_2$–ribose (H, H, H, H, OH, OH)–Cytosine

Cytidine diphosphoethanolamine

Figure 14-8 Biosynthesis of CDP-ethanolamine from phosphoethanolamine.

Biosynthesis of Phosphatidyl Choline

Animal cells can synthesize phosphatidyl choline by a reaction sequence beginning with the phosphorylation of choline by ATP in a reaction catalyzed by **choline kinase**:

$$(CH_3)_3N^+{-}CH_2CH_2{-}OH + ATP \rightarrow \underset{\text{Phosphocholine}}{(CH_3)_3N^+{-}CH_2CH_2{-}O{-}PO_3^{2-}} + ADP$$

CDP-choline is next formed by reaction of phosphocholine with CTP. The enzyme involved is **phosphocholine cytidylyltransferase**:

$$\text{phosphocholine} + \text{CTP} \rightarrow \text{CDP-choline} + PP_i$$

Finally, **phosphocholine transferase** catalyzes the formation of phosphatidyl choline from CDP-choline and diacylglycerol:

$$\text{CDP-choline} + \text{diacylglycerol} \rightarrow \text{phosphatidyl choline} + \text{CMP}$$

Alternatively, phosphatidyl choline may be synthesized by direct methylation of the amino group of phosphatidyl ethanolamine by S-adenosylmethionine in a reaction catalyzed by **phosphatidyl ethanolamine methyltransferase**:

$$\text{phosphatidyl ethanolamine} + 3\text{S-adenosylmethionine} \rightarrow \text{phosphatidyl choline} + 3\text{S-adenosylhomocysteine}$$

Biosynthesis of Phosphatidyl Serine

Phosphatidyl serine is formed in animal cells by the displacement of the ethanolamine group of phosphatidyl ethanolamine by serine:

$$\text{phosphatidyl ethanolamine} + \text{serine} \rightarrow \text{phosphatidyl serine} + \text{ethanolamine}$$

Biosynthesis of Phosphatidyl Inositol

The formation of phosphatidyl inositol involves the interaction of CDP-choline and phosphatidic acid to form a CDP-diacylglycerol, followed by the transfer of a phosphatidic acid unit to inositol:

$$\begin{array}{l} \quad\quad H_2C{-}O{-}CO{-}R_1 \\ R_2{-}CO{-}O{-}CH \\ \quad\quad H_2C{-}O{-}PO_3H_2 \end{array} + \text{CDP-choline} \longrightarrow \text{phosphocholine} + \begin{array}{l} \quad\quad H_2C{-}O{-}CO{-}R_1 \\ R_2{-}CO{-}O{-}CH \\ \quad\quad H_2C{-}O{-}CDP \end{array}$$

CDP—diacylglycerol

$$\begin{array}{l} \quad\quad H_2C{-}O{-}CO{-}R_1 \\ R_2{-}CO{-}O{-}CH \\ \quad\quad H_2C{-}O{-}CDP \end{array} + \text{inositol} \longrightarrow \text{CMP} + \begin{array}{l} \quad\quad H_2C{-}O{-}CO{-}R_1 \\ R_2{-}CO{-}O{-}CH \quad\quad O \\ \quad\quad H_2C{-}O{-}\overset{\|}{P}{-}O{-}\text{inositol} \\ \quad\quad\quad\quad\quad\quad\; OH \end{array}$$

Phosphatidyl inositol

The latter reaction is catalyzed by **CDP-diacylglycerol inositol phosphatidyl-transferase**.

Biosynthesis of Sphingomyelin

The two terminal carbon atoms at the hydroxyl end of sphingosine are derived from L-serine, while the balance of the molecule is obtained from palmitoyl CoA:

$$\underset{\text{Palmitoyl-CoA}}{R{-}CO{-}S{-}CoA} + \underset{\text{L-Serine}}{\begin{array}{c} CH_2OH \\ | \\ HC{-}NH_2 \\ | \\ COOH \end{array}} \longrightarrow \underset{\text{3-Dehydrosphinganine}}{R{-}CO{-}\overset{NH_2}{\overset{|}{C}H}{-}CH_2OH} + CoA + CO_2$$

$$R{-}CO{-}\overset{NH_2}{\overset{|}{C}H}{-}CH_2OH + NADPH + H^+ \longrightarrow NADP^+ + \underset{\text{Sphinganine}}{R{-}CHOH{-}\overset{NH_2}{\overset{|}{C}H}{-}CH_2OH}$$

$$R{-}CHOH{-}\overset{NH_2}{\overset{|}{C}H}{-}CH_2OH + FAD \longrightarrow FADH_2 + CH_3(CH_2)_{12}{-}CH{=}CH{-}\overset{OH}{\overset{|}{C}H}{-}\overset{NH_2}{\overset{|}{C}H}{-}H_2COH$$

Sphingosine

In the next step of the reaction sequence leading to sphingomyelin, the amino group of sphingosine is acylated by a fatty acyl CoA to form *N*-acyl-sphingosine or **ceramide** in a reaction catalyzed by **sphingosine acyltransferase**:

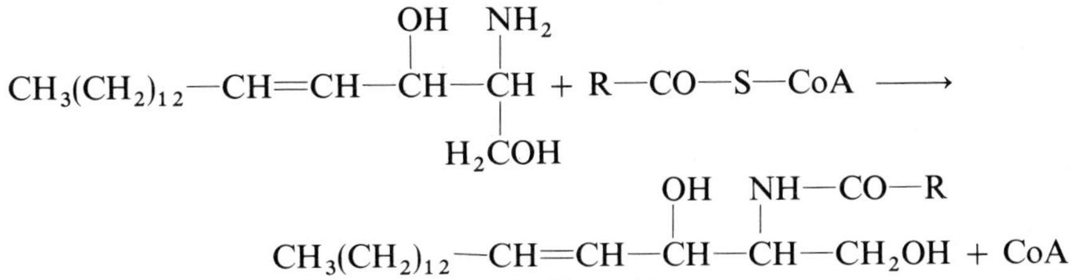

$$CH_3(CH_2)_{12}-CH{=}CH-\overset{OH}{\overset{|}{C}}H-\overset{NH_2}{\overset{|}{\underset{\underset{H_2COH}{|}}{C}}}H + R-CO-S-CoA \longrightarrow$$

$$CH_3(CH_2)_{12}-CH{=}CH-\overset{OH}{\overset{|}{C}}H-\overset{NH-CO-R}{\overset{|}{C}}H-CH_2OH + CoA$$

Ceramide

Finally, sphingomyelin is formed by reaction of a ceramide with CDP-choline. The enzyme involved is **ceramide cholinephosphotransferase**:

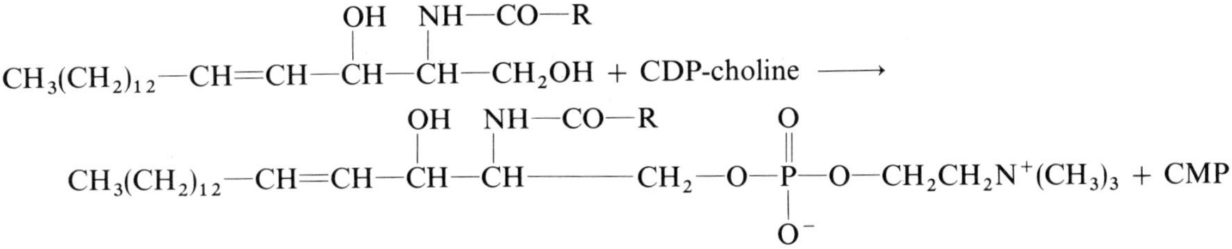

$$CH_3(CH_2)_{12}-CH{=}CH-\overset{OH}{\overset{|}{C}}H-\overset{NH-CO-R}{\overset{|}{C}}H-CH_2OH + \text{CDP-choline} \longrightarrow$$

$$CH_3(CH_2)_{12}-CH{=}CH-\overset{OH}{\overset{|}{C}}H-\overset{NH-CO-R}{\overset{|}{C}}H-CH_2-O-\overset{O}{\overset{\|}{\underset{\underset{O^-}{|}}{P}}}-O-CH_2CH_2N^+(CH_3)_3 + CMP$$

Sphingomyelin

SUMMARY

While the **phospholipids** resemble the triacylglycerols in being fatty acid derivatives, their primary biological function is to serve as **membrane components** rather than fuels. The phospholipids are **diesters** of **phosphoric** acid of the general type:

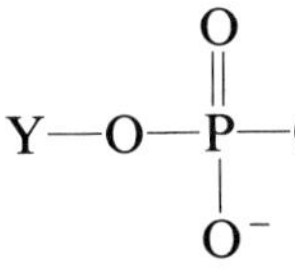

$$Y-O-\overset{O}{\overset{\|}{\underset{\underset{O^-}{|}}{P}}}-O-X$$

In the **glycerophosphatide** series, Y is a diacylglycerol; in the sphingolipids, it is sphingosine or a derivative. X is **choline, ethanolamine**, or **serine**, or the polyhydroxy compound **inositol**. The glycerophosphatides are esters of **phosphatidic acid** with a nitrogenous alcohol or inositol. In the **plasmalogens**, the fatty acid at the terminal position of glycerol is replaced by a *cis* α-β unsaturated ether; the central position is esterified by an unsaturated fatty acid, while the base is ethanolamine or serine.

The phosphatides differ chemically from the triacylglycerols in possessing a charged site. Since they contain both a *polar* head group and *hydrocarbon* tails, they tend to form bilayer aggregates in solution, in which the heads are exposed to solvent and the tails are sequestered inside so as to be in mutual contact.

In the **sphingomyelins**, which occur in brain and nervous tissue, a fatty acid residue is in amide linkage to the amino group of **sphingosine**. The **cerebrosides** are structurally related to the sphingomyelins, but contain no phosphorus and are uncharged. The phosphorylcholine group of the sphingomyelins is replaced by a monosaccharide unit, usually D-galactose. The **gangliosides** are carbohydrate-containing lipids of complex structure that, like the cerebrosides, contain no phosphorus. Their components include fatty acids, glucose, galactose, hexosamine, sphingosine, and *N*-acetylneuraminic acid.

The biosynthesis of phosphatidic acid may occur by several routes. One of these is via the reaction of **L-glycerol-3-phosphate** with acyl CoA. Alternatively, acyl CoA may esterify the hydroxyl group of **dihydroxyacetone phosphate**; this is followed by reduction of the carbonyl to a hydroxyl group by NADPH and esterification of the hydroxyl by a second molecule of acyl CoA.

The formation of **phosphatidyl ethanolamine** begins with the phosphorylation of ethanolamine by ATP to form **phosphoethanolamine**. This reacts with CTP to yield **CDP-ethanolamine**. Finally, the phosphoethanol portion of CDP-ethanolamine is transferred to diacylglycerol to form phosphatidyl ethanolamine.

Phosphatidyl choline may be formed either by direct methylation of the amino group of **phosphatidyl ethanolamine** by **S-adenosyl methionine** or by a reaction sequence beginning with the phosphorylation of **choline** by ATP to form **phosphocholine**. This reacts with CTP to yield **CDP-choline**, from which the phosphocholine portion is transferred to diacylglycerol to form phosphatidyl choline.

Phosphatidyl serine is formed in animal cells by the displacement of the ethanolamine group of phosphatidyl ethanolamine by serine. The formation of **phosphatidyl inositol** involves the reaction of **CDP-choline** and phosphatidic acid to form a **CDP-diacylglycerol**, followed by the transfer of a phosphatidic acid unit to **inositol**.

The biosynthesis of **sphingosine** begins with the transfer of a **palmitoyl** group from **palmitoyl CoA** to **serine**, forming **3-dehydrosphinganine** which is reduced by NADPH to **sphinganine**. This is oxidized to sphingosine by FAD. In the subsequent formation of **sphingomyelin**, the amino group of sphingosine is acylated by a fatty acyl CoA to form **N-acyl-sphingosine** or **ceramide**. Sphingomyelin arises by the reaction of a ceramide with CDP-choline.

REFERENCES

The following code is used to classify references. I: particularly useful as an introduction to the subject; R: useful primarily as a reference text; A: an advanced account of the material; H: a publication of historical importance.

General G. M. Ansell, J. N. Hawthorne, and R. M. C. Dawson, *Form and Function of Phospholipids*, Elsevier, New York (1973). (A)

D. M. Greenberg, ed., *Metabolic Pathways*, 3rd. ed., vol. 2, Academic, New York (1968). (R)

W.C. McMurray and W.L. Magee, "Phospholipid Metabolism," *Ann. Rev. Biochem.* 42, 61 (1973). (R)

S. Wakil, ed., *Lipid Metabolism*, Academic, New York (1970). (A)

Glycerophosphatides

H. Van Den Bosch, "Phosphoglyceride Metabolism," *Ann. Rev. Biochem.* 43, 243 (1974). (R)

Sphingolipids

W. Stoffel, "Sphingolipids," *Ann. Rev. Biochem.* 40, 57 (1971). (R)

REVIEW QUESTIONS

Questions marked with an asterisk are of a high level of difficulty.

*14-1 Cells of *E. coli* (and other organisms) regulate the fatty acid composition of their membrane phospholipids to maintain membrane fluidity at a *constant* level at a given temperature. *E. coli* has both saturated and unsaturated fatty acid side chains in its membrane phospholipids. A particular mutant of *E. coli* has lost the ability to synthesize its own unsaturated fatty acids; it can synthesize palmitic acid (saturated $C_{16:0}$) but requires an unsaturated fatty acid supplement to grow. This requirement can be satisfied by adding any one of four unsaturated fatty acids to the growth medium:

(a) *cis*-Δ^9-$C_{14:1}$ (a *cis*-unsaturated 14-carbon fatty acid with one double bond between C-9 and C-10)
(b) *cis*-Δ^9-$C_{18:1}$
(c) *cis*-Δ^9-$C_{16:1}$
(d) *cis,cis*-$\Delta^{9,12}$-$C_{14:2}$

Cultures of the mutant are grown on each of the four unsaturated fatty acids at 30°C, and then the ratio of palmitate to unsaturated fatty acid in membrane phospholipid is determined. The four results, in random order, are 53:47, 79:21, 46:54, and 70:30. Match each of the four unsaturated fatty acids, (a), (b), (c), and (d), with its most likely result. Rationalize your answers.

14-2 Paper electrophoresis is carried out at pH 7.0 for the compounds listed below. Which would move toward the cathode, which toward the anode, and which would remain stationary?

(a) phosphatidyl serine
(b) phosphatidyl choline
(c) phosphatidyl ethanolamine
(d) phosphatidyl inositol

14-3 What products are formed by *complete* hydrolysis of the following?

(a) phosphatidyl choline, containing 1 oleic and 2 palmitic residues
(b) phosphatidyl serine, containing 1 stearic and 2 oleic residues

*14-4 State how many high energy phosphate bonds are needed for the synthesis of dipalmityl phosphatidyl ethanolamine, starting from palmitic acid, glucose, and ethanolamine.

*14-5 Suggest a biological reason why phosphatides generally contain one saturated and one unsaturated fatty acid.

*14-6 Calorimetric measurements indicate the occurrence of processes resembling phase transitions in phospholipid bilayers as the temperature is increased. What events at the molecular level might these reflect?

15

Biological Membranes

15-1 STRUCTURE AND COMPOSITION OF MEMBRANES

We have already had numerous occasions to mention biological membranes, which have been encountered as boundaries of eucaryotic cells and organelles and as *selective* barriers to the free diffusion of small ions and molecules. It is this selective permeability that is the basis for the indispensable function of membranes in regulating the composition of the intracellular medium, in controlling the transport of metabolites, and in governing the flow of information between cells and their environment. Membranes are not merely passive barriers but highly organized structures that contain specific molecular gates and pumps. The latter require metabolic energy to function and are said to engage in **active transport**. Membranes also contain specific receptors that allow a response of cells to external stimuli.

While membranes are highly diverse in structure, they all share certain structural features. All membranes are made up of proteins and lipids in proportions that vary widely for different membranes (Table 15-1). These are organized into thin sheets, about 6–10 nm thick. The proteins are responsible for the specific functions of membranes, including selective permeability, active transport, and energy transduction. The proteins and lipids that comprise natural membranes are held together by noncovalent interactions.

The lipids present in membranes are chiefly polar. The phospholipids are generally the most abundant species, with some contribution from the sphingolipids. Some types of membranes contain cholesterol (chapter 16). While the proportions of lipid types are genetically controlled and characteristic of the membrane, the fatty acid compositions of individual lipids are variable and subject to nutritional influence.

Table 15-1
Composition of Some Membranes

COMPOUND	PERCENTAGE OF MEMBRANE		
	Human Red Blood Cell	Mitochondrial	E. coli
Protein	60	76	75
Phosphatidyl choline	7	9	
Phosphatidyl ethanolamine	7	8	18
Phosphatidyl serine	3		
Phosphatidyl inositol	0.3	0.8	
Sphingomyelin	7		
Cholesterol	9	0.2	
Cardiolipin		4	3

Membrane proteins may be structurally classified into **extrinsic** proteins, which are only loosely attached to the membrane surface, and **intrinsic** proteins, which are integrated into the membrane structure and tightly combined with lipid. Extrinsic proteins are readily extracted under mild conditions; intrinsic proteins require relatively drastic treatment with denaturing solvents.

Physical measurements have revealed that the phospholipids of natural membranes have a high degree of **lateral mobility**; that is, they can move freely in the plane of the membrane, but do not cross readily from one side to the other. From this and other findings there has been developed the **fluid mosaic** model of membrane structure. According to this model, which has found wide acceptance, the phospholipids of membranes form a bilayer, in which the polar heads are oriented outwards toward the surface, while the hydrocarbon tails are oriented inward to form a quasifluid hydrocarbon phase (Fig. 15-1). Some of the globular membrane proteins are completely immersed in the membrane, others are only partially buried. The degree of penetration is governed by the proportion of hydrophobic amino acids on the protein surface. The membrane proteins thereby generate a mosaic-type structure in the phospholipid bilayer.

Direct evidence for the presence of intrinsic proteins in the interior of membranes has been provided by **freeze-etching** electron microscopy. In

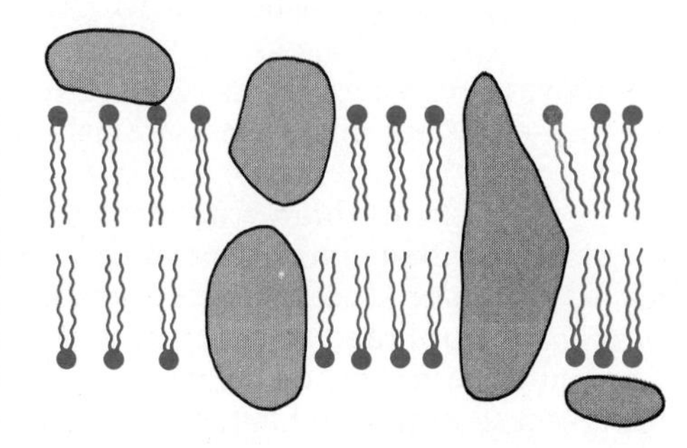

Figure 15-1 Schematic representation of the fluid-mosaic model of a membrane, showing imbedded protein molecules. Three intrinsic and two extrinsic proteins are shown.

this technique the interior of a frozen membrane is exposed by fracturing with a microtome blade. The membrane is usually cleaved along a plane in the middle of the bilayer, thereby exposing the interior of the bilayer. The ice covering a membrane surface is removed by sublimation. Electron microscopic observation permits direct visualization of the membrane interior. In this way, completely buried intrinsic proteins have been observed directly.

While it is not possible to generalize much further, the fluid-mosaic model for membrane structure is able to account for most of the properties of natural membranes to a satisfactory degree and provides a basis for understanding the detailed properties of specific membranes.

One additional structural feature of natural membranes is of importance. Natural membranes are intrinsically *asymmetric*. The two surfaces of biological membranes differ in composition and properties. The active transport systems to be considered in section 15-3 have a definite orientation with respect to the membrane surfaces. Membrane carbohydrates, which, in the form of glycoproteins and glycolipids, comprise up to 5% of eucaryotic membranes, are localized on the external surface.

15-2 PHYSICAL PROPERTIES OF MEMBRANES

Permeability

Permeability studies upon synthetic phospholipid bilayers have indicated that the ease with which they are penetrated by small molecules is very dependent upon the characteristics of the latter. Charged and polar molecules diffuse through these synthetic membranes at very slow rates; the incorporation of a charged group into a nonpolar molecule typically depresses its rate of diffusion by orders of magnitude. Water is an exception: it readily diffuses across these membranes.

This general property is a consequence of the structure of the bilayers, which have interiors composed of nonpolar hydrocarbons. In order to penetrate a bilayer, a polar molecule must leave an aqueous environment, where it is extensively solvated, and then enter a hydrophobic environment, first shedding its solvation shell. This process is thermodynamically highly unfavorable and tends to occur at a slow rate unless the molecule can combine with a specific acceptor within the bilayer. The selective permeability to small molecules arises from either **passive facilitated diffusion** mediated by a specific carrier protein or else from **active transport** by a specific molecular pump, which requires a supply of metabolic energy.

Electrical Properties

Natural membranes are characterized by a high electrical resistance. This property is conferred by their hydrocarbon interiors, which impede the transport of ions. Because of this resistance to current flow, a difference in electrical potential may exist across a membrane.

Active transport can produce a biased distribution of a particular solute on the two sides of a membrane. In general, the free energy change per mole accompanying the transfer of one mole of an uncharged solute from a solution where its concentration is c_1, to one where its concentration is c_2

is given by

$$\Delta G = RT \ln \frac{c_2}{c_1}$$

If the solute is electrically charged, there will be superimposed upon the above concentration gradient a gradient of electrical potential:

$$\Delta G = RT \ln \frac{c_2}{c_1} + zF\,\Delta\phi$$

where z is the number of charges on the solute molecule, F is the faraday, and $\Delta\phi$ is the difference in volts of electrical potential between the two solutions separated by the membrane, or the **membrane potential**. In nerve and muscle cells, in which it has a major functional role, the membrane potential can be measured directly.

15-3 ACTIVE TRANSPORT

Small molecules and ions can cross a biological membrane by passive diffusion. In this case the *driving force* responsible for mass flow is the gradient of solute concentration across the membrane; the solute diffuses in the direction of decreasing concentration. When the concentrations (more precisely, *activities*) of solute are the same on both sides of the membrane, this driving force vanishes. While diffusion can abolish a concentration gradient, it cannot establish such a gradient and cannot develop an excess of solute on one side of the membrane.

In **passive facilitated diffusion** the movement of a solute across a membrane is promoted by its combination with a specific carrier molecule that is compatible with the membrane interior and is mobile within it. This form of transport always occurs *down* a concentration gradient and may proceed in either direction across a membrane, depending upon the relative concentrations of solute in the two compartments. No metabolic energy is required and the presence of inhibitors of metabolism does not block the process. Both passive facilitated diffusion and active transport display **saturation**; that is, the rate of mass transport does not increase indefinitely as the concentration of solute in one compartment increases, but rather approaches a plateau corresponding to saturation of the carrier system with the solute (Fig. 15-2).

An example of passive facilitated diffusion is the transport of glucose across erythrocyte membranes. The carrier system is inactive toward many other mono- and disaccharides, may occur in either direction, and is not dependent upon a supply of metabolic energy. The carrier itself is a protein.

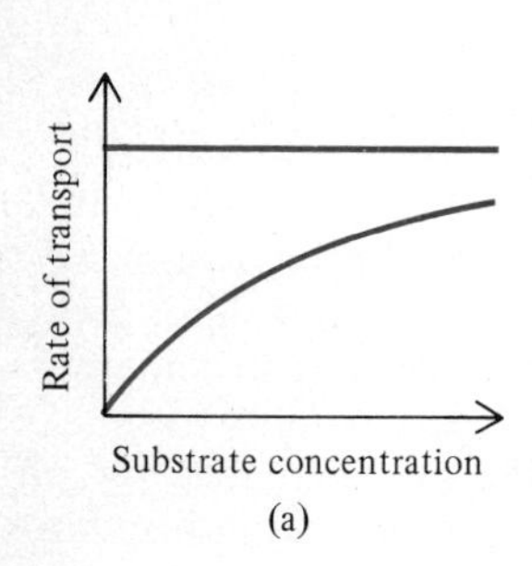

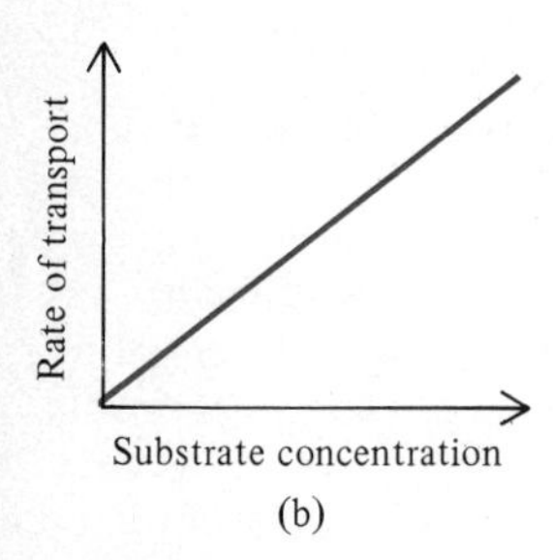

Figure 15-2
Saturation effects in membrane transport. (a) When transport across the membrane occurs by ordinary diffusion, the rate increases linearly with concentration. (b) When transport involves a carrier, the rate approaches a plateau with increasing concentration.

In contrast, active transport is absolutely dependent upon metabolic energy. It moreover normally occurs in only one direction and may operate against a concentration gradient. A highly biased distribution of solute across a membrane may be developed in this way.

Active Transport Systems

Among the best characterized systems for active transport is the Na^+ and K^+ pump, which transfers K^+ into cells and Na^+ out of them. The K^+ concentration is high within nearly all cells, while that of Na^+ is low. The operation of the ion pump has been observed directly in many types of cell, including erythrocytes and the giant axons of squid nerves.

Ion transport in intact cells is blocked by such inhibitors of oxidative metabolism as cyanide, or more specifically by the steroid **ouabain**, which binds to the outer surface of cells and thereby blocks the action of the pump. In the case of erythrocytes there are 100–200 ouabain-binding sites per cell, which provides an index of the number of pumping sites. Other experiments have shown that the transport of K^+ and Na^+ is always accompanied by the hydrolysis of ATP.

Observations of this kind have led to the proposal of **Na^+-K^+ activated ATPase**, which is an essential component of the ion pump. Within the cell, the activation of this enzyme system requires the presence of K^+ on one side of the membrane and Na^+ on the other. Such an enzyme has in fact been isolated from membranes and shown to catalyze the hydrolysis of ATP in the presence of Na^+, K^+, and Mg^{2+}. SDS gel electrophoresis indicates that it consists of a large polypeptide of molecular weight $\sim$100,000 and a smaller glycoprotein of molecular weight $\sim$50,000.

The stoichiometry of the ion pump is somewhat surprising. Two K^+ ions are transported into the inside of the cell and three Na^+ ions to the outside for each molecule of ATP hydrolyzed. An excess of negative charge thus develops within the cell and is reflected by an electrical membrane potential. Because of the significant "leakiness" of the membrane for K^+, a passive diffusion of K^+ occurs in the opposite direction, leading to a steady state in which active transport and passive diffusion are exactly balanced.

While the details of the operation of the ion pump are uncertain, one possible mechanism explains the process in terms of two conformations of the protein, one of which binds Na^+ selectively while the other binds K^+. The hydrolysis of ATP proceeds via an initial phosphorylation of a group on the protein which induces a transition between the two conformations. The conformational change is correlated with the closing of a channel to the inside and the opening of one to the outside. The hydrolysis of the phosphoryl group completes the cycle. The sequential steps are (Fig. 15-3):

(1) The protein is in the Na^+-binding conformation. A channel is open that leads from the inside solution to a cavity containing the ion-binding sites.
(2) Three Na^+ ions diffuse through the channel to the cavity and combine with the Na^+ sites.
(3) A group on the protein is phosphorylated by ATP. The protein shifts to the K^+-binding conformation. The first channel closes and a second channel opens leading from the cavity to the outside solution.

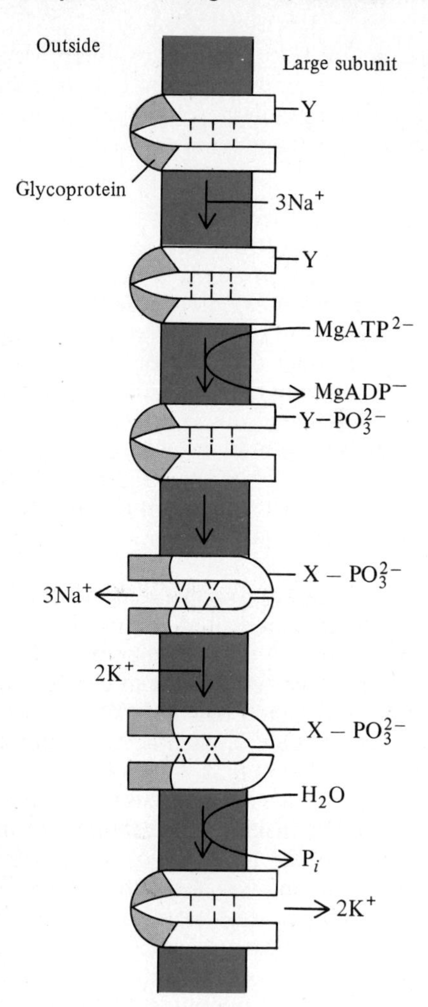

Figure 15-3
One possible model for the action of the Na^+-K^+ pump. (Adapted from D. Metzler, *Biochemistry*, Academic, 1977.)

(4) The Na^+ ions are released and diffuse through the second channel to the outside solution. Two K^+ ions diffuse into the cavity from the outside solution and combine with their binding sites.
(5) The phosphorylated group is hydrolyzed restoring the protein to the initial Na^+-binding conformation. The channel to the outside solution closes and that to the inside solution opens.
(6) K^+ ions are released and diffuse through the channel to the inside solution.

A number of peptide antibiotics are known that can combine with metal ions and facilitate their diffusion through membranes. An important example is **valinomycin** (Fig. 15-4), which binds K^+ and enormously accelerates its diffusion through natural or synthetic membranes. Valinomycin is a highly efficient agent for the passive facilitated transport of K^+ and effectively

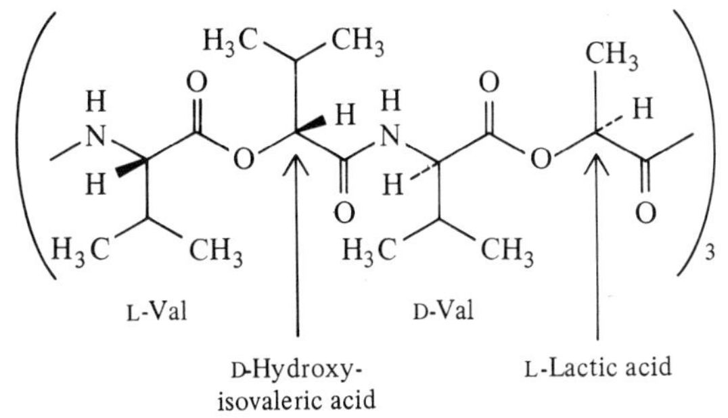

Figure 15-4 Structure of valinomycin. This is a cyclic peptide that folds about a K^+ ion.

counters the action of the ion pump. This is presumably the basis for its antibiotic activity.

Calcium ions are also pumped out of most cells by the action of a system whose properties resemble those of the Na^+-K^+ ATPase. There are many other ions for which selective pumping systems do not appear to exist. Chloride ions do not undergo active transport and diffuse relatively readily through membranes.

It has already been mentioned that glucose can be transported across some membranes by passive facilitated diffusion. However, active transport systems for glucose also exist. The transport of glucose against a concentration gradient occurs in the epithelial cell layers of the small intestine and the kidney tubules. The latter process contributes to the conservation of glucose in blood plasma and normally prevents its excretion in the urine.

Active transport is also important in the absorption of amino acids from the intestine and in the reclaiming of amino acids from the kidney tubules. Specific transport systems have been identified for different classes of amino acids.

SUMMARY

Membranes enclose the cytoplasms of all living cells; eukaryotic cells also contain membrane-bounded organelles within the cytoplasm as well as a nuclear membrane. A very important property of biological membranes is their **selective permeability**, which allows a discriminate passage of molecules. A molecule or ion may cross a membrane by ordinary diffusion, by passive **facilitated diffusion** via combination with a carrier molecule, or by **active transport**, which involves the expenditure of metabolic energy. An active transport system may be regarded as a **biological pump**, whereby a net transport of a substance may occur against a concentration gradient. Facilitated diffusion and active transport are characterized by saturation effects of the solute.

All membranes are composed of proteins and lipids in proportions that vary widely for different membranes. These form thin sheets, about 6–10 nm thick. The protein and lipid components of natural membranes are held together by noncovalent interactions. The protein component is responsible for the specific functions of membranes, including selective permeability, active transport, and energy transduction. The lipids commonly present include **phosphatidyl choline, phosphatidyl ethanolamine, phosphatidyl serine, phosphatidyl inositol, cholesterol, cardiolipin**, and **sphingomyelin**.

Membrane proteins may be either **extrinsic** proteins, which are only loosely attached to the membrane surface, or **intrinsic** proteins, which are integrated into the membrane structure and tightly combined with lipid.

According to the widely accepted **fluid mosaic** model, the phospholipids of membranes form a bilayer, in which the polar heads are oriented outwards toward the surface, while the hydrocarbon tails are oriented inward to form a quasifluid hydrocarbon phase. Physical measurements have revealed that membrane phospholipids have a high degree of **lateral mobility**, being able to move freely in the plane of the membrane but not able to cross from one side to the other. Natural membranes are intrinsically *asymmetric*; the two surfaces differ in composition and properties. Active transport systems are oriented with respect to the membrane surfaces. Membrane **carbohydrates** are localized on the external surface.

Membranes are characterized by a high electrical resistance. Because of this resistance to current flow, a difference in electrical potential may exist across a membrane. A biased distribution of ions, arising from active transport, may generate a **membrane potential**.

A well-characterized active transport system is the Na^+ and K^+ pump which transfers K^+ into cells and Na^+ out of them. A Na^+-K^+ activated ATPase is an essential component of the ion pump. A plausible model for the process postulates that two conformations of the protein exist; one binds Na^+ selectively, while the other binds K^+. The hydrolysis of ATP proceeds via an initial phosphorylation of a group on the protein, which induces a transition between the two conformations. The conformational change is correlated with the closing of a channel to the inside and the opening of one to the outside. The hydrolysis of the phosphoryl group completes the process.

REFERENCES

The following code is used to classify references. I: particularly useful as an introduction to the subject; R: useful primarily as a reference text; A: an advanced account of the material; H: a publication of historical importance.

General

M. S. Bretscher, "Membrane Structure: Some General Principles," *Science* 181, 622 (1973). (I)

L. I. Rothfield, ed., *Structure and Function of Biological Membranes*, Academic, New York (1971). (A)

G. Rouser, G. J. Nelson, S. Fleischer, and G. Simon, *Biological Membranes*, D. Chapman, ed., Academic, New York (1968). (A)

P. J. Quinn, *Molecular Biology of Cell Membranes*, University Park, Baltimore (1976). (A)

S. J. Singer and G. L. Nicolson, "The Fluid Mosaic Model of the Structure of Membranes," *Science* 175, 720 (1972). (H)

H. T. Tien, *Bilayer Lipid Membranes*, Dekker, New York (1974). (A)

Membrane Transport

W. Boos, "Bacterial Transport," *Ann. Rev. Biochem.* 43, 123 (1974). (R)

H. N. Christensen, *Biological Transport*, Benjamin, Menlo Park, California (1974). (I)

J. L. Dahl and L. E. Hokin, "The Sodium-Potassium ATPase," *Ann. Rev. Biochem.* 43, 327 (1974). (R)

D. R. Greenberg, ed., *Metabolic Transport*, vol. 6 of *Metabolic Pathways*, 3rd. ed., Academic, New York (1972). (R)

D. A. Hayden and S. B. Hladky, "Ion Transport across Thin Lipid Membranes," *Q. Rev. Biophys.* 5, 187 (1972). (A)

L. E. Hokin, ed., *Metabolic Transport*, Academic, New York (1972). (A)

D. L. Oxender, "Membrane Transport," *Ann. Rev. Biochem.* 41, 777 (1972). (R)

REVIEW QUESTIONS

Questions marked with an asterisk are of a high level of difficulty.

15-1 Which of the following observations support the fluid mosaic membrane model?
- (a) low melting points of phospholipids composing the lipid bilayer
- (b) membrane composed of 50% lipid and 50% protein
- (c) membrane proteins dissociated by detergent but not by high salt concentrations
- (d) phospholipid head groups on the outside of lipid bilayer near water solvent
- (e) none of the above

15-2 An animal membrane consists of 45% by weight of phospholipid and 55% protein. What is the mole ratio of the two if the average molecular weight of the phospholipids is 800 and that of the proteins is 80,000?

15-3 The rate of transport (mole/cm^2/sec) of a substance across a membrane increases linearly with increasing initial concentration of the substance up to the highest concentration attainable. What is the probable mechanism of transport?

15-4 The rate of transport of a substance across a membrane approaches a limiting value as the initial concentration of the substance increases to a high value. If the same concentration of the substance is present on both sides of the membrane, no net transport occurs. What is the probable mechanism of transport?

15-6 What does the finding that many pairs of different membranes can fuse into one continuous structure suggest?

15-7 How could you test whether a particular substance crossed a membrane by active transport?

16

Terpenes and Steroids

16-1 BIOLOGICAL FUNCTIONS

In this chapter we shall, primarily for convenience, consider together a collection of compounds that have diverse biological roles and whose mutual structural relationship is confined to the recurrence of a particular structural unit, the **isoprene** grouping, in each compound or a precursor. For reasons that remain obscure, living cells have found it advantageous to put this structural unit to multiple uses.

Among the members of this family of compounds that we shall consider in detail are the important vitamins A and D, which have central functions in vision and bone metabolism, respectively. In the latter part of the chapter we shall discuss the **steroids**, whose parent compound, **cholesterol**, is synthesized via precursors which contain the isoprene grouping.

The biological significance of steroids is two-fold. Cholesterol itself occurs with phospholipids as a constituent of many membranes. Its metabolic derivatives include the **steroid hormones**. Of the latter, the hormones of the **adrenal cortex** are important metabolic regulators, while the male and female **sex hormones** (androgens and estrogens, respectively) are determinants of sexual characteristics.

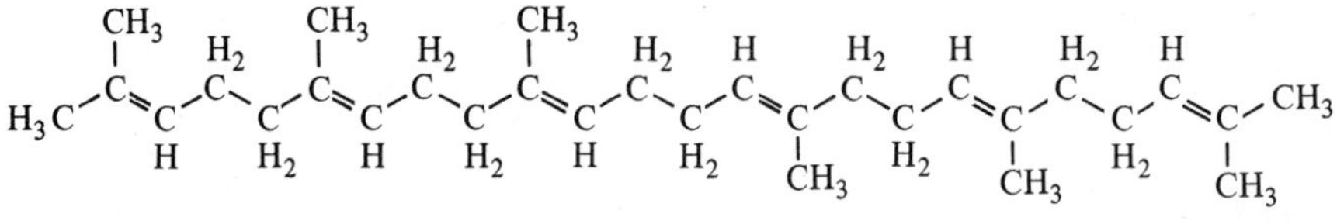

Figure 16-1
Structure of the linear terpene **squalene**, which is found in shark-liver oil. Squalene is an intermediate in cholesterol biosynthesis.

16-2 TERPENES

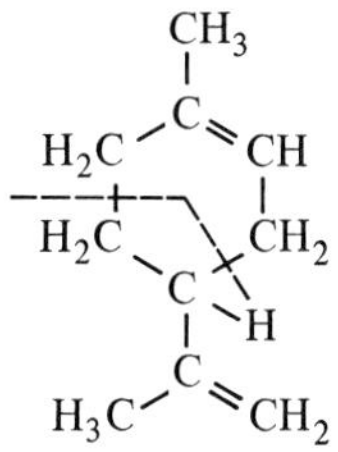

The biological world contains many compounds with a structural relationship to **isoprene** (2-methyl butadiene):

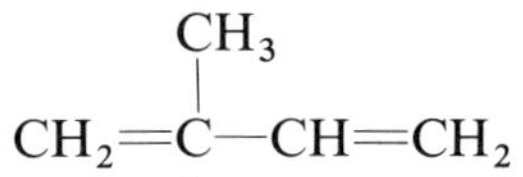

$$CH_2{=}C(CH_3){-}CH{=}CH_2$$

Isoprene

This group of compounds, called **terpenes**, includes a wide range of natural products of diverse functions and properties. Among these are natural rubber, the essential oils, such as camphor, citral, and geraniol, and a number of plant pigments, including lycopene and the carotenoids. Terpenes may be linear, like squalene (Fig. 16-1), or cyclic, like limonene (Fig. 16-2). The molecular structures of these materials can formally be dissected into isoprenelike subunits.

Figure 16-2
Structure of the cyclic terpene, **limonene**.

Carotenoids

While the biosynthesis of the carotenoids takes place in plants, they are of considerable biochemical importance in animals as well. All are isoprene derivatives with a high degree of unsaturation (Fig. 16-3). The presence of an extensive system of conjugated double bonds renders them strongly colored, usually red or yellow. The configuration about the double bonds is generally *trans*.

The carotenoids include **vitamin A_1** (Fig. 16-4), which is one of the fat-soluble vitamins and an essential growth factor in animals. While a diet

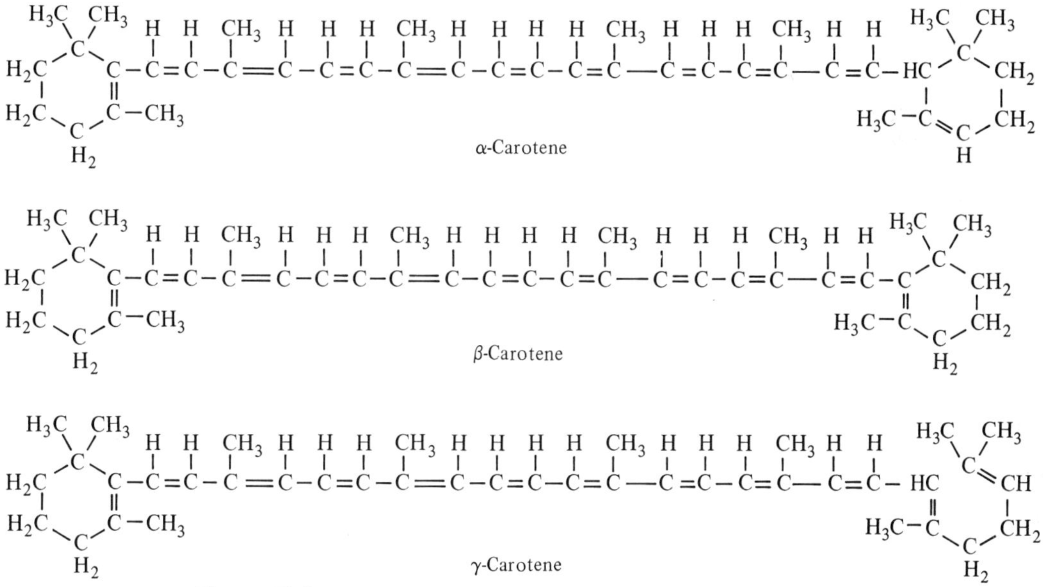

α-Carotene

β-Carotene

γ-Carotene

Figure 16-3
Structures of three carotenoids.

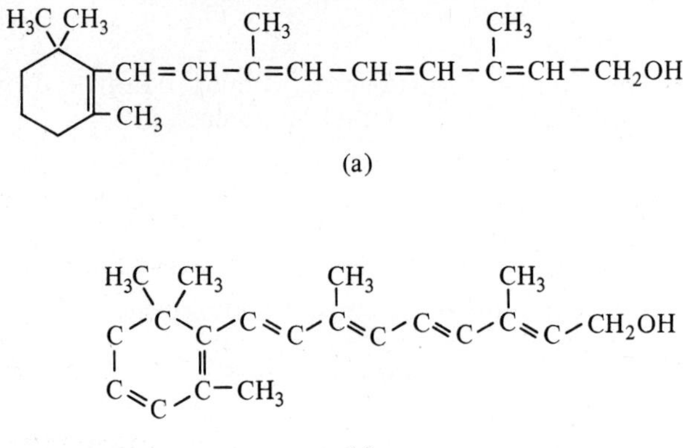

(a)

(b)

Figure 16-4
Structures of (a) vitamin A_1 and of (b) vitamin A_2.

deficient in vitamin A_1 has adverse effects upon all tissues, the best-characterized changes occur in the eyes. The earliest symptom is a loss of night vision, followed by an increasing loss of general visual function, and finally by total blindness. The related compound **vitamin A_2** (Fig. 16-4) is abundant in fish. This differs from vitamin A_1 only in possessing a second double bond between carbon atoms 3 and 4 in the six-membered ring. The configuration of all double bonds in the side-chains of both compounds is *trans*.

Vitamin A activity in animals is also bestowed by the carotenoids, which are converted into vitamin A by enzymic processes occurring in the liver. β-Carotene (Fig. 16-3) may be split in the middle to form two molecules of vitamin A_1.

Two isomeric aldehyde derivatives of vitamin A_1 are of central importance in mammalian vision. These are ***trans*-retinal**, all of the double bonds of whose side-chain are in the *trans* configuration, and **11-cis-retinal**, whose side-chain contains a *cis* double bond at the 11,12 position (Fig. 16-5). The two compounds are otherwise identical.

The Role of Vitamin A in Vision

The mammalian retina contains two kinds of light-sensitive cells. The **cone cells** are concerned with acute perception at high light intensities, while the **rod cells** are adapted to responding to low light levels. The general outline

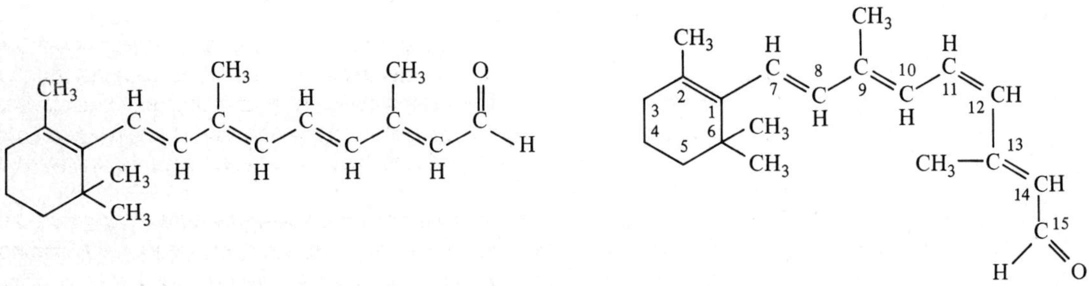

trans-Retinal

11-*cis*-Retinal

Figure 16-5
Structures of *trans*-retinal and of 11-*cis*-retinal.

of the photochemical processes occurring in rod cells is now fairly well understood.

The rod cells of the retina contain numerous disklike membrane vesicles, which are stacked parallel to the surface. These comprise the photosensitive elements of these cells. The membranes of these vesicles are rich in the protein **rhodopsin**, of molecular weight 28,000, which consists of the protein **opsin**, plus strongly bound 11-*cis*-retinal.

The exposure of rhodopsin to light results in the isomerization of 11-*cis*-retinal to the *trans* form by a purely photochemical reaction in which no enzymes are involved. The close similarity of the visible absorption spectrum of rhodopsin to the wavelength dependence of visual sensitivity in weak light suggests that rhodopsin is the primary light sensitive component of rod cells. The isomerization of retinal is followed by a series of other molecular changes, terminating in the decomposition of rhodopsin into opsin and *trans*-retinal.

The resynthesis of rhodopsin must occur at a sufficient rate to replace that lost by photodecomposition. Exposure of the retina to light of high intensity results in extensive decomposition of rhodopsin. This is responsible for the time lag required for visual adaptation to subsequent dim illumination. During this period of dark adaptation rhodopsin is reformed, with a progressive improvement in vision.

According to the model developed by Wald only 11-*cis*-retinal can combine with opsin. This may be formed by oxidation of the 11-*cis* isomer of vitamin A_1, which may in turn be produced by enzymic isomerization of the *trans* form of the vitamin. This process appears to occur in a sequence of reactions catalyzed by two different enzymes:

$$\textit{trans}\text{-retinal} + \text{NADH} + \text{H}^+ \rightleftarrows \textit{trans}\text{-vitamin A}_1 + \text{NAD}^+$$

$$\textit{trans}\text{-vitamin A}_1 \rightleftarrows 11\text{-}\textit{cis}\text{-vitamin A}_1$$

$$11\text{-}\textit{cis}\text{-vitamin A}_1 + \text{NAD}^+ \rightleftarrows 11\text{-}\textit{cis}\text{-retinal} + \text{NADH} + \text{H}^+$$

The first and third of the above reactions are catalyzed by **retinal reductase** and the second by **retinal isomerase**. In summary

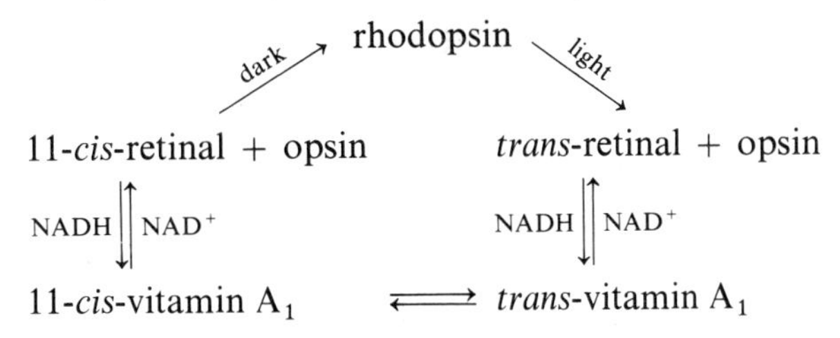

The photodecomposition of rhodopsin initiates a nerve impulse in the retina, so that the illumination of the retina is registered by the brain. This appears to reflect a change in membrane permeability resulting from a conformational change in rhodopsin accompanying photodecomposition. This change in permeability allows an efflux of Ca^{2+} from the rhodopsin-containing vesicles, thereby triggering the nerve impulse.

Vitamin E (Tocopherol)

The tocopherols are widely distributed in plant tissues and occur in some animal tissues. Several related tocopherols have been identified in plants;

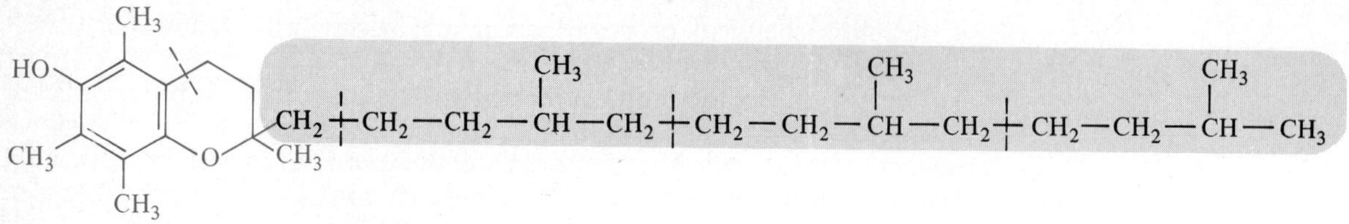

Figure 16-6
Structure of α-tocopherol. The fused ring structure on the left (exclusive of substituents) is a chromane group.

the most abundant of these is α-tocopherol (Fig. 16-6). A deficiency in tocopherol results in sterility in male and female rats, as well as numerous other pathological symptoms, including deterioration of the liver, kidneys, and skeletal muscles.

The tocopherols consist of a substituted **chromane** residue, plus a side-chain containing three hydrogenated isoprene units (Fig. 16-6). The β- and γ-tocopherols are similar in structure to α-tocopherol, except that the 7-methyl substituent is missing for β-tocopherol and the 5-methyl for γ-tocopherol.

The mechanism of action of the tocopherols is still uncertain. There is some evidence that they may function as antioxidants and thereby protect unsaturated fatty acids present in membranes from oxidation by molecular oxygen.

Vitamin K

At least two forms of vitamin K (Fig. 16-7) are of natural occurrence; vitamin K_2 is probably the active form. A deficiency in this vitamin is associated with an impairment of the blood-clotting process in mammals. This hemorrhagic disease appears to reflect a lowered **prothrombin** level in the blood.

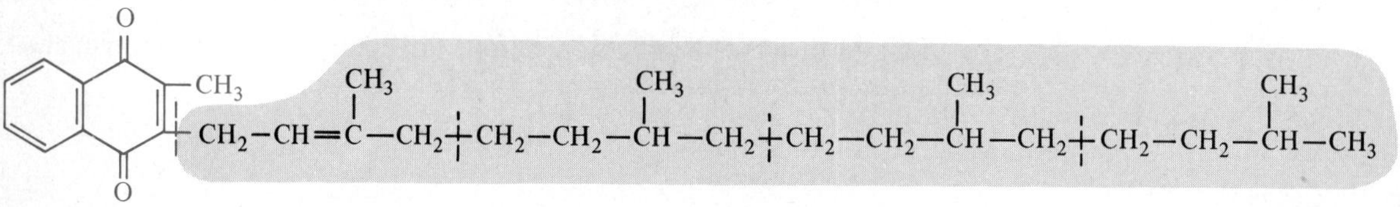

Vitamin K_1

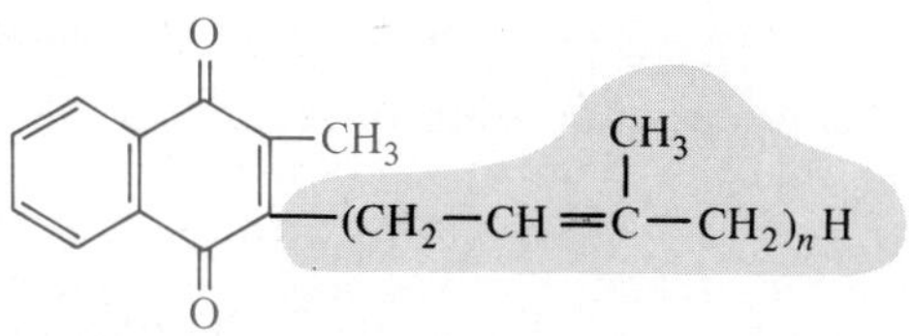

Vitamin K_2

Figure 16-7
Structures of vitamin K_1 and K_2. The value of n for the latter may be 6–10, depending on the species.

Prothrombin is the precursor of **thrombin**, the active enzyme responsible for the conversion of the blood protein **fibrinogen** into the polymeric network **fibrin**, which comprises the structural framework of blood clots. A deficiency in vitamin K appears to interfere with the biosynthesis of prothrombin.

Since vitamin K is synthesized by many biological organisms, including microorganisms and plants, which lack a blood clotting system, there is a possibility that it may possess a second, more general biological activity. Since it is a quinone and is readily reducible, it has been proposed that it may function as an electron carrier in an alternate path of electron transport.

16-3 STEROIDS

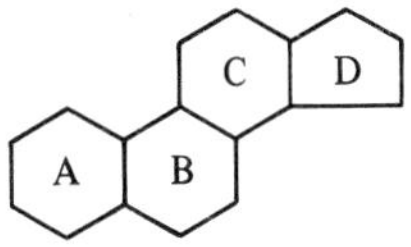

While the steroids are, by the operational criteria mentioned in section 13-1, usually classified as lipids, their chemical nature is altogether different from that of the other representatives of this class. The steroids may be regarded as derivatives of the parent hydrocarbon **sterane**, or **perhydrocyclopentanophenanthrene** (Fig. 16-8), which consists of four fused rings, of which three are 6-membered and the fourth is 5-membered.

The number of biologically active variants of the basic structure is large; however, the naturally occurring derivatives form a definite pattern. A hydrocarbon side-chain is often present at the 17 position, while methyl groups may occur at the 10 and 13 positions (Figs. 16-9 and 16-10). A hydroxyl or carbonyl group is often found at the 3 position.

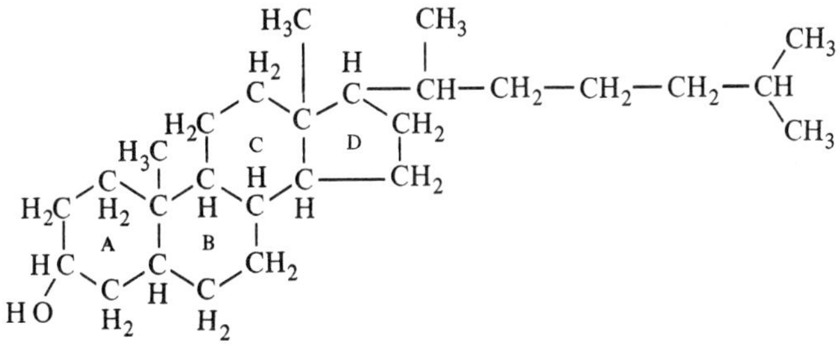

(a)

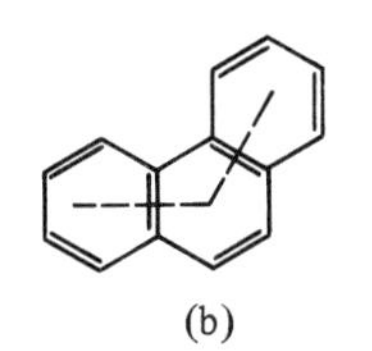

(b)

Figure 16-8
Structures of (a) the parent hydrocarbon **perhydrocyclopentanophenanthrene** and of (b) phenanthrene.

Figure 16-9
Structure of the sterol **cholestanol**, written out in full.

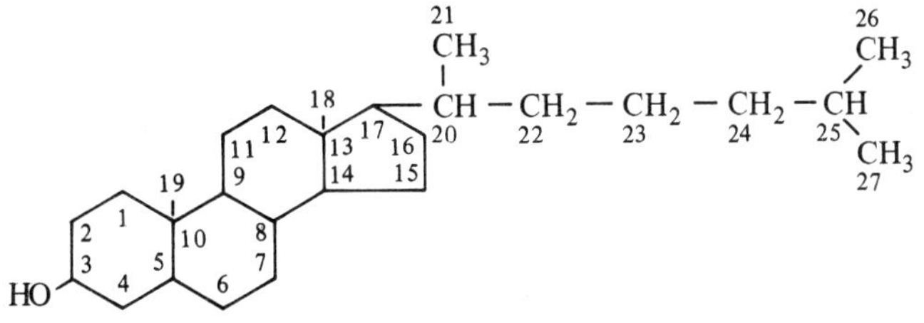

Figure 16-10
Structure of cholestanol in abbreviated form, showing the conventional scheme of numbering positions. Methyl groups at the C-10 and C-13 positions are indicated by short vertical lines.

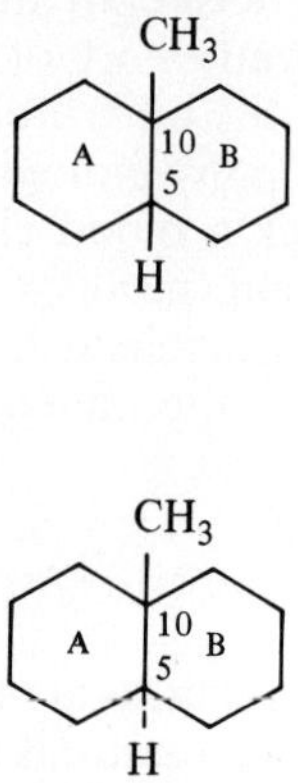

Figure 16-11
Cis and *trans* configurations of the fused A and B rings.

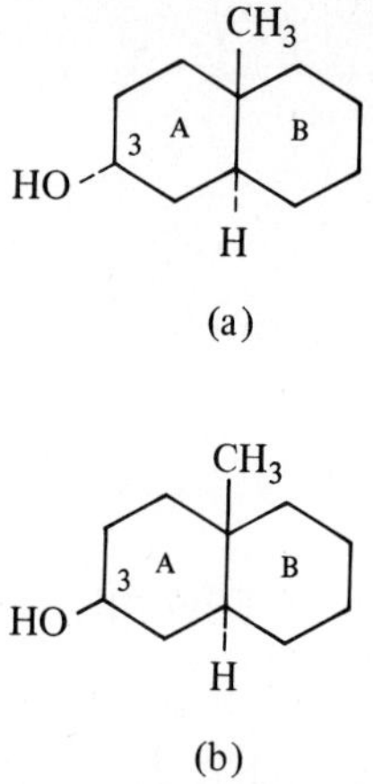

Figure 16-12
(a) α-orientation and (b) β-orientation of the 3-hydroxyl.

Configuration

The parent hydrocarbon contains six chiral centers (positions 5,8,9,10,13, and 14), permitting a total of 64 stereoisomers. If a side-chain is present at the 17 position, a seventh chiral carbon arises, so that the total number of possible isomers rises to 128. With a hydroxyl group at the 3 position, 256 isomers become possible. It is hardly surprising that the determination of the absolute configuration of a steroid is a formidable problem.

Two saturated rings can be fused so that their mutual configuration is either *cis* or *trans*. In the case of the pair of 6-membered rings designated as A and B (Fig. 16-10), the substituents on the 10- and 5-carbons may lie on either the same side of the plane of the molecule in the *cis* configuration, or on opposite sides in the *trans* configuration. The corresponding series of compounds are termed *normal* and *allo*, respectively (Fig. 16-11).

In the conventional representation of steroids, a substituent which projects out of the plane of the paper toward the viewer is designated by a solid line; one which extends behind the paper and away from the viewer is represented by a broken line. The latter is termed **α-oriented** and the former **β-oriented**. For example, the hydroxyl group on the 3-carbon may be α- or β-oriented (Fig. 16-12).

16-4 STEROLS

Cholesterol

The **sterols** are steroid alcohols containing a hydroxyl group at the 3-position. Cholesterol (Fig. 16-13), which was first isolated in the eighteenth century, is widely distributed in animals, occurring in all vertebrates and many

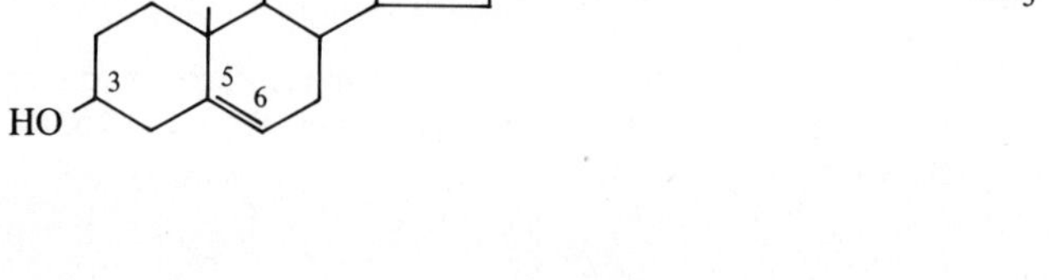

Figure 16-13
Structure of cholesterol.

invertebrates. In the higher animals it is found in all tissues, occurring in highest concentrations in brain, liver, skin, and adrenal glands. Cholesterol is a component of the membranes of many animal cells and also occurs in the lipoproteins of blood plasma.

Cholesterol belongs to the allo series of steroids. The hydroxyl at the 3-carbon is in the β-configuration. A double bond is present at the 5,6 position. A saturated hydrocarbon side-chain is attached to the 17-carbon (Fig. 16-13). The 3-hydroxyl and 17-hydrocarbon are characteristic of the sterols.

In addition to serving as a structural element, along with the phosphatides, of many membranes, cholesterol is also important as a precursor of many other biologically active steroids, including the bile acids and numerous hormones.

Cholesterol also appears in a less benign biological role. It and other lipids may be deposited on the inner walls of blood vessels, in a condition termed **atherosclerosis**, which appears to be associated with a blocking of blood vessels, sometimes leading to heart attacks or strokes.

Other Sterols

The reduction of the double bond of cholesterol produces two compounds differing in configuration. **Cholestanol** belongs to the allo series, while **coprostanol** is normal. In both molecules the 3-hydroxyl is in the β-orientation (Figs. 16-11 and 16-12). Cholestanol and coprostanol are formed from cholesterol in mammals by the action of bacteria in the intestine. Both are excreted in feces.

It is instructive to compare the conformations of the two molecules. The 6-carbon cyclohexane rings are not really planar hexagons, but assume the **chair** form (section 9-1), as shown in Fig. 16-14. When depicted in this way, cholestanol and coprostanol are seen to have very different molecular shapes. However, a clear visualization of their geometry requires the use of molecular models.

7-Dehydrocholesterol, which contains a second double bond at the 7,8 position (Fig. 16-15), is common in animal tissues and is an intermediate in steroid biosynthesis.

Ergosterol (Fig. 16-16) possesses a third double bond in the side-chain. It occurs in yeast and, like 7-dehydrocholesterol, is a precursor of vitamin D.

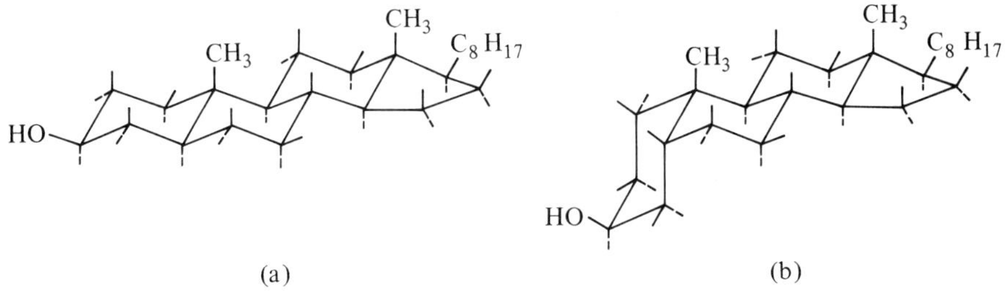

HO, CH_3, CH_3, C_8H_{17} (a)

HO, CH_3, CH_3, C_8H_{17} (b)

Figure 16-14
Conformation of (a) **cholestanol** and of (b) **coprostanol**.

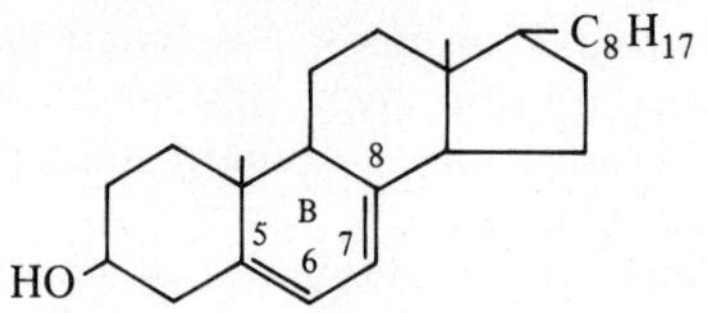

Figure 16-15 Structure of **7-dehydrocholesterol.**

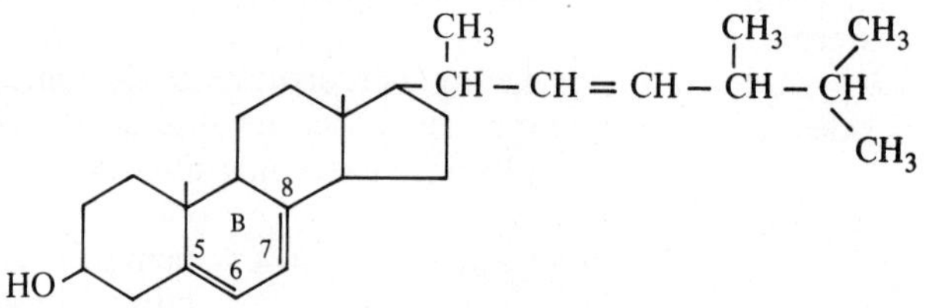

Figure 16-16 Structure of **ergosterol.**

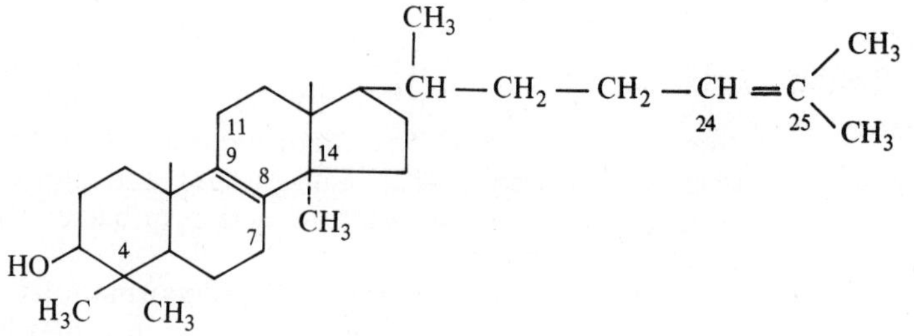

Figure 16-17 Structure of **lanosterol.**

Another important sterol is **lanosterol** (Fig. 16-17), which has been found in wool. This differs from the sterols described earlier in the dimethyl group on the 14-carbon. Double bonds occur in the side-chain and in the 8,9 position. Lanosterol is an intermediate in the biosynthesis of cholesterol.

Vitamin D

While the related substances bearing this designation are not true steroids, as they lack the complete four-ring system, they have a close structural relationship with the steroids, from which they are derived.

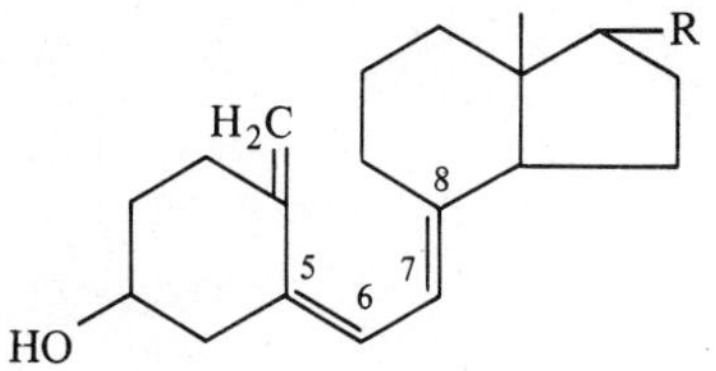

$R = C_8H_{17}$ in vitamin D_3.

$R = C_9H_{17}$ in vitamin D_2.

Figure 16-18 Structure of vitamin D.

The clinical syndrome associated with vitamin D deficiency is characterized by a softening of the bones, which is a consequence of inadequate calcification. Vitamin D assists in the absorption of Ca^{2+} from the gastrointestinal tract and promotes its incorporation into bone.

Vitamins D_2 and D_3 (Fig. 16-18) arise from ergosterol and 7-dehydrocholesterol, respectively, via ultraviolet irradiation, which ruptures the B ring. The two differ only in the side-chain.

The mechanism of formation of vitamin D_2 (ergocalciferol) is summarized in Fig. 16-19. Vitamin D_3 or **cholecalciferol**, the form normally found in mammals, arises from the irradiation of 7-dehydrocholesterol. The normal precursor of cholecalciferol in humans is the 7-dehydrocholesterol occurring in the skin. The formation of cholecalciferol involves the action of solar radiation upon the skin. Since, except for fish liver oils, the normal human diet is almost devoid of vitamin D, the above mechanism is the primary source of this vitamin.

The primary function of vitamin D is the regulation of calcium metabolism. This is achieved by the action of hydroxylated derivatives. The

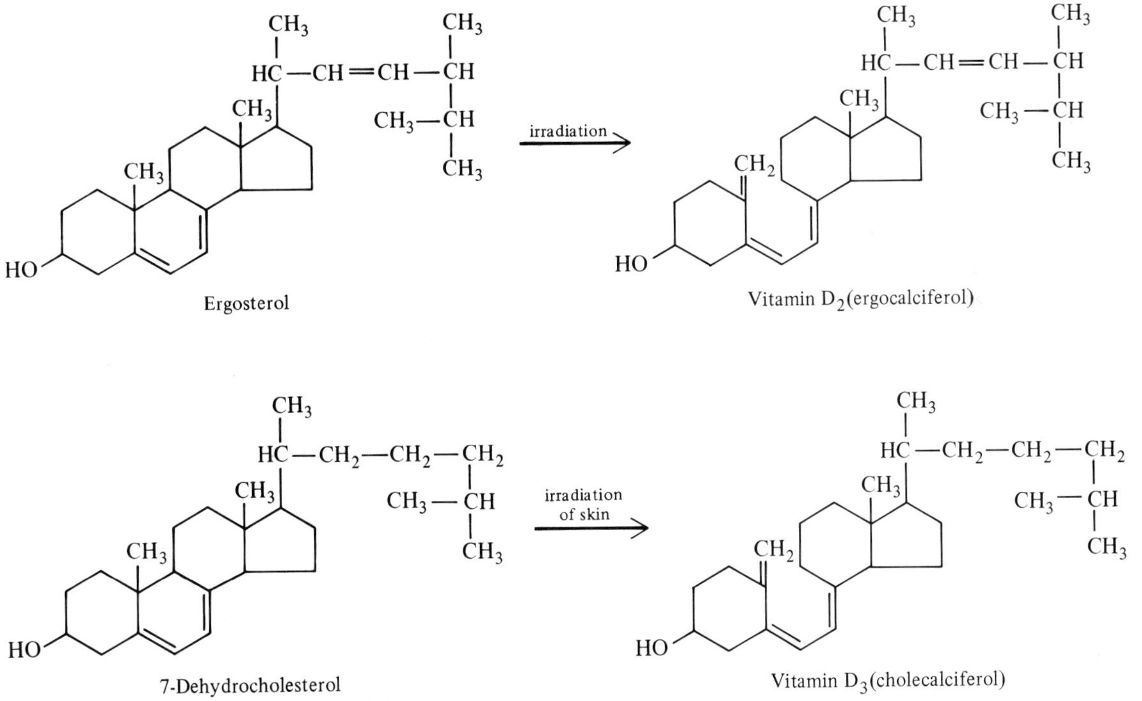

Ergosterol —irradiation→ Vitamin D_2(ergocalciferol)

7-Dehydrocholesterol —irradiation of skin→ Vitamin D_3(cholecalciferol)

Figure 16-19

Mechanism of formation of vitamin D_2 by the ultraviolet irradiation of ergosterol. The mechanism of formation of cholecalciferol from 7-dehydrocholesterol is analogous.

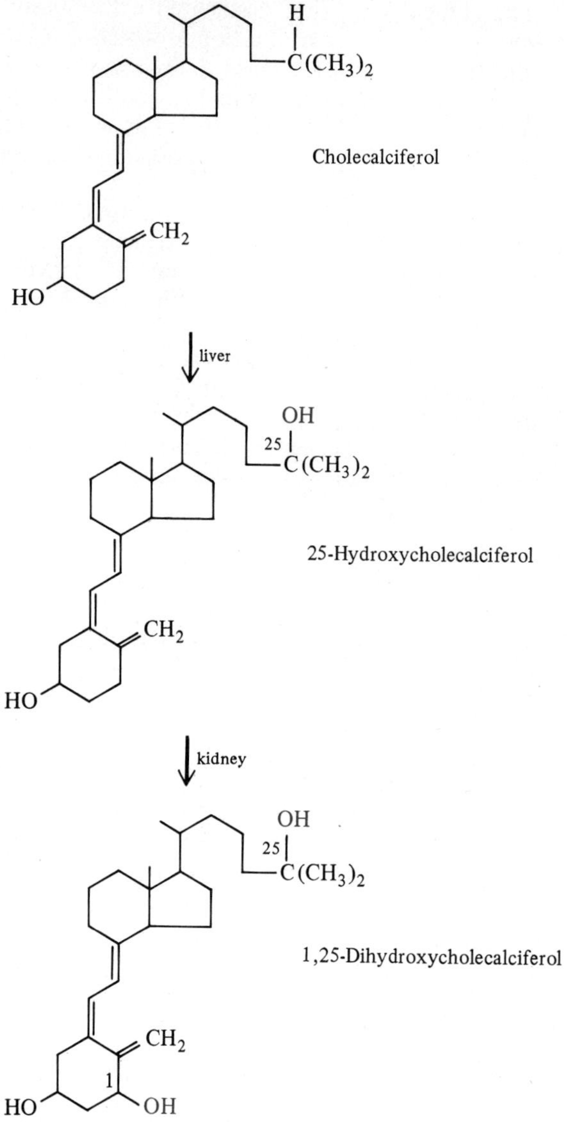

Cholecalciferol

liver

25-Hydroxycholecalciferol

kidney

1,25-Dihydroxycholecalciferol

Figure 16-20 Conversion of cholecalciferol into its hydroxylated derivatives.

initial hydroxylation to form **25-hydroxycholecalciferol** (Fig. 16-20) occurs in the liver, while the subsequent hydroxylations which yield ultimately 1,24,25-trihydroxy-cholecalciferol take place in the kidneys (Fig. 16-20). Both the 1,25-dihydroxy and the 1,24-25-trihydroxy derivatives facilitate the uptake of Ca^{2+} by the intestinal mucosa. These compounds appear to elicit an increase in the level of calcium-binding proteins in the body and may function by regulating gene transcription.

16-5 BILE SALTS

The metabolic derivatives of cholesterol are diverse and include the steroid hormones. The most abundant products are the **bile acids** (Figs. 16-21 and 16-22), which undergo combination with the amino acids glycine and taurine to form the **bile salts**. These are powerful emulsifying agents, whose detergent properties enable them to assist in the emulsification of fats in the intestine, thereby promoting their digestion and absorption. Like the fatty acid detergents, the bile salts tend to form micelles in aqueous solution.

Four different bile salts have been isolated from human bile (Fig. 16-21), of which cholic acid (3α, 7α, 12α-trihydroxycholanic acid) is the most abundant. All have 5-carbon side-chains, terminating in a carboxyl group, at the 17 position. The A and B rings are in the normal configuration.

In bile all of the bile acids exist as glycine or taurine conjugates. Formally, these arise as follows:

$$\underset{\text{Bile acid}}{R{-}COOH} + \underset{\text{Taurine}}{H_2N{-}CH_2CH_2{-}SO_3^-} \longrightarrow \underset{\text{Bile salt}}{R{-}CONH{-}CH_2CH_2{-}SO_3^-} + H_2O$$

$$\underset{\text{Bile acid}}{R{-}COOH} + \underset{\text{Glycine}}{H_2NCH_2COOH} \longrightarrow \underset{\text{Bile salt}}{RCONH{-}CH_2COOH} + H_2O$$

The bile salts are ionized under physiological conditions. The detergent properties of bile salts arise from the addition of a charged site to the nonpolar steroid.

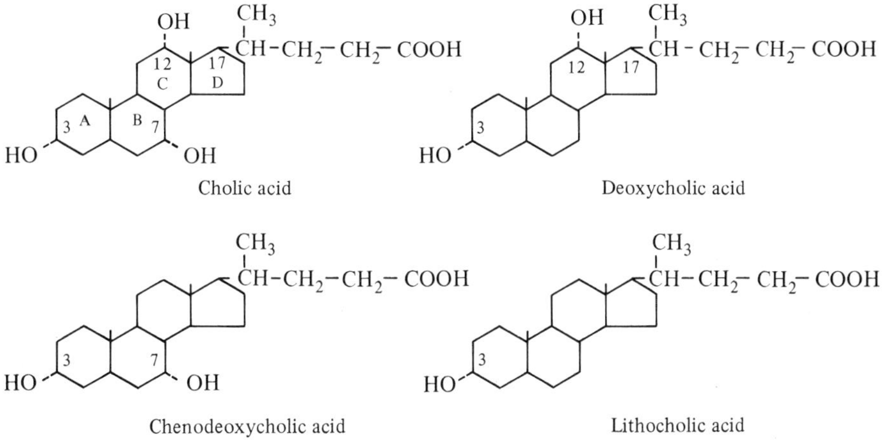

Figure 16-21
Structures of four bile acids.

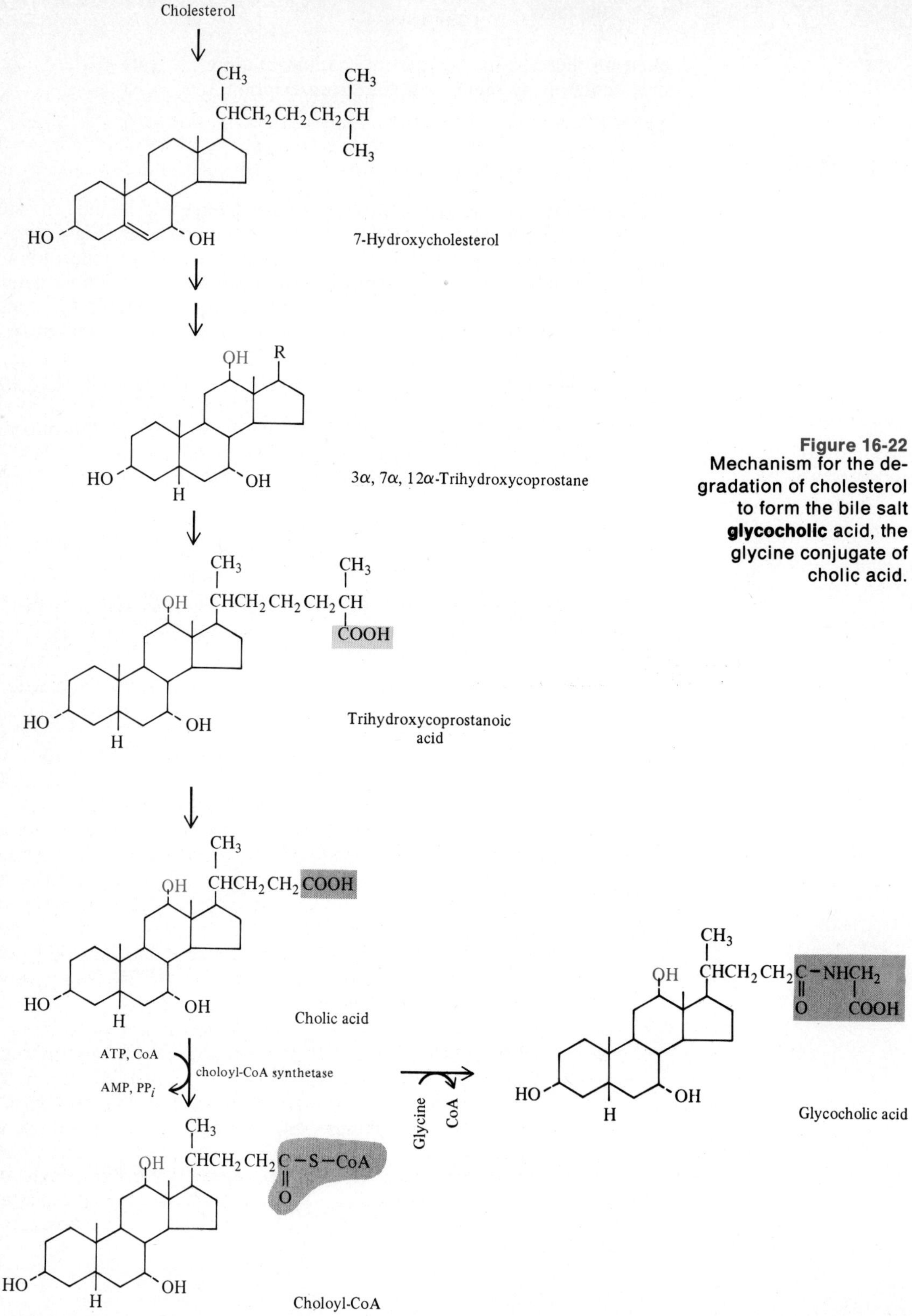

Cholesterol

CH_3 CH_3
$CHCH_2CH_2CH_2CH$
CH_3

HO OH

7-Hydroxycholesterol

OH R

HO OH
H

3α, 7α, 12α-Trihydroxycoprostane

CH_3 CH_3
OH $CHCH_2CH_2CH_2CH$
COOH

HO OH
H

Trihydroxycoprostanoic acid

CH_3
OH $CHCH_2CH_2COOH$

HO OH
H

Cholic acid

ATP, CoA
choloyl-CoA synthetase
AMP, PP_i

CH_3
OH $CHCH_2CH_2C-S-CoA$
O

HO OH
H

Choloyl-CoA

Glycine
CoA

CH_3
OH $CHCH_2CH_2C-NHCH_2$
O COOH

HO OH
H

Glycocholic acid

Figure 16-22
Mechanism for the degradation of cholesterol to form the bile salt **glycocholic** acid, the glycine conjugate of cholic acid.

The biosynthesis of the bile salts occurs in the liver, from which they flow into the bile duct and the small intestine. A major fraction is subsequently reabsorbed in the duodenum and returned to the liver for reuse. Their formation from cholesterol involves the removal of the double bond in the B ring, the conversion of the β-hydroxyl at the 3-position to an α-hydroxyl, and the oxidative shortening of the side-chain (Fig. 16-22). The acyl CoA derivatives of the bile acids are then linked with glycine or taurine.

16-6 BIOSYNTHESIS OF CHOLESTEROL

The carbon skeleton of cholesterol, which is synthesized *de novo*, arises entirely from *acetyl CoA*. The initial step is the condensation of three molecules of acetyl CoA to form β-hydroxy-β-methylglutaryl CoA, in a reaction catalyzed by **hydroxymethylglutaryl-CoA synthase** (Fig. 16-23):

$$\text{acetyl CoA} + \text{acetoacetyl CoA} \rightarrow \beta\text{-hydroxy-}\beta\text{-methylglutaryl-CoA} + \text{CoA} + CO_2$$

The β-hydroxy-β-methylglutaryl-CoA subsequently undergoes reduction of one of its carboxyl groups to an alcohol by the action of **hydroxymethylglutaryl-CoA reductase** to form mevalonate, with an accompanying loss of CoA (Fig. 16-23):

$$\beta\text{-hydroxy-}\beta\text{-methylglutaryl-CoA} + 2\text{NADPH} + 2\text{H}^+ \rightarrow \text{mevalonate} + 2\text{NADP}^+ + \text{CoA}$$

Mevalonate is next converted to squalene in a complex series of reactions (Fig. 16-24). The process begins with the sequential phosphorylation of mevalonate to form first the 5-monophosphate and the 5-pyrophosphate, in reactions catalyzed by **mevalonate kinase** and by **phosphomevalonate kinase**, respectively. The latter derivative is further phosphorylated by **pyrophosphomevalonate kinase** to yield an unstable intermediate which simultaneously decarboxylates and loses phosphoric acid to form **3-isopentenyl pyrophosphate**. This isomerized by the action of **isopentenyl pyrophosphate isomerase** to yield **3,3-dimethylallyl pyrophosphate**.

The isopentenyl and dimethylallyl pyrophosphates are combined next, with release of pyrophosphate, to yield the terpene geranyl pyrophosphate, in a reaction catalyzed by **dimethylallyl transferase** (Fig. 16-25). A further condensation then occurs with a second molecule of isopentenyl pyrophosphate to form **farnesyl pyrophosphate** in a reaction also mediated by dimethylallyl transferase.

The condensation of two molecules of farnesyl pyrophosphate by the action of **presqualene synthase** next yields **presqualene pyrophosphate** (Fig. 16-25). This is reduced by NADPH to form squalene, in a reaction catalyzed by **squalene synthase**.

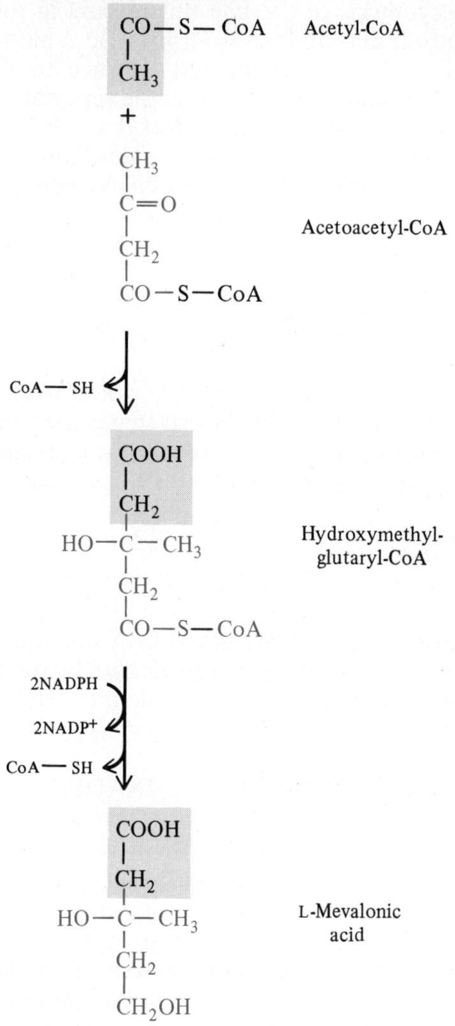

$CO-S-CoA$
CH_3 Acetyl-CoA

+

CH_3
$C=O$
CH_2
$CO-S-CoA$ Acetoacetyl-CoA

$CoA-SH$ ↓

COOH
CH_2
$HO-C-CH_3$
CH_2
$CO-S-CoA$ Hydroxymethyl-glutaryl-CoA

2NADPH
$2NADP^+$
$CoA-SH$ ↓

COOH
CH_2
$HO-C-CH_3$
CH_2
CH_2OH L-Mevalonic acid

Figure 16-23 Biosynthetic pathway leading to **mevalonic acid** from acetyl CoA.

In the final stage of cholesterol biosynthesis (Fig. 16-25), squalene reacts with molecular oxygen to yield squalene 2,3-epoxide, in a reaction mediated by **squalene monooxygenase**. The squalene 2,3-epoxide next undergoes ring closure to form lanosterol, a true sterol. The enzyme involved is **squalene epoxide lanosterol-cyclase**.

The formation of cholesterol from lanosterol requires a loss of three methyl groups, a shift in position of the double bond in the B ring, and a hydrogenation of the side-chain double bond (Fig. 16-26).

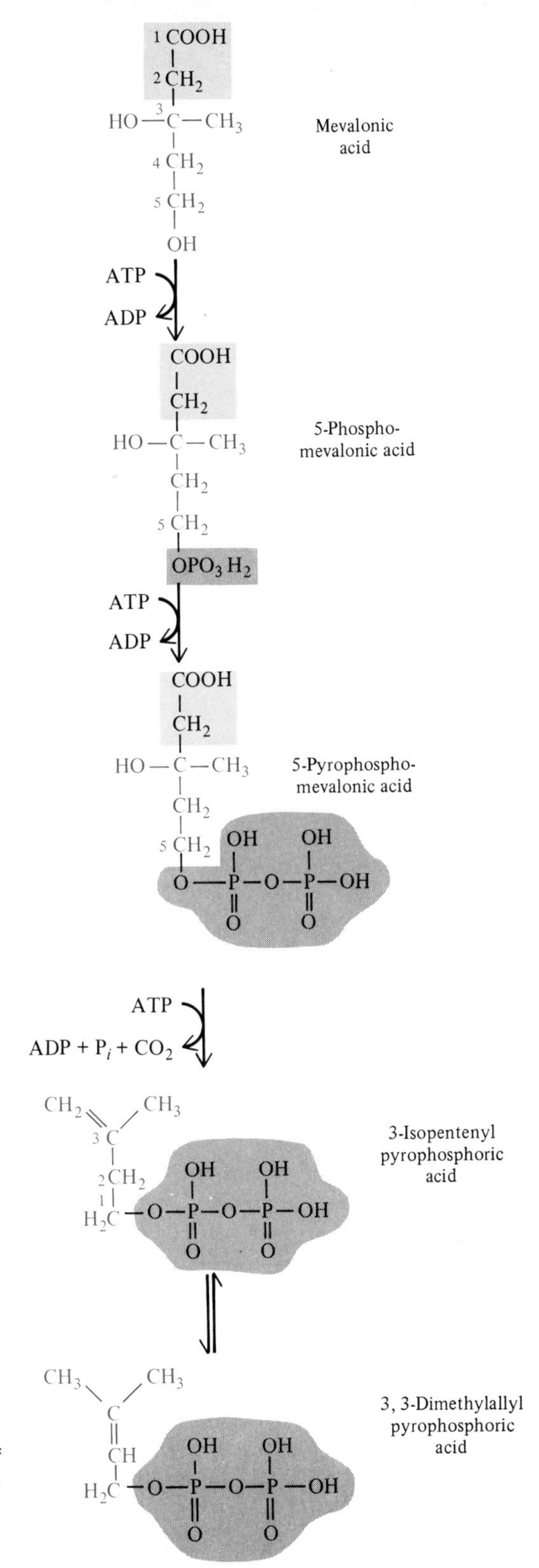

Figure 16-24 Formation of **3-isopentenyl pyrophosphate** from mevalonic acid.

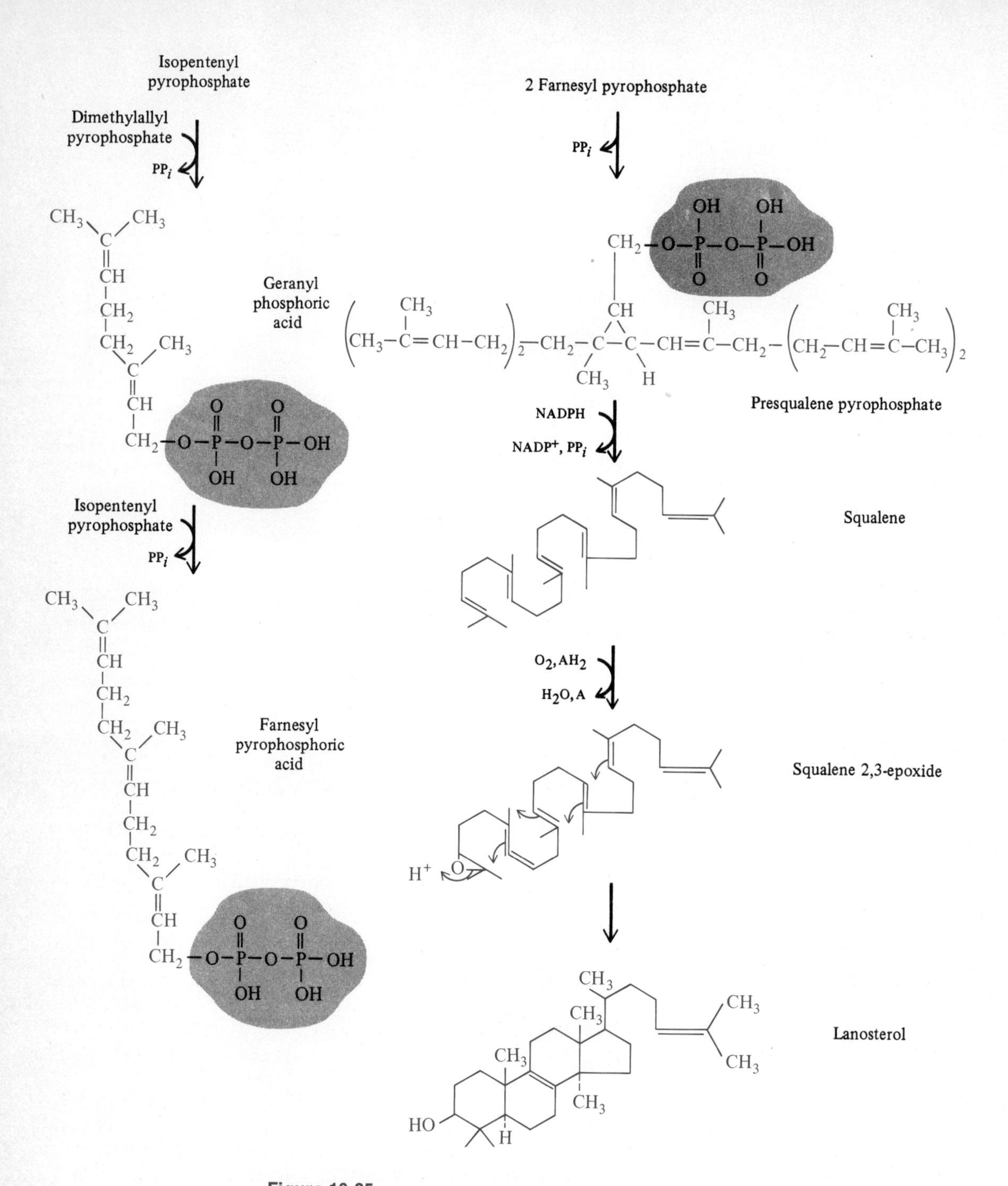

Figure 16-25
Formation of **squalene** and its conversion to lanosterol. In the next to last step AH_2 is an unidentified reducing agent.

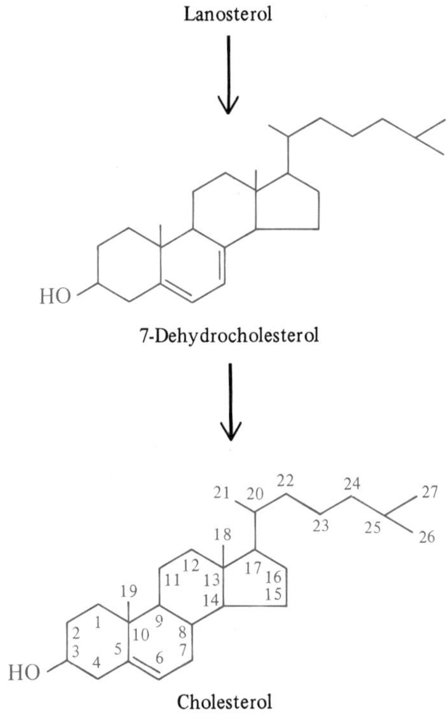

Lanosterol

7-Dehydrocholesterol

Cholesterol

Figure 16-26 Formation of cholesterol from lanosterol. While this is the most important metabolic route, alternative pathways exist.

16-7 STEROID HORMONES

In general hormones (chapter 22) are regulatory substances that are secreted by specialized glands, usually in trace quantities. Unlike the vitamins, they are biosynthesized by the organism. Some hormones are steroids. These are, for the most part, synthesized from cholesterol or a derivative in specialized tissues, including the testes, ovaries, adrenal cortex, and corpus luteum.

Hormones of the Adrenal Cortex

The adrenocortical hormones are of primary physiological importance; an animal does not survive removal of the adrenal gland for long. The adverse effects include a severe disturbance of electrolyte balance, reflected by a loss of Na^+ by excretion. This group of hormones also influences the metabolism of glucose.

The three principal hormones of the adrenal cortex are **cortisol, corticosterone,** and **aldosterone** (Fig. 16-27). All contain 21 carbon atoms. The other common structural features include a keto group at the 3-position, a double bond at the 4,5-position, a hydroxyl group at the 11-position, and a side-chain at the 17-position consisting of a —$COCH_2OH$ group.

Cortisol is secreted by the adrenal gland in milligram quantities in a human adult. It promotes gluconeogenesis and the accumulation of glycogen

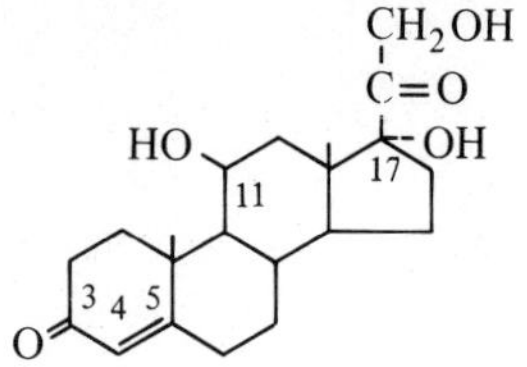

Figure 16-27 Structures of the steroid hormones of the adrenal cortex: (a) cortisol, (b) corticosterone, and (c) aldosterone.

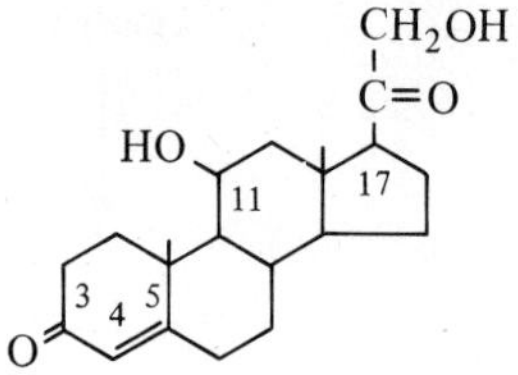

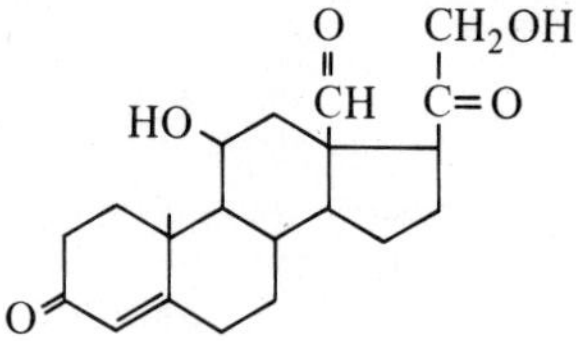

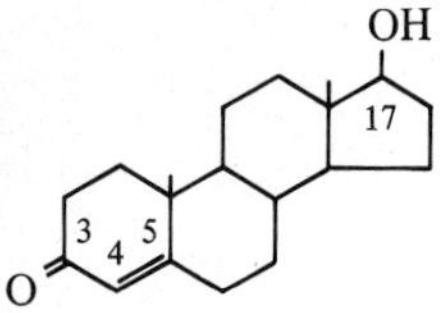

in the liver. Its other biological effects include the inhibition of protein synthesis in muscle and other tissues and the stimulation of fat breakdown to fatty acids in adipose tissue. Cortisol is also important as an anti-inflammatory agent.

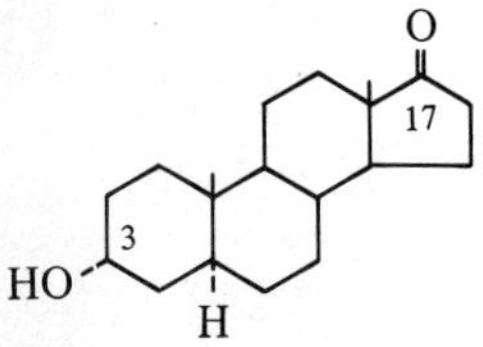

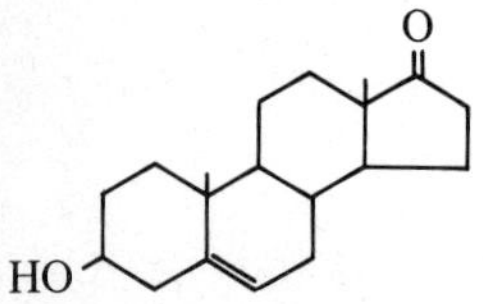

Aldosterone is primarily concerned with mineral metabolism and the regulation of electrolyte levels. It promotes the reabsorption of Na^+ ions by the kidneys and thereby governs water and electrolyte metabolism.

The consequences of a loss of adrenal function, resulting from surgical removal of the gland or from Addison's disease, include the following:

(1) Lowered concentrations of Na^+, Cl^-, and HCO_3^- in the plasma and increased levels of K^+.
(2) Loss of ability to excrete ingested water.
(3) Decreased excretion of K^+ and increased excretion of Na^+, Cl^-, and HCO_3^-.
(4) Increased content of K^+ and H_2O and decreased content of Na^+ in muscle.
(5) Decreased levels of liver and muscle glycogen.
(6) Lowered level of glucose in plasma.

Figure 16-28 Structures of several androgens: (a) testosterone, (b) androsterone, and (c) dihydroepiandrosterone.

The adrenal cortex is itself subject to hormonal control. The peptide hormone ACTH, which is released by the pituitary gland, acts upon the adrenal cortex so as to stimulate the production and secretion of the adrenocortical hormones. The release of ACTH is in turn influenced by neurohormones released by the hypothalamus of the brain. The pituitary gland is also controlled by a feedback system, whereby its activity is modified by the level of adrenocortical hormones in the circulation (chapter 22).

Androgens

The androgens or male sex hormones are essential for the development of many male characteristics by the growing animal. Their administration serves to prevent or even to reverse the effects of castration. In the adult male the continuous biosynthesis of these hormones is essential for the maturation of sperm and the functioning of the accessory glands of the genital tract.

In addition to their specific influence upon the genital tract, the androgens have a generalized effect upon metabolism. They stimulate protein synthesis and bone growth and enhance nitrogen retention. This effect is retained by several hormone derivatives that lack androgenic properties.

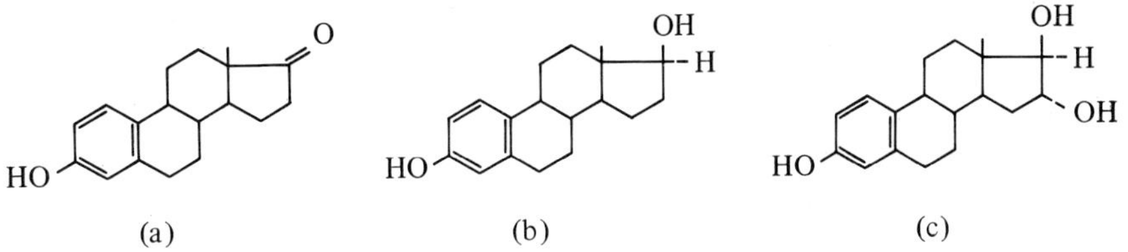

Figure 16-29
Structures of the estrogens: (a) estrone, (b) estradiol, and (c) estriol.

The principal androgen is testosterone (Fig. 16-28), which occurs in testes, together with the less active steroids androsterone and dehydro-epiandrosterone. In all of these the usual side-chain at the 17-position is replaced by a keto or hydroxyl group.

Estrogens

The estrogens or female sex hormones are different from all the steroids described previously in that the A ring has aromatic character (Fig. 16-29). As a consequence, there is no methyl group at the 10-position and the 3-hydroxyl has phenolic properties. The major estrogens include estrone, estradiol, and estriol.

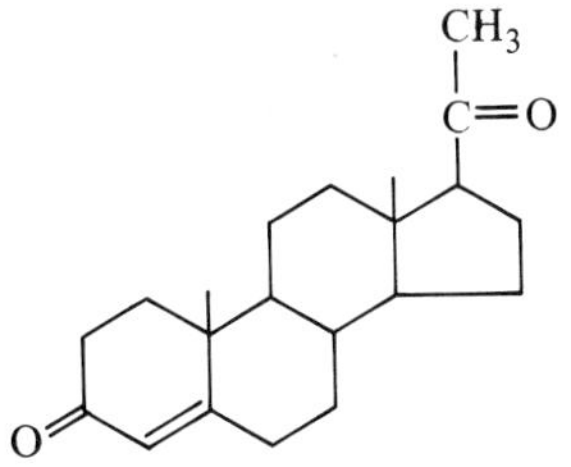

The estrogens are responsible for the female genital cycle, including the menstrual cycle in man and monkeys and the estrus cycles in other animals. The principal estrogen is estradiol (Fig. 16-29), which is formed by oxidative removal of C-19 of testosterone and subsequent aromatization of the A ring. The estrogens are formed largely in the ovary and also in the placenta during pregnancy. However, their biosynthesis also occurs in the testes.

The estrogens have a pronounced influence upon the metabolism of both inorganic and organic substances, including bone, fat, and protein. They appear to be responsible for the distribution of body fat characteristic of human females. The administration of estrogens produces an elevation of serum calcium levels and a redistribution of calcium within the bone structure.

Figure 16-30
Structure of progesterone.

Progestins

Progesterone (Fig. 16-30) is the primary hormone produced by the corpus luteum, the gland which develops in the ovarian follicle after release of an ovum. Other sites of formation are the adrenals, testes, and placenta. The combined action of progesterone and estradiol govern the menstrual cycle. Progesterone action is also the basis of the oral contraceptive.

16-8 BIOSYNTHESIS OF THE STEROID HORMONES

The sex hormones, the progestins, and the adrenal cortical hormones are all metabolic products of cholesterol (Fig. 16-31). The cholesterol side-chain is initially shortened to two carbon atoms by oxidative cleavage to form the intermediate pregnenolone. Subsequent oxidation of the 3-hydroxyl group of pregnenolone to a carbonyl, followed by a double bond shift, leads to progesterone (Fig. 16-31).

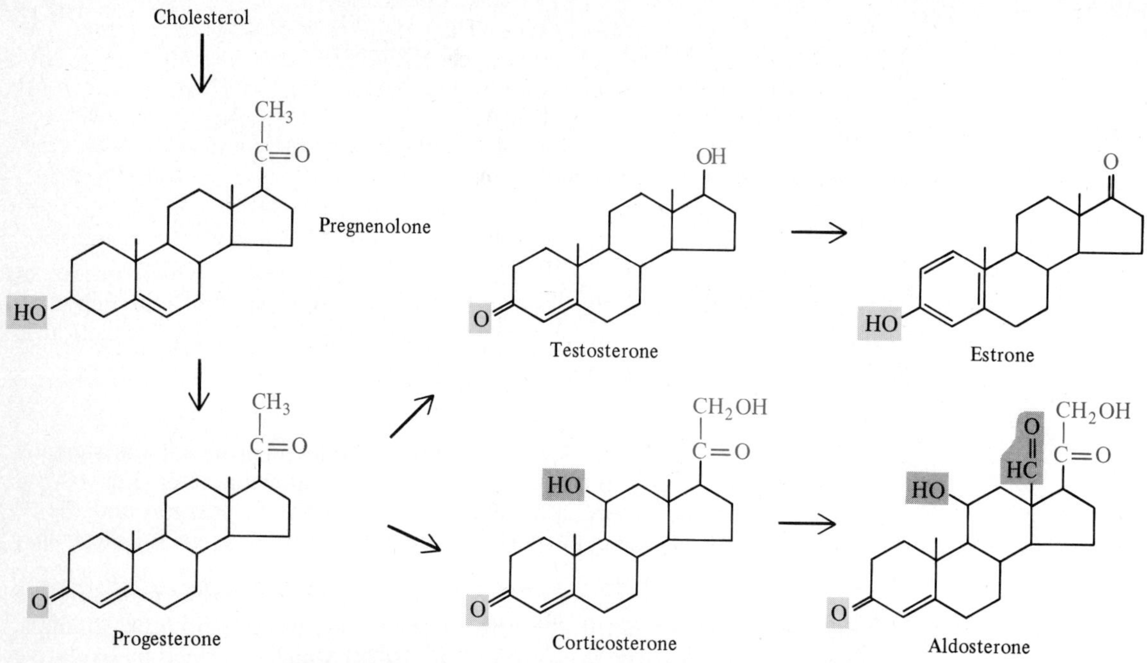

Cholesterol

Pregnenolone

Testosterone

Estrone

Progesterone

Corticosterone

Aldosterone

Figure 16-31
Biosynthesis of some steroid hormones from cholesterol.

Progesterone is the metabolic precursor of the adrenal cortical hormones cortisol and corticosterone, which arise by hydroxylation steps. Corticosterone, in turn, is the precursor of aldosterone.

Testosterone is formed from progesterone by removal of the side-chain at the C-17 position (Fig. 16-31). Oxidative removal of the C-19 of a testosterone derivative, followed by aromatization of the A ring, yields estrone.

SUMMARY

A number of biologically important molecules called **terpenes** are equivalent to combinations of **isoprene** $(\mathrm{C{-}C{-}\overset{\overset{\displaystyle C}{|}}{C}{-}C{-}})$ units. Among the most important of these are **vitamin A** and its precursor ***β*-carotene**. Vitamin A has a central role in vision as the precursor of **retinal**, which is complexed with the protein **opsin** to form the visual pigment **rhodopsin**, the light-sensitive component of the rod cells of the eye. Retinal can exist as ***cis*** or ***trans*** stereoisomers. Upon illumination of rhodopsin, a *cis* → *trans* conversion occurs for retinal, resulting in a dissociation of rhodopsin to yield *trans*-retinal and free opsin, thereby inducing a nerve impulse to the brain.

The **steroids** are derivatives of a parent tetracyclic hydrocarbon. The steroid alcohols or **sterols** include **cholesterol**, which is abundant in animal tissues and is a component of many membranes. Cholesterol, whose ultimate metabolic precursor is acetyl CoA, is itself a precursor of the sex and **adrenocortical** hormones, as well as of **vitamin D**. The male and female sex hormones, which are termed **androgens** and **estrogens**, respectively, act on the sex accessory organs and function as secondary sexual determinants. The adrenocortical hormones have profound effects upon the metabolism of carbohydrate and protein in many tissues.

Cholecalciferol or vitamin D_3 is essential for the proper development of bone in mammals. It is formed from the cholesterol derivative **7-dehydrocholesterol** by ultraviolet irradiation and is converted to the more active forms **25-hydroxy-** and **1,25-dihydroxycholecalciferol** in the liver and kidney, respectively.

REFERENCES

The following code is used to classify references. I: particularly useful as an introduction to the subject; R: useful primarily as a reference text; A: an advanced account of the material; H: a publication of historical importance.

General

M. H. Briggs and J. Brotherton, *Steroid Biochemistry and Pharmacology*, Academic, New York (1970). (R)

L. F. Fieser and M. Fieser, *Steroids*, Reinhold, New York (1959). (R)

T. W. Goodwin, ed., *Aspects of Terpenoid Chemistry and Biochemistry*, Academic, New York (1971). (A)

E. Heftmann, *Steroid Biochemistry*, Academic, New York (1969). (A)

W. Templeton, *An Introduction to the Chemistry of Terpenoids and Steroids*, Butterworths, London (1969). (I)

Steroid Hormones

R. I. Dorfman and F. Ungar, *Metabolism of Sex Hormones*, Academic, New York (1965). (R)

K. W. McKerns, *Steroid Hormones and Metabolism*, Appleton, New York (1969). (R)

Bile Salts

G. A. D. Haslewood, *Bile Salts*, Methuen, London (1967). (A)

Cholesterol Biosynthesis

K. Bloch, "The Biological Synthesis of Cholesterol," *Science* 150, 19 (1956). (A)

Vitamins

H. F. Deluca and J. W. Suttie, eds., *The Fat-Soluble Vitamins*, University of Wisconsin, Madison (1970). (A)

REVIEW QUESTIONS

Questions marked with an asterisk are of a high level of difficulty.

*16-1 Write a balanced equation for the synthesis of cholesterol from acetyl CoA.

*16-2 Speculate as to the biological significance of the recurrence of terpene-type structures in compounds of diverse function.

16-3 Which of the following statements is more likely to be correct? Why?
(a) The beneficial effects of vitamin D increase monotonically with increasing dosage.
(b) No further benefit occurs upon increasing the dosage beyond an optimal level.

16-4 Acetate, both of whose carbon atoms are isotopically labeled, is administered to rats. Which of the carbon atoms of cholesterol would be labeled?

*16-5 If only the methyl carbon in the experiment in 16-4 were labeled, what would the result be?

17

Porphyrins, Hemes, and Heme Proteins

17-1 GENERAL PROPERTIES OF THE PORPHYRINS

Figure 17-1
Structure of porphin ($C_{20}H_{14}N_4$).

This chapter will discuss the class of compounds termed porphyrins, which have essential functions in at least three processes of central importance in biology: the use of light by plants to synthesize carbohydrates, the transport of electrons in the respiratory chain, and the transport of molecular oxygen in animals. Without the latter, higher species could not exist. Without the first two, there could be no life at all. That chemically very similar molecules can mediate such diverse processes is a striking example of efficiency in biochemical design.

The porphyrins are derivatives of the parent compound **porphin** (Fig. 17-1). This consists of four **pyrrole** rings joined by four CH bridges to form a large ring-shaped molecule. The inner ring of porphin contains 12 carbon and four nitrogen atoms (Fig. 17-1). Because of resonance, the positions of the double bonds should not be regarded as fixed.

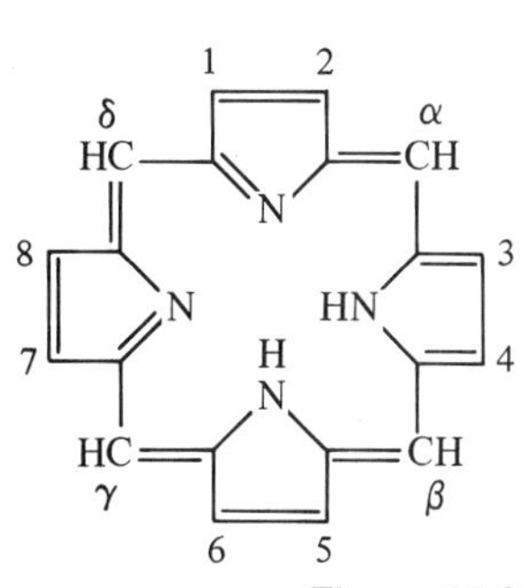

Figure 17-2
Simplified representation of the porphin ring.

The naturally occurring porphyrins are derivatives of porphin in which the eight pyrrole CH groups are partially replaced by various substituents (Fig. 17-2). It is convenient for many purposes to indicate the positions of the substituents with the aid of the following abbreviated formula:

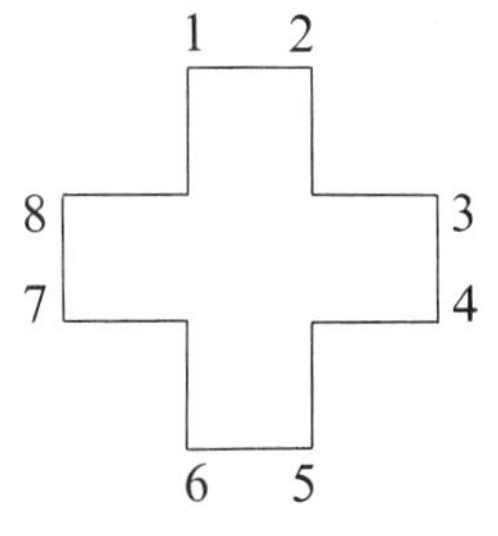

Figure 17-3 Structure of ferroprotoporphyrin or heme.

Protoporphyrin, the porphyrin component of some heme groups, contains four methyl, two vinyl (—CH═CH$_2$), and two propionic acid (—CH$_2$—CH$_2$—COOH) residues (Fig. 17-3). Only one of the possible isomers, protoporphyrin IX, is of major biological importance:

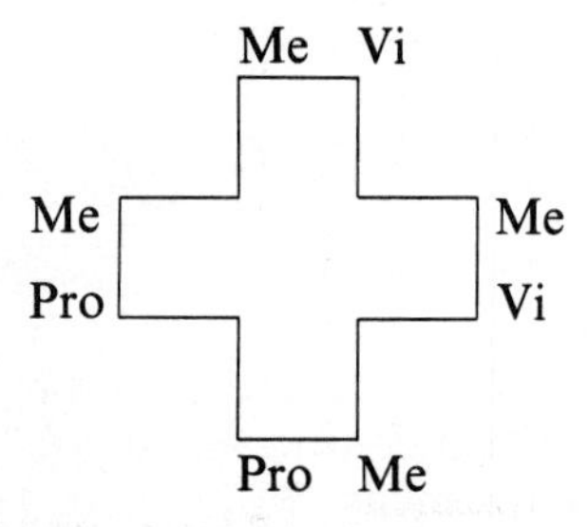

The porphyrins are intensely colored compounds, possessing sharply banded absorption spectra in the visible wavelengths. In the case of protoporphyrin dissolved in a mixture of ether and acetic acid, the absorption maxima are at 633, 576, 537, 502, and 396 nm. In organic solvents porphyrins also display a characteristic red fluorescence, which may be used for analytical purposes.

The presence of tertiary nitrogens in two of the four pyrrole rings endows the porphyrins with weakly basic properties:

$$+ H^+ \rightleftharpoons \quad N\text{-}H^+$$

17-2 HEMES

An important property of the porphyrins is their ability to form complexes with many metallic cations. The complexes formed with iron and with magnesium are of particular biological importance.

The porphyrin ring is planar and rigid. The dimensions of the central hole are such that an iron atom can be inserted readily, to form coordinated linkages with the four pyrrole nitrogens, displacing the hydrogens of the two NH groups as protons (Fig. 17-3). The complexes with protoporphyrin IX are known as **hemes**.

The iron may be in either the ferrous [Fe(II)] or ferric [Fe(III)] states. The corresponding complexes with protoporphyrin IX are termed **ferrohemes** and **ferrihemes**, respectively. Ferroheme has no net charge because of the loss of two protons. Ferriheme or hemin has a single net positive charge and may form salts, being usually prepared as the chloride. Since iron atoms normally have a coordination number of six, two other ligands may be linked to the iron from the two open sides of the heme. In heme complexes with proteins, one of these ligands is often the **imidazole** group of histidine.

Heme occurs in virtually all living organisms. Some of the most familiar hemoproteins are **hemoglobin** and **myoglobin**, which combine reversibly with molecular oxygen, the **cytochromes**, which figure in mitochondrial electron transport, and the **peroxidases**, which catalyze the decomposition of hydrogen peroxide. Hemoproteins have an intense absorption band near 400 nm, which is called the **Soret band**.

17-3 HEMOGLOBIN

Hemoglobin is the principal agent for the transport of molecular oxygen by the mammalian bloodstream. It occurs in the red blood cells or **erythrocytes**, accounting for a major fraction of their mass. This protein has the remarkable property of binding molecular oxygen in a readily reversible attachment.

Mammalian hemoglobin is a conjugated protein of molecular weight 67,000, containing four heme groups. The polypeptide portion of the molecule, exclusive of the heme groups, has been named **globin**. The heme and globin portions of hemoglobin may be separated by acid treatment. In the presence of hydrochloric acid plus acetone, the globin is precipitated:

$$\text{hemoglobin} \xrightarrow{\text{HCl}} \text{globin} + \text{ferroheme}$$

The ferroheme remaining in solution is rapidly oxidized by atmospheric oxygen to ferriheme.

At neutral pH globin combines with ferroheme to form hemoglobin, or with ferriheme to yield **methemoglobin**:

$$\text{globin} + \text{ferroheme} \rightarrow \text{hemoglobin}$$
$$\text{globin} + \text{ferriheme} \rightarrow \text{methemoglobin}$$

Methemoglobin lacks the capacity to combine reversibly with oxygen. It is also formed by the action upon hemoglobin of oxidizing agents, including peroxides and ferricyanide.

Oxyhemoglobin

The distinctive property of hemoglobin is its ability to form a stable oxygen complex while the iron atom of the heme group remains in the Fe(II) state. Free ferroheme and its derivatives can also bind oxygen, but the iron is

X
N N
II
Fe
N N
X

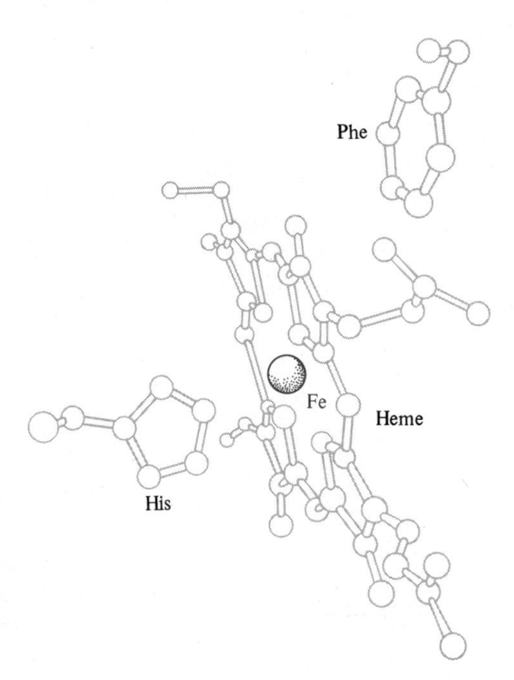

Figure 17-4
Environment of the iron atom in hemoglobin. The proximal histidine is shown.

rapidly oxidized to the Fe(III) state. The *reversible* character of oxygen complex formation by hemoglobin reflects the specific protein environment of the heme group, which is embedded in a hydrophobic region of the molecule.

In hemoglobin the imidazole group of a histidine (termed the **proximal** histidine) lies near the iron atom of each heme group (Fig. 17-4). Imidazole occupies the *fifth* coordinate position of the iron; in oxyhemoglobin the *sixth* position is occupied by oxygen, probably in exchange for H_2O. The noncovalent attachment of the heme group to a histidine residue recurs in many heme proteins.

Carbon monoxide (CO) can replace oxygen in its complex with hemoglobin. Its binding is so strong that minute levels in the atmosphere can block oxygen transport.

The binding of oxygen to the heme iron is believed to occur as a result of donation of an electron pair of oxygen to the iron atom. In the deoxygenated state, four of the five 3*d* orbitals of the iron atom each contain one *unpaired* electron. The binding of oxygen causes a transition to a state in which all of the electrons are paired; this is accompanied by a loss of the paramagnetism characteristic of deoxygenated hemoglobin. The radius of the iron atom is sufficiently large for the deoxygenated state so that it cannot fit exactly into the center of the porphyrin ring but is displaced about 0.06 nm toward the proximal imidazole ring. Upon oxygenation, the iron atom moves into the center of the porphyrin ring, pulling the proximal histidine with it. The resultant conformational change is of central importance for the allosteric properties of hemoglobin.

Primary Structure

The globin portion of all mammalian hemoglobins consists of four polypeptide chains. Hemoglobin contains no cystine and is not cross-linked. Each polypeptide chain is combined with a heme group.

Normal *adult* human hemoglobin (HbA) is constructed of two equivalent half molecules, each of which contains an α-chain and a distinct, though similar, β-chain (Fig. 17-5); it is often designated in abbreviated form as $\alpha_2^A\beta_2^A$. In *fetal* hemoglobin (HbF) the two β chains are replaced by two γ chains and this form is accordingly designated as $\alpha_2^A\gamma_2^A$. The primary structures of the α, β, and γ chains are shown in Fig. 17-6. The three chains are synthesized by distinct genes.

A number of known hereditary diseases are characterized by alterations in the amino acid sequence. One of these, sickle cell hemoglobin (HbS), has already been mentioned (section 4-6). Some others are cited in Table 17-1. In each case the variant protein arises from a point mutation in the gene responsible for the α or β polypeptide. This is inherited according to the rules of Mendelian genetics.

Three-Dimensional Structure

Hemoglobin shares with myoglobin the distinction of being the first protein whose complete structure was determined by x-ray crystallography. This classical achievement was the culmination of over 20 years work by Perutz and his associates.

The overall shape of the molecule is that of a compact spheroid with very little open space in the interior. The greater part of the polypeptide chains is incorporated into a set of short α-helical segments separated by nonhelical bends, which contain the proline residues (Fig. 17-7–17-9). The total α-helical content is about 70%.

The interior of the molecule consists of nonpolar residues in a very close-packed array. Almost all the polar side-chains lie on the surface in contact with water. Within each subunit the polypeptide chain is folded about the planar heme group in a characteristic pattern, which is similar for both the α and β chains of all hemoglobins.

Within each hemoglobin tetramer each α chain is closely paired with a β unit. The pairs are designated $\alpha_1\beta_1$ and $\alpha_2\beta_2$. The number of interchain contacts within each of the two $\alpha\beta$ units is substantially greater than that between different $\alpha\beta$ units. In dilute solution oxyhemoglobin dissociates into two $\alpha\beta$ submolecules; little or no further dissociation into separate α and β chains occurs. This reflects the greater cohesion within the $\alpha\beta$ pairs, which arises primarily from hydrophobic bonding.

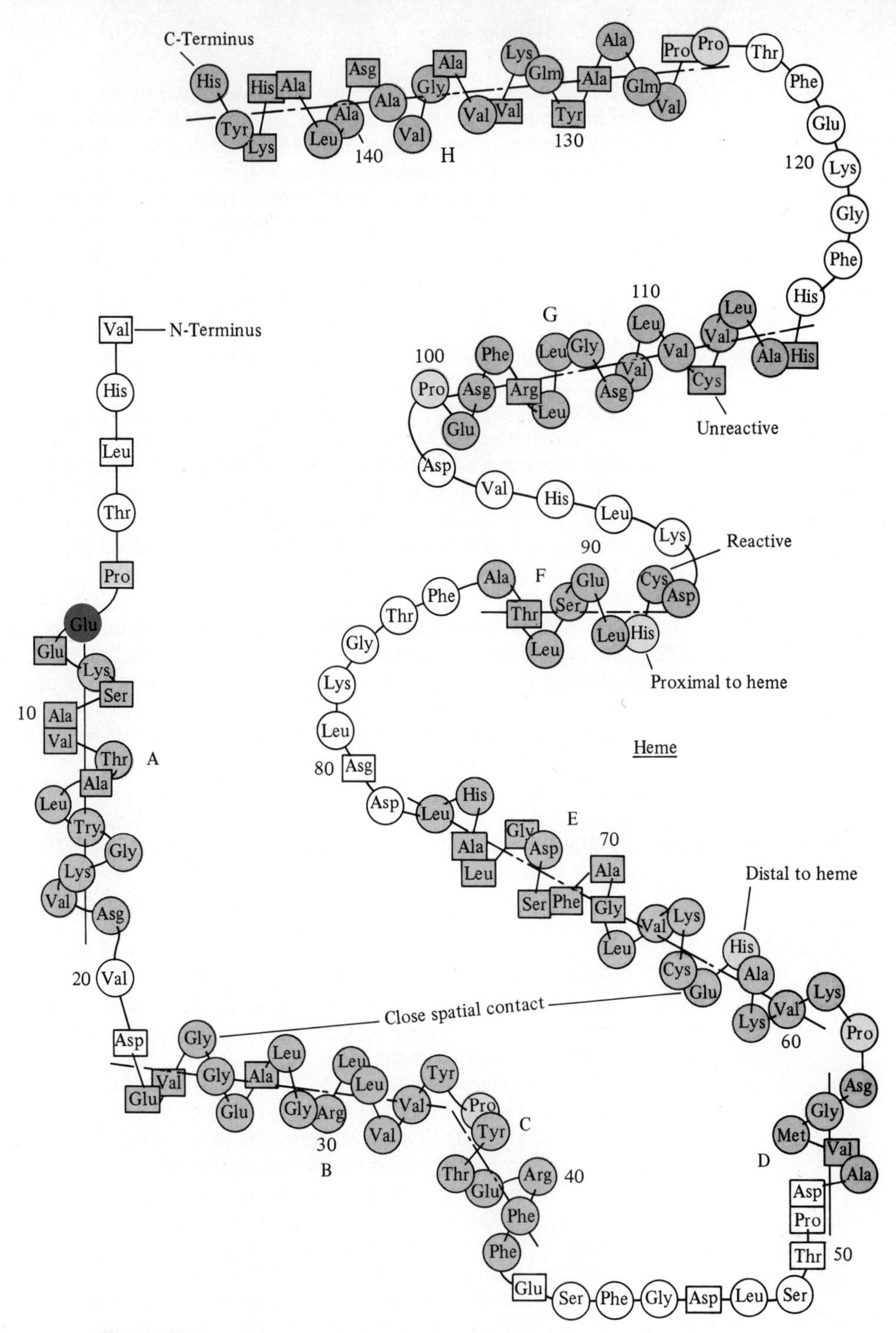

Figure 17-5
Conformation of the β-chain of human hemoglobin, shown in planar projection. A–H represent α-helical segments.

10 20
α Val·Leu·Ser·Pro·Ala·Asp·Lys·Thr·Asg·Val·Lys·Ala·Ala·Try·Gly·Lys·Val·Gly·Ala·His·Ala·Gly·Glu·Tyr·
β Val·His·Leu·Thr·Pro·Glu·Glu·Lys·Ser·Ala·Val·Thr·Ala·Leu·Try·Gly·Lys·Val·Asn·Val· Asp·Glu·Val·
γ Gly·His·Phe·Thr·Glu·Glu·Asp·Lys·Ala·Thr·Ile·Thr·Ser·Leu·Try·Gly·Lys·Val·Asn·Val· Glu·Asp·Ala·
10 20

30 40
α Gly·Ala·Glu·Ala·Leu·Glu·Arg·Met·Phe·Leu·Ser·Phe·Pro·Thr·Thr· Lys·Thr·Tyr·Phe·Pro·His·Phe·Asp·Leu·
β Gly·Gly·Glu·Ala·Leu Gly·Arg·Leu·Leu·Val·Val·Tyr·Pro·Try·Thr·Gln·Arg·Phe·Phe·Glu·Ser·Phe·Gly·Asp·Leu·
γ Gly·Gly·Glu·Thr·Leu·Gly·Arg·Leu·Leu·Val·Val·Tyr·Pro·Try·Thr·Gln·Arg·Phe·Phe·Asp·Ser·Phe·Gly·Asg·Leu·
30 40

50 60
α Ser·His·Gly·Ser·Ala· Gln·Val·Lys·Gly·His·Gly·Lys·Lys·Val·Ala·Asp·Ala·Leu·Thr·Asg·
β Ser·Thr·Pro·Asp·Ala·Val·Met·Gly·Asn·Pro·Lys·Val·Lys·Ala·His·Gly·Lys·Lys·Val·Leu·Gly·Ala·Phe·Ser·Asp·
γ Ser·Ser·Ala·Ser·Ala·Ile·Met·Gly·Asn·Pro·Lys·Val·Lys·Ala·His·Gly·Lys·Lys·Val·Leu·Thr·Ser·Leu·Gly·Asp·
50 60 ↑ 70

70 80 90
α Ala·Val·Ala·His·Val·Asp·Asp·Met·Pro·Asg·Ala·Leu·Ser·Ala·Leu·Ser·Asp·Leu·His·Ala·His·Lys·Leu·Arg·Val·
β Gly·Leu·Ala·His·Leu·Asp·Asp·Leu·Lys·Gly·Thr·Phe·Ala·Thr·Leu·Ser·Gln·Leu·His·Cys·Asp·Lys·Leu·His·Val·
γ Ala·Ile·Lys·His·Leu·Asp·Asp·Leu·Lys·Gly·Thr·Phe·Ala·Gln·Leu·Ser·Glu·Leu·His·Cys·Asp·Lys·Leu·His·Val·
80 90 ↑

100 110
α Asp·Pro·Val·Asn·Phe·Lys·Leu·Leu·Ser·His·Cys·Leu·Leu·Val·Thr·Leu·Ala·Ala·His·Leu·Pro·Ala·Glu·Phe·Thr·
γ Asp·Pro·Gln·Asp·Phe·Arg·Leu·Leu·Gly·Asg·Val·Leu·Val·Cys·Val·Leu·Ala·His·His·Phe·Gly·Lys·Glu·Phe·Thr·
β Asp·Pro·Glu·Asn·Phe·Lys·Leu·Leu·Gly·Asg·Val·Leu·Val·Thr·Val·Leu·Ala·Ile·His·Phe·Gly·Lys·Glu·Phe·Thr·
100 110 120

120 130 141
α Pro·Ala·Val·His·Ala Ser·Leu·Asp·Lys·Phe·Leu·Ala·Ser·Val·Ser·Thr·Val·Leu·Thr·Ser·Lys·Tyr·Arg
β Pro·Pro·Val·Gln·Ala·Ala·Tyr·Gln·Lys·Val·Val·Ala·Gly·Val·Ala·Asp·Ala·Leu·Ala·His·Lys·Tyr·His
γ Pro·Glu·Val·Gln·Ala·Ser·Try·Gln·Lys Met·Val·Thr·Gly·Val·Ala·Ser·Ala·Leu·Ser·Ser·Arg·Tyr·His
130 140 146

Figure 17-6
Amino acid sequences of the three polypeptide chains which occur in human hemoglobin.

Table 17-1
Some Abnormal Hemoglobins

Type	Position	Residue	Residue in Mutant
S	$\beta 6$	Glu	Val
I	$\alpha 16$	Lys	Asp
$G_{Honolulu}$	$\alpha 30$	Glu	$GluNH_2$
M_{Boston}	$\alpha 58$	His	Tyr
$G_{Philadelphia}$	$\alpha 68$	$AspNH_2$	Lys
C	$\beta 6$	Glu	Lys
E	$\beta 26$	Glu	Lys
Q_{Arabia}	$\beta 121$	Glu	Lys
$O_{Indonesia}$	$\alpha 116$	Glu	Lys

Figure 17-7
Model of hemoglobin with one chain removed. (Courtesy of National Naval Medical Center, Bethesda, Maryland.)

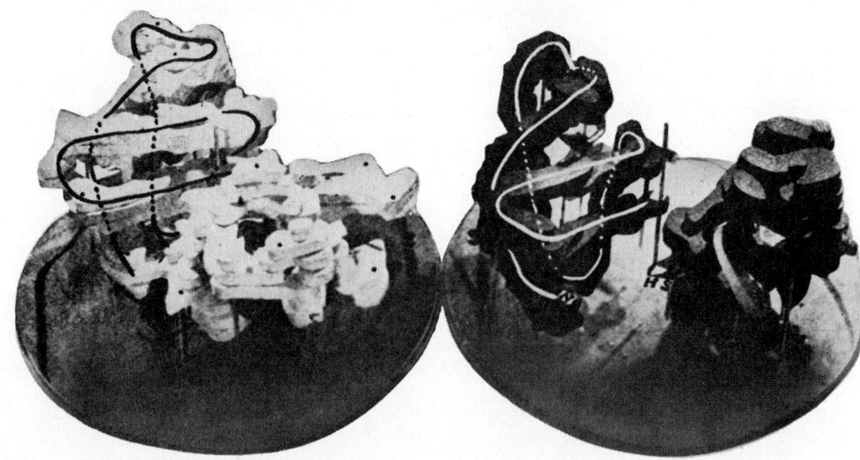

Figure 17-8
Partially dissembled model of hemoglobin. (Courtesy of National Naval Medical Center, Bethesda, Maryland.)

Erythrocytes

The erythrocytes or red cells of the blood are biconcave disks about 8 μm in diameter and 2 μm thick at the periphery. About 35% of their weight is solids, of which 33% is hemoglobin. A human adult possesses about 10^{12} erythrocytes, containing about 950 g of hemoglobin.

An erythrocyte consists of an external sac that surrounds and contains the hemoglobin solution. A semipermeable membrane is present at the surface. In contrast to plasma, the principal cation of the interior is K^+. No

Figure 17-9
Complete model of hemoglobin, showing $\alpha_1\beta_1$ and $\alpha_2\beta_2$ units. (Courtesy of National Naval Medical Center, Bethesda, Maryland.)

nucleus is present. Various chemical and physical treatments, including exposure to a solvent of low electrolyte content and attack by certain snake venoms, cause rupture of the external sac and release of the hemoglobin. The process is termed **hemolysis** and the resultant preparations of hemoglobin-free sacs are called **ghosts**.

The erythrocyte membrane is extensively involved in both passive facilitated and active transport. Glucose and several other sugars enter the cells by facilitated diffusion, utilizing a specific protein carrier. An active transport system exists for the transport of K^+ into and Na^+ out of the red cells.

The erythrocytes are formed in the red bone marrow. A series of cellular precursors are formed. In contrast to the mature erythrocytes, the earlier precursors are nucleated. The number of erythrocytes present in the circulation is roughly constant, formation and destruction being approximately balanced. The greater part of erythrocyte breakdown occurs in the spleen and to a lesser extent in the bone marrow.

The Transport of Oxygen by Erythrocytes

The oxygen requirements of mammals are considerable, amounting to as much as 30 mole/day for a human adult, about 80% of which is eliminated as CO_2. The active metabolism of mammalian tissues, which is based upon oxidation, requires the removal of oxygen from the air, its rapid transport to the most remote parts of the body, and its delivery to the tissues at a pressure close to that at which it is present in the atmosphere.

The transport of oxygen by the blood occurs both by ordinary solution and by combination with the hemoglobin of erythrocytes. Because of the limited solubility of molecular oxygen in water, the former mechanism is of minor importance, accounting for only a small fraction of the total oxygen transported. Hemoglobin is thus an indispensable factor in the maintenance of an adequate oxygen supply.

Inhaled air is mixed with the gases already present in the respiratory tract. A part of this mixture enters the alveolar sacs of the lungs, where it makes contact with the pulmonary capillaries. Molecular oxygen diffuses across the capillary walls into the circulating blood, where most of it combines with the hemoglobin of the erythrocytes. Carbon dioxide moves in the opposite direction. In expiration, a part of the gas of the alveolar sacs is forced into the upper respiratory tract, where it is mixed with the gases already present, a fraction of which leave as expired air. Under normal conditions the rates of entry and loss of CO_2 and O_2 in the alveolar sacs are balanced, so that their gas composition remains constant.

The oxygen combined with the erythrocyte hemoglobin is delivered by the arterial blood to the capillary systems of the various tissues. The partial pressure of O_2 in the interstitial fluid surrounding the capillaries is only about one-third of that of arterial blood. As a consequence, O_2 is released from the erythrocytes in the capillaries and diffuses through plasma into the interstitial fluid and then into the tissue cells.

Simultaneously with this process, the CO_2 formed by oxidative metabolism diffuses from the tissues into the termini of the circulatory system. Since a rise in the CO_2 level reduces the affinity of hemoglobin for oxygen, the release of oxygen from oxyhemoglobin is thereby stimulated.

The CO_2 is transported by the venous blood back to the lungs, where it is released. It is transported both as bicarbonate (HCO_3^-) and as a complex formed with hemoglobin.

Bicarbonate ion is formed by the reaction of CO_2 with H_2O:

$$CO_2 + H_2O \rightleftarrows H_2CO_3 \rightleftarrows HCO_3^- + H^+$$

In the absence of a catalyst this reaction attains equilibrium rather slowly. The enzyme **carbonic anhydrase**, which is present in erythrocytes, accelerates the reaction sufficiently to meet the requirements for CO_2 transport.

The delivery of oxygen to tissues and the removal of CO_2 complement each other to a remarkable degree. The two processes may be summarized as in Table 17-2.

The Oxygen Equilibria of Hemoglobin

The transport of molecular oxygen by the hemoglobin of mammalian erythrocytes occurs through the formation of a dissociable hemoglobin-oxygen complex:

$$Hb + O_2 \rightleftarrows HbO_2$$

Each of the four heme groups may combine with a single oxygen molecule:

$$Hb_4 \overset{O_2}{\rightleftarrows} Hb_4O_2 \overset{O_2}{\rightleftarrows} Hb_4O_4 \overset{O_2}{\rightleftarrows} Hb_4O_6 \overset{O_2}{\rightleftarrows} Hb_4O_8$$

Table 17-2
Oxygen Transport and Removal of Carbon Dioxide

O_2 *Transport*	O_2 *Tension* (mm Hg)	*Reaction*
Inhaled air	158	
↓		
Alveolar air	100	
↓		
Arterial blood	90	$Hb + O_2 \rightarrow HbO_2$
↓		
Capillaries	40	$HbO_2 \rightarrow Hb + O_2$
↓		
Interstitial fluid	30	
↓		
Cell interiors	10	
CO_2 *Removal*	CO_2 *Tension* (mm Hg)	
Tissues	50	
↓		
Venous blood	46	
↓		
Alveolar air	40	
↓		
Exhaled air	32	
↓		
Atmosphere	0.3	

The combination is very rapid, being complete within 0.1 sec under physiological conditions.

In accordance with the principle of mass action, the degree of oxygenation of hemoglobin is a function of the partial pressure of oxygen and increases with increasing oxygen tension (Fig. 6-15). A maximum of four oxygen molecules can be bound by a single molecule of hemoglobin. However, the binding equilibria of the four heme groups are interdependent, so that the affinity of a particular heme for oxygen is dependent upon the state of oxygenation of the other hemes. This interdependence is a consequence of the conformational change accompanying oxygenation and is an example of **allosterism**.

The binding affinity for O_2 of deoxyhemoglobin is relatively low, but after one or more of the subunits have combined with O_2 the affinity of the remaining subunits increases by a factor of several hundred under physiological conditions. This **cooperativity** is of major importance for the physiological role of hemoglobin in that it greatly sharpens its response to changes in oxygen partial pressure. This is an important contributing factor to the ability of the hemoglobin of erythrocytes to become saturated with oxygen in the lungs and to release nearly all of its oxygen upon encountering the 20-fold reduction in partial pressure characteristic of the tissue capillaries.

The cooperativity in oxygen binding displayed by hemoglobin can be explained semiquantitatively by postulating that each α or β subunit may

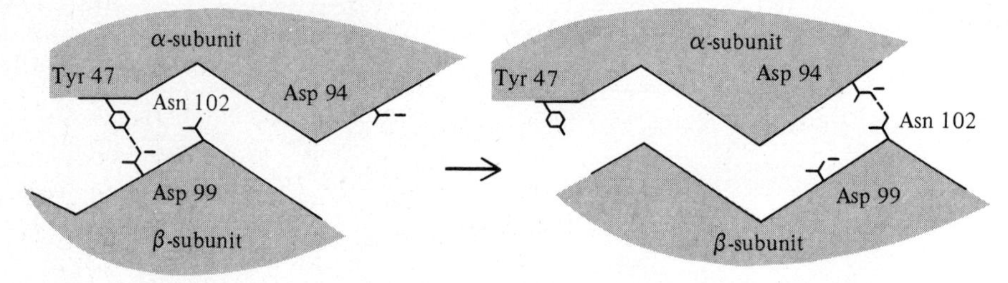

Figure 17-10
Structural alterations accompanying the oxygenation of hemoglobin. The change in hydrogen bonding in part of the α_1-β_2 contact region of the interface between two $\alpha\beta$ units is shown. Other changes are described in the text.

exist in either of two conformations, one of which, the T conformation, has a low affinity for O_2, and the other, the R conformation, has a much higher affinity. Since the presence of intermolecular contacts between R and T subunits is energetically unfavored, a hemoglobin tetramer tends to be either all T(T_4) or all R(R_4). In the absence of O_2 binding an equilibrium exists between R_4 and T_4 units which is biased greatly in favor of the latter. Upon the binding of an O_2 molecule by a subunit initially in the T conformation, that subunit is converted to the R conformation and pulls the other subunits with it into that conformation, so that their affinity for O_2 increases. This simplified picture corresponds to the symmetry model for allosteric systems (section 6-8). However, the actual mechanism is almost certainly more complex.

X-ray crystallography has detected small but distinct differences in the conformations of the subunits of deoxy- and oxyhemoglobin, as well as in their mutual arrangement. The heme groups of the two β subunits move about 0.07 nm closer together in the oxy than in the deoxy form. No major change is observed within the $\alpha_1\beta_1$ contact region. However, important changes do occur within the $\alpha_1\beta_2$ contact, with an alteration in the hydrogen bonding pattern (Fig. 17-10). In addition to the tyrosine-aspartate linkage shown in Fig. 17-10, deoxyhemoglobin also contains hydrogen bonds involving the COOH-terminal arginine of each α-chain, whose carboxyl is linked to a lysine of the other α-chain, and the COOH-terminal histidine of each β-chain, whose carboxyl is hydrogen bonded to a lysine of the opposite α-chain. The imidazole group of the same COOH-terminal histidine is also hydrogen bonded to an aspartate within the same β-chain. Oxygenation, which results in a shift in the mutual positions of the α_1 and β_2 subunits (Fig. 17-10), ruptures the above linkages.

Under physiological conditions the oxygen-binding equilibria of hemoglobin are influenced to an important degree by the binding of the effector **2,3-diphosphoglycerate** (DPG), which reduces the oxygen affinity. DPG, which is present in erythrocytes in a roughly 1:1 mole ratio to hemoglobin, is bound preferentially by deoxyhemoglobin. One molecule of DPG binds to a single hemoglobin tetramer between the two β chains. The reduction in O_2 affinity by DPG is of physiological importance in that it permits a more complete release of O_2 in the tissue capillaries.

The Bohr Effect

The affinity of hemoglobin for O_2 is also dependent upon the pH. Under physiological conditions the oxygenation of hemoglobin is accompanied by an appreciable dissociation of hydrogen ions. Conversely, increasing protonation of hemoglobin reduces its affinity for O_2, so that a drop in pH results in a dissociation of O_2:

$$HbO_2 + nH^+ \rightleftarrows Hb(H^+)_n + O_2$$

This **Bohr effect** contributes to the efficiency of O_2 release in tissue capillaries, where the O_2 partial pressure is low and the level of CO_2 may be high, with a consequent acidification.

The Bohr effect arises from a rupture of electrostatic bonds or **salt bridges** at the termini of the hemoglobin subunits upon oxygenation. In deoxyhemoglobin both the NH_2-terminal valines of the α-subunits and the histidine-143 groups of the β-subunits are involved in salt bridges. As a result, the protonated forms of these groups are stabilized and their pK values are raised (section 3-6). Upon oxygenation, the groups are freed, the pK values decrease, and protons are released. A decrease in hydrogen ion concentration at constant oxygen pressure favors deprotonation of these groups and increases the affinity for O_2.

17-4 MYOGLOBIN

Myoglobin, which is found in the muscles of all vertebrates and invertebrates, is a hemoprotein capable of reversibly binding O_2. Its affinity for O_2 is higher than that of hemoglobin and it functions as a storage agent.

The single polypeptide chain of myoglobin, whose molecular weight is 17,000, is combined with a single heme group. The mode of attachment of the latter is similar to that of the heme groups of hemoglobin. The spatial arrangement of the polypeptide chain also closely resembles those of the four hemoglobin chains, although there is little correlation in amino acid sequence (Fig. 17-11).

Much of what has been said of the structure of hemoglobin holds also for myoglobin. The molecule is quite compact with very little internal space accessible to solvent. All of the polar side-chains are located on the surface, while the interior contains most of the nonpolar side-chains. The polypeptide backbone consists of eight short α-helical segments, which are separated by bends, to which the proline residues are confined. The overall conformation of myoglobin is invariant for all of the species thus far examined, although important differences in amino acid content occur. The invariant amino acids appear sufficient to maintain the characteristic structure.

As myoglobin contains only a single heme, and hence only one binding site for O_2, it displays no cooperativity of binding (Fig. 6-15). This is consistent with its function as a storage protein, which does not require a sharpened response to changes in oxygen partial pressure.

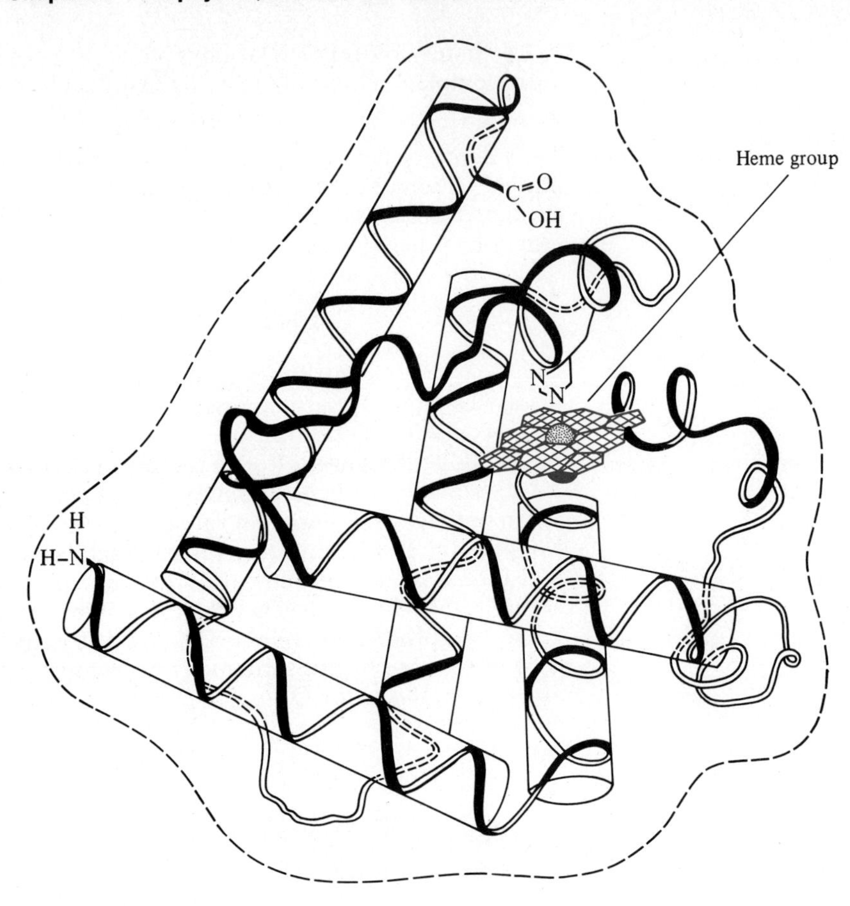

Figure 17-11
Conformation of myoglobin, showing α-helical regions.

17-5 OTHER HEMOPROTEINS

Cytochromes

The cytochromes are a group of mitochondrial heme-containing proteins, which are of central importance as electron carriers in the respiratory chain (section 12-6). The best characterized of these is cytochrome c, whose complete three-dimensional structure is known.

The molecular weight of cytochrome c is close to 12,400. It contains one heme group per molecule. The protein component consists of a single polypeptide chain devoid of cystine cross-links (Fig. 17-12). The heme group is attached by covalent **thioether** ($-CH_2-S-\overset{|}{C}H-$) bonds from the two cysteine residues to the *reduced* vinyl groups of the protoporphyrin ring (Fig. 17-13).

In contrast to hemoglobin and myoglobin, the heme group of cytochrome c is completely enveloped by the polypeptide chain and virtually inaccessible to solvent. The heme may exist in ferrous and ferric forms; in both forms the fifth and sixth coordination positions of the iron are occupied by a methionine and a histidine. The reduced form of cytochrome c does not form complexes with molecular O_2 or CO.

Acetyl·Gly·Asp·Val·Glu·Lys·Gly·Lys·Lys·Ile·Phe·
1 10

Ile·Met·Lys·Cys·Ser·GluNH$_2$·CyS·His·Thr·Val·Glu·Lys·
11 12 15 20
Heme

Gly·Gly·Lys·His·Lys·Thr·Gly·Pro·AspNH$_2$·Leu·His·Gly·
30

Leu·Phe·Gly·Arg·Lys·Thr·Gly·GluNH$_2$·Ala·Pro·Gly·
40

Tyr·Ser·Tyr·Thr·Ala·Ala AspNH$_2$·Lys·AspNH$_2$·Lys·Gly·
46 47 50

Ile·Ile·Try·Gly·Glu·Asp·Thr·Leu·Met·Glu·Tyr·Leu·Glu·
58 60 62

AspNH$_2$·Pro·Lys·Lys·Tyr·Ile·Pro·Gly·Thr·Lys·Met·Ile·
70 80

Phe·Val·Gly·Ile·Lys·Lys·Lys·Glu·Glu·Arg·Ala·Asp·Leu·
83 89 90 92

Ile·Ala·Tyr·Leu·Lys·Lys·Ala·Thr·AspNH$_2$·GluCOOH
100 104

Figure 17-12 Amino acid sequence of human heart cytochrome c.

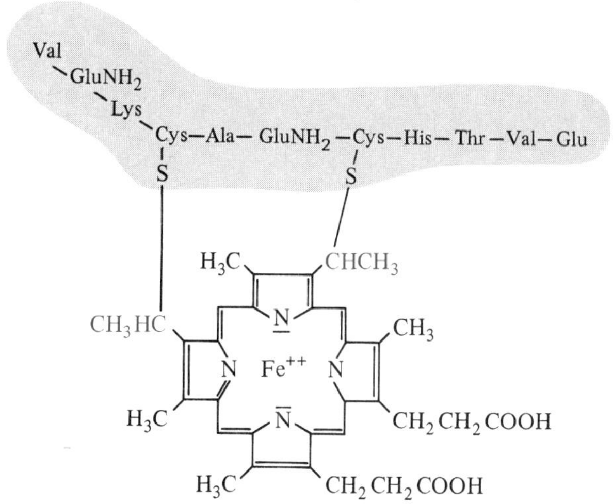

Val
GluNH$_2$
Lys
Cys—Ala— GluNH$_2$—Cys—His— Thr—Val— Glu
S S
H_3C CHCH$_3$
CH$_3$HC N CH$_3$
N Fe^{++} N
H_3C N CH$_2$CH$_2$COOH
H_3C CH$_2$CH$_2$COOH

Figure 17-13 Attachment of the heme group in cytochrome c.

A number of other cytochromes have been identified. These include cytochromes b, b$_5$, a, and a$_3$. The structure of cytochrome b$_5$, which occurs in the *outer* mitochondrial membrane, is totally unlike that of cytochrome c. The heme group is not covalently bonded to the protein but is coordinated with two histidine side-chains. As in the case of cytochrome c, the heme is completely buried in the interior of the protein.

Cytochromes a, b, and c all function in electron transport in the respiratory chain, being alternately reduced and reoxidized. The mechanism whereby electrons reach the buried heme is still uncertain.

Cytochrome a (cytochrome oxidase), the terminal cytochrome in the respiratory chain, differs from the above in containing two hemes. Unlike the other cytochromes it combines with molecular O_2, which is then rapidly reduced to H_2O. A variant of cytochrome a also combines with CO; it is designated a_3.

Catalase and Peroxidase

This group of heme-containing enzymes catalyzes reactions of hydrogen peroxide H_2O_2, rather than O_2. The **peroxidases** occur widely in plant tissues and to some extent in animal tissues. They catalyze reactions of the type

$$H_2O_2 + AH_2 \rightarrow 2H_2O + A$$

Catalase, a tetramer of molecular weight 250,000, is of general occurrence in aerobic cells. It catalyzes the decomposition of hydrogen peroxide, whose accumulation in tissues would be toxic.

$$2H_2O_2 \rightarrow 2H_2O + O_2$$

Catalase is one of the most efficient enzymes known. Each catalytic center can convert over 10^5 H_2O_2 molecules/sec.

17-6 THE METABOLISM OF HEME GROUPS

Biosynthesis of Heme

The immediate metabolic precursor of heme and related porphyrins is **porphobilinogen**. The synthesis of this intermediate begins with the combination of glycine and succinyl CoA to form α-amino-β-ketoadipic acid; succinyl CoA itself is supplied by way of the TCA cycle (section 12-3). Succinyl CoA condenses with glycine within the mitochondrial matrix of animal cells:

$$\text{succinic acid} \longrightarrow \text{succinyl CoA} \xrightarrow{\text{glycine}} \mathrm{HOOC{-}CH_2{-}CH_2{-}\overset{\displaystyle O}{\overset{\|}{C}}{-}\underset{}{\overset{\displaystyle NH_2}{\overset{|}{C}H}}{-}COOH}$$

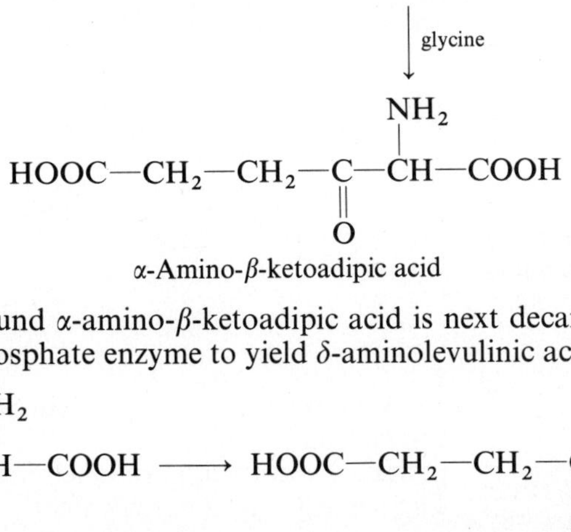

α-Amino-β-ketoadipic acid

The enzyme-bound α-amino-β-ketoadipic acid is next decarboxylated by a pyridoxal-5′-phosphate enzyme to yield δ-aminolevulinic acid:

$$\mathrm{HOOC{-}CH_2{-}CH_2{-}C({=}O){-}CH(NH_2){-}COOH} \longrightarrow \mathrm{HOOC{-}CH_2{-}CH_2{-}C({=}O){-}CH_2{-}NH_2}$$

δ-Aminolevulinic acid

Two molecules of δ-aminolevulinic acid then combine to form porphobilinogen in a reaction catalyzed by **δ-aminolevulinate dehydratase** (Fig. 17-14).

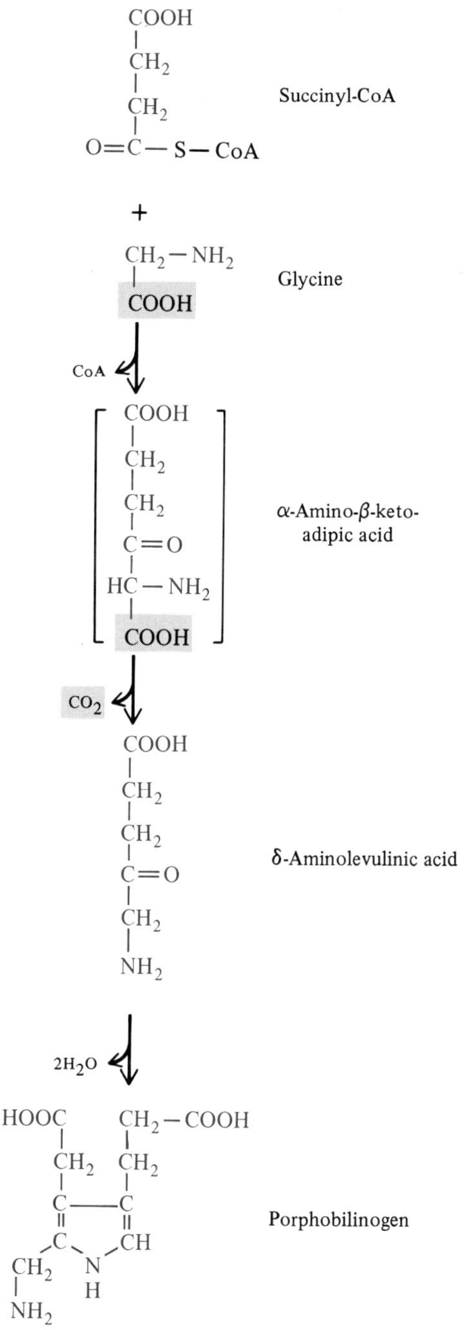

$$\begin{array}{c} COOH \\ | \\ CH_2 \\ | \\ CH_2 \\ | \\ O{=}C{-}S{-}CoA \end{array}$$ Succinyl-CoA

+

$$\begin{array}{c} CH_2{-}NH_2 \\ | \\ COOH \end{array}$$ Glycine

↓ CoA

$$\left[\begin{array}{c} COOH \\ | \\ CH_2 \\ | \\ CH_2 \\ | \\ C{=}O \\ | \\ HC{-}NH_2 \\ | \\ COOH \end{array}\right]$$ α-Amino-β-keto-adipic acid

↓ CO_2

$$\begin{array}{c} COOH \\ | \\ CH_2 \\ | \\ CH_2 \\ | \\ C{=}O \\ | \\ CH_2 \\ | \\ NH_2 \end{array}$$ δ-Aminolevulinic acid

↓ $2H_2O$

HOOC–CH_2–C, CH_2–CH_2–COOH, C=C, CH, N–H, CH_2–NH_2 Porphobilinogen

Figure 17-14 Biosynthesis of porphobilinogen from succinyl CoA.

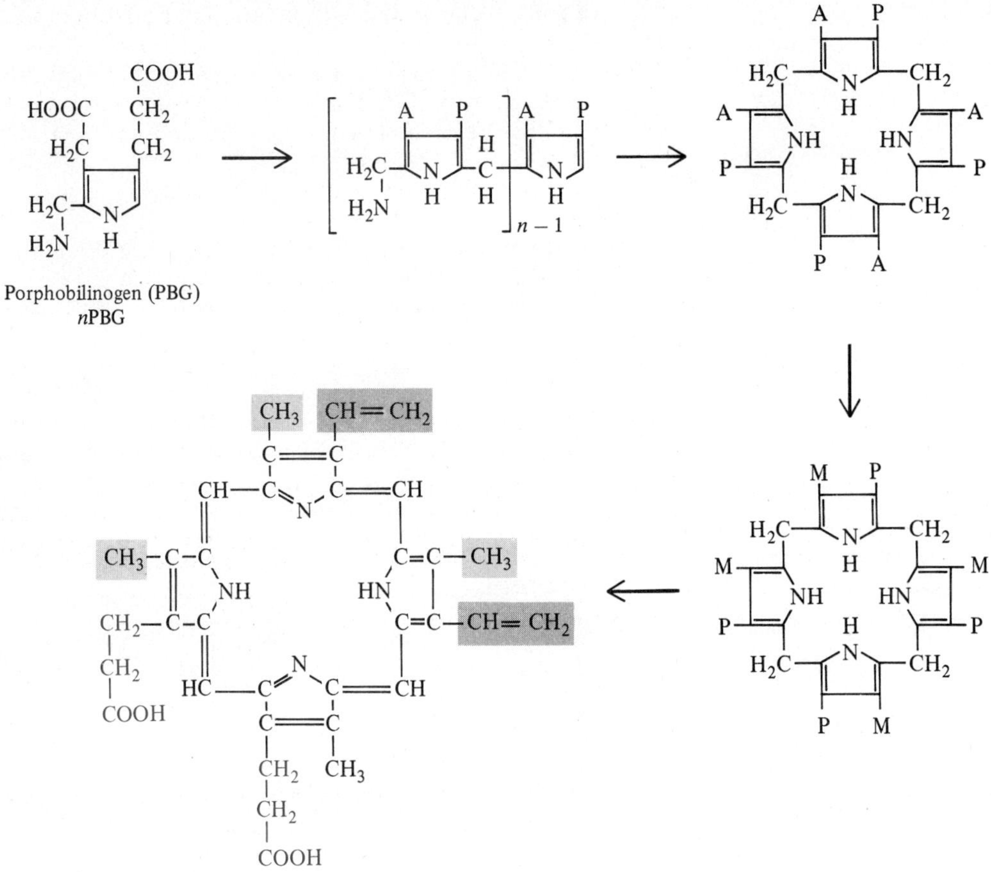

Porphobilinogen (PBG)
nPBG

Protoporphyrin IX

Figure 17-15
Biosynthetic pathway leading from porphobilinogen to protoporphyrin IX. (M stands for $—CH_3$; P for $—CH_2—CH_2—COOH$; A for $—CH_2—COOH$.)

The formation of protoporphyrin involves the condensation of four molecules of porphobilinogen to yield ultimately protoporphyrin IX (Fig. 17-15).

Iron incorporation occurs at the level of protoporphyrin. The active enzyme, **ferrochelatase**, is present in mitochondria. Another enzyme, **heme synthetase**, catalyzes the formation of hemoglobin from protoporphyrin, Fe^{2+}, and globin.

SUMMARY

The **porphyrins** are planar nitrogen-containing cyclic compounds, of which **protoporphyrin IX** is important as the functional group of several biologically active proteins. Protoporphyrin IX forms a chelate complex with Fe(II) known as a **heme.**

The four heme groups of the oxygen transport protein **hemoglobin** have the unusual property of combining reversibly with molecular oxygen without oxidation of their Fe(II). Adult human hemoglobin consists of two α and two β polypeptide chains, each of which binds a heme group. Each α is closely paired with a β, so that hemoglobin resembles a dimer of two $\alpha\beta$ units.

Hemoglobin is an **allosteric** protein that undergoes conformational changes upon combining with consecutive oxygen molecules. These changes result in **cooperativity** of oxygen binding. The binding of one oxygen increases the affinity of hemoglobin for binding additional oxygens. This cooperativity sharpens the response of hemoglobin to differences in oxygen pressure and increases its efficiency in transporting oxygen from the lungs to the tissues.

Myoglobin, a heme-containing oxygen storage protein, contains only one polypeptide chain and does not display cooperativity of oxygen binding. The three-dimensional structure of myoglobin resembles that of each of the four polypeptide chains of hemoglobin (which have similar conformations).

The **cytochromes** are heme-containing proteins that are important components of the mitochondrial respiratory chain. In contrast to hemoglobin, their heme groups are buried within the molecule and their Fe atoms are readily oxidized and reduced. The cytochromes function as electron carriers.

REFERENCES

The following code is used to classify references. I: particularly useful as an introduction to the subject; R: useful primarily as a reference text; A: an advanced account of the material; H: a publication of historical importance.

General

R. E. Dickerson and I. Geis, *The Structure and Action of Proteins*, Harper, New York (1969). (I)

G. L. Eichhorn, ed., *Inorganic Biochemistry*, Elsevier, Amsterdam (1973). (R)

Hemoglobin

E. Antonini and M. Brunori, *Hemoglobin and Myoglobin in Their Reactions with Ligands*, North-Holland, Amsterdam (1971). (R)

J. N. Baldwin, "Structure and Function of Hemoglobin," *Prog. Biophys. Mol. Biol.* 29, 225 (1975). (A)

S. J. Edelstein, "Cooperative Interactions of Hemoglobin," *Ann. Rev. Biochem.* 44, 209 (1975). (R)

M. F. Perutz, "Stereochemistry of Cooperative Effects in Hemoglobin," *Nature* 228, 726 (1970). (H)

M. F. Perutz and L. F. Ten Eyck, "Stereochemistry of Cooperative Effects in Hemoglobin," *Cold Spring Harbor Symp. Quant. Biol.* 36, 295 (1971). (H)

W. A. Schroeder, "The Hemoglobins," *Ann. Rev. Biochem.* 32, 301 (1963). (R)

Cytochrome c E. Keilin, *The History of Cell Respiration and Cytochromes*, Cambridge University, New York (1966). (I)

R. Lemberg and J. Barrett, *Cytochromes*, Academic, New York (1973). (A)

REVIEW QUESTIONS

Questions marked with an asterisk are of a high level of difficulty.

17-1 Which of the following statements is true for the O_2 carrier and storage proteins in the blood and tissues?
(a) Myoglobin has four Fe atoms/molecule.
(b) Hemoglobin is composed of four *identical* polypeptide chains.
(c) Hemoglobin binds O_2 more tightly at pH 7.2 than at pH 7.6.
(d) The hemoglobin molecule changes its shape when it binds O_2.
(e) none of these

17-2 Suggest a biological reason why myoglobin has only one subunit.

*17-3 In sickle-cell disease, the HbS molecules form elongated aggregates when deoxygenated. Suggest possible reasons for the increased tendency of HbS to aggregate.

17-4 About what fraction of the amino acids of the hemoglobin β chains occur in α-helices? What amino acids occur between helical regions?

*17-5 Deoxy HbA has a greatly reduced tendency to dissociate into $\alpha\beta$ units than the oxygenated form. Suggest reasons for this.

*17-6 How are the differences in the microenvironments of the heme groups of hemoglobin and cytochrome c related to their functions?

17-7 Would a mixture of native oxy- and deoxyhemoglobin display a redox potential? Why or why not?

17-8 Certain oxygen-binding fluorocarbons are coming to be used clinically as blood substitutes. Suggest some possible long-term disadvantages in this practice.

17-9 Although many mutant forms of human hemoglobin A have been identified, no replacement of the β-92 histidine has ever been observed. Why? Cite some other amino acids likely to be invariant.

*17-10 Speculate as to the mechanism of electron transfer between cytochromes.

18

Nucleotides and Nucleic Acids

18-1 THE BIOLOGICAL FUNCTION OF THE NUCLEIC ACIDS

Together with the proteins, the nucleic acids are inseparable from life as we know it. The simplest known biological objects, such as **tobacco mosaic virus**, which have some of the properties of living systems, consist of only protein plus nucleic acid.

The basic biological function of the nucleic acids may be tersely stated. They are responsible for the storage, in coded form, of the genetic information required for the synthesis of the enzymes and other proteins needed by the living cell, and for the direction of the process that translates this stored information into amino acid sequences.

One of the two kinds of nucleic acid, deoxyribonucleic acid (DNA), is found most abundantly in the nucleus, where in eukaryotic systems it occurs in the chromosomes as complexes with protein. It is also present in the nucleolus and in some cytoplasmic organelles, including the mitochondria. The other form, ribonucleic acid (RNA), occurs primarily in the cytoplasm, but it is also present in the nucleus.

Both forms of nucleic acid are linear polymers of fundamental subunits called **nucleotides**. The more common nucleotides are much fewer in number than the amino acids; only four different kinds are of frequent occurrence for each type of nucleic acid.

The earliest evidence for the central genetic function of DNA was indirect and implied. The cellular content of DNA depends upon the function of the cells. Ordinary somatic cells (chapter 1), which are not concerned with reproduction, contain twice the amount of DNA present in the haploid germ cells. This is just the distribution expected for a genetic determinant, since the germ cells contain only single copies of genetic information, instead of the two copies present in diploid somatic cells. Moreover, many mutagenic physical and chemical agents alter DNA, which indicates that a modification of its structure produces an alteration of its genetic message.

According to what was once facetiously referred to as the "central dogma," the mechanism for the guided synthesis of proteins is as follows:

(1) The necessary genetic information is stored in DNA as specific sequences of nucleotides, similar to a message in Morse code determined by sequences of dots and dashes.
(2) DNA does not itself engage directly in the mediation of protein synthesis. Instead, it directs the formation of molecules of a certain type of RNA, **messenger RNA**, which serve as secondary carriers of genetic information in which the original message is preserved.
(3) Messenger RNA in turn guides the assembly of amino acids into the polypeptide chains of proteins, with the assistance of the synthetic apparatus present in the cytoplasm.

Although, as we shall see, the above scheme requires amplification and modification in some respects, time has proven it to be basically correct and it may serve as a reasonable guide to the more complete account that follows.

18-2 NUCLEOTIDES

Both forms of nucleic acid consist of linear polymers of fundamental repeating units called **nucleotides**. Each nucleotide consists of:

(1) A nitrogenous heterocyclic ring, called the **base**, which may be a **purine** or **pyrimidine** derivative (Figs. 18-1 and 18-2).

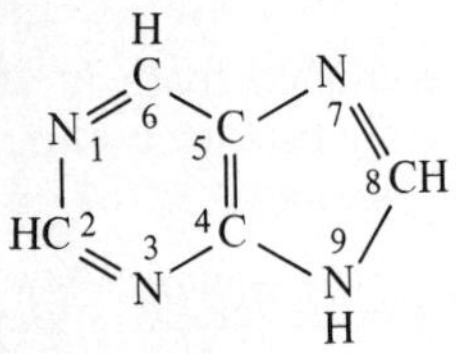

Purine

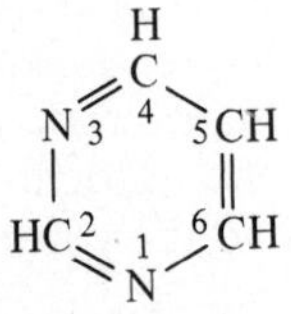

Pyrimidine

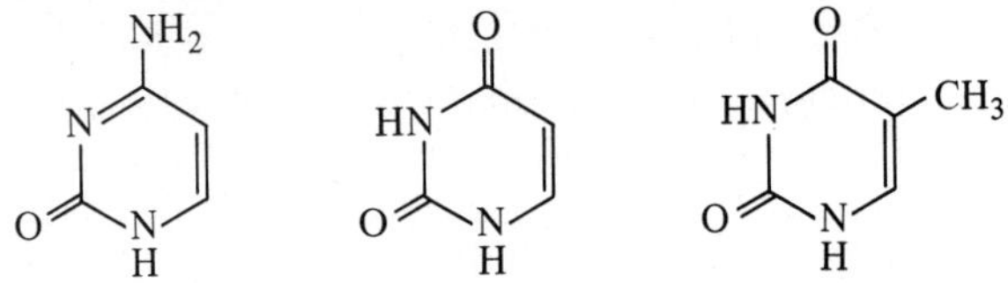

Cytosine (C)

Uracil (U)

Thymine (T)

Pyrimidines

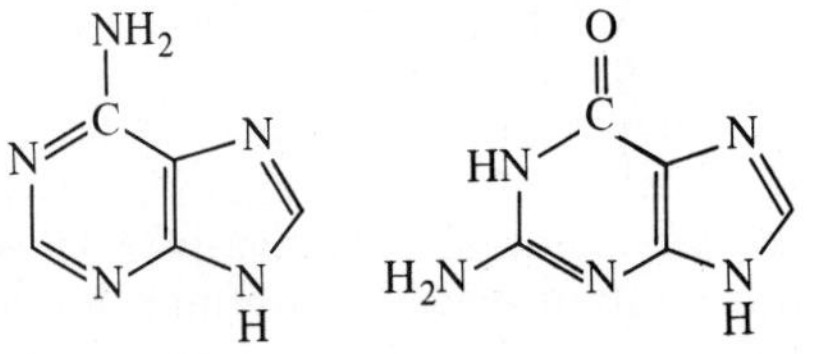

Adenine (A)

Guanine (G)

Purines

Figure 18-1 Skeletal structures of purine and pyrimidine bases and the designation of ring positions.

Figure 18-2 Structures of the purine and pyrimidine bases that commonly occur in nucleic acids.

(2) A 5-carbon sugar, which is D-ribose for RNA and 2-deoxy-D-ribose for DNA (Figs. 18-3 and 18-4). It is conventional to number the carbon atoms in sequence, as shown in Figs. 8-1 and 8-2. (When the sugar is part of a nucleotide the numbers are primed in order to distinguish them from positions on the base.) The two sugars differ in that the 2-hydroxyl of D-ribose is replaced by a hydrogen atom in 2-deoxy-D-ribose.

(3) A phosphate group, attached by an ester linkage at the 2′-, 3′-, or 5′-position of the sugar (Figs. 8-1, 8-2, or 18-5). The nucleotide is accordingly designated as a 2′-, 3′-, or 5′-nucleotide.

There are five bases of frequent occurrence. The purine bases **adenine** and **guanine**, which consist of fused heterocyclic 5- and 6-membered rings,

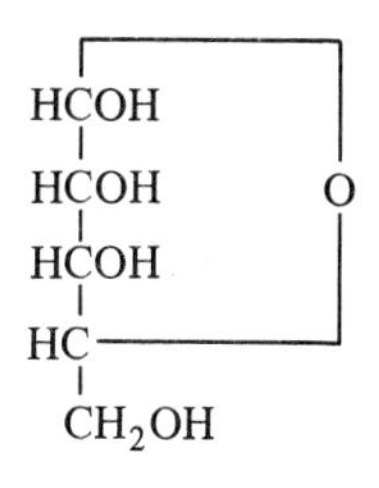

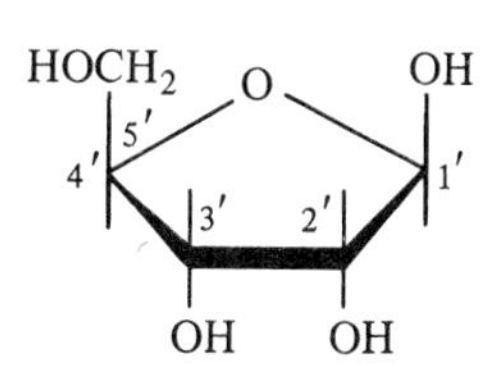

Figure 18-3 Structure of D-ribose, represented by Fischer and Haworth formulas.

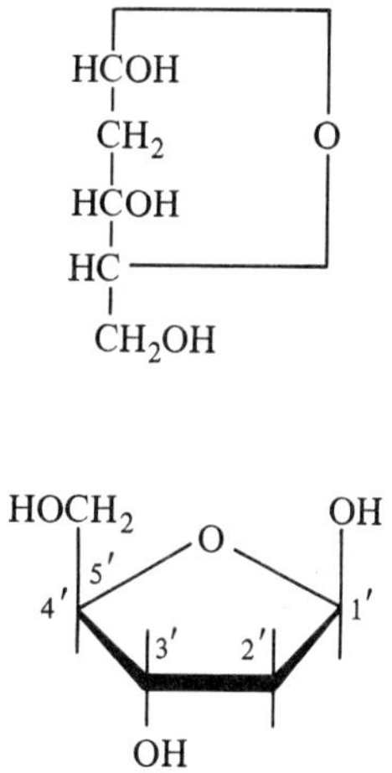

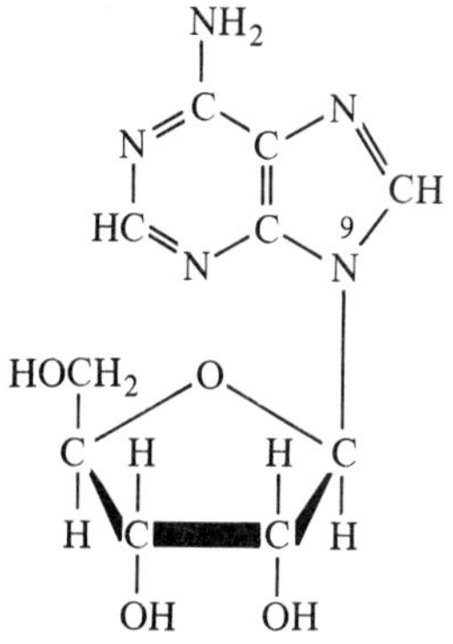

Adenosine

2′-Deoxyadenosine

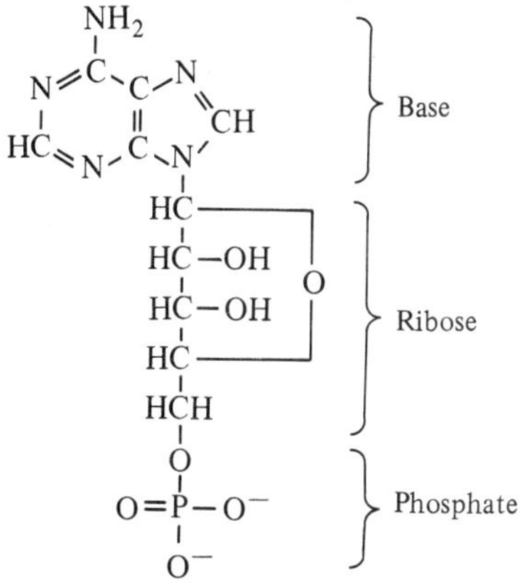

Figure 18-4 Structure of 2-deoxy-D-ribose.

Base

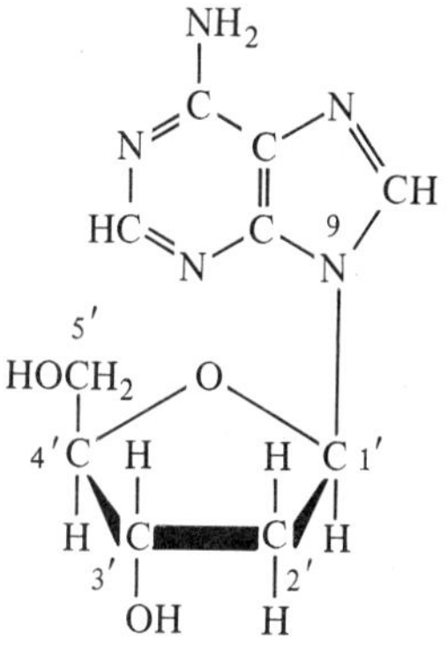

Ribose

Phosphate

(a)

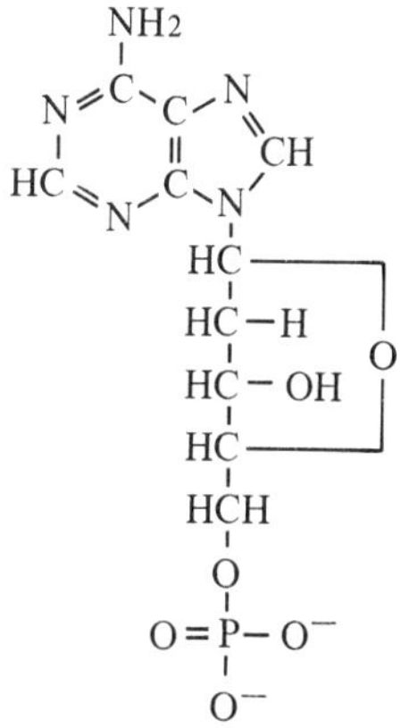

(b)

Figure 18-5 Comparison of the structures of (a) 5′-AMP and of (b) 5′-dAMP and of the corresponding nucleosides.

2-Methylguanine

5-Hydroxymethylcytosine

6-Methyladenine

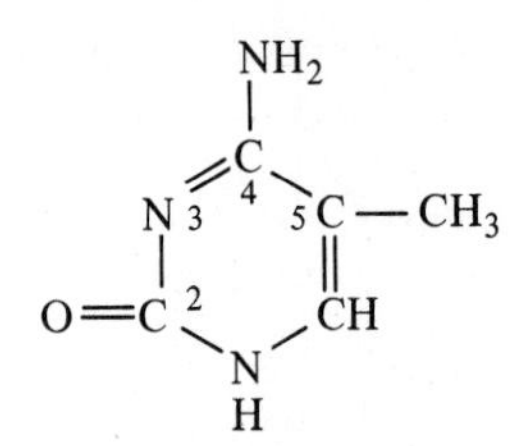

5-Methylcytosine

Figure 18-6 Some less common bases occurring in DNA.

are common to DNA and RNA. The pyrimidine bases **uracil** and **cytosine**, which are single 6-membered rings, are found in RNA (Fig. 18-2). While cytosine also occurs in DNA, uracil is replaced by its 5-methyl derivative **thymine**. (The conventional system of designating ring positions by numbers is indicated in Fig. 18-1.)

In addition to the five most common bases, others occur with reduced frequency or in special cases. A few unusual bases occur in natural DNA. In the T-even bacteriophages of *Escherichia coli*, cytosine is entirely replaced by 5-hydroxymethylcytosine (Fig. 18-6). Methylated derivatives of the common bases, such as those in Fig. 18-6, also occur in DNA.

All of the bases share definite aromatic characteristics, including a **planar** conformation and resistance to oxidation.

In the nucleotides the purine and pyrimidine bases are joined to the 1′-carbon of ribose (or deoxyribose) by N_9-$C_{1'}$ or N_1-$C_{1'}$ glycosidic bonds, respectively. The special configuration about the glycosidic linkage is shown in Figs. 8-1 and 8-2 and is termed the β-configuration.

A **nucleoside** is the part of a nucleotide that consists of the base plus the sugar without the phosphate. The conventional nomenclature of the nucleosides and nucleotides is summarized in Table 18-1.

The presence of the bases endows the nucleotides with intense absorption of light in the near ultraviolet region (Fig. 18-7). The position of maximum absorption is in the range 250–280 nm, depending upon the base and its state of ionization.

The nucleotides containing a diphosphate group, such as ADP, or a triphosphate group, such as ATP, have already been discussed in section 8-2.

Table 18-1

Names and Abbreviations of the Principal Nucleotides and Nucleosides

Name	*Abbreviation*	*Name*	*Abbreviation*
Adenine		Deoxyribonucleotide series	
adenosine	A	deoxyadenosine	dA
2′-adenylic acid		3′-deoxyadenylic acid	
adenosine-2′-phosphate	2′-AMP	deoxyadenosine-3′-phosphate	3′-dAMP
2′-adenosine monophosphate		3′-deoxyadenosine monophosphate	
3′-adenylic acid		5′-deoxyadenylic acid, etc.	5′-dAMP
adenosine-3′-phosphate	3′-AMP; Ap	deoxyadenosine-5′-diphosphate	5′-dADP
3′-adenosine monophosphate		deoxyadenosine-5′-triphosphate	5′-dATP
5′-adenylic acid		deoxycytidine	dC
adenosine-5′-phosphate	5′-AMP; pA	3′-deoxycytidylic acid, etc.	3′-dCMP
5′-adenosine monophosphate		5′-deoxycytidylic acid, etc.	5′-dCMP
adenosine-5′-diphosphate	5′-ADP; ppA	deoxycytidine-5′-diphosphate	5′-dCDP
adenosine-5′-triphosphate	5′-ATP; pppA	deoxycytidine-5′-triphosphate	5′-dCTP
Uracil		thymidine	T
uridine	U	3′-thymidylic acid, etc.	3′-TMP
2′-uridylic acid		5′-thymidylic acid, etc.	5′-TMP
uridine-2′-phosphate	2′-UMP	thymidine-5′-diphosphate	5′-TDP
2′-uridine monophosphate		thymidine-5′-triphosphate	5′-TTP
3′-uridylic acid, etc.	3′-UMP; Up	deoxyguanosine	dG
5′-uridylic acid, etc.	5′-UMP; pU	3′-deoxyguanylic acid, etc.	3′-dGMP
uridine-5′-diphosphate	5′-UDP; ppU	5′-deoxyguanylic acid, etc.	5′-dGMP
uridine-5′-triphosphate	5′-UTP; pppU	deoxyguanosine-5′-diphosphate	5′-dGDP
		deoxyguanosine-5′-triphosphate	5′-dGTP
Cytosine			
cytidine	C		
2′-cytidylic acid			
cytidine-2′-phosphate	2′-CMP		
2′-cytidine monophosphate			
3′-cytidylic acid, etc.	3′-CMP; Cp		
5′-cytidylic acid, etc.	5′-CMP; pC		
cytidine-5′-diphosphate	5′-CDP; ppC		
cytidine-5′-triphosphate	5′-CTP; pppC		
Guanine			
guanosine	G		
2′-guanylic acid			
guanosine-2′-phosphate	2′-GMP		
2′-guanosine-monophosphate			
3′-guanylic acid, etc.	3′-GMP; Gp		
5′-guanylic acid, etc.	5′-GMP; pG		
guanosine-5′-diphosphate	5′-GDP; ppG		
guanosine-5′-triphosphate	5′-GTP; pppG		

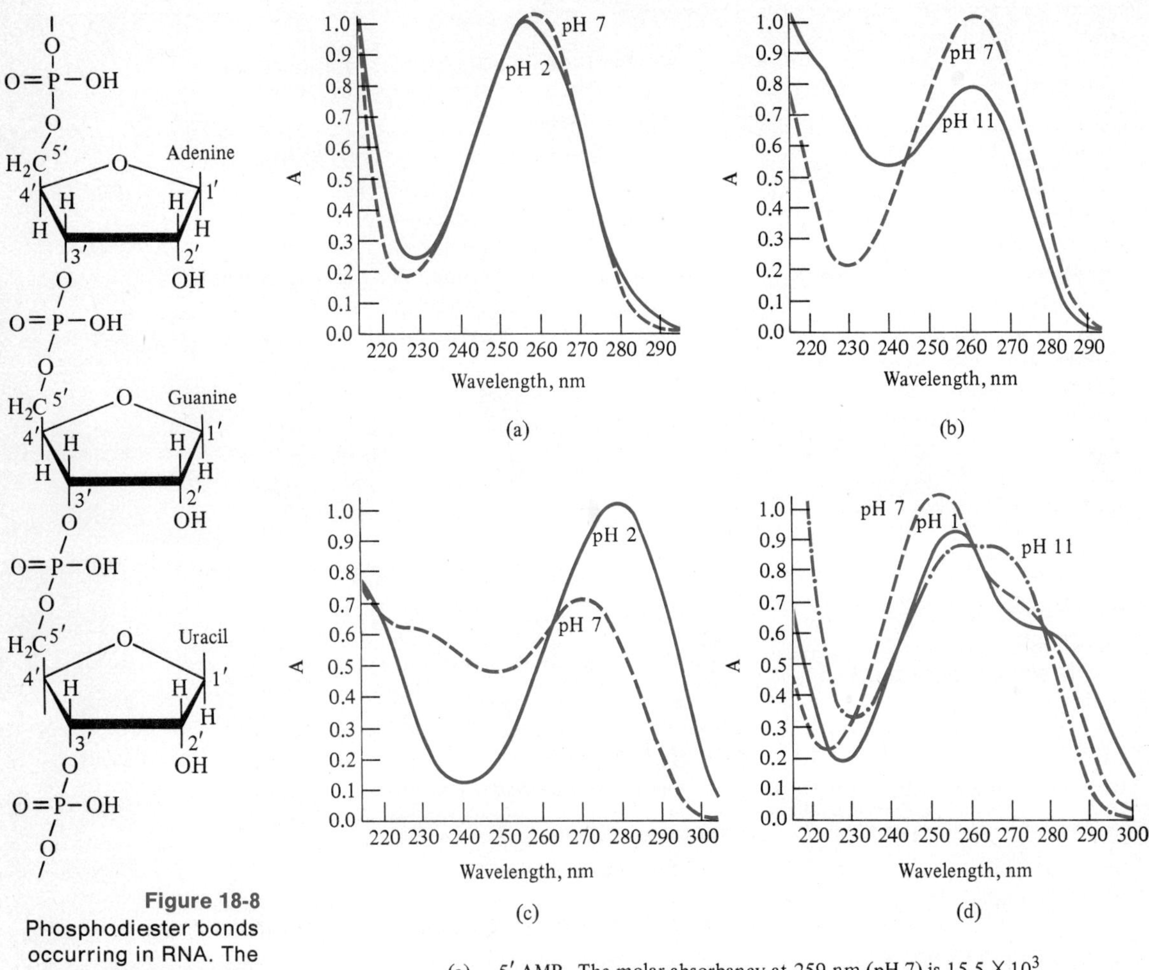

Figure 18-8
Phosphodiester bonds occurring in RNA. The same type of bond is formed in DNA. In both cases the 3′- and 5′-positions of the sugars are joined.

(a) 5′-AMP. The molar absorbancy at 259 nm (pH 7) is 15.5×10^3.
(b) 5′-UMP. The molar absorbancy at 262 nm (pH 7) is 10.0×10^3.
(c) 5′-CMP. The molar absorbancy at 271 nm (pH 7) is 9.0×10^3.
(d) 5′-GMP. The molar absorbancy at 252 nm (pH 7) is 13.7×10^3.

Figure 18-7
Ultraviolet absorption of the four principal ribonucleotides: (a) 5′-AMP, (b) 5′-UMP, (c) 5′-CMP, (d) 5′-GMP. Their molar absorbancies ($\times 10^{-3}$) at pH 7 are 15.5, 10.0, 9.0, and 13.7, respectively.

18-3 CHEMICAL STRUCTURE OF THE NUCLEIC ACIDS

Both DNA and RNA consist of linear polymers of nucleotides united by **phosphodiester** bonds joining the 3′ and 5′ carbons of adjacent nucleotides (Fig. 18-8). The formation of the second ester linkage abolishes the secondary phosphate ionization, leaving only the primary. At neutral pH nucleic acids are negatively charged polyelectrolytes with each phosphate group bearing a single negative charge. The molecular weights that have been reported

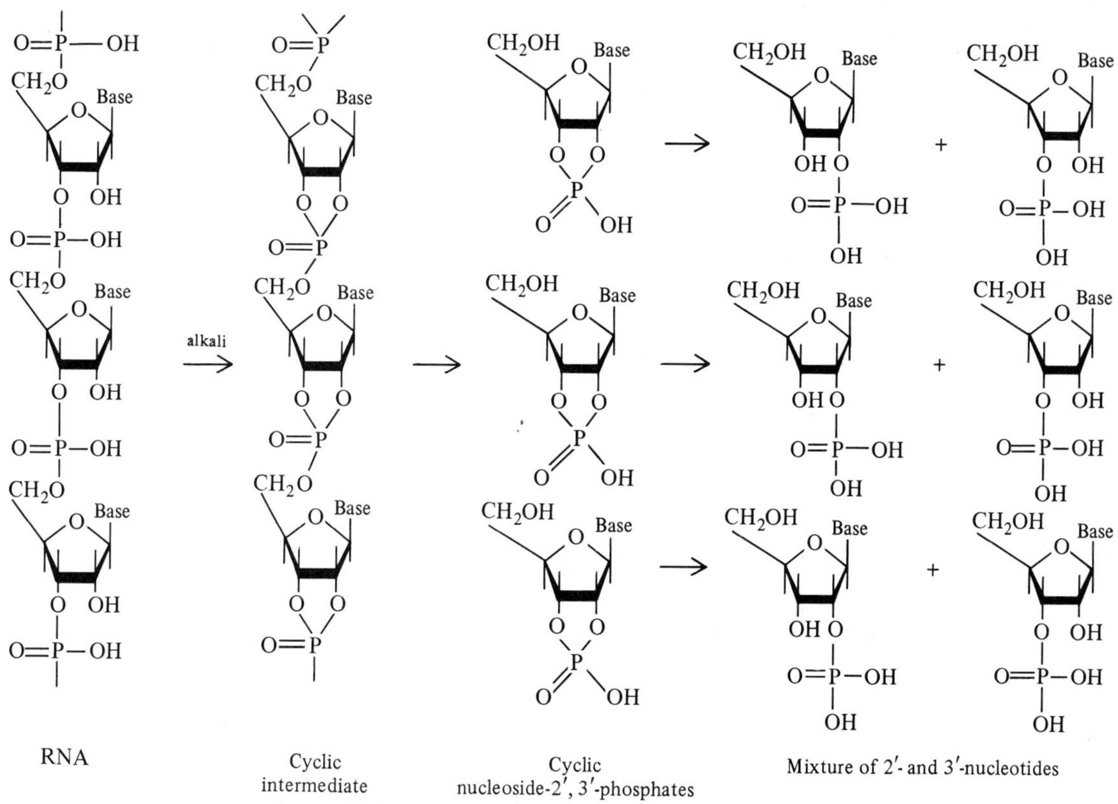

Figure 18-9
Alkaline hydrolysis of RNA. Hydrolysis by pancreatic ribonuclease involves the same cyclic intermediate.

for native viral and bacterial DNA's range from 10^6 to 10^9 or higher. (The values for eukaryotic DNA's may be much higher.)

The absence of the 2′-hydroxyl group in DNA causes it to differ substantially from RNA with respect to susceptibility to alkaline hydrolysis. In alkaline media RNA is readily hydrolyzed to a mixture of 2′- and 3′-nucleotides. The formation of 2′-nucleotides is a consequence of the mechanism of alkaline hydrolysis, which proceeds by way of a cyclic phosphate intermediate in which the 2′- and 3′-hydroxyls of ribose are esterified by a single phosphate group (Fig. 18-9). Random hydrolysis of the cyclic intermediate produces the observed mixture of 2′- and 3′-nucleotides. DNA, which lacks a 2′-hydroxyl and cannot form the cyclic intermediate, is resistant to alkaline hydrolysis.

The enzyme pancreatic **ribonuclease** catalyzes the hydrolysis of RNA phosphodiester bonds involving pyrimidine nucleosides. A cyclic intermediate is formed, as in alkaline hydrolysis, but the intermediate is exclusively cleaved so as to leave the phosphate in the 3′-position of the pyrimidine nucleosides. Pancreatic **deoxyribonuclease** hydrolyzes DNA so as to leave the phosphate in the 5′-position.

18-4 PHYSICAL STRUCTURE OF THE NUCLEIC ACIDS

The Watson-Crick Structure

In 1953 biology turned an important corner with the proposal by Watson and Crick of a bihelical structure for DNA. X-ray diffraction had earlier provided evidence that the molecular architecture of DNA was based on some form of helix. Moreover, careful studies by Chargaff of the base compositions of DNA's from a wide range of sources had uncovered the following general relationship:

$$\text{adenine} = \text{thymine}$$
$$\text{guanine} = \text{cytosine}$$

This was found to hold irrespective of the overall base composition.

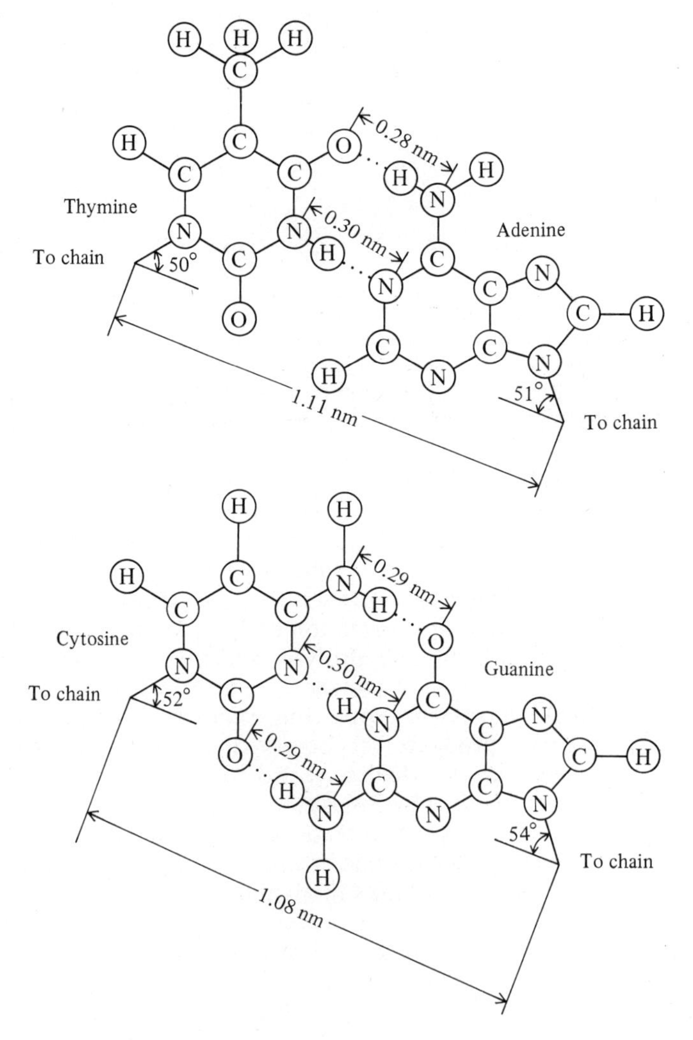

Figure 18-10
Hydrogen-bonded base pairs occurring in DNA

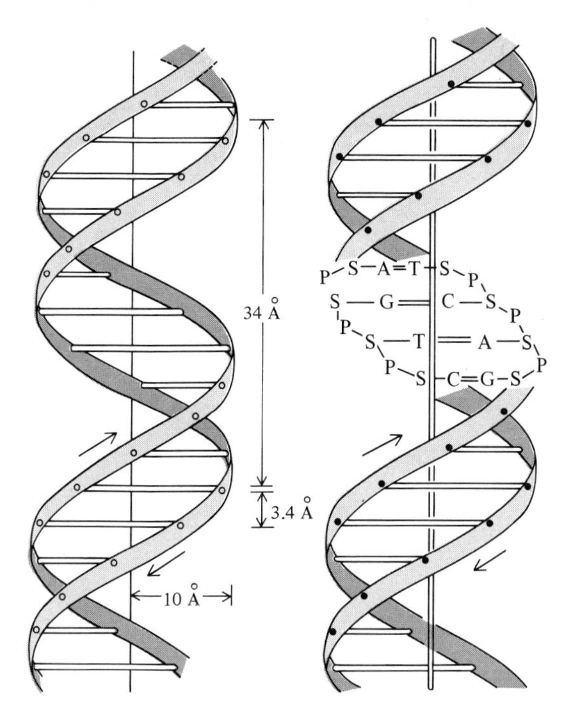

Figure 18-11
Schematic representation of the DNA helix.

The Watson-Crick model consists of two polynucleotide strands wound into a double stranded helix about 2 nm in diameter. The strands are antiparallel, that is, the directions of their phosphodiester bonds are opposed, so that they run in opposite directions.* The helix is right handed. The planar bases are placed in the core of the helix with the phosphate groups on the periphery. Each adenine is paired with a thymine and each guanine with a cytosine. The members of each base pair are linked by hydrogen bonds of the N—H· · ·N or N—H· · ·O types (Fig. 18-10). The base pairs are roughly perpendicular to the helical axis, being separated by about 0.34 nm in the direction of the axis, with 10 base pairs for each turn of the helix. There is considerable overlap or **stacking** of adjacent base pairs (Figs. 18-11 and 18-12). The bihelical structure contains one deep and one shallow helical curve.

The forces stabilizing the DNA helix are now believed to be predominantly the van der Waals interactions of the parallel array of "stacked" bases. The hydrogen bonding of the base pairs makes only a secondary contribution. Indeed the hydrogen bonding may function primarily as a discriminator, to avoid an energetic penalty arising from lost hydrogen bonds with water. There is sufficient space between the stacked bases to permit the insertion or **intercalation** of a planar aromatic molecule, such as acridine or its derivatives.

* Each strand of noncircular DNA has a free 3′-hydroxyl at one end (the 3′-end) and a free 5′-hydroxyl at the other (the 5′-end). The two 3′-ends lie at opposite termini of the DNA duplex.

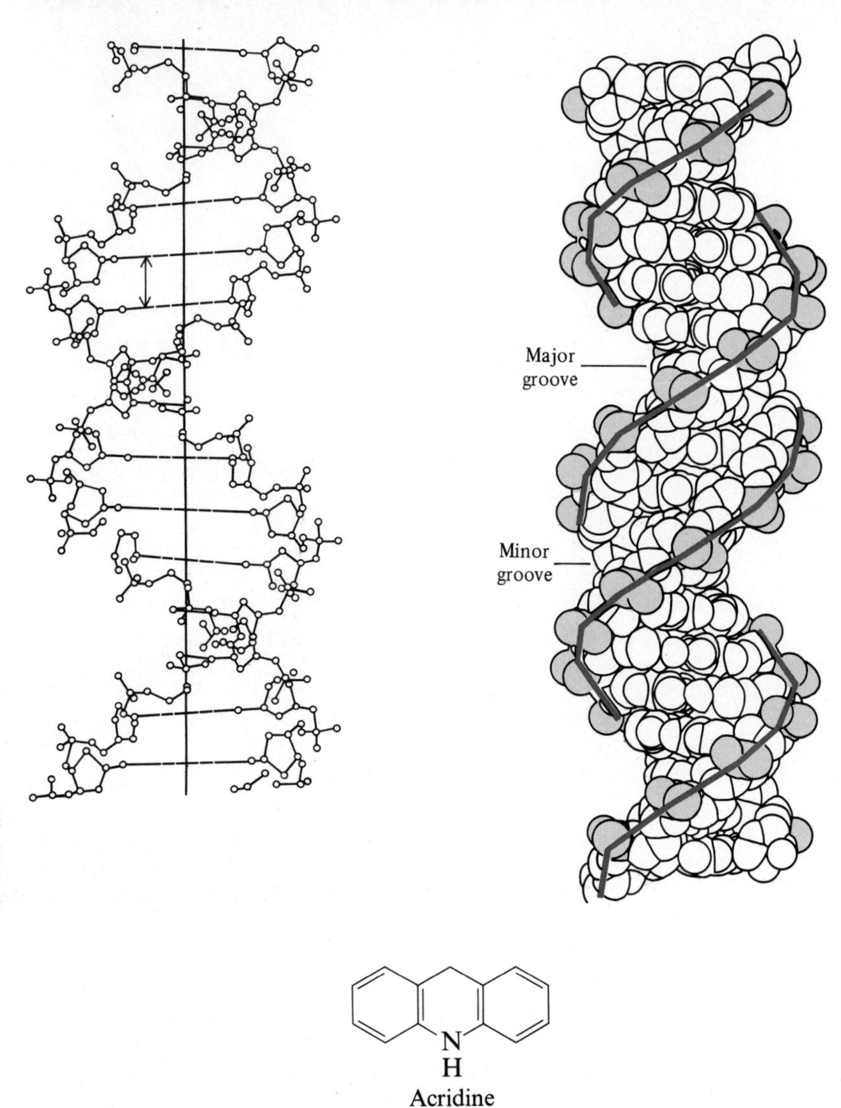

Figure 18-12
Models of the B form of DNA, which is the form occurring in biological systems.

Intercalating molecules are often mutagenic.

Denaturation

The helical organization of DNA in aqueous solution can be disrupted by thermal stress. This process, which also results in major changes in hydrodynamic properties, is most easily monitored by observations of the ultraviolet absorbance. The absorbance per mole of nucleotide at 260 nm is much less for native DNA than that predicted for an equivalent mixture of nucleotides. This effect, known as **hypochromism**, arises from the mutual interaction of the electron systems of the stacked bases.

When, as a consequence of thermal denaturation, the ordered structure of DNA is lost, a pronounced rise in absorbance occurs. The transition is

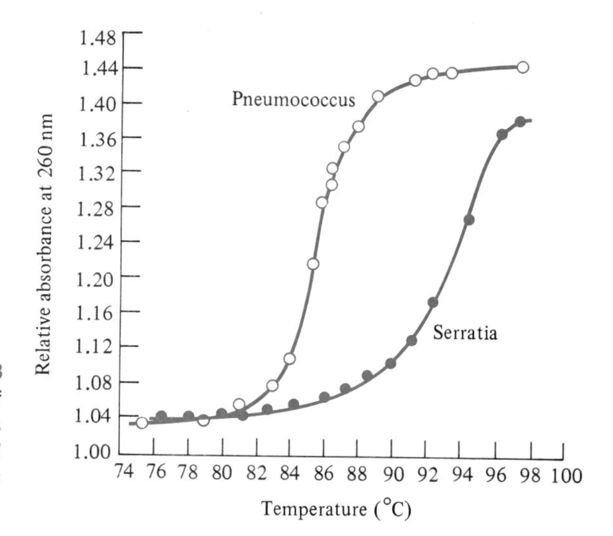

Figure 18-13 Thermal transition of DNA from two species at neutral pH and 0.1-*M* KCl, as monitored by ultraviolet absorbance.

remarkably sharp for DNA of high molecular weight, generally attaining completion over a temperature range of 5°C (Figs. 18-13–18-15). The midpoint of the thermal profile is often designated as the **melting point**, T_m, and provides an index of the stability of the particular DNA.

The value of T_m is dependent upon the base composition, increasing linearly with increasing guanine-cytosine content (Fig. 18-16). This empirical relationship may be used to estimate the base composition of an unknown DNA.

Denaturation also occurs at extremes of pH, where the bases are ionized. Denaturation, whether induced by heat or pH, results in a separation of base pairs, a collapse of the ordered helical structure, and a separation of the strands.

The thermal denaturation of DNA is reversible, provided that cooling to room temperature is carried out *slowly*, with extended periods at intermediate temperatures. This process is often termed **annealing**. The annealing temperature must be high enough so that hydrogen-bonded base pairs may be broken up and reformed at a relatively rapid rate. Under these conditions the strands possess sufficient lateral mobility to permit the competitive exploration of different base-pairing schemes, until that of maximum stability (the native form) is selected. Irregularities and zones of mismatching may be progressively eliminated in this way. This property permits the formation of hybrid duplexes.

The hybridization technique can be used to form DNA molecules whose strands are derived from different genetic variants of the same organism. The two strands have nucleotide sequences that are complementary over much of their length, but are noncomplementary in certain regions, usually as the result of addition or deletion of one or more nucleotides. The complementary regions form normal Watson-Crick double helices, while single stranded loops occur in regions where additions or deletions interfere with complementary base pairing.

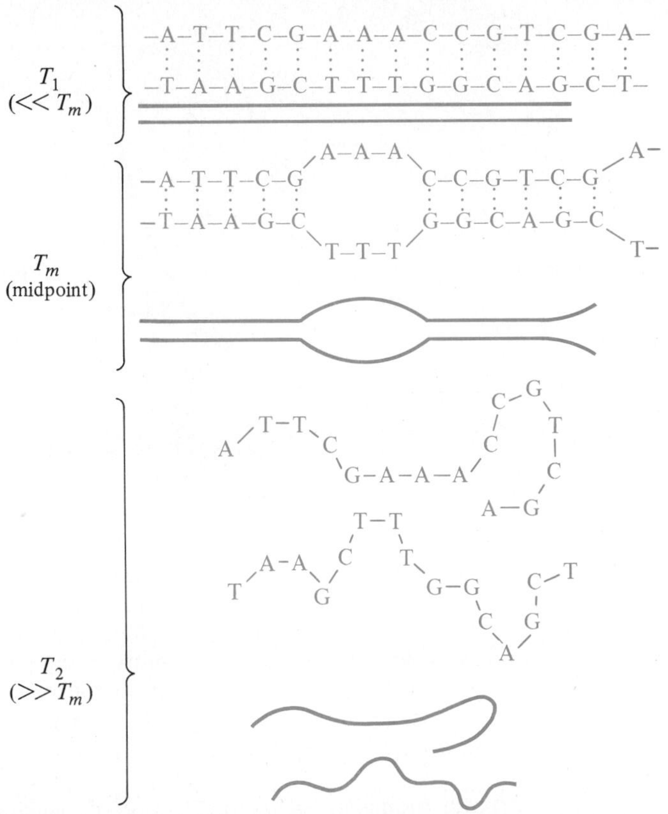

T_1 ($<< T_m$)

—A—T—T—C—G—A—A—A—C—C—G—T—C—G—A—

—T—A—A—G—C—T—T—T—G—G—C—A—G—C—T—

T_m (midpoint)

—A—T—T—C—G A—A—A C—C—G—T—C—G A—

—T—A—A—G—C T—T—T G—G—C—A—G—C T—

T_2 ($>> T_m$)

A—T—T—C—G—A—A—A—C—C—G—T—C—G—A

T—A—A—G—C—T—T—T—G—G—C—A—G—C—T

Figure 18-14
Schematic version of the thermal denaturation of DNA.

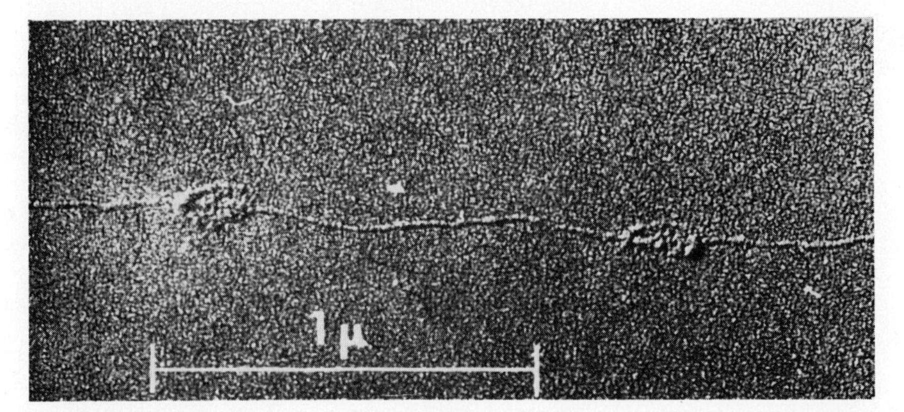

Figure 18-15
Electron microscope photograph of a strand of DNA from T-2 bacterophage after heating to a temperature near the midpoint of the thermal transition. Note the contrast between the linear filament of the native portion and the featureless denatured portion. (Courtesy of M. Beer and C. A. Thomas, J. Mol. Biol. 3, 699, 1961.)

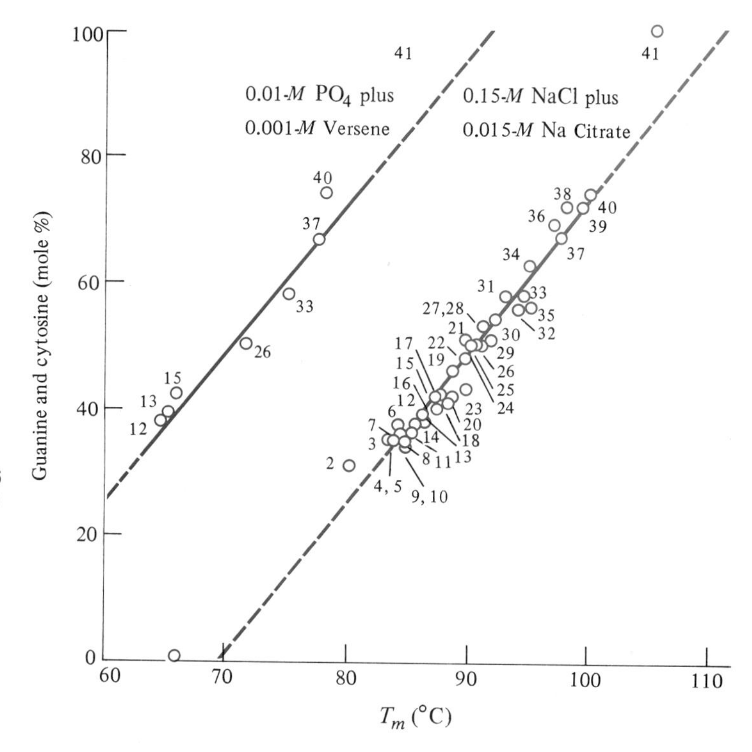

Figure 18-16 Dependence of midpoint of the thermal transition upon the base composition for DNA's from various sources. The midpoint increases with increasing guanine and cytosine content.

Hybridization experiments have also been used to study **homology** in DNA's from different species. In this case DNA from a given species is fractured into smaller pieces by sonic treatment, denatured, and mixed with denatured DNA from a second species. The extent of hybridization is a measure of the occurrence of complementary nucleotide sequences in the two DNA's and hence of their phylogenetic separation.

Circular DNA

While DNA molecules may exist as linear structures, the two ends are often covalently joined. Circular DNA molecules occur in viruses, mitochondria, and in most bacteria. The Watson-Crick double helix of circular DNA is often twisted further to form **supercoils**. The sense of the additional turns may be the same as the double helix (positive) or may be opposite (negative). If one imagines that an uncoiled duplex is twisted to form the supercoiled molecule the total number of turns made, α, is given by

$$\alpha = \beta + \tau$$

when β is the number of Watson-Crick turns and τ, which may be positive or negative, is the number of superhelical turns. The **superhelix density**, σ, is defined as the number of superhelical turns per 10 base pairs. The value of σ is frequently negative, being often in the range -0.05.

The occurrence of superhelices in circular DNA molecules can be detected by their influence upon the sedimentation coefficient (section 5-5), which is substantially increased over the value for a molecule of the same molecular weight, but without superhelices. If one of the strands of supercoiled DNA is "nicked" by short exposure to a hydrolytic enzyme, the resultant "relaxed" form of the molecule sediments more slowly.

Palindromes

Palindromes are inverted repetitive base pair sequences of DNA which read the same in the forward and backward directions. Sequences of this kind, which have been found in many bacterial DNA's, may be represented by

$$\begin{array}{c} \qquad\qquad\qquad\quad\;\downarrow \\ \cdots \mathrm{A}-\mathrm{B}-\mathrm{C}-\mathrm{D}-\mathrm{E}-\mathrm{E'}-\mathrm{D'}-\mathrm{C'}-\mathrm{B'}-\mathrm{A'}\cdots \\ \cdots \mathrm{A'}-\mathrm{B'}-\mathrm{C'}-\mathrm{D'}-\mathrm{E'}-\mathrm{E}-\mathrm{D}-\mathrm{C}-\mathrm{B}-\mathrm{A}\cdots \end{array}$$

where the arrow indicates the point of inversion. The palindrome sequences occurring in natural DNA's may be quite long and involve hundreds of base pairs. They were originally detected by denaturing fragments of eukaryotic DNA and observing their renaturation by electron microscopy. The renatured DNA displayed extensive intrachain base pairing to form hairpin-shaped structures, which arise from the inverted repetitive base sequences. Palindromes often mark specific sites in DNA molecules and may be important in specific DNA-protein interactions.

Satellite DNA

If the DNA from higher organisms is fragmented, thermally denatured, and then allowed to renature, it is often found that the renaturation of the resultant single-stranded fragments occurs in a stepwise manner. A portion of the material reforms helical duplexes rapidly, while other material renatures slowly.

The rapidly renaturing fraction often has a different base composition, and hence a different density, than the bulk of the DNA. If the technique of **density gradient centrifugation** (section 18-6) is employed to obtain the density distribution of the DNA, one obtains a main peak, plus one or more smaller ("satellite") peaks of distinctly different density.

This satellite DNA consists of highly repetitive short sequences. For example, that of the crab has the repetitive sequence —A—T—A—T—A—T—A—T—. Sequences of this kind occur in blocks of 10^2 residues and may recur as many as 10^7 times within the cell.

The function of satellite DNA is uncertain. It is often associated with chromosomal regions that do not unravel during the telophase period of mitosis, as does the bulk of the DNA.

18-5 MOLECULAR GENETICS

Transformation

The earliest convincing evidence for the central role of DNA as a carrier of genetic information came from the pivotal experiments of Avery and his colleagues in 1944, which many view as the birth of molecular biology. DNA, which heretofore had been regarded as only one of a confusing multitude of cellular components of uncertain function, was thereby elevated to a position of supreme importance, which it has never relinquished.

Transformation had been known for many years as a puzzling property of pneumococcus bacteria. When pneumococci (*Streptococcus pneumoniae*) are obtained directly from the sputum of a person suffering from pneumonia, microscopic examination shows them to be enveloped by polysaccharide capsules. When grown on agar plates, colonies of encapsulated pneumococci have a characteristic smooth appearance and are referred to as S colonies.

Pneumococci have been classified into *types* on the basis of characteristic chemical differences in their capsular polysaccharides. The different types have been designated by Roman numerals as I, II, III, etc. The type of a pneumococcus colony is a hereditary property subject to genetic control. Interconversions of different types do not ordinarily occur in cultures.

If an S strain is grown outside of a host in a nutritive medium for many generations, spontaneous mutations result in the appearance of altered cells without capsules that have little capacity to cause disease. When grown on agar plates, colonies of such cells have a rough appearance and are called R cells.

In 1928 Griffith observed that the injection into mice of a *mixture* of living R cells with heat-killed and noninfectious S cells of type I produced the symptoms of pneumonia, despite the ineffectiveness of either when injected alone. Live S cells of type I were recovered from the infected animals. If heat-killed cells of type II were injected, together with R cells, the encapsulated pneumococci were of type II.

It was subsequently found that the R to S transition could be induced *outside* the host by adding *cell-free* extracts of killed S cells to growing cultures of R cells. The new S cells were invariably of the same type as those from which the extract was prepared. The unknown substance responsible was called **transforming principle**.

In 1944 Avery and co-workers succeeded in purifying transforming principle and in identifying it with DNA. Transforming DNA has two properties normally associated with genes:

(1) It can direct a particular cell function, namely, the synthesis of capsular polysaccharide.
(2) It can induce its own replication. A culture of transformed cells may in turn be extracted to yield more transforming DNA than was required to cause their own formation (Fig. 18-17).

In bacterial transformation, DNA from the donor cells is able to penetrate the recipient cells directly without the intervention of the donor cells. In both cases the added DNA is incorporated into the recipient chromosome by some kind of recombination process. Transformation has been shown to occur in numerous bacterial species and the transfer of many different properties, including resistance to antibiotics, has been demonstrated. In

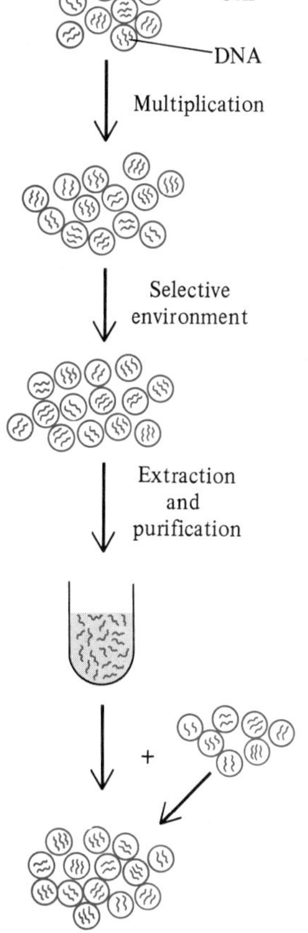

Figure 18-17
Experimental procedure for isolation and demonstration of transforming DNA. If the transforming DNA confers resistance to a drug, then the selective environment would be a nutrient medium containing the drug. Isolation of the DNA of a strain grown on the selective medium provides transforming principle, which may be used to transform a normal strain, conveying the property of drug resistance.

1952 Hershey and Chase confirmed the genetic role of DNA by showing with radioisotope labeling experiments that only the DNA of a bacterial virus entered the host bacterial cell and initiated infection.

Transduction

A second mechanism for the transfer of genes between bacterial cells involves their passive transport or **transduction** by a class of viruses parasitic upon bacteria and collectively known as **bacteriophages** or **phages**.

Certain phages (transducing phage) are able to undergo accidental genetic interchange with the chromosome of the host cell so as to incorporate a fraction of its chromosome. If the virus now carrying a composite genetic message is released from the host cell and enters a new host, a fragment of the original bacterial chromosome is injected into the second cell. It may then engage in crossing over with the new host chromosome, so that the latter acquires one or more of the transported genes.

Transduction has proved to be useful in genetic mapping of bacteria. Since the number of genes carried by a single transducing particle is small, usually less than 1% of the bacterial chromosome, the frequency with which groups of genes are transported by the same phage is an accurate index of their physical separation on the DNA molecule.

Conjugation

Cells of *E. coli* have the property of transferring genetic material by a sexual mechanism. In sexual conjugation, some or all of the chromosome of a *male* or (+) cell is transferred to a *female* or (−) cell through a hollow filament. About 1.5 h is required for complete transfer of the (+) chromosome. The process may be interrupted at any time by the shearing action produced by agitating the cell suspension in a blender, so that (−) cells can be obtained that carry varying amounts of male chromosomal material. This may be incorporated into the (−) chromosome. From the number and kind of distinct and identifiable traits that are governed by the acquired segments of the (+) chromosome and appear in the progeny of the (−) cell, as a function of the time of transfer, the relative position of the genes governing these traits can be deduced (Fig. 18-18).

The *E. Coli* Chromosome

The intestinal bacterial species, *E. coli*, is perhaps the most extensively studied of all microorganisms and its genetic structure has been characterized in considerable detail. The chromosome is a linear array of the individual genes, each of which carries genetic information in the form of a coded sequence of nucleotides. Its chemical structure is that of a circular DNA molecule containing 3.8×10^6 nucleotides.

Several hundred distinct genes have been identified in the *E. coli* chromosome. Most of these have been recognized through the formation of defective mutants by ultraviolet irradiation or treatment with mutagenic compounds. A mutation of this kind usually results in the loss of the capability to synthesize an essential metabolic intermediate, which must therefore be supplied from external sources if the cells are to grow and replicate. If a viable colony of a mutant strain can be maintained only in a medium containing a nutrient that is not required by the original *wild-type* strain, it may

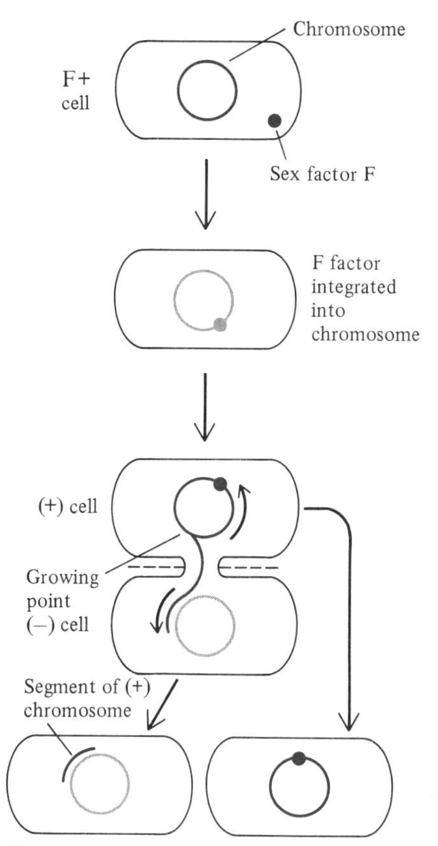

Figure 18-18
Schematic depiction of bacterial conjugation and chromosome transfer as they occur in *E. coli*. F^+ or "male" cells contain a sex factor F which is separate from the chromosome. It may become integrated into the chromosome. A portion of a chromosomal strand of the F^+ cell is transferred to the F^- cell, which lacks the F factor, carrying a number of genes with it.

be concluded that the mutation has inactivated a gene essential for the synthesis of this substance. By experiments of this kind a series of genes may be identified, each gene being recognized and named according to the metabolic factor for whose synthesis the intact gene is essential.

In practice, ordinary wild-type *E. coli* cells, which can grow on a *minimal* medium containing a carbon compound as an energy source together with some inorganic nutrients, are exposed to mutagenic conditions and then "plated out" on agar gel media containing a wide range of nutrients. Distinct colonies are allowed to develop by replication of individual bacteria. Each colony may consist of wild-type cells or of any one of a large number of possible mutant forms. Particular mutations may be recognized and selected by the transfer of cells from individual colonies to nutrient agar plates containing the minimal medium with or without specific additives.

Identified genes may be transported to other bacterial cells by transduction or conjugation and may become incorporated into the chromosome of the recipient cell by recombination. The presence of an incorporated gene may be recognized by the altered nutrient requirements it confers.

Genes may be transferred in blocks of varying size. The probability that two genes will be transferred together is inversely proportional to their linear separation on the chromosome, as the likelihood of their being separated

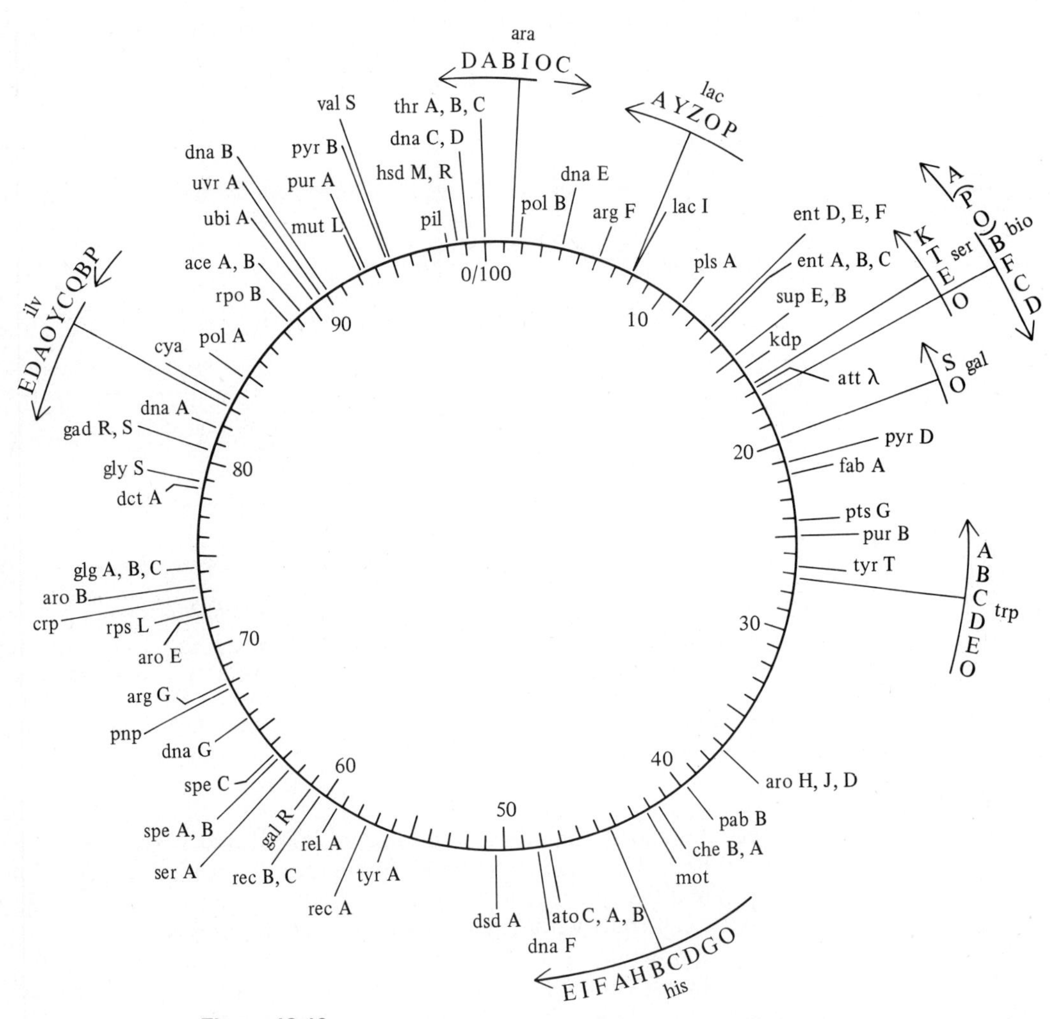

Figure 18-19
Genetic map of *E. coli* strain K-12. The genetic symbols stand for identified genes, each of which is responsible for a particular enzymic, or other, function. For example, *cya* (position 83) stands for adenylate cyclase, while *glg A* (position 74) represents glycogen synthetase. The numbers within the circle indicate the order of transfer in interrupted conjugation experiments. (The position of the origin is arbitrary.)

by a break in the chromosome increases with increasing distance between them. In this way, determinations of the relative frequency of simultaneous transfer permit the arrangement of the identified genes into a one-dimensional genetic map, which, in the case of *E. coli*, has the form of a closed circle (Fig. 18-19).

Since a gene consists of a sequence of nucleotides, a mutation may occur at more than one point within the gene. For each gene there may exist a collection of different mutations. Two strains containing different mutations of the same gene may, by chromosomal transfer and recombination, yield a form in which a normal gene is reformed by recombination of the normal portions of each altered gene (Fig. 18-20).

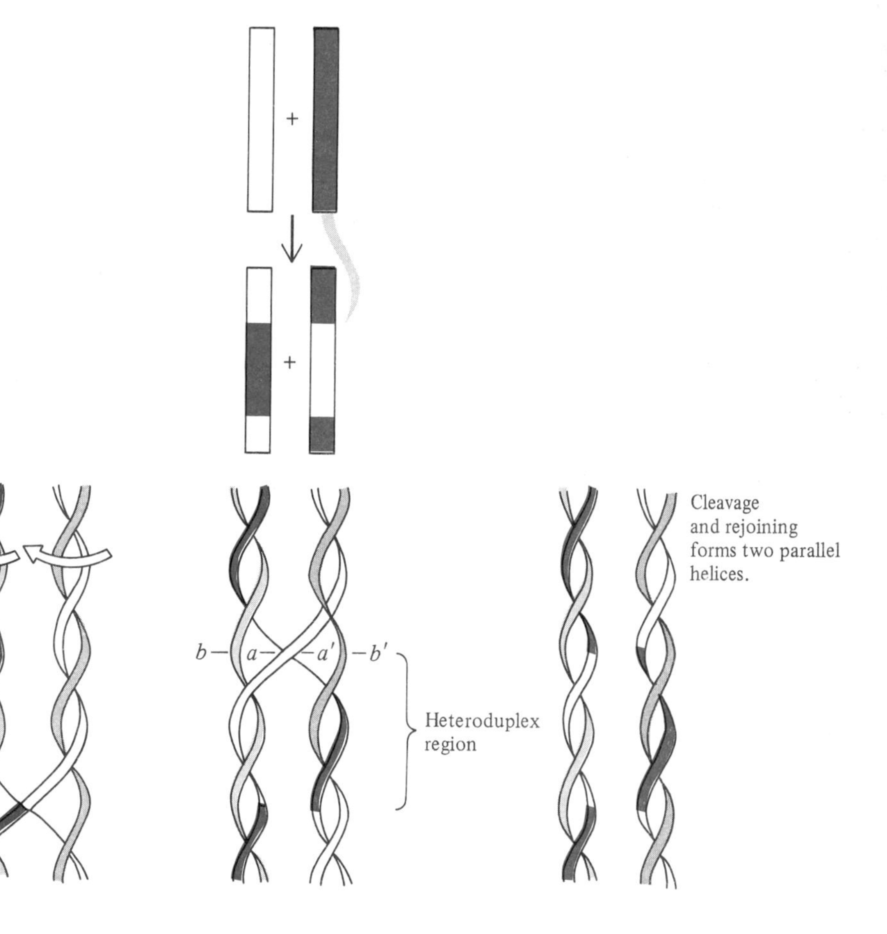

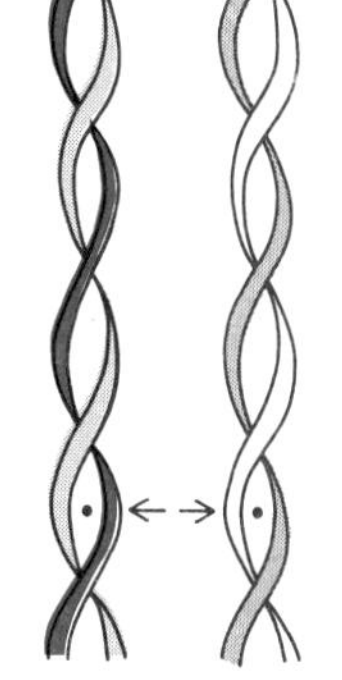

Figure 18-20
(a) Schematic depiction of DNA recombination showing the exchange of homologous regions in two chromosomes. (b) One possible mechanism for recombination, involving cleavage and rejoining. (Adapted from D. E. Metzler, *Biochemistry*, Academic, 1977.)

18-6 REPLICATION OF DNA

Cytological studies established early that the DNA content of cells approximately doubles prior to cell division. While it was thereby clear that both progeny cells must receive one or more identical molecules of DNA, it remained uncertain whether a replica of the original double-stranded DNA molecule was formed, or whether the two strands separated and were replicated individually. The issue was decided by one of the classic experiments of molecular biology, which deserves discussion in detail.

The Meselson-Stahl Experiment

By growing *E. coli* for many generations in a medium containing only the ^{15}N isotope of nitrogen, it was possible to obtain a culture of cells all of whose nitrogen was present as this isotope. The density of its DNA was significantly higher than that of ordinary *E. coli*, which contains predominantly the common ^{14}N isotope. The culture was transferred abruptly to a medium containing only ^{14}N and allowed to replicate. Samples of cells were removed at successive intervals after transfer, and their DNA was isolated.

The density of the DNA of successive generations was measured by an elegant technique called **density gradient centrifugation** (Figs. 18-21 and 18-22). When an 8-*M* solution of cesium chloride is centrifuged at high speed in an analytical ultracentrifuge cell, some degree of sedimentation of the solute (CsCl) occurs so that a gradient of solute concentration, and hence of density, is formed between the meniscus and the outer periphery of the solution. If the range of solvent densities spans that of the DNA, then **sedimentation** of DNA will occur from regions of lower density and **flotation** from regions of higher density, so that all of the DNA is ultimately concentrated in a narrow band at the point of the cell where its density is equal to that of the solvent. If two species of different density are present, two bands will be formed at the corresponding points of the cell. The process may be observed by measurements of the ultraviolet absorption, using either the darkening of a photographic plate or a photoelectric device for direct scanning.

Application of this technique to the *E. coli* system described above showed that the DNA obtained prior to transfer formed a single band at the position

Figure 18-21
Sedimentation of a protein in water, as monitored by absorption of ultraviolet light. The time of sedimentation increases from left to right. The light areas correspond to regions from which the protein has sedimented, leaving behind transparent solvent. (Courtesy of National Naval Medical Center, Bethesda, Maryland.)

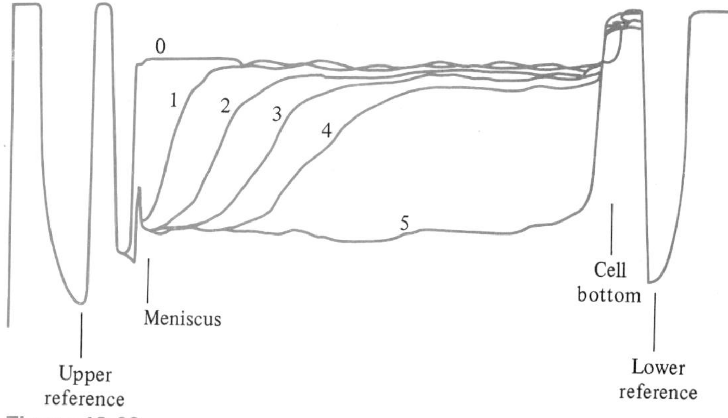

Figure 18-22

Densitometer tracings of the photographs of Fig. 18-21, showing the degree of reduction of light transmission as a function of position within the cell. Curves 1–5 correspond to progressive stages of sedimentation.

expected for ^{15}N-labeled DNA. For the *first* generation after transfer this band almost disappeared, being replaced by a new band of density intermediate to that of ^{14}N- and ^{15}N-containing DNA. For the *second* generation this band, plus a new band corresponding to ^{14}N-containing DNA, were present, with no material in the interband region (Fig. 18-23). With increasing time the ^{14}N band was progressively enlarged at the expense of the intermediate band.

These observations could be simply interpreted at the molecular level. The density of the DNA obtained from the first generation is that predicted for a hybrid molecule, one of whose strands contains ^{14}N and the other ^{15}N. It is clear that the replication process cannot involve any extensive fragmentation or dispersal of the original strands, since otherwise a range of species of different densities should be present. The hybrid species contains one strand of parental DNA and one new strand synthesized from the new medium. The replication scheme thus involves a separation of the strands of the original DNA, whose *chemical* structure is preserved, but the *physical* structure is not. In later replications the hybrid species itself undergoes strand separation, to give rise to a new hybrid molecule and to one containing only ^{14}N-labeled strands (Fig. 18-24).

It is plausible to postulate that replication and unwinding may be simultaneous events. Each individual strand may guide the synthesis of its complementary strand by base pairings of the Watson-Crick type. For example, a sequence —ACAATGT— would elicit the sequence —TGTTACA— in the new strand.

DNA Polymerases

The first enzyme with DNA-synthesizing capability to be identified and characterized was the **DNA polymerase I** isolated by Kornberg and coworkers. This enzyme requires both a **template** strand and a short **primer**

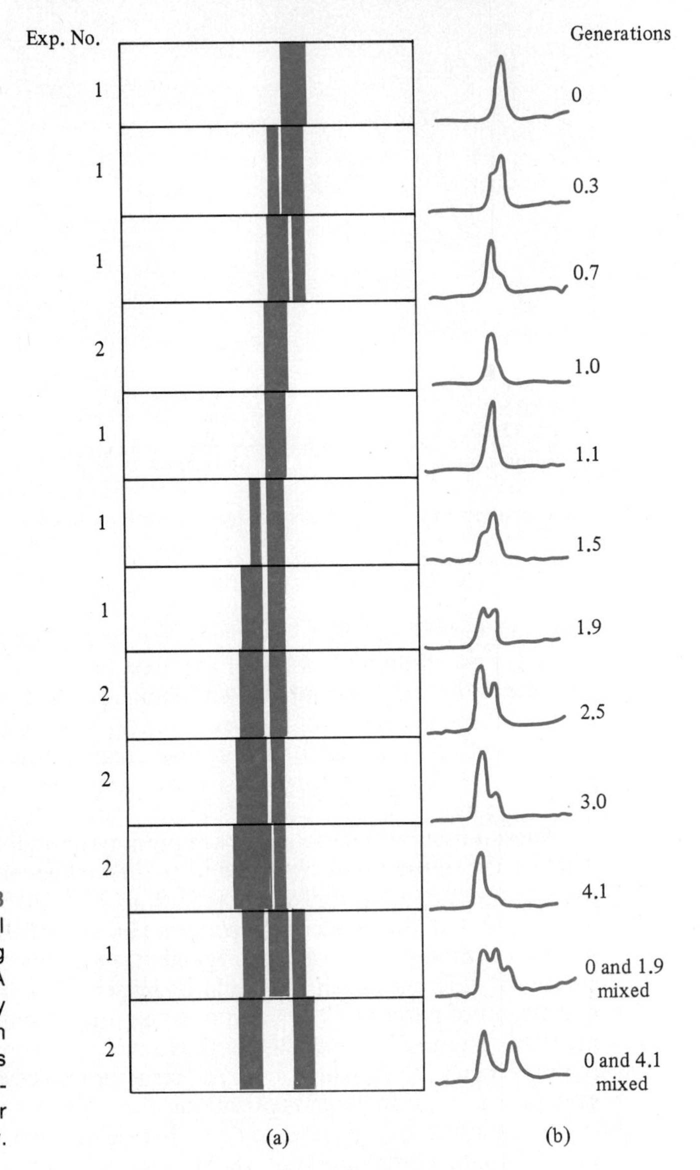

Figure 18-23
The Meselson-Stahl experiment, showing the appearance of DNA of intermediate density in the first generation after transfer and its subsequent replacement by DNA of higher density.

strand. The enzyme recognizes the 3′-end of the primer strand and binds the appropriate **deoxynucleoside triphosphate** (dATP, dGTP, dCTP, or dTTP) to pair with the next base in the template strand (Fig. 18-25). The new nucleotide is joined to the primer strand by formation of a phosphodiester bond, while pyrophosphate is released. In this way, a single stranded template DNA directs the synthesis of a strand with complementary base sequence. The overall reaction is

$$n\text{XTP} \rightarrow (\text{XP})_n + \text{PP}_i$$

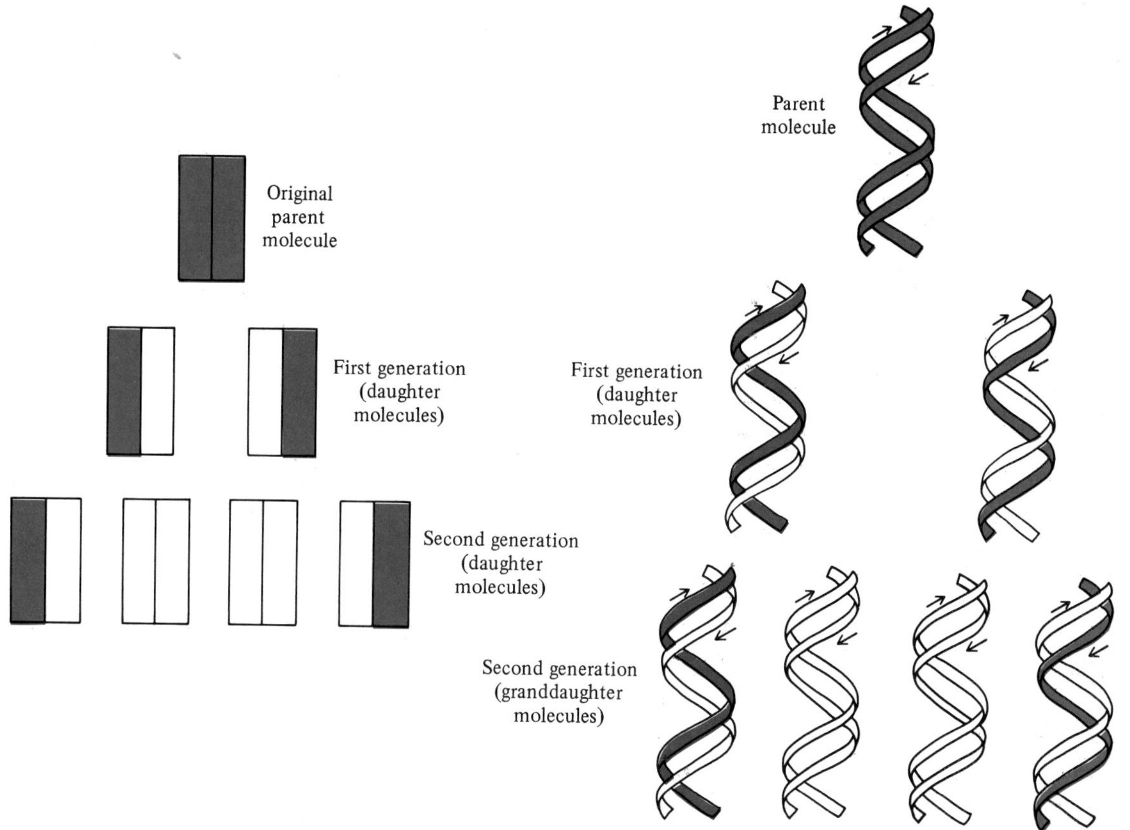

Figure 18-24
Schematic version of guided replication of DNA in *E. coli*, as suggested by Meselson-Stahl experiment.

In addition to catalyzing the growth of a DNA strand from the 3′ end of the primer strand, DNA polymerase I also catalyzes the hydrolytic removal of nucleotides from both the 3′ and the 5′ ends. The two hydrolytic activities involve distinct and different sites.

Since mutant forms of *E. coli* have been found that lack polymerase I, but are nevertheless able to carry out DNA synthesis, it is unlikely that this enzyme is the sole biological agent for DNA chain elongation. It is currently believed that its primary function within the cell may be to correct errors in replication by removal of any 3′-terminal nucleotide that is incorrectly paired. It may also have the ancillary function of filling in the gaps between freshly synthesized DNA fragments.

Two other DNA polymerases, II and III, have been identified in *E. coli.* Neither of these catalyze hydrolysis from the 5′ end. Genetic evidence suggests strongly that polymerase III is essential for DNA replication and may be the primary biological agent for chain elongation. It adds new nucleotide units only at the 3′ ends of the growing chain.

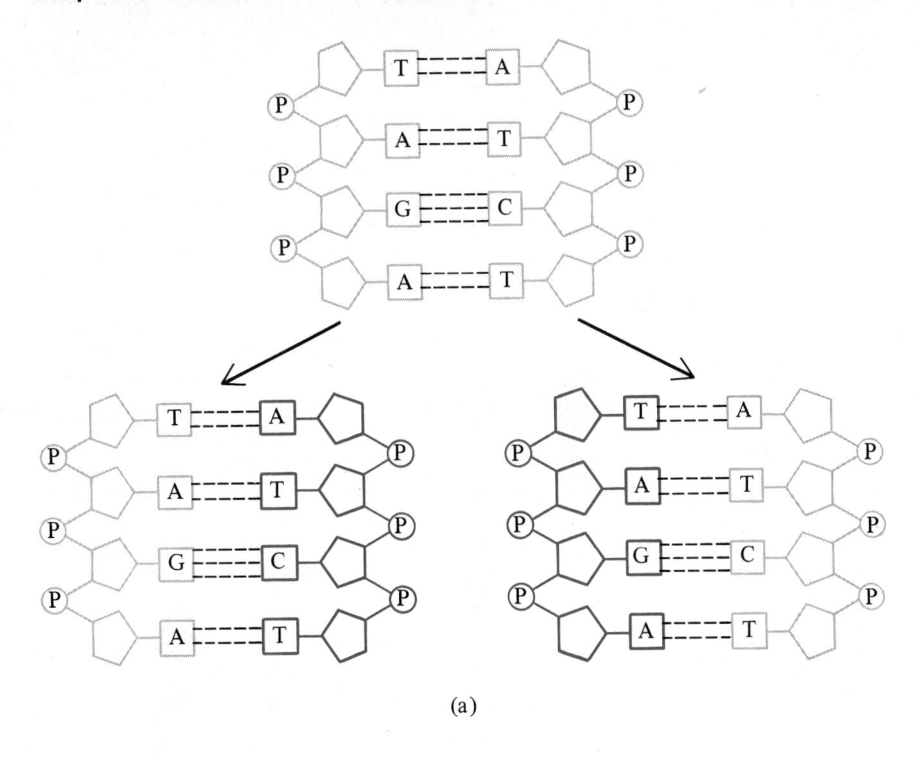

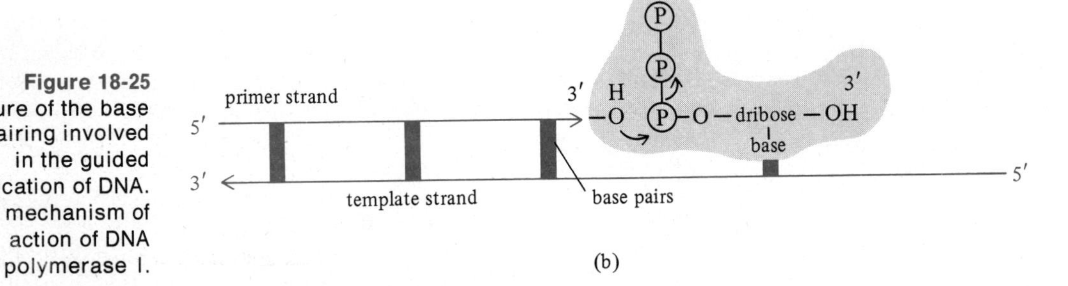

Figure 18-25
(a) Nature of the base pairing involved in the guided replication of DNA.
(b) The mechanism of action of DNA polymerase I.

DNA Ligase

A definitive picture of DNA replication in bacteria has begun to emerge in recent years. According to the currently popular model, a double-stranded DNA molecule is first opened up locally, probably through the mediation of an **unwinding protein**. Unwinding and replication take place at a **replication fork** (Fig. 18-26). In order to account for the observed rate of replication of the *E. coli* chromosome the entire molecule must spin at a rate of several hundred revolutions per second. A short RNA **primer** is formed at a specific primer region and is paired with a DNA strand (Fig. 18-26). The RNA primer is elongated by a DNA polymerase using deoxyribonucleoside triphosphates as substrates. In this way both strands of the parental DNA

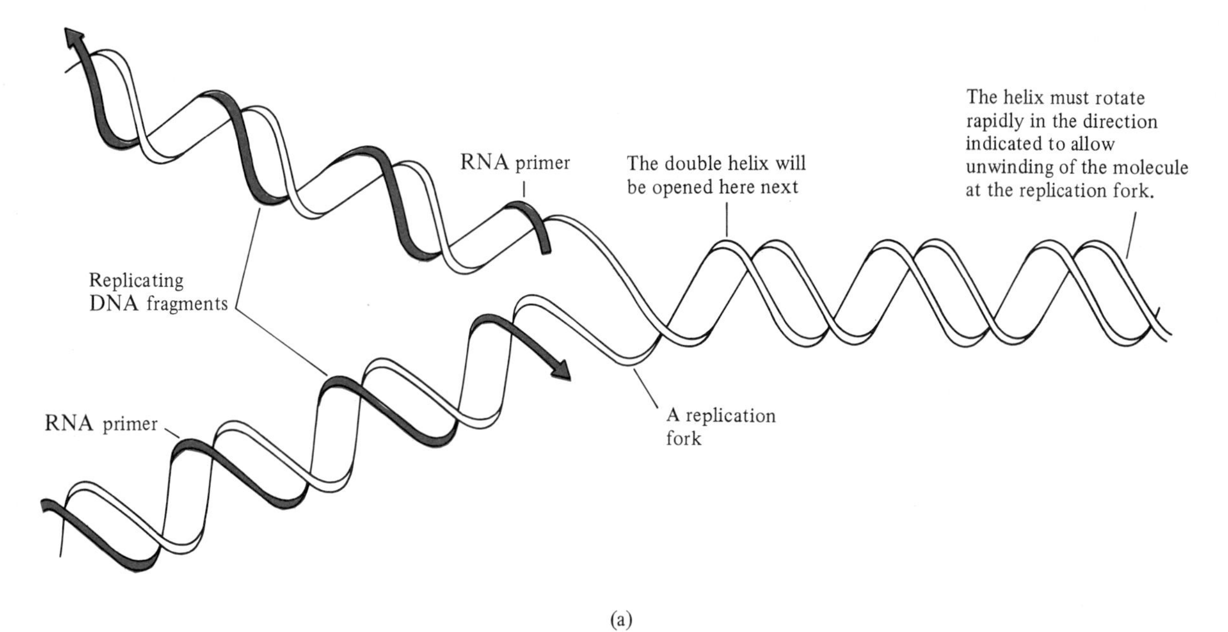

(a)

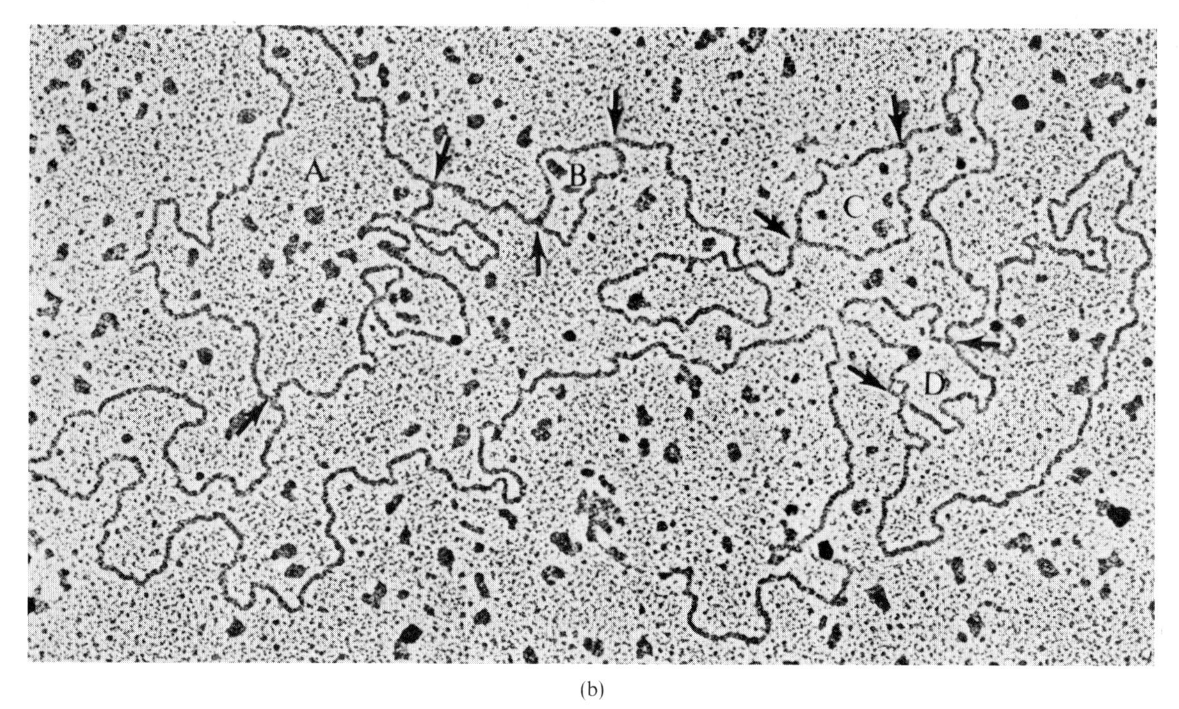

(b)

Figure 18-26
(a) Schematic version of a model for DNA replication. (Adapted from D. E. Metzler, *Biochemistry*, Academic Press, New York, 1977.) (b) Electron microscope photograph of replicating regions of eukaryotic DNA. (Courtesy of D. R. Wolstenholme, *Chromosoma* 43, 1, 1973.)

molecule are replicated. The RNA primer ends are later removed by a hydrolytic enzyme.

Since the known DNA polymerases can add nucleotides only to the 3′-end of a DNA chain, at least one of the new chains must be formed as fragments, which must subsequently be joined. Such fragments, which are called **Okazaki fragments**, have been identified in bacterial cells undergoing replication. The stitching together of these fragments to form a complete DNA strand is achieved by the action of the enzyme **DNA ligase**, whose specialty is the repair of "nicked" DNA.

An interrupted DNA strand has a 3′-hydroxyl group and a 5′-phosphate, which must be joined to form an intact strand. The mechanism of action of DNA ligase is unusual in that activation of the phosphate group occurs by transfer of an AMP group from NAD^+ (Fig. 18-27). The new phosphodiester bond is then formed, with the simultaneous removal of AMP.

In this way, the two new DNA strands are synthesized in synchrony with the unwinding of the parental double helix. Synthesis is generally bidirectional and occurs at one or more replication forks. The swiveling action necessary for unwinding to occur at a replication fork requires that a nick be present in one strand.

The unwinding of the original DNA molecule may require the mediation of an unwinding protein. Examples of these have been identified in *E. coli* cells infected with the bacterial virus T4 bacteriophage, as well as in some eukaryotic cells. These proteins bind preferentially to single-stranded DNA

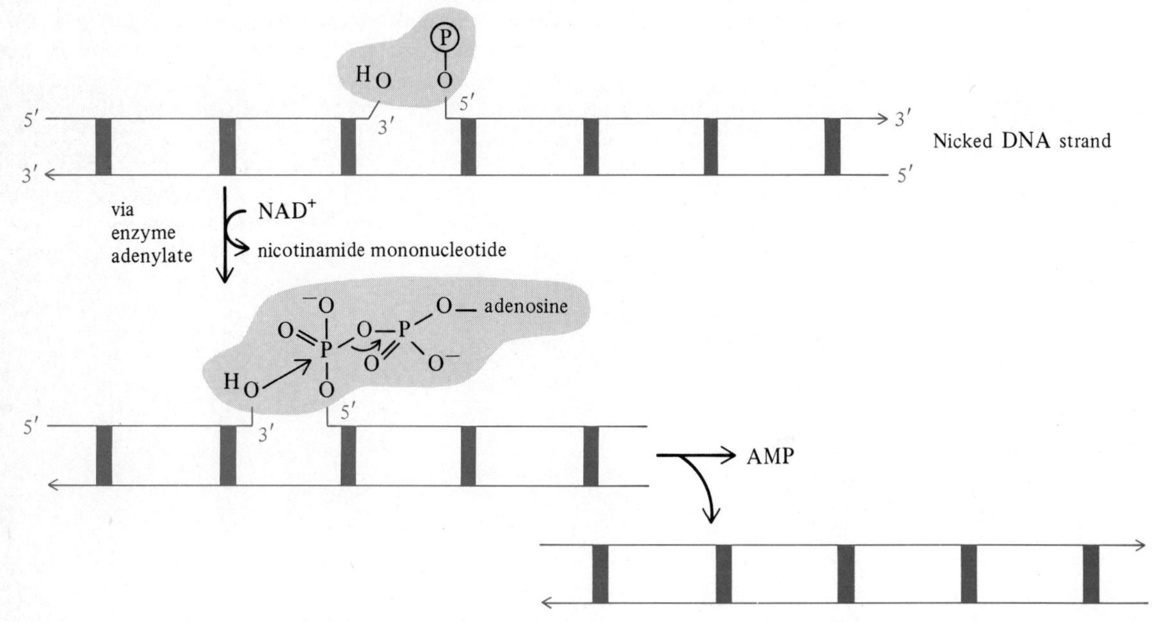

Figure 18-27
Mechanism of action of DNA ligase. The reaction mechanism involves the transient addition of an AMP unit from NAD^+ to the 5′-end of a fragment of a nicked DNA strand.

and consequently cause local unwinding of the double-stranded DNA, with exposure of the purine and pyrimidine bases.

18-7 RESTRICTION AND MODIFICATION OF DNA

Many bacteria are subject to invasion by bacterial viruses, which may result in the lysis and destruction of the bacterial cells. Many bacterial species have developed a defensive mechanism against invading viruses which depends upon the digestion of their DNA by specific hydrolytic enzymes which do not attack the bacterial DNA. This phenomenon is termed **restriction** and the enzymes involved are called **restriction endonucleases**. Restriction enzymes have a high degree of specificity and cleave DNA chains only at a few points characterized by unique base sequences that are recognized by the enzymes.

DNA molecules that are resistant to the action of restriction enzymes, such as the bacterial DNA itself and the DNA of viruses capable of successfully replicating within the host, are modified in such a way as to block the action of the enzymes. This modification consists of the methylation of specific bases within the critical nucleotide sequence (Fig. 18-28). The usual mechanism for methylation involves transfer of a methyl group from *S*-adenosyl methionine.

The high degree of selectivity displayed by the restriction enzymes, which cleave DNA only at specific sequences that may recur only a few times within the DNA molecule, renders them useful for the determination of DNA sequences and the manipulation of isolated genes. By converting a DNA molecule of high molecular weight into a limited number of segments, they greatly simplify the problem of nucleotide sequence analysis.

In recent years advances in molecular biology have made possible the exciting development of **genetic engineering**. Restriction enzymes may be used to excise particular genes from the chromosomal DNA of a given organism. The excised genes may be incorporated by various means into the DNA of a second organism. In this way the cells of the second organism acquire foreign genes with capabilities not present in the native cells. Although genetic engineering is still in its infancy, it is evolving rapidly and will undoubtedly be of vast future importance.

$$
\begin{array}{l}
\qquad\qquad\quad\downarrow\quad\ \ * \\
5'\text{–(T or A)–G–A–}\overset{}{\text{A}}\text{–T–T–C–(A or T)–}3' \\
3'\text{–(A or T)–C–T–T–}\underset{*}{\text{A}}\text{–A–G–(T or A)–}5' \\
\qquad\qquad\qquad\qquad\qquad\ \ \uparrow
\end{array}
$$

Figure 18-28
Action of methylase and the restriction endonuclease *E. coli* RI (EcoRI) upon a specific DNA region. Most susceptible regions have an axis of symmetry. The asterisks indicate methylated bases and the arrows designate points attacked by the endonuclease in an unmethylated foreign DNA.

18-8 REPLICATION OF A VIRAL DNA

ϕX174

The search for relatively simple systems to study the replication of DNA has led many investigators to turn to the smaller DNA-containing bacterial viruses. These share the advantage of having extensive genetic information available, as well as that of permitting easy DNA replication in cell extracts.

Among the simplest and best understood of these systems is the bacterial virus ϕX174, which infects *E. coli* cells. This virus is unusual in that the infectious particle contains only a *single-stranded* circular DNA molecule, whose molecular weight is only 1.8×10^6, corresponding to 5×10^3 nucleotides and to nine identified genes.

Few findings in molecular biology are more startling than the observation that the limited information coded into this minuscule chromosome is sufficient to seize control of the comparatively vast *E. coli* cell and to subvert its synthetic apparatus for viral replication. The initial step in replication is the conversion of the single stranded DNA of the infecting particle into a circular double-stranded **replicative** form (RF). This form subsequently undergoes several replications to yield additional RF circles that function as templates for the synthesis of numerous single strands of viral DNA, which are incorporated into mature viruses.

The initial conversion of the single-stranded infecting DNA to the double-stranded RF form requires the action of a DNA polymerase synthesized by the host cell. Formation of the RF circles also requires an enzyme, **RNA polymerase**, that catalyzes the DNA-directed synthesis of RNA (section 18-11). This finding was part of the original evidence leading to the postulated involvement of primer RNA in DNA replication.

The replication of the RF form requires a gene product of the specific gene A of ϕX174. This gene codes for a specific nuclease of molecular weight 56,000, which nicks one strand of the double-stranded RF form and thereby initiates replication. The short sequence of primer RNA is probably synthesized shortly after formation of the nick.

Unlike the DNA of more complex systems, whose replication generally occurs in a bidirectional manner, that of ϕX174 is believed to be unidirectional and to occur by a **rolling circle** mechanism (Fig. 18-29). As a new viral strand is synthesized along the complementary template strand, the original strand is progressively displaced in the form of a single-stranded tail of the new strand (Fig. 18-29).

A complementary strand to the single-stranded tail is next synthesized, probably in the form of segments that are subsequently joined by a ligase. The final step is the cleavage by a suitable endonuclease of the tail and the formation of closed circles by the action of a ligase.

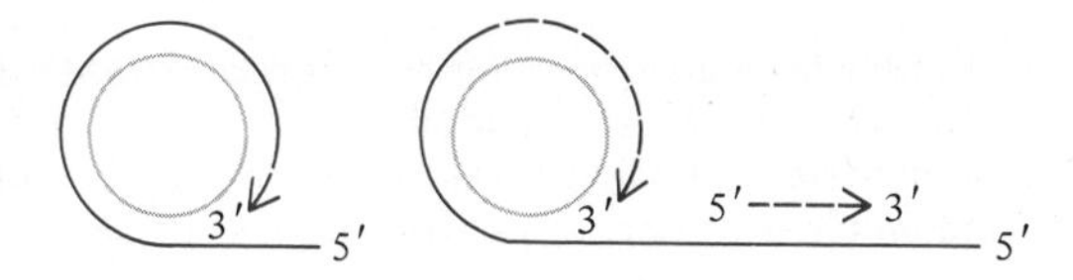

Figure 18-29 Rolling circle model for the replication of ϕX174 DNA.

18-9 NUCLEOPROTEINS

The chromosomal DNA of eukaryotic organisms exists as complexes with positively charged proteins. Electrostatic interactions of the oppositely charged components probably make a major contribution to the stabilization of the complexes.

The protein components of chromosomal nucleoproteins have been classified into two major groups, **protamines** and **histones**. Protamines, which are the simpler in structure, occur in fish and mammalian sperm cells. Their molecular weights are relatively low, ranging from 2×10^3 to 2×10^4.

The protamines are very basic proteins, being positively charged at pH's below 12. This is largely a consequence of their high arginine content. In the case of **clupeine**, a protamine isolated from herring sperm, over 70% of the molecule consists of this residue. The aromatic amino acids are generally absent.

Within somatic cells, DNA exists as electrostatic complexes with the group of positively charged proteins termed the histones. Five different classes of histone have been identified, which range in molecular weight from 1.1×10^4 to 2.1×10^4. The different classes of histone are rich in lysine or arginine.

The histones have not figured prominently in the discussion of the biological role of DNA. This is not because they have no function, but rather because this function is poorly understood. It was once proposed that these proteins might serve as gene repressors; however, no supporting evidence has appeared. Electron microscopy has shown that many chromatin fibers display a regular repeating structure, in which DNA-histone nodules of about an 8-nm diameter are spaced along a DNA molecule like beads on a wire, separated by varying lengths of DNA. The nodules or **nucleosomes** appear to contain DNA that is folded into a relatively compact shape and complexed with a set of histones. These observations have led to suggestions that the role of histones is primarily structural and that they function as determinants of DNA folding.

In addition to the histones, a large number of other proteins occur in eukaryotic nuclei. These include enzymes, gene repressors, and hormone-binding proteins, most of which have uncertain functions.

18-10 DIFFERENT KINDS OF RNA

The biological role of RNA in the directed biosynthesis of proteins has already been mentioned briefly and will be discussed at length in chapter 19. In this section we shall describe the various different forms of RNA, each of which represents a **transcription** of a nucleotide sequence of chromosomal DNA and each of which has a specialized function.

Ribosomal RNA

This form of RNA is confined to the **ribosomes** (chapter 1), accounting for about half their mass, the balance being largely protein. The ribosomes are particles of roughly spherical shape which are distributed throughout the cytoplasm. In eukaryotic cells they occur in association with the complex system of internal membranes, termed the **endoplasmic reticulum**. The most detailed studies have been done for the ribosomes of *E. coli*, which have a molecular weight close to 2.6×10^6 and a sedimentation coefficient of about 70 svedbergs or 70 S (section 5-5), and are usually designated as the 70S species. A 70S ribosome is composed of two smaller particles, the 50S and 30S ribosomes. The association of these subunits to form the 70S particle is reversible and is sensitive to the level of Mg^{2+}.

The 50S ribosome of *E. coli* contains two molecules of RNA, which have sedimentation coefficients of 23S and 5S, plus about 30–40 protein molecules; the total molecular weight is about 1.8×10^6. The 30S ribosome contains a single molecule of RNA with a sedimentation coefficient of 16S and about 20 molecules of protein, corresponding to a total molecular weight of 8×10^5.

Eukaryotic ribosomes, which are less completely characterized than those of *E. coli*, are substantially larger. Plant and animal ribosomes usually have sedimentation coefficients close to 80S, being composed of 60S and 40S subunits.

Ribosomal RNA (rRNA) is quantitatively the dominant form of RNA, accounting for about 90% of cellular RNA. There is evidence that about six regions coding for rRNA occur in the *E. coli* chromosome. Each region contains genes coding for 16S, 23S, and 5S rRNA. A single transcript (section 18-11) for the entire region is formed and subsequently cleaved enzymically into pieces corresponding to the three forms of rRNA.

Messenger RNA

The rRNA's of each species are quite homogeneous in size and nucleotide sequence and thus are unlikely to be able to code for the enormous variety of proteins synthesized by a cell. This is the function of a different form of RNA, termed **messenger RNA** (mRNA), whose nucleotide sequence is directly translated into polypeptides. It consists of transcripts of specific nucleotide sequences of chromosomal DNA. These subsequently migrate to the cytoplasm, where, together with the synthetic machinery of the cell, they direct the synthesis of protein. Since mRNA guides the formation of many different polypeptide chains, whose size and amino acid sequence vary considerably, it is not surprising that it is heterogeneous with respect to both size and base sequence. Since, if protein synthesis is to occur in a controlled manner, mRNA cannot persist for long periods, it is relatively unstable and short live .

A mRNA molecule may undergo further processing before becoming involved in protein synthesis. Instances are known where the original transcript may contain the messages for several genes. Enzymic cleavage splits such large precursors into the functional mRNA molecules.

Many bacterial species, including *E. coli*, which have been grown on glucose as the sole energy source are initially unable to grow when abruptly transferred to lactose. However, after a period of several minutes, new enzymes required for the utilization of lactose begin to be synthesized. Upon depletion of the lactose, the level of these enzymes declines rapidly. These findings illustrate the principle that the mRNA required for the synthesis of these **inducible** enzymes is formed rapidly when needed and

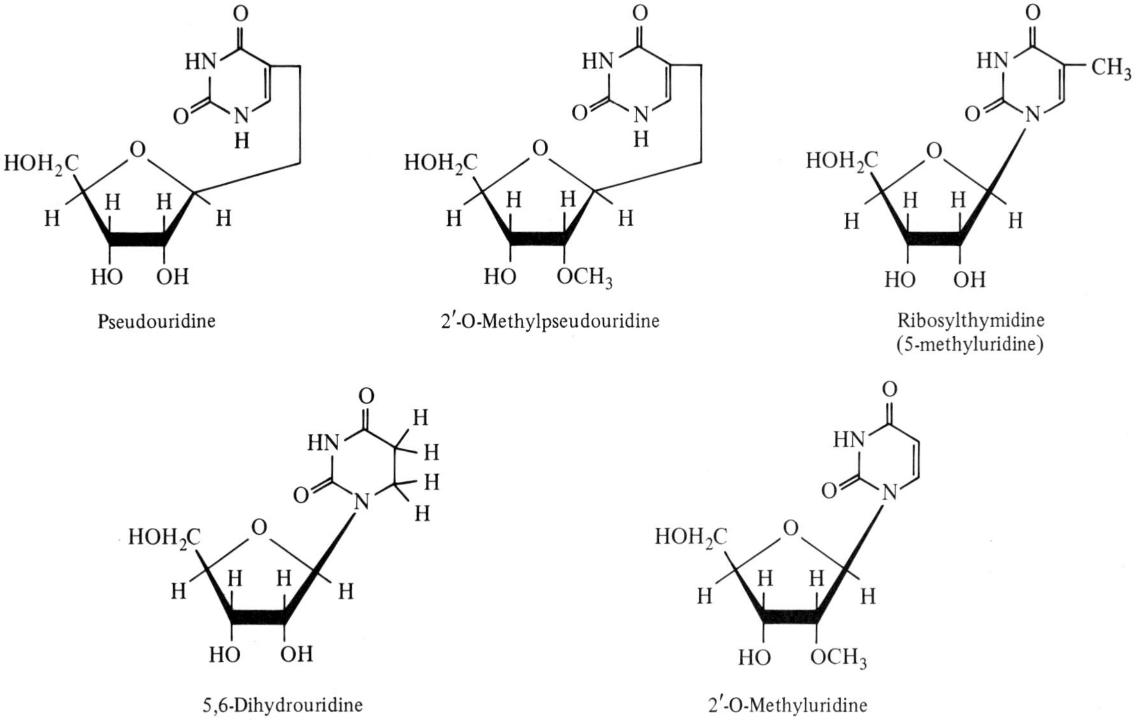

Pseudouridine

2′-O-Methylpseudouridine

Ribosylthymidine (5-methyluridine)

5,6-Dihydrouridine

2′-O-Methyluridine

Figure 18-30
Structures of some unusual nucleosides occurring in tRNA.

disappears when no longer required. The adaptation to a lactose medium is governed by a portion of the *E. coli* chromosome called the ***lac* operon** (section 18-11).

Transfer RNA

Transfer RNA's (tRNA) are not incorporated into any subcellular fraction, but occur in the free state in the cytoplasm. They consist of a collection of species of similar weights (23,000–28,000) but different nucleotide sequences. All biologically active forms, however, possess the same terminal trinucleotide sequence, which consists of two cytidylic residues followed by an adenosine (—CCA). tRNA contains a somewhat higher proportion of uncommon bases than the other types of RNA (Fig. 18-30). Many of these are methylated derivatives of the more common bases.

The three-dimensional structures of several tRNA's have been determined by x-ray crystallography. A prominent structural feature is the presence of hairpinlike helical segments formed by the bending back upon itself of the single polynucleotide strand and stabilized by base pairings of the Watson-Crick type (Fig. 18-31).

The biological function of tRNA is to serve as a carrier for amino acids in their directed assembly into polypeptide chains of specific sequence. This will be discussed in detail in chapter 19.

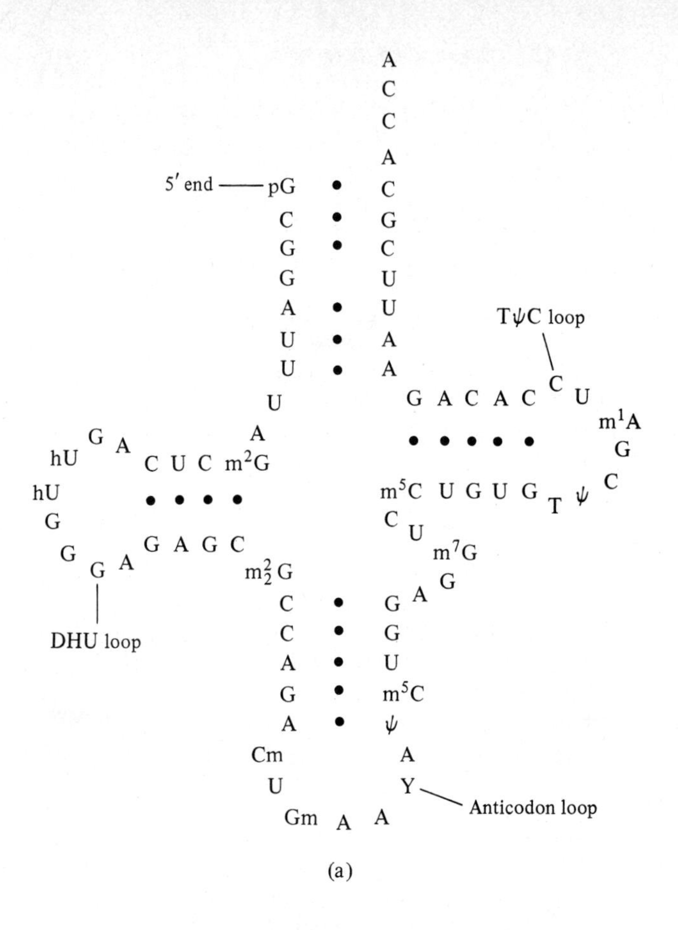

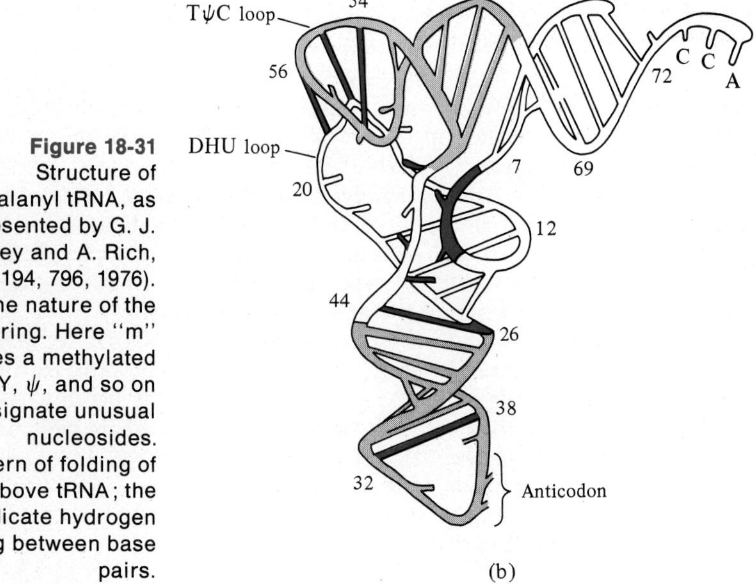

Figure 18-31 Structure of phenylalanyl tRNA, as presented by G. J. Quigley and A. Rich, *Science* 194, 796, 1976). (a) The nature of the bair pairing. Here "m" indicates a methylated base; Y, ψ, and so on designate unusual nucleosides. (b) Pattern of folding of the above tRNA; the bars indicate hydrogen bonding between base pairs.

In both bacterial and eukaryotic cells, the genes for tRNA molecules occur as clusters which are **transcribed** (section 18-11) as long RNA sequences; these may contain more than one type of tRNA. The large precursor RNA molecules are subsequently cleaved and trimmed enzymically to form the tRNA molecules. In addition to the cleavage and trimming process, extensive alteration of the purine and pyrimidine bases is required to generate the final form of a tRNA molecule. Over 50 enzymic modification reactions are known; the nature and extent of modification depends upon the tRNA species. The loops in which the hairpin-shaped helical segments terminate contain a high proportion of modified bases. The majority of the modification reactions are methylations; however, more complicated conversions also occur, such as the rearrangement of uridine to form pseudouridine (Fig. 18-30).

18-11 DNA TRANSCRIPTION

RNA Polymerase

The probable biological mechanism for the formation of RNA involves the action of the enzyme **RNA polymerase**, which is somewhat analogous to the DNA polymerases in its specificity. In the presence of Mg^{2+} and a DNA template it catalyzes the polymerization of *ribo*nucleoside 5′-*tri*phosphates to form biosynthetic RNA whose base sequence is complementary to that of a strand of the template DNA. Pyrophosphate is split out in the reaction

$$n\text{XTP} \rightarrow (\text{XP})_n + \text{PP}_i$$

DNA is required for the reaction to occur; the presence of all four ribonucleoside triphosphates is necessary. RNA polymerase catalyzes the sequential addition of nucleotides to the 3′-hydroxyl end of an RNA chain, which is synthesized in the 5′ → 3′ direction.

Several different RNA polymerases have been identified. RNA polymerase from *E. coli* has a molecular weight of 5.0×10^5 and contains four different kinds of subunits, designated α, β, β', and σ. The composition of the *E. coli* enzyme is $\alpha_2\beta\beta'\sigma$. The σ subunit is not essential for the actual catalytic action of RNA polymerase, but is necessary for the correct selection of the initiation site on the DNA template.

Only *one* strand of bihelical DNA is normally transcribed within cells. This is illustrated by the case of the double-stranded RF form of ϕX174 viral DNA, which is readily transcribed within infected *E. coli*. The mRNA isolated from infected cells readily forms a hybrid molecule with denatured RF DNA, but does not hybridize with the single-stranded DNA occurring in native virus particles. If the latter is designated as the *plus* strand, then it is clear that the mRNA transcript is also a plus strand and that it was transcribed from the complementary *minus* strand. It also follows that the action of the polymerase involves the transient separation of the two strands and the copying of only one.

In eukaryotic cells at least three distinct RNA polymerases are present, which are responsible for the transcription of rRNA genes (polymerase I), mRNA genes (polymerase II), and tRNA genes (polymerase III).

The Operon

The **operon** model was originally proposed by Jacob and Monod to account for the regulation of the transcription of DNA into mRNA. An operon is a linear cluster of genes subject to simultaneous control. The controlling region is a segment of the DNA molecule located at the 3′ end of the operon. The initial part of this control region is termed the **promoter** (Fig. 18-32). It is the site of the initial attachment of the RNA polymerase to DNA. The rates of binding and of initiation may be profoundly influenced by various regulatory proteins.

A second control segment, called the **operator**, is immediately adjacent to the promoter. It is a binding site for a **repressor** molecule. If the operator is unoccupied, transcription may be initiated and proceed unimpeded through the operator region and the genes that code for specific proteins. If a repressor molecule is bound by the operator, transcription is blocked.

The repressor protein molecule may itself be inactivated by combination with an **inducer**, which produces a conformational change leading to a loss in affinity of the repressor protein for the operator site. In the presence of inducer the operator is not blocked and transcription may occur.

The synthesis of repressor protein is coded for by a **regulatory** gene. This gene may be in proximity to the operon or at some distance from it. The transcription of regulatory genes normally occurs continuously at a slow rate.

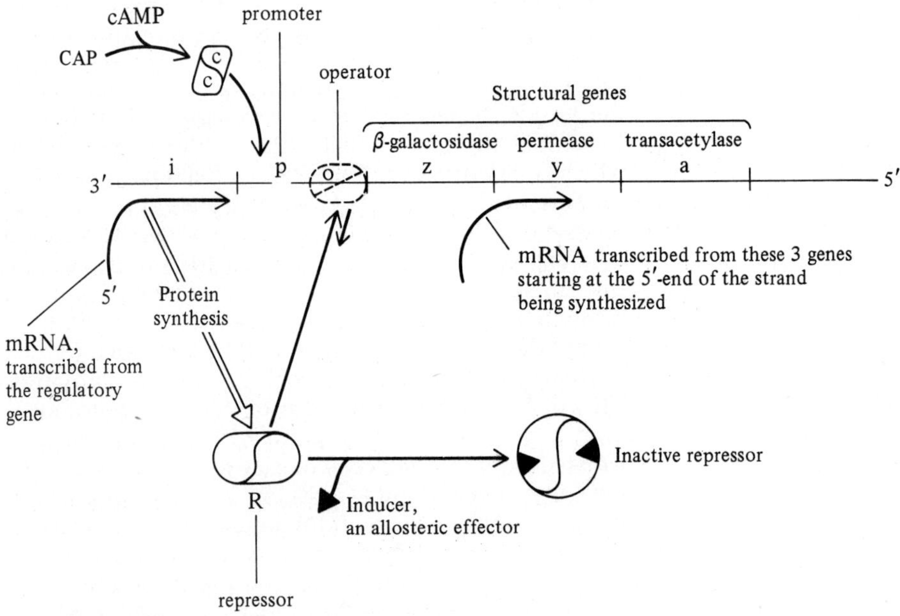

Figure 18-32
Schematic depiction of the *lac* operon of *E. coli*. If inducer is absent, the repressor, coded for by the i gene, combines with the operator, o, and blocks formation of mRNA coding for the three indicated enzymes. The promoter, p, is adjacent to the operator.

The control of transcription may be interfered with by a mutation in the regulatory gene, which produces a defective repressor unable to combine with the operator or by a mutation in the operator resulting in a loss of ability to bind repressor. Such **constitutive** mutations result in uncontrolled transcription of the operon and in the rapid synthesis of mRNA.

Perhaps the best-studied example of an operon is the ***lac* operon** of *E. coli*, which governs the adaptability of *E. coli* to growth in a lactose-containing medium. This contains a promoter, an operator, and three **structural** genes, which code for specific enzyme molecules required for the utilization of lactose (Fig. 18-32). Since the three genes are subject to synchronous control, it is likely that they form a single transcriptional unit and that the entire unit is transcribed as a single piece of mRNA.

The promoter site is controlled by an allosteric **catabolite activator protein** (CAP), itself regulated by the modifier cAMP (section 8-8). Upon binding cAMP the CAP molecule is able to combine with the promoter site and thereby enhance the rate of initiation of transcription.

The *lac* operator is immediately adjacent to the promotor. It is normally combined with repressor, so that transcription of the operon is blocked. The *lac* repressor is a protein consisting of four identical subunits, each of molecular weight 37,200. Each subunit has a binding site for the operator and another for an inducer.

Upon binding an inducer molecule, the *lac* repressor undergoes a structural transition that eliminates its affinity for the operator site, thereby freeing the operator and allowing transcription to occur. The natural inducer is **allolactose** (β-D-galactose-1,6-D-glucose). The transcription of the *lac* operon forms the mRNA, which directs the synthesis of the three enzymes required for the growth of *E. coli* on a lactose medium.

Initiation and Termination of Transcription

The specific association of RNA polymerase with the promoter region implies that the promoter contains a specific sequence of base pairs that is recognized by the polymerase. While the initial complex presumably involves the intact bihelical form of the DNA segment, the beginning of transcription is probably preceded by a separation of the strands to form an **open complex**, which is ready to begin mRNA synthesis.

An mRNA chain is begun by the reaction of ATP or GTP with a second ribonucleoside triphosphate to form a dinucleotide with a triphosphate group at the 5′ end. The chain is extended by the progressive addition of nucleotide units to the 3′ end (Fig. 18-33). While the base sequence of the mRNA is complementary to that of the DNA strand that is transcribed, the RNA polymerase may assist in achieving correct pairing by virtue of the nature of its active site, which may be incompatible with incorrect base pairings between the DNA template strand and the nascent mRNA. In eukaryotic mRNA, there is an initial sequence of adenylate residues preceding the transcript of the actual message.

The transcribed DNA contains not only initiating promoter regions, but also **termination signals** that end the synthesis of the mRNA chain. In some cases specific proteins such as the **rho factor** of *E. coli* may also be involved in terminating the message.

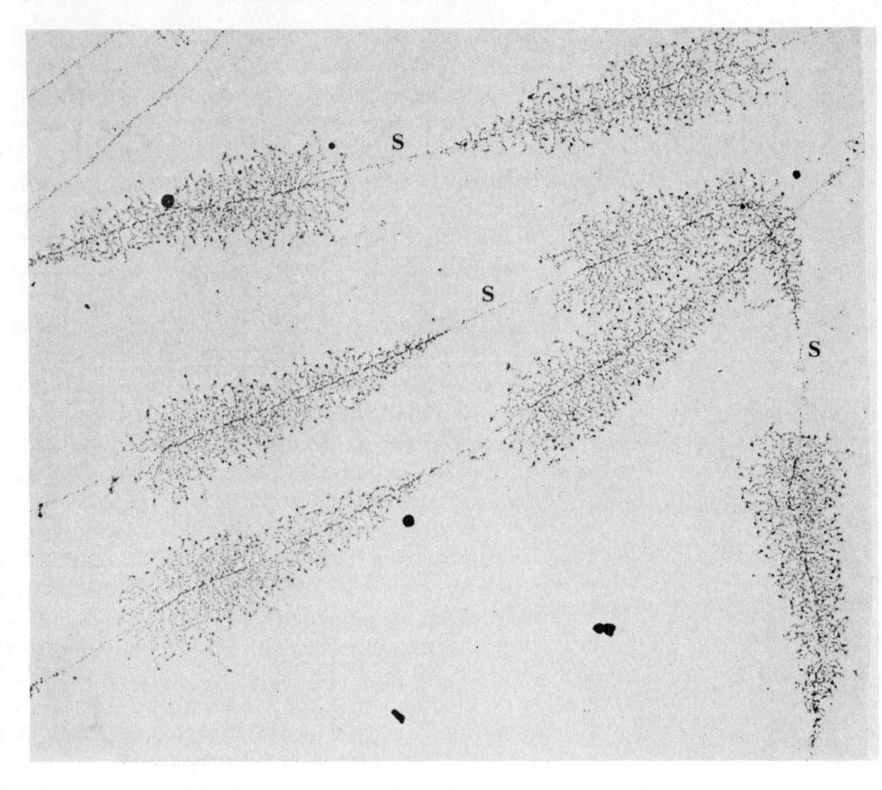

Figure 18-33
Electron microscope photograph of transcribing DNA from oocyte nucleoli of the newt. (Courtesy of O. L. Miller and B. Beatty, *Science*. 164, 955, 1969.)

A number of antibiotics are known that inhibit transcription. These include **rifamycin**, which binds to the β subunit of RNA polymerase and competitively blocks the binding of the initial ribonucleoside triphosphate, and **actinomycin D**, which is intercalated between DNA bases and inhibits RNA polymerase.

SUMMARY

The **nucleic acids** are linear polymers of **nucleotides**, which consist of a purine or pyrimidine base joined by an N-glycosidic linkage to a pentose sugar, which is esterified in the 3′ or 5′ position with phosphoric acid. In **ribonucleic acid** (RNA), the pentose is D-ribose, while in **deoxyribonucleic acid** (DNA), it is 2-deoxy-D-ribose. The mode of linkage joining the nucleotides of DNA or RNA is a phosphodiester bond between the 3′ and 5′ positions of the pentose.

DNA has a **two-stranded helical structure** stabilized by hydrogen-bonded base pairings between **adenine** and **thymine** and between **guanine** and **cytosine**. The bases occur in the core of the double helix, while the sugar-phosphate groups lie on the periphery. The two polynucleotide strands comprising the double helix have opposite **polarities**; that is, the directions of the 3′-5′ phosphodiester bonds are opposite for the two. A stable bihelical structure requires that the bases of the two strands be in register; the two nucleotide sequences are **complementary**. Thermal **denaturation** results in the collapse of the bihelical structure and in the separation of the strands. The process may be reversed by slow cooling, which allows **annealing** and **renaturation**. By thermally denaturing a mixture of two DNA's from related species whose base sequences are similar, and then renaturing them, a hybrid DNA may be formed, whose two strands are derived from different species.

DNA functions as the repository of the **genetic information** that guides the synthesis of cellular proteins. This genetic information is encoded into the DNA in the form of specific nucleotide sequences or **genes**. The DNA of prokaryotic cells normally occurs as a single, often circular, molecule comprising the bacterial or viral chromosome. In eukaryotic cells, the DNA occurs in the nuclear chromosomes as complexes with the basic proteins termed **histones** and **protamines**, as well as in mitochondria.

In the replication of DNA, the parental strands are progressively unwound as each strand directs the synthesis of its complementary strand. The net result is the formation of two new DNA molecules, each of which contains one strand of the original DNA molecule plus a newly synthesized complementary strand. The initiation of DNA replication involves the pairing of a short segment of **primer RNA** with one of the strands.

DNA polymerase III, the enzyme which catalyzes DNA replication, utilizes *deoxy*nucleoside *tri*phosphates as substrates. Since it can only synthesize DNA strands in one direction, one of the two new strands must be synthesized in fragments, which are joined by a second enzyme, **DNA ligase**. The unwinding of the original DNA duplix is mediated by one or more **unwinding proteins.**

In contrast to DNA, RNA is generally **single-stranded**; however, segments of an RNA molecule may fold to form *hairpinlike* bihelical regions with complementary segments within the same molecule. The translation of the genetic message encoded in chromosomal DNA into specific amino acid sequences of proteins involves the **transcription** of nucleotide sequences of DNA into complementary nucleotide sequences of molecules of **messenger RNA**. Only one DNA strand is transcribed. The process is catalyzed by **RNA polymerase**. The strand of messenger RNA in turn guides the **translation** of specific nucleotide sequence into specific amino acid sequences of proteins. This requires a preliminary combination of each amino acid with a molecule of **transfer RNA**, which serves as an **adaptor** to guide the correct positioning of its attached amino acid along the messenger RNA strand.

The regulation of messenger RNA transcription is achieved by the organization of chromosomal DNA nucleotide sequences or **structural genes**, which are responsible for the biosynthesis of particular proteins, into **operons**, which also contain **regulatory genes**. Each operon contains an **operator** region, which must be unblocked for transcription to proceed. A regulatory gene directs the synthesis of a **repressor** molecule that combines with the operator and thereby blocks transcription. The repressor may be inactivated by combination with an **inducer** molecule, which frees the operator and allows transcription to occur. An essential feature of regulation is the short persistence of messenger RNA.

REFERENCES

The following code is used to classify references. I: particularly useful as an introduction to the subject; R: useful primarily as a reference text; A: an advanced account of the material; H: a publication of historical importance.

General

E. Chargaff, "Preface to a Grammar of Biology: A Hundred Years of Nucleic Acid Research," *Science* 172, 637 (1971). (I)

J. N. Davidson, *The Biochemistry of the Nucleic Acids*, Academic, New York (1972). (A)

A. M. Michelson, *The Chemistry of Nucleosides and Nucleotides*, Academic, New York (1963). (R)

A. E. Mirsky, "The Discovery of DNA," *Sci. Am.* 218, 78 (1968). (I)

J. D. Watson, *Molecular Biology of the Gene*, Benjamin, Menlo Park (1975). (I)

DNA Replication

A. Kornberg, *DNA Synthesis*, Freeman, San Francisco (1974). (A)

I. R. Lehman, "DNA Ligase: Structure, Mechanism, and Function," *Science* 186, 790 (1974). (H)

B. Lewin, *Gene Expression*, Wiley, New York (1972). (A)

M. Meselson and F. W. Stahl, "The Replication of DNA in *E. coli*," *Proc. Natl. Acad. Sci.* U.S. 44, 671 (1958). (H)

R. Schekman, A. Weiner, and A. Kornberg, "Multienzyme Systems of DNA Replication," *Science* 186, 987 (1974). (A)

A. Sugino, S. Hiross, and R. Okazaki, "RNA-Linked Nascent DNA Fragments in *E. coli*," *Proc. Natl. Acad. Sci. U.S.* 69, 1863 (1972). (H)

G. L. Zubay and J. Marmur, eds., *Papers in Biochemical Genetics*, Holt, New York (1973). (H)

DNA Transcription

M. J. Chamberlin, "The Selectivity of Transcription," *Ann. Rev. Biochem.* 43, 721 (1974). (A)

W. Gilbert, N. Maizels, and A. Maxam, "Sequences of Controlling Regions of the Lactose Operon," *Cold Spring Harbor Symp. Quant. Biol.* 38, 845 (1973). (A)

F. Gros, "Control of Gene Expression in Prokaryotic Systems," *FEBS Lett.* 40, S19 (1974). (A)

F. Jacob and J. Monod, "Genetic Regulatory Mechanisms in the Synthesis of Proteins," *J. Mol. Biol.* 3, 318 (1961). (H)

A. D. Riggs and S. Bourgeois, "On the Assay, Isolation, and Characterization of the *lac* Repressor," *J. Mol. Biol.* 34, 361 (1968). (A)

J. L. Sirlin, *Biology of RNA*, Academic, New York (1972). (A)

REVIEW QUESTIONS

Questions marked with an asterisk are of a high level of difficulty.

18-1 It is known that a considerable number (up to 100/molecule) of the phosphodiester bonds of DNA may be hydrolyzed by deoxyribonuclease with only a minor (20%) drop in molecular weight. Suggest a reason.

*18-2 What factors would be expected to influence the rate of separation of the two strands of denatured DNA?

18-3 Why is the problem of determining nucleotide sequence in the ribonucleic acids more difficult than that of determining amino acid sequence in the proteins?

*18-4 Compute a complete hydrogen ion titration curve for adenylic acid, using pK values of 4.1 for adenine and 2.0 and 6.6 for phosphate.

*18-5 Suppose that one is confronted with the problem of determining the composition of a mixture of 5′-AMP and 5′-UMP in unknown proportion. How might this be accomplished by measurement of ultraviolet absorption alone?

*18-6 How might the molecular weight of a ribonucleic acid of low molecular weight (<10 nucleotide units) be determined from its hydrogen ion titration curve?

18-7 Compare and contrast RNA and DNA with respect to
(a) chemical composition
(b) secondary structure
(c) location within living cells
(d) number of different "types" or "species"
(e) biological functions

18-8 Draw the structural formula and give the name of
(a) a purine riboside
(b) a purine deoxyriboside-5′-diphosphate
(c) a pyrimidine riboside-3′,5′-diphosphate
(d) a purine riboside-2′,3′-diphosphate
(e) a purine deoxyriboside-3′-5′-cyclic monophosphate

18-9 Draw the basic structure of a portion of the DNA molecule. Indicate the cleavage positions that would yield
(a) predominately nucleoside-3′-phosphates
(b) predominately nucleoside-5′-phosphates

18-10 Select the statement(s) from the column on the right that might apply to the role of the nucleic acid components on the left in DNA structure.

(a) pentose sugar	(1) involved in hydrophobic interactions
(b) phosphate group	(2) hydrogen bonded
(c) purines and pyrimidines	(3) forms cyclic intermediate during cleavage of chain at pH 11–12

(4) strong interaction with water
(5) joined by O-glycoside bonds to purines and pyrimidines
(6) only L isomers found in nature

18-11 A sample of RNA contains equimolar amounts of U, C, A, and G nucleotides, yet is resistant to digestion with pancreatic ribonuclease. What can you deduce about the structure of this RNA? Show your reasoning.

*18-12 (a) An organism's DNA contains 35% A, 23% T, 27% G, and 15% C. It is partly degraded by a single-strand specific endonuclease. If the DNA shows ultraviolet hyperchromicity upon an increase in temperature, what can you conclude about the secondary structure of the material?
(b) How would the percent hyperchromicity and shape of the melting profile of this DNA compare with that of *E. coli* DNA?
(c) Above T_m, how does the A_{260} of this DNA compare to that of the same concentration of *E. coli* DNA above its T_m?
(d) Compare the kinetics of reassociation of this organism's DNA with that of *E. coli* DNA. Assume that the total number of bases in this DNA is the same as that of *E. coli* DNA.

*18-13 Suggest a biological advantage in the fact that only one DNA strand is transcribed into RNA.

19

The Biosynthesis of Proteins

19-1 ORIGINS OF THE CURRENTLY ACCEPTED MODEL FOR PROTEIN SYNTHESIS

Early Evidence for The Role of RNA

We now turn our attention to the later phases of the directed biosynthesis of proteins. In chapter 18 we mentioned that one form of RNA (mRNA) acts as a secondary carrier of genetic information, functioning as an intermediary in the process whereby specific nucleotide sequences in chromosomal DNA are translated into amino acid sequences of proteins. Other types of RNA (rRNA and tRNA) also have essential roles in protein synthesis.

The earliest suggestive evidence for the involvement of RNA in protein synthesis was indirect. The activities in protein synthesis of the cells of many tissues were found to be roughly correlated with their RNA contents. Tissues intensely engaged in the production of proteins, such as liver, were found to be rich in RNA, while the RNA content of tissues with only limited activity in protein synthesis, such as nerve, was much lower.

Since protein synthesis was found to persist for significant periods in cells from which the nuclei had been removed, it was unlikely that DNA itself could engage directly in the guided assembly of amino acids. Experiments with disrupted cells retaining some synthetic capability showed that enzymic destruction of RNA reduced or abolished this activity.

Evidence of another kind was provided by studies with **tobacco mosaic virus**, the infective agent of a characteristic disease of tobacco plants. This virus consists of only a single strand of RNA of molecular weight 2×10^6, plus about 2,300 identical coat protein subunits of molecular weight 17,000.

Earlier work upon the virus and its components had developed methods for removing the protein and isolating the purified viral RNA.

In 1955 Fraenkel-Conrat and his colleagues found that such protein-free preparations of viral RNA possessed some residual infectivity and could produce the symptoms of the disease when introduced into tobacco plants. Tobacco mosaic virus that was indistinguishable from the original strain could be isolated from the sap of the diseased plants.

Since the protein content of the new virus was identical to that of the original, this experiment proved conclusively that viral RNA carried all the specifications required for the production of normal protein and could utilize the synthetic machinery of the plant cells for this purpose.

Translation of Base Sequence into Amino Acid Sequence

The early results outlined above provided a persuasive argument for the direct involvement of RNA in protein synthesis. However, two serious obstacles remained to the acceptance of RNA as a surrogate for DNA in the transmission of genetic information. One of these—the requirement for an explicit mechanism for the transcription into RNA base sequences of the genetic message encoded in the base sequences of DNA—was removed by the discovery of RNA polymerase.

The remaining difficulty concerned the mechanism of the translation of RNA base sequences into the amino acid sequences of polypeptide chains. It was at first assumed that the amino acids interacted directly with the RNA messenger. Since there are 20 amino acids and only four nucleotide bases, ambiguity can be avoided only if the coding units consist of **combinations** of nucleotides each of which recognizes and positions a particular amino acid. Each coding unit or **codon** must consist of at least three nucleotides if a specific codon is to exist for each amino acid. A nucleotide triplet is too large to permit simultaneous contact of most amino acids with all three bases.

Although ingenious versions of the triplet code were proposed that circumvented this difficulty, these were all shown to be infeasible on other grounds. Moreover, there was, and is, no evidence for any specific interaction of amino acids and nucleotide bases.

A model was finally proposed which avoided both the geometrical and chemical difficulties by postulating that a preliminary combination occurred of each amino acid with an **adaptor** molecule, which could in turn recognize and become attached to the triplet codons of mRNA. The combination of the amino acids to form polypeptides would be accompanied by their detachment from the adaptors.

A large body of experimental investigations subsequently verified the adaptor hypothesis and identified transfer RNA as the adaptor molecules. Corresponding to each amino acid are one or more tRNA molecules which can serve as adaptors. A tRNA molecule carrying its amino acid interacts specifically with the coding triplets of mRNA.

Figures 19-1 and 19-2 depict in schematic form the general mechanism for the mRNA-directed synthesis of proteins.

Direction of Protein Synthesis

After the general idea that the assembly of amino acids into the polypeptide chains of proteins was guided somehow by a mRNA template attained wide

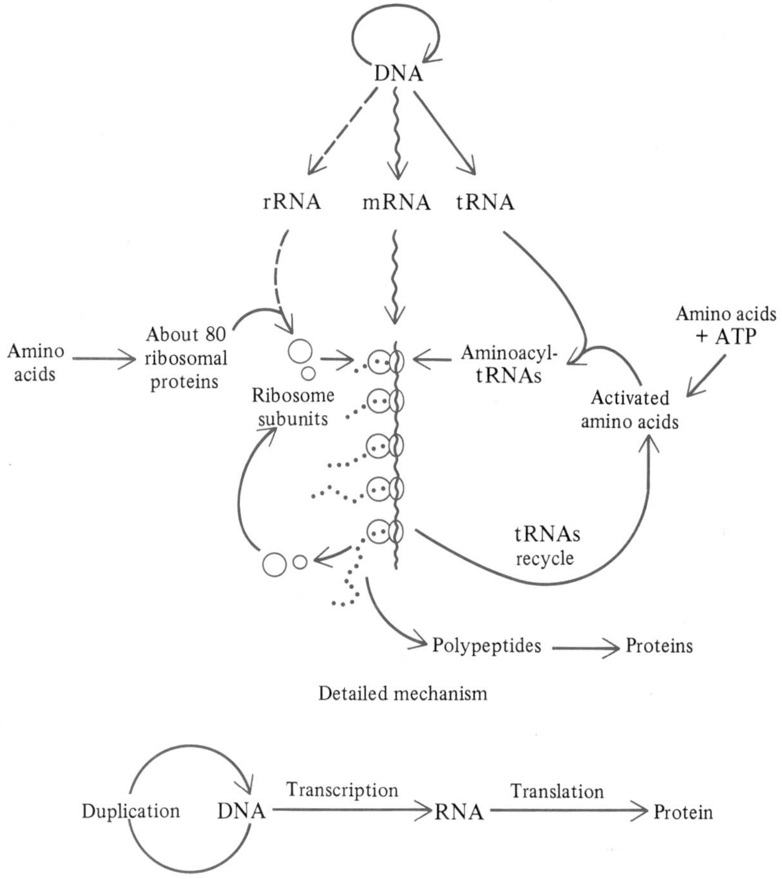

Detailed mechanism

Simplified mechanism

Figure 19-1
Overall mechanism for the direction of protein synthesis by DNA, showing the specific activities of mRNA, tRNA, and ribosomes.

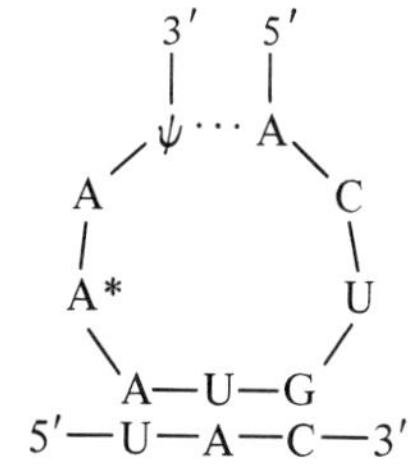

Transfer RNA^{tyrosine} from *E. coli*

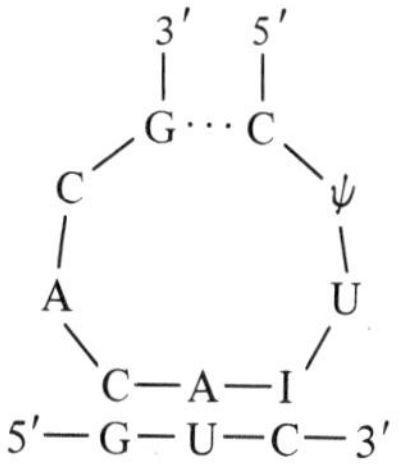

Transfer RNA^{valine} from yeast

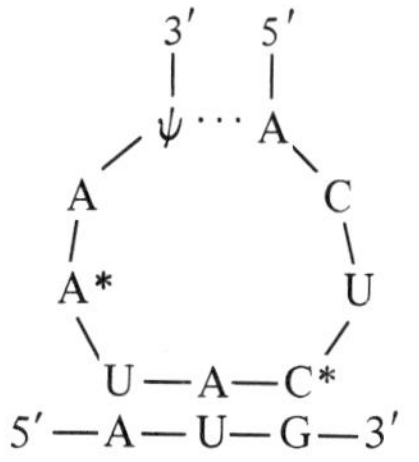

Transfer RNA^{methionine} from *E. coli*

Figure 19-2
Base pairing involved in the recognition of specific codons by the corresponding tRNA's. The asterisks indicate modified (usually methylated) bases, while ψ stands for pseudouridine.

acceptance, there remained a fundamental uncertainty as to the basic mechanism. A number of mechanisms could easily be postulated, between which it was difficult to make a choice *a priori.* The formation of peptide bonds might begin at one end of the chain and proceed to the other in an orderly, stepwise manner. Alternatively, peptide bond formation might occur simultaneously between all neighboring amino acids on a loaded template. It was also conceivable that bond formation might be initiated at several different points of the chain and that the resultant peptide fragments subsequently coalesce to form the complete chain.

This question was resolved by Dintzis, by examining the radioactive labeling of freshly synthesized hemoglobin. Rabbit **reticulocytes** were selected as the synthesizing system. These immature red cells accumulate in large numbers in the blood of rabbits made anemic by phenylhydrazine injections;

their protein synthesis is largely confined to the production of hemoglobin. The synthetic activity of the reticulocytes survives their isolation from blood and they continue to form hemoglobin for many hours after transfer to an incubation medium containing the essential amino acids.

The approach of Dintzis was to add radioactive leucine to the suspension of reticulocytes and observe the distribution of radioactivity in different parts of the polypeptide chains of freshly synthesized hemoglobin. The isolated α- and β-chains of hemoglobin were split by trypsin to a series of peptide fragments that were separated by paper electrophoresis and paper chromatography. The fragments were identified from their positions on the chromatogram, designated arbitrarily by numbers, and assayed for radioactivity. Each of the polypeptide chains of hemoglobin may be represented by

$$H_2N\text{-}A_1A_2A_3 \cdots A_k/B_1B_2B_3 \cdots B_m/, \ldots, /G_1G_2G_3 \cdots G_n/, \ldots$$

where the diagonals indicate the points of attack by trypsin and $A_1A_2A_3 \cdots A_k$ are the amino acids of the tryptic fragment originating from the NH_2-*terminal* end of the chain, and so forth. Or, in abbreviated form, the chain may be represented by

$$\mathrm{A\ B\ C\ D\ E\ F\ G}$$

where the letters stand for tryptic fragments. If it is assumed that the linear assembly begins at A_1 and proceeds *sequentially* to the end of the chain, the peptides would be synthesized in the order A,B,C,D,E,F,G. This **sequential** model predicts a very uneven labeling of polypeptide chains formed at *short times* after the introduction of radioactive leucine.

At zero time, immediately prior to the addition of labeled leucine, the unfinished chains already present would, according to this model, include all stages of synthesis:

A
AB
⋮
ABCDEFG
⋮

All peptides formed after the introduction of radioactive leucine will be labeled. The completed chains corresponding to the above cases may be represented by

A B* C* D* E* F* G*
A B C* D* E* F* G*
A B C D* E* F* G*
⋮
A B C D E F G

where the asterisks indicate the presence of radioactivity. Since the probability that a particular fragment is already incorporated at zero time into an incomplete chain decreases from A to G, it would be expected on the sequential model that, for hemoglobin chains formed at short time intervals after addition of labeled leucine, the degree of labeling of the fragments would increase progressively from A to G.

With increasing time, the fraction of the completed hemoglobin that originates from chains that were partially synthesized at zero time decreases progressively. Since chains that are *initiated* at times *subsequent* to the addition of labeled leucine will have a uniform distribution of radioactivity, it may be predicted that the gradient of radioactivity from peptide A to peptide G will decrease with time until uniform labeling is ultimately approached.

The above predicted behavior corresponds entirely with the actual findings of Dintzis. For short incubation times the degrees of labeling of the peptides of each chain followed a definite order, which was preserved at subsequent times. The degree of labeling was least for the NH_2-terminal peptide and increased as the COOH-terminus was approached. With increasing time, the gradient of radioactivity progressively declined until, at very long times, all peptides were labeled to the same extent.

These experiments established that the polypeptide chains of proteins are formed by the *stepwise* addition of amino acids, starting from the NH_2-terminus. This finding provided an indispensable foundation for the more detailed studies which followed.

19-2 THE PROTEIN-SYNTHESIZING SYSTEM OF THE CELL

Much of the spectacular progress in our understanding of protein biosynthesis has stemmed from studies with **cell-free** systems which retained synthetic capability. Perhaps the most widely used has been that derived from *E. coli*. The minimal constituents of the active system include the following:

(1) amino acids
(2) ribosomes
(3) an amino acid activating enzyme system
(4) ATP and GTP
(5) mRNA
(6) tRNA
(7) Mg^{2+}
(8) enzymes and other proteins responsible for the initiation and elongation of the polypeptide chain and its release from the ribosomes

Amino Acid Activation

The first step in protein synthesis is the **activation** of amino acids. Each amino acid combines with a molecule of ATP to form an active **acyl adenylate** derivative (Fig. 19-3). The reaction is catalyzed by an activating enzyme (aminoacyl-tRNA synthetase), which is specific for the particular amino acid and for its corresponding tRNA. The acyl adenylate then condenses with a molecule of tRNA through formation of an ester linkage with the 2′ or 3′ ribose hydroxyls at the 3′ end of the tRNA molecule, which bears the terminal CCA sequence (Fig. 19-3). AMP is simultaneously split out. The

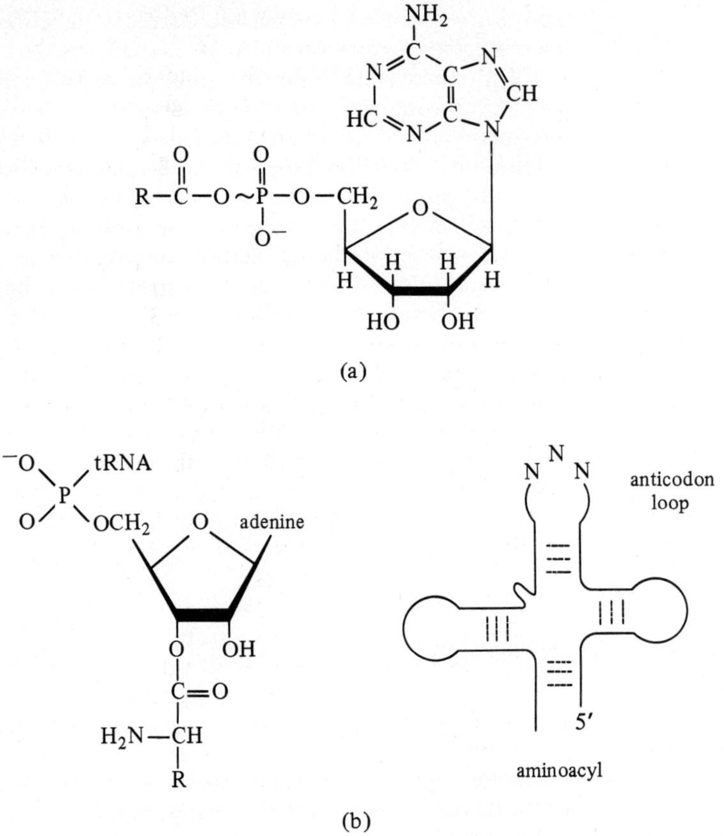

(a)

(b)

Figure 19-3
(a) Generalized structure of an aminoacyl adenylate. (b) The generalized structure of an aminoacyl tRNA. The amino acid is attached by an acyl linkage to the 2′ or 3′ ribose hydroxyl of the adenosine at the 3′-end of the tRNA.

reaction of the acyl adenylate with tRNA is catalyzed by the same enzyme, without dissociation of the initial enzyme-substrate complex:

$$\text{aa} + \text{ATP} \rightarrow \text{aa-AMP} + \text{PP}_i$$

$$\text{aa-AMP} + \text{tRNA} \rightarrow \text{aa-tRNA} + \text{AMP}$$

In the initial acyl adenylate intermediate the carboxyl group of the amino acid is linked by an *acid anhydride* bond with the 5′-phosphate group of the AMP; the high-energy anhydride bond effectively activates the carboxyl group for condensation with the terminal adenylate group of the tRNA. Since the pyrophosphate formed in the first reaction is hydrolyzed to phosphate by a phosphatase, the overall reaction is essentially irreversible.

The aminoacyl-tRNA synthetases are quite specific for the amino acid and the tRNA. This is vital for the accuracy of mRNA translation, since the aminoacyl-tRNA is recognized and positioned only by virtue of a specific **anticodon** base sequence in the tRNA and not by the amino acid itself.

Most of the synthetases of the *E. coli* system have been obtained in purified form. Their molecular weights are generally in the range $1\text{–}2 \times 10^5$. They differ widely in structure, but share a requirement for Mg^{2+} and the presence of one or more sulfhydryl groups essential for activity.

Initiation of Synthesis

According to the model which is accepted today, a strand of mRNA may be visualized as marked off into nonoverlapping triplet codons, each of which codes for an amino acid. The strand is read in linear sequence, proceeding from near the 5′ end to near the 3′ end.

The incorporation of amino acids into polypeptide occurs on a mRNA strand that may be associated with several ribosomes to form a **polysome** (Fig. 19-4). One of the several hairpinlike loops present in each tRNA molecule contains the anticodon nucleotide triplet which recognizes and forms a base-paired complex with a complementary codon triplet in the mRNA. One of the three bases in the anticodon is less specific in its interaction with its partner in the codon than the other two and is called the **wobble** base. The attachment of a tRNA molecule to its mRNA anticodon and the transfer of its amino acid to the growing polypeptide occur at the complex between a segment of a mRNA and a ribosome (Fig. 19-5).

In *E. coli* and other prokaryotic organisms (but not in eukaryotic organisms) the formation of a polypeptide chain begins with the modified amino acid N-formylmethionine. N-formylmethionyl-tRNA (fMet-tRNA) is formed by an enzymic reaction in which a formyl group is transferred from N^{10}-formyltetrahydrofolate to the α-amino group of methionyl-tRNA (Met-tRNA). Of the two known species of Met-tRNA, only one is capable of participating in this reaction. The reaction is

$$N^{10}\text{-formyltetrahydrofolate} + \text{Met-tRNA} \rightarrow \text{tetrahydrofolate} + \text{fMet-tRNA}$$

In order for synthesis to begin, the specific initiation codon (AUG) on the mRNA must be correctly positioned on the ribosome. This is achieved by base-paired complex formation between an **initiator** base sequence in the

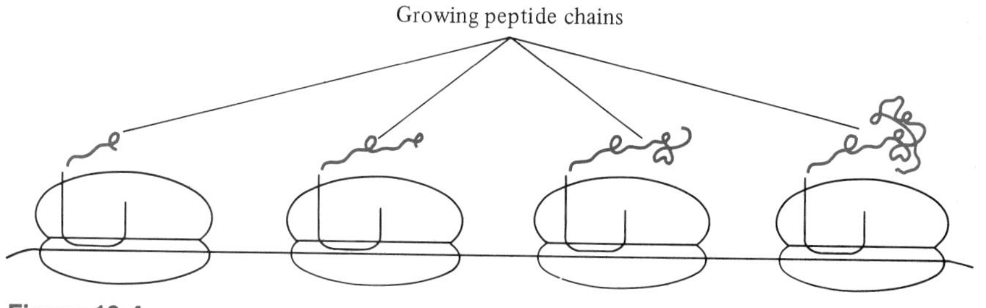

Figure 19-4
Simplified version of the simultaneous action of several ribosomes utilizing a single strand of mRNA. Each ribosome moves down the strand, whose coded information is read in sequence. tRNA molecules, each bearing its amino acid, become attached transiently to the mRNA-ribosome system, deliver their amino acids to the growing polypeptide chain (dark wavy line) and are then detached.

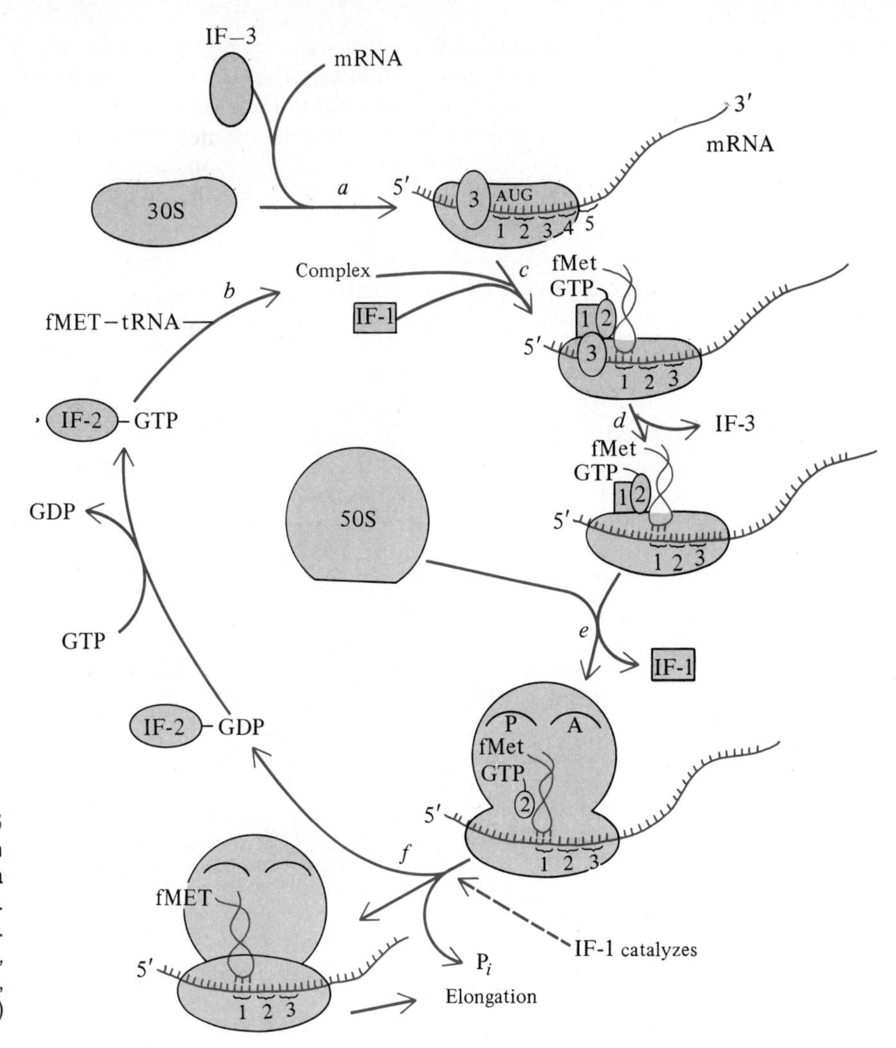

Figure 19-5
Initiation of protein synthesis in a eukaryotic system. (Adapted from D. E. Metzler, *Biochemistry*, Academic, New York, 1977.)

mRNA and a complementary sequence at the 3′-end of the 16S RNA of a 30S ribosome. A ribosomal protein S 1 and a nonribosomal protein initiation factor IF-3 are also involved in the stabilization of the initiation complex of mRNA and the 30S ribosome (Fig. 19-5).

A second protein initiation factor IF-2 becomes complexed with GTP and with a molecule of fMet-tRNA. This complex next combines with the 30S ribosome-mRNA complex in such a way that base pairing occurs between the initiator codon (AUG) of the mRNA and the corresponding anticodon of the tRNA. The process is assisted by a third protein initiation factor IF-1. A 50S ribosome is now added to form the complete 70S ribosome. The initiation factors IF-3 and IF-1 leave the complex at this time.

The 50S ribosome contains two binding sites for tRNA. These have been named the peptidyl (P) site and the aminoacyl (A) site. The tRNA portion of the fMet-tRNA is bound by the P site with the subsequent hydrolysis of GTP and release of IF-2 as a complex with GDP.

At this stage the mRNA is attached to the 30S subunit of a 70S ribosome. The anticodon of a tRNA molecule carrying a formylmethionine residue is base paired with the initiation codon of the mRNA, while the tRNA molecule is also bound by the P site of the 50S subunit. The initiation sequence is now complete.

Elongation of Polypeptide Chains

The placing of the initiating fMet-tRNA on its site on the 50S ribosome allows polypeptide synthesis to begin. The sequential process is as follows (Fig. 19-6):

(1) A tRNA molecule carrying the next amino acid combines with the A site on the 50S ribosome in such a way that its anticodon pairs with the mRNA codon next to that for fMet-tRNA. A protein elongation

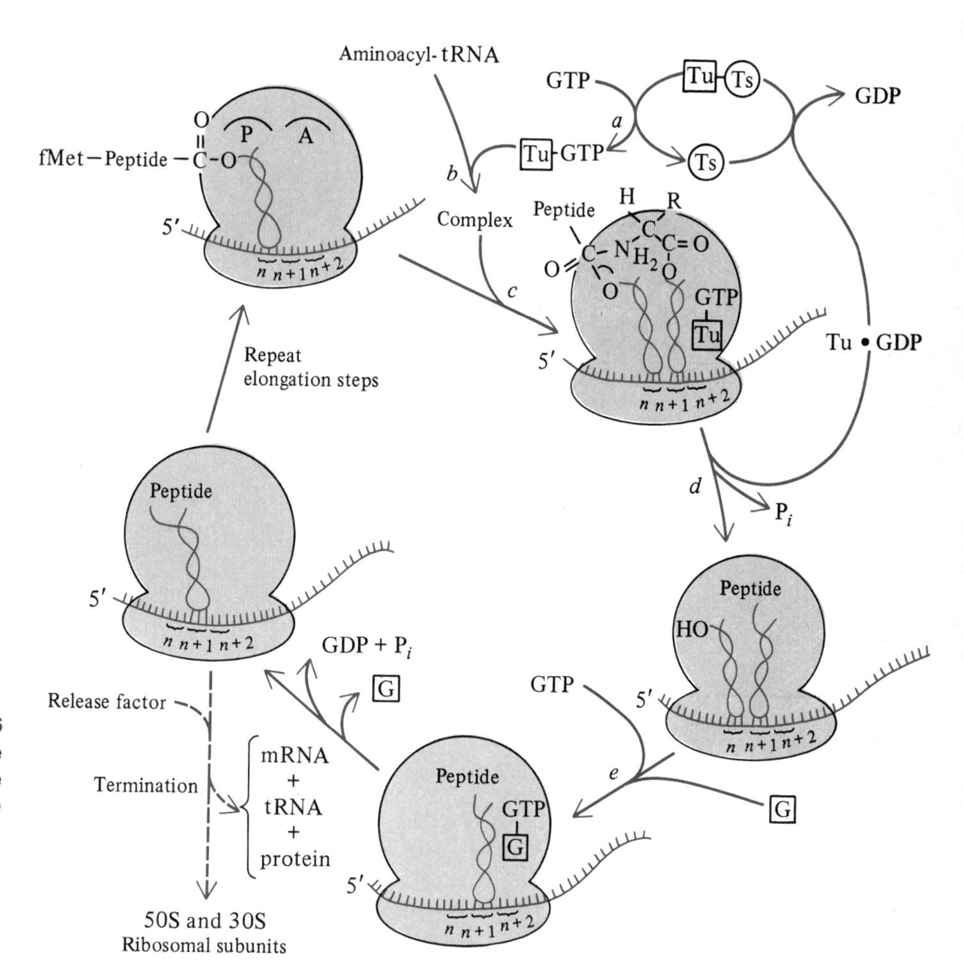

Figure 19-6 Elongation of the nascent polypeptide chain. G represents the elongation factor EF-G. (Adapted from D. E. Metzler, *Biochemistry*, Academic, New York, 1977.)

factor EF-T is involved in this step. This factor consists of two proteins Ts and Tu, which form a complex.

The Tu-Ts complex reacts with GTP to form a GTP-Tu complex and liberate Ts. The GTP-Tu complex next combines with the ribosome and the charged molecule of tRNA.

(2) A peptide bond is formed between formylmethionine and the new amino acid. The tRNA on the P site is released and the nascent polypeptide is transferred to the tRNA on the A site. The bound GTP of the GTP-Tu complex is hydrolyzed with the release of P_i and the GDP-Tu complex. The GDP-Tu complex reacts with free Ts to yield free GDP and reform the Tu-Ts complex. The net result of this step is the hydrolysis of a molecule of GTP and the formation of a peptide linkage.

(3) The remaining tRNA, which bears the growing polypeptide chain, is **translocated** from the A site to the P site. A shift of the mRNA occurs which brings the next codon into place on the A site. The energy required by this process is supplied by the hydrolysis of GTP. This step requires the participation of another elongation factor, EF-G (translocase), which catalyzes the hydrolysis of GTP. The used tRNA which was released in step (2) does not leave the P site until EF-G is bound.

The continuation of the elongation consists of the repetition of the above steps, with each mRNA codon being translated in turn and the corresponding amino acid being added to the polypeptide chain. The process continues until a stop codon is reached. A termination factor, RF-1 or RF-2, then blocks further synthesis, probably by binding directly to the stop codon. The termination factors also catalyze the hydrolytic separation of the polypeptide from the final tRNA. Release of the polypeptide and subsequent dissociation of the tRNA and mRNA from the ribosome now occur.

In eukaryotic cells the basic mechanism of protein synthesis is similar to that found in prokaryotic organisms. However, methionyl tRNA, without a formyl group, is responsible for the initiation of synthesis.

Further Processing of The Polypeptide

The completed polypeptide may require chemical modification before it can assume its biologically active form. The N-formyl group of the terminal fMet residue is removed enzymically. The terminal methionine residue as well as a few other amino acids near the NH_2-terminal end may be split off by a peptidase.

The biologically active three-dimensional structure appears to be formed spontaneously (chapter 5), without the mediation of other cellular elements. As a consequence, the formation of disulfide bonds by the correct pairing of particular —SH groups also proceeds spontaneously. The actual oxidation of cysteine —SH groups to form cystine disulfide bridges is catalyzed enzymically. The information essential for both the structural organization of the protein and the proper positioning of disulfide cross-links is implicit in the amino acid sequence of the newly synthesized protein.

A number of other covalent modifications may also occur after the completion of translation. In particular instances these include methylation, hydroxylation, or phosphorylation, as well as the linkage with prosthetic groups.

Antibiotics That Inhibit Protein Synthesis

Many antibiotics function by interfering with a particular stage of protein synthesis. **Puromycin** has a structure that resembles that of an aminoacyl derivative of the terminal adenylate residue of a tRNA molecule. It can bind at the A site of a ribosome when the P site is occupied by a tRNA molecule carrying a growing polypeptide chain. A covalent derivative of the peptide and puromycin is formed that cannot be elongated and subsequently dissociates from the ribosome. The **tetracycline** antibiotics, which include **chloramphenicol** and **lincomycin**, act on the 50S ribosomal subunit of prokaryotic organisms and are inhibitors of peptide bond formation. **Streptomycin** binds to a 30S ribosomal protein and causes ribosomes to misread the code (Fig. 19-7).

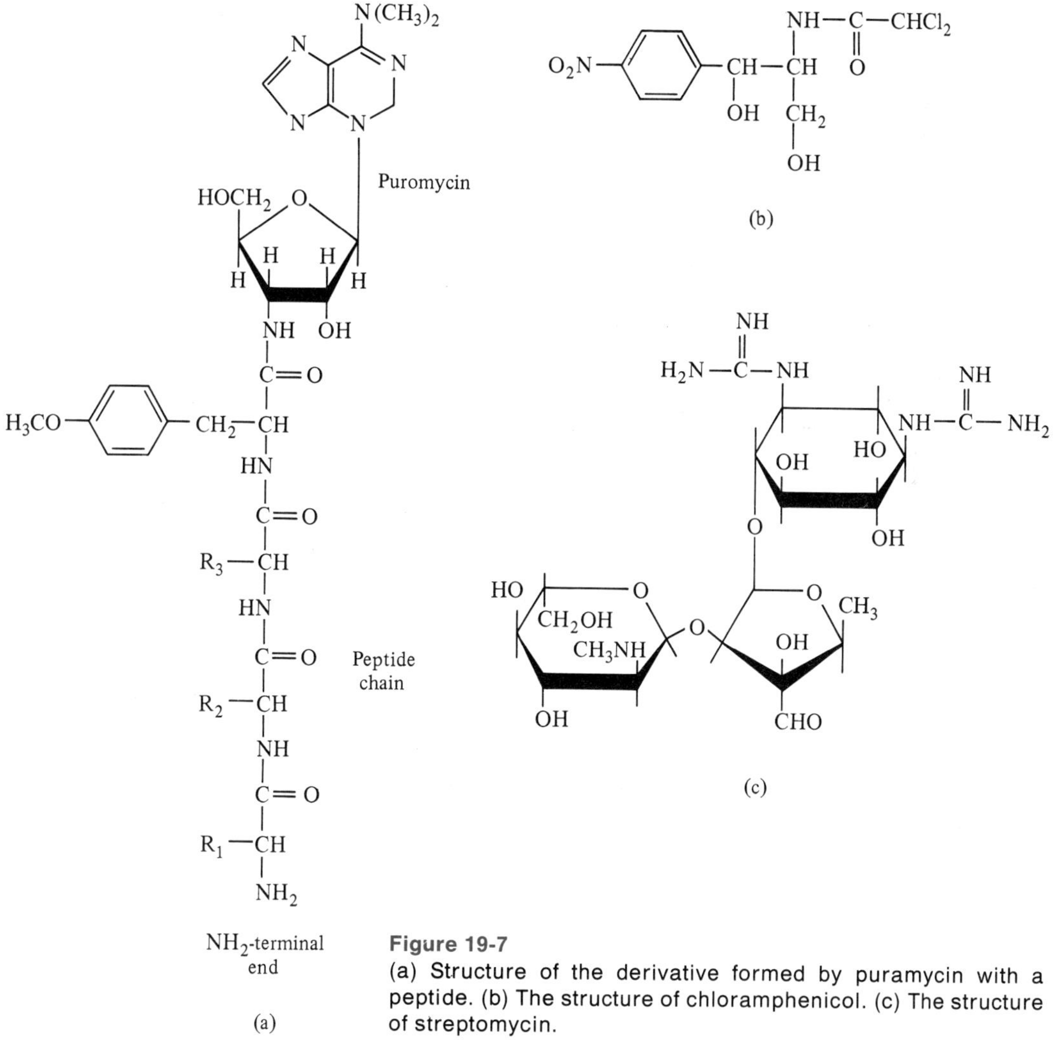

Figure 19-7
(a) Structure of the derivative formed by puramycin with a peptide. (b) The structure of chloramphenicol. (c) The structure of streptomycin.

19-3 THE GENETIC CODE

The crucial step of identifying the trinucleotide codons corresponding to each amino acid was achieved by the parallel work of Nirenberg and of Ochoa. Their experimental approach involved the use of **biosynthetic** mRNA's of *random* base sequence prepared by the action of the enzyme **polyribonucleotide phosphorylase**. This enzyme, which has been isolated from the cells of several bacterial species, including *M. lysodeikticus*, catalyzes the polymerization of *ribo*nucleoside *di*phosphates, with the splitting out of inorganic phosphate.

$$n\text{XDP} \xrightleftharpoons{\text{Mg}^{2+}} (\text{XP})_n + n\text{P}_i$$

In contrast to RNA polymerase, no template is required. If a mixture of different ribonucleoside diphosphates is employed, the nucleotide composition of the polymeric product reflects the composition of the reactant mixture and the base sequence is essentially random and governed primarily by statistical factors.

The initial observation was that a polymer containing only uridylic acid (polyribouridylic acid), when made available to the cell-free protein-synthesizing system from *E. coli*, resulted in the incorporation of radioisotope-labeled phenylalanine into a polypeptide containing only this amino acid. Other amino acids were not incorporated. The implication was that the triplet coding for phenylalanine was U-U-U.

The dictionary of code words was extended by the use of nucleotide copolymers. Because of the random sequence of the polynucleotides synthesized by polyribonucleotide phosphorylase, the relative frequency of occurrence of any nucleotide triplet could be predicted on statistical grounds if the overall base composition was known. For example, a copolymer consisting of 25% adenylic acid and 75% uridylic acid would contain the following trinucleotides in the cited relative frequencies:

Trinucleotide	*Frequency*
AAA	$1/4 \times 1/4 \times 1/4 = 1/64$
AUA	$1/4 \times 3/4 \times 1/4 = 3/64$
AAU or UAA	$1/4 \times 3/4 \times 1/4 = 3/64$
AUU or UUA	$1/4 \times 3/4 \times 3/4 = 9/64$
UUU	$3/4 \times 3/4 \times 3/4 = 27/64$
UAU	$3/4 \times 1/4 \times 3/4 = 9/64$

By correlation of the relative efficiency with which a particular copolymer promoted the incorporation of different amino acids with the relative frequency of occurrence of its various trinucleotide sequences, it was possible to compile the base compositions of a list of codons without regard to their actual nucleotide sequence. The establishment of base sequences within the codons required a different approach based upon comparing the ability of trinucleotides of known sequence to promote the binding of specific tRNA molecules by ribosomes. The present status of the genetic code is indicated in Table 19-1.

Table 19-1

The Genetic Code

Codon*	Amino Acid	Codon	Amino Acid
UUU, UUC	Phe	UCU, UCC, UCA, UCG, AGU, AGC	Ser
UUA, UUG, CUU, CUC	Leu	CCU, CCC, CCA, CCG	Pro
AUU, AUC, AUA	Ile	ACU, ACC, ACA, ACG	Thr
AUG	Met	GAA, GAG	Glu
GUU, GUC, GUA, GUG	Val	UGU, UGC	Cys
GCU, GCC, GCA, GCG	Ala	UGG	Trp
UAU, UAC	Tyr	CGU, CGC, CGA, CGG, AGA, AGG	Arg
CAU, CAC	His	GGU, GGC, GGA, GGG	Gly
CAA, CAG	Glu-HN_2		
AAU, AAC	Asp-NH_2		
AAA, AAG	Lys		
GAU, GAC	Asp		

*The convention is to place the 5′ end of the codon at the left and the 3′ end at the right.

Khorana was able to confirm the codon assignments by an independent approach that involved the use of synthetic mRNA's containing base sequences that repeated in a regular way. For example, the synthetic polynucleotide AGAGAGAGAGAG ··· contains the alternating nucleotide triplets AGA and GAG that code for arginine and glutamic acid, respectively. When used as an mRNA this polynucleotide leads to the formation of the polypeptide Arg-Glu-Arg-Glu-Arg-Glu- ··· . A number of other nucleotide copolymers were synthesized which in each case gave rise to polypeptides whose composition was in accord with the genetic code.

Several features of the genetic code deserve comment.

(1) There is no punctuation separating codons. A series of codons must therefore be read correctly from first to last, or the translation will contain only nonsense. This underlines the importance of an initiator signal to assure the correct setting of the reading frame at the beginning of the translation of a mRNA molecule. The reading frame is then moved sequentially from one triplet to the next; without this arrangement, all of the codons would be out of register.
(2) The code is **degenerate**; there is more than one codon for all amino acids except tryptophan and methionine. The degeneracy varies widely for different amino acids; for example leucine has six different codons, while tyrosine has only two.
(3) Only three of the 64 possible codons—UAG, UAA, and UGA—do not code for any amino acid. These function as termination codons which signal the end of translation. AUG functions as the initiation codon.
(4) The code is basically universal, although occasional species variations may occur.

19-4 CONTROL OF PROTEIN SYNTHESIS

A question of central importance in protein biosynthesis is how the rates of formation of particular proteins are regulated so as to conform to the needs of each cell. The need for such a regulatory mechanism is obvious when one considers the diverse functions of the cells of a complex organism. For example, the genes for hemoglobin must be present in every cell of the mammalian body, yet most mammalian cells produce no hemoglobin whatsoever.

In the discussion of DNA transcription and the operon model (section 18-11) we have already described one of the most important mechanisms for the regulation of protein biosynthesis. Many examples are known of the repression or induction of enzyme synthesis by the presence of a product or a substrate of the enzyme. A classical example of enzyme induction is the inactivation of the *lac* repressor by a lactose derivative and the resultant freeing of the *lac* operon for the synthesis of the enzymes required for lactose metabolism (section 18-11).

Examples of the converse phenomenon are also known. *E. coli* cells grown in a medium devoid of amino acids contain enzymes for the biosynthesis of all 20 amino acids. If the amino acids are present in the medium, the corresponding enzymes are not produced. In this case the products of the action of the enzymes serve to activate the repressors of enzyme synthesis and are termed **corepressors.**

Gradations in both induction and repression occur. At intermediate levels of inducer or repressor, enzyme synthesis can proceed at an intermediate rate. Similar variations can occur in the cellular content of ribosomes. Under conditions of optimal growth the ribosome content may be greater by a factor of five or more than that for a slowly growing culture.

The control of enzyme activity by the repressor mechanism has the disadvantage of a relatively slow response to an abrupt change in conditions.

In the case of a repressible enzyme the supply of the product of its action causes synthesis of the enzyme to cease. However, in the absence of a supplementary control system the *pre-existing* enzyme would continue its now unnecessary activity.

This wasteful synthesis may be avoided by means of a **feedback** mechanism for control, by which an enzyme is allosterically inhibited by high levels of a product formed by its action. For example, *E. coli* cells growing in a medium lacking isoleucine can synthesize this amino acid. If isoleucine is suddenly supplied the production of the enzymes needed for its formation is halted. In addition, the existing biosynthetic pathway is blocked through inhibition by isoleucine of one of its enzymes.

19-5 VIRUSES AS MODEL SYSTEMS

Tobacco Mosaic Virus

The inherent simplicity of viral systems, together with their rapid rate of multiplication, has rendered them very popular subjects for studies of molecular genetics (section 19-1).

A combination of electron microscopy and physicochemical measurements has revealed that the virus particles have the shape of rigid cylinders about 300 nm long (Fig. 19-8). The single strand of RNA is wound into a helical conformation and surrounded by the protein subunits, which are arranged in a helical array (Fig. 19-9).

It has already been mentioned (section 19-1) that purified preparations of viral RNA retain infectious properties. The complete virus isolated from plants infected in this way is always of the same strain as that from which the RNA was obtained.

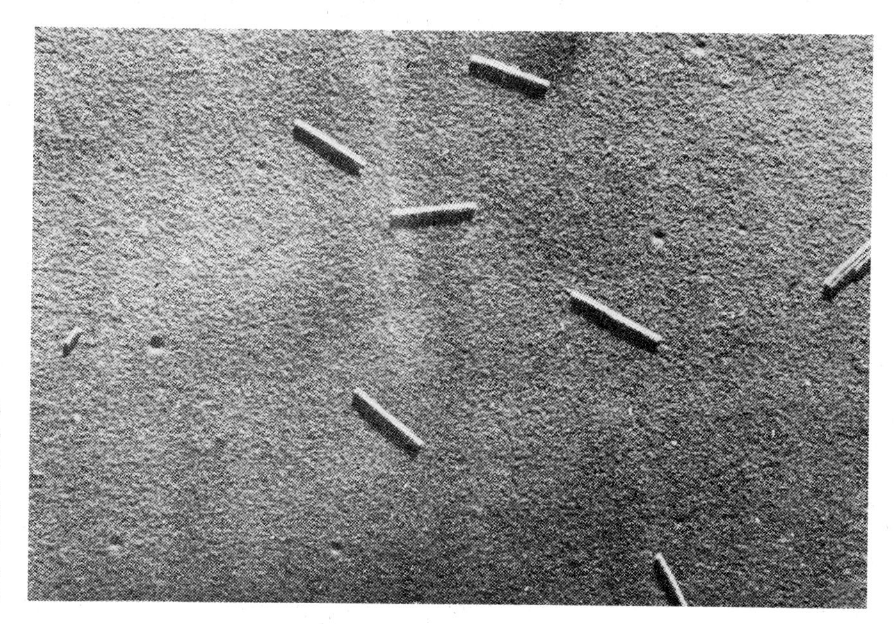

Figure 19-8
Electron microscope photograph of individual tobacco mosaic virus particles. (Courtesy of National Naval Medical Center, Bethesda, Maryland.)

Figure 19-9 Mutual arrangement of RNA and protein subunits of tobacco mosaic virus.

The RNA from tobacco mosaic virus (TMV) may be regarded as a preformed mRNA, which contains all the information needed for its own replication and the formation of its protein subunit, as well as any necessary enzymes not normally present in the host.

TMV is subject to mutations, which can result in changes in disease symptoms and host specificity. Mutations may occur naturally or may be induced by treatment with chemical or physical agents. Among the best-known mutagenic agents is nitrous acid (HNO_2), which deaminates the adenine, cytosine, and guanine bases of RNA, converting them to hypoxanthine, uracil, and xanthine, respectively (Fig. 19-10). While most mutations are lethal, limited treatment with nitrous acid can produce viable mutants of significantly different properties, which can be isolated and studied.

An alteration in the nucleotide sequence of the viral RNA results in a modified amino acid sequence in its protein subunit. Since the amino acid sequence of the subunit is known, it is possible to utilize the "fingerprinting" technique described in section 3-7 to pinpoint the altered regions.

A number of such mutations have been identified and correlated with changes in amino acid sequence. All correspond to replacement of one amino acid by another, rather than to the deletion or interchange of amino acids. Moreover, all of the observed changes (Table 19-2) are consistent both with

Table 19-2
Single Amino Acid Mutations in Tobacco Mosaic Virus Protein

Base Change	*Amino Acid Change*
C → U	Thr → Met
C → U	Ser → Phe
C → U	Ser → Leu
A → G	Thr → Ala
A → G	Ser → Gly
A → G	Arg → Gly

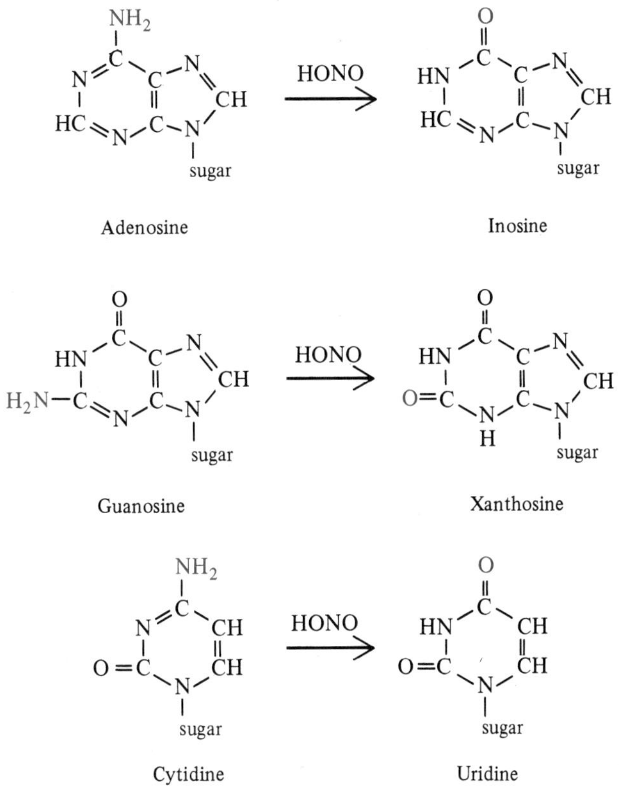

Adenosine $\xrightarrow{HONO}$ Inosine

Guanosine $\xrightarrow{HONO}$ Xanthosine

Cytidine $\xrightarrow{HONO}$ Uridine

Figure 19-10 Action of nitrous acid upon three nucleosides. Inosine resembles guanine in base-pairing characteristics.

the genetic code and with the expected chemical alteration of the codon by nitrous acid. For example, serine, whose codon is UCU, is replaced by phenylalanine, whose codon is UUU, reflecting the conversion of cytosine to uracil by nitrous acid treatment. The inverse transformation, from phenylalanine to serine, is never observed. The molecular genetic studies upon TMV are thus in basic harmony with the conclusions drawn from studies with other systems.

T-Even Bacteriophage

Perhaps an even more interesting viral system is the group of bacterial viruses or **bacteriophages**. The best characterized of these are the set of **T-even** bacteriophages which invade cells of *E. coli*. (The forms designated as T2, T4, and T6 proved quite fortuitously to be similar in properties and are customarily termed T-even.)

The T-even bacteriophages are tadpole-shaped organisms of particle weight about 500×10^6 (Fig. 19-11). They consist of a single molecule of DNA, of molecular weight close to 1.2×10^8, plus several species of protein, including "head" and "tail" proteins, as well as a few minor constituents of low molecular weight. The DNA is imbedded in the interior of the head and surrounded by a protein coat. The tail terminates in a flat plate, from which several fibers trail.

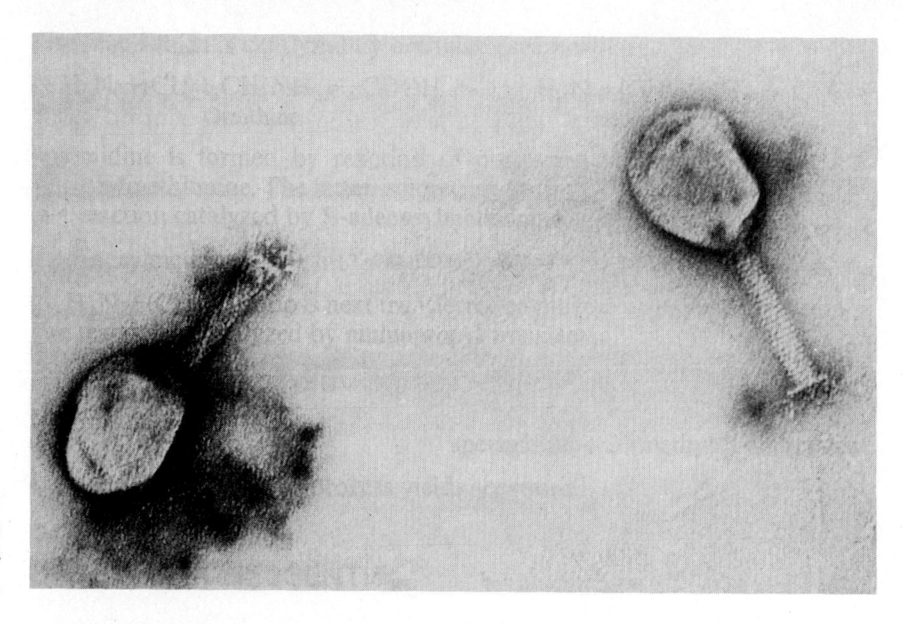

Figure 19-11 Electron microscope photograph of T4 phages.

During the infection process the virus particles become attached to the surfaces of *E coli* cells by their tails, which subsequently undergo contraction. The viral DNA is extruded through the tail into the host cell, the phage particle acting somewhat like a minuscule syringe. Little or no protein enters the cell and there is no evidence of the participation of phage protein in the subsequent events.

The normal metabolism of the *E. coli* cell virtually ceases soon after infection. The injected genetic material assumes dictatorial powers over the metabolic system of the host, diverting it entirely to the formation of fresh virus.

After a latent period of several minutes, the synthesis of phage DNA becomes detectable, followed by the appearance of phage protein and of intact virus. After about 20 min, lysis of the cell occurs, with the release of bacteriophage capable of initiating further cycles of infection.

The genetic makeup of the T-even bacteriophages is much more complex than that of TMV. The T-4 virus contains over 100 identifiable genes, as compared with only a few in TMV. The genes of T-4 must direct the synthesis of not only the viral proteins, but also of enzymes needed for the replication of its DNA and for the rupture of the cell wall to permit release of mature virus particles.

While no synthesis of the normal RNA of *E. coli* occurs in the infected cell, new RNA of different base composition appears. This has been shown to form molecular hybrids with phage DNA, indicating that it represents transcripts of regions of the viral chromosome. The newly formed RNA undoubtedly corresponds to mRNA concerned with the synthesis of viral protein, as well as any necessary enzymes normally absent from *E. coli* cells.

Additional enzymes are required for the replication of viral DNA. The DNA of the T-even bacteriophages is unusual in that cytosine is replaced entirely by 5-hydroxymethylcytosine, which forms a similar Watson-Crick base pair with guanine. Since this base is not a normal constituent of *E. coli*, it must be synthesized from available material.

Several new enzymes arise in infected cells whose function is the synthesis of 5-hydroxymethylcytosine and its conversion to a nucleotide suitable for incorporation into DNA. One of these converts deoxycytidylic acid (dCMP) into its 5-hydroxymethyl derivative. A second enzyme converts the latter to the triphosphate, which is incorporated into phage DNA. Both enzymes are presumably coded for by the phage chromosome, since neither has been detected in noninfected *E. coli*.

19-6 MUTATIONS AND THE GENETIC CODE

Not all mutations reflect the alteration of a single nucleotide. Natural mutations also occur that correspond to the addition or removal of one or more nucleotides. The consequences of mutations of this class are much more drastic than for those which leave the number of nucleotides unchanged.

This qualitative difference stems from the fact that the reading of mRNA begins at a fixed point and proceeds in sequential blocks of three nucleotides. For example, if the normal base sequence is GTA-ACG-CAA · · · , then the replacement of T by C yields GCA-ACG-CAA · · · , in which only one codon is changed. However, if a nucleotide (G) is inserted after T, the trinucleotide groupings become GTG-AAC-GCA-A · · · . All codons following the change are altered and the reading frame is completely disrupted. Mutations of this kind are called **frame-shift mutations**. A nucleotide deletion has similar results. In neither case is an active protein likely to be formed.

In contrast, the replacement of a single nucleotide alters only one codon. Because of the degeneracy of the genetic code, the altered codon will probably code for *some* amino acid. In this way an altered, but still biologically active protein may be synthesized.

A mutation which replaces a single codon specific for a given amino acid by another codon specific for a second amino acid is termed a **missense** mutation, as opposed to a **nonsense** mutation, which produces a codon which does not code for any amino acid. The extensive degeneracy of the genetic code insures that most mutations are likely to be missense rather than nonsense. As has already been discussed, nonsense codons function as "periods" in the reading of mRNA and terminate protein synthesis at the points where they occur.

Suppressor Genes

Many examples are known of the reversal of the effects of a harmful mutation by a second change at a *different* point on the chromosome. Such **suppressor** mutations may occur either within the original gene (**intragenic suppression**) or in another gene (**intergenic suppression**). The corresponding mechanisms are altogether different. A gene which suppresses a mutation in a second gene is called a **suppressor** gene.

Intragenic suppression may arise from a second mutation leading to an amino acid replacement whose structural effects cancel those of the replacement resulting from the initial mutation. As an example, if the first mutation replaces an amino acid with a small side-chain by one with a large side-chain, a suppressor mutation in a nearby codon might have the converse result, thereby restoring the ability of the protein to fold correctly. Alternatively, the effects of a mutation involving the insertion or deletion of a single nucleotide might be reversed by a second mutation that deletes or inserts a nucleotide in an adjacent location, thereby restoring a correct reading of the mRNA (except for a disordered region between the two changes).

In contrast to the above, the action of suppressor genes does not involve a change in nucleotide sequence of the gene that produces the affected mRNA. Instead, the manner of reading the mRNA is changed.

A number of distinct suppressor genes have been identified in strains of *E. coli*. Each alters the reading of a specific codon and can reverse the effects of a single nucleotide change which converts a codon to a nonsense codon. For example, many otherwise lethal mutations of bacteriophage can grow in mutant strains of *E. coli* that possess appropriate suppressor genes, but not in the normal B strain. Suppressor genes have been identified that cancel the effects of mutation to each of the three nonsense codons UAG, UAA, and UGA.

Most suppressor genes function by directing the synthesis of atypical tRNA molecules whose anticodons can pair with nonsense codons and permit reading of the mRNA to proceed to completion, although with one amino acid substitution. Since suppression is usually relatively inefficient, it permits most protein chains to be terminated normally, while allowing production of enough of the missing enzyme for survival. In the absence of the suppressor gene the enzyme would not be synthesized at all.

Suppressor genes are of course ordinarily liabilities, which tend to be eliminated by evolution, unless a specific harmful mutation must be compensated for. In the latter case, enough copies of the active protein may be synthesized to permit normal growth.

SUMMARY

Prior to incorporation into a polypeptide each amino acid is linked to the terminal adenosine of a specific tRNA molecule to form an **aminoacyl** tRNA. This requires an initial **activation** of the amino acid by reaction with ATP to form an **aminoacyl adenylate**; the subsequent reaction that yields aminoacyl tRNA is catalyzed by the same enzyme, which is specific for the amino acid.

In bacteria, polypeptide chains are always initiated with the amino acid **N-formylmethionine**. Protein synthesis begins with the binding of a tRNA molecule carrying this amino acid by an **initiation codon** of tRNA.

The assembly of polypeptides occurs on the **mRNA-ribosome** system. The **ribosomes** of *E. coli* are globular cytoplasmic particles composed of protein and RNA in roughly equal quantities and designated by their sedimentation coefficients as **50S** and **30S** ribosomes; these may interact to form a **70S** particle.

The initial binding of N-formylmethionine occurs to an mRNA molecule complexed with a 30S ribosome. The initiation process requires **initiation**

factors and involves the hydrolysis of GTP. A 50S ribosome is added to the complex after binding of formylmethionine-tRNA.

The growth of the polypeptide begins with the binding of a second tRNA carrying an amino acid to the adjacent codon of the mRNA. The first and second tRNA molecules are also combined with adjacent P and A binding sites, respectively, on the 50S ribosome. Formation of the peptide bond occurs, followed by the release of the first tRNA and by the **translocation** of the second tRNA, which bears the peptide, from the A to the P site; a shift of the mRNA occurs to bring the next codon into position. A protein **elongation factor** and GTP are involved. Elongation of the polypeptide continues until a termination codon is reached and release of the polypeptide occurs.

A molecule of mRNA is marked off sequentially into a series of trinucleotide codons, each of which codes for a specific amino acid and combines with the trinucleotide **anticodon** of its tRNA. By studies with synthetic mRNA's it has proved possible to identify the codons corresponding to each amino acid and thereby to establish the **genetic code**. The code is degenerate; more than one codon exists for nearly all amino acids. Three of the 64 possible trinucleotides do not code for any amino acid; these function as termination codons.

The principles of protein synthesis are illustrated by several viral systems which can utilize the synthetic machinery of the host cell to form enzymes and other proteins required for their own replication.

REFERENCES

The following code is used to classify references. I: particularly useful as an introduction to the subject; R: useful primarily as a reference text; A: an advanced account of the material; H: a publication of historical importance.

General

J. N. Davidson, *The Biochemistry of the Nucleic Acids*, 7th ed., Academic, New York (1972). (A)

R. Haselkorn and L. B. Rothman-Denes, "Protein Synthesis," *Ann. Rev. Biochem.* 42, 397 (1973). (R)

B. Lewin, *Gene Expression*, Wiley, New York (1976). (A)

A. M. Michelson, *The Chemistry of Nucleosides and Nucleotides*, Academic, New York (1964). (R)

J. D. Watson, *Molecular Biology of the Gene*, 3rd ed., Benjamin, New York (1976). (I)

Initiation of Protein Synthesis

"Initiation of Protein Synthesis in Prokaryotic and Eukaryotic Systems," a summary of an international workshop, *FEBS Letters* 48, 1 (1974). (A)

Amino Acid Activation

L. L. Kisselev and O. O. Favorova, "Aminoacyl-tRNA Synthetases: Some Recent Results and Achievements," *Adv. Enzymol.* 40, 141 (1974). (R)

M. Sprinzl and F. Cramer, "Accepting Site for Aminoacylation of $tRNA_{Phe}$ from Yeast," *Nature New Biol.* 245, 3 (1973). (A)

Ribosomal Function

C. G. Kurland, "Ribosome Structure and Function Emergent," *Science* 169, 1171 (1970). (A)

O. Pongs, K. H. Nierhaus, V. A. Erdmann, and H. G. Wittmann, "Active Sites in *E. coli* Ribosomes," *FEBS Letters* 40, S28 (1974). (A)

The Genetic Code

F. H. C. Crick, "The Origin of the Genetic Code," *J. Mol. Biol.* 38, 367 (1968). (I)

M. W. Nirenberg and J. H. Matthaei, "The Dependence of Cell-Free Protein Synthesis in *E. coli* upon Naturally Occurring or Synthetic Polyribonucleotides," *Proc. Natl. Acad. Sci. U.S.* 47, 1588 (1961). (H)

G. S. Stent, *Molecular Genetics*, Freeman, San Francisco (1971). (I)

C. R. Woese, *The Genetic Code*, Harper and Row, New York (1967). (A)

M. Ycas, *The Biological Code*, North-Holland, New York (1967). (I)

REVIEW QUESTIONS

Questions marked with an asterisk are of a high level of difficulty.

*19-1 Predict the size (number of residues relative to template) and sequence (or composition—specify if sequence is fixed or random) of the peptides synthesized in a cell free protein synthesizing system using the following polynucleotides as templates:

(a) poly(UC)
(b) poly(AAUU)
(c) poly(IU)
(d) poly(GAU)
(e) poly(A, C) (50% A, 50% C)

Polymer (e) has a random sequence; the others are regular and repetitive, that is, (a) is UCUCUC· · ·, and so on.

*19-2 List three good reasons why it is believed that all tRNA's must possess the same overall folding scheme.

*19-3 Why is there speculation that tRNA may have to undergo a conformational change at the ribosome during protein synthesis?

19-4 What are the roles of GTP during protein synthesis?

*19-5 Suggest some biological reasons why tRNA has a high proportion of unusual bases and mRNA does not.

19-6 What are the biological advantages of the short life of mRNA?

*19-7 How many high energy phosphate bonds are required to translate the mRNA of a protein containing 50 amino acids? Assume that the process begins with free amino acids, tRNA's, GTP, ATP, ribosomes, and necessary enzymes and protein cofactors.

19-8 If only three different bases occurred in mRNA, how many different amino acids could be coded for, assuming a triplet code?

20

Amino Acid Metabolism

20-1 METABOLIC CONVERSIONS OF AMINO ACIDS

Since there are substantial differences in the metabolic processing of amino acids by different species, some degree of selectivity in the discussion will be necessary. Our frame of reference in this chapter will normally be *mammals*.

Only a fraction of the amino acids ingested in the mammalian diet are normally incorporated into protein. The remainder enter degradative pathways for the most part and supplement fats and carbohydrates as a major biological fuel. When diverted to this purpose, amino acids lose their amino groups, whose nitrogen is excreted in the urine as urea or some other degradation product. Their carbon skeletons may be either converted to glucose via gluconeogenesis or oxidized to CO_2 via the TCA cycle. In addition to the dietary amino acids, tissue proteins undergo a continuous metabolic turnover, which contributes to the pool of available amino acids.

Amino acids are also important precursors in many biosynthetic pathways leading to hormones, purines, porphyrins, and other biomolecules.

This chapter will outline the various pathways by which the carbon skeletons of amino acids enter the mainstream of metabolic degradation, as well as the common pathway for the disposal of excess nitrogen as waste products. We shall also describe various biosynthetic pathways for the formation of amino acids and for their conversion into other important biological molecules.

Amino acid metabolism has a particular interest for the physician since a number of hereditary diseases result in aberrant metabolic patterns; several of these produce mental retardation.

20-2 ESSENTIAL AND NONESSENTIAL AMINO ACIDS

While the primary metabolic function of amino acids is to serve as precursors of proteins, their metabolism is also related to the oxidative pathways discussed in chapter 12, so as to provide a supplementary energy source. The chief source of amino acids for mammals is dietary protein, which is hydrolyzed to amino acids by the digestive process. In addition, the mammalian body can synthesize many of the amino acids from other compounds. The capacity to synthesize the remaining amino acids has been lost in the course of evolutionary time, so that they must be supplied in the diet. The two classes of amino acids have accordingly been termed nonessential and essential (section 8-4 and Table 8-1). The deficiency of one or more of the essential amino acids in the diet blocks normal development of the young mammal.

A basic difference exists between mammals and plants or microorganisms in the nature of their controls over the utilization of amino acids. Since the mammals have a relatively restricted ability to synthesize amino acids *de novo*, they are more dependent upon an external supply.

In mammals amino acids are normally absorbed in excess of needs. Some mechanism for the disposal of surplus amino acids is thus necessary. Amino nitrogen is removed by formation and excretion of urea. The carbon chain is eliminated through entry into the common terminal pathways of oxidative metabolism.

20-3 DIGESTION OF PROTEIN

In mammals, dietary protein is largely converted to amino acids by hydrolysis in the gastrointestinal tract. Hydrolysis is catalyzed by proteolytic enzymes, which are secreted in the gastric or pancreatic juice, or are present in the mucosa of the small intestine. With the exception of certain intestinal peptidases, the proteolytic enzymes are synthesized and secreted as zymogens, which are subsequently converted to the active form.

The ingested proteins first encounter the enzyme pepsin in the gastric juices of the stomach, whose pH of 2–3 is optimal for its action. The products of the action of pepsin are a mixture of short polypeptides, which subsequently undergo further hydrolysis in the small intestine.

The pancreatic enzymes present in the small intestine include trypsin, chymotrypsin, elastase, and carboxypeptidase. In addition to these, the intestinal mucosa contains a group of aminopeptidases and dipeptidases. The ultimate product of their combined action is a mixture of free amino acids.

The amino acids are absorbed very rapidly, chiefly in the small intestine. The absorption of amino acids occurs by active transport and requires an energy supply. Some degree of structural specificity is present, the neutral or polar amino acids being absorbed more rapidly than the basic or hydrophobic.

The absorbed amino acids enter the circulation via the portal blood. Oligopeptides may also be absorbed from the small intestine, as may even intact protein under certain circumstances.

20-4 PATHWAYS OF AMINO ACID CONVERSION

The synthesis of proteins from amino acids was discussed in chapter 19. In the balance of the present chapter we shall be concerned with the alternative metabolic fates of the amino acids. These include:

(1) Oxidative deamination to α-keto acids
(2) Transamination to α-keto acids
(3) Decarboxylation
(4) Alteration of the side-chains.

Oxidative Deamination

This type of reaction provides one route to the formation of α-keto acids. The process is equivalent to the removal of two hydrogen atoms from the amino group, followed by its hydrolysis to ammonia and the keto acid:

$$\begin{array}{c} COOH \\ | \\ HC{-}NH_2 \\ | \\ R \end{array} \longrightarrow \begin{array}{c} COOH \\ | \\ C{=}NH \\ | \\ R \end{array} \xrightarrow{H_2O} \begin{array}{c} COOH \\ | \\ C{=}O \\ | \\ R \end{array} + NH_3$$

An important mechanism for oxidative deamination is the reaction catalyzed by the NAD^+-dependent enzyme **glutamate dehydrogenase**, which occurs in the cytoplasm and mitochondria of liver cells.

$$\underset{\text{Glutamic acid}}{\begin{array}{l} COOH \\ | \\ CH_2 \\ | \\ CH_2 \\ | \\ CHNH_2 \\ | \\ COOH \end{array}} + NAD^+ + H_2O \longrightarrow \underset{\alpha\text{-Ketoglutaric acid}}{\begin{array}{l} COOH \\ | \\ CH_2 \\ | \\ CH_2 \\ | \\ C{=}O \\ | \\ COOH \end{array}} + NADH + NH_3$$

Since glutamate dehydrogenase is the only enzyme of high activity involved in amino acid deamination, it is of central importance in this process. The glutamate dehydrogenase of beef liver consists of six identical subunits with a total molecular weight of 336,000.

A mechanism of secondary importance for oxidative deamination is provided by the **amino acid oxidases**, which have flavin coenzymes as prosthetic groups. An FMN-dependent **L-amino acid oxidase** is present in the endoplasmic reticulum of kidneys and liver.

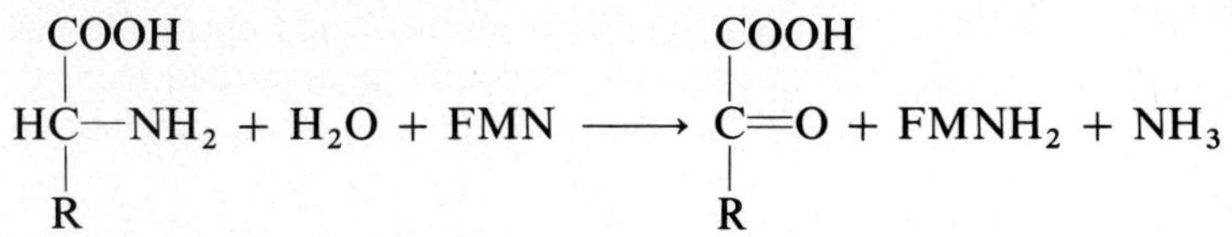

$$\underset{\substack{| \\ R}}{\overset{\substack{COOH \\ |}}{HC}}-NH_2 + H_2O + FMN \longrightarrow \underset{\substack{| \\ R}}{\overset{\substack{COOH \\ |}}{C}}=O + FMNH_2 + NH_3$$

A D-amino acid oxidase, with FAD as prosthetic group, is also present in kidney and liver.

The reduced forms of both enzymes can react with molecular oxygen to form hydrogen peroxide and regenerate the oxidized forms of the enzymes:

$$FMNH_2 + O_2 \rightarrow FMN + H_2O_2$$

$$FADH_2 + O_2 \rightarrow FAD + H_2O_2$$

The hydrogen peroxide is decomposed by **catalase** to H_2O and O_2.

$$H_2O_2 \rightarrow H_2O + \tfrac{1}{2}O_2$$

Transamination

This is the most important pathway for the interconversion and synthesis of the nonessential amino acids. A widely distributed group of enzymes called **transaminases** catalyze reactions of the general type:

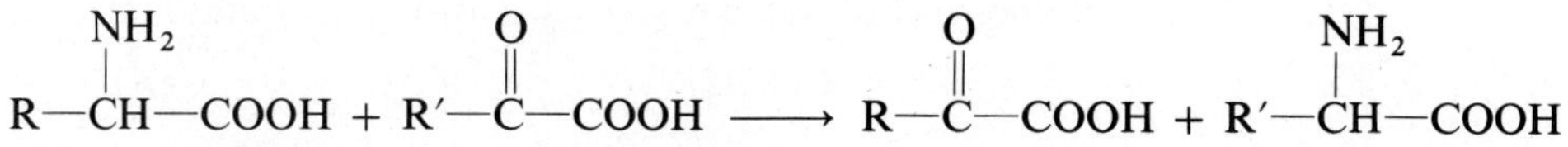

$$R-\overset{\substack{NH_2 \\ |}}{CH}-COOH + R'-\overset{\substack{O \\ \|}}{C}-COOH \longrightarrow R-\overset{\substack{O \\ \|}}{C}-COOH + R'-\overset{\substack{NH_2 \\ |}}{CH}-COOH$$

Two examples of such a reaction are the transfer of an amino group from glutamic acid to oxaloacetic acid and to pyruvic acid, both of which are intermediates of oxidative metabolism:

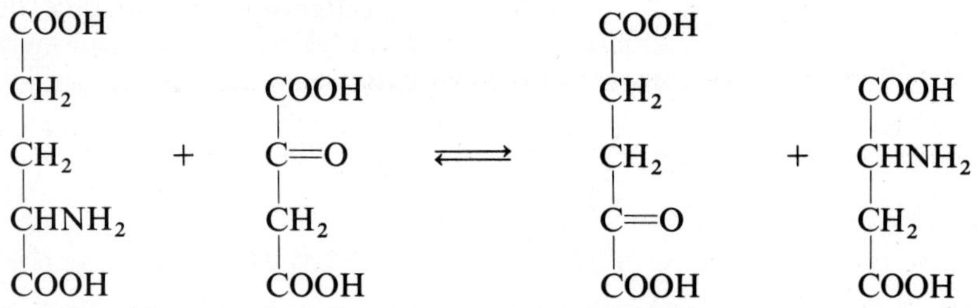

$$\underset{\text{Glutamic acid}}{\begin{matrix} COOH \\ | \\ CH_2 \\ | \\ CH_2 \\ | \\ CHNH_2 \\ | \\ COOH \end{matrix}} + \underset{\text{Oxaloacetic acid}}{\begin{matrix} COOH \\ | \\ C=O \\ | \\ CH_2 \\ | \\ COOH \end{matrix}} \rightleftharpoons \underset{\alpha\text{-Ketoglutaric acid}}{\begin{matrix} COOH \\ | \\ CH_2 \\ | \\ CH_2 \\ | \\ C=O \\ | \\ COOH \end{matrix}} + \underset{\text{Aspartic acid}}{\begin{matrix} COOH \\ | \\ CHNH_2 \\ | \\ CH_2 \\ | \\ COOH \end{matrix}}$$

This reaction is catalyzed by **aspartate transaminase**.

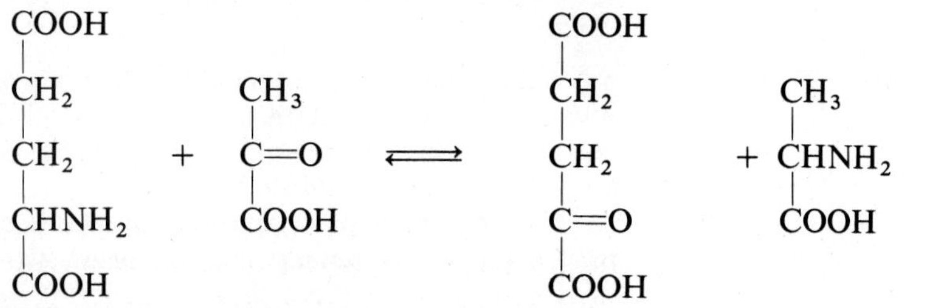

$$\underset{\text{Glutamic acid}}{\begin{matrix} COOH \\ | \\ CH_2 \\ | \\ CH_2 \\ | \\ CHNH_2 \\ | \\ COOH \end{matrix}} + \underset{\text{Pyruvic acid}}{\begin{matrix} CH_3 \\ | \\ C=O \\ | \\ COOH \end{matrix}} \rightleftharpoons \underset{\alpha\text{-Ketoglutaric acid}}{\begin{matrix} COOH \\ | \\ CH_2 \\ | \\ CH_2 \\ | \\ C=O \\ | \\ COOH \end{matrix}} + \underset{\text{Alanine}}{\begin{matrix} CH_3 \\ | \\ CHNH_2 \\ | \\ COOH \end{matrix}}$$

The enzyme involved is **alanine transaminase**.

The transaminases present in liver are capable of transaminating the α-keto acids corresponding to most of the nonessential amino acids. In addition to providing a means for amino acid synthesis, transamination also serves to redistribute nitrogen among amino acids and to correct any imbalance in availability of ingested amino acids. Most of the transaminases require α-ketoglutarate as an amino group acceptor. The specificity for the other substrate is somewhat blurred, although usually one amino acid is preferred.

The α-keto acids are widely available, often, as in the cases of pyruvate and oxaloacetate, arising from carbohydrate oxidation. Because of this and the reversibility of the transamination reactions, the amino acids may to some extent be regarded as forming a common metabolic pool. The administration to rats of an isotopically labeled amino acid is followed by the appearance of the label in all other amino acids.

All of the transaminases utilize **pyridoxal phosphate** (PLP) as a prosthetic group (section 7-2). The reaction involves the formation of a Schiff's base between the aldehyde group of pyridoxal phosphate and the α-amino group of the amino acid:

$$\underset{\text{Amino acid 1}}{R_1\text{-}CH(NH_2)\text{-}COOH} + \underset{\text{PLP}}{O{=}CH\text{-}X} \rightleftharpoons R_1\text{-}CH(COOH)\text{-}N{=}CH\text{-}X \rightleftharpoons R_1\text{-}C(COOH){=}N\text{-}CH_2\text{-}X \underset{}{\overset{H_2O}{\rightleftharpoons}} \underset{\alpha\text{-Keto acid 1}}{R_1\text{-}C({=}O)\text{-}COOH} + \underset{\text{Pyridoxamine}}{H_2N\text{-}CH_2\text{-}X}$$

$$\underset{\alpha\text{-Keto acid 2}}{R_2\text{-}C({=}O)\text{-}COOH} + H_2N\text{-}CH_2\text{-}X \rightleftharpoons R_2\text{-}C(COOH){=}N\text{-}CH_2\text{-}X \rightleftharpoons R_2\text{-}CH(COOH)\text{-}N{=}CH\text{-}X \overset{H_2O}{\rightleftharpoons} \underset{\text{Amino acid 2}}{R_2\text{-}CH(NH_2)\text{-}COOH} + O{=}CH\text{-}X$$

The deamination of amino acid 1 converts pyridoxal phosphate to pyridoxamine. Pyridoxal phosphate is reformed by reaction with an α-keto acid, remaining attached to the same enzyme throughout the process.

The amide derivatives glutamine and asparagine may also participate in transamination:

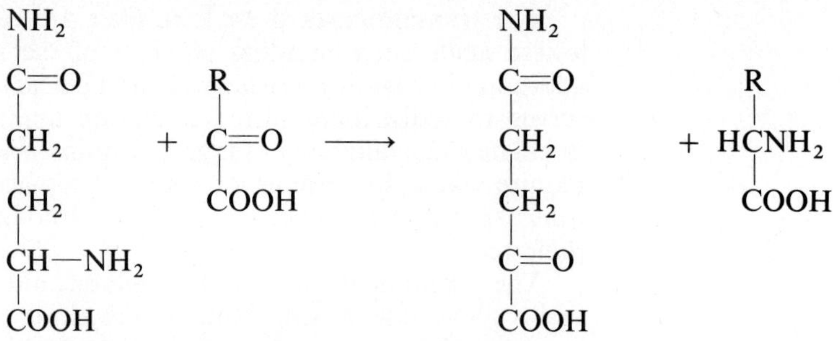

$$\underset{\text{Glutamine}}{\mathrm{H_2N{-}C(=O){-}CH_2{-}CH_2{-}CH(NH_2){-}COOH}} + \underset{\text{Keto acid}}{\mathrm{R{-}C(=O){-}COOH}} \longrightarrow \underset{\alpha\text{-Ketoglutaramic acid}}{\mathrm{H_2N{-}C(=O){-}CH_2{-}CH_2{-}C(=O){-}COOH}} + \mathrm{R{-}CH(NH_2){-}COOH}$$

The ω-amide, α-ketoglutaramic acid, is subsequently hydrolyzed to α-ketoglutaric acid by a **transaminase-amidase**:

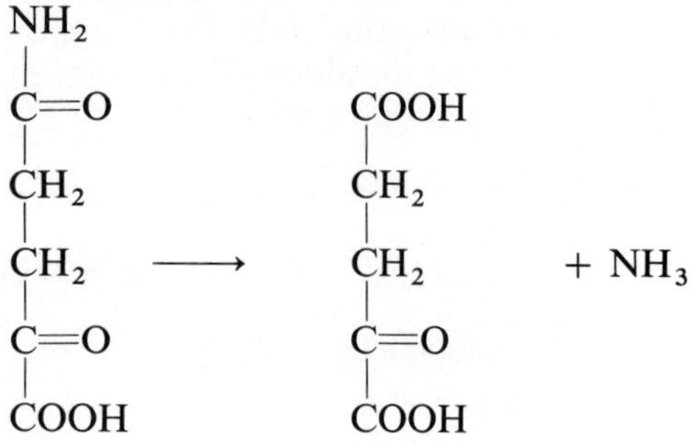

$$\mathrm{H_2N{-}C(=O){-}CH_2{-}CH_2{-}C(=O){-}COOH} \longrightarrow \underset{\alpha\text{-Ketoglutaric acid}}{\mathrm{HOOC{-}CH_2{-}CH_2{-}C(=O){-}COOH}} + \mathrm{NH_3}$$

Decarboxylation Several amino acids, including arginine, histidine, and tryptophan may undergo decarboxylation in animal tissues or microorganisms. The processes are catalyzed by specific **decarboxylases**, which, like the transaminases, contain PLP as a coenzyme. The reaction again probably proceeds by way of Schiff's base formation between the aldehyde group of PLP and the amino group. This is followed by decarboxylation to yield finally a primary amine:

$$\mathrm{R{-}CH_2{-}CHNH_2{-}COOH \rightarrow R{-}CH_2{-}CH_2NH_2 + CO_2}$$

Alteration of the Side-Chain This pathway is important only for certain amino acids, which will be discussed individually.

20-5 THE UREA CYCLE

The most efficient of the deamination reactions, that of L-glutamic acid as catalyzed by glutamate dehydrogenase, yields ammonia as a product. Because of the dynamic equilibrium between different amino acids, this reaction provides a potential mechanism for the elimination of surplus amino nitrogen.

However, ammonia is not a suitable form for the excretion of nitrogen by mammals. Since ammonia is a cellular poison even at low concentrations, some mechanism for its removal is necessary.

In mammals, ammonia is converted to urea by a cyclic process. The overall reaction, which is endergonic, is equivalent to the combination of two molecules of ammonia with a molecule of carbon dioxide to yield urea plus H_2O:

$$CO_2 + 2NH_3 \longrightarrow O{=}C(NH_2)_2 + H_2O$$

For physiological concentrations, ΔG is close to 14 kcal. Such a reaction could never occur spontaneously for an isolated system and must proceed via coupling with one or more exergonic processes.

The entry of ammonia into the urea cycle (Fig. 20-1) occurs via formation of the unstable high energy compound **carbamoyl phosphate**. In mammalian liver the reaction, which consumes two molecules of ATP, is catalyzed by **carbamoyl phosphate synthetase**.

$$H_2O + CO_2 + NH_3 + 2ATP \rightarrow H_2N{-}CO{-}OPO_3^{2-} + 2ADP + P_i + H^+$$

The reaction as written is composite, representing the sum of two coupled processes, one of which is the hydrolysis of ATP. The ammonia is generated by the oxidative deamination of glutamic acid (section 20-3) in liver mitochondria. The formation of carbamoyl phosphate occurs in the mitochondrial matrix. The enzyme has an absolute requirement for N-acetylglutamic acid, which is an allosteric activator:

$$HOOC{-}CH_2{-}CH_2{-}CH(NH{-}CO{-}CH_3){-}COOH$$

N-Acetylglutamic acid

Carbamoyl phosphate reacts with the δ-amino group of ornithine to form **citrulline** (Fig. 20-1), in a reaction catalyzed by **ornithine carbamoyltransferase** of the mitochondrial matrix.

$$\underset{\text{Carbamoyl phosphate}}{H_2N{-}CO{-}OPO_3^{2-}} + \underset{\text{Ornithine}}{H_2N{-}(CH_2)_3{-}CH(NH_2){-}COOH} \longrightarrow$$

$$\underset{\text{Citrulline}}{H_2N{-}CO{-}NH{-}(CH_2)_3{-}CH(NH_2){-}COOH} + P_i$$

Citrulline is next transformed to arginine in two steps. Prior to these, citrulline leaves the mitochondria and enters the surrounding cytoplasm, where the remaining reactions of the urea cycle occur. In the first reaction, citrulline is condensed with aspartic acid, with the consumption of a mole of ATP, to yield **argininosuccinic acid** (Fig. 20-2), in a reaction catalyzed by

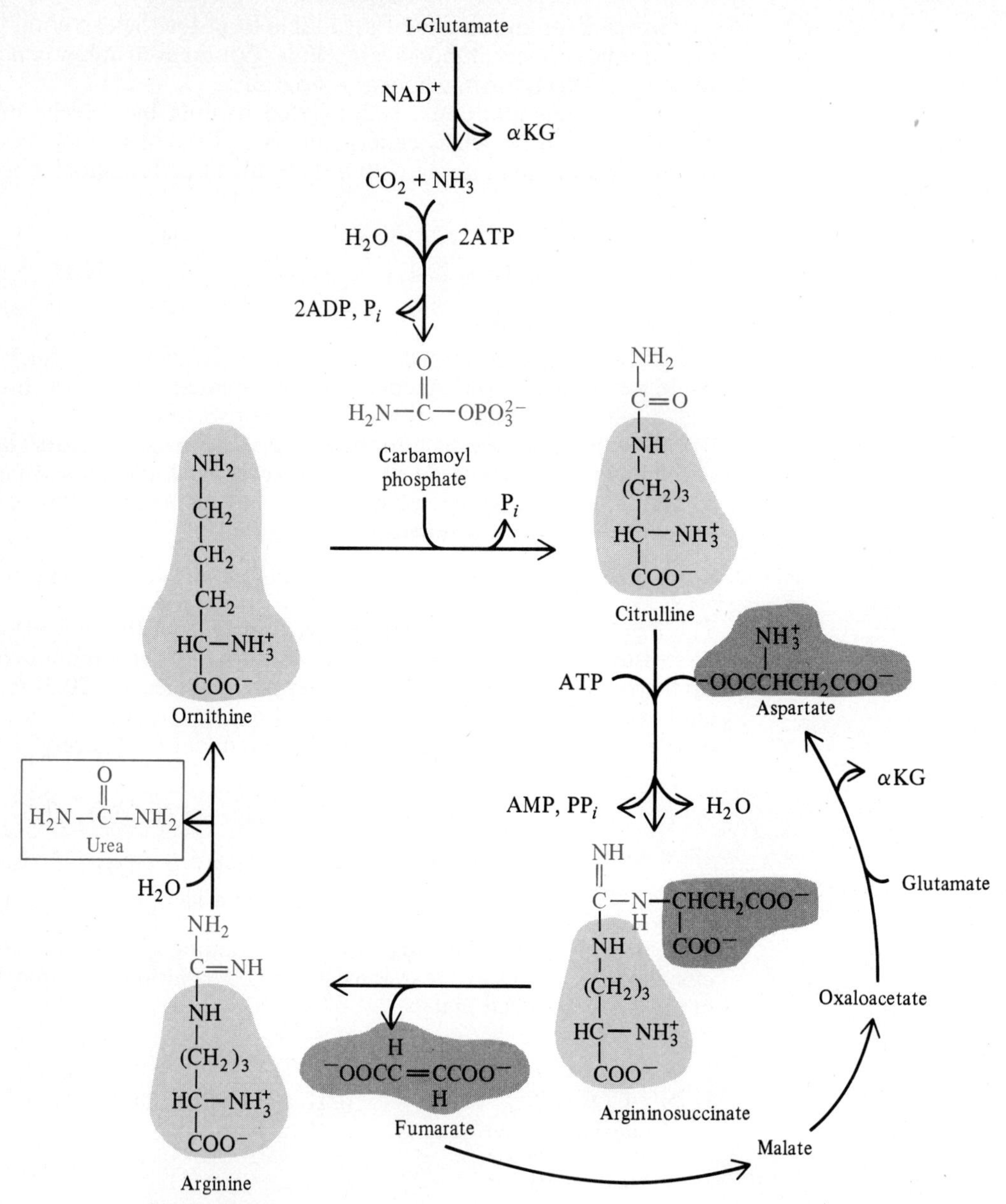

Figure 20-1
Mechanism of the urea cycle.

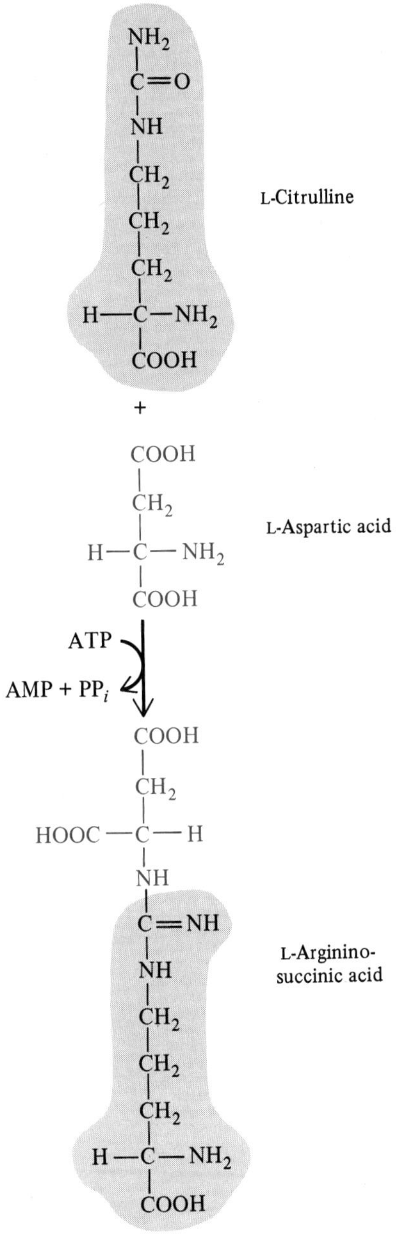

$$\begin{array}{c} NH_2 \\ | \\ C{=}O \\ | \\ NH \\ | \\ CH_2 \\ | \\ CH_2 \\ | \\ CH_2 \\ | \\ H{-}C{-}NH_2 \\ | \\ COOH \end{array}$$

L-Citrulline

+

$$\begin{array}{c} COOH \\ | \\ CH_2 \\ | \\ H{-}C{-}NH_2 \\ | \\ COOH \end{array}$$

L-Aspartic acid

ATP → AMP + PP_i

$$\begin{array}{c} COOH \\ | \\ CH_2 \\ | \\ HOOC{-}C{-}H \\ | \\ NH \\ | \\ C{=}NH \\ | \\ NH \\ | \\ CH_2 \\ | \\ CH_2 \\ | \\ CH_2 \\ | \\ H{-}C{-}NH_2 \\ | \\ COOH \end{array}$$

L-Argininosuccinic acid

Figure 20-2
Condensation of citrulline with aspartic acid.

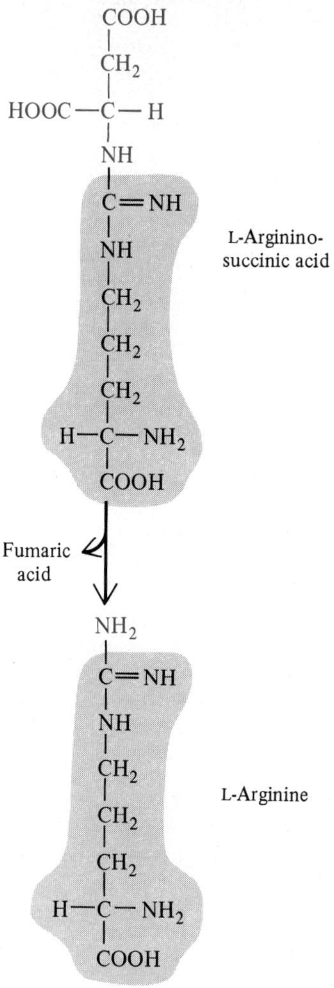

L-Arginino-succinic acid

Fumaric acid

L-Arginine

Figure 20-3
Formation of arginine.

argininosuccinate synthetase:

$$\text{citrulline} + \text{aspartate} + \text{ATP} \rightleftarrows \text{argininosuccinate} + \text{AMP} + \text{PP}_i$$

The pyrophosphate is hydrolyzed by a pyrophosphatase to inorganic phosphate, thereby displacing the reaction in favor of products.

In the second reaction, argininosuccinate is decomposed to arginine and fumarate by the action of **argininosuccinate lyase** to form free arginine and fumarate (Fig. 20-3).

$$\text{argininosuccinate} \rightarrow \text{arginine} + \text{fumarate}$$

Urea arises by the hydrolysis of arginine as catalyzed by **arginase**; ornithine, a second product of the reaction, is itself a precursor of arginine, thereby completing the cycle:

$$H_2N-\overset{\overset{\displaystyle NH}{\|}}{C}-NH-(CH_2)_3-\overset{\overset{\displaystyle NH_2}{|}}{C}H-COOH + H_2O \longrightarrow$$
Arginine

$$H_2N-(CH_2)_3-\overset{\overset{\displaystyle NH_2}{|}}{C}H-COOH + O{=}C\begin{matrix} \diagup NH_2 \\ \diagdown NH_2 \end{matrix}$$
Ornithine Urea

The synthesis of urea is "paid for" by the hydrolysis of 3 moles of ATP. The overall reaction is

$$2NH_3 + CO_2 + 3ATP + 3H_2O \rightarrow \text{urea} + 2ADP + AMP + 4P_i$$

The liver is the most important site of urea formation, although minor quantities are synthesized in other organs.

Not all animals eliminate waste nitrogen as urea. Birds and many reptiles use uric acid as a vehicle for disposal. Other species, including the teleost fishes, dispose of excess nitrogen as ammonia in the excreta.

20-6 METABOLIC DEGRADATION OF INDIVIDUAL AMINO ACIDS

The carbon skeletons of the amino acids enter the general pathways of oxidative metabolism by way of their α-keto derivatives arising from transamination. As such they are ultimately burned to CO_2 and H_2O. The present discussion follows the conversions of amino acids only to the point where intermediates common to the metabolism of all foodstuffs are formed.

Acetyl-CoA-Forming Amino Acids

The oxidative degradation pathways of 11 amino acids converge in acetyl-CoA, which is the common point of entry into the TCA cycle. Two of these—threonine and leucine—yield acetyl-CoA directly; five amino acids—leucine, lysine, phenylalanine, trytophan, and tyrosine—yield acetoacetyl-CoA as an intermediate; while five amino acids—alanine, cysteine, glycine, serine, and threonine—are converted to acetyl-CoA by way of pyruvate.

Threonine and Leucine

The principal degradative pathway for threonine involves its cleavage to acetaldehyde and glycine by **serine hydroxymethyltransferase**. The acetaldehyde is converted to acetyl-CoA and the glycine enters the common pathways for this amino acid.

$$\underset{\text{Threonine}}{CH_2CHOHCHNH_2COOH} \longrightarrow \underset{\text{Glycine}}{CH_2NH_2COOH} + CH_3CHO \xrightarrow{NAD^+,\ CoA} CH_3CO-S-CoA$$

Two of the six carbon atoms of leucine yield acetyl-CoA directly; the remaining four carbon atoms are converted to acetoacetic acid (Fig. 20-4). Transamination of the α-amino group, followed by oxidative decarboxylation of the resultant α-keto acid, yields isovaleryl-CoA. Dehydrogenation

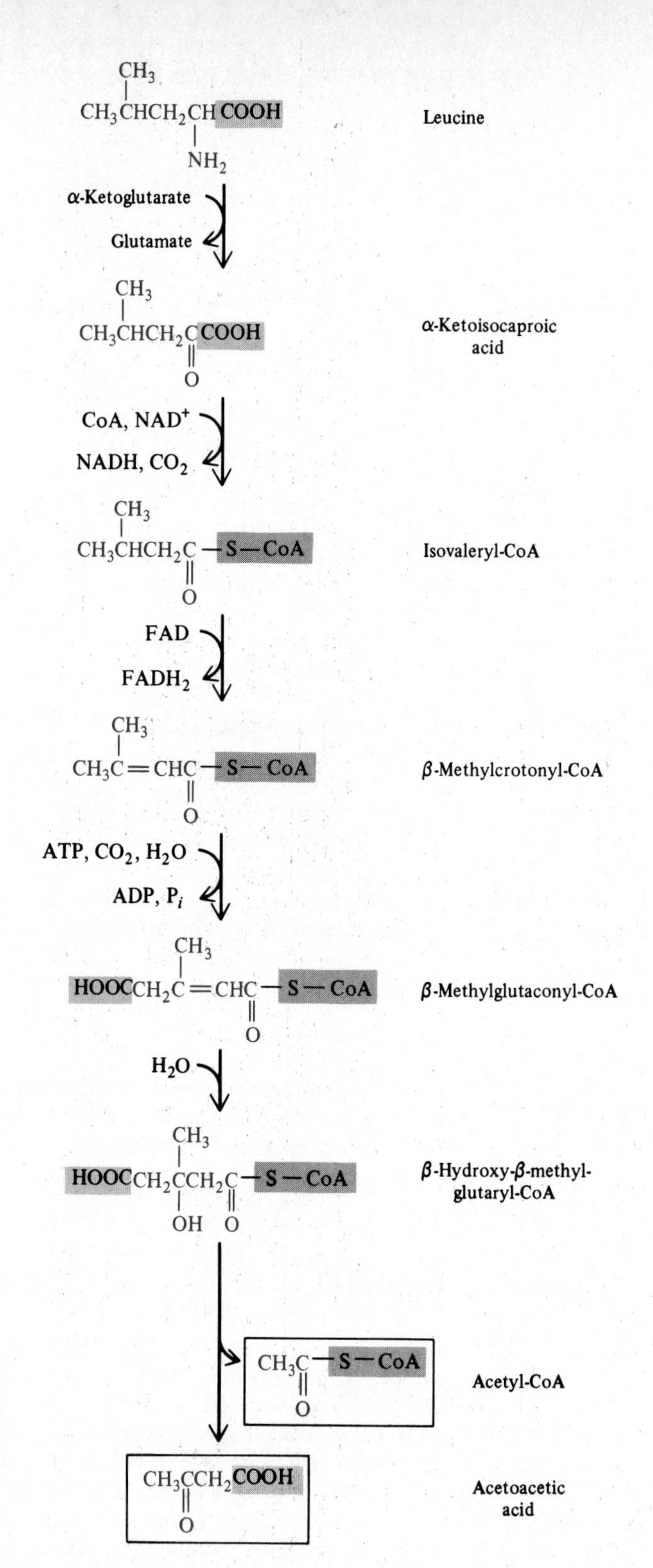

Figure 20-4 Degradation of leucine to acetyl CoA and acetoacetic acid. The acetoacetic acid subsequently reacts with succinyl CoA to form acetoacetyl CoA.

and subsequent carboxylation of the latter forms the 6-carbon intermediate **β-hydroxy-β-methylglutaryl-CoA**. This is in turn split to produce acetyl-CoA and acetoacetate; the latter is converted to acetoacetyl-CoA by reaction with succinyl-CoA, as catalyzed by 3-ketoacid-CoA transferase:

$$
\begin{array}{ccccccc}
\mathrm{COOH} & & \mathrm{CH_3} & & \mathrm{COOH} & & \mathrm{CH_3} \\
| & & | & & | & & | \\
\mathrm{CH_2} & + & \mathrm{C{=}O} & \longrightarrow & \mathrm{CH_2} & + & \mathrm{C{=}O} \\
| & & | & & | & & | \\
\mathrm{CH_2} & & \mathrm{CH_2} & & \mathrm{CH_2} & & \mathrm{CH_2} \\
| & & | & & | & & | \\
\mathrm{C{-}S{-}CoA} & & \mathrm{COOH} & & \mathrm{COOH} & & \mathrm{C{=}O} \\
\| & & & & & & | \\
\mathrm{O} & & & & & & \mathrm{S{-}CoA} \\
\text{Succinyl CoA} & & \text{Acetoacetic acid} & & \text{Succinic acid} & & \text{Acetoacetyl-CoA}
\end{array}
$$

The β-hydroxy-β-methylglutaryl-CoA is an intermediate in the biosynthesis of cholesterol (section 16-6).

A rare hereditary disease arises from a biochemical defect in the enzyme catalyzing the oxidative decarboxylation of the branched chain α-keto acids (Fig. 20-4) arising from the transamination of leucine, as well as from parallel reactions for valine and isoleucine (Fig. 20-16). This condition, which is called the **maple syrup disease** from the characteristic odor of the urine and perspiration of its victims, is fatal early in life if untreated. A low protein diet, supplemented with essential amino acids, may permit survival of patients. The maple syrup odor stems from decomposition products of the α-keto acids, which accumulate and are present in high levels in the urine.

A second exotic disease arising from interference with leucine metabolism is the **Jamaican vomiting sickness**, which is caused by eating unripe ackee fruit. The latter contains a toxin which inhibits isovaleryl-CoA dehydrogenase and causes an accumulation of isovaleric acid in the blood. Death from this very lethal illness is believed to come from a resultant drastic drop in blood glucose level.

Lysine, Phenylalanine, Tryptophan, and Tyrosine

All of these amino acids are degraded to acetoacetate, which may be converted to acetoacetyl-CoA, as described above. Lysine, which does not undergo transamination, is first converted to L-α-aminoadipic semialdehyde by either of two alternative pathways (Fig. 20-5); the latter is transformed to acetoacetyl-CoA by a series of reactions.

The degradation pathways of tyrosine and phenylalanine converge as the latter is initially converted to tyrosine (Figs. 20-6 and 20-7). The remaining steps of the pathway convert the rest of the molecule into acetoacetate and fumarate, which lies on the TCA cycle.

The conversion of phenylalanine to tyrosine involves the incorporation of an oxygen atom from molecular oxygen in a reaction catalyzed by **phenylalanine-4-monooxygenase**. When this enzyme is absent as a result of a genetic defect, phenylalanine is degraded by an alternative, normally unused pathway to yield **phenylpyruvate** via transamination.

This proved to be the origin of a cruel and baffling hereditary disorder afflicting newborn children. The affected child at first appears normal but rapidly develops the symptoms of mental retardation. This disease, which

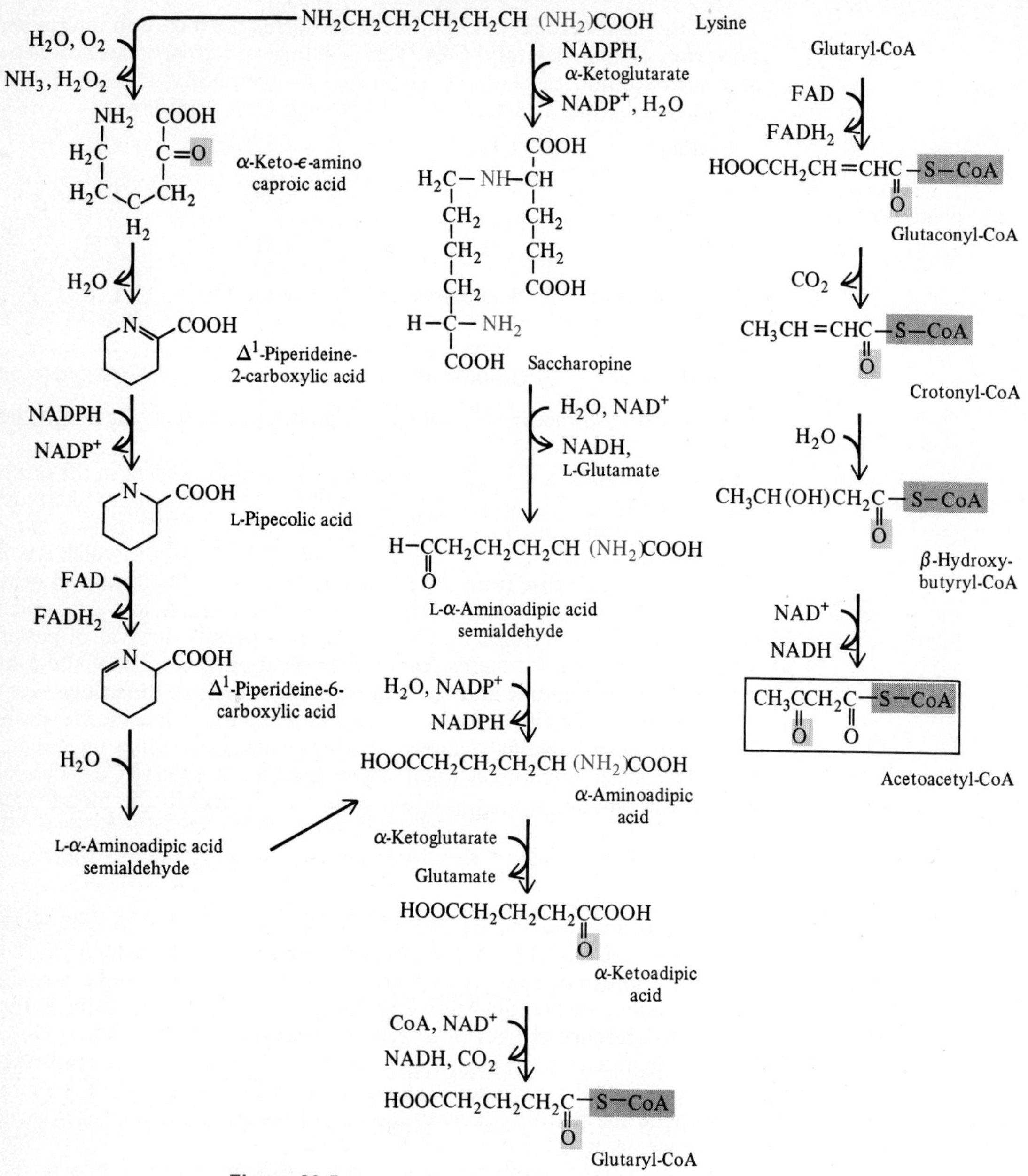

Figure 20-5
Degradation of lysine to acetoacetyl CoA.

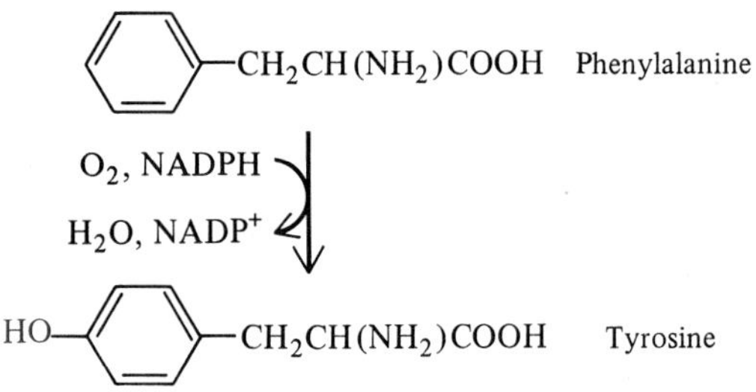

$$\text{C}_6\text{H}_5\text{—CH}_2\text{CH(NH}_2\text{)COOH} \quad \text{Phenylalanine}$$

$$\xrightarrow[\text{H}_2\text{O, NADP}^+]{\text{O}_2\text{, NADPH}}$$

$$\text{HO—C}_6\text{H}_4\text{—CH}_2\text{CH(NH}_2\text{)COOH} \quad \text{Tyrosine}$$

Figure 20-6 Conversion of phenylalanine to tyrosine.

is now known as **phenylketonuria** (PKU), arises from the accumulation of phenylpyruvate to excessive levels, causing brain damage. It has been treated by rearing children on a special low-phenylalanine diet.

The oxidative degradation of tryptophan yields one molecule each of acetyl-CoA and acetoacetyl-CoA; the remaining carbon atoms appear as four molecules of CO_2 and one of formate (Fig. 20-8). The initial step is the opening of the indole ring by reaction with molecular oxygen, in a process catalyzed by **tryptophan pyrrolase**. A genetic deficiency that abolishes or impairs the activity of this enzyme can result in mental retardation.

The α-ketoadipic acid formed in the degradation of tryptophan (Fig. 20-8) is also an intermediate in the lysine pathway; the two pathways therefore converge in the later stages.

Alanine, Cysteine, Glycine, and Serine

These four amino acids are degraded to pyruvate; this may either be converted to acetyl-CoA and enter the TCA cycle or may be converted to glucose (section 12-5) (Fig. 20-9).

Alanine is converted to pyruvate directly by transamination with α-ketoglutarate:

$$\text{alanine} + \alpha\text{-ketoglutarate} \rightleftharpoons \text{pyruvate} + \text{glutamate}$$

The end products of cysteine degradation are pyruvate and inorganic sulfate. In one major pathway cysteine is oxidized to cysteine sulfinic acid in a reaction catalyzed by **cysteine dioxygenase**; molecular oxygen is involved.

$$\underset{\text{Cysteine}}{\text{HS—CH}_2\text{—}\overset{\displaystyle\text{NH}_2}{\overset{|}{\text{CH}}}\text{—COOH}} \xrightarrow[\text{NAD}^+]{\text{O}_2} \underset{\text{Cysteine sulfinic acid}}{\text{HO}_2\text{S—CH}_2\text{—}\overset{\displaystyle\text{NH}_2}{\overset{|}{\text{CH}}}\text{—COOH}}$$

Cysteine sulfinic acid is converted by transamination with α-ketoglutarate to β-sulfinylpyruvate; enzymic removal of the sulfinyl group yields pyruvate and inorganic sulfite; the latter is further oxidized to sulfate, which is excreted.

$$\text{HO}_2\text{S—CH}_2\text{—}\overset{\displaystyle\text{NH}_2}{\overset{|}{\text{CH}}}\text{—COOH} \xrightarrow{\text{transamination}}$$

$$\underset{\beta\text{-Sulfinyl pyruvate}}{\text{HO}_2\text{S—CH}_2\text{—}\overset{\displaystyle\text{O}}{\overset{\|}{\text{C}}}\text{—COOH}} \longrightarrow \text{SO}_3^{2-} + \text{CH}_3\text{COCOOH} \longrightarrow \text{SO}_4^{2-}$$

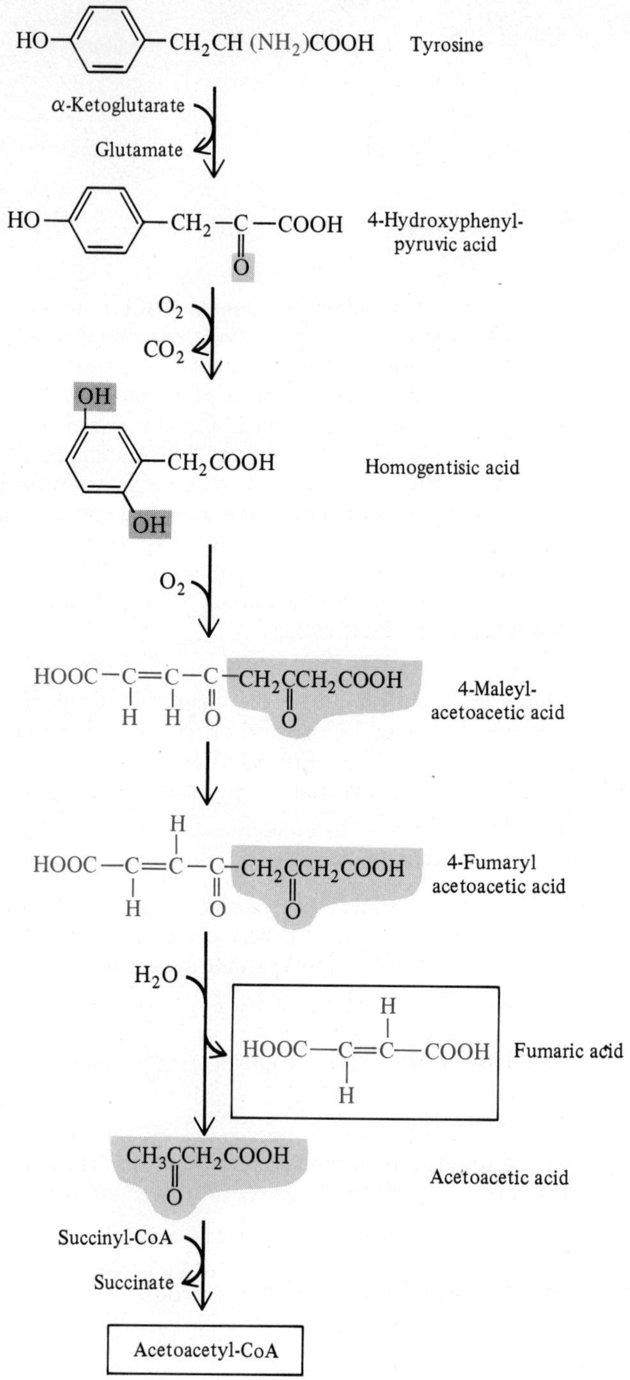

Figure 20-7
Degradation of tyrosine to fumaric and acetoacetic acids.

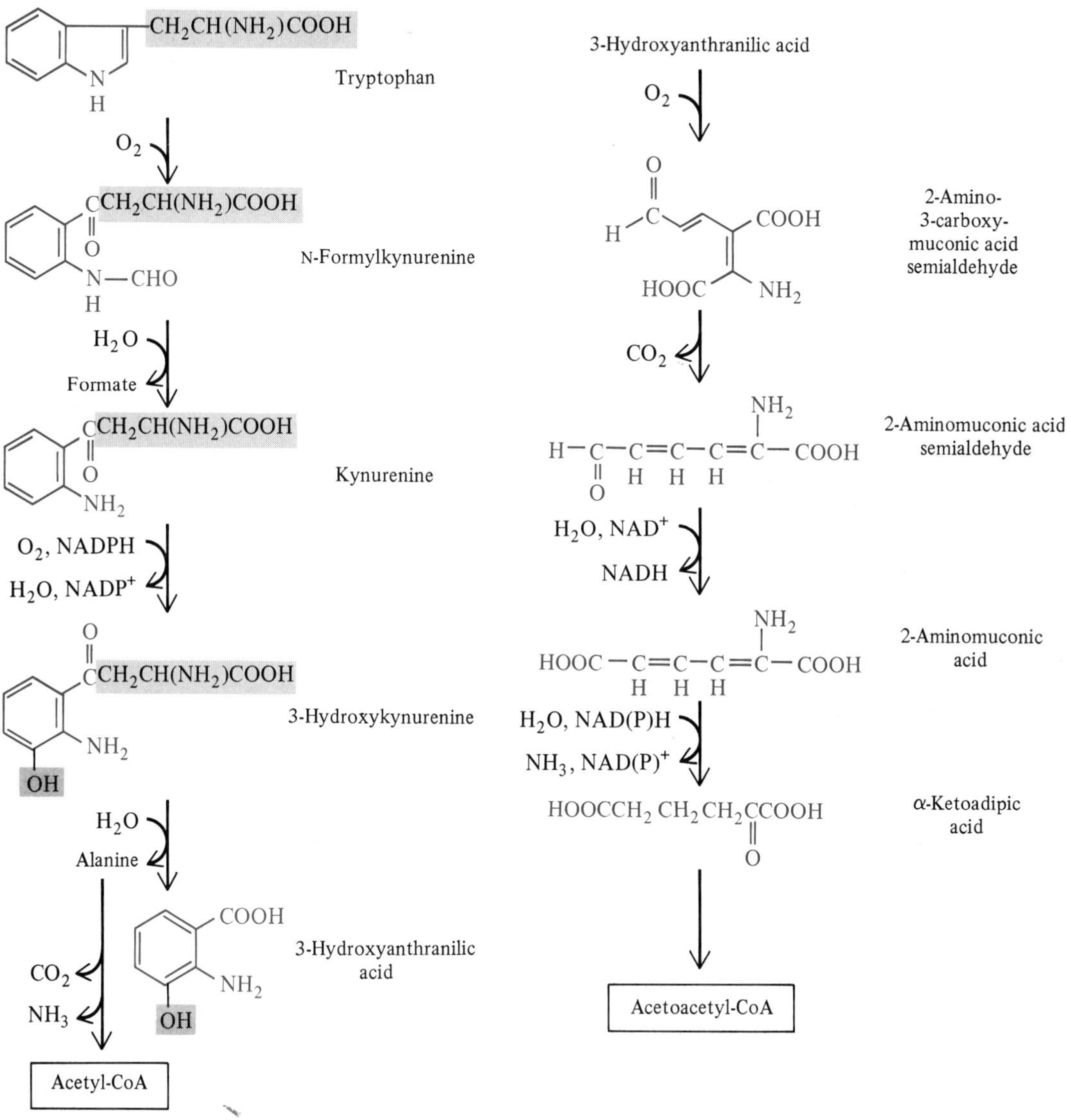

Tryptophan (indole ring–$CH_2CH(NH_2)COOH$)

O_2 ↓

N-Formylkynurenine ($C(=O)CH_2CH(NH_2)COOH$; N—CHO, H)

H_2O ↓ Formate

Kynurenine ($C(=O)CH_2CH(NH_2)COOH$; NH_2)

O_2, NADPH; H_2O, $NADP^+$ ↓

3-Hydroxykynurenine ($C(=O)CH_2CH(NH_2)COOH$; NH_2; OH)

H_2O ↓ Alanine

3-Hydroxyanthranilic acid (COOH; NH_2; OH) — (CO_2, NH_3) → Acetyl-CoA

3-Hydroxyanthranilic acid

O_2 ↓

2-Amino-3-carboxy-muconic acid semialdehyde (H—C(=O)—CH=CH—C(COOH)=C(NH_2)—COOH: HOOC, COOH, NH_2)

CO_2 ↓

2-Aminomuconic acid semialdehyde

$$\mathrm{H{-}C(=O){-}CH{=}CH{-}CH{=}C(NH_2){-}COOH}$$

H_2O, NAD^+ ↓ NADH

2-Aminomuconic acid

$$\mathrm{HOOC{-}CH{=}CH{-}CH{=}C(NH_2){-}COOH}$$

H_2O, NAD(P)H ↓ NH_3, $NAD(P)^+$

α-Ketoadipic acid

$$\mathrm{HOOCCH_2\,CH_2CH_2CCOOH}\ (\mathrm{C{=}O})$$

↓

Acetoacetyl-CoA

Figure 20-8
Degradation of tryptophan to acetyl CoA and acetoacetyl CoA.

Serine is converted by consecutive dehydration and deamination to pyruvate; the reaction is catalyzed by **serine hydratase**, which contains pyridoxal phosphate as a prosthetic group.

$$\underset{\text{Serine}}{\mathrm{HOCH_2{-}CH(NH_2){-}COOH}} \longrightarrow \mathrm{CH_2COCOOH + NH_3}$$

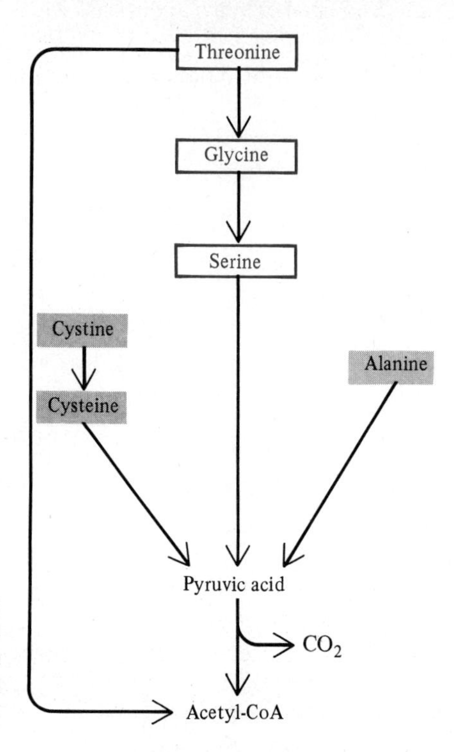

Figure 20-9 Convergence of the degradative pathways of several amino acids in pyruvic acid.

The degradation of glycine is partially channeled through serine, with which it is metabolically interconvertible (Fig. 20-10), by the action of **serine hydroxymethyltransferase**. The major pathway is by oxidative cleavage to form CO_2, NH_3, and N^5,N^{10}-methylenetetrahydrofolate (section 7-4); the enzyme is **glycine synthase**.

$$CH_2NH_2COOH + FH_4 + NAD^+ \rightleftharpoons N^5,N^{10}\text{-methylene-}FH_4 + CO_2 + NH_3 + NADH + H^+$$

where FH_4 stands for tetrahydrofolate.

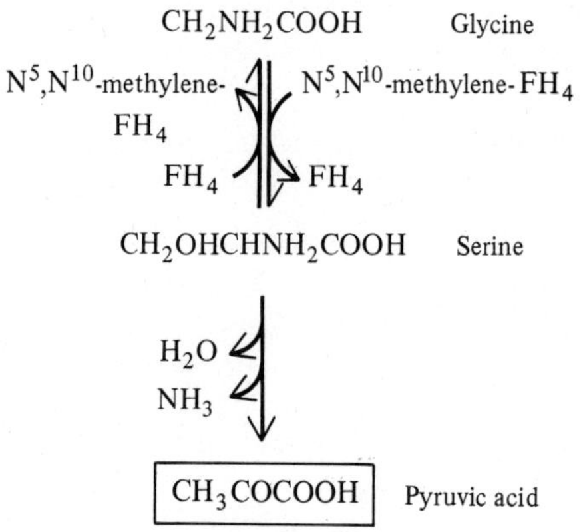

CH_2NH_2COOH Glycine

N^5,N^{10}-methylene-FH_4 / N^5,N^{10}-methylene-FH_4

FH_4 / FH_4

$CH_2OHCHNH_2COOH$ Serine

H_2O

NH_3

$CH_3COCOOH$ Pyruvic acid

Figure 20-10 Conversion of glycine to serine and the subsequent conversion of the latter to pyruvate.

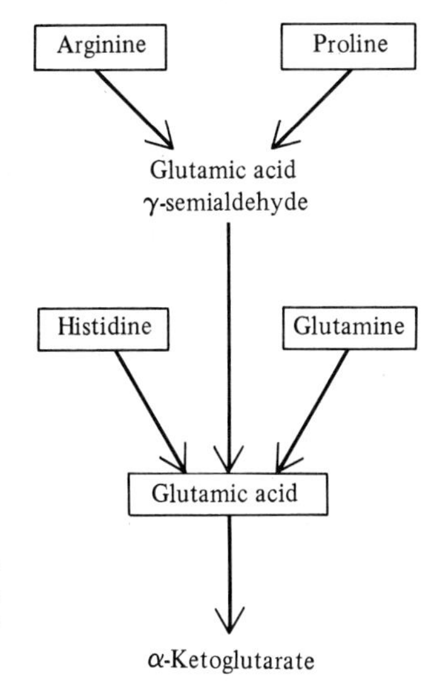

Figure 20-11 Degradation of five amino acids to α-ketoglutarate by way of glutamic acid.

α-Ketoglutarate-Forming Amino Acids

Five amino acids—arginine, glutamic acid, glutamine, histidine, and proline—enter the TCA cycle by way of α-ketoglutarate, their common degradation product. All five are converted to α-ketoglutarate by way of glutamate (Fig. 20-11).

In mammalian liver arginine is converted to ornithine and urea by the action of arginase (section 20-4); this reaction is a part of the urea cycle. Ornithine is next transaminated in the delta (δ) position to glutamic acid semialdehyde, which is subsequently transformed to glutamic acid, with whose pathway it merges (Fig. 20-11 and 20-12).

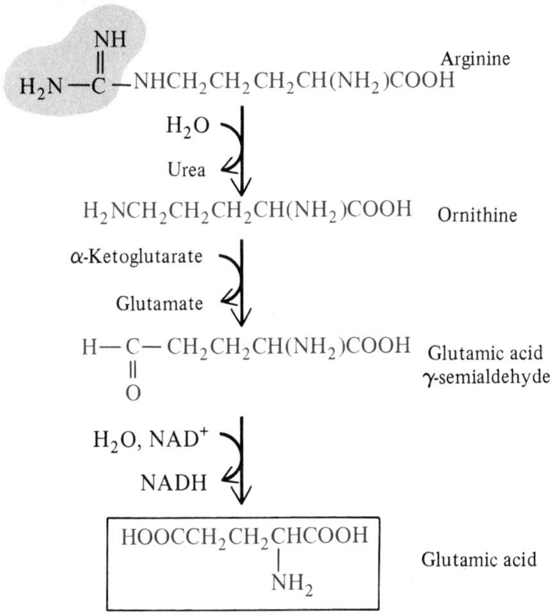

$$H_2N{-}\overset{\overset{NH}{\|}}{C}{-}NHCH_2CH_2CH_2CH(NH_2)COOH \quad \text{Arginine}$$

$$\downarrow \; H_2O \rightarrow \text{Urea}$$

$$H_2NCH_2CH_2CH_2CH(NH_2)COOH \quad \text{Ornithine}$$

$$\downarrow \; \alpha\text{-Ketoglutarate} \rightarrow \text{Glutamate}$$

$$H{-}\underset{\underset{O}{\|}}{C}{-}CH_2CH_2CH(NH_2)COOH \quad \text{Glutamic acid } \gamma\text{-semialdehyde}$$

$$\downarrow \; H_2O, NAD^+ \rightarrow NADH$$

$$HOOCCH_2CH_2\underset{\underset{NH_2}{|}}{C}HCOOH \quad \text{Glutamic acid}$$

Figure 20-12 Degradation of arginine to glutamic acid.

In the kidney, glutamine is converted by hydrolysis to glutamic acid and ammonia in a reaction catalyzed by **glutaminase**:

$$\text{glutamine} + H_2O \rightarrow \text{glutamate} + NH_3$$

An alternative mechanism is via the action of the NADPH-dependent enzyme **glutamate synthase**.

$$\text{glutamine} + \alpha\text{-ketoglutarate} + \text{NADPH} + H^+ \rightarrow 2\ \text{glutamate} + \text{NADP}^+$$

The glutamic acid formed in this way enters the normal pathway for this amino acid, being converted to α-ketoglutarate by the action of glutamate dehydrogenase.

Proline undergoes an initial dehydrogenation, followed by a spontaneous ring opening to yield glutamic acid γ-semialdehyde, which is converted to glutamic acid (Fig. 20-13).

The degradation of histidine occurs by an opening of the imidazole ring to yield N-formiminoglutamic acid (Fig. 20-14). The formimino (HN=CH—) group is removed enzymically and transferred to tetrahydrofolate, which acts as a 1-carbon group acceptor to form formimino-tetrahydrofolate.

Succinyl-CoA-Forming Amino Acids

The carbon skeletons of three amino acids—isoleucine, methionine, and valine—enter the TCA cycle by way of succinyl-CoA, which is deacylated to yield succinate (Fig. 20-15).

Isoleucine and valine follow similar pathways of degradation. Both are initially transaminated to the corresponding α-keto acids, which subsequently undergo oxidative decarboxylation (Fig. 20-16) with the loss of CO_2. Further

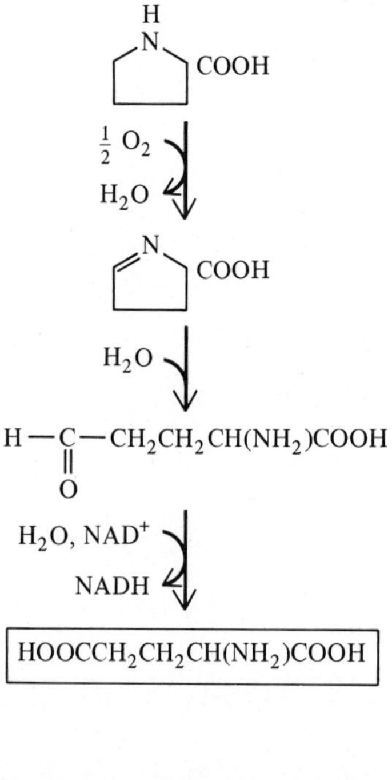

H N COOH — Proline

$\frac{1}{2} O_2$ ↓ H_2O

N COOH — Δ^1-Pyrroline 5-carboxylic acid

H_2O ↓

$H-\underset{\underset{O}{\|}}{C}-CH_2CH_2CH(NH_2)COOH$ — Glutamic acid γ-semialdehyde

H_2O, NAD^+ ↓ NADH

$HOOCCH_2CH_2CH(NH_2)COOH$ — Glutamic acid

Figure 20-13 Conversion of proline to glutamic acid.

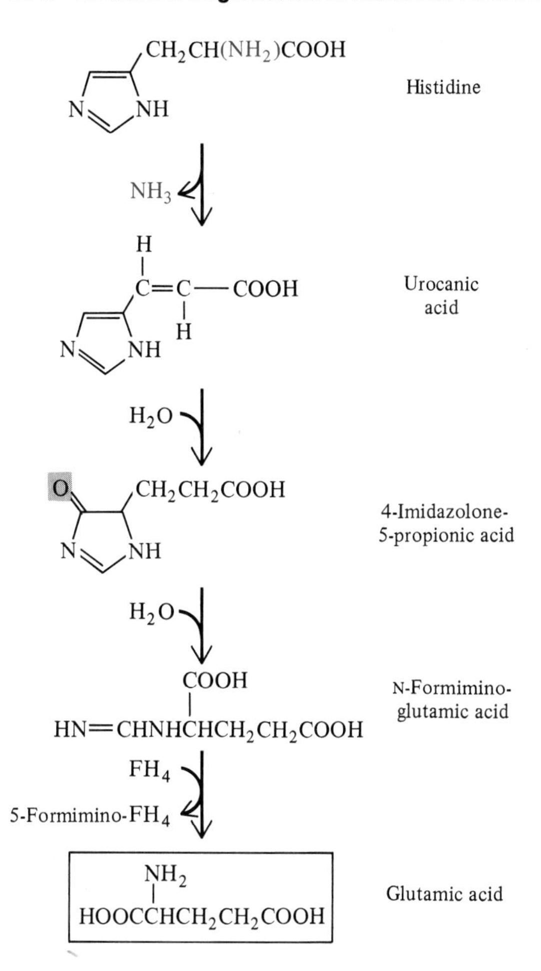

Figure 20-14
Degradation of histidine to glutamic acid.

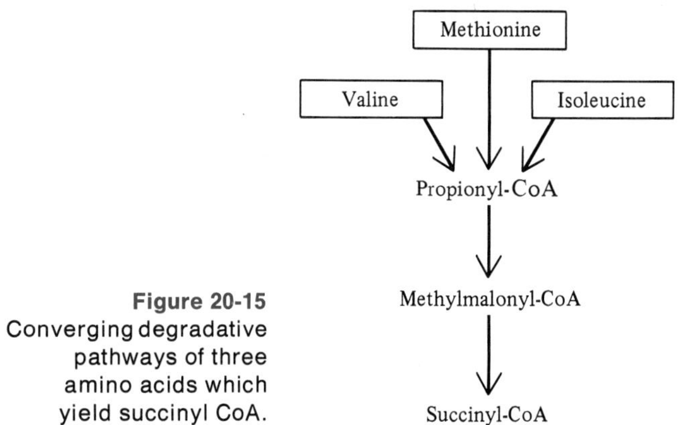

Figure 20-15
Converging degradative pathways of three amino acids which yield succinyl CoA.

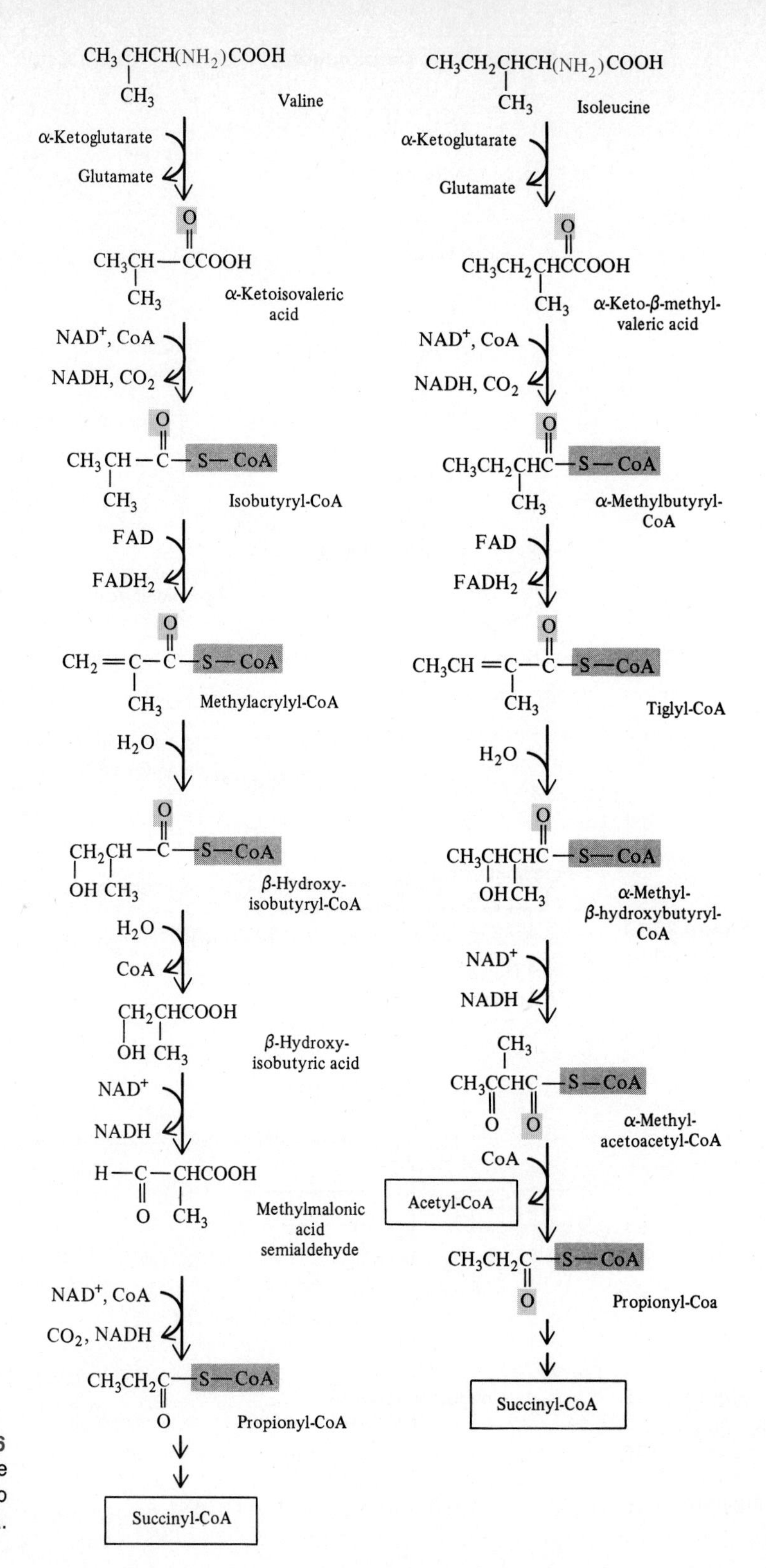

Figure 20-16
Degradation of valine and isoleucine to succinyl CoA.

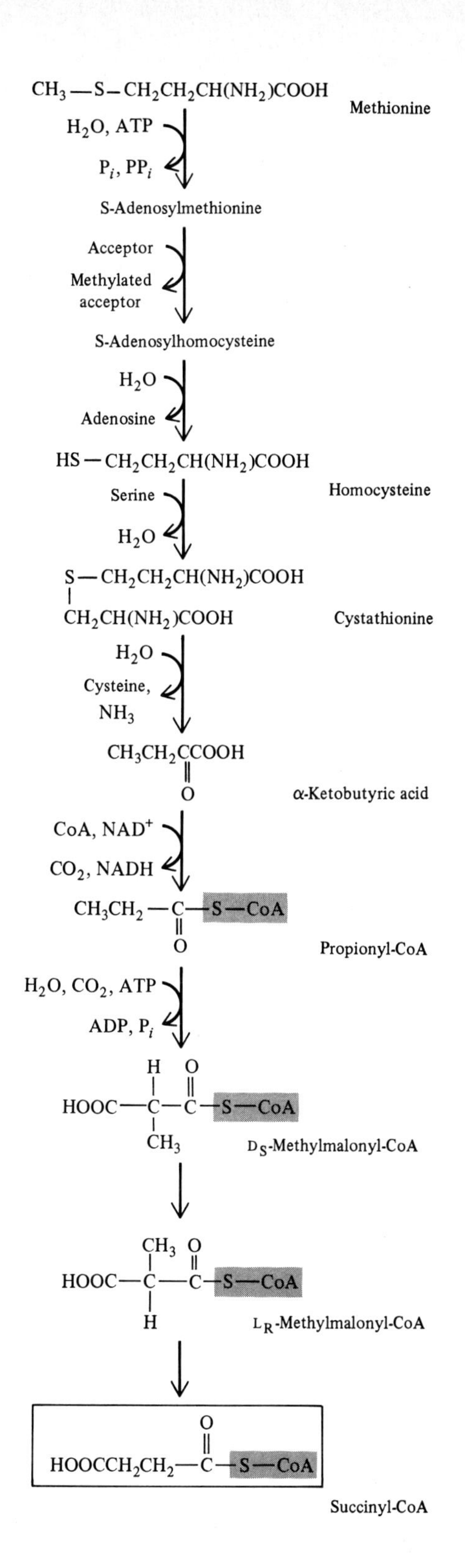

Figure 20-17 Degradation of methionine to succinyl CoA.

degradation of the side-chains yields ultimately propionyl-CoA, in which the pathways of this group of amino acids merge. Propionyl-CoA is carboxylated to form methylmalonyl-CoA (section 13-6), which is converted to succinyl-CoA.

Methionine is converted initially to **homocysteine** (Fig. 20-17) by way of S-adenosylmethionine (section 7-2) as intermediate. Homocysteine combines with serine to form **cystathionine**, which is decomposed to yield α-ketobutyrate, ammonia, and cysteine. Oxidation decarboxylation of α-ketobutyrate produces propionyl-CoA.

The initial phase of methionine degradation is important because of its involvement in many biological methylation reactions. Methionine is activated by reaction with ATP to form S-adenosylmethionine (section 7-2); this contains a high energy methylsulfonium group. The reaction is catalyzed by **methionine adenosyl transferase**.

$$\text{ATP} + \text{L-methionine} \rightarrow \text{S-adenosylmethionine} + \text{PP}_i + \text{P}_i$$

S-adenosylmethionine is a methyl group donor for a large number of methylation reactions involving a diverse group of acceptors. These reactions are catalyzed by specific **methyltransferases**.

Oxaloacetate-Forming Amino Acids

Asparagine and aspartic acid are converted to oxaloacetate, an intermediate of the TCA cycle. Asparagine is hydrolyzed to aspartic acid plus NH_3 in a reaction catalyzed by **asparaginase**.

$$\text{asparagine} + H_2O \rightarrow \text{aspartic acid} + NH_3$$

Aspartic acid is converted to oxaloacetic acid by transamination with α-ketoglutarate:

$$\text{aspartic acid} + \alpha\text{-ketoglutaric acid} \rightarrow \text{oxaloacetic acid} + \text{glutamic acid}$$

20-7 BIOSYNTHETIC REACTIONS OF THE AMINO ACIDS

In addition to their roles as structural elements of proteins and contributors to oxidative metabolism, amino acids also function as precursors of biologically important compounds. A few of these biosynthetic pathways will be considered in this section.

Creatine

Creatine, whose derivative creatine phosphate is an important energy reservoir in muscle, is synthesized by the mammalian body via a pathway which begins with the transfer of the guanidine group of arginine to glycine; the reaction is catalyzed by **glycine amidinotransferase**.

$$H_2N-\underset{\underset{NH}{\|}}{C}-NH-(CH_2)_3-\overset{\overset{NH_2}{|}}{CH}-COOH + H_2N-CH_2-COOH \longrightarrow$$

Arginine Glycine

$$H_2N-(CH_2)_3-\overset{\overset{NH_2}{|}}{CH}-COOH + H_2N-\underset{\underset{NH}{\|}}{C}-NH-CH_2-COOH$$

Ornithine Guanidoacetic acid

Methionine is the ultimate source of methyl groups for the formation of creatine from guanidoacetic acid. The initial reaction is the formation of S-adenosylmethionine (section 20-6), which functions as the actual methylating agent. The methyl group of methionine in S-adenosylmethionine is subsequently transferred to guanidoacetic acid to form creatine.

$$\text{S-adenosylmethionine} + HN=\overset{\overset{NH_2}{|}}{C}-NH-CH_2-COOH \longrightarrow$$

$$\text{S-adenosylhomocysteine} + HN=\overset{\overset{NH_2}{|}}{C}-\underset{\underset{CH_3}{|}}{N}-CH_2-COOH$$

Creatine

Nonprotein Peptide Bond Formation

A few small peptides are formed in cells quite independently of the protein-synthesizing system. One of the most abundant of these is glutathione (**γ-glutamylcysteinylglycine**). The biosynthesis of glutathione proceeds in two independent stages (Fig. 20-18).

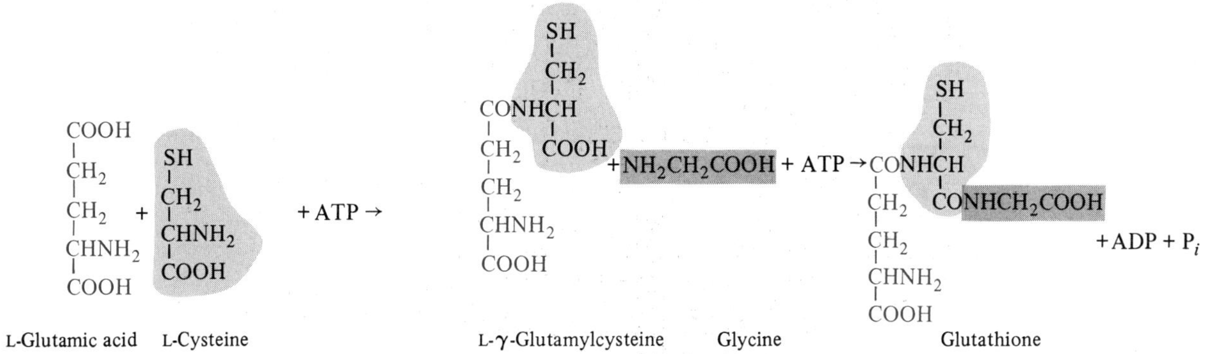

$$\begin{matrix} COOH \\ | \\ CH_2 \\ | \\ CH_2 \\ | \\ CHNH_2 \\ | \\ COOH \end{matrix} + \begin{matrix} SH \\ | \\ CH_2 \\ | \\ CHNH_2 \\ | \\ COOH \end{matrix} + ATP \rightarrow \begin{matrix} & & SH \\ & & | \\ & & CH_2 \\ & & | \\ CONHCH \\ | & & | \\ CH_2 & & COOH \\ | \\ CH_2 \\ | \\ CHNH_2 \\ | \\ COOH \end{matrix} + NH_2CH_2COOH + ATP \rightarrow \begin{matrix} & & SH \\ & & | \\ & & CH_2 \\ & & | \\ CONHCH \\ | & & | \\ CH_2 & & CONHCH_2COOH \\ | \\ CH_2 \\ | \\ CHNH_2 \\ | \\ COOH \end{matrix} + ADP + P_i$$

L-Glutamic acid L-Cysteine L-γ-Glutamylcysteine Glycine Glutathione

Figure 20-18
Biosynthesis of glutathione.

Glutamic acid is initially combined with cysteine, with the splitting of ATP to ADP, in a reaction catalyzed by **γ-glutamylcysteine synthetase**.

$$\text{glutamate} + \text{cysteine} + \text{ATP} \rightarrow \gamma\text{-glutamylcysteine} + \text{ADP} + P_i$$

A second ATP-consuming reaction, catalyzed by **glutathione synthetase**, adds glycine to γ-glutamylcysteine:

$$\gamma\text{-glutamylcysteine} + \text{glycine} + \text{ATP} \rightarrow \text{glutathione} + \text{ADP} + P_i$$

Bacteria may synthesize relatively complex polypeptides, such as the antibiotic gramacidin S (section 4-5), by stepwise enzyme action, without the involvement of mRNA or ribosomes.

Decarboxylation

Amino acid decarboxylation is catalyzed by a group of enzymes termed **amino acid decarboxylases**; these contain pyridoxal phosphate as a prosthetic group. The mechanism probably involves formation of a Schiff's base between pyridoxal phosphate and the amino acid, followed by decarboxylation:

$$R{-}CH_2{-}CHNH_2{-}COO^- + H^+ \rightarrow R{-}CH_2{-}CH_2{-}NH_2 + CO_2$$

In this way histidine is converted to **histamine**, a potent vasodilator. The enzyme responsible, **histidine decarboxylase**, is present in many tissues, including lungs and liver. Similarly, glutamic acid is decarboxylated to γ-aminobutyric acid by an enzyme present in brain:

$$HOOC{-}CH_2{-}CH_2{-}CHNH_2{-}COOH \rightarrow HOOC{-}CH_2{-}CH_2{-}CH_2{-}NH_2 + CO_2$$

γ-Aminobutyric acid may act as an inhibitory neurotransmitter.

Tyrosine Derivatives

Several biosynthetic pathways lead from tyrosine. In one of the most important of these, tyrosine is initially converted to 3,4-dihydroxy-L-phenylalanine (DOPA), which is converted via the amine **dopamine** to the important regulatory hormones **epinephrine** (adrenalin) and **norepinephrine** (section 22-3). DOPA has been found to be effective in treating Parkinson's disease, whose symptoms reflect a deficiency of dopamine in certain regions of the brain. DOPA is also a precursor of the dark pigment **melanin**.

Tryptophan Derivatives

Tryptophan is a precursor of **serotonin**, a powerful vasoconstrictor occurring in blood platelets, brain, and intestinal tissue. Tryptophan is first hydroxylated to form 5-hydroxytryptophan. Decarboxylation yields serotonin (Fig. 20-19). Alternatively, indolepyruvic acid may arise by transamination Fig. 20-19); oxidative decarboxylation of this molecule yields indoleacetic acid. In plants the latter compound functions as a growth hormone.

Polyamines

Three polyamines, **putrescine**, **spermidine**, and **spermine** (Fig. 20-20), occur in a wide variety of cells. They are components of bacterial membranes, ribosomes, and many chromosomes. Both spermidine and spermine are derived from **putrescine**, which is formed in mammals by the decarboxylation of

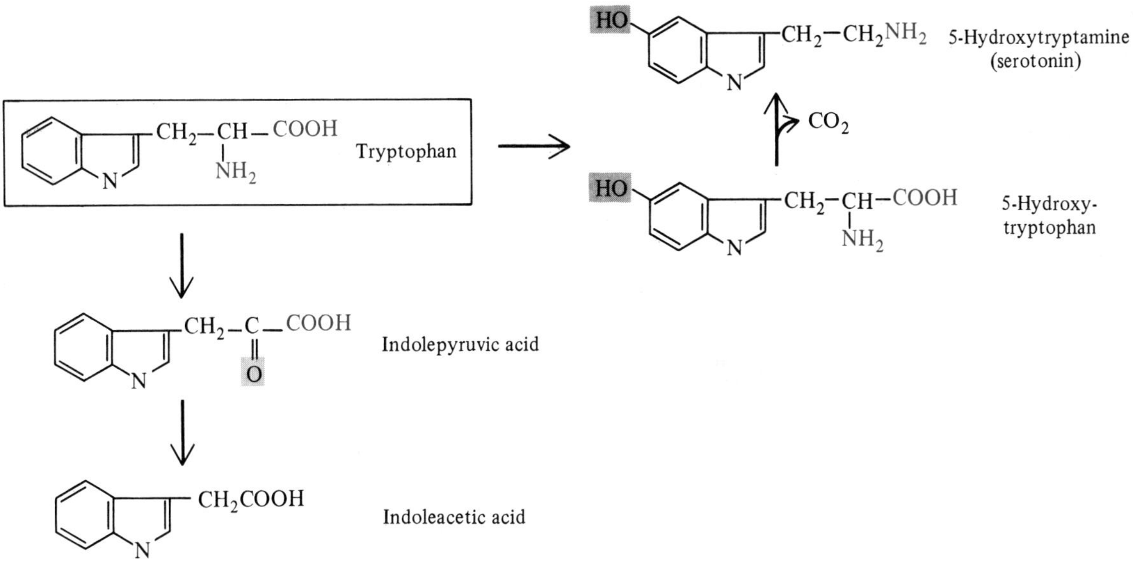
HO
$CH_2-CH_2NH_2$
5-Hydroxytryptamine (serotonin)
CO_2
$CH_2-CH-COOH$
NH_2
Tryptophan
HO
$CH_2-CH-COOH$
NH_2
5-Hydroxy-tryptophan
$CH_2-C-COOH$
O
Indolepyruvic acid
CH_2COOH
Indoleacetic acid

Figure 20-19
Alternative conversions of tryptophan.

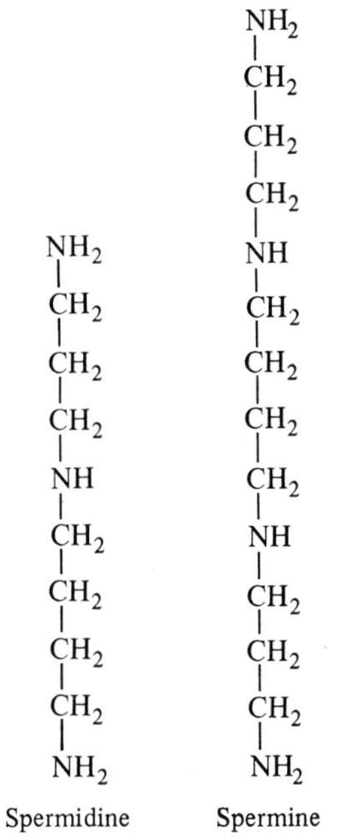
Spermidine
Spermine

Figure 20-20
Structures of spermidine and spermine.

ornithine, which is catalyzed by **ornithine decarboxylase**.

$$\underset{\text{Ornithine}}{H_2N{-}(CH_2)_3CH(NH_2){-}COOH} \rightarrow \underset{\text{Putrescine}}{H_2N{-}(CH_2)_4NH_2} + CO_2$$

Spermidine is formed by reaction of putrescine with a derivative of S-adenosylmethionine. The latter is converted to an amine by decarboxylation in a reaction catalyzed by **S-adenosylmethionine decarboxylase**.

S-adenosylmethionine → S-(5′-adenosyl)-3-methylmercaptopropylamine + CO_2

An $H_2N{-}(CH_2)_3$ group is next transferred to putrescine to yield spermidine. The reaction is catalyzed by **aminopropyl transferase**.

S-(5′-adenosyl)-3-methylmercaptopropylamine + putrescine →

spermidine + 5′-methylthioadenosine

A repetition of the above process yields spermine.

20-8 BIOSYNTHESIS OF THE NONESSENTIAL AMINO ACIDS BY MAMMALS

Glutamic Acid and Glutamine

Glutamic acid, which is of central importance in the biosynthesis of other amino acids, may be derived from α-ketoglutaric acid by the action of glutamate dehydrogenase (section 20-3). The level in tissues of free glutamic acid is high in comparison with other amino acids.

The synthesis of glutamine from glutamic acid is catalyzed by **glutamine synthetase**.

$$\text{glutamic acid} + NH_3 + \text{ATP} \rightleftarrows \text{glutamine} + \text{ADP} + P_i$$

Aspartic Acid and Asparagine

Aspartic acid is formed from oxaloacetic acid by transamination with glutamic acid:

$$HOOC{-}CO{-}CH_2COOH + HOOC{-}(CH_2)_2CH(NH_2)COOH \rightarrow$$
$$HOOC{-}CH(NH_2){-}CH_2COOH + HOOC{-}(CH_2)_2{-}CO{-}COOH$$

Asparagine is formed from aspartic acid by the action of **asparagine synthetase**, whose mechanism is analogous to that of glutamine synthetase.

Alanine

The biosynthesis of alanine occurs via transamination of pyruvic acid with glutamic acid:

glutamic acid + pyruvate → α-ketoglutarate + alanine

Tyrosine

Tyrosine is derived from the essential amino acid phenylalanine by a hydroxylation catalyzed by **phenylalanine 4-monooxygenase**:

$$\text{phenylalanine} + \text{NADPH} + H^+ + O_2 \rightarrow \text{tyrosine} + \text{NADP}^+ + H_2O$$

This reaction also figures in the degradation of phenylalanine (section 20-5).

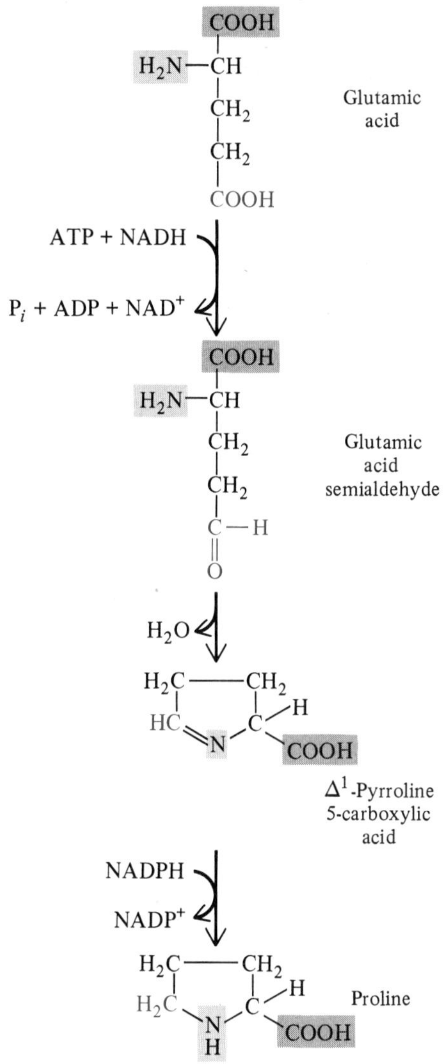

COOH
H_2N—CH
CH_2
CH_2
COOH

Glutamic acid

ATP + NADH

P_i + ADP + NAD^+

COOH
H_2N—CH
CH_2
CH_2
C—H
‖
O

Glutamic acid semialdehyde

H_2O

H_2C——CH_2
HC   C—H
  N   COOH

Δ^1-Pyrroline 5-carboxylic acid

NADPH

$NADP^+$

H_2C——CH_2
H_2C C—H
  N   COOH
  H

Proline

Figure 20-21
Biosynthesis of proline in mammals.

Proline and Hydroxyproline

Both of these amino acids are derived from glutamic acid by way of γ-glutamic semialdehyde, which is formed by the enzymic reduction of glutamic acid by NADH, as catalyzed by **glutamate kinase** and **dehydrogenase** (Fig. 20-21). The reaction involves an initial phosphorylation by ATP of the γ-carboxyl of glutamic acid, followed by reduction by NADH. γ-Glutamic semialdehyde undergoes a spontaneous cyclization to yield Δ'-pyrroline-5-carboxylic acid (Fig. 20-21); reduction of the latter by NADPH, as catalyzed by **pyrroline-5-carboxylate** reductase, yields proline, which is an allosteric inhibitor of the initial reaction of the sequence.

4-Hydroxyproline, which occurs in collagen, is formed by the action of **proline-4-monooxygenase**, which converts only proline residues which have been incorporated into a polypeptide chain.

$$\text{proline} + O_2 + \alpha\text{-ketoglutarate} + \text{CoA} \rightarrow \text{4-hydroxyproline} + \text{succinyl-CoA} + CO_2 + H_2O$$

The reaction involves Fe^{3+} and ascorbic acid as cofactors.

Serine and Glycine

The main pathway for the biosynthesis of serine begins with 3-phosphoglycerate, an intermediate of glycolysis (section 10-4). 3-Phosphoglycerate is oxidized by **phosphoglycerate dehydrogenase**, an NAD^+-dependent enzyme, to form **3-phosphohydroxypyruvate** (Fig. 20-22). Transamination from glutamate produces **3-phosphoserine**, in a reaction catalyzed by **phosphoserine transferase**. Finally, hydrolysis by **phosphoserine phosphatase** yields serine.

A second pathway involves the hydrolysis of 3-phosphoglycerate to glyceric acid, followed by oxidation by NAD^+ to hydroxypyruvic acid and transamination of the latter with glycine or alanine to yield serine (Fig. 20-22).

Serine is also a metabolic precursor of glycine. The formation of glycine occurs by removal of the terminal-CH_2OH group of serine. Tetrahydrofolate acts as an acceptor for the hydroxymethyl group; the reaction is catalyzed by

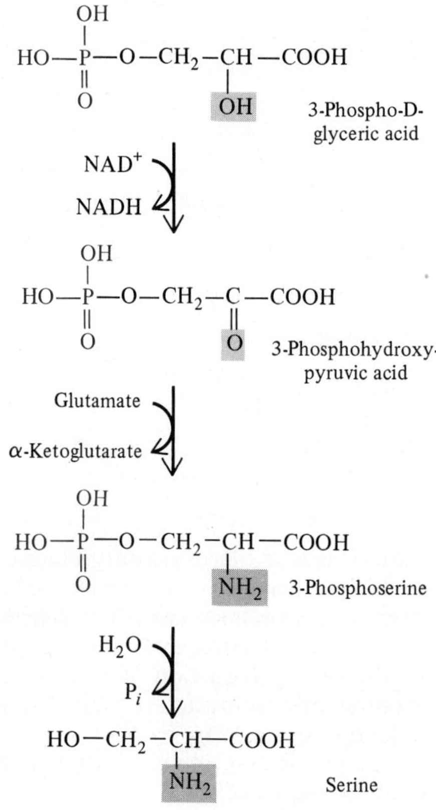

$$\mathrm{HO{-}P(OH)(=O){-}O{-}CH_2{-}CH(OH){-}COOH} \quad \text{3-Phospho-D-glyceric acid}$$

$$\downarrow \; NAD^+ \rightarrow NADH$$

$$\mathrm{HO{-}P(OH)(=O){-}O{-}CH_2{-}C(=O){-}COOH} \quad \text{3-Phosphohydroxypyruvic acid}$$

$$\downarrow \; \text{Glutamate} \rightarrow \alpha\text{-Ketoglutarate}$$

$$\mathrm{HO{-}P(OH)(=O){-}O{-}CH_2{-}CH(NH_2){-}COOH} \quad \text{3-Phosphoserine}$$

$$\downarrow \; H_2O \rightarrow P_i$$

$$\mathrm{HO{-}CH_2{-}CH(NH_2){-}COOH} \quad \text{Serine}$$

Figure 20-22
Pathway leading from 3-phosphoglycerate to serine.

serine hydroxymethyltransferase, a pyridoxal phosphate-containing enzyme (Fig. 20-10).

$$\underset{\text{Serine}}{HO{-}CH_2{-}\overset{\overset{NH_2}{|}}{C}H{-}COOH} + \text{tetrahydrofolic acid} \rightleftharpoons$$

$$\underset{\text{Glycine}}{H_2N{-}CH_2{-}COOH} + N^5,N^{10}\text{-methylenetetrahydrofolic acid}$$

This reaction is also of importance as a source of single carbon fragments at the oxidation level of CH_2OH, HCHO, or HCOOH. An example is provided by an alternative mechanism for the biosynthesis of glycine from CO_2 and NH_3 by **glycine synthase**, a pyridoxal phosphate enzyme.

$$CO_2 + NH_3 + N^5,N^{10}\text{-methylenetetrahydrofolic acid} + NADH + H^+ \rightarrow$$
$$\text{glycine} + \text{tetrahydrofolic acid} + NAD^+$$

This reaction is the primary pathway in mammalian liver.

Cysteine and Cystine

Although cysteine is not classified as an essential amino acid, its precursor in mammals is the essential amino acid methionine. The mechanism is equivalent to the replacement of the hydroxyl oxygen of serine by a sulfur atom derived from methionine, thereby converting serine to cysteine.

The sequence begins with the conversion of methionine to S-adenosylmethionine by **methionine adenosyltransferase** via reaction with ATP (Fig. 20-23).

$$\text{methionine} + ATP \rightarrow \text{S-adenosylmethionine} + PP_i + P_i$$

S-Adenosylmethionine is important as a donor of methyl groups, as in the biosynthesis of creatine (section 20-7). The methyl group may be transferred to any of a number of acceptors, yielding S-adenosylhomocysteine (Fig. 20-23), which is converted to homocysteine by **adenosylhomocysteinase**.

In the next step, homocysteine is condensed with serine by **cystathionine-β-synthase** to form cystathionine:

$$\underset{\text{Homocysteine}}{HOOC{-}CH(NH_2){-}CH_2{-}CH_2{-}SH} + \underset{\text{Serine}}{HOCH_2{-}CH(NH_2){-}COOH} \rightarrow$$

$$\underset{\text{Cystathionine}}{HOOC{-}CH(NH_2){-}CH_2{-}S{-}CH_2{-}CH_2{-}CH(NH_2){-}COOH}$$

Finally, cystathionine is decomposed to yield free cysteine by the action of **cystathionine-γ-lyase**:

$$\text{cystathionine} \rightarrow \text{cysteine} + \alpha\text{-ketobutyrate} + NH_3$$

Cystine does not occur in the free state in significant quantities. The cystine cross-links of proteins arise from the oxidation of cysteine groups incorporated into the polypeptide chains.

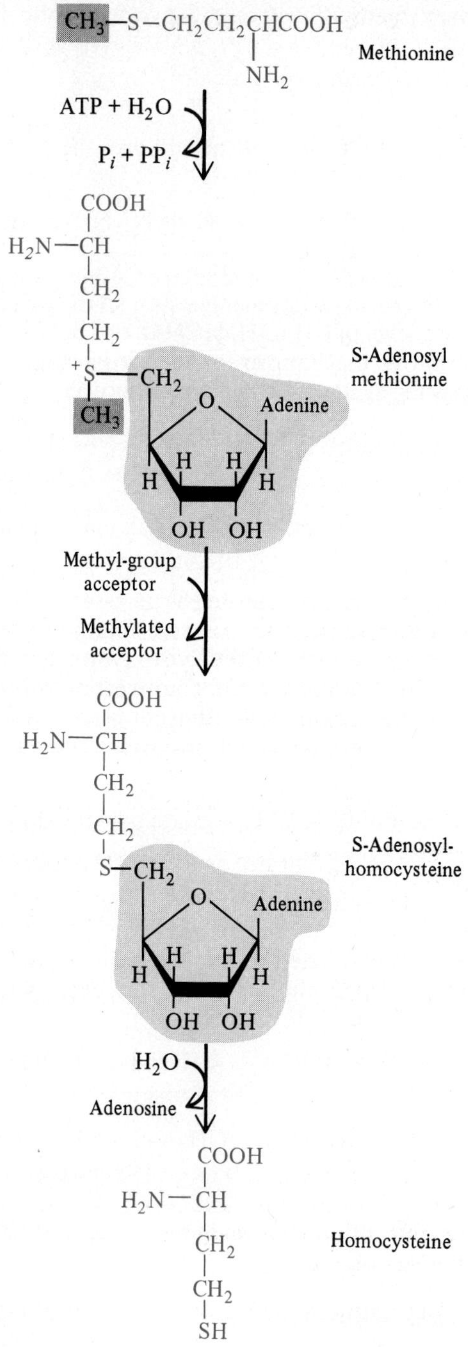

$CH_3-S-CH_2CH_2CHCOOH$ (with NH_2 on CH)   Methionine

$ATP + H_2O$ → $P_i + PP_i$

COOH
H_2N-CH
CH_2
CH_2
$^+S-CH_2$ (ribose)   Adenine
CH_3
H H H H
OH OH

S-Adenosyl methionine

Methyl-group acceptor → Methylated acceptor

COOH
H_2N-CH
CH_2
CH_2
$S-CH_2$ (ribose)   Adenine
H H H H
OH OH

S-Adenosyl-homocysteine

H_2O → Adenosine

COOH
H_2N-CH
CH_2
CH_2
SH

Homocysteine

Figure 20-23 Conversion of methionine to homocysteine.

Arginine

Arginine is formed in the urea cycle (section 20-5). The rate of arginine production is not sufficient to supply the normal needs of mammals, so that an external source is required, especially for young humans.

SUMMARY

The digestion of proteins by animals converts them enzymically to a mixture of amino acids. These are absorbed from the intestine and transported by the blood to the liver, where most of the degradation of amino acids occurs.

The carbon skeletons of the amino acids are converted by oxidative degradation to intermediates of the TCA cycle. The different possible intermediates are acetyl CoA, α-ketoglutarate, succinate, fumarate, and oxaloacetate. The amino groups of the majority of the amino acids undergo transamination with α-ketoglutarate to form glutamate, which is subsequently converted to α-ketoglutarate and ammonia by the action of **glutamate dehydrogenase**. The ammonia is used to form **carbamoyl phosphate** by an ATP-dependent reaction.

Carbamoyl phosphate reacts with **ornithine** to form **citrulline**. Ornithine and citrulline are intermediates of the **urea cycle**, by which the body disposes of excess nitrogen by incorporation into urea, which is excreted.

One of the amino groups of urea arrives as ammonia; the second enters the urea cycle as the amino group of aspartate, which reacts with citrulline to form **argininosuccinate** in an ATP-dependent reaction. Argininosuccinate is the precursor of arginine, which is hydrolyzed to urea and ornithine. The urea cycle occurs in the liver.

Mammals can synthesize one-half of the amino acids required for protein biosynthesis; the remainder must be supplied in the diet. Glutamic acid is formed from α-ketoglutarate by the action of **glutamate dehydrogenase**; glutamic acid is itself the precursor of glutamine and proline. Aspartic acid and alanine arise by transamination of oxaloacetic acid and pyruvic acid, respectively. Serine, a precursor of glycine, is formed from 3-phosphoglycerate. Cysteine arises from methionine and serine by a complex process. Tyrosine is synthesized from phenylalanine.

REFERENCES

The following code is used to classify references. I: particularly useful as an introduction to the subject; R: useful primarily as a reference text; A: an advanced account of the material; H: a publication of historical importance.

General

D. M. Greenberg, *Metabolic Pathways*, Academic, New York (1975). (R)

A. Meister, *Biochemistry of the Amino Acids*, Academic, New York (1965). (A)

W. L. Nyham, *Heritable Disorders of Amino Acid Metabolism*, Wiley, New York (1974). (R)

Amino Acid Degradation

A. E. Braunstein, "Amino Group Transfer," in *The Enzymes*, P. D. Boyer, ed., 3rd ed., vol. 9, part B, page 379, Academic, New York (1973). (R)

H. Eisenberg, "Glutamate Dehydrogenase: Anatomy of a Regulatory Enzyme," *Acc. Chem. Res.* 4, 379 (1971). (I)

S. Ratner, "Enzymes of Arginine and Urea Synthesis," *Adv. Enzymol.* 39, 1 (1973). (R)

Amino Acid Biosynthesis

S. Dagley and D. E. Nicholson, *An Introduction to Metabolic Pathways*, Wiley, New York (1970). (I)

S. Prusiner and E. R. Stadtman, *The Enzymes of Glutamine Metabolism*, Academic, New York (1973). (A)

REVIEW QUESTIONS

Questions marked with an asterisk are of a high level of difficulty.

20-1 Which of the following statements are *true about all nonessential* amino acids?
(a) All are synthesized from essential amino acids.
(b) All are synthesized from intermediates in the TCA cycle.
(c) All are able to obtain amino groups by direct reaction with NH_3 in the blood.
(d) None are required in the human diet.
(e) All of the above are *true* for all nonessential amino acids.

20-2 Which of the following is a possible fate of *some* amino acids in the cells?
(a) become part of newly-synthesized proteins
(b) loss of amino groups in transamination reactions
(c) burned to CO_2 and H_2O via the TCA cycle
(d) only (a)
(e) (a), (b), and (c)

20-3 Glucose labeled with the isotope ^{14}C in all six carbon atoms is injected into a rat. The rat is sacrified 30 min later. In which compounds below would you expect to find labeled carbon atoms?
(a) L-alanine isolated from liver
(b) HCO_3^- in the liver
(c) liver glycogen
(d) liver fatty acids
(e) all of the above

20-4 The fatty acid $CH_3(CH_2)_{16}COOH$ labeled with the isotope ^{14}C *in all its carbon atoms* is injected into a rat. The rat is sacrificed 30 min later. In which compound below, isolated from its liver, would you *not expect* to find labeled carbon atoms?

(a) 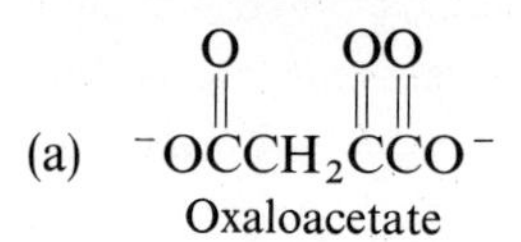

Oxaloacetate

(b) 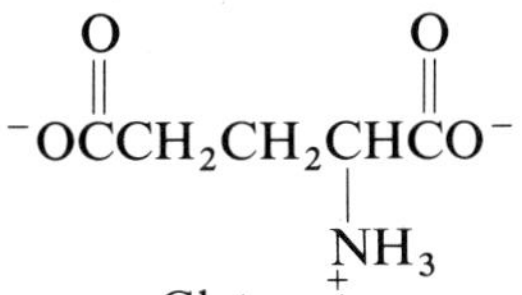 $^{-}O\overset{\overset{O}{\|}}{C}CH_2CH_2\underset{\underset{\overset{+}{N}H_3}{|}}{C}H\overset{\overset{O}{\|}}{C}O^{-}$

Glutamate

(c) 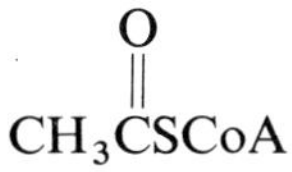 $CH_3\overset{\overset{O}{\|}}{C}SCoA$

Acetyl CoA

(d) 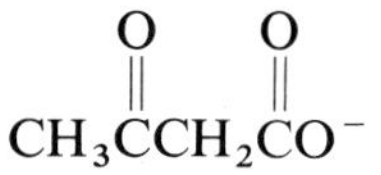 $CH_3\overset{\overset{O}{\|}}{C}CH_2\overset{\overset{O}{\|}}{C}O^{-}$

Acetoacetate

(e) 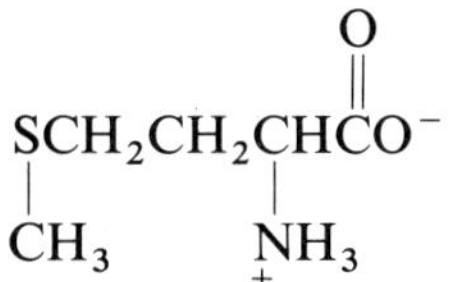 $\underset{\underset{CH_3}{|}}{S}CH_2CH_2\underset{\underset{\overset{+}{N}H_3}{|}}{C}H\overset{\overset{O}{\|}}{C}O^{-}$

Methionine

*20-5 Write a balanced equation for the complete oxidative degradation of tyrosine by a mammal. Include all activation steps.

*20-6 In a hereditary disease, isovaleric acid accumulated in the blood. The metabolism of which amino acid is likely to be affected? At what step?

*20-7 How many ATP molecules are generated by the complete oxidation of isoleucine to H_2O, CO_2, and urea?

*20-8 Write a balanced equation for the synthesis of serine starting with glucose as the sole carbon source.

*20-9 What is the number of high energy phosphate bonds needed for the formation of one molecule of arginine from CO_2, NH_3, and glutamate?

21

The Metabolism of Nucleotides and Their Components

21-1 BIOSYNTHESIS OF PYRIMIDINE RIBONUCLEOTIDES

None of the purine and pyrimidine bases present in nucleic acids are essential components of the mammalian diet, with the implication that they can be synthesized *in vivo*. If *free* pyrimidine bases are consumed, they are rapidly metabolized, their nitrogen being largely excreted as urea.

In contrast, the ingestion of pyrimidine **nucleotides** labeled with radioisotopes, or of intact nucleic acids, results in significant incorporation of the intact labeled pyrimidine bases, suggesting that the nucleotides, but not the free bases themselves, may function as nucleic acid precursors.

Orotic acid (6-carboxyluracil) is the key substance in the biosynthesis of pyrimidines (Fig. 21-1). The pathways are similar for animals and microorganisms.

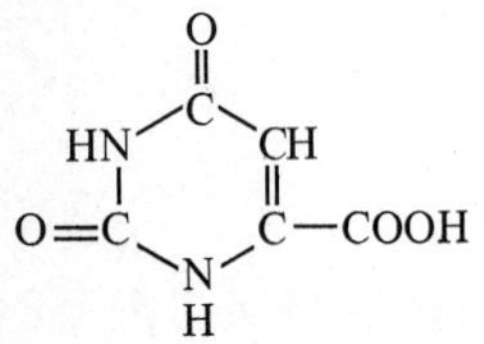

The biosynthesis of orotic acid begins with the formation of **ureidosuccinic** acid (or **N-carbamoylaspartic acid**) by the condensation of carbamoyl phosphate (section 20-3) and aspartic acid (Fig. 21-2); the reaction is catalyzed by **aspartate carbamoyl transferase** (or **aspartate transcarbamoylase**).

Ring closure of ureidosuccinic acid to form dihydroorotic acid is catalyzed by **dihydroorotase** (Fig. 21-2). Dihydroorotic acid is subsequently oxidized to orotic acid by **dihydroorotate dehydrogenase**, with NAD^+ as cofactor.

```
        O
        ‖
        C
      /   \
   HN       CH
    |       ‖
 O=C        C—COOH
      \   /
        N
        H
```

Figure 21-1
Structure of orotic acid.

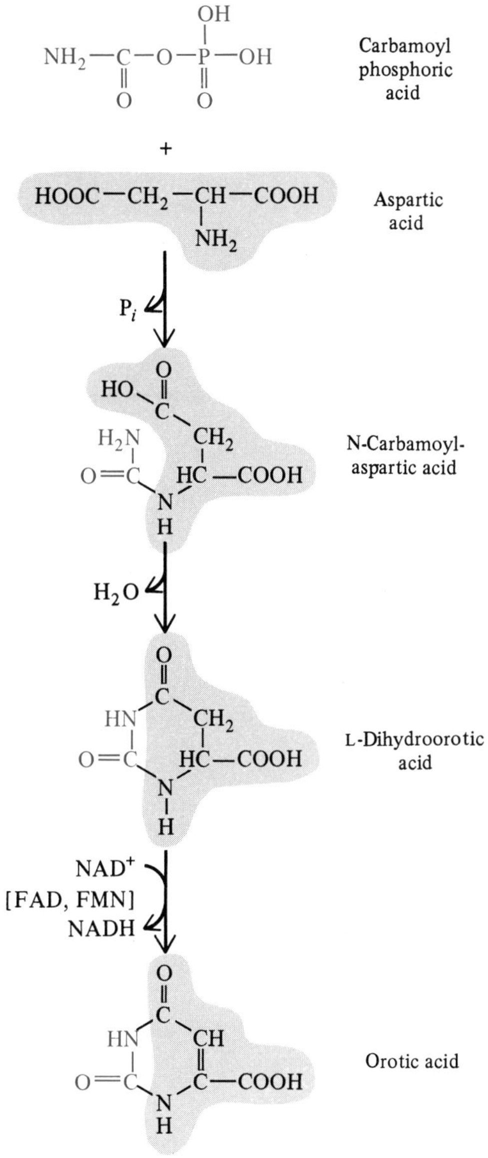

$$NH_2-\underset{\underset{O}{\|}}{C}-O-\overset{\overset{OH}{|}}{\underset{\underset{O}{\|}}{P}}-OH$$ Carbamoyl phosphoric acid

+

$$HOOC-CH_2-\underset{\underset{NH_2}{|}}{CH}-COOH$$ Aspartic acid

↓ P_i

N-Carbamoyl-aspartic acid

↓ H_2O

L-Dihydroorotic acid

↓ NAD^+ [FAD, FMN] NADH

Orotic acid

Figure 21-2
Metabolic pathway leading to orotic acid.

The parent pyrimidine nucleotide arises via the combination of orotic acid with **5-phosphoribose-1-pyrophosphate**. The latter compound is formed from ribose-5-phosphate and ATP in a reaction catalyzed by **ribosephosphate pyrophosphatase** (Fig. 21-3).

The coupling reaction, whereby **orotidine-5′-phosphate** is formed (Fig. 21-4) is catalyzed by **orotate phosphoribosyl transferase**. Finally, orotidine-5′-phosphate is decarboxylated to yield uridine-5′-phosphate (UMP) (Fig. 21-5).

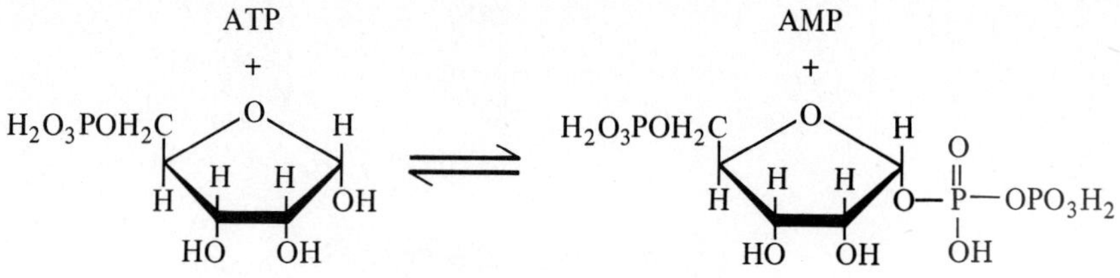

ATP + [ribose 5-phosphate structure] ⇌ AMP + [PRPP structure]

Figure 21-3
Formation of 5-phosphoribose-1-pyrophosphate (PRPP).

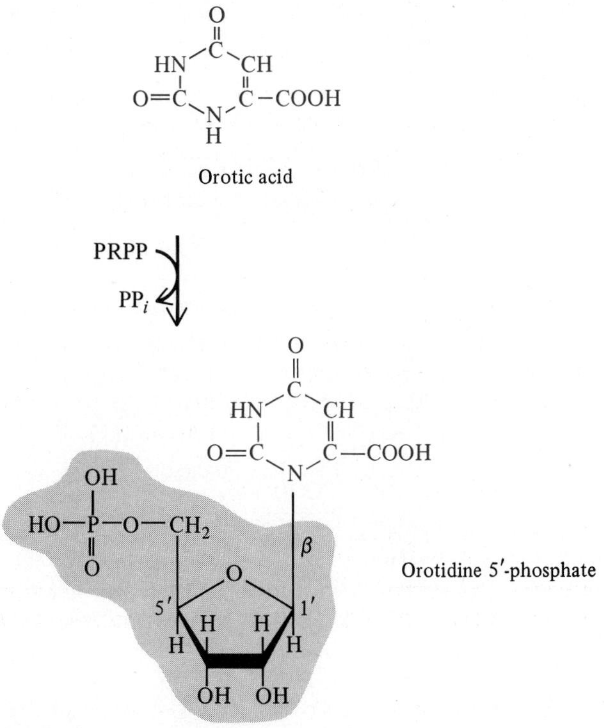

Orotic acid

PRPP → PP_i

Orotidine 5′-phosphate

Figure 21-4
Formation of orotidine-5′-phosphate from orotic acid. Orotidine-5′-phosphate is the parent pyrimidine nucleotide from which other pyrimidine nucleotides are derived.

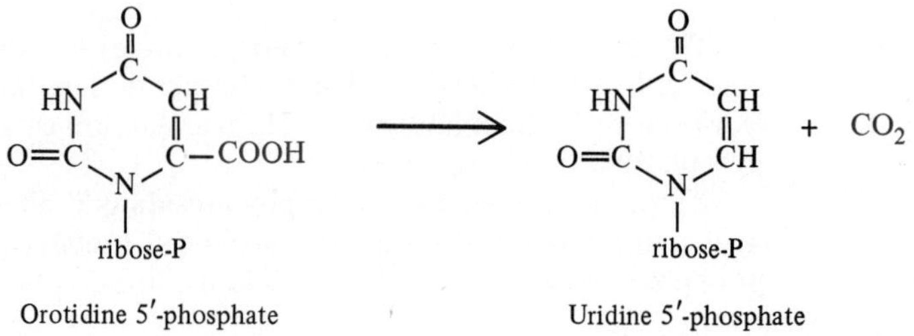

Orotidine 5′-phosphate → Uridine 5′-phosphate + CO_2

Figure 21-5
Formation of UMP by decarboxylation of orotidine-5′-phosphate.

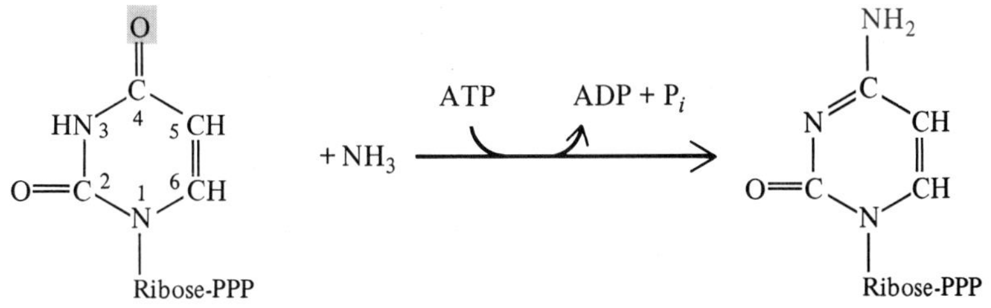

$$\text{Uridine triphosphate (UTP)} + NH_3 \xrightarrow{ATP \;\; ADP + P_i} \text{Cytidine triphosphate (CTP)}$$

Figure 21-6 Formation of CTP from UTP in bacteria.

The conversion of UMP to UTP, an immediate precursor of nucleic acids, occurs through two consecutive transfers of phosphate from ATP, catalyzed by a nucleoside monophosphate kinase:

$$UMP + ATP \rightarrow UDP + ADP$$
$$UDP + ATP \rightarrow UTP + ADP$$

Cytidine triphosphate arises in bacteria by the reaction of UTP with NH_3 in an ATP-driven reaction catalyzed by **CTP-synthetase** (Fig. 21-6). In animals, NH_3 is replaced by glutamine.

The carbamoyl phosphate involved in the reaction that initiates this biosynthetic sequence (Fig. 21-2) is generated from glutamine by the reaction

$$2ATP + \text{glutamine} + CO_2 + H_2O \rightarrow 2ADP + P_i + \text{glutamate} + \text{carbamoyl phosphate}$$

This reaction, which occurs *outside* the mitochondria in the *cytoplasm*, is not the same as the mitochondrial process forming part of the urea cycle (section 20-3). The corresponding enzymes are termed **carbamoyl phosphate synthase (glutamine)** and **carbamoyl phosphate synthase (ammonia)**, respectively.

Control of Nucleotide Biosynthesis

The ultimate products of the pyrimidine biosynthetic pathway include CTP. This has a regulatory influence upon the rate of pyrimidine biosynthesis and hence upon the rate of its own formation, which is thereby adjusted to the needs of the organism.

The intervention of CTP in the pyrimidine biosynthetic pathway occurs at the step leading to ureidosuccinic acid, which is catalyzed by aspartate carbamoyltransferase.

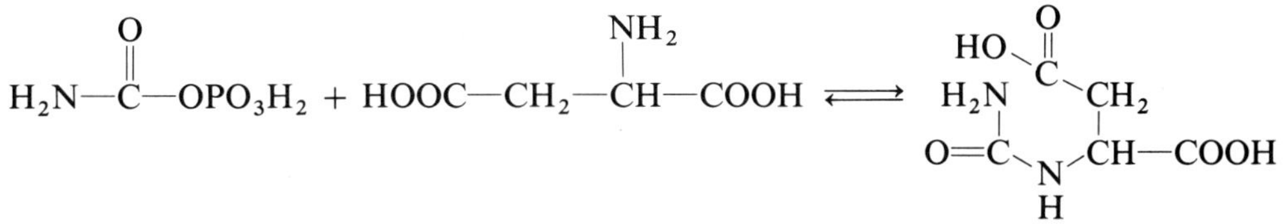

$$H_2N{-}\overset{O}{\overset{\|}{C}}{-}OPO_3H_2 + HOOC{-}CH_2{-}\overset{NH_2}{\overset{|}{C}H}{-}COOH \rightleftharpoons \text{Ureidosuccinic acid}$$

Carbamoyl phosphate — Aspartic acid — Ureidosuccinic acid (N-carbamoyl aspartate)

Aspartate carbamoyltransferase, whose molecular weight is 310,000, consists of two catalytic subunits of molecular weight 100,000 each and three regulatory subunits of molecular weight 34,000, which contain the allosteric sites. CTP acts as an allosteric inhibitor for this enzyme in *E. coli*, so that a rise in the level of CTP partially blocks the formation of ureidosuccinic acid and reduces the rate of pyrimidine biosynthesis.

This process is a classic example of the feedback regulation of a metabolic reaction sequence through inhibition by an end product. The mechanism of inhibition involves a structural change induced by the binding of CTP at an **allosteric** site, which is distinct from the catalytic site.

21-2 BIOSYNTHESIS OF PURINE RIBONUCLEOTIDES

Like the pyrimidines, the purine bases can be synthesized from other compounds by mammalian tissues. The mechanism of purine biosynthesis, which appears to be similar for mammals and bacteria, consists of a stepwise construction of the purine ring upon the C-1 carbon of ribose-5-phosphate.

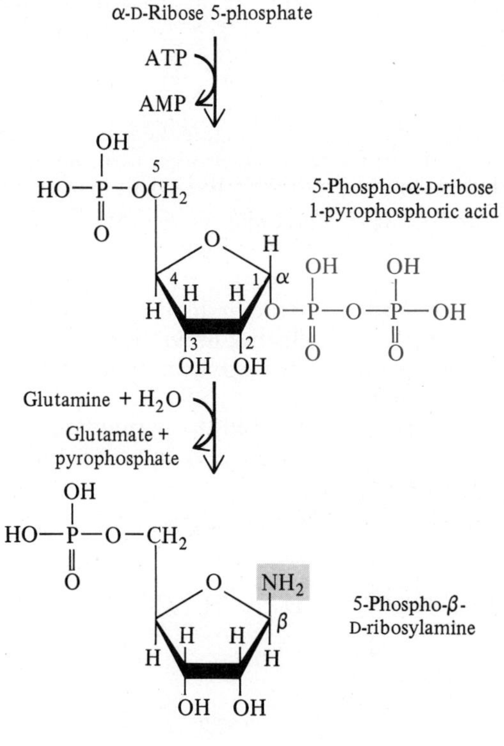

α-D-Ribose 5-phosphate

ATP → AMP

5-Phospho-α-D-ribose 1-pyrophosphoric acid

Glutamine + H_2O → Glutamate + pyrophosphate

5-Phospho-β-D-ribosylamine

Figure 21-7 Formation of 5-phospho-β-D-ribosylamine.

Ribose-5-phosphate is converted to 5-phosphoribose-1-pyrophosphate by reaction with ATP (Fig. 21-3). Reaction with glutamine yields the amino sugar **5-phospho-β-ribosylamine** in a reaction catalyzed by **amidophosphoribosyl transferase** (Fig. 21-7).

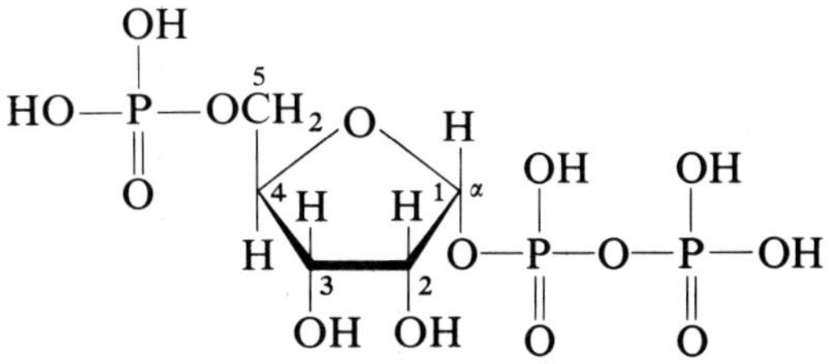

5-Phospho-α-D-ribose-1-pyrophosphate + glutamine $\longrightarrow$

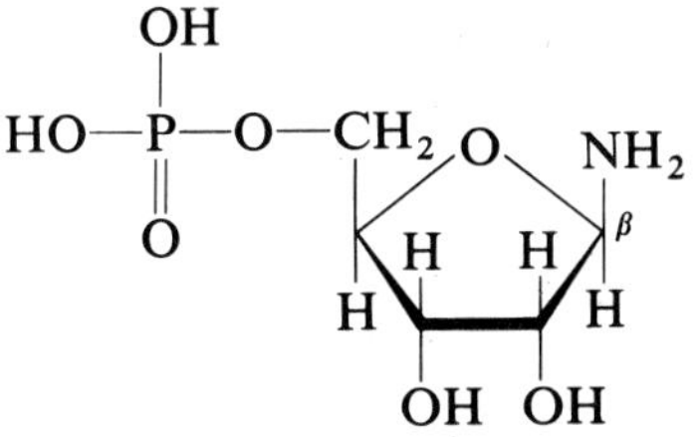

5-Phospho-β-D-ribosylamine + glutamate + PP_i

The formation of the amine is accompanied by an inversion of the spatial conformation at the C-1 position, which is α for the pyrophosphate and β for the amine (section 9-2).

The next step in the biosynthesis consists of the condensation of glycine with 5-phospho-β-ribosylamine, in a reaction catalyzed by **phosphoribosylglycinamide synthetase** (Fig. 21-8):

glycine + ATP + 5-phospho-β-ribosylamine $\longrightarrow$

ADP + P_i + 5′-Phosphoribosylglycinamide

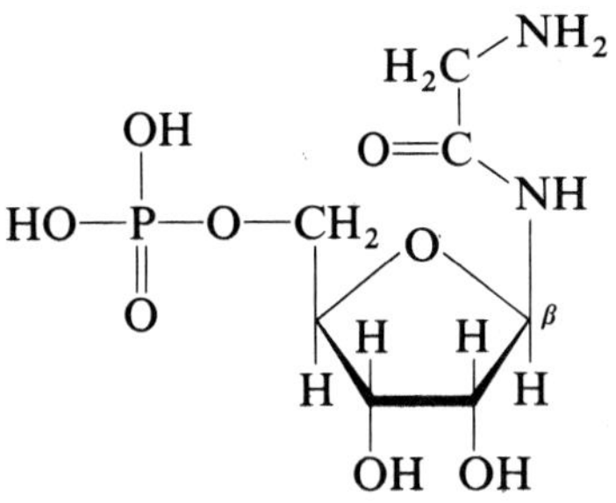

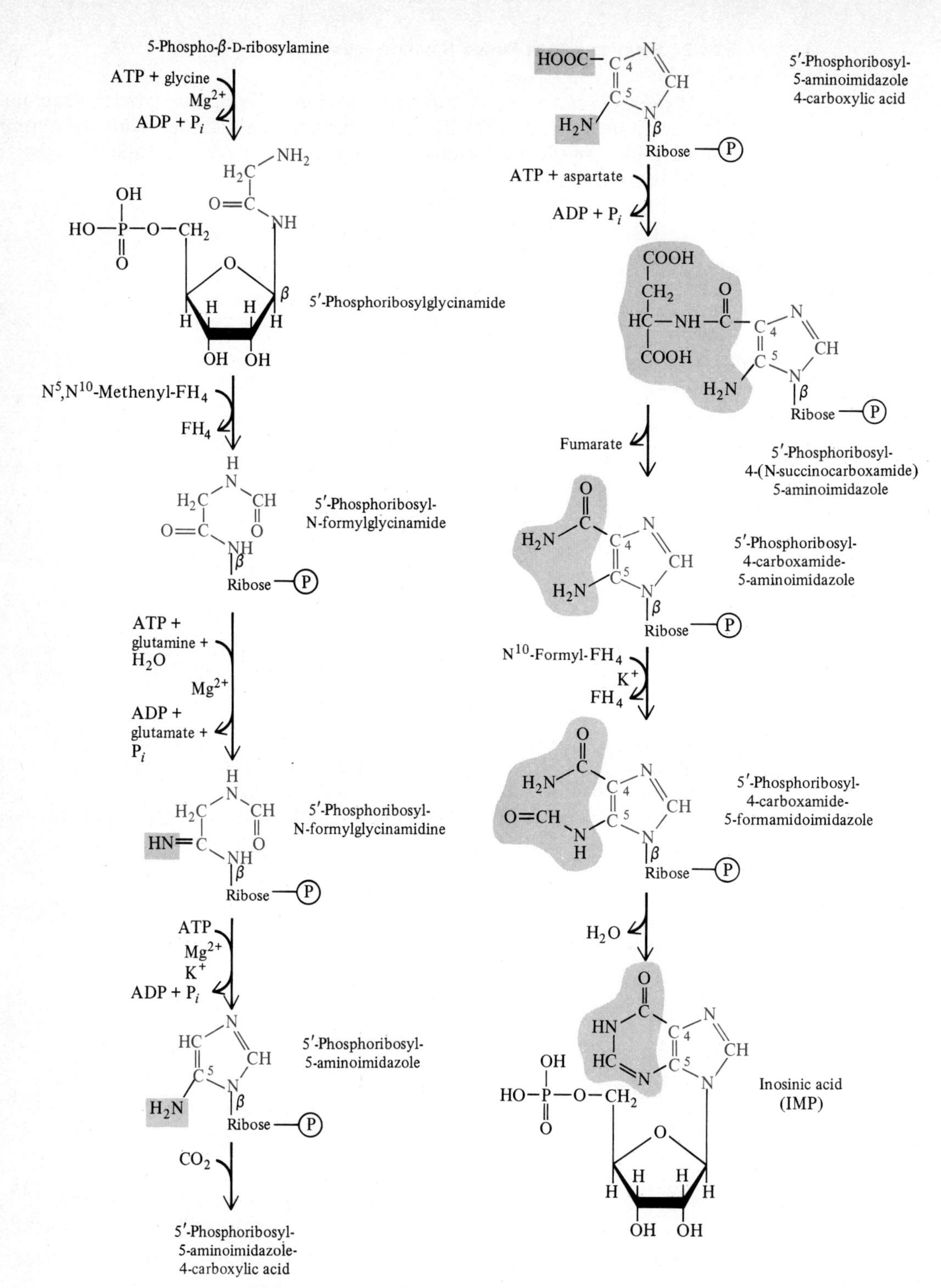

Figure 21-8
Metabolic pathway leading to IMP from 5-phospho-β-D-ribosylamine.

5′-Phosphoribosylglycinamide is next formylated by transfer from the formyl folic acid derivative in a reaction catalyzed by **phosphoribosylglycinamide formyltransferase**:

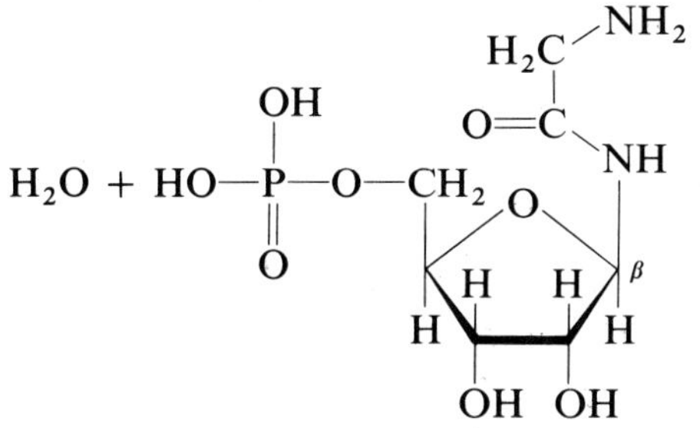

H_2O + 5′-Phosphoribosylglycinamide + N^5,N^{10}-anhydroformyltetrahydrofolate ⟶

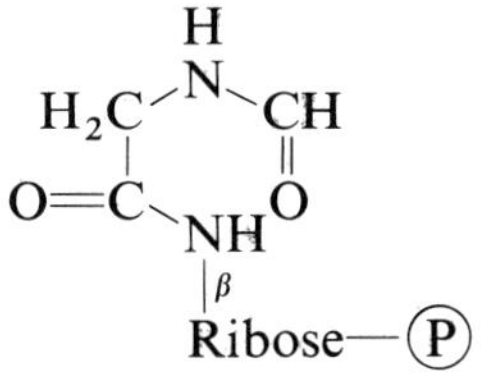

5′-Phosphoribosyl-N-formylglycinamide + tetrahydrofolate

An amino group is transferred from glutamine to 5′-phosphoribosyl-N-formylglycinamide to form the amidine derivative (Fig. 21-8); the enzyme is **phosphoribosylformylglycinamide synthetase**.

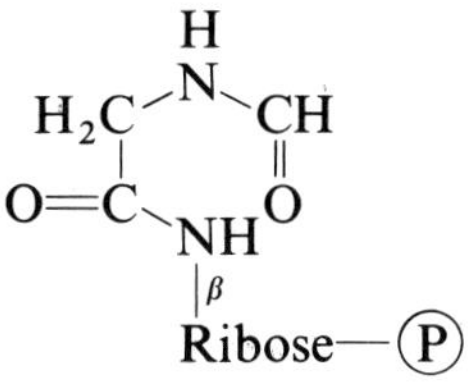

5′-Phosphoribosyl-N-formylglycinamide + glutamine + H_2O + ATP ⟶

glutamic acid + ADP + 5′-Phosphoribosyl-N-formylglycinamidine + P_i

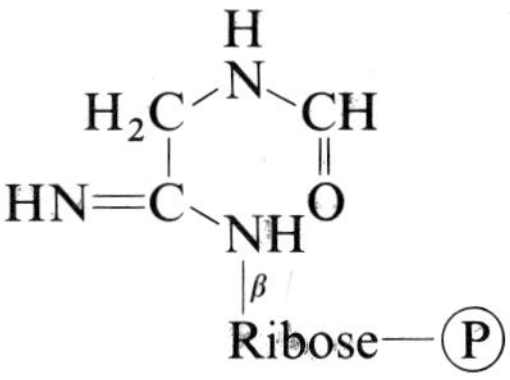

Ring closure converts the amidine to an imidazole derivative (Fig. 21-8); the reaction is catalyzed by **phosphoribosylaminoimidazole synthetase**.

ATP + 5′-phosphoribosyl-N-formylglycinamidine ⟶

P_i + ADP + 5′-Phosphoribosyl-5-aminoimidazole

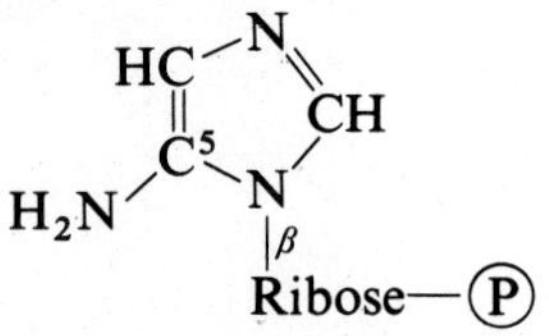

The imidazole derivative is next carboxylated by **phosphoribosyl aminoimidazole carboxylase**:

5′-phosphoribosyl-5-aminoimidazole + CO_2 ⟶ 5′-Phosphoribosyl-5-aminoimidazole-4-carboxylic acid

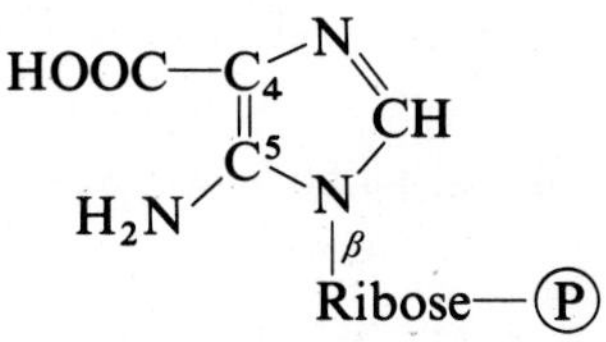

The above intermediate is condensed with aspartic acid by **phosphoribosyl aminoimidazole-succinocarboxamide synthetase.**

5′-phosphoribosyl-5-aminoimidazole-4-carboxylic acid + aspartic acid + ATP ⟶

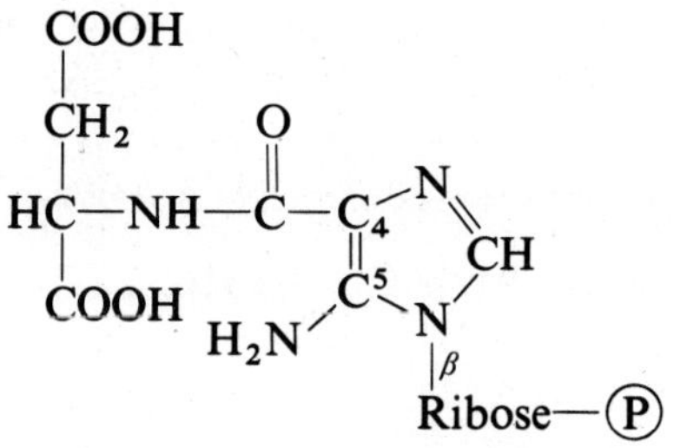

5′-Phosphoribosyl-4-(N-succinocarboxamide)-5-aminoimidazole + ADP + P_i

The latter derivative is cleaved by **adenylosuccinate lyase**:

5′-phosphoribosyl-4-(N-succinocarboxamide)-5-aminoimidazole ⟶

fumarate + 5′-Phosphoribosyl-4-carboxamide-5-aminoimidazole

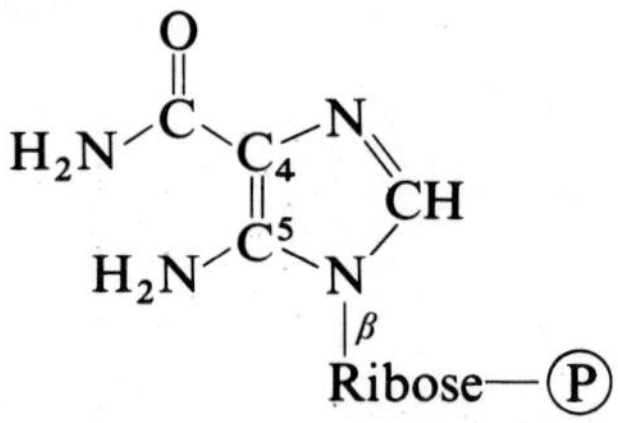

The next reaction is a formylation. The donor is N^{10}-formyltetrahydrofolate, the enzyme is **phosphoribosyl-aminoimidazole-carboxamide formyl transferase.**

5′-phosphoribosyl-4-carboxamide-5-aminoimidazole + N^{10}-formyltetrahydrofolate ⟶

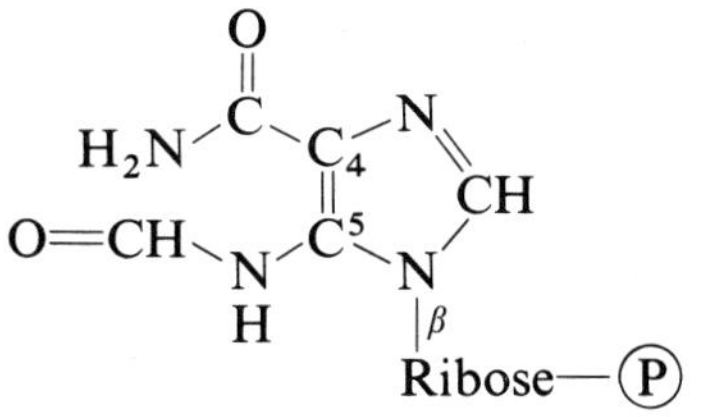

```
        O
        ‖
   H2N–C–C4–N
         ‖    ⟍
         ‖     CH       + tetrahydrofolate
 O=CH–N–C5–N⟋
      H     |β
            Ribose—Ⓟ
```

5′-Phosphoribosyl-4-carboxamide-5-formamidoimidazole

Closure of the second ring by **IMP-cyclohydrolase** produces inosinic acid (IMP), a ribonucleotide with a complete purine ring structure

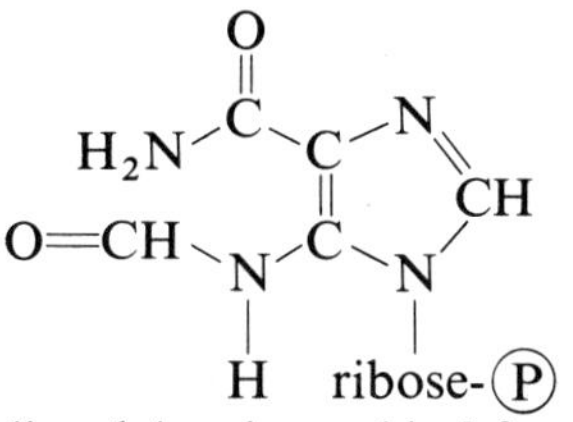

```
        O
        ‖
   H2N–C–C–N
         ‖   ⟍
         ‖    CH        ⟶ inosinic acid
 O=CH–N–C–N⟋
      |    |
      H    ribose-Ⓟ
```

5′-Phosphoribosyl-4-carboxamide-5-formamidoimidazole

Inosinic acid is the point of departure for the synthesis of the other ribonucleotides. Adenylic acid (AMP) arises by condensation with aspartic acid, with participation of GTP, to form adenylosuccinic acid, which is subsequently cleaved to yield adenylic acid (Fig. 21-9). The enzymes involved in the two successive steps are **adenylosuccinate synthetase** and **adenylosuccinate lyase**, respectively:

inosinic acid + GTP + aspartic acid → GDP + P_i + adenylosuccinic acid →
adenylic acid + fumaric acid

Guanylic acid (GMP) is synthesized from inosinic acid via oxidation by NAD^+ to xanthylic acid (catalyzed by **IMP dehydrogenase**), followed by acquisition of an amino group from glutamine (Fig. 21-10).

```
                                 O
                                 ‖
                             HN–C–C–N
                             |    ‖   ⟍
inosinic acid + NAD+ ⟶       |    ‖    CH + NADH
                           O=C–N–C–N⟋
                               H    |
                                    ribose-Ⓟ
```

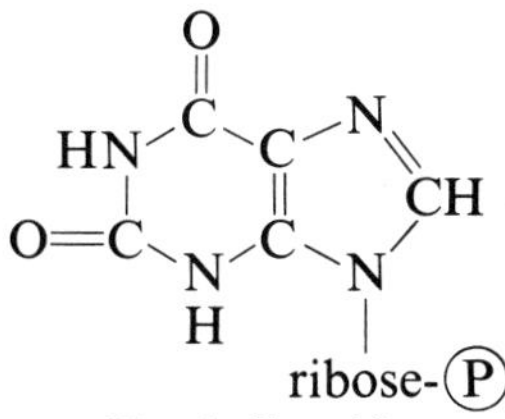

Xanthylic acid

xanthylic acid + ATP + glutamine →
guanylic acid + glutamic acid + AMP + PP_i

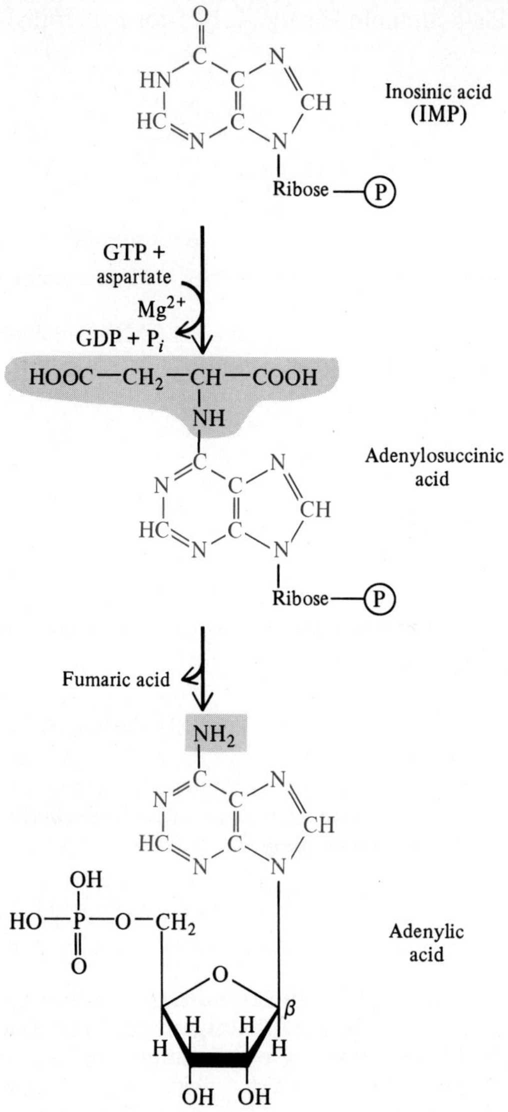

Inosinic acid (IMP)
Ribose—(P)
GTP + aspartate
Mg2+
GDP + Pi
HOOC—CH2—CH—COOH
NH
Adenylosuccinic acid
Ribose—(P)
Fumaric acid
NH2
Adenylic acid
OH
HO—P—O—CH2
O
β
H H
H H
OH OH

Figure 21-9

Regulation of Purine Nucleotide Biosynthesis

In *E. coli*, control of inosinic acid formation, which governs the rate of synthesis of the purine nucleotides, is achieved at the step corresponding to the transfer of an amino group from glutamine to 5-phosphoribose-1-pyrophosphate. The enzyme **amidophosphoribosyl transferase**, which catalyzes this reaction, has allosteric properties and is inhibited both by AMP, ADP, or ATP and by GMP, GDP, or GTP. Each of the two classes of inhibitor binds to its own distinct allosteric site. In this way, a rise in the level of any of the purine nucleotides tends to depress the rate of inosinic acid formation, and hence to retard that of the nucleotides derived from it.

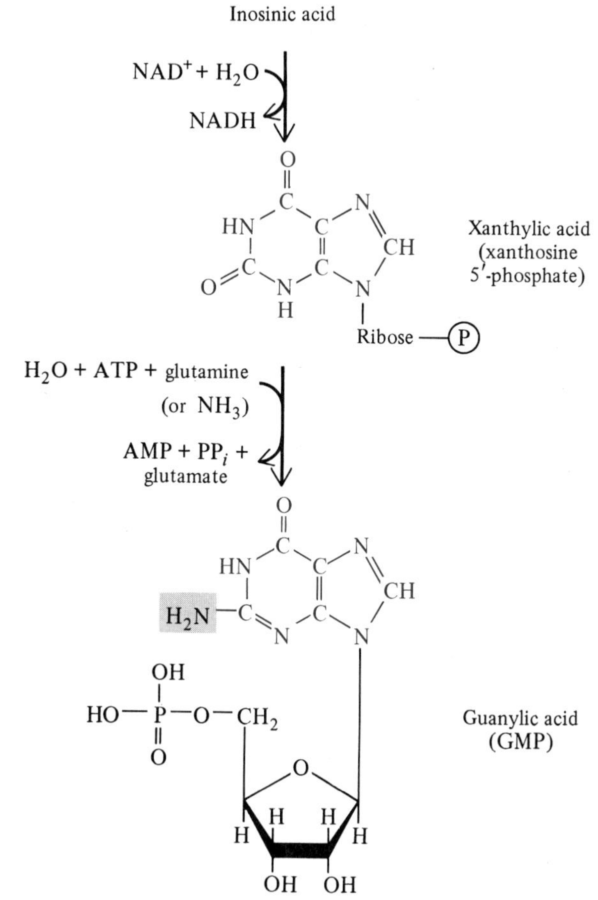

Inosinic acid

$NAD^+ + H_2O$

NADH

Xanthylic acid (xanthosine 5′-phosphate)

H_2O + ATP + glutamine (or NH_3)

AMP + PP_i + glutamate

Guanylic acid (GMP)

Figure 21-10 Formation of GMP from IMP.

21-3 FORMATION OF THE DEOXYRIBONUCLEOTIDES

The deoxyribonucleotides are formed from the corresponding ribonucleotides by deoxygenation of the sugar, thereby converting ribose to 2-deoxy-D-ribose. Two pathways have been identified for the conversion of ribonucleotides, depending upon the species. In *E. coli* all four of the *ribo*nucleoside *di*phosphates (ADP, GDP, UDP, and CDP) are deoxygenated to the corresponding *deoxy*ribonucleoside diphosphates (dADP, dGDP, dUDP, and dCDP) by a complex multienzyme system.

The system includes a protein **thioredoxin**, which contains two free cysteine-SH groups that may be reversibly oxidized to a cystine disulfide. The disulfide form of thioredoxin is reduced to the sulfhydryl (—SH) form by NADPH in a reaction catalyzed by **thioredoxin reductase**, an FAD enzyme:

$$\text{thioredoxin}(\text{—S—S—}) + \text{NADPH} + \text{H}^+ \rightleftharpoons \text{thioredoxin}\begin{matrix}\text{SH}\\ \text{SH}\end{matrix} + \text{NADP}^+$$

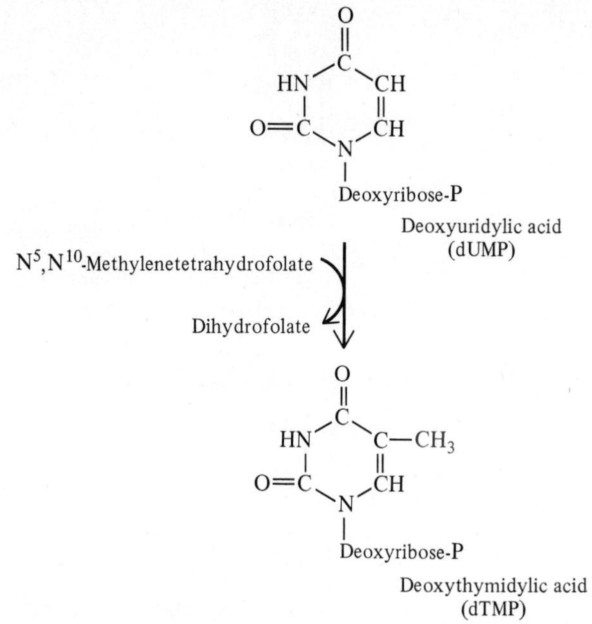

Deoxyribose-P

Deoxyuridylic acid (dUMP)

N^5,N^{10}-Methylenetetrahydrofolate

Dihydrofolate

Deoxyribose-P

Deoxythymidylic acid (dTMP)

Figure 21-1
Formation of dTMP from dUMP.

The reduced form of thioredoxin next removes an oxygen from the ribonucleoside diphosphate (NDP) in a reaction catalyzed by **ribonucleoside diphosphate reductase**.

$$\text{thioredoxin}\begin{matrix}\diagup SH \\ \diagdown SH\end{matrix} + \text{NDP} \longrightarrow \text{thioredoxin}(\text{—S—S—}) + \text{dNDP} + H_2O$$

Each of the four ribonucleoside diphosphates may be reduced in this way to form the analogous 2-deoxy-D-ribose derivative.

In some microorganisms there exists another pathway which utilizes the ribonucleoside *tri*phosphates rather than the diphosphates. Either thioredoxin or dihydrolipoic acid may function as reducing agents.

Biosynthesis of Deoxythymidylic Acid

The precursor of deoxythymidylic acid (dTMP) is deoxyuridylic acid (dUMP). A methyl group is transferred from N^5,N^{10}-methylenetetrahydrofolate (section 7-2) in a reaction catalyzed by **thymidylate synthetase** (Fig. 21-11):

$$N^5,N^{10}\text{-methylene tetrahydrofolate} + \text{dUMP} \rightarrow \text{dTMP} + \text{dihydrofolate}$$

21-4 DEGRADATION OF NUCLEOTIDE BASES

Purines

In mammals the nucleotides produced by the hydrolysis of nucleic acids by nucleases are usually further hydrolyzed to yield the free purine and pyrimidine bases as the ultimate products. The free bases may either re-enter biosynthetic pathways leading to nucleotides and nucleic acids, or may be further degraded and excreted. Most of the nitrogen of the purine bases consumed by mammals is ultimately excreted in the urine as **uric acid** or **allantoin**. The ring structures of the bases are thereby preserved (Fig. 21-12).

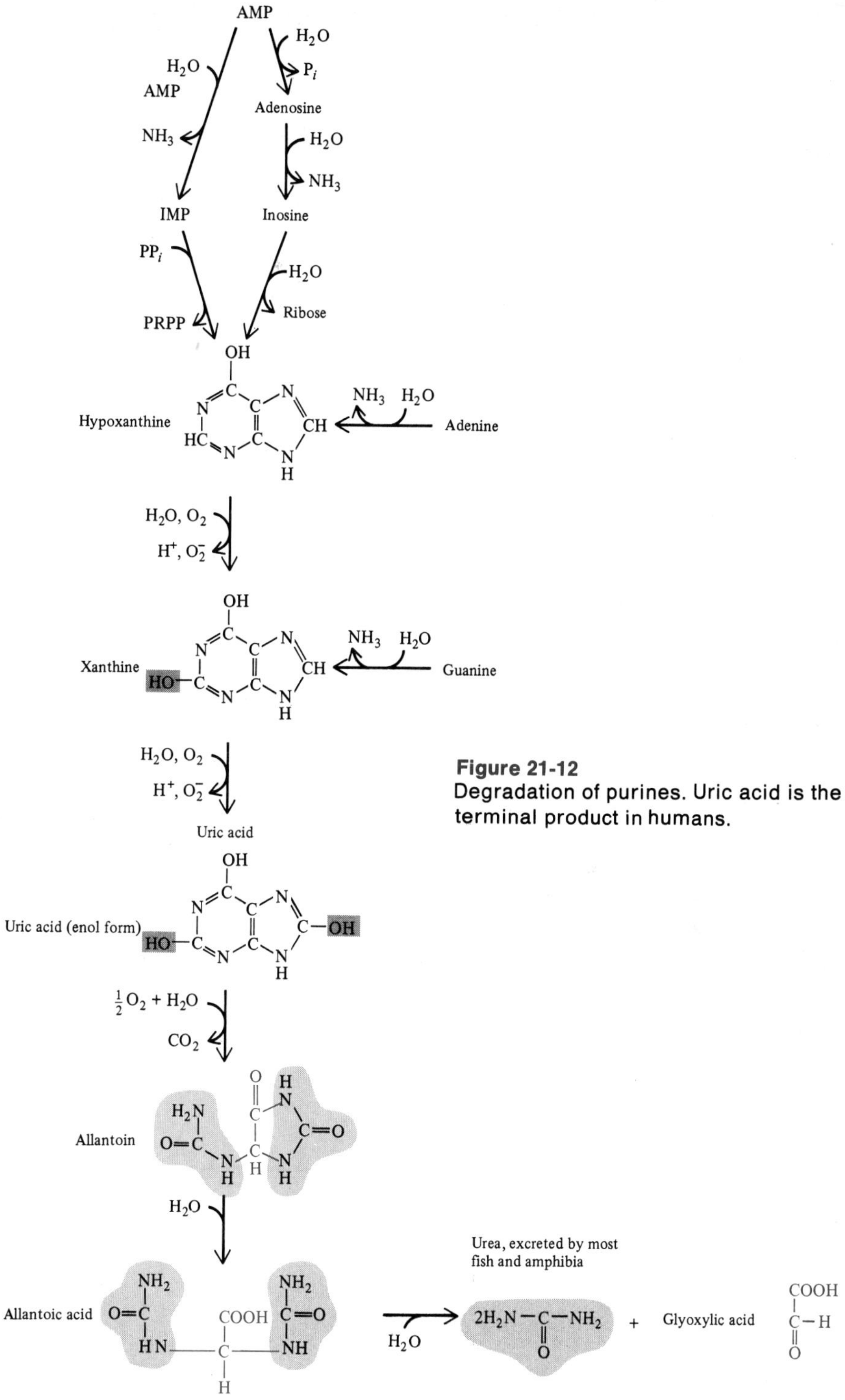

AMP

H_2O, AMP, NH_3 → IMP

H_2O, P_i → Adenosine

H_2O, NH_3 → Inosine

IMP: PP_i, PRPP →

Inosine: H_2O, Ribose →

Hypoxanthine

Adenine → (H_2O, NH_3) → Hypoxanthine

H_2O, O_2 ; H^+, O_2^-

Xanthine

Guanine → (H_2O, NH_3) → Xanthine

H_2O, O_2 ; H^+, O_2^-

Uric acid

Uric acid (enol form)

$\frac{1}{2}O_2 + H_2O$; CO_2

Allantoin

H_2O

Allantoic acid

H_2O → $2H_2N-C(=O)-NH_2$ + Glyoxylic acid (COOH–C(=O)–H)

Urea, excreted by most fish and amphibia

Figure 21-12
Degradation of purines. Uric acid is the terminal product in humans.

In man, both adenine and guanine are first converted to **xanthine** (Fig. 21-12); this is then oxidized to uric acid by the flavoprotein **xanthine oxidase**.

$$\text{xanthine} + H_2O + O_2 \rightarrow \text{uric acid} + O_2^-$$

In some species, including turtles and mollusks, uric acid is further oxidized to allantoin by **urate oxidase** (Fig. 21-12).

$$\text{uric acid} + \tfrac{1}{2}O_2 + H_2O \rightarrow \text{allantoin} + CO_2$$

In fishes allantoin is degraded to allantoic acid by **allantoinase** (Fig. 21-12). In some fishes allantoic acid is further broken down to urea by **allantoicase**.

In man, the terminal product of purine degradation, uric acid, may accumulate to excessive levels, resulting in supersaturation and subsequent precipitation in tissues. Precipitation of uric acid in cartilaginous tissues causes the disease **gout**.

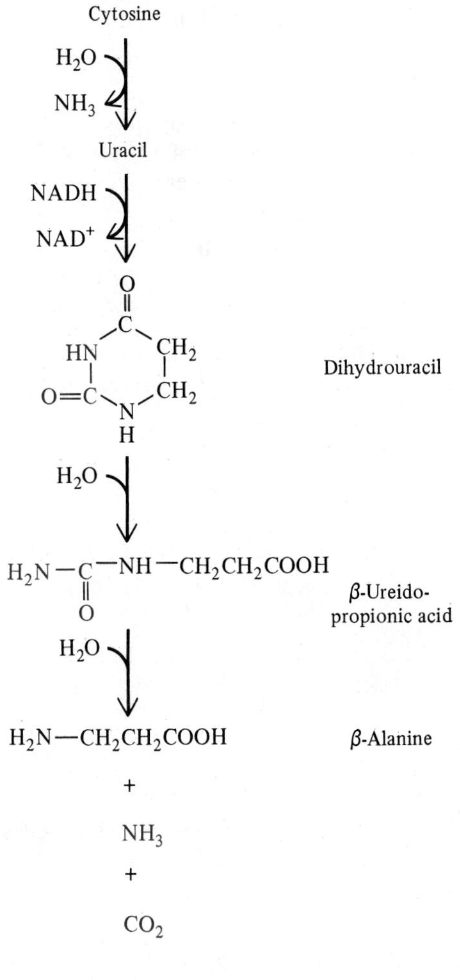

Cytosine

$\downarrow$ (H_2O in, NH_3 out)

Uracil

$\downarrow$ (NADH in, NAD^+ out)

Dihydrouracil (ring: O=C, HN, CH_2, CH_2, N–H, C=O)

$\downarrow$ (H_2O)

$H_2N{-}C({=}O){-}NH{-}CH_2CH_2COOH$ β-Ureido-propionic acid

$\downarrow$ (H_2O)

$H_2N{-}CH_2CH_2COOH$ β-Alanine

+

NH_3

+

CO_2

Figure 21-13
Degradation of pyrimidines.

Pyrimidines

The metabolic degradation of the amino pyrimidines, cytosine and methylcytosine, is channeled through uracil and thymine, respectively, to which they are converted by deamination (Fig. 21-13).

The initial step in the metabolism of uracil and thymine is reduction to yield the dihydro derivatives, dihydrouracil and dihydrothymine. These are hydrolyzed to the corresponding β-ureido compounds, which are subsequently hydrolyzed further to the β-amino acids (Fig. 21-13). β-Alanine, the degradation product of uracil, is a precursor of coenzyme A.

21-5 BIOSYNTHESIS OF THE NUCLEOTIDE COENZYMES

Flavin Nucleotides

Riboflavin (section 7-3) is an essential component of the mammalian diet. Riboflavin-5′-phosphate, also called flavin mononucleotide (FMN), is synthesized from riboflavin and ATP by **riboflavin kinase**:

$$\text{riboflavin} + \text{ATP} \rightarrow \text{riboflavin-5'-phosphate} + \text{ADP}$$

Flavin adenine dinucleotide (FAD) is formed by further combination of riboflavin-5′-phosphate with ATP in a reaction catalyzed by **FMN adenylyltransferase**.

$$\text{riboflavin-5'-phosphate} + \text{ATP} \rightarrow \text{FAD} + \text{PP}_i$$

Pyridine Nucleotides

Nicotinic acid (section 7-3) may be supplied in the diet, or may be synthesized from tryptophan. The sequence of reactions involved in the synthesis of NAD^+ from nicotinic acid in bacteria is as follows:

(1) Nicotinic acid reacts with 5-phosphoribose-1-pyrophosphate to form **nicotinic acid mononucleotide**:

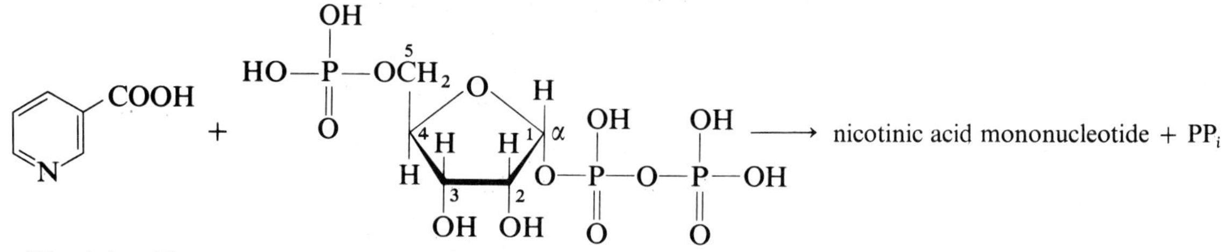

Nicotinic acid + 5-Phosphoribose-1-pyrophosphate ⟶ nicotinic acid mononucleotide + PP_i

This is a 5′-ribonucleotide, in which the base, nicotinic acid, is joined to ribose by an N-C glycosidic bond (Fig. 7-13).

(2) Nicotinic acid mononucleotide next condenses with ATP to yield desamido-NAD^+:

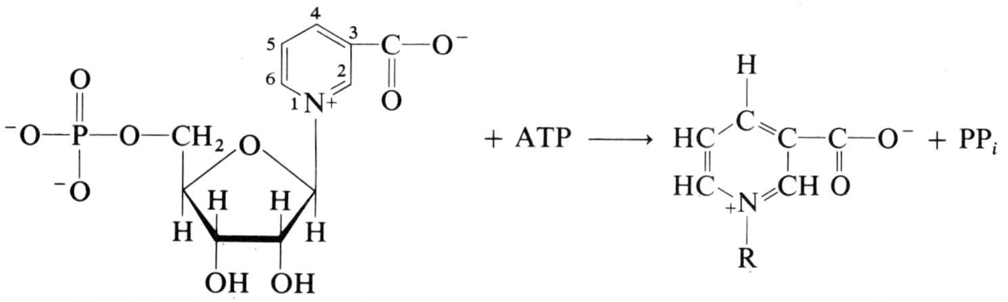

Nicotinic acid mononucleotide + ATP ⟶ Desamido-NAD^+ + PP_i

Desamido-NAD^+ is identical to NAD^+, except for the absence of the amide group on the carboxyl of nicotinic acid.

(3) Finally, in a reaction with glutamine and ATP, desamido-NAD^+ is converted to NAD^+:

$$\text{desamido-NAD}^+ + \text{glutamine} + \text{ATP} \longrightarrow \underset{\text{NAD}^+}{\text{(pyridinium ring: HC=CH, C(H), C–C(=O)–NH}_2\text{, CH=N}^+\text{–R)}} + \text{glutamate} + \text{PP}_i + \text{AMP}$$

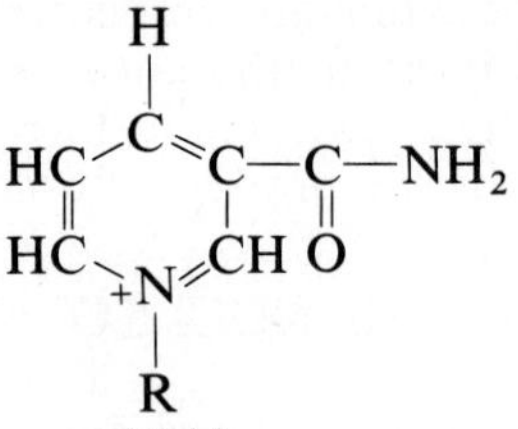

Coenzyme A

The pantothenic acid portion of coenzyme A is an essential nutritional factor for mammals. In mammalian liver and some microorganisms, the pathway for the synthesis of CoA is as follows (Fig. 21-14).

Pantothenic acid is initially phosphorylated by ATP to form 4′-phosphopantothenic acid:

$$\text{pantothenic acid} + \text{ATP} \longrightarrow \underset{\text{4}'\text{-Phosphopantothenic acid}}{H_2O_3PO\text{–}CH_2\text{–}C(CH_3)_2\text{–}CH(OH)\text{–}C(=O)\text{–}NHCH_2CH_2COOH} + \text{ADP}$$

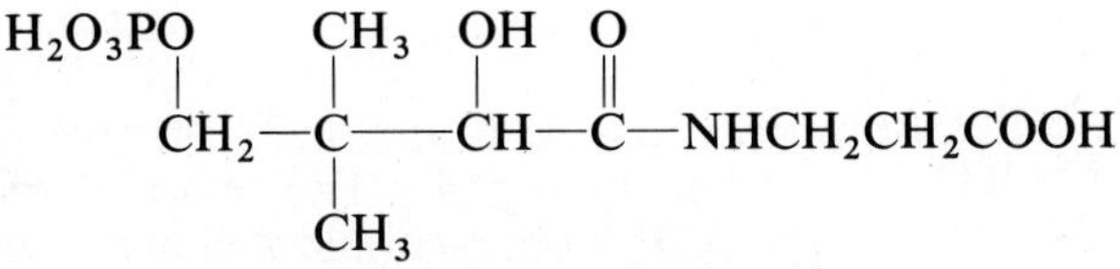

The latter compound combines with cysteine to form 4′-phosphopantothenoylcysteine in a CTP-driven reaction (Fig. 21-14):

$$\text{4}'\text{-phosphopantothenic acid} + \text{cysteine} + \text{CTP} \rightarrow \text{4}'\text{-phosphopantothenoylcysteine} + \text{CDP}$$

4′-Phosphopantothenoylcysteine is next decarboxylated to yield 4′-phosphopantetheine, which is then combined with ATP to form dephospho CoA (Fig. 21-14):

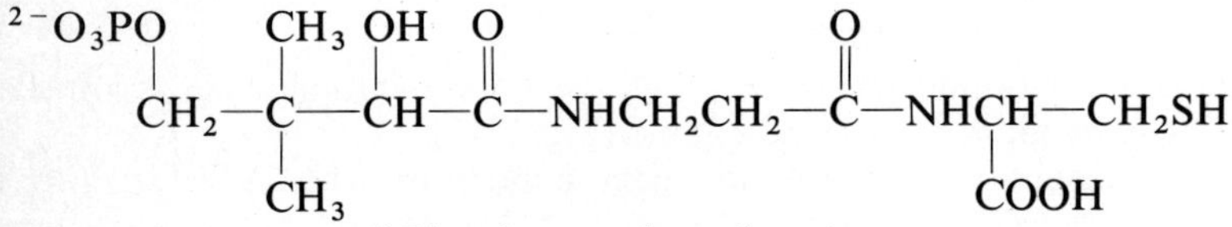

$$\underset{\text{4}'\text{-Phosphopantothenoylcysteine}}{{}^{2-}O_3PO\text{–}CH_2\text{–}C(CH_3)_2\text{–}CH(OH)\text{–}C(=O)\text{–}NHCH_2CH_2\text{–}C(=O)\text{–}NHCH(COOH)\text{–}CH_2SH} \longrightarrow \text{4}'\text{-phosphopantetheine} + CO_2$$

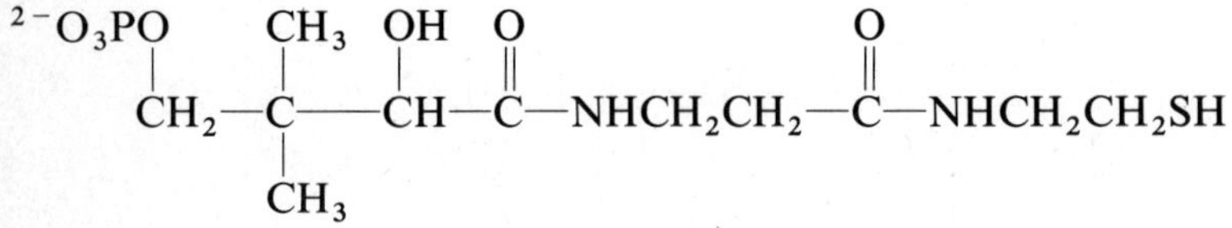

$$\underset{\text{4}'\text{-Phosphopantetheine}}{{}^{2-}O_3PO\text{–}CH_2\text{–}C(CH_3)_2\text{–}CH(OH)\text{–}C(=O)\text{–}NHCH_2CH_2\text{–}C(=O)\text{–}NHCH_2CH_2SH} + \text{ATP} \longrightarrow \text{dephospho CoA} + \text{PP}_i$$

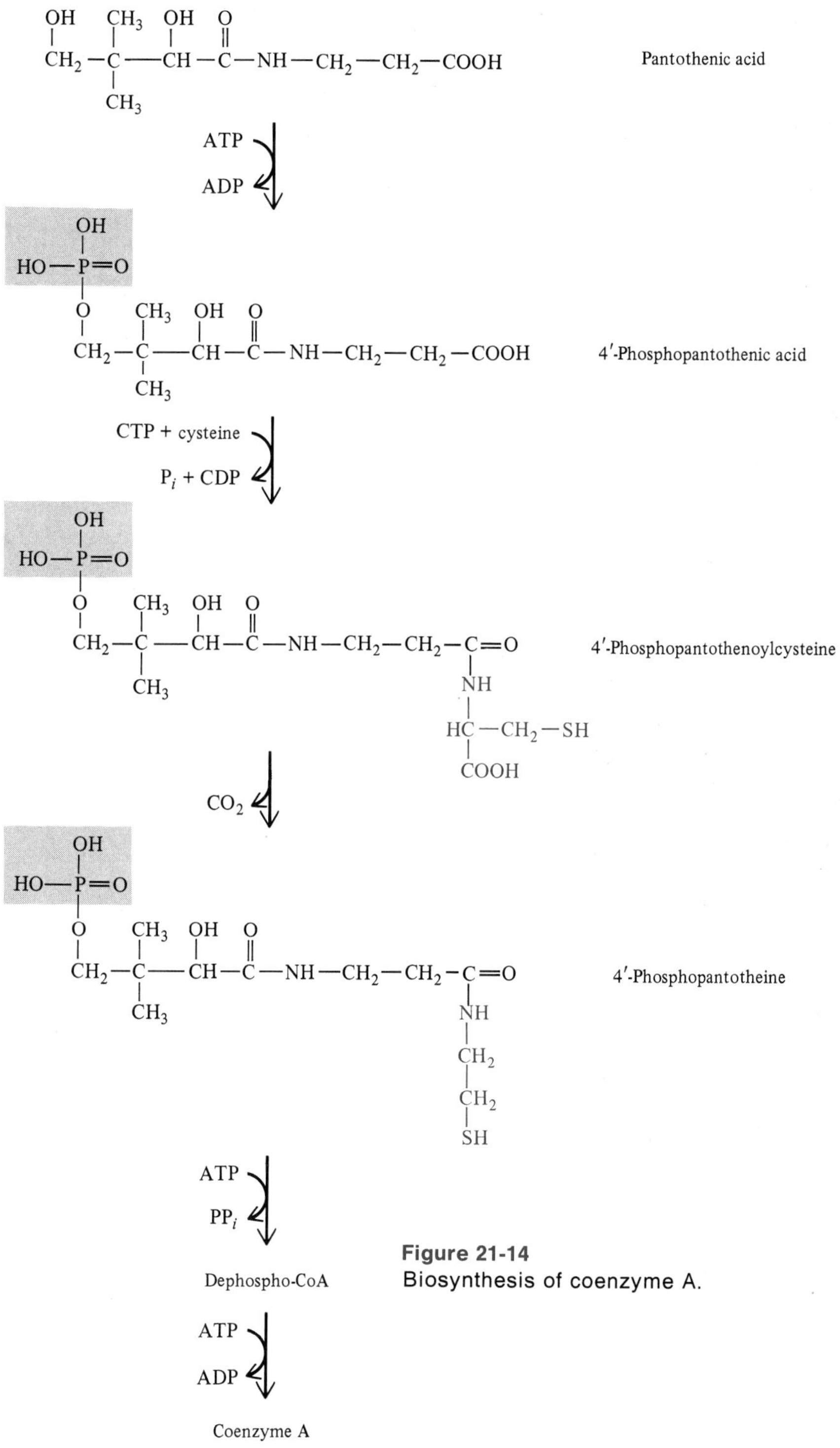

$$HOCH_2-C(CH_3)_2-CH(OH)-C(=O)-NH-CH_2-CH_2-COOH$$

Pantothenic acid

ATP → ADP

$$(HO)_2P(=O)-O-CH_2-C(CH_3)_2-CH(OH)-C(=O)-NH-CH_2-CH_2-COOH$$

4′-Phosphopantothenic acid

CTP + cysteine → P_i + CDP

$$(HO)_2P(=O)-O-CH_2-C(CH_3)_2-CH(OH)-C(=O)-NH-CH_2-CH_2-C(=O)-NH-CH(COOH)-CH_2-SH$$

4′-Phosphopantothenoylcysteine

→ CO_2

$$(HO)_2P(=O)-O-CH_2-C(CH_3)_2-CH(OH)-C(=O)-NH-CH_2-CH_2-C(=O)-NH-CH_2-CH_2-SH$$

4′-Phosphopantotheine

ATP → PP_i

Dephospho-CoA

ATP → ADP

Coenzyme A

Figure 21-14
Biosynthesis of coenzyme A.

Dephospho CoA lacks only the 3′-phosphate of the adenosine portion of CoA. It is converted to CoA by phosphorylation by ATP

$$\text{dephospho CoA} + \text{ATP} \longrightarrow \text{CoA} + \text{ADP}$$

SUMMARY

The pyrimidine ribonucleotides are synthesized by a pathway proceeding via **orotic acid**, which is formed by a reaction sequence beginning with the condensation of **aspartic acid** and **carbamoyl phosphate**. A ribose phosphate moiety is donated by **5-phosphoribose-1-pyrophosphate** to form **orotidine-5′-phosphate**, which is decarboxylated to yield UMP. UDP and UTP are formed by consecutive phosphorylations of UMP catalyzed by a kinase.

UTP is the precursor of CTP, to which it is converted by amination. CTP is an **allosteric** inhibitor of **aspartate transcarbamoylase**, which catalyzes the initial condensation of aspartate and carbamoyl phosphate. CTP is thus the key compound in the regulation of the rate of pyrimidine ribonucleotide synthesis.

Purine ribonucleotide biosynthesis begins with **ribose-5-phosphate**, which is activated by conversion to **5-phosphoribose-1-pyrophosphate**. After conversion to the amine **5-phospho-β-D-ribosylamine** by reaction with glutamine, a condensation occurs with glycine to yield **5′-phosphoribosyl-glycinamide**, which contains four atoms of the purine ring. The remaining atoms of the purine ring are introduced one by one until the process culminates in the formation of **inosine-5′-phosphate** (IMP). Amination of IMP at the 6-position yields AMP; conversion of IMP to xanthylic acid, followed by amination, forms GMP. AMP and GMP are converted to di- and triphosphates by consecutive kinase-catalyzed phosphorylations.

Regulation of IMP biosynthesis is achieved at the early step whereby **5-phospho-β-ribosylamine** is formed. The enzyme **amidophosphoribosyl transferase**, which catalyzes this reaction, is inhibited by ATP, ADP, or AMP, as well as by GTP, GDP, or GMP; the two classes of nucleotides bind to distinct allosteric sites.

The deoxyribonucleoside diphosphates are formed from the corresponding ribonucleoside diphosphates via reduction by **thioredoxin**.

The purines and pyrimidines formed by degradation of nucleic acids may be reused for nucleotide synthesis or may be degraded and excreted. The purines are ultimately degraded to uric acid in human beings, to allantoin in turtles and mollusks, and to allantoic acid or urea in fish.

REFERENCES

The following code is used to classify references. I: particularly useful as an introduction to the subject; R: useful primarily as a reference text; A: an advanced account of the material; H: a publication of historical importance.

General

S. Dagley and D. E. Nicholson, *An Introduction to Metabolic Pathways*, Wiley, New York (1970). (I)

J. N. Davidson, *The Biochemistry of Nucleic Acids*, Academic, New York (1972). (A)

J. F. Henderson and A. R. P. Paterson, *Nucleotide Metabolism*, Academic, New York (1973). (A)

I. G. Leder, *Metabolic Pathways*, 3rd ed., Academic, New York (1975). (R)

J. B. Stanbury, J. B. Wyngaarden, and D. S. Fredrickson, *The Molecular Basis of Inherited Disease*, 3rd ed., McGraw-Hill, New York (1972). (A)

Regulation

J. F. Henderson, "Regulation of Purine Biosynthesis," *Am. Chem. Soc. Monog.* 170, Washington, D.C. (1972). (A)

Formation of Deoxyribonucleotides

A. Larsson and P. Reichard, "Enzymatic Reduction of Ribonucleotides," *Prog. Nucleic Acid Res. Mol. Biol.* 7, 303 (1967). (A)

REVIEW QUESTIONS

Questions marked with an asterisk are of a high level of difficulty.

*21-1 Write a balanced equation for the biosynthesis of UTP from CO_2, NH_3, ATP, oxaloacetate, and ribose-5-phosphate by *E. coli.*

*21-2 How many high energy phosphate bonds are required for the synthesis of one molecule of UTP?

21-3 Which of the atoms of AMP are derived directly from glycine?

21-4 Which of the atoms of AMP are derived directly from aspartic acid?

*21-5 The greater part of the free adenine produced by vertebrates is recycled by the reaction

$$\text{adenine} + \text{5-phosphoribose-1-pyrophosphate} \rightarrow \text{AMP} + \text{PP}_i$$

If the AMP thereby salvaged is reused for nucleic acid synthesis, how many high energy phosphate bonds are conserved?

*21-6 Speculate on the biological significance of the fact that 5-phosphoribose-1-pyrophosphate is a precursor of both the purine and pyrimidine nucleotides.

22

Hormonal Regulation

22-1 GENERAL ASPECTS

Nothing in biology is more impressive than the elaborate system of controls that governs the multitude of biochemical reactions occurring in the various tissues of a complex organism. Within the cells of all organisms, regulation is achieved by the induction or repression of protein synthesis, as well as by metabolic feedback, whereby a product or intermediate modifies the kinetics of an enzyme placed earlier in the pathway.

In higher organisms an additional dimension of regulation is provided by hormonal control. In this case the active substances are synthesized by specialized tissues or **glands** and secreted into the circulation.

Hormones are compounds synthesized and secreted by one tissue which act on a different group of cells, often called the **target tissue**.

Hormones are particularly remarkable for the extremely low concentrations at which they are effective. The minute quantities present in tissues have presented major problems of isolation and purification.

One cannot generalize as to the mechanisms of hormonal action. Some hormones interact directly with key enzymes and alter their kinetics. Others influence the permeability of membranes. In some cases the hormone may intervene in protein synthesis by interacting with specific gene loci, thereby stimulating or blocking the formation of mRNA and the synthesis of specific enzymes.

Efficient regulation requires some mechanism for the removal of excess hormones from the circulation; the organism would otherwise soon become saturated with hormones.

With respect to chemical nature, the three principal classes of hormones are amino acid derivatives, peptides or proteins, and steroids. Table 22-1 lists the principal known hormones and their physiological effects.

Table 22-1

A: Major Peptide Hormones

Hormone	Site of Synthesis	Site of Action	Major Physiological Role
Thyrotropin (TSH)	Anterior pituitary	Thyroid gland	Stimulates synthesis and secretion of thyroid hormone
Adrenocorticotropin (ACTH)	Anterior pituitary	Adrenal medulla	Stimulates synthesis of adrenal steriods
Growth Hormone (Somatotropin)	Anterior Pituitary	Many tissues	Promotes protein synthesis and skeletal growth
Luteinizing Hormone (LH)	Anterior pituitary	Gonads	Stimulates synthesis of progesterone in ovary and testosterone in testes; causes ovulation
Follicle Stimulating Hormone (FSH)	Anterior pituitary	Gonads	Stimulates growth of ovarian follicle and of sertoli cells of testes
Prolactin	Anterior pituitary	Breast	Synthesis of milk proteins and growth of breast
Insulin	Pancreas	Many tissues	Promotes transport of glucose and amino acids into certain cells; synthesis of fatty acids in adipose cells and in liver; glycolysis of glucose; protein synthesis
Glucagon	Pancreas	Liver	Stimulates glycogenolysis and gluconeogenesis in liver
Antidiuretic Hormone (vasopressin)	Hypothalamus	Kidney	Prevents loss of water and NaCl and controls blood pressure
Oxytocin	Hypothalamus	Milk glands	Secretion of milk and uterine contractions
Parathyroid Hormone	Parathyroid gland	Bone, kidney	Acts to increase Ca^{2+} in blood
Calcitonin	Thyroid C cells	Bone, kidney	Acts to decrease Ca^{2+} in blood

B: Major Steroid Hormones

Hormone	Site of Synthesis	Site of Action	Major Physiological Role
Cortisol	Adrenal cortex	Liver and peripheral tissue	Stimulates glycogenolysis and synthesis of certain liver proteins; promotes protein breakdown in peripheral tissues
Aldosterone	Adrenal cortex	Kidney	NaCl retention
Estradiol	Ovary	Uterus, breast	Secondary female sex characteristics
Testosterone	Testis	Spermatogonia	Sperm synthesis; secondary male sex characteristics
Progesterone	Ovary, placenta	Uterus, breast	Preservation of pregnancy
Vitamin D_3	Skin	Bone, kidney	Stimulates Ca^{2+} transport by small intestine; acts on bone to increase Ca^{2+} in blood

Table 22-1 *(continued)*

C: Hormones Related to Amino Acids

Hormone	*Site of Synthesis*	*Site of Action*	*Major Physiological Role*
Epinephrine	Adrenal medulla	Liver, heart, adipose tissue	Stimulates glycogenolysis and breakdown of fats; increases cardiac output
Norepinephrine	Adrenal medulla, sympathetic nerves, central nervous system	Heart, adipose tissue; synaptic cleft	Increases blood pressure; neurotransmitter
Melatonin	Pineal	Gonads in rodents	Regulates sexual cycle
Thyroid hormones (tetraiodothyronine and triiodothyronine)	Thyroid gland	Many tissues	Increases respiration; required for nervous system growth in fetus and young animal; stimulates synthesis of certain enzymes

22-2 THYROID HORMONES

Thyroxine, the principal hormone of the thyroid gland, was one of the first hormones to be recognized and isolated. It is an iodinated derivative of the aromatic amino acid thyronine (Fig. 22-1). An unusual feature of its structure is the presence of a diphenyl ether group. The di- and triiodo derivatives of thyronine are also physiologically active.

The thyroid gland has a generalized influence upon the overall metabolism of the organism and especially upon oxygen consumption. A perturbation of thyroid function is reflected by an alteration in the **basal metabolic rate** or rate of oxygen consumption, which is elevated in **hyperthyroidism** and depressed in **hypothyroidism**. The administration of thyroxine relieves the symptoms of hypothyroidism.

Thyroxine produces an increased level of several proteins involved in respiration; it also augments the rate of respiration of mitochondria and induces mitochondrial swelling. There is some evidence that it may influence gene expression.

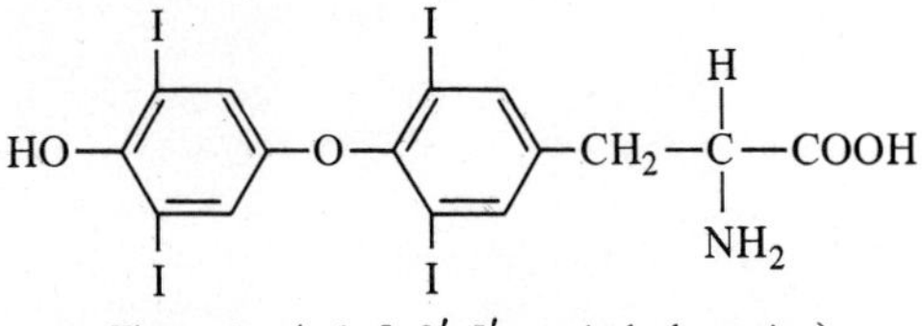

Thyroxine (L-3, 5, 3′, 5′-tetraiodothyronine)

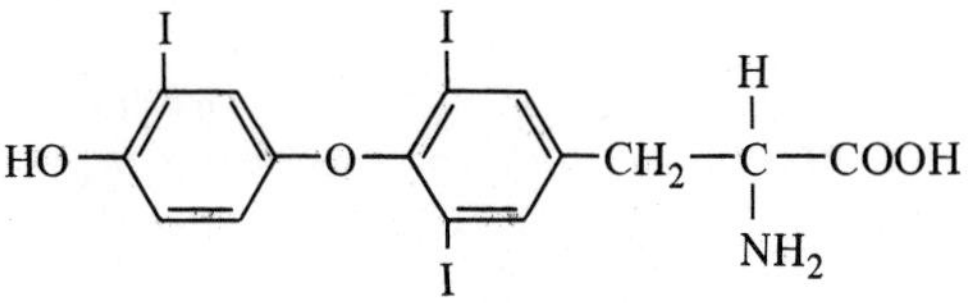

Triiodothyronine (L-3, 5, 3′-triiodothyronine)

Figure 22-1 Structures of thyroxine and triiodothyronine.

The gross physiological effects of severe thyroxine deficiency in mammals include a slowing or cessation of growth, a loss of reproductive capacity, and changes in the hair and skin. Lack of thyroxine adversely affects the functioning of all tissues and glandular organs. Hypothyroidism in early childhood may result in severe mental retardation.

At the other extreme of thyroid malfunction, hyperthyroidism produces accelerated metabolism, loss of weight, and serious psychological disturbances. A frequent symptom in human patients is a pronounced bulging of the eyeballs, giving the impression of surprise or panic.

Thyroxine is synthesized in the thyroid gland from the tyrosine residues of the protein **thyroglobulin**. Iodide is withdrawn from the circulation, oxidized to iodine, and introduced into the tyrosine ring. The combination of two iodinated tyrosines yields thyroxine. Proteolytic degradation of thyroglobulin releases free thyroxine into the circulation, where it is bound and transported by several protein carriers, of which the most important is **thyroxine binding globulin**. Like the other amino acids it is ultimately degraded.

The biosynthesis of thyroxine is of course dependent upon an adequate iodine supply. This is rarely a problem except in a few restricted geographical areas.

22-3 PANCREATIC HORMONES

The pancreas has already been mentioned (section 6-4) as the site of formation of the zymogens of a set of proteolytic enzymes. It is also the site of biosynthesis of two important hormones, **insulin** and **glucagon**, which are produced by specialized cells termed **islet cells**. Insulin is synthesized and secreted by the β-type of islet cells and **glucagon**, by the **α-type**.

Insulin

Insulin, a globular protein of molecular weight 5,700, consists of two polypeptide chains joined by disulfide bonds (section 4-4). Two precursors of insulin have been identified, neither of which possesses any hormonal activity. **Prepropinsulin**, a single polypeptide of molecular weight 11,500, is probably the translation product of mRNA. The cleavage of a sequence of 23 amino acids from the NH_2-terminal end of the molecule yields **proinsulin**; the latter is incorporated into granules formed from the Golgi apparatus (chapter 1), which subsequently enter the cytosol. Folding of the molecule and formation of disulfide bonds occur. A sequence of 30 amino acids, termed the C-peptide, is excised by a proteolytic enzyme to yield insulin (Fig. 22-2).

Regulation of Insulin Secretion

The major regulator of the secretion of insulin by the pancreas is the concentration of blood glucose. An elevation of blood glucose stimulates the secretion of insulin; a deficiency has the opposite effect. Insulin secretion is also stimulated by the ingestion of proteins and by the intravenous administration of arginine and leucine, as well as by several hormones, including glucagon and growth hormone.

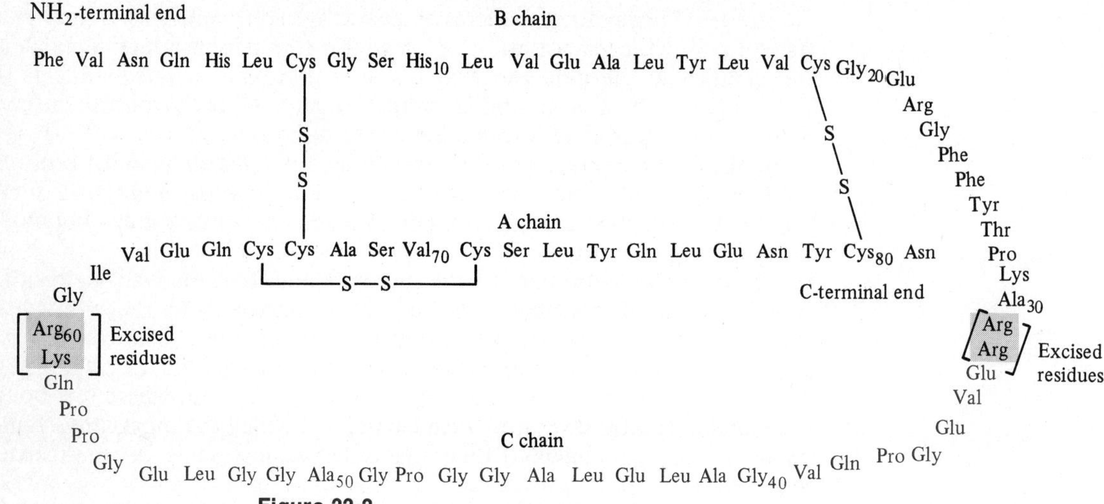

Figure 22-2
Formation of insulin from its precursor.

Secretion of insulin involves a migration of the insulin-containing granules to the plasma membrane of the β-cell and their fusion with it. Secretion occurs via an extrusion of insulin from the cell in a reaction to stimulation.

The Action of Insulin

Many biological effects have been ascribed to insulin. There is strong evidence that insulin acts by first binding to a specific glycoprotein receptor in the plasma membrane of a target cell. Beyond this, it is not as yet possible to cite any explicit mechanisms for the manifold actions of this hormone upon carbohydrate and lipid metabolism.

The role of insulin in carbohydrate metabolism includes its stimulation of the transport of glucose into skeletal and heart muscle cells. At the usual (fasting) levels of blood glucose (70–90 mg/100 ml), the presence of insulin is required for the normal rate of uptake of glucose by these tissues. If there is a deficiency of insulin, as in diabetes, there is a reduction in the rate of transport of glucose which is partially compensated for by an elevated level of blood glucose. However, insulin has little effect upon the glucose transport systems of many other tissues, including brain, erythrocytes, liver, and kidney.

Insulin also influences glucose utilization by promoting its conversion to glycogen in liver and muscle, by enhancing glycolysis (section 10-4), and by inhibiting gluconeogenesis (section 12-5) in liver. Insulin stimulates both glycogen synthesis and glycolysis by increasing the level of glucokinase and thereby producing a higher concentration of glucose-6-phosphate, a substrate of glycolysis and a precursor of glycogen. An additional stimulation of glycogen synthesis arises from the enhancement by insulin of the activity of glycogen synthetase.

Insulin affects protein metabolism in several ways: it stimulates the transport of amino acids into cells and also enhances the incorporation of amino acids into proteins, particularly in muscle.

Fat metabolism is also influenced by insulin. Triacylglycerol synthesis is stimulated in all tissues, but especially in liver and adipose tissue. Insulin is required for the optimal synthesis of the enzymes required for fatty acid formation in liver, including acetyl CoA carboxylase and the fatty acid synthetase complex. These enzymes are reduced in concentration in the livers of diabetic animals and are increased by the administration of insulin. In adipose cells insulin stimulates the accumulation of triacylglycerols in two ways: it increases the rate of transport of glucose into the cells, thereby increasing the rates of glycolysis and production of α-glycerophosphate, a precursor of triacylglycerols; it also acts to retard the lipolysis of triacylglycerols.

The disease **diabetes** results from a *relative* insufficiency of insulin. Two forms of the disease have been distinguished: the **juvenile**, in which insulin is essentially absent, and the **maturity onset**, in which insulin is present in the pancreas and in blood, but is either released without adequate control or not bound efficiently to cells. The major symptom of diabetes is **hyperglycemia**, stemming from a reduced efficiency of glucose transport into muscle and fat cells. A secondary symptom is the appearance in the blood of the "ketone bodies" acetoacetate and β-hydroxybutyrate. These are accompanied by hunger, thirst, and weight loss.

Glucagon

This hormone is a polypeptide of molecular weight 3,500, consisting of 29 amino acids. Glucose is the most important regulator of glucagon secretion; elevated glucose levels suppress glucagon secretion, while low glucose levels stimulate it. Many amino acids, including alanine and arginine, enhance secretion.

Glucagon has two major effects in the liver, both of which tend to increase the level of blood glucose. It increases the rate of phosphorylysis of glycogen (section 10-9), while curtailing the action of glycogen synthetase; it also stimulates gluconeogenesis. The first of these is the result of the interaction of glucagon with its receptor sites, which produces an increase in the level of cyclic AMP. Cyclic AMP figures prominently in the cascade of reactions (section 10-9) that culminate in the activation of glycogen phosphorylase and in the deactivation of glycogen synthetase. The mechanism of the stimulation of gluconeogenesis is uncertain.

22-4 HORMONES OF THE ADRENAL MEDULLA

The adrenal gland is composite in structure, consisting of the adrenal medulla and the adrenal cortex. The steroid hormones produced by the adrenal cortex have already been discussed in chapter 16.

The three principal hormones synthesized in the adrenal medulla are **epinephrine**, **norepinephrine**, and **dopamine** (section 20-6). Dopamine, a metabolic precursor of the other two, also has a hormonal function of its own, acting as a neurotransmitter for specific nerves in the central nervous

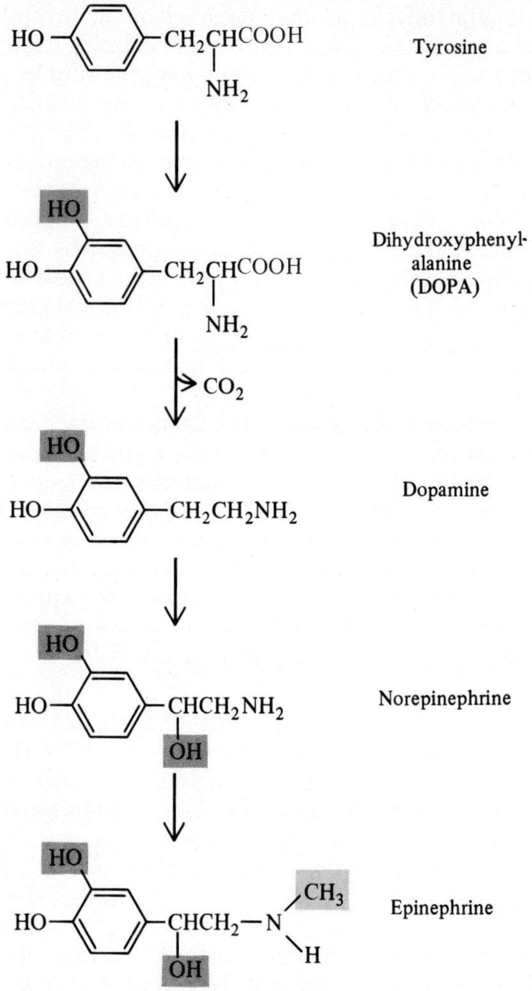

Tyrosine

Dihydroxyphenylalanine (DOPA)

CO_2

Dopamine

Norepinephrine

Epinephrine

Figure 22-3 Biosynthesis of epinephrine and norepinephrine.

system; it occurs in storage granules in sympathetic nerves and in the central nervous system. Norepinephrine and epinephrine (Fig. 22-3) are stored in medullary **chromaffin** granules and are released into the blood upon appropriate stimulus. Like dopamine, norepinephrine is also stored in granules in sympathetic nerves and in the central nervous system and functions as a neurotransmitter. While epinephrine is largely confined to the adrenal medulla, the other two are widely distributed in body tissues. These three hormones are often referred to as **catecholamines**.

Biosynthesis of the Catecholamines

The biosynthetic pathway for the formation of dopamine, epinephrine, and norepinephrine is depicted in Fig. 22-3. The four enzymes involved do not have the same cellular location, so that there is transport of the compounds into and out of the chromaffin granules.

Conversion of tyrosine to dopa (3,4-dihydroxyphenylalanine) is catalyzed by **tyrosine hydroxylase**. This is the rate-limiting step in the pathway and, as such, is probably the main point of allosteric regulation. Tyrosine hydroxylase is inhibited by dopamine, norepinephrine, and epinephrine, which probably act as feedback inhibitors.

Tyrosine hydroxylation takes place in the cytoplasm of the cell, as does the next step, the decarboxylation of dopa to yield dopamine, which is catalyzed by **aromatic amino acid decarboxylase**, a pyridoxal phosphate enzyme. As the name implies, the enzyme is nonspecific and is able to catalyze the decarboxylation of several aromatic amino acids. While tyrosine hydroxylase is restricted to those cells which synthesize catecholamines, the decarboxylase is widely distributed.

The conversion of dopamine to norepinephrine is catalyzed by **dopamine-β-hydroxylase**, which requires a reducing agent, such as ascorbic acid, as a cofactor. This enzyme is largely bound by the chromaffin granules, so that dopamine must enter these granules for conversion to norepinephrine.

The final step in the pathway is the N-methylation of norepinephrine, which is catalyzed by **phenylethanolamine-N-methyl-transferase**, with S-adenosylmethionine as the methyl group donor. Since the transferase is present in the cytoplasm, the reaction requires a diffusion of norepinephrine from the granules.

The human adrenal medulla contains about 8 mg of epinephrine and 2 mg of norepinephrine within the chromaffin granules. Similar granules are found at sympathetic nerve endings and in nerve terminals in the brain.

Biological Roles of the Catecholamines

Norepinephrine and dopamine have life-essential functions as neurotransmitters for their respective neurons. A deficiency in the tyrosine hydroxylase system required for their biosynthesis is associated with severe neurological disorders.

The roles of epinephrine are quite different and concern the glycogen enzyme system of the liver and the hormone-sensitive lipase of adipose tissue. The rate of release of epinephrine from the adrenal medulla is normally quite low, but is greatly augmented under conditions of stress. The resultant elevated level of epinephrine in the blood is sufficient to cause physiological effects. Epinephrine interacts with receptors on the surfaces of sensitive cells; this interaction results in an activation of **adenylate cyclase**, which leads to an increase in cyclic AMP (section 10-10). Cyclic AMP is the activator for **cAMP-dependent protein kinase**. The latter, in turn, phosphorylates and activates **glycogen phosphorylase**. Cyclic AMP-dependent protein kinase also phosphorylates and **inactivates** glycogen synthetase (section 10-10). The combined effects produce a breakdown of liver glycogen and a formation of glucose-1-phosphate, which is converted to glucose-6-phosphate and then to glucose, causing an increase in the level of blood glucose.

In adipose cells, the rise in cyclic AMP level results in the activation of a hormone-sensitive lipase, causing an increase in the formation of free fatty acids and their secretion into the blood. A number of other hormones also have this effect.

Epinephrine also increases cardiac output; this effect may also be mediated by cyclic AMP. The mechanism remains obscure.

As long as epinephrine is secreted into the blood by the adrenal medulla, the liver adenylate cyclase system continues to be active, thereby maintaining a high level of cyclic AMP. When epinephrine secretion ceases, the bound epinephrine on the liver cell membrane dissociates. Cyclic AMP ceases to be formed and the remainder is decomposed by a phosphodiesterase.

22-5 PITUITARY HORMONES

The pituitary gland or **hypophysis** is located immediately below an area of the brain called the **hypothalamus** and is composed of two or three elements, depending upon the species. The functions of the **anterior pituitary** are regulated by the hypothalamus. This close relationship is mediated by the blood supply of the anterior pituitary, most of which flows to the gland via long portal vessels from the hypothalamus. The hypothalamus contains small peptide hormones which travel to the anterior pituitary and there regulate the release of six or seven hormones.

The **posterior pituitary** or **neurohypophysis** has neural connections with the hypothalamus. The important peptide hormones of the posterior pituitary, **oxytocin** and **antidiuretic hormone** (ADH), are synthesized in the hypothalamus and are then transported to the posterior pituitary for storage.

In lower animals, there exists a third element of the pituitary, called the **intermediate lobe**, which synthesizes **α- and β-melanocyte stimulating hormones** (α- and β-MSH).

Anterior Pituitary Hormones

There are at least six peptide hormones secreted by the anterior pituitary which have important biological roles; each of these is synthesized by a specific cell type.

Adrenocorticotropin (ACTH), a peptide of 39 amino acid residues (section 4-4), stimulates the synthesis of steroid hormones from the adrenal cortex (section 16-6). There is evidence that a larger precursor of ACTH is synthesized initially and subsequently degraded to the ACTH secreted by the pituitary. This is a common pattern for peptide hormones. The adrenal cortex is the target tissue of ACTH; there it stimulates the conversion of cholesterol to pregnenolone, thereby accelerating the synthesis of several steroid hormones (section 16-6). ACTH also seems to be required for maintaining the integrity of those specialized adrenal cells which synthesize the steroids; these cells atrophy in animals whose pituitaries have been removed.

Thyrotropin, **luteinizing hormone** (LH), and **follicle stimulating hormone** (FSH) are structurally related, altthough different in function. All are glycoproteins and consist of two subunits, one of which, the **α-subunit**, is common to all three. Thyrotropin stimulates the thyroid gland to incorporate iodide into the thyroid hormones. LH acts on the gonads to stimulate the synthesis of sex steroids (androgens in males, progesterone and estrogens in females) from cholesterol. FSH furthers growth and development of the gonads: in the ovaries the follicles are enlarged; in the testes sperm formation is augmented. FSH has also a dominant function in the initial stages of the menstrual cycle. Each of the above three hormones may be dissociated and separated into its two subunits, neither of which possesses hormonal activity; activity is restored upon recombination. Specificity is conferred by the ***β*-subunit**; if the α-subunit from thyrotropin is recombined with the β-subunit from LH, the recombinant has LH activity.

Prolactin stimulates milk secretion by the mammary gland. The stimulation occurs only after complete development of the gland through the action of the ovarian hormones.

Somatotropin or growth hormone stimulates protein synthesis, fat metabolism, and bone growth. Its administration produces a characteristic weight gain, which has been used as a biological assay. Somatotropin exerts its action indirectly by raising the level of another peptide, **somatomedin**, which is probably synthesized in the liver.

Regulaton of the Anterior Pituitary Hormones

The release of the anterior pituitary hormones is controlled by the hypothalamus. This control is exercised by means of a set of peptide hormones, whose target tissue is the anterior pituitary. Three of the peptide hormones (Fig. 22-4) from the hypothalamus that regulate the secretion of particular

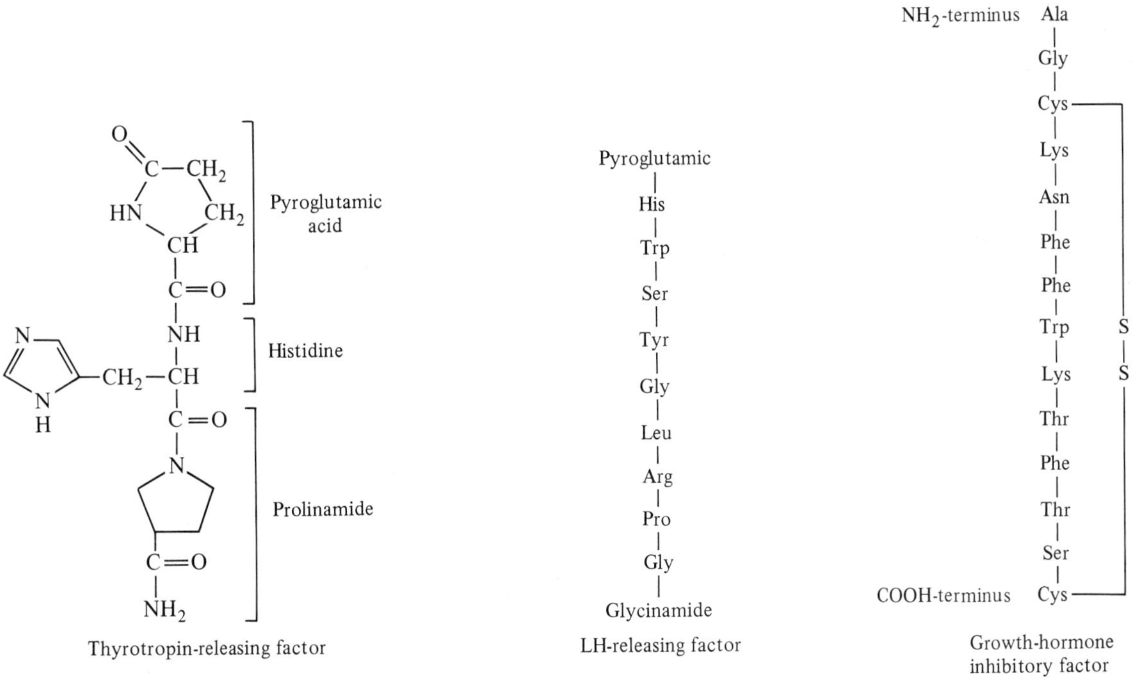

Pyroglutamic acid
Histidine
Prolinamide
Thyrotropin-releasing factor

Pyroglutamic—His—Trp—Ser—Tyr—Gly—Leu—Arg—Pro—Gly—Glycinamide
LH-releasing factor

NH_2-terminus Ala—Gly—Cys—Lys—Asn—Phe—Phe—Trp—Lys—Thr—Phe—Thr—Ser—Cys COOH-terminus (S—S bridge between the two Cys)
Growth-hormone inhibitory factor

Figure 22-4
Peptide hormones secreted by the hypothalamus. Growth-hormone inhibitory factor is also called somatotropin release inhibitor hormone.

anterior pituitary hormones have been isolated and characterized: **thyrotropin releasing hormone** (TRH), **LH releasing hormone** (LRH), and **somatotropin release inhibitor hormone** (SRIH).

TRH stimulates the secretion of both thyrotropin and prolactin; LRH mediates the release of both luteinizing hormone and follicle stimulating hormone; SRIH inhibits the release of both thyrotropin and somatotropin. There is also evidence for a **corticotropin releasing factor** (CRF), which governs the secretion of ACTH, and for a second prolactin releasing factor. However, the hypothalamus appears to have a predominantly inhibiting effect upon the secretion of prolactin; dopamine is the primary inhibiting factor.

It is of interest that TRH, LRH, and SRIH have been detected in other regions of the brain and that SRIH has been found in the pancreas; this suggests that these hormones may have a broader function than that outlined above.

The major elements involved in the control of ACTH secretion by the anterior pituitary are CRF and **cortisol** (section 16-6), a steroid hormone of the adrenal cortex whose synthesis is stimulated by ACTH. An excessive level of cortisol tends to inhibit the release of CRF from the hypothalamus and thereby to block its own formation:

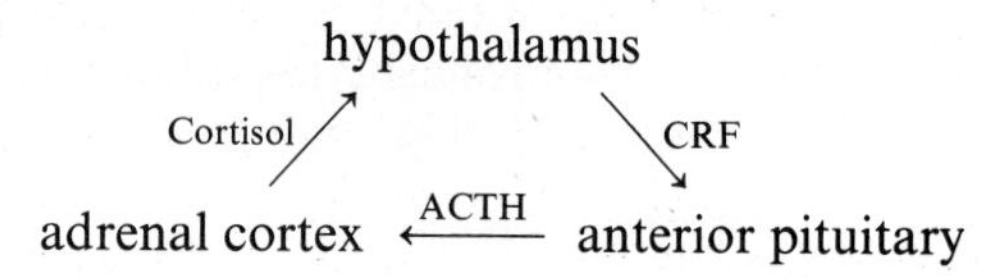

In some animals cortisol inhibits the release of ACTH from the anterior pituitary by direct action on this gland.

Posterior Pituitary Hormones

The hormones **oxytocin** and **vasopressin** (Fig. 22-5) are synthesized in the hypothalamus as parts of larger precursor polypeptides termed **neurophysins**. The active hormones are split from their neurophysin precursors by enzymic

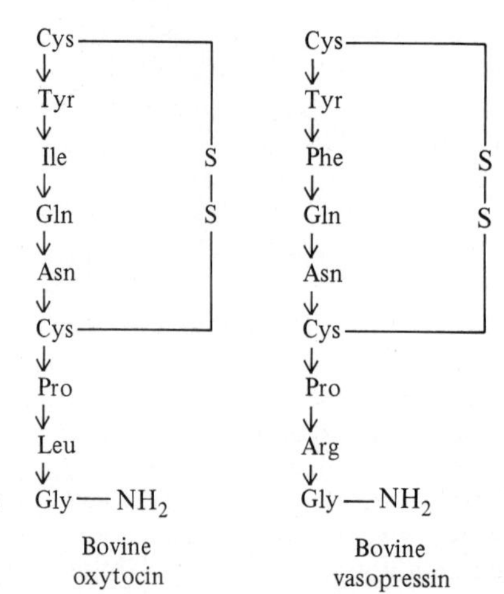

Figure 22-5 Structures of oxytocin and vasopressin.

proteolysis and then form molecular complexes with their specific (altered) neurophysins. The complexes are transported from the hypothalamus to the posterior pituitary, where they are stored. Both the hormones and the neurophysins are secreted from the gland, but the hormones are dissociated from the neurophysins while circulating in the blood.

Suckling of the breast by the young is a stimulus for the secretion of oxytocin. In turn, oxytocin causes rapid release of milk by the gland. This hormone also causes uterine contractions during parturition and has been used in obstetrics to induce labor.

Vasopressin acts on the distal convoluted tubules and collecting ducts of the kidney to promote the rapid reabsorption of water. In humans, a dose of 0.1 μg produces a maximum effect. The stimuli for the secretion of vasopressin include a fall in blood pressure and a loss of plasma volume. The action of vasopressin differs from that of aldosterone (section 16-6) in that the reabsorption of water, rather than Na^+, is stimulated. However, both produce the same ultimate effect, namely, the retention of water in the blood.

The Mechanism of Action of the Peptide Hormones

The biological action of most peptide hormones arises from their interaction with specific receptor sites located in the plasma membranes of the target cells. Binding of the hormone activates an adjacent membrane-bound molecule of **adenylate cyclase**, causing a rapid increase in the rate of formation of cyclic AMP. The latter is the "second messenger" which mediates the biological function of the hormone. The subsequent steps are, however, well understood only for the case of the synthesis and degradation of glycogen (section 10-10).

The action of the hormone may be terminated by the hydrolysis of cyclic AMP by a phosphodiesterase and by dissociation of the hormone from the receptor site, followed by enzymic degradation of the hormone.

22-6 PARATHYROID HORMONES

The parathyroid glands, four small organs adjacent to the thyroid, release **parathyroid hormone** into the blood whenever the Ca^{2+} level of the blood falls below normal values. The target tissue of this hormone, a polypeptide containing 84 amino acid residues, is the kidney, where it stimulates adenylate cyclase activity. The resultant increase in the cyclic AMP level causes the excretion of more phosphate via the urine, thereby reducing the phosphate level in the blood.

Parathyroid hormone also stimulates the release of Ca^{2+} from bone. It moreover stimulates the synthesis of 1,25-dihydroxycholecalciferol in the kidneys. 1,25-Dihydroxycholecalciferol is itself a hormone which promotes the absorption of Ca^{2+} from the intestine.

A second parathyroid hormone is **calcitonin**; it is a polypeptide containing 32 amino acids. Calcitonin, which counters the action of parathyroid hormone, inhibits the loss of Ca^{2+} from bone to blood.

22-7 CONTROL OF BLOOD GLUCOSE

The level of glucose in the blood is maintained at a fairly constant value, about 0.1%, despite the intermittent and irregular nature of the supply of dietary carbohydrate. This is very important for the nutrition of individual tissues, some of which, such as the brain, do not possess an adequate supply of reserve carbohydrate.

The most important reservoir of carbohydrate is liver glycogen. This may be drawn upon by the action of glycogen phosphorylase to release glucose for oxidation or biosynthetic reactions. If the concentration of blood glucose is to be constant, the withdrawal of glucose from the glycogen reservoir must be balanced by the glucose-consuming processes.

The attainment of this balance is achieved by the regulatory action of a set of hormones. The elevation of blood glucose is promoted by epinephrine and glucagon, both of which act by stimulating adenylate cyclase activity and thereby indirectly instigating the mobilization of the glycogen reserve.

A second important factor in raising blood glucose levels is the stimulation of glucose synthesis from amino acids by the adrenocortical hormones, especially cortisol. This effect is countered by growth hormone, which favors protein synthesis over gluconeogenesis.

The action of the above hormones is opposed by insulin, which accelerates the entry of glucose into the peripheral tissues, where it is consumed. In addition, the utilization of glucose by muscle and liver is enhanced.

A complex feedback system correlates the actions of the various hormones. The secretion of insulin is augmented by a rise in glucose level. Epinephrine comes into action under emergency conditions when there is an urgent need for glucose. The mobilization of glucose by epinephrine is under the control of the nervous system.

SUMMARY

The hormones secreted by the mammalian endocrine system provide an additional dimension of biochemical regulation. Hormones in general act upon a tissue which is different from that in which they are synthesized; the former is termed the **target tissue**. In many cases a hormone interacts with a specific **receptor site**. Hormones often stimulate the secretion of a second substance which exerts the actual physiological effect. In this way the influence of the original hormone may be amplified.

The thyroid gland produces the two hormones **thyroxine** and **triiodothyronine**. These have a generalized effect upon metabolism; they stimulate respiration, promote nervous system growth in young animals, and augment the synthesis of certain enzymes.

The hypothalamus secretes a set of peptide hormones, each of which stimulates or inhibits the release of one of the peptide hormones from the anterior pituitary gland; these include **thyrotropin**, **prolactin**, **adrenocorticotropin luteinizing hormone**, and **follicle stimulating hormone**. The latter two act on the gonads; adrenocorticotropin stimulates the synthesis of the steroid hormones formed by the adrenal cortex; prolactin promotes the growth of the breast and the synthesis of milk proteins; and the thyrotropin stimulates the synthesis and secretion of the thyroid hormones.

The posterior pituitary gland secretes the peptide hormones **vasopressin**, which acts on the kidney to control the loss of water and NaCl, and **oxytocin**, which stimulates the secretion of milk. **Parathyroid hormone**, which is synthesized in the parathyroid gland, acts in bone and kidney to increase the Ca^{2+} level in blood; **calcitonin**, which is formed by thyroid C cells, acts in the same tissues to lower the blood Ca^{2+} level.

Three **catecholamine** hormones, **dopamine, norepinephrine**, and **epinephrine**, are synthesized by the adrenal medulla gland. While dopamine and norepinephrine function primarily as neurotransmitters, epinephrine acts on liver, heart, and adipose tissue, where it stimulates the activity of **adenylate cyclase**, thereby enhancing the synthesis of cyclic AMP and ultimately stimulating glycogenolysis and the breakdown of fats.

The polypeptide hormones **insulin** and **glucagon** are synthesized by specialized cells in the pancreas as inactive precursors which are subsequently activated by cleavage by a proteolytic enzyme. Insulin acts on many tissues, promoting the transport of glucose into muscle and fat cells, enhancing the synthesis of fatty acids in adipose tissue and liver, stimulating glycogen synthesis, and promoting protein synthesis. The administration of insulin reduces the blood glucose level. Glucagon stimulates glycogenolysis and gluconeogenesis, acting to increase the blood glucose level.

REFERENCES

The following code is used to classify references. I: particularly useful as an introduction to the subject; R: useful primarily as a reference text; A: an advanced account of the material; H: a publication of historical importance.

General

E. Frieden and H. Lipner, *Biochemical Endocrinology of the Vertebrates*, Prentice-Hall, Englewood Cliffs, New Jersey (1971). (R)

E. A. Newsholme and C. Start, *Regulation in Metabolism*, Wiley, New York (1973). (A)

J. B. Stanbury, J. B. Wyngaarden, and D. S. Fredrickson, *The Metabolic Basis of Inherited Disease*, 3rd ed., McGraw-Hill, New York (1978). (A)

A. P. White, P. Handler, E. L. Smith, R. L. Hill, and I. R. Lehman, *Principles of Biochemistry*, 6th ed., McGraw-Hill, New York (1978). (R)

Insulin and Glucagon

P. Cuatrecasas, "Interaction of Insulin with the Cell Membrane: The Primary Action of Insulin," *Proc. Natl. Acad. Sci. U.S.* 63, 450 (1969). (H)

P. Cuatrecasas, "Insulin Receptor of Liver and Fat Cell Membranes," *Fed. Proc.* 32, 1838 (1973). (H)

M. P. Czech, "Molecular Basis of Insulin Action," *Ann. Rev. Biochem.* 46, 359 (1977). (R)

P. J. Lefebre and R. H. Vager, *Glucagon: Molecular Physiology, Clinical and Therapeutic Implications*, Pergamon, New York (1972). (A)

A. C. Maurer, "Therapy of Diabetes," *Amer. Scientist* 67, 422 (1979). (I)

Cyclic AMP

P. Greengard and J. W. Kebabian, "Role of Cyclic AMP in Synaptic Transmission in the Peripheral Nervous System," *Fed. Proc.* 33, 1054 (1974). (A)

I. Pastan, "Cyclic AMP," *Sci. Amer.* 227, 97 (1972). (I)

Hormones of the Hypothalamus

S. Reichlin, R. Saperstein, I. M. D. Jackson, A. E. Boyd, and Y. Patel, "Hypothalamic Hormones," *Ann. Rev. Physiol.* 38, 389 (1976). (R)

A. V. Schally, A. Arimura, and A. J. Kastin, "Hypothalamic Regulatory Hormones," *Science* 179, 341 (1973). (A)

Pituitary Hormones

M. J. Brownstein, J. T. Russell, and H. Gainer, "Synthesis, Transport, and Release of Posterior Pituitary Hormones," *Science* 207, 373 (1980). (A)

J. M. Tanner, "Human Growth Hormone," *Nature* 237, 433 (1972). (A)

REVIEW QUESTIONS

Questions marked with an asterisk are of a high level of difficulty.

22-1 Suggest a biological advantage for the synthesis of insulin as an inactive precursor.

22-2 In general, hormones are active at much lower concentrations than other biochemical agents. Cite some factors which contribute to their high efficiency.

*22-3 In contrast to many proteins, insulin is not spontaneously regenerated by reoxidation of its reduced and separated polypeptide chains. Cite a logical reason. How does this suggest the existence of a precursor?

22-4 What biological advantage is conferred by a stepwise regulatory mechanism, such as the indirectly transmitted effect of epinephrine, upon the blood glucose level?

22-5 List all the hormones involved in the regulation of blood glucose level and cite their modes of action.

23

Plasma

23-1 MULTIPLE FUNCTIONS OF BLOOD PLASMA

The survival of higher organisms is impossible without some means for the transport of oxygen and metabolic intermediates between tissues. The transport of oxygen by the red cells of the blood has already been described and the transport of metabolic intermediates in the blood plasma has been alluded to.

While the transport properties of plasma are very important and will be discussed in this chapter, they by no means exhaust the list of biological functions of this fluid. The self-sealing properties of blood, as reflected by the **clotting** process, arise through a very elaborate and sensitively balanced mechanism that shall be described in some detail.

A major element in the resistance of the body to infection by microorganisms is the **antibody** response, which results in the formation of specialized proteins circulating in the plasma.

23-2 COMPOSITION OF PLASMA

If the red blood cells are separated from whole mammalian blood by centrifugation, there remains a clear supernatant liquid, called the **plasma**. Untreated plasma gels or **clots** spontaneously. The gel may be removed by further centrifugation, leaving the **serum**, which differs from plasma in the absence of the clotting protein **fibrinogen**.

Whole plasma contains about 7-8% protein. The total number of protein components is large, but only a few of these occur in quantity (Table 23-1). If normal human serum is analyzed by electrophoresis at pH 8.6, at least five species are usually observed, all of which migrate toward the anode. In order

Table 23-1
The Major Plasma Proteins

Protein	Molecular Weight	Amount in Human Plasma (Percentage of Total Protein)
Plasma albumin	69,000	55
Fibrinogen	340,000	7
α_1-Globulin		5
α_2-Globulin		9
β-Globulin	93,000	13
γ-Globulin	160,000	11

of decreasing mobility, these are **serum albumin**, followed by the α_1-, α_2-, β-, and γ-globulins. The latter four components are themselves heterogeneous.

The electrophoretic pattern of a pathological serum can show important differences. These can be of diagnostic value.

The separation of plasma proteins was originally achieved by fractional precipitation with ammonium sulfate. This approach was superseded by alcohol fractionation under carefully controlled conditions, as developed by Cohn and co-workers. More recently, chromatographic methods have come to be widely used.

23-3 THE BLOOD CLOTTING SYSTEM

The self-sealing properties of blood arise from its property of spontaneous gelation upon traumatic injury to a blood vessel. The sensitive and delicately controlled clotting system involves an elaborate hierarchy of enzymes and other factors.

The central event at the molecular level is the action of the proteolytic enzyme **thrombin** upon **fibrinogen**. Fibrinogen is thereby converted to an altered form capable of spontaneous polymerization. The polymerization of activated fibrinogen produces a three-dimensional network of thin strands (Fig. 23-1) called **fibrin**, whose external properties are those of a stiff gel.

Thrombin exists in circulating blood as the inactive precursor **prothrombin**. The initial phase of blood clotting is the conversion of prothrombin to thrombin.

The principal stages involved in the clotting mechanism (Fig. 23-2) can be summarized as follows:

(1) prothrombin $\rightarrow$ thrombin
(2) fibrinogen $\xrightarrow{\text{thrombin}}$ activated fibrinogen
(3) activated fibrinogen $\rightarrow$ fibrin gel

The Activation of Prothrombin

The initial phase of blood clotting is by far the most complex. Numerous protein factors are involved, in addition to particulate bodies called **platelets**. Platelets are small cell-like particles whose diameters are 2–3 μm.

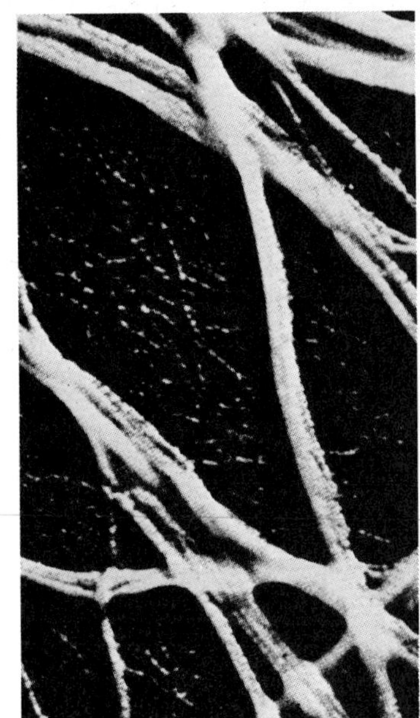

Figure 23-1
Electron microscope photograph of a fibrin network. (Courtesy of National Naval Medical Center, Bethesda, Maryland.)

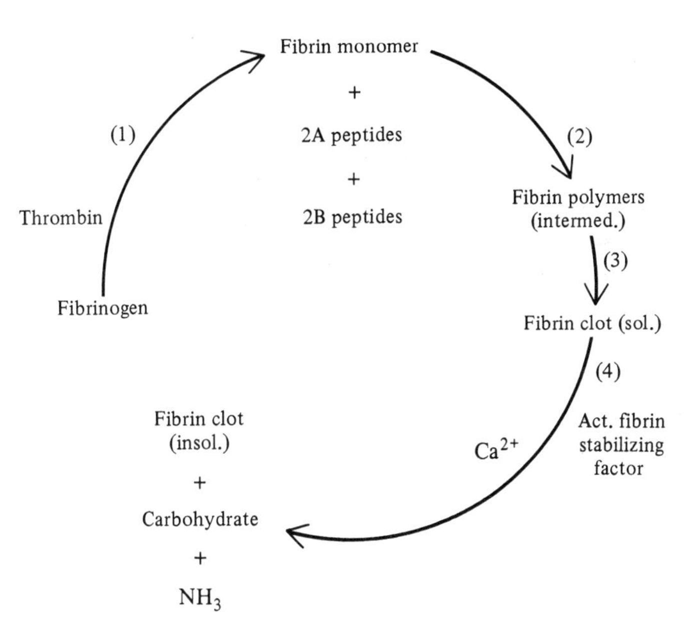

Figure 23-2
The major steps in the clotting of fibrinogen.

One of the two principal clotting pathways requires only factors present in plasma and is termed **intrinsic**. A mechanism has been proposed which is shown in Fig. 23-3. The various components, which are incompletely characterized, are designated by their common names and by Roman numerals.

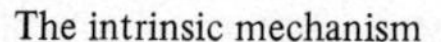

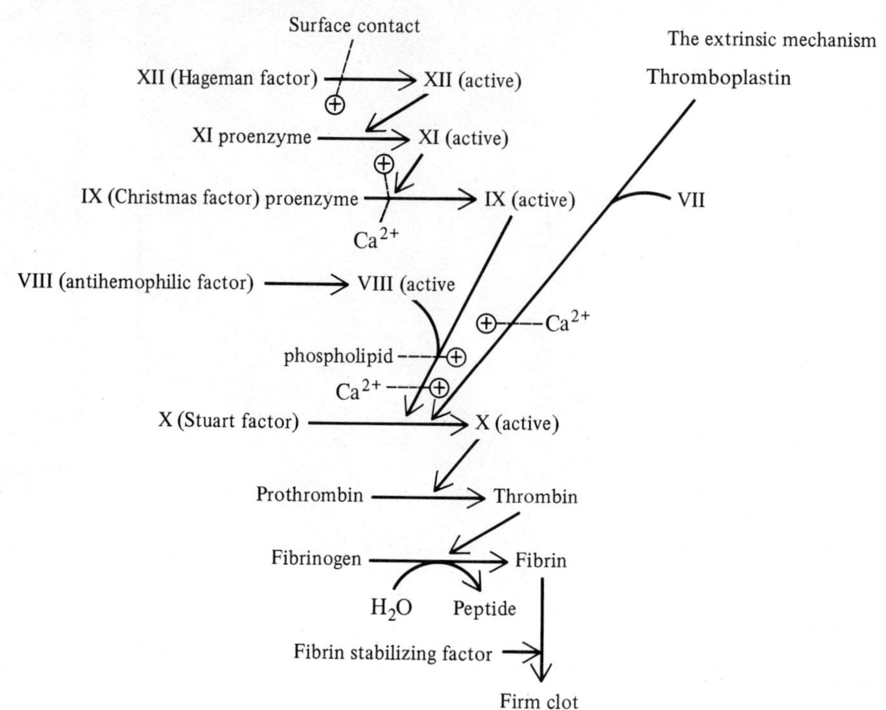

Figure 23-3
Intrinsic and extrinsic clotting pathways.

The process is stepwise. Each component is activated by the factor preceding it in the chain and, in turn, activates that which follows it. Most, or all, of the reactions are presumably enzymic. The activation of antihemophilic factor (VIII) involves the participation of phospholipids present in the platelets.

Hemophilia, a hereditary disease that has figured prominently among European royalty, represents a deficiency in the antihemophilic factor (VIII). It can be countered by the transfusion of normal plasma.

The intrinsic mechanism is triggered by surface contact; the **Hageman factor** is activated by contact with some foreign surfaces, including glass. There exists also an **extrinsic** mechanism, which is initiated by the release of the lipoprotein **thromboplastin** from injured tissues. Both pathways culminate in the conversion of prothrombin to thrombin.

Prothrombin is a protein of molecular weight 63,000 which migrates with the α_2-globulin fraction. Thrombin, the product of its proteolytic activation, has a molecular weight of 8,000–13,000, but exists in solution as aggregates of this basic unit. Like many other proteolytic enzymes, it is stoichiometrically inhibited by diisopropylfluorophosphate (DFP). Its active center probably contains a serine residue.

The Activation of Fibrinogen

Fibrinogen is one of the larger plasma proteins, having a molecular weight close to 340,000. Bovine fibrinogen consists of three pairs of polypeptide

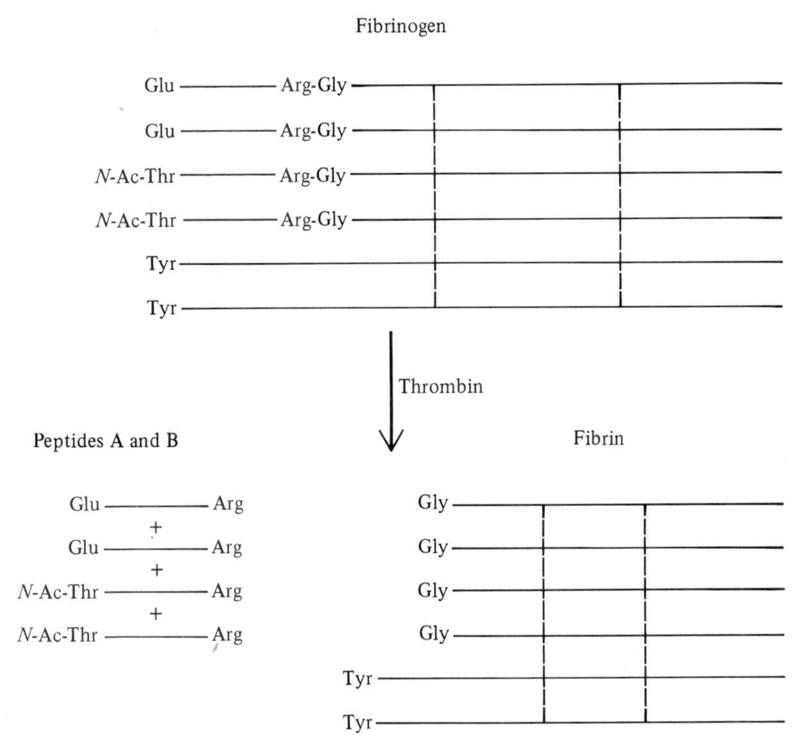

Figure 23-4 Proteolytic activation of fibrinogen by thrombin. The polypeptide chains are shown schematically with the disulfide cross-links represented by broken lines.

chains, whose NH_2-terminal groups are tyrosine, glutamic acid, and O-acetyl threonine. Electron microscopy shows the molecule to have a somewhat extended and asymmetric shape, consisting of globular nodules joined by a thin polypeptide thread. The molecular length is about 50 nm.

The proteolytic action of thrombin upon fibrinogen liberates two peptides, A and B, from the NH_2-terminal ends of two of the three pairs of chains (Fig. 23-4). In the case of bovine fibrinogen, the two glutamic and two N-acetyl threonine NH_2-terminal groups are replaced by four NH_2-terminal glycines. The modified fibrinogen has the property of spontaneous polymerization.

Since the peptides which are split off account for only about 3% of the molecule, activated fibrinogen does not differ greatly from the native molecule in size and shape.

The Polymerization of Activated Fibrinogen

Electron microscopic examination of the intermediate polymer formed initially indicates that it arises largely by end-to-end combination of the rodlike fibrinogen units (Fig. 23-5). At later stages of the polymerization, considerable lateral association occurs to build up thicker strands. The linear polymers increase progressively in length until gelation occurs. Occasional contacts between strands produce a three-dimensional network.

The fibrin gel formed by the action of thrombin upon *purified* fibrinogen can be dispersed by the action of such denaturing solvents as 6-*M* urea to yield a solution of particles of size and shape similar to those of fibrinogen

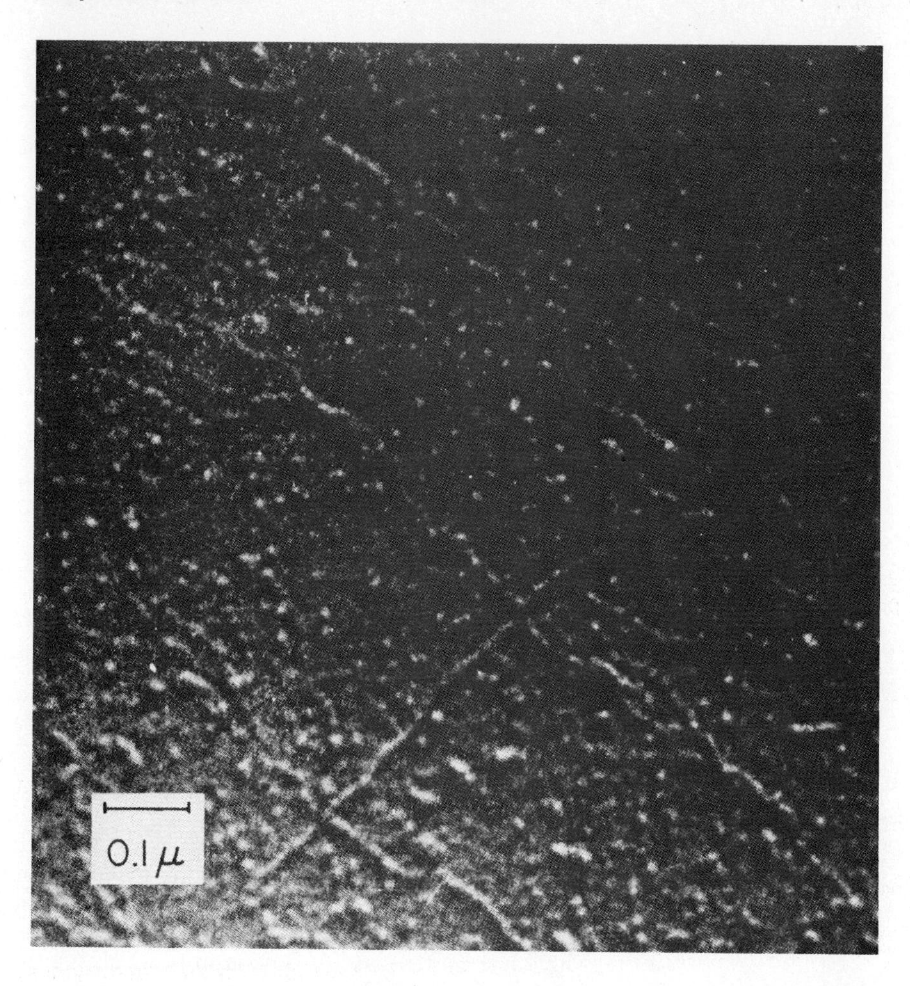

Figure 23-5
Electron microscope photograph of the intermediate polymer formed at an early stage of the polymerization of fibrinogen.

in the same medium. The intermolecular bonds formed are thus reversible and noncovalent.

In the presence of Ca^{2+}, plus a **fibrin stabilizing factor** present in plasma, covalent links are formed between fibrinogen units, so that the gel becomes insoluble in urea.

23-4 PLASMA ALBUMIN

Plasma albumin is by far the most abundant protein in plasma (Table 23-1). As a consequence of its high concentration (6–8%) and relatively low molecular weight (69,000), plasma albumin accounts for most of the osmotic pressure of blood. The maintenance of osmotic pressure is perhaps its primary function.

The molecule consists of a single polypeptide chain cross-linked by 18 disulfide bonds. About 70% of the molecules of a bovine or human albumin preparation contain a single free cysteine sulfhydryl and are designated as **mercaptalbumin**.

A distinctive property of plasma albumin is its capacity to bind a wide variety of organic anions reversibly. No other protein approaches it in the generalized character of its binding ability. Albumin may therefore serve as an agent for the transport of many biological materials, such as fatty acids, by the blood.

23-5 IMMUNOGLOBULINS

The bodies of higher animals can mobilize a plural system of defenses against the invasion of pathogenic microorganisms. In addition to the relatively nonspecific activity of such agents as interferon and the phagocytes, there is the powerful and selective weapon of the antibody response.

An **antibody** is a protein that is synthesized by the body in response to the introduction into its circulation of a foreign molecule or **antigen**, and has the property of combining specifically with the antigen that stimulates its formation. The criteria for antigenicity are amazingly unrestrictive. The only general requirement is that the substance be foreign to the circulation of the animal, although the most effective antigens are usually of high molecular weight. Proteins, nucleic acids, lipids, carbohydrates, and a variety of synthetic materials have been found to elicit antibody synthesis. Most cellular membranes, including those of bacteria, contain antigenic material.

The appearance of antibodies in the circulation can occur within a few days after introduction of the antigen, which may be introduced deliberately by injection, or accidentally by bacterial or viral infection. In the latter case, the combination of antibodies with the invading organism facilitates its destruction by other defensive elements in the circulation.

An antibody or **immunoglobulin** is formed by the **plasma cells** of the lymphoid system. These arise from specialized cells termed **lymphocytes**, which are present in bone marrow and thymus gland. These occur in a very large number of cell lines, each of which makes only a single type of immunoglobulin with a characteristic amino acid sequence. If the antigen is absent, each cell line exists only in small numbers and makes only small quantities of its particular antibody, to which it is committed in advance. Exposure to a specific antigen stimulates the proliferation of the corresponding cell line, with an accompanying increase in the amount of specific antibody formed by the cell line. The lymphocytes that are stimulated are thereby transformed into the plasma cells, which actually form the antibody.

The Structure of Antibodies

Five different types of immunoglobulins have been identified in human plasma (IgG, IgA, IgM, IgD, and IgE). Of these, the IgG are by far the most common and the most extensively studied. The IgG immunoglobulins consist of two equivalent *heavy* (H) chains with 430 amino acids and two identical *light* (L) chains with 214 amino acids; the total molecular weight is 150,000. The chains are joined by cystine disulfide bridges (Fig. 23-6) to form a Y-shaped structure.

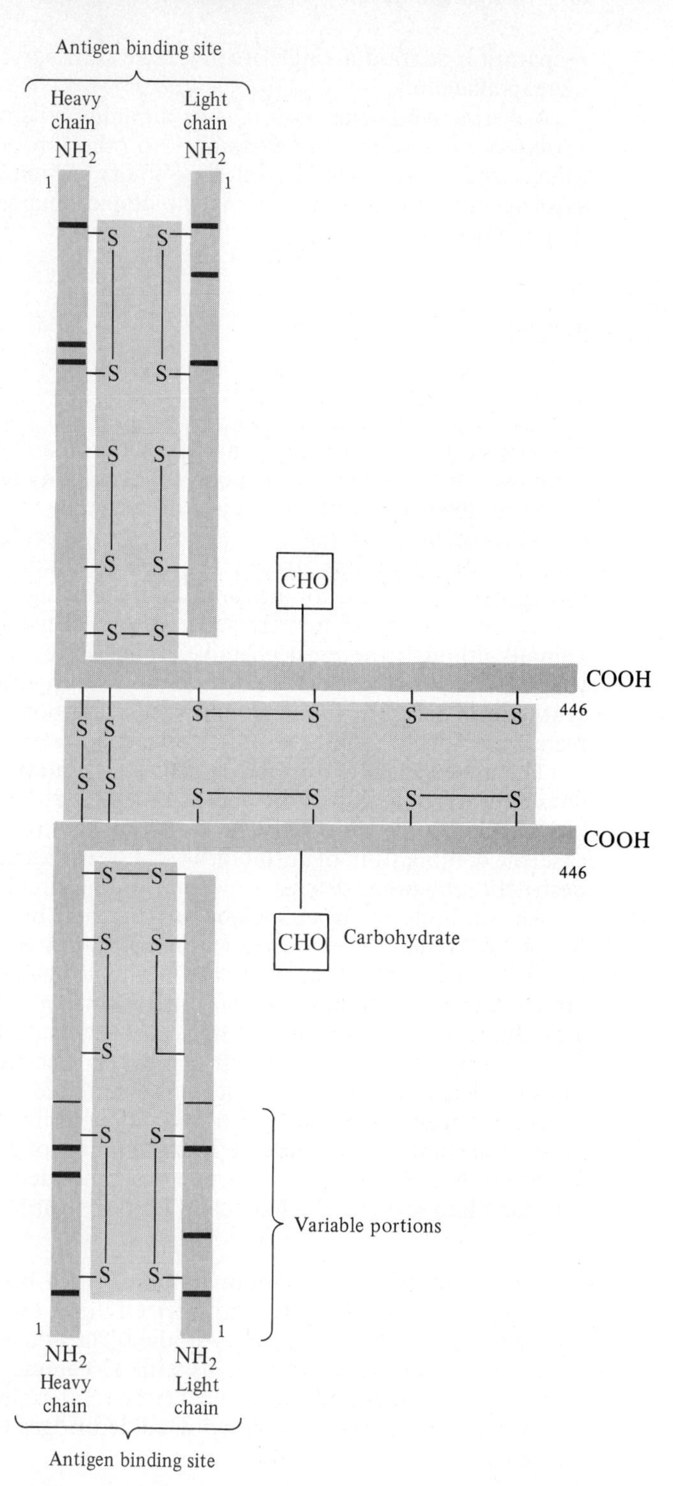

Figure 23-6 Structure of the IgG immunoglobulins.

Table 23-2
The Human Immunoglobulins

Symbol	Molecular Weight	Formula
IgG	150,000	$\kappa_2\gamma_2$ or $\lambda_2\gamma_2$
IgM	950,000	$(\kappa_2\mu_2)_5$ or $(\lambda_2\mu_2)_5$
IgA	300,000 or more	$(\kappa_2\alpha_2)_n$ or $(\lambda_2\alpha_2)_n$
IgD	160,000	$\kappa_2\delta_2$ or $\lambda_2\delta_2$
IgE	190,000	$\kappa_2\epsilon_2$ or $\lambda_2\epsilon_2$

Each chain contains a zone where the amino acid sequence is invariant and a region where it varies with the specific antibody. The variable regions occur in the arms of the Y, where the two antigen-binding sites are present. The arms can be removed by the action of papain to form fragments termed $F_{ab'}$.

Two types of L chain, κ and λ, are found in human immunoglobulins; five kinds of H chain, γ, μ, α, δ, and ε, have been identified. The compositions of the different classes of human antibodies are cited in Table 23-2. Two of these, IgM and IgA, consist of disulfide-linked polymers of basic four-chain units.

The Antigen-Antibody Reaction

The immunoglobulins contain two equivalent combining sites in each basic four-chain unit. In the case of protein antigens, the antigenic sites are probably complex and may consist of patches of the protein surface containing several amino acid residues. The valency of a natural protein antigen depends primarily upon its size and ranges from three for ribonuclease (molecular weight 13,000) to about 40 for thyroglobulin (molecular weight 700,000).

The complex equilibria occurring upon mixing solutions of a protein antigen and its antibody can be interpreted formally as the reversible condensation of a multivalent with a bivalent species. For example, the reaction of bovine serum albumin (valency ~ 6) with its rabbit antibody (Fig. 23-7) may be monitored by ultracentrifugal analysis, which indicates that the nature of the complex series depends upon the mole ratio of antigen (G) to antibody (A).

At very high values of the G/A ratio (> 10) the antibody sites are saturated with antigen, so that the only species present are free antigen and the AG_2 complex.

As the G/A ratio is lowered, the antibody sites are no longer saturated, so that equilibria occur of the types:

$$A + G \rightleftarrows AG$$
$$AG + A \rightleftarrows A_2G$$
$$A_2G + G \rightleftarrows A_2G_2, \text{ and so on}$$

The relative concentration of species containing more than one antibody increases progressively as the G/A ratio decreases, at the expense of the AG_2 complex. As the G/A ratio approaches unity, very large complexes are formed

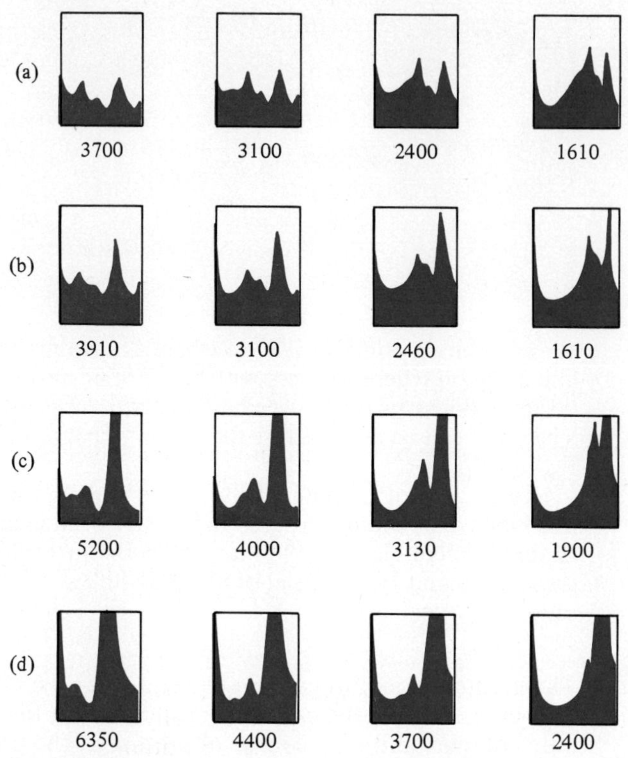

Figure 23-7
Sedimentation diagrams (section 5-5) of mixtures of bovine plasma albumin and its rabbit antibody as a function of antibody: antigen (A:G) weight ratio. Sedimentation proceeds to the left. The peaks observed, beginning at the right, are free antigen, AG_2 complex, A_2G_2 (or A_2G), and higher complex species. The numbers indicate sedimentation times. (a) A:G = 1:0.56; (b) A:G = 1:1; (c) A:G = 1:1.6; (d) A:G = 1.59. (Courtesy of National Naval Medical Center, Bethesda, Maryland.)

of the general composition A_xG_y, resulting ultimately in the appearance of a precipitate, which, in the **equivalence zone**, incorporates all of the antigen and antibody (Fig. 23-8). This is the basis of the **precipitin** reaction, which is the classical means for the recognition and estimation of antibodies.

As G/A decreases further, the proportion of antibody in the specific precipitate increases. In the limit, as G/A becomes very small, the antigen

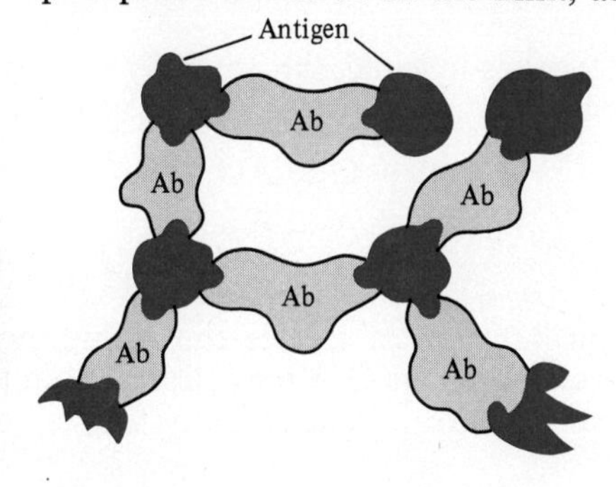

Figure 23-8
Schematic representation of the formation of a specific precipitate by the combination of a protein antigen with bivalent antibody.

becomes saturated with antibody, so that complexes of the type GA_n are formed, where n is the valency of the antigen. Here differences arise for different species. In the case of rabbit antibodies these complexes are of such low solubility that precipitation persists in the region of large antibody excess. For horse antibodies the solubility of the GA_n is much higher, so that the formation of a specific precipitate is confined to the equivalence zone.

The Antibody Site

While small molecules are ordinarily rather poor antigens, they may function as potent antigenic determinants when chemically coupled to a carrier protein. Such groups are termed **haptens**.

The injection of a protein labeled with a specific hapten residue can result in a formation of antibodies directed specifically against the hapten, as well as antibodies directed against the protein carrier. If the carrier is a plasma protein from the same species, the antibodies may be directed exclusively against the hapten. In this case a precipitin reaction may occur with a second, unrelated protein that is labeled with the same hapten.

Antibodies to a given hapten often show a cross-reaction with a second, structurally similar hapten, indicating that antigenic specificity is not unlimited (Figs. 23-9 and 23-10). A large number of such cross-reactions have been examined and have permitted several generalizations as to the factors governing specificity.

The most effective artificial haptens are usually aromatic groups containing one or more charged sites. The nature of the charged site appears to be critical and cross-reaction usually does not occur for haptens differing in this respect. The geometry of substitution is also important and a change from *para* to *ortho* substitution generally produces a diminution, or loss of, cross-reaction.

It is not possible to generalize as to the detailed nature of hapten-antibody interactions. It is likely that only secondary forces are involved; hydrogen bonding and hydrophobic interactions, and possibly electrostatic forces, may play a role in particular instances. An instructive concept has been the "lock and key" model of Pauling, which represents the antibody site as analogous to a cavity complementary in shape to the antigenic determinant (Fig. 23-11). The strength of interaction will, in this model, depend on the closeness of fit, which should be sensitive to minor changes in the hapten.

In any event the combining sites must be complex in nature, presumably containing several residues whose proper orientation is maintained by the

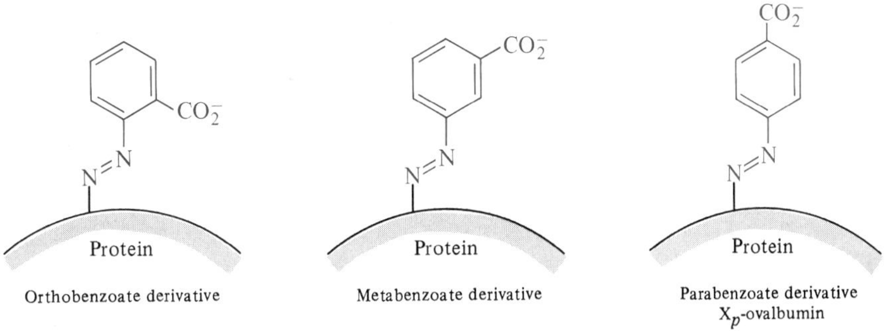

CO_2^- N=N Protein — Orthobenzoate derivative

CO_2^- N=N Protein — Metabenzoate derivative

CO_2^- N=N Protein — Parabenzoate derivative X_p-ovalbumin

Figure 23-9 Hapten groups coupled to a protein carrier by diazotization.

Antisera made with	Tested against antigens made with					
	NH_2	NH_2 COOH	NH_2 COOH	NH_2 Cl COOH	NH_2 CH_3 COOH	NH_2 SO_2H
NH_2 COOH	0	+++	0	++++	+++	+
NH_2 SO_2H	0	0	0		0	++++

0, no reaction; +, positive reaction; ++++, very strong reaction.

Figure 23-10
Cross-reaction between antibodies to related haptens. Note that the nature and position of the acidic group are critical.

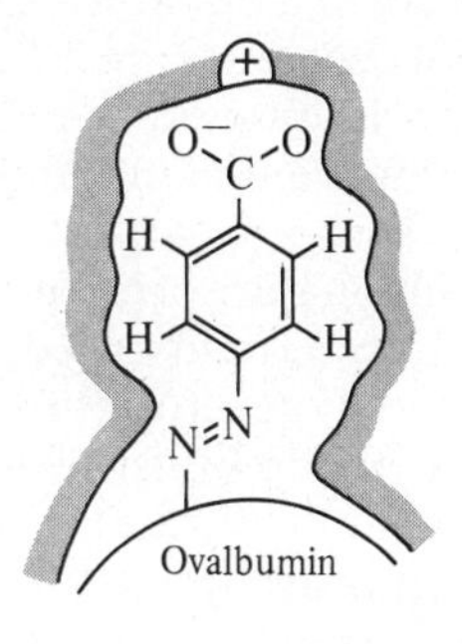

Interaction of X*p*-ovalbumin with anti-X*p*-globulin

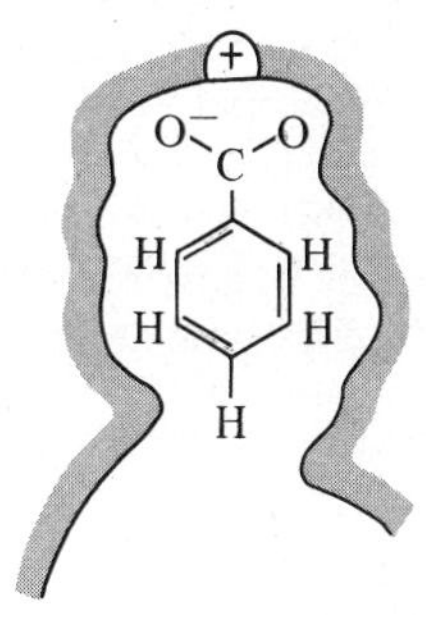

Interaction of benzoate ion with anti-X*p*-globulin

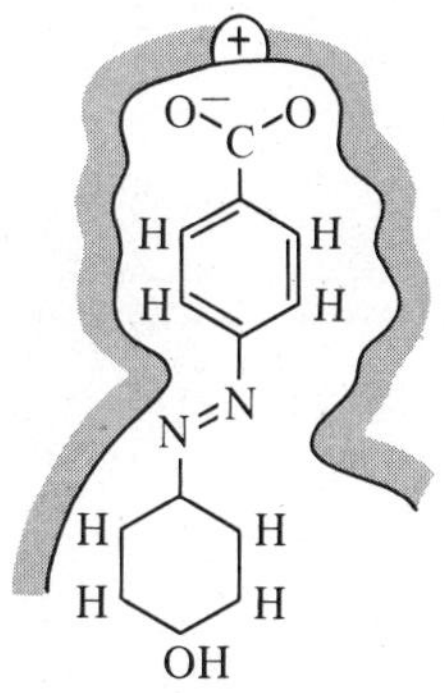

Interaction of *p*-hydroxyphenylazobenzoate with anti-X*p*-globulin

Figure 23-11
Simplified depiction of the "lock and key" model for the combination of a hapten group with an antibody site.

tertiary structure of the antibody. If antibodies are converted to a structureless state by treatment with a denaturant, removal of the denaturing agent is followed by recovery of activity. The implication is that those aspects of the tertiary structure responsible for the formation of the active site are governed entirely by the amino acid sequence, just as in the cases of other proteins (section 4-6), and that direct contact with the antigenic determinant is not necessary to assure correct folding.

23-6 OTHER PLASMA PROTEINS

Lipoproteins

The plasma lipoproteins are of central importance for the transport of lipids, including triacylglycerols, phospholipids, and cholesterol. In recent years they have attracted intense clinical interest because of their probable role in atherosclerosis and coronary disease.

There are three principal classes of human plasma lipoproteins. These are high density lipoprotein (HDL), low density lipoprotein (LDL), and very low density lipoprotein (VLDL), whose lipid contents increase in that order, ranging from 40–70% for HDL to 95% or more for VLDL. The molecular weights range from 10^5 to 10^7; their chemical composition is variable with respect to the nature and relative quantities of the lipid components. All the lipoproteins contain triacylglycerols, phospholipids, cholesterol, and fatty acids.

The lability of the lipoproteins prevents their isolation by the usual preparative techniques. The most useful methods for fractionation have utilized ultracentrifugation and have depended upon their low and variable densities. By suitable adjustment of the solvent density it is possible to separate the plasma lipoproteins into fractions of different density by differential flotation.

Transferrin

The principal agent for the transport of iron to and from its storage depots in bone marrow, spleen, and liver is the protein **transferrin**. This is a glycoprotein of molecular weight 80,000 containing two Fe^{3+}-binding sites; the carbohydrate content is 6%, consisting of hexose, hexosamine, and sialic acid. The iron-transferrin complex is stable in the pH range 7.5–10, but dissociates at pH 4. The combination with iron results in significant changes in absorption spectrum and optical rotation.

Over 12 different genetic variants of transferrin have been identified, which differ in electrophoretic mobility; none of these is associated with a hereditary disease.

Ceruloplasmin

This is a blue protein of molecular weight 150,000 which contains $8Cu^{+}$ and $8Cu^{2+}$ ions; it is the principal copper-containing protein of plasma, containing over 90% of plasma copper and 3% of total body copper. In **Wilson's disease** the ceruloplasmin content is abnormally low.

Haptoglobin

This is a glycoprotein occurring in the α-globulin fraction of many mammalian plasmas. Its ability to form specific complexes with hemoglobin may help to avoid loss by urinary excretion.

SUMMARY

Mammalian blood **plasma**, or the fluid supernatant remaining after the sedimentation of erythrocytes and other particulate bodies, is a very complex solution containing a large number of different proteins. The self-sealing properties of blood arise from the **blood-clotting system**. Traumatic injury to the tissues results in the initiation of a series of reactions involving a number of protein factors, each of which is activated by the preceding element in the series and, in turn, activates the following element. The process culminates in the formation of the proteolytic enzyme **thrombin** from its inactive precursor **prothrombin**. Thrombin cleaves two short peptides from the protein **fibrinogen**, thereby converting it to an activated form that spontaneously polymerizes. Extended linear aggregates of fibrinogen are formed which interact to produce a three-dimensional network, manifested externally by **gelation**.

The most abundant protein in plasma is **plasma albumin**, which has a dual function in maintaining osmotic pressure and as an agent for the binding and transport of many small molecules, including fatty acids.

The **antibodies** or **immunoglobulins** present in plasma are an important element in the body's defenses against invasion by foreign organisms. Antibodies arise in response to the introduction of an **antigen** foreign to the animal's circulation and are produced by specific **plasma cells**, whose proliferation is stimulated by exposure to a particular antigen. Antibodies have the property of combining specifically with the antigen that elicited them; the antigen may be a protein or other biopolymer, or a virus or bacterial cell. Antibodies may also be obtained which are directed against a synthetic **hapten** conjugated with a protein.

The best characterized antibodies are the IgG immunoglobulins. These consist of two *light* (L) and two *heavy* (H) polypeptides; both types of chain contain invariant portions and regions whose amino acid sequences are characteristic of the specific antibody. Antibodies are **bivalent**; this enables them to form specific precipitates with multivalent protein antigens.

Plasma also contains a set of **lipoproteins**, which consist of a polypeptide core, plus a high proportion of various lipids. The lipoproteins are important as carriers of lipids.

REFERENCES

The following code is used to classify references. I: particularly useful as an introduction to the subject; R: useful primarily as a reference text; A: an advanced account of the material; H: a publication of historical importance.

General

J. T. Edsall and J. Wyman, *Biophysical Chemistry*, Academic, New York (1958). (A)

R. Montgomery, R. L. Dryer, T. W. Conway, and A. Spector, *Human Biochemistry*, Mosby, St. Louis (1974). (R)

F. W. Putnam, *The Plasma Proteins*, 2nd ed., Academic, New York (1975). (A)

A. P. White, P. Handler, and E. L. Smith, *Principles of Biochemistry*, McGraw-Hill, New York (1973). (H)

Immunoglobulins

G. Edelman, "Antibody Structure and Molecular Immunology," *Science* 180, 830 (1973). (I)

G. Edelman, B. A. Cunningham, W. E. Gall, P. D. Gottlieb, U. Ruttishauser, and M. J. Waxdal, "The Covalent Structure of an Entire IgG Immunoglobulin Molecule," *Proc. Natl. Acad. Sci. U.S.* 63, 78 (1969). (H)

H. N. Eisen, *Immunology*, Harper, New York (1974). (A)

A. Nisonoff, J. E. Hopper, and S. B. Spring, *The Antibody Molecule*, Academic, New York (1975). (A)

REVIEW QUESTIONS

Questions marked with an asterisk are of a high level of difficulty.

23-1 If you added 10^{-5} *moles* of OH^- to 10 *liters* of blood in the body, you would obtain which result(s) below?
- (a) increase in final blood pH to pH 8
- (b) conversion of more dissolved CO_2 to HCO_3^- in blood
- (c) addition of protons to protein side-chains like *glutamate* and *aspartate*
- (d) increased binding of O_2 to hemoglobin
- (e) (b) and (d)

23-2 The buffering effect of the carbonate-bicarbonate system in blood is due to:
- (a) the fact that the pK of the first dissociable proton of H_2CO_3 is near 7.0.
- (b) the fact that the pK of the second dissociable proton of H_2CO_3 is near 7.0.
- (c) a mass action effect involving dissolved CO_2.
- (d) the fact that the pH of blood is near 9.0.
- (e) all of the above

24-3 If all antibodies were univalent, how would the nature of the reaction with antigen be changed?

23-4 Gels of purified fibrinogen are soluble in concentrated urea solutions. Upon removal of the urea by dialysis, the gel reforms. What does this suggest as to the nature of the bonds joining fibrinogen units in the gel?

*23-5 Why does blood not ordinarily clot in the body?

*23-6 Speculate as to why antibodies to homologous plasma proteins are not formed.

*23-7 Plasma albumin has a single binding site for compound X. If the equilibrium constant for binding X is 10^7 and it is independently found that 50% of the plasma albumin is combined with X, what is the concentration of free X in plasma?

*23-8 There is evidence that a flexible "hinge" exists between the portions of an immunoglobin containing the combining sites and the balance of the molecule. What are some possible biological advantages of this structural feature?

24

Specialized Tissues

24-1 CONNECTIVE TISSUE

Connective tissue occurs throughout the mammalian body in ligaments, tendons, bone, cartilage, and the walls of blood vessels. It has a primarily structural function, conferring mechanical strength and providing support and protection for softer tissues. The role of connective tissue becomes more important with increasing size and weight of the animal and accounts for an increasing fraction of its total protein.

Collagen Fibers

Together with elastin, collagen is a principal fibrous constituent of connective tissue; it ranks with the most abundant proteins known. In biological systems, collagen occurs as bundles of linear fibrils (Fig. 24-1). These may be aligned in a parallel manner, as in tendon, or randomly arranged, as in skin. The fibers have a remarkable tensile strength, approaching that of steel wire.

Collagen fibers from many sources show a characteristic banded pattern when observed with the electron microscope (Fig. 24-1). The band spacing is quite regular, being close to 70 nm.

The distinctive properties of macroscopic collagen include an abrupt contraction which occurs upon heating to a critical range of temperature. Like the phase transitions which it resembles in many respects, this process has a remarkably sharp thermal profile, usually attaining completion over a rather narrow temperature range, whose midpoint is close to 60°C for many mammalian collagens. The shrunken fibers, which may be shortened from $\frac{1}{3}$ to $\frac{1}{4}$ of the original length, have greatly altered mechanical properties, including a reduced tensile strength and rubberlike elasticity.

Soluble Collagen

A detailed study of collagen at the molecular level became possible with the discovery that the normally insoluble fibrils of some tissues can be dissolved

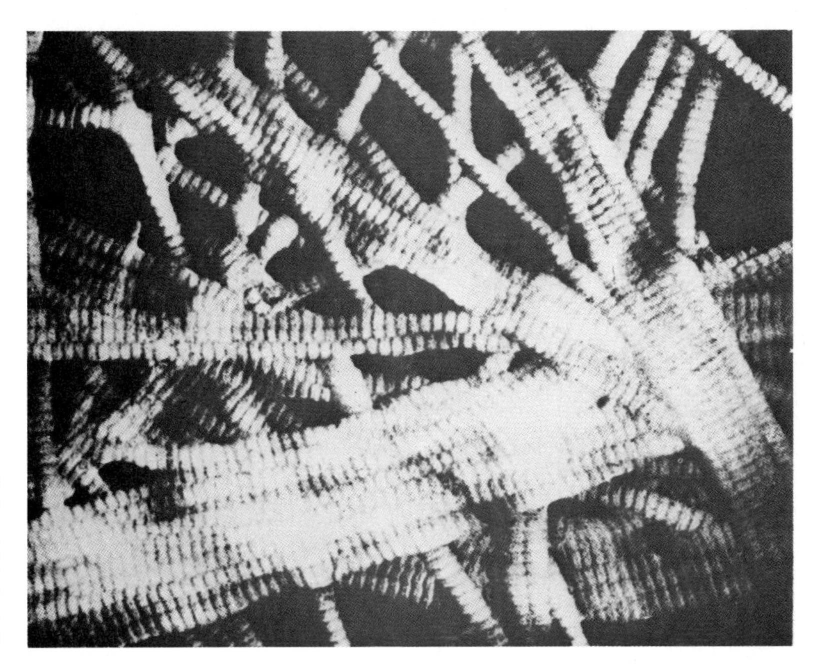

Figure 24-1 Electron microscope photograph of native collagen fibrils from skin. (Courtesy of National Naval Medical Center, Bethesda, Maryland.)

in certain solvents, such as dilute acetic acid. The resultant solutions contain, in addition to aggregates of varying size, individual collagen molecular units known as **tropocollagen** or **collagen monomer**. Removal of the aggregates by high speed centrifugation permits physical studies upon the residual molecularly dispersed material.

Many of the early studies that established the molecular characteristics of tropocollagen were made using **ichthyocol**, the collagen of carp bladder. Tropocollagens from other sources appear to be basically similar to ichthyocol in structure and properties.

From electron microscopy and other physical techniques, the gross molecular shape of tropocollagen is known to be that of a thin rigid rod of close to 300 nm in length and 1.5 nm in diameter. The molecular weight is about 350,000.

By the use of x-ray diffraction a fairly detailed picture of the structure of tropocollagen has been obtained. The molecule consists of three intertwined polypeptide chains twisted into a helical conformation quite different from the α-helix described earlier (section 5-6) (Figs. 24-2 and 24-3). The cablelike structure is itself twisted further into a right-handed superhelix. The three strands are linked by interchain hydrogen bonds.

Solutions of ichthyocol undergo a thermal denaturation upon heating that results in a splitting into the constituent polypeptide chains and a collapse of the helical structure. The physical properties of the denatured protein revert to those expected for a system of unorganized random coils (Fig. 24-4). This process is undoubtedly related to the thermal transition observed for collagen fibers.

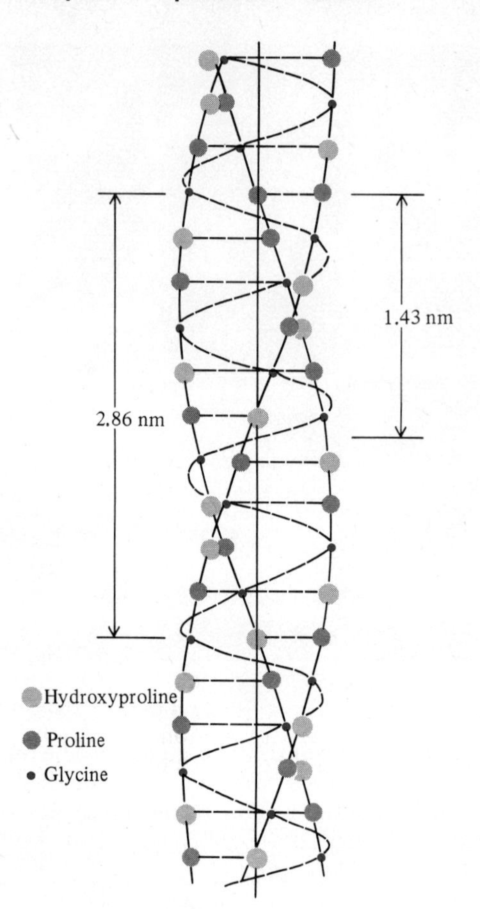

Figure 24-2 Molecular model of tropocollagen.

The explanation for the unusual structural characteristics of tropocollagen may be found in its amino acid composition. All known collagens are characterized by very high contents of glycine (about 35%), proline (about 12%), and hydroxyproline (about 9%). The latter two residues cannot be incorporated into an α-helix, as the *single* hydrogen atom on their α-nitrogen is lost upon formation of a peptide bond so that it is impossible for them to form hydrogen bonds of the type N—H · · · O. Theoretical considerations and parallel studies with polymers of proline have reinforced the conclusion that a structure of the multistranded helical type is the most stable for this system.

Collagen is unusual in containing the two amino acids hydroxyproline and hydroxylysine. These are formed from proline and lysine residues in the growing polypeptide chain by the action of specific hydroxylases.

The principal form of collagen in the tissues of most species (collagen I) consists of two equivalent α1(I) chains and one α2 chain and is designated α1(I)α2. The complete collagen fibers are stabilized by the formation of covalent cross-links; these often involve lysine or hydroxylysine residues whose amino groups have been enzymically oxidized to aldehydes and may con-

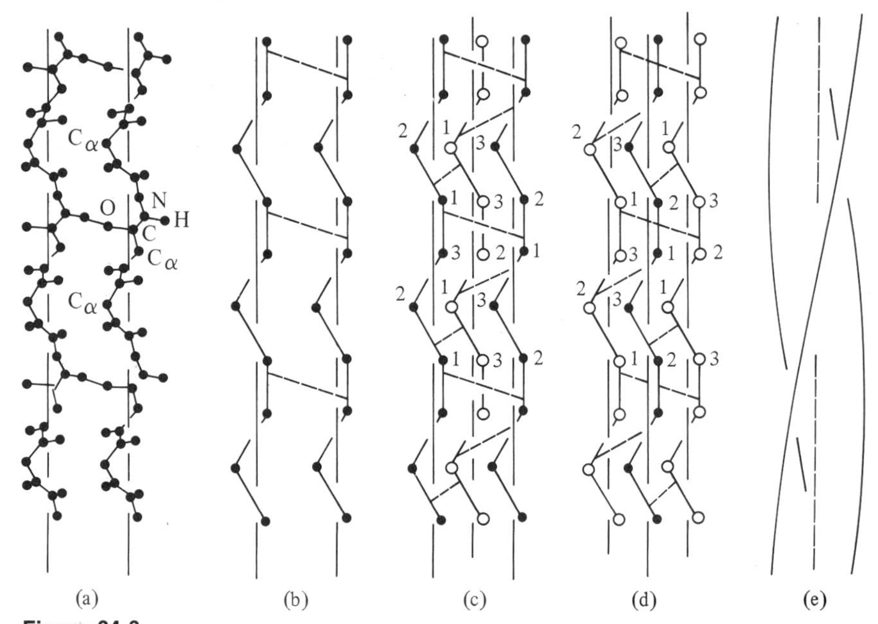

Figure 24-3
Assembly of the molecular model of tropocollagen: (a) and (b) show two polypeptide chains lying side by side; (c) and (d) show two alternative ways of combining three chains to form a triple-stranded structure; (e) shows the deformation of the triple-stranded structure to form tropocollagen.

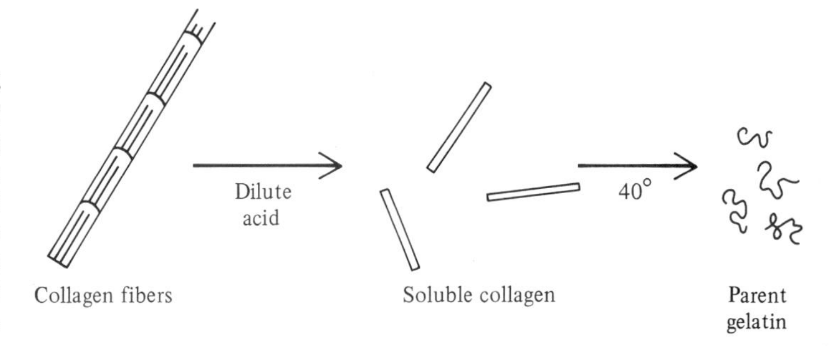

Figure 24-4
Denaturation of collagen. Parent gelatin is the mixture of separated polypeptide chains formed by the breakup of tropocollagen (soluble collagen).

dense with a lysine amino group via Schiff's base formation or form other types of covalent linkage.

Reformation of Collagen Fibrils

Collagen fibrils may be regenerated by the neutralization of acid solutions of tropocollagen. Fibrils reformed in this way resemble those of native collagen when examined with the electron microscope (Fig. 24-5). In particular, they have the same banded appearance with the characteristic 70-nm spacing.

Fibrils of somewhat different properties may be obtained by altering the conditions of reformation. The addition of certain glycoproteins results in fibrils for which the 70-nm spacing is replaced by a 280-nm spacing. Regenerated fibrils of this type have been designated as the **fibrous long spacing**

Figure 24-5 Regenerated collagen fibrils having the native structure. (Courtesy of National Naval Medical Center, Bethesda, Maryland.)

Figure 24-6 Regenerated collagen fibrils having the FLS form. (Courtesy of National Naval Medical Center, Bethesda, Maryland.)

(FLS) form (Fig. 24-6). The similarly of the spacing to the long dimension of tropocollagen is of course suggestive of a model in which tropocollagen units are arranged in parallel without overlapping.

Fibrils of different structural characteristics can be prepared by carrying out regeneration in other media. In the presence of ATP, the **segment long spacing** (SLS) form appears. This consists of short polarized segments 260 nm in length, in which no end-to-end association of tropocollagen units occurs. Under other conditions fibrils devoid of regular periodicity may be formed.

It seems clear that the banded appearance of native and regenerated collagen fibrils arises from the alignment of particular structural features. If the two ends of the tropocollagen unit are distinguishable from the balance of the molecule and each other, then their arrangement in a staggered, overlapping configuration could account for the appearance of native collagen with the 70-nm band spacing. In the SLS and FLS forms they are presumably aligned with their ends in register (Fig. 24-7).

Elastin

As the name implies, **elastin**, which is the second major protein component of connective tissue, predominates in such elastic structures as ligaments and the walls of blood vessels. In amino acid composition it is significantly different from collagen, containing a higher proportion of the higher aliphatic

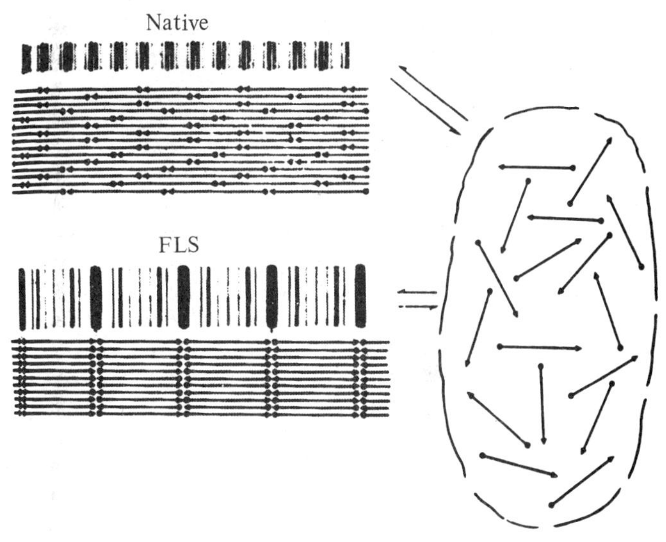

Figure 24-7
Schematic model for the formation of different kinds of regenerated collagen fibrils. A staggered alignment of the (distinguishable) ends could account for the 70-nm spacing of the bands of the native fibrils. (Courtesy of National Naval Medical Center, Bethesda, Maryland.)

hydrocarbon side-chains and many fewer acidic, basic, and aliphatic hydroxyl groups. Like collagen, it is rich in proline and hydroxyproline. There is evidence that its polypeptide chains are covalently linked to form an elastic two-dimensional sheet. Elastin is resistant to attack by trypsin and chymotrypsin, but may be hydrolyzed by elastase.

Other Components of Connective Tissue

In addition to the protein components already described, the connective tissues contain varying amounts of several mucopolysaccharides, which may be complexed with protein. These include hyaluronic acid and chondroitin sulfate. A number of glycoproteins are also present.

24-2 MYOFIBRIL

Contractile Tissues

The property of contractility is widespread in biological systems and is by no means confined to tissues ordinarily classed as muscle; some random examples include amoeba movement, the contraction of blood platelets, and the motion of bacterial flagella. However, the discussion here will be confined to classical muscle, which is itself histologically quite diverse.

Smooth muscle cells in mammals are found in the walls of blood vessels, in the skin, in the uterus, and in the walls of the digestive tract. Elsewhere, smooth muscle occurs in the invertebrates, including the snail and some mollusks. Mammalian smooth muscle cells are generally spindle shaped and up to 500 nm in length. They have no distinct enclosing sheath and are not visibly striated.

Figure 24-8
Electron microscope photograph of myofibril, showing striations. (Courtesy of National Naval Medical Center, Bethesda, Maryland.)

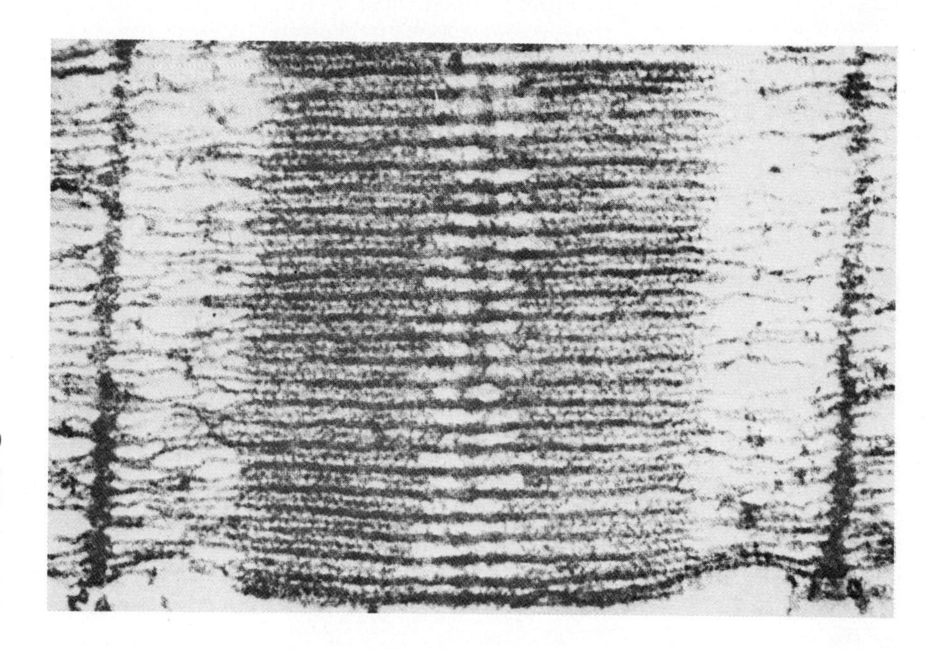

Figure 24-9
Electron microscope photograph of striated muscle filaments. (Courtesy of National Naval Medical Center, Bethesda, Maryland.)

Skeletal muscle is by far the most abundant form of muscle in mammals, occuring in the limbs, chest, and shoulders and accounting for almost half the mass of the human body. It displays characteristic striations and is often referred to as **striated** muscle. These striations arise from alternating bands of different refractive index (Figs. 24-8 and 24-9).

A mammalian muscle is generally enclosed in a sheath, called the **perimysium**, and may contain fat deposits and connective tissue. The strictly contractile tissue consists of fibers of diameter 10–100 μm that are linked ultimately with the tendon fibrils. Each fiber is contained in a membrane, the **sarcolemma**, which is itself of complex structure. A fiber is a single cell with multiple nuclei.

Each fiber is actually a bundle of parallel **fibrils** about 1 μm in diameter. While the fibrils are at the limit of resolution of ordinary light microscopy, the electron microscope shows them to consist, at the molecular level, of thinner **filaments.**

Cardiac muscle, of which the heart is largely composed, is intermediate in properties to smooth and striated muscle. Transverse striations are present which are less distinct than those of skeletal muscle.

Fine Structure of Muscle Fibrils

By far the most detailed information is available for skeletal muscle, upon which the present discussion will be focused. The striated character of muscle fibrils or **myofibrils** has proved to be directly related to their structural organization. Two broad bands, the A and I bands, are prominent. When viewed with polarized light the (isotropic) I band appears dark and the A band light. Both the A and I bands are bisected by narrower bands, called H in the former case and Z in the latter (Fig. 24-10).

A repeating structural pattern occurs along the myofibrils. The repeating units, termed the **sarcomeres**, are the actual contractile units. In striated muscle, the sarcomeres of many parallel myofibrils are aligned in register, thereby producing the characteristic striations.

Electron microscopy has shown the myofibril to consist of thick and thin filaments (Fig. 24-10). Only thin filaments occur in the I bands; these have a diameter close to 6 nm. The thin filaments begin at the Z band and extend throughout the I band and part of the A band, terminating at the edge of the H zone (Fig. 24-10). The thick filaments, of diameter about 15 nm, extend across the entire length of the A band. Observations of cross-sections of muscle have shown that the thick filaments are arranged in a hexagonal array; each thick filament is surrounded by six thin filaments (Fig. 24-11).

Within the A bands projections extend at regular intervals from the thick filaments to the thin (Fig. 24-11), thereby forming cross-links between the two sets of filaments. The thick and thin filaments are otherwise not connected.

Contraction of The Myofibril

The primary biological activity of muscle is of course the performance of external work by virtue of its property of contractility. The mechanical tension developed *in vivo* is considerable—up to 4 kg/cm^2.

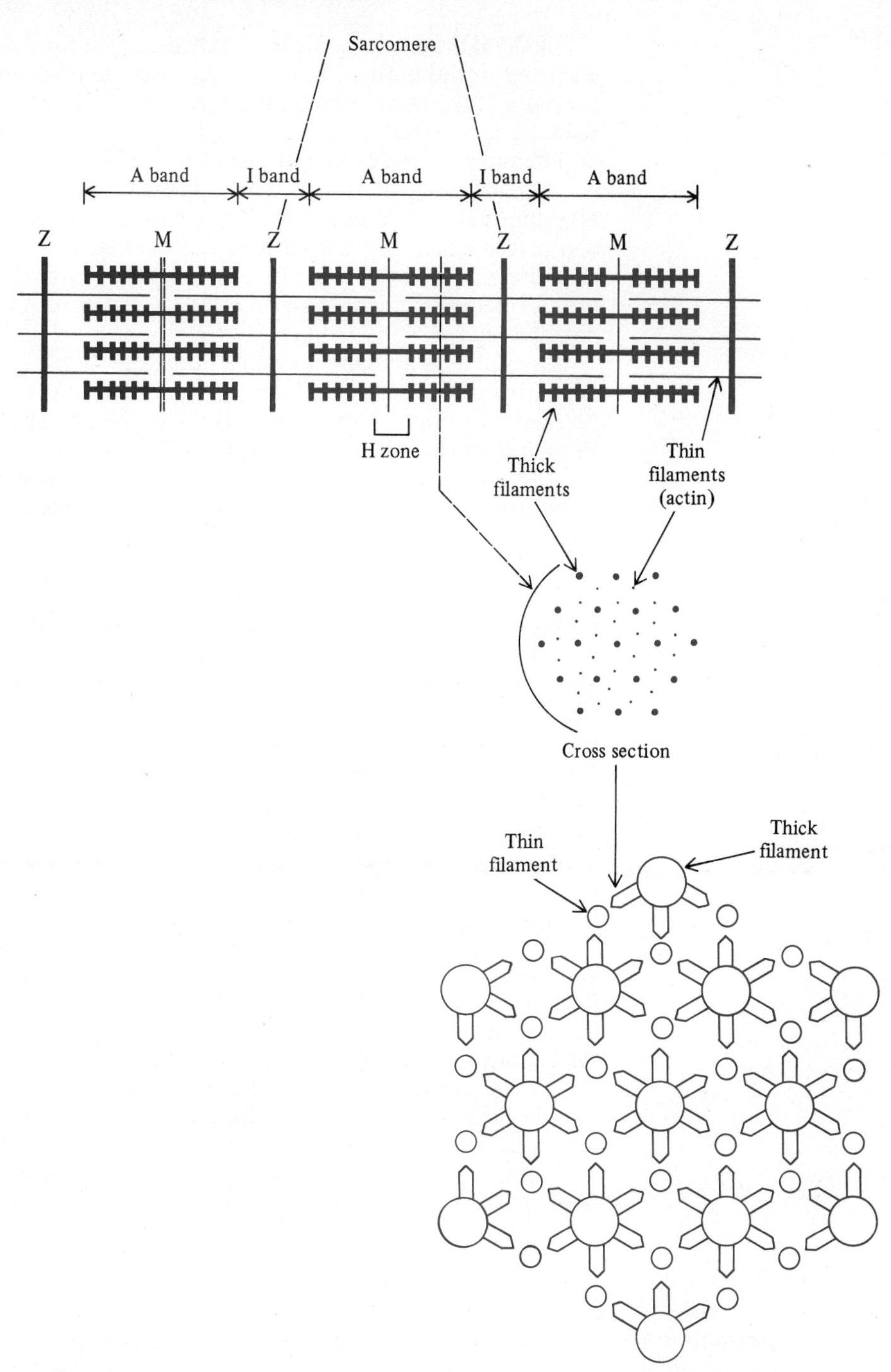

Figure 24-10
Model for the structure of the myofibril.

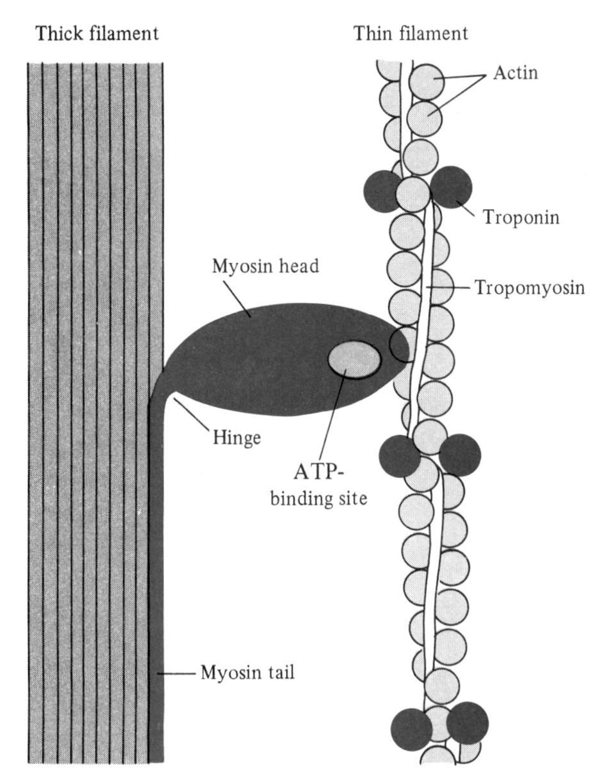

Figure 24-11 Organization of the thick and thin filaments.

When the sarcomeres are maximally contracted, they are shortened by up to 50%. Electron microscopic studies have revealed that the contractile process does not involve any change in length of the thick or thin filaments, but rather a mutual *sliding* of the filaments.

During contraction the thin filaments slide past the thick filaments so as to be drawn into the A band until the H zone is filled. Further shortening may involve some overlapping of the ends of the thin filaments (Fig. 24-12).

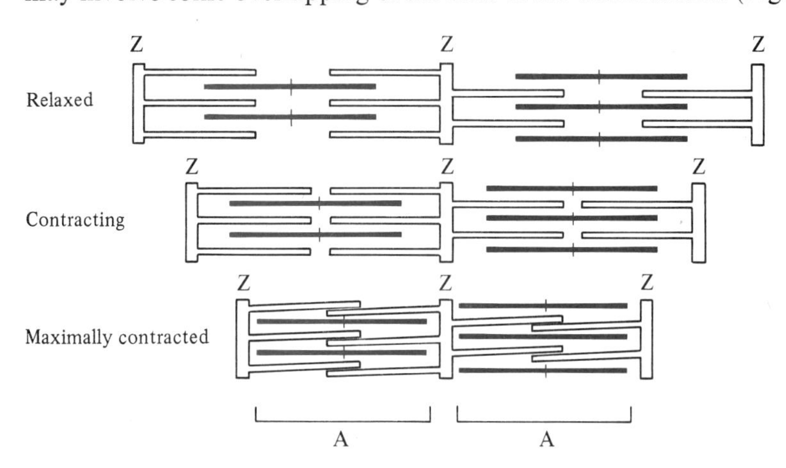

Figure 24-12 Model for the contraction of the myofibril.

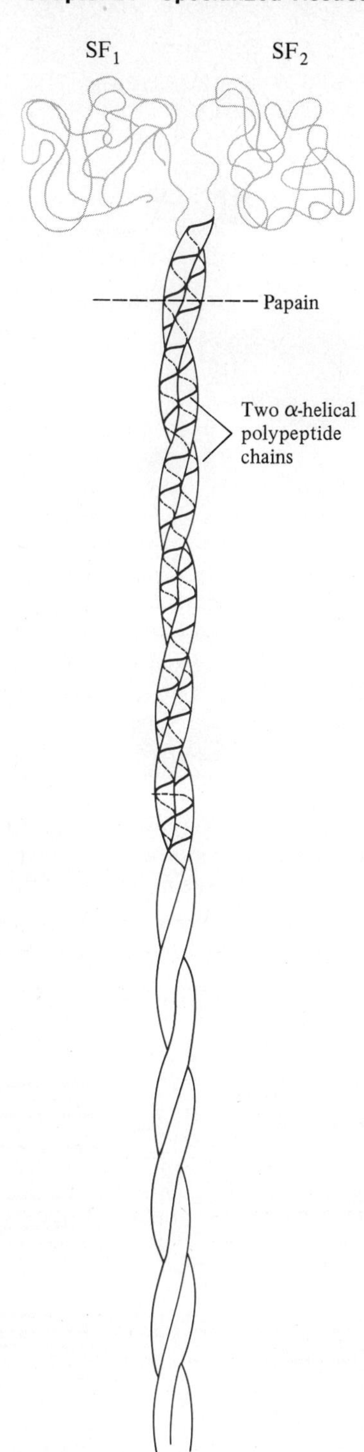

Figure 24-13
Structure of myosin.

Muscle Proteins

The intracellular fluid of muscle cells or **sarcoplasm**, which surrounds the myofibrils, contains numerous water-soluble proteins, including glycolytic enzymes and glycogen phosphorylase. The proteins of the myofibril are insoluble in water, but can be extracted with 0.6-*M* KCl.

Myosin, the most abundant of the muscle proteins, is an asymmetric protein that has a molecular weight of 460,000 and a length of 160 nm. It contains two equivalent long polypeptide chains of molecular weight close to 200,000; these are termed *heavy* chains. These are α-helical for most of their length (Fig. 24-13). The two α-helices are wound about each other and together constitute the linear *tail* of the myosin molecule; the balance of the heavy chains are folded into a globular *head* at one end. The head also contains four smaller, or *light* polypeptide chains.

Myosin is an enzyme and catalyzes the hydrolysis of ATP to ADP and inorganic phosphate. The reaction is stimulated by Ca^{2+} and inhibited by Mg^{2+}.

Proteolysis by papain cleaves the myosin head from the tail and further divides the head into two equivalent fragments, called the **SF fragments**, which have been isolated and purified. These fragments account for all the ATPase activity of myosin. The catalytic activity of myosin is thus confined to the head. There are two catalytic sites, one on each SF unit.

Actin, the second major protein of the myofibril, can be isolated as a globular protein, **G-actin**. Its molecular weight is 46,000. At neutral pH G-actin is stable only in the absence of added electrolyte. In the presence of salts polymerization occurs to form extended filaments, **F-actin**, whose lengths may be several micrometers.

Each G-actin molecule contains tightly bound molecules of ATP and Ca^{2+}. During polymerization the ATP is hydrolyzed to ADP, which remains bound by the actin.

A third important muscle protein is **tropomyosin**, which consists of two α-helical polypeptide chains of total molecular weight 70,000. These are twisted about each other to form a double coil of length about 40 nm.

The myofibril also contains a globular protein, **troponin**, which has an important regulatory function. It consists of three subunits: a Ca^{2+}-binding subunit of molecular weight 18,000; an **inhibitory** subunit of molecular weight 23,000, which contains an actin-binding site; and a tropomyosin-binding subunit.

Structural Organization of the Myofibril

The application of electron microscopy and other techniques has led to a definitive model for the myofibril. The thick filaments consist of parallel bundles of myosin molecules, in a staggered overlapping arrangement. The orientation is such that the myosin heads project laterally out of the bundle in a regular array. A single thick filament contains several hundred myosin molecules.

The thin filaments consist of two mutually coiled strands of F-actin (Fig. 24-14) plus tropomyosin and troponin. The linear tropomyosin molecules are placed in the grooves of the F-actin coil in an end-to-end arrangement; each tropomyosin is in contact with only one F-actin strand. The tropomyosin molecules can move along the grooves. The troponin molecules

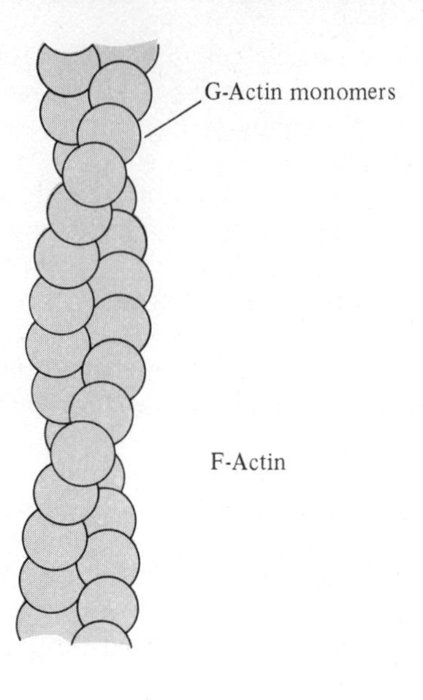

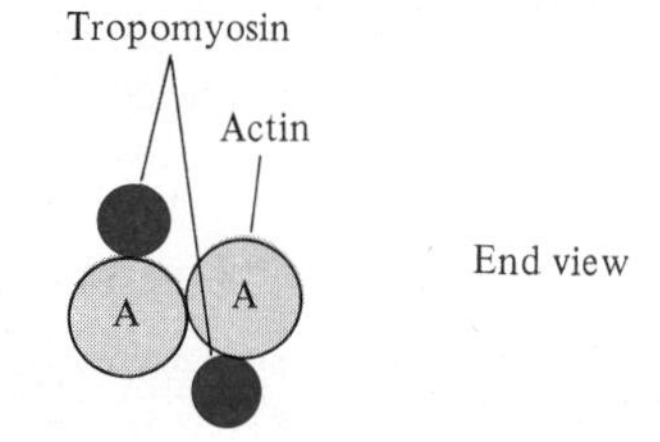

Figure 24-14 Structure of the thin filaments.

are linked to the thin filaments via their actin- and tropomyosin-binding sites; the former linkage may be broken and reformed in response to the binding of Ca^{2+}. One molecule each of tropomyosin and troponin occur for every seven G-actin monomers.

The myosin heads protruding from the thick filaments account for the cross-bridges between the two sets of filaments.

The Contractile Process

The basic source of energy for the contractile process is the hydrolysis of ATP. The involvement of ATP in muscle contraction initially came to be recognized, in part, from studies with glycerol-extracted fibers. This treatment removes the soluble components, leaving only the contractile elements. Such preparations undergo a dramatic contraction in the presence of ATP, supporting the idea that ATP interacts directly with the contractile proteins.

Evidence of another kind was provided by studies upon artificial complexes of myosin and F-actin. Such complexes, called **actomyosin**, are formed upon mixing purified myosin and F-actin in solution. Actomyosin threads may be formed by extrusion of actomyosin solutions into suitable solvents. Such threads contract when exposed to ATP. The addition of ATP to *solutions* of actomyosin results in dissociation into actin and myosin.

The broad outlines of the mechanism of contraction of the myofibril are now becoming partially visible in a culmination of the early studies cited above. The arrival of a nerve signal causes the release of Ca^{2+} into the sarcoplasm; the Ca^{2+} is at first only bound to the Ca^{2+}-binding sites of troponin. This, in turn, produces a structural change in troponin so as to expose a previously shielded myosin-binding site on the G-actin monomer. The latter now combines with the protruding myosin head, which has previously bound two molecules of ATP.

A conformational change, accompanied by the hydrolysis of ATP, next occurs in the myosin head. The orientation of the cross-bridge with respect to the thick filament changes in such a way as to slide the thin filament along the thick filament. It is at this step that the contraction of muscle fibrils is translated into external work. It is not yet possible to supply details as to how the free energy of hydrolysis of ATP is utilized for the development of contractile force.

The return of muscle to the relaxed state is achieved by the removal of Ca^{2+} from the sarcoplasm.

Skeletal muscle possesses an energy reserve in **phosphocreatine**, which can regenerate ATP from ADP in a reaction catalyzed by **creatine kinase**:

$$\underset{\text{Phosphocreatine}}{HOOC-CH_2-\overset{\overset{\displaystyle CH_3}{|}}{N}-\underset{\underset{\displaystyle NH}{\|}}{C}-NH-PO_3^{2-}} + ADP \longrightarrow \underset{\text{Creatine}}{HOOC-CH_2-\overset{\overset{\displaystyle CH_3}{|}}{N}-\underset{\underset{\displaystyle NH}{\|}}{C}-NH_2} + ATP$$

Since phosphocreatine is present in muscle in considerable excess and the equilibrium of the above reaction lies well to the right, the ATP hydrolyzed during a muscle contraction is quickly replaced, so that no net decrease in ATP level occurs.

24-3 BONE TISSUE

Composition

The discussion in this section will apply primarily to long bones, for which the most definitive information is available. Intact bone consists of roughly equal weights of mineral and protein components. The former may be selectively removed by treatment with dilute acid, leaving the protein portion as a flexible residue retaining the original shape of the bone.

The inorganic fraction of bone consists primarily of a crystalline mineral whose approximate composition is that of **hydroxyapatite**, $Ca_{10}(PO_4)_6(OH)_2$. Small amounts of Mg^{2+} and F^- replace some of the Ca^{2+} and OH^-, respectively.

The protein portion, which consists largely of collagen, provides a matrix in which hydroxyapatite is distributed. The mineral components endow bone with its characteristic hardness. The collagen fibers present in bone are basically similar to those of connective tissue.

Formation

In the development of an animal embryo, the formation of bone begins with the introduction of collagen fibrils synthesized by specialized cells called fibroblasts. At this initial stage of bone formation considerable quantities of mucopolysaccharides are present. Despite the widespread occurrence of both collagen and mucopolysaccharide in the animal body, the deposition of calcium phosphate is confined to the regions destined to become bone.

The mineralization of bone appears to result from the removal of calcium phosphate from a supersaturated fluid medium via some form of nucleation induced by the collagen fibrils. The mucopolysaccharide initially present disappears as bone is formed.

The mineral content of bone is in dynamic equilibrium with other tissues and the administration of radioactive P^{32} is followed by its early appearance in bone.

A deficiency of vitamin D, which promotes Ca^{2+} absorption from the intestine, produces the symptoms of rachitis in young animals. Excessive levels of vitamin D are toxic, resulting in withdrawal of calcium and phosphate from bone, so that their concentrations in plasma become high enough to produce renal calculi.

Another important factor in the regulation of the metabolism of calcium and phosphate is the action of parathyroid hormone and calcitonin. Parathyroid hormone stimulates Ca^{2+} transport from bone and also promotes Ca^{2+} reabsorption from kidney tubules. The net effect is to increase the Ca^{2+} level in plasma. Calcitonin, which is secreted when the Ca^{2+} level rises, stimulates the deposition of Ca^{2+} in bone. The opposed actions of the two hormones provides a means of controlling the incorporation of Ca^{2+} in bone by a feedback mechanism.

SUMMARY

The fibrous protein **collagen** has an important structural function in animals, occurring in tendon, skin, and bone. The basic molecular unit, **tropocollagen**, is a rodlike molecule consisting of three helically wound polypeptide chains. The unusual helical structure of collagen probably arises from its high content of **proline** and **hydroxyproline**, which cannot fit into an α-helix. Thermal denaturation of collagen destroys the helical structure.

Muscle fibrils contain sets of interdigitating *thick* and *thin* filaments, linked by regularly spaced **cross-bridges**. The thick filaments consist of **myosin**, while the thin filaments are composed of **F-actin** complexed with **tropomyosin** and **troponin**. Myosin is a rod-shaped molecule consisting of an elongated helical *tail* and a globular *head*; the head portions account for the cross-bridges observed in muscle fibrils. The myosin head contains two enzymic sites which catalyze the hydrolysis of ATP.

F-actin is an elongated polymer of globular **G-actin** units. The thin filaments consist of two helically intertwined F-actin strands complexed with tropomyosin and troponin.

The arrival of a nerve pulse at a muscle cell causes a release of bound Ca^{2+} and an increase in the level of free Ca^{2+}. Ca^{2+} is bound by a troponin site, producing a change in the mutual arrangement of myosin and F-actin, so that the latter interacts directly with the myosin head. ATP is simultaneously hydrolyzed. The net result is a sliding of the F-actin filaments past the myosin filaments, resulting in contraction of the muscle fibers.

Bone consists of a protein component, which is primarily collagen, and a mineral component, which is largely **hydroxyapatite** $[Ca_{10}(PO_4)_6(OH)_2]$. The collagen component provides a matrix in which the mineral component is deposited. The calcium of bone is in dynamic equilibrium with that present in body fluids; the process is subject to hormonal control.

REFERENCES

The following code is used to classify references. I: particularly useful as an introduction to the subject; R: useful primarily as a reference text; A: an advanced account of the material; H: a publication of historical importance.

General

A. L. Lehninger, *Biochemistry*, Worth, New York (1975). (A)

D. E. Metzler, *Biochemistry*, Academic, New York (1977). (R)

A. White, P. Handler, and E. L. Smith, *Principles of Biochemistry*, McGraw-Hill, New York (1973). (R)

Muscle

The Mechanism of Muscle Contraction," *Cold Spring Harbor Symp. Quant. Biol.* 37 (1972). (A)

H. E. Huxley, "The Mechanism of Muscular Contraction," *Science* 164, 1356 (1969). (H)

H. E. Huxley, "Muscular Contraction and Cell Motility," *Nature* 243, 445 (1973). (A)

A. Weber and J. M. Murray, "Molecular Control Mechanism in Muscle Contraction," *Physiol. Rev.* 53, 612 (1973). (A)

Collagen

R. E. Dickerson and I. Geis, *The Structure and Action of Proteins*, pp. 33–43, Benjamin, New York (1969). (I)

REVIEW QUESTIONS

Questions marked with an asterisk are of a high level of difficulty.

24-1 Which of the following statements is true about the structure of collagen?
- (a) It has H-bonds between adjacent peptide groups on the same chain.
- (b) It has H-bonds between almost all of the peptide groups on one chain with peptide groups on the other chains.
- (c) Each chain follows a helical path in space.
- (d) Rotation angles around the C—C (carbonyl) and N—C bonds are nearly the same for every residue.

(e) Chains are easily extended to about twice normal length.
(f) Collagen has a high proline content.

24-2 There is evidence that a flexible "hinge" exists between the head and tail of myosin. How does this fit into the model for muscle contraction?

24-3 The standard free energy, $\Delta G^{\circ\prime}$, of hydrolysis of phosphocreatine is -10.3 kcal at 25°C and pH 7.0.
(a) Calculate the equilibrium constant for the hydrolysis of phosphocreatine.
(b) Calculate the equilibrium constant for the reaction catalyzed by phosphocreatine kinase.

24-4 About 0.03 moles of phosphocreatine and 5×10^{-3} moles of ATP are present per kilogram of skeletal muscle. How much work per kilogram of muscle could be carried out utilizing *only* the existing supply of high energy phosphate? How does this compare with the work required to lift a 10-kg weight 1 m?

*24-5 How much glucose would have to be degraded by anaerobic glycolysis for each kilocalorie of external work done by muscle?

Glossary

acetylcholine A molecule involved in nerve conduction.

acetylcholinesterase An enzyme that hydrolyzes acetylcholine to choline and acetate.

acetyl CoA The acetylated form of coenzyme A.

acetyl CoA carboxylase An enzyme that catalyzes the carboxylation of acetyl CoA to form malonyl CoA.

ACP-acyltransferase An enzyme that catalyzes the transfer of an acetyl group from acetyl CoA to an acyl carrier protein (ACP).

ACP-malonyl transferase An enzyme that catalyzes the transfer of a malonyl group from malonyl CoA to ACP.

actin A major protein component of the myofibril.

actinomycin An antibiotic that inhibits transcription.

action potential The transient change in electric potential accompanying the passage of a nerve impulse.

active acetaldehyde The complex of acetaldehyde and thiamine pyrophosphate.

active carbon dioxide The complex of CO_2 and biotin.

active formate The formyl derivative of tetrahydrofolate.

active site A restricted region in a protein, in which biological activity is centered.

active transport The transport of a substance across a biological membrane at the expense of metabolically derived energy.

actomyosin A complex of myosin and F-actin.

acyl carrier protein (ACP) A component of the fatty-acid synthetase system to which the growing fatty acid chain is attached.

acyl-CoA dehydrogenase An enzyme that catalyzes the reduction of a fatty acyl-CoA to form an α, β-unsaturated derivative.

acyl-CoA synthetase An enzyme that catalyzes the conversion of fatty acids to fatty acyl-CoA.

adenine A purine base occurring in nucleic acids.

adenohypophysis The anterior pituitary.

adenosine diphosphate (ADP) An adenine nucleotide containing a 5′-pyrophosphate group.

adenosine monophosphate (AMP) An adenine nucleotide containing a 5′-phosphate group.

adenosine triphosphate (ATP) A major high energy compound, whose hydrolysis to ADP drives many biochemical processes.

adenosylhomocysteinase An enzyme that hydrolyzes *S*-adenosylhomocysteine to homocysteine.

***S*-adenosyl methionine decarboxylase** An enzyme that converts *S*-adenosyl methionine to the corresponding amine by decarboxylation.

adenylate cyclase An enzyme that forms cAMP from ATP.

adenylosuccinate lyase An enzyme that cleaves 5′-phosphoribosyl-4-(N-succinocarboxamide)-5-aminoimidazole to form fumarate and 5′-phosphoribosyl-4-carboxamide-5-aminoimidazole.

adenylosuccinate synthetase An enzyme that catalyzes

adenylosuccinate synthetase (*continued*) the formation of adenylosuccinate by the condensation of aspartic acid with inosinic acid.

adipose cells Storage cells for fat.

adrenal cortex A gland that secretes a number of steroid hormones.

adrenocorticotropin (ACTH) A peptide pituitary hormone that stimulates the synthesis of steroid hormones by the adrenal cortex.

aerobic cells Cells whose metabolism uses molecular oxygen as the ultimate electron acceptor.

affinity chromatography A form of column chromatography in which a small molecule is linked chemically to the chromatographic material and selectively binds the protein.

alanine transaminase An enzyme that catalyzes the transfer of an amino group between alanine and α-ketoglutarate.

alcohol dehydrogenase An enzyme that catalyzes the reduction of acetaldehyde to ethanol.

aldaric acid A dicarboxylic sugar acid formed by oxidation of the aldehyde group and the terminal alcohol group.

aldonic acid A sugar acid formed by oxidation of an aldehyde group.

aldose A member of a family of sugars containing aldehyde groups, whose simplest prototype is glyceraldehyde.

aldosterone A steroid hormone secreted by the adrenal cortex that regulates mineral metabolism and electrolyte levels.

algae A class of eukaryotic microorganisms.

aliphatic group A hydrocarbon other than benzene or its derivatives.

allantoin A degradation product of purines in some higher species.

allantoicase An enzyme that catalyzes the degradation of allantoic acid to urea in some fish.

allantoinase An enzyme that catalyzes the degradation of allantoin to allantoic acid in fish.

allolactose The natural inducer that combines with the *lac* repressor.

allosterism The process whereby the activity of an enzyme is altered by a structural change resulting from the binding of a modifier.

amidophosphoribosyl transferase An enzyme that catalyzes the formation of 5-phosphoribosylamine.

amino acid decarboxylases Enzymes that decarboxylate amino acids to form the corresponding amines.

amino acid oxidase A flavin enzyme that oxidizes amino acids to α-carbonyl derivatives.

amino acids The constituents of proteins.

α-amino group The primary amino group attached to the α-carbon of an amino acid.

δ-aminolevulinate dehydratase An enzyme that catalyzes the formation of porphobilinogen from δ-aminolevulinate.

aminopropyl transferase An enzyme that forms spermidine by transfer of an NH_2-$(CH_2)_3$-group to putrescine.

α-amylase, β-amylase Enzymes that hydrolyze the glycosidic bonds of starch.

anaerobic cells Cells whose metabolism uses some molecule other than molecular oxygen as the ultimate electron acceptor.

anaphase A phase of mitosis during which separation of the chromatids occurs.

androgens The male sex hormones.

anion exchangers Resins used for the separation of anions.

annealing The restoration of the native helical structure of a thermally denatured DNA by slow and controlled cooling.

anterior pituitary A lobe of the pituitary that secretes a set of peptide hormones and whose blood supply flows from the hypothalamus.

antibody A protein molecule formed in response to the presence of a foreign molecule in the circulation.

anticodon A nucleotide triplet in tRNA, which pairs with a mRNA codon.

antigen A foreign molecule that elicits the formation of an antibody from an organism's immune system

apoferritin The protein component of ferritin.

arginase An enzyme that hydrolyzes arginine to ornithine plus urea.

argininosuccinate lyase An enzyme that catalyzes the decomposition of argininosuccinate to arginine plus fumarate.

argininosuccinate synthetase An enzyme that catalyzes the formation of argininosuccinate.

argininosuccinic acid An intermediate of the urea cycle formed by the condensation of citrulline with aspartate.

aromatic amino acid decarboxylase An enzyme that decarboxylates dopa to form dopamine.

aromatic group A group containing a benzene ring or a derivative thereof.

A site The site on a ribosome with which a tRNA carrying an amino acid initially combines.

asparaginase An enzyme that hydrolyzes asparagine to aspartate.

asparagine synthetase An enzyme that forms asparagine from aspartate.

aspartate transaminase An enzyme that catalyzes the transfer of an amino group between aspartate and α-ketoglutarate.

atherosclerosis The condition resulting from the deposition of cholesterol on the inner walls of blood vessels.

ATP-citrate lyase An enzyme that converts citrate to acetyl CoA plus oxaloacetate.

autotrophic cells Cells that obtain their energy requirements from sunlight.

axoplasm The interior of a nerve fiber.

bacteria A class of prokaryotic microorganisms.

bacteriophages Viruses that infect bacterial cells.

bilayer A double layer of phospholipids.

bile salts Steroid derivatives important for the emulsification and digestion of fats in the intestine.

biliverdin A degradation product of heme.

basal metabolism rate The rate of oxygen consumption by an organism.

Bohr effect The reduction of the oxygen affinity of hemoglobin as a result of protonation.

buffer A solution containing a weak acid and its corresponding salt, which minimizes the pH changes resulting from the addition of acid or base.

Bragg equation An equation describing the diffraction of x rays by regularly spaced layers of atoms.

Brownian motion The thermal motion of a dissolved solute.

bundle-sheath cells Specialized cells occurring in C_4 plants, in which the decarboxylation of malate occurs.

calcitonin A parathyroid hormone that inhibits the loss of Ca^{2+} from bone to blood.

cAMP-dependent protein kinase An enzyme that catalyzes the phosphorylation and activation of phosphorylase kinase and is activated by binding cAMP.

carbamoyl phosphate A high energy intermediate of the urea cycle that is formed by the reaction of CO_2, NH_3, and ATP.

carbamoyl phosphate synthetase An enzyme that catalyzes the formation of carbamoyl phosphate.

carbohydrates The class of organic molecules that includes the sugars, their derivatives, and their polymers.

α-carbon The carbon atom to which the amino and carboxyl groups of an amino acid are attached.

carbonic anhydrase An enzyme that catalyzes the formation of carbonic acid from dissolved CO_2.

α-carboxyl group The carboxyl group attached to the α-carbon of an amino acid.

cardiac muscle A striated, involuntary form of muscle found in heart tissue.

carnitine A molecule that functions as a carrier for fatty acyl CoA across the inner mitochondrial membrane.

carnitine acyltransferase An enzyme that catalyzes the combination of carnitine and fatty acyl CoA.

carotenes A class of photosynthetic pigments.

catabolite activator protein (CAP) An allosteric protein that combines with a promoter and facilitates the initiation of transcription.

catalase An enzyme that decomposes hydrogen peroxide.

catecholamines The hormones epinephrine, norepinephrine, and dopamine.

cation exchangers Resins used for the separation of cations.

CDP-diacylglycerol inositol phosphatidyl-transferase An enzyme that catalyzes the formation of phosphatidyl inositol from a CDP-diacylglycerol and inositol.

cell The fundamental structural unit of living systems.

cellulases Enzymes that hydrolyze cellulose.

cellulose The principal structural carbohydrate of plants.

centrioles Cellular granules involved in cell replication.

centromere A chromosomal region that is visible during mitosis and meiosis.

ceramide The sphingosine–fatty acid portion of a cerebroside.

ceramide cholinephosphotransferase An enzyme that catalyzes the formation of sphingomyelin from ceramide and CDP-choline.

cerebrosides Compounds structurally related to the sphingomyelins, in which the phosphorylcholine group of the sphingomyelins is replaced by a monosaccharide.

chemical coupling theory A proposed mechanism for oxidative phosphorylation that postulates the participation of a high energy intermediate.

chemiosmotic hypothesis A proposed mechanism for oxidative phosphorylation that postulates that the driving force is a gradient of hydrogen ions across the mitochondrial membrane.

chemolithotrophs Organisms whose carbon source is atmospheric CO_2 and whose energy is derived from oxidation-reduction reactions involving inorganic compounds.

chemoorganotrophs Organisms that derive both their carbon and their energy from the metabolism of consumed organic compounds.

chirality An intrinsic asymmetry of a molecule resulting in optical activity.

chloroplasts Chlorophyll-containing cytoplasmic particles occurring in green plants.

cholecalciferol A form of vitamin D.

cholestanol A compound arising from the reduction of the double bond of cholesterol, differing in configuration from coprostanol.

cholesterol An important steroid that occurs in many membranes.

choline kinase An enzyme that catalyzes the phosphorylation of choline by ATP to form phosphocholine.

chromaffin granules Granules of the adrenal medulla that store and release catecholamines.

chromane The ring structure occurring in the tocopherols.

chromatids The partially separated chromosome pairs appearing during mitosis.

chromosomes DNA-containing particles present in cell nuclei that carry the genetic instructions for the organism.

chylomicrons Plasma particles composed of lipid plus a small amount of protein.

chymotrypsin A proteolytic enzyme formed in the pancreas.

cilia External cellular filaments involved in cellular motility.

circular dichroism The differential absorption of left- and right-handed circularly polarized light by an optically active substance.

circularly polarized light A form of polarized light for which the electric vector rotates about the direction of propagation of the beam.

cisternae A set of flattened sacs within the endoplasmic reticulum.

***cis-trans* isomerism** A type of spatial isomerism about a double bond.

citric synthase An enzyme of the TCA cycle that catalyzes the combination of acetyl CoA with oxaloacetic acid.

citrulline An intermediate of the urea cycle that is formed by the reaction of carbamoyl phosphate with ornithine.

clotting The process whereby blood seals itself by gelation.

clupein A protamine of high arginine content that occurs in herring sperm.

codon A nucleotide triplet in mRNA that directs the incorporation of an amino acid into protein.

coenzyme A molecule that participates in an enzyme-catalyzed reaction with a substrate.

cofactor A second molecule that is essential for the activity of an enzyme.

collagen A protein of connective tissue.

competitive inhibitor An enzyme inhibitor that interferes directly with the binding of substrate.

cone cells Cells of the mammalian retina that are concerned with vision at high light intensities.

conjugate base The ion or molecule remaining after dissociation of a hydrogen ion by a weak electrolyte.

conjugated protein A protein containing a structural element not composed of, or derived from, amino acids.

conjugation The direct transfer of a portion of a chromosome between two bacterial cells.

constitutive mutations Mutations that block repression and allow uncontrolled transcription of an operon.

COOH-terminal end The end of a polypeptide chain with a free carboxyl group.

cooperativity The process whereby the binding of a molecule by a protein enhances the affinity of the protein for binding additional molecules.

coprostanol A compound arising from the reduction of cholesterol, differing in configuration from cholestanol.

corticosterone A steroid hormone secreted by the adrenal cortex.

corticotropin releasing factor A peptide hormone secreted by the hypothalamus that controls the secretion of ACTH.

cortisol A steroid hormone secreted by the adrenal cortex that promotes gluconeogenesis.

C_4 plants Plants whose CO_2 fixation proceeds by incorporation into phosphoenolpyruvate.

creatine kinase An enzyme that catalyzes the regeneration of ATP from ADP by phosphocreatine.

cristae Characteristic foldings of the inner mitochondrial membrane.

critical micelle concentration The concentration above which soaps form micelles.

crossing over An exchange of chromosomal segments between maternal and paternal chromosomes that occurs during meiosis.

crystal A macroscopic system of completely ordered three-dimensional structure.

cyclic AMP (cAMP) A nucleotide important in biological regulation that contains an internal diester of phosphoric acid.

cyclic photophosphorylation A photosynthetic pathway occurring in chloroplasts, which generates ATP.

cystathionine γ-lyase An enzyme that decomposes cystathionine to form cysteine and α-ketobutyrate.

cystathionine β-synthase An enzyme that condenses homocysteine with serine to form cystathionine.

cysteine dioxygenase An enzyme that converts cysteine to cysteine sulfinic acid.

cytochromes Heme-containing proteins occurring in photosynthetic systems and in the mitochondrial respiratory chain.

cytoplasm The portion of a cell exclusive of the nucleus and the external membrane.

dansyl chloride A fluorescent reagent that reacts with amino groups.

decarboxylases PLP-dependent enzymes that decarboxylate α-amino acids to form amines.

7-dehydrocholesterol A derivative of cholesterol containing a second double bond at the 7, 8 position.

denaturation The loss of organized structure by a protein.

density gradient centrifugation An ultracentrifugal technique that resolves solutes of different densities using a gradient of CsCl concentration.

2-deoxy-D-ribose The sugar component of DNA

deoxyribonuclease An enzyme that hydrolyzes phosphodiester bonds of DNA.

deoxyribonucleic acid (DNA) A class of nucleic acid containing deoxyribose that acts as the cellular repository of genetic information.

diacylglycerol acyltransferase An enzyme that catalyzes the formation of a triacylglycerol from a diacylglycerol and fatty acyl CoA.

diffusion The net transport of a solute by thermal motion from a region of high concentration to one of low concentration.

diffusion coefficient The parameter characterizing the diffusion of a solute.

dihydroorotase An enzyme that closes the ring of ureidosuccinic acid to form dihydroorotic acid.

dihydroxyacetone phosphate acyltransferase An enzyme that catalyzes the transfer of a fatty acyl group from fatty acyl CoA to dihydroxyacetone phosphate.

dimethylallyl transferase An enzyme that catalyzes the formation of geranyl pyrophosphate from the combination of isopentenyl and dimethylallyl pyrophosphates.

2, 3-diphosphoglycerate An effector for hemoglobin that reduces the oxygen affinity.

diploid cells Cells containing a double set of chromosomes.

disc gel electrophoresis A form of gel electrophoresis in which a pH gradient is used to obtain very thin bands.

dispersion The wavelength dependence of optical rotation.

DNA ligase An enzyme that repairs nicked DNA.

DNA polymerase I, II, and III Enzymes capable of synthesizing DNA from nucleoside triphosphates in the presence of a DNA template.

Donnan equilibrium The contribution to osmotic pressure of a biased distribution of small ions.

dopamine An intermediate in tyrosine degradation and a precursor of epinephrine.

dopamine β-hydroxylase An enzyme that converts dopamine to norepinephrine.

double displacement reactions Enzymic reactions in which one substrate is bound and one product is released prior to the binding of a second substrate and the release of a second product.

Edman degradation A method for the determination of amino acid sequence that utilizes reaction with phenylisothiocyanate.

EF-T An elongation factor involved in protein synthesis.

elution volume The volume of solvent required to elute a solute from a gel filtration column.

enantiomorphs Optical isomers that are mirror images.

endergonic reaction A reaction for which the predicted free energy change is positive.

endoplasmic reticulum A system of membranous channels within the cytoplasm.

enolization The migration of a hydrogen atom from a carbon atom to a carbonyl group.

enoyl-ACP hydratase An enzyme that dehydrates D-β-hydroxybutyryl ACP to form crotonyl ACP.

enoyl-ACP reductase An enzyme that reduces crotonyl ACP to butyryl ACP.

enoyl-CoA hydratase An enzyme that converts an α, β-unsaturated fatty acyl CoA to a β-hydroxy derivative.

enthalpy The heat content of a system.

entropy A thermodynamic quantity that is a measure of the randomness or disorder of a system.

enzymes Proteins with catalytic capabilities.

epimers Two sugars differing solely in the configuration of a single carbon atom.

epinephrine A regulatory hormone derived from tyrosine.

equivalence zone The ratio of antigen and antibody at which both are incorporated entirely into a specific precipitate.

ergocalciferol A form of vitamin D.

erythrocytes Red blood cells.

essential amino acids Amino acids that the body cannot synthesize and that must be supplied in the diet.

estrogens The female sex hormones.

ethanolamine kinase An enzyme that catalyzes the phosphorylation of ethanolamine to form phosphoethanolamine.

etiophorphyrins Porphyrin derivatives containing four methyl and four ethyl groups.

eukaryotic cell A cell characteristic of plants, animals, fungi, protozoa, and most algae that contains a membrane-bounded nucleus.

evolutionary convergence The approach of unrelated species to similar characteristics as a result of the evolutionary process.

exciton A quantum of excitation energy.

exergonic reaction A reaction for which the predicted free energy change is negative.

extrinsic membrane proteins Proteins that are only loosely attached to a membrane surface.

Fabry's disease A hereditary disease that produces high levels of sphingolipid.

F-actin The polymeric form of actin.

fat-soluble vitamins Vitamins soluble in nonpolar media, which are vitamins A, D, E, and K.

fatty acids Molecules consisting of linear hydrocarbon chains terminating in a carboxyl group.

fatty acid synthetase An organized enzyme system that mediates the biosynthesis of fatty acids.

feedback control A regulatory mechanism in which a product of an enzymic reaction sequence modifies the action of an enzyme occurring earlier in the sequence.

ferredoxin An iron-sulfur protein whose reduced form transfers an electron to $NADP^+$ in a photosynthetic pathway.

ferredoxin-$NADP^+$ reductase An enzyme that catalyzes the transfer of an electron from ferredoxin to $NADP^+$.

ferredoxin reducing substance An iron-sulfur protein that accepts an electron and transfers it to ferredoxin in a photosynthetic pathway.

ferriheme A heme whose iron atom is in the Fe(III) state.

ferritin An iron-storage protein.

ferrochelatase An enzyme that catalyzes the incorporation of iron into protoporphyrin to form heme.

ferroheme A heme whose iron atom is in the Fe(II) state.

fibrin The three-dimensional network of polymerized fibrinogen.

fibrinogen The plasma protein whose polymerization is responsible for blood clotting.

fibrous long spacing fibrils A form of regenerated collagen fibrils with 280-nm spacings.

fibrous protein A protein whose geometrical shape has a high axial ratio.

Fick's laws The equations describing diffusion.

flagella External filaments responsible for the cellular motility of some microorganisms.

flavin adenine dinucleotide (FAD) A coenzyme important in biological oxidation-reduction processes, whose structure consists of riboflavin and adenosine-5′-diphosphate.

flavin mononucleotide (FMN) Riboflavin-5′-phosphate, a coenzyme important in biological oxidation-reduction processes.

fluid mosaic model A model for membrane structure that postulates a high degree of lateral mobility for the membrane phospholipids.

FMN adenylyltransferase An enzyme that forms FAD from FMN and ATP.

follicle stimulating hormone (FSH) A protein hormone secreted by the pituitary that furthers growth and development of the gonads.

free energy The usable portion of the total energy of a system.

freeze-etching An electron microscopic technique that involves fracturing of a frozen sample and partial removal of ice by sublimation.

frictional coefficient A parameter characterizing the frictional drag that impedes the motion of a particle through a viscous medium.

frictional ratio The ratio of the frictional coefficient of a particle to that predicted for an anhydrous sphere of the same molecular weight.

fructokinase An enzyme that catalyzes the phosphorylation of fructose.

fructose A ketohexose sugar.

fructose diphosphatase An enzyme that hydrolyzes fructose-1,6-diphosphate to fructose-6-phosphate.

fructose diphosphate aldolase An enzyme that cleaves fructose-1,-6-diphosphate to D-glyceraldehyde-3-phosphate and dihydroxyacetone phosphate.

fumarase An enzyme of the TCA cycle that converts fumarate to malate.

fungi A class of eukaryotic microorganisms.

furanose A sugar containing a five-membered hemiacetal ring.

G-actin The globular monomeric form of actin.

galactokinase An enzyme that catalyzes the phosphorylation of galactose to form galactose-1-phosphate.

galactosamine An amino derivative of galactose.

galactosyl transferase The enzyme component of lactose synthase.

gametes Cells specializing in reproduction.

ganglioside A carbohydrate-containing lipid of complex structure.

Gaucher's disease A hereditary disease resulting in extensive deposition of cerebrosides.

gel filtration A form of column chromatography that separates on the basis of molecular size.

genes The chromosomal subelements that are the functional units of heredity.

genetic map A diagram showing the mutual locations of genes within a chromosome.

genome A chromosome carrying the complete haploid set of genetic information for an organism.

ghost The hemoglobin-free sac of an erythrocyte after release of its contents.

glands Specialized tissues that secrete hormones.

glass electrode A standard electrode containing a glass membrane that responds to differences in hydrogen ion level.

globin The protein portion of hemoglobin.

globular protein A protein whose geometrical shape has a low axial ratio.

glomeruli The kidney structures through which plasma is filtered.

glucagon A pancreatic polypeptide hormone that is involved in the regulation of carbohydrate metabolism.

glucokinase An enzyme that catalyzes the phosphorylation of glucose.

gluconeogenesis The biosynthesis of glucose.

glucosamine An amino derivative of glucose.

glucose An aldohexose sugar.

glucose-1-phosphate, glucose-6-phosphate Biologically important phosphoric acid esters of glucose.

glucose-6-phosphate dehydrogenase An enzyme that catalyzes the oxidation of glucose-6-phosphate to 6-phosphogluconate.

glucose-phosphate isomerase An enzyme that catalyzes the interconversion of glucose-6-phosphate and fructose-6-phosphate.

glucose-1-phosphate uridyl transferase An enzyme that catalyzes the formation of UDP-glucose from glucose-1-phosphate and UTP.

α,1,6-glucosidase An enzyme that hydrolyzes the α,1,6 linkages of amylopectin.

glucoside An acetal derivative formed by glucose.

glutamate dehydrogenase An NAD^+-dependent enzyme that converts glutamate to α-ketoglutarate.

glutamate kinase and dehydrogenase An enzyme that reduces glutamate to γ-glutamic semialdehyde.

glutamate synthase An enzyme that converts glutamine to glutamate.

glutaminase An enzyme that hydrolyzes glutamine to glutamate.

glutamine synthetase An enzyme that forms glutamine from glutamate.

γ-glutamylcysteine synthetase An enzyme that catalyzes the condensation of glutamate with cysteine.

γ-glutamylcysteinylglycine Glutathione, a naturally occurring tripeptide.

glutathione synthetase An enzyme that catalyzes the combination of γ-glutamylcysteine with glycine to form glutathione.

glyceraldehydephosphate dehydrogenase An enzyme that catalyzes the conversion of glyceraldehyde-3-phosphate to 3-phosphoglyceroyl phosphate.

glycerol kinase An enzyme that phosphorylates glycerol to L-glycerol-3-phosphate.

glycerol-3-phosphate dehydrogenase An enzyme that catalyzes the reduction by NAD^+ of L-glycerol-3-phosphate to dihydroxyacetone phosphate.

glycerophosphatide A phospholipid containing a diacylglycerol esterified with phosphoric acid.

glycine amidinotransferase An enzyme that catalyzes the transfer of the guanidine group of arginine to glycine.

glycine synthase An enzyme that catalyzes the synthesis of glycine from CO_2, NH_3, and N^5, N^{10}-methylene-FH_4.

glycogen The principal reserve carbohydrate of animals.

glycogen phosphorylase An enzyme that catalyzes the phosphorolysis of glycogen.

glycolate oxidase An enzyme that catalyzes the oxidation of glycolic acid to glyoxylic acid.

glycolysis A set of metabolic reactions by which glucose-6-phosphate is converted to lactate.

glycoproteins Proteins containing short oligosaccharide chains, which are covalently attached.

glycosidic bond An acetal bond involving a sugar.

glycosyl-1-phosphate nucleotidyltransferase An enzyme that catalyzes the formation of UDP-glucose.

glyoxylate cycle A metabolic pathway occurring in plants and microorganisms that forms succinate from acetyl CoA.

glyoxysomes Cytoplasmic particles containing the enzymes isocitrate lyase and malate synthase.

Golgi apparatus A region of the endoplasmic reticulum that processes proteins for excretion by the cell.

gout A disease arising from the precipitation of uric acid in cartilaginous tissues.

grana Parallel arrays of flat sacs formed by chloroplast lamellae.

guanine A purine base occurring in nucleic acids.

Hageman factor A component of the blood clotting system.

haploid cells Cells containing a single set of chromosomes.

hapten A group chemically coupled to a carrier protein that controls its antigenic specificity.

α-helix A form of polypeptide conformation in which the chain is arranged in a helical structure stabilized by hydrogen bonding.

heme A porphyrin molecule complexed with iron that functions as a prosthetic group in several proteins.

heme synthetase An enzyme that catalyzes the formation of hemoglobin from its constituents.

hemolysis The rupture of the external sac of an erythrocyte and the release of its contents.

hemosiderin An iron-storage protein.

Henderson-Hasselbalch equation An equation characterizing the ionization of a weak electrolyte.

heparin A mucopolysaccharide that inhibits blood clotting and reduces plasma lipid content.

heterotrophic cells Cells that require a supply of chemical energy from an external source.

heterozygous organism An organism with two different alleles of a particular gene.

hexokinase An enzyme that catalyzes the phosphorylation of a hexose sugar by ATP.

hexose-1-phosphate uridyl transferase An enzyme that catalyzes the transfer of sugars between UDP-glucose and galactose-1-phosphate.

high energy compounds Molecules, such as ATP, whose chemical transformation releases substantial amounts of free energy.

high voltage paper electrophoresis A form of paper electrophoresis employing electric fields of several thousand volts.

histamine The amine formed by the decarboxylation of histidine.

histidine decarboxylase An enzyme that decarboxylates histidine to form histamine.

histones Basic proteins occurring in the nuclei of eukaryotic cells as complexes with DNA.

homocysteine The amino acid that differs from cysteine in having an additional methylene group in the side-chain.

homozygous organism An organism with two identical alleles of a particular gene.

hormones Compounds secreted by specialized glands that function as metabolic regulators.

hydrogen bonding A noncovalent bond in which a hydrogen atom links two electronegative atoms.

hydrogen electrode A standard electrode that consists of a platinum electrode equilibrated with hydrogen gas at a known pressure and temperature.

hydrophobic Nonpolar character, as for a hydrocarbon molecule or group.

3-hydroxyacyl CoA dehydrogenase An enzyme that converts a β-hydroxy fatty acyl CoA to a β-keto derivative.

hydroxyapatite The mineral component of bone.

D-β-hydroxybutyric acid dehydrogenase An enzyme that reduces acetoacetate to D-β-hydroxybutyric acid.

25-hydroxycholecalciferol A derivative of cholecalciferol formed in the liver.

β-hydroxy-β-methylglutaryl CoA An intermediate in the degradation of leucine. The intermediate is formed from isovaleryl CoA.

hydroxymethylglutaryl-CoA reductase An enzyme that catalyzes the formation of mevalonate from β-hydroxy-β-methylglutaryl CoA.

hydroxymethylglutaryl-CoA synthase An enzyme that catalyzes the formation of β-hydroxy-β-methylglutaryl CoA from acetyl CoA.

hyperglycemia An elevation of the blood glucose level that is characteristic of diabetes.

hyperthyroidism Excessive activity of the thyroid gland.

hypochromism The decrease in absorbance of the nucleotides of helical nucleic acids that arises from base stacking.

hypophysis The pituitary gland.

hypothalamus The area of the brain that regulates the pituitary gland.

hypothyroidism Depressed activity of the thyroid gland.

ichthyocol The collagen of carp bladder.

α-imino acid A molecule similar to an amino acid, except that its side-chain forms a closed ring including the α-nitrogen.

immunoglobulin An antibody molecule.

IMP-cyclohydrolase An enzyme that forms inosinic acid by closure of a ring of 5′-phosphoribosyl-4-carboxamide-5-formamidoimidazole.

IMP-dehydrogenase An enzyme that forms xanthylic acid from inosinic acid by NAD^+ oxidation.

inducer A molecule that combines with and inactivates a repressor, thereby allowing transcription to occur.

inducible enzymes Enzymes whose biosynthesis is elicited by the presence of a substrate.

inositol phosphatide (inositide) A glycerophosphatide in which phosphoric acid is esterified with inositol.

insulin A polypeptide hormone secreted by the pancreas that is involved in the regulation of carbohydrate metabolism.

interference optics A technique for monitoring the transport of a solute that depends upon the formation of interference fringes by rays of light traversing the system.

intergenic suppression The cancellation of the effects of a mutation by a second mutation in a different gene.

intermediate lobe A lobe of the pituitary present in lower animals that secretes the melanocyte stimulating hormones.

intragenic suppression The cancellation of the effects of a mutation by a second mutation within the same gene.

intrinsic membrane proteins Proteins that are integrated into a membrane structure.

inversion The change in sign of the optical rotation of sucrose that accompanies its hydrolysis to glucose and fructose.

invertase An enzyme catalyzing the hydrolysis of sucrose to glucose and fructose.

ion-exchange A form of column chromatography that separates on the basis of charge.

ionization constant The equilibrium constant characterizing the dissociation of a weak electrolyte.

ionophores A class of antibiotics that block oxidative phosphorylation in the presence of monovalent cations.

ion-product The product of the molar concentrations of the H^+ and OH^- ions into which water dissociates.

irreversible inhibition The stoichiometric abolition of enzymic activity as a result of the reaction with a small molecule.

islet cells Specialized cells of the pancreas that synthesize insulin or glucagon.

isocitrate lyase An enzyme that cleaves isocitrate to form succinate and glyoxylate.

isopentenyl pyrophosphate isomerase An enzyme that forms 3,3-dimethylallyl pyrophosphate from 3-isopentenyl pyrophosphate.

isoprene group A hydrocarbon group of the form C—C—C—C (with a C branching from the second carbon), which recurs in compounds of the terpene class.

isozymes Distinct forms of the same enzyme arising from different combinations of subunits.

β-ketoacyl-ACP reductase An enzyme that reduces acetoacetyl ACP to form D-β-hydroxybutyrl ACP

β-ketoacyl-ACP synthase An enzyme that catalyzes the removal of an acetyl group from acetyl ACP.

ketose A member of a family of sugars containing keto groups, whose simplest prototype is dihydroxyacetone.

kinetics The factors involved in governing the velocity of an enzymic reaction.

kinetic stability The stability of a molecule conferred by the slow rate of its decomposition.

lac operon A portion of the *E. coli* chromosome that governs the adaptation to a lactose medium.

lactate dehydrogenase An enzyme that interconverts lactate and pyruvate.

lactone A cyclic ester formed by a sugar acid.

lactose synthase An enzyme system that forms lactose from UDP-galactose and glucose.

lamellae Foldings of the inner membrane of a chloroplast.

lanosterol A sterol found in wool.

lecithin (phosphatidyl choline) A glycerophosphatide in which phosphoric acid is esterified with choline.

LH releasing hormone (LRH) A peptide hormone secreted by the hypothalamus that mediates the release of LH and FSH.

lipases Enzymes that hydrolyze fatty acid esters.

lipids Biological molecules containing long hydrocarbon chains.

lipoamide dehydrogenase A component of the pyruvate dehydrogenase system that reoxidizes dihydrolipoic acid to lipoic acid.

lipoate acetyltransferase A component of the pyruvate dehydrogenase system that catalyzes the conversion of active acetaldehyde to acetyl CoA.

lipoproteins Proteins containing a high proportion of lipids.

liposome A spherical phospholipid bilayer.

luteinizing hormone (LH) A protein hormone secreted by the pituitary that stimulates the synthesis of sex hormones by the gonads.

lymphocytes Percursors of plasma cells.

lysosomes Membrane-bounded cellular compartments that contain high concentrations of digestive enzymes.

malate dehydrogenase An enzyme that catalyzes the oxidation of malate to oxaloacetate.

malate synthase An enzyme that catalyzes the combination of glyoxylate and acetyl CoA to form malate.

malic enzyme An enzyme that catalyzes the carboxylation of pyruvate to malate.

meiosis Nuclear division of the cell resulting in the formation of haploid gametes.

melanin A dark pigment derived from DOPA.

α- and β-melanocyte stimulating hormones (α- and β-MSH) Peptide hormones that are formed by the intermediate lobe of the pituitary and influence pigmentation.

melting point The midpoint of the thermal denaturation profile of a nucleic acid.

membrane A film composed of protein and phospholipid that surrounds a cell or subcellular organelle.

membrane potential The electrical potential between two solutions separated by a membrane.

mercaptalbumin A component of serum albumin containing a reactive sulfhydryl group.

mesophyll cells Specialized cells occurring in C_4 plants in which CO_2 fixation occurs.

messenger RNA (mRNA) That form of RNA that codes directly for the amino acid sequence of polypeptides.

metabolism The set of chemical processes carried out by living systems.

metallo-enzyme An enzyme in which a metal atom is an essential component of the active site.

metaphase A phase of mitosis during which each chromosome appears as a replicated pair of chromatids.

methemoglobin The form of hemoglobin containing Fe(III).

methionine adenosyl transferase An enzyme which catalyzes the formation of *S*-adenosyl methionine from ATP and methionine.

methylmalonyl CoA mutase An enzyme that converts methylmalonyl CoA to succinyl CoA.

methylmalonyl CoA racemase An enzyme that converts a stereoisomer of methylmalonyl CoA to a mixture of two stereoisomers.

methyltransferases Enzymes that catalyze the transfer of a methyl group from *S*-adenosyl methionine to an acceptor.

mevalonate kinase An enzyme that phosphorylates mevalonate to the 5-monophosphate.

microorganisms Living systems of microscopic size.

missense mutation A mutation that produces a codon specific for a different amino acid.

mitochrondria Organelles occurring in the cytoplasm of eukaryotic cells, where the processes of respiration and oxidative phosphorylation take place.

mitochondrial matrix The space enclosed by the inner mitochondrial membrane.

mitosis Nuclear division of the somatic cells resulting in the formation of two diploid cells.

monosaccharide A monomeric sugar.

mucins Glycoproteins of high carbohydrate content.

mucopolysaccharides Polymeric carbohydrates containing derivatives of amino sugars.

muscle fibrils The elongated structures, about 1 μm in diameter, which comprise muscle fibers.

muscle filaments The actin and myosin polymers that comprise muscle fibrils.

mutarotation A time dependence of optical rotation for a sugar, reflecting the slow attainment of an equilibrium mixture of α and β forms.

mutase An enzyme that catalyzes the transfer of a phosphate group between two positions on the same molecule.

mutation An alteration in the base sequence of chromosomal DNA, resulting in a protein of altered amino acid sequence.

myelin The substance comprising the sheaths of nerve fibers.

myofibrils Muscle fibrils.

myoglobin A heme-containing oxygen storage protein occurring in muscle.

NADH dehydrogenase A flavoprotein enzyme that reoxidizes the NADH produced by the TCA cycle.

Na^+-K^+ activated ATPase An enzyme involved in the active transport of Na^+ and K^+ across a membrane.

negative staining A technique used to enhance contrast in electron microscopy by darkening the background

nephrons The functional units of the kidney.

neurophysins Polypeptide precursors of the posterior pituitary hormones.

neurohypophysis The posterior pituitary.

NH_2-terminal end The end of a polypeptide chain with a free α-amino group.

nicotinamide adenine dinucleotide (NAD^+) A coenzyme, important in biological oxidation-reduction processes, that has a dinucleotide structure whose bases are nicotinamide and adenine.

nicotinamide adenine dinucleotide phosphate ($NADP^+$) A coenzyme, important in biological oxidation-reduction processes, whose structure is NAD^+ with the adenosine portion esterified with an additional phosphate.

nicotinic acid mononucleotide An intermediate in the biosynthesis of NAD^+ that is formed by the reaction of nicotinic acid and 5-phosphoribose-1-pyrophosphate.

Niemann-Pick disease A hereditary disease that produces high levels of sphingomyelin.

noncompetitive inhibitor An enzyme inhibitor that interferes with the conversion, but not the binding, of a substrate.

nonsense mutation A mutation that produces a codon that does not code for any amino acid.

nonessential amino acids Amino acids that the body can synthesize and that do not need to be supplied in the diet.

norepinephrine A hormone secreted by the adrenal medulla that functions as a neurotransmitter.

normal electrode potential The potential developed by a standard half-cell with unit activities of all ions with respect to a hydrogen electrode.

nucleic acids Linear polymers of nucleotides that mediate the storage and expression of genetic information.

nucleoli Particles present in cell nuclei that mediate the synthesis of ribosome precursors.

nucleoplasm That part of the cell nucleus exclusive of chromosomes, nucleoli, and nuclear membrane.

nucleoside The portion of a nucleotide that consists of the base plus the sugar, without the phosphate.

nucleotide A compound consisting of a purine or pyrimidine base, five-carbon sugar, and phosphoric acid.

nucleus The part of a cell that contains the genetic determinants (chromosomes). It is membrane bounded in eukaryotic cells.

Okazaki fragments Freshly synthesized segments of a DNA strand, which are subsequently joined to form the complete strand.

oligosaccharases Enzymes that hydrolyze disaccharide substrates to monosaccharides.

operator A portion of the control region of an operon at which the repressor is bound.

operon A cluster of genes subject to simultaneous control.

opsin The protein component of rhodopsin.

ornithine carbamoyltransferase An enzyme that catalyzes the formation of citrulline.

ornithine decarboxylase An enzyme that decarboxylates ornithine to putrescine.

orotate phosphoribosyl transferase An enzyme that catalyzes the formation of orotidine-5′-phosphate.

osazone A derivative formed by a sugar with phenylhydrazine.

osmotic pressure An increase in hydrostatic pressure across a membrane arising from a difference in solute concentration on the two sides of the membrane.

oxidation A chemical transformation involving a loss of electrons.

oxidative phosphorylation The set of reactions that phosphorylate ADP to ATP and are coupled to the transport of electrons down the respiratory chain.

oxytocin A peptide hormone secreted by the pituitary that controls release of milk by the mammary gland and causes uterine contractions during parturition.

palindrome A base pair sequence of DNA that reads the same in the forward and backward directions.

parathyroid hormone A peptide hormone secreted by the parathyroid glands that enhances excretion of phosphate in the urine.

partial specific volume The volume increment per gram of solute.

partition chromatography A form of chromatography dependent upon the distribution of solute between stationary and mobile phases.

passive facilitated diffusion The diffusion of a solute across a membrane in combination with a specific carrier molecule.

peptide bond An amide linkage formed between the amino and carboxyl groups of two amino acids.

peptidoglycan A network composed of polysaccharide chains covalently cross-linked by short polypeptides.

perimysium A sheath enclosing a mammalian muscle.

peroxidase A heme-containing enzyme that decomposes hydrogen peroxide.

phenylalanine-4-monooxygenase An enzyme that converts phenylalanine to tyrosine.

phenylethanolamine N-methyl-transferase An enzyme that catalyzes the methylation of norepinephrine to form epinephrine.

phosphatidate phosphatase An enzyme that hydrolyzes phosphatidic acid to form a diacylglycerol.

phosphatidic acid An ester of phosphoric acid with a diacylglycerol.

phosphatidyl ethanolamine A glycerophosphatide in which phosphoric acid is esterified with ethanolamine.

phosphatidyl ethanolamine methyltransferase An enzyme that catalyzes the methylation of phosphatidyl ethanolamine by *S*-adenosyl methionine to form phosphatidyl choline.

phosphatidyl serine A glycerophosphatide in which phosphoric acid is esterified with serine.

phosphocholine cytidylyltransferase An enzyme that catalyzes the reaction of phosphocholine with CTP to form CDP-choline.

phosphocholine transferase An enzyme that catalyzes the formation of phosphatidylcholine from CDP-choline and diacylglycerol.

phosphocreatine An energy reserve molecule in muscle that can regenerate ATP from ADP.

phosphodiester bonds Bonds occurring in nucleic acids that consist of doubly esterified phosphoric acid.

phosphoglucomutase An enzyme that catalyzes the interconversion of glucose-1-phosphate and glucose-6-phosphate.

6-phosphogluconate dehydrogenase An enzyme that catalyzes the oxidation of 6-phosphogluconate to D-ribulose-5-phosphate.

phosphogluconate oxidative pathway A metabolic pathway that oxidizes glucose, forms pentose sugars, and generates NADPH.

phosphoglycerate dehydrogenase An enzyme that converts 3-phosphoglycerate to 3-phosphohydroxypyruvate.

phosphoglyceromutase An enzyme that catalyzes the interconversion of 3- and 2-phosphoglycerate.

3-phosphohydroxypyruvate An intermediate in the biosynthesis of serine.

phospholipid A diester of phosphoric acid with sphingosine or a diacylglycerol and with an alcohol.

phosphomevalonate kinase An enzyme that converts phosphomevalonate-5-phosphate to the corresponding pyrophosphate.

4-phosphopantetheine The prosthetic group of acyl carrier protein.

phosphopyruvate hydratase (enolase) An enzyme that interconverts 2-phosphoglycerate and phosphoenolpyruvate.

5-phosphoribose-1-pyrophosphate An intermediate in nucleotide biosynthesis that is formed from ribose-5-phosphate and ATP.

5-phospho-β-ribosylamine An intermediate in the biosynthesis of purine nucleotides that is formed by the reaction of 5-phosphoribose-1-pyrophosphate with glutamine.

phosphoribosyl-aminoimidazole-carboxamide formyl transferase An enzyme that formylates 5′-phosphoribosyl-4-carboxamide-5-formamidoimidazole.

phosphoribosylaminoimidazole carboxylase An enzyme that carboxylates 5′-phosphoribosyl-5-aminoimidazole to form 5′-phosphoribosyl-5-aminoimidazole-4-carboxylic acid.

phosphoribosylaminoimidazole-succinocarboxamide synthetase An enzyme that condenses 5′-phosphoribosyl-5-aminoimidazole-4-carboxylic acid with aspartic acid.

phosphoribosylaminoimidazole synthetase An enzyme that converts 5′-phosphoribosyl-N-formylglycinamidine to the corresponding imidazole derivative by ring closure.

phosphoribosylformylglycinamidine synthetase An enzyme that catalyzes the formation of 5′-phosphoribosyl-N-formyl-glycinamidine.

phosphoribosylglycinamide formyltransferase An enzyme that formylates 5′-phosphoribosylglycinamide.

phosphoribosylglycinamide synthetase An enzyme that catalyzes the formation of 5′-phosphoribosylglycinamide from glycine, ATP, and 5-phospho-β-ribosylamine.

phosphoribulokinase An enzyme that phosphorylates ribulose-5-phosphate to ribulose-1,5-diphosphate.

phosphorylase a The active, phosphorylated form of glycogen phosphorylase.

phosphorylase b The inactive form of glycogen phosphorylase.

phosphorylase kinase The enzyme that phosphorylates phosphorylase b to form phosphorylase a.

phosphorylase phosphatase The enzyme that hydrolyzes the serine phosphate of phosphorylase a, converting it to phosphorylase b.

3-phosphoserine An intermediate in the biosynthesis of serine that is formed by the transamination of 3-phosphohydroxypyruvate from glutamate.

phosphoserine phosphatase An enzyme that hydrolyzes 3-phosphoserine to serine.

phosphoserine transferase An enzyme that catalyzes the formation of 3-phosphoserine.

photolithotrophs Organisms that obtain their carbon from atmospheric CO_2 and their energy from sunlight.

photoorganotrophs Organisms that obtain their carbon from organic compounds and their energy from sunlight.

photosynthetic organisms Organisms that obtain a major fraction of their energy requirements from the radiant energy of sunlight.

photosystem I, photosystem II Photosynthetic units containing chlorophyll molecules with exceptional properties, which supply electrons to reaction chains and thereby reduce $NADP^+$ and split H_2O.

phytol The alcohol constituent of chlorophyll.

pitch The spacing between successive turns of a helix.

plane-polarized light Light whose electric vector is confined to a single plane.

plasma cells Cells responsible for the formation of antibodies.

plasmalemma A membrane enclosing a living cell.

plasmalogen A glycerophosphatide in which a fatty acid at the terminal position of glycerol is replaced by a *cis* α-β unsaturated ether.

plastids Cytoplasmic particles occurring in plants.

plastocyanin A copper-containing protein occurring in photosynthetic systems.

platelets Particulate bodies occurring in blood plasma and involved in blood clotting.

β-pleated sheet A polypeptide conformation in which a set of side-by-side chains are stretched to their maximum extension and linked by hydrogen bonds.

polyenoic fatty acids Multiply unsaturated fatty acids.

polypeptides Linear polymers of amino acids

polyribonucleotide phosphorylase An enzyme that catalyzes the formation of polynucleotides from ribonucleoside diphosphates.

polysaccharide A polymer of one or more sugars.

polysome A complex of several ribosomes with a strand of tRNA.

pores Microscopic openings in cell membranes.

porphin A ring-shaped molecule consisting of four pyrrole rings that is the parent compound of the porphyrins.

porphobilinogen A metabolic precursor of heme and related porphyrins.

portal vein A blood vessel that supplies the liver.

posterior pituitary A lobe of the pituitary with neural connections to the hypothalamus that secretes oxytocin and several other peptide hormones.

precipitin reaction The formation of a specific precipitate by antigen and antibody.

preproinsulin A precursor of proinsulin and insulin.

presqualene synthase An enzyme that catalyzes the formation of presqualene from farnesyl pyrophosphate.

primary structure The amino acid sequence of a polypeptide.

primer A DNA or RNA strand that is essential for the biosynthesis of new DNA and is incorporated into the product.

progesterone A hormone produced by the corpus luteum that is involved in governing the menstrual cycle.

proinsulin The precursor of insulin that is formed from preproinsulin.

prokaryotic cell A cell characteristic of bacteria and blue-green algae that lacks a membrane-bounded nucleus.

prolactin A protein hormone secreted by the pituitary that stimulates milk secretion by the mammary gland.

proline-4-monooxygenase An enzyme that converts proline to 4-hydroxyproline.

prometaphase A phase of mitosis during which the chromosomes migrate toward the equatorial plane of the spindle and the nuclear membrane disappears.

promoter A portion of the control region of an operon at which RNA polymerase is initially attached.

prophase An initial phase of mitosis.

propionyl CoA carboxylase An enzyme that converts propionyl CoA to methylmalonyl CoA.

prostaglandins Hormonelike compounds derived from fatty acids.

prosthetic group A nonamino acid constituent of a protein that is essential for its biological activity.

protamines Basic proteins occurring in sperm cells as complexes with DNA.

proteins Linear polymers of amino acids that are folded into organized structures.

proteolytic enzymes Enzymes that catalyze the hydrolysis of peptide bonds.

prothrombin The inactive precursor of thrombin.

protomer A subunit of an allosteric enzyme.

protozoa A class of eukaryotic microorganisms.

purines Constituents of some nucleotides, which consist of two fused heterocyclic nitrogenous rings.

putrescine An amine formed by the decarboxylation of ornithine.

pyranose A sugar containing a six-membered hemiacetal ring.

pyrimidines Constituents of some nucleotides, which consist of heterocyclic nitrogenous rings.

pyrrole A five-membered nitrogenous ring that occurs in porphyrins.

pyrroline-5-carboxylate reductase An enzyme that reduces Δ'-pyrroline-5-carboxylate to form proline.

pyruvate carboxylase A biotin enzyme that carboxylates pyruvate to oxaloacetate.

pyruvate decarboxylase An enzyme that catalyzes the conversion of pyruvate to acetaldehyde.

pyruvate dehydrogenase An enzyme complex that converts pyruvate to acetyl CoA.

pyruvate dehydrogenase kinase An enzyme that catalyzes the phosphorylation by ATP of pyruvate dehydrogenase.

pyruvate kinase An enzyme that catalyzes the phosphorylation of ADP by phosphoenolpyruvate to yield ATP and pyruvate.

racemic mixture An optically inactive mixture of equal quantities of two enantiomorphs.

redox potential The electrical potential developed by a half-cell with respect to a hydrogen electrode.

reduction A chemical transformation involving a gain of electrons.

regulatory gene A gene that codes for a repressor protein.

release factors Proteins that mediate the conclusion of polypeptide synthesis and the release of the polypeptide-tRNA from the ribosome.

renal tubules Kidney tubules from which water and many solutes are reabsorbed.

replication fork A Y-shaped region corresponding to the junction between the replicated and original portions of a replicating DNA molecule.

replicative form (RF) The double-stranded form of the DNA of ϕX174 virus, which is formed in the initial step of replication.

repressor A molecule that inhibits the synthesis of mRNA by binding at an operator site.

resolving power The smallest separation of two objects that permits their differentiation.

respiration The reoxidation of reduced coenzymes by the transfer of electrons to molecular oxygen.

respiratory chain A set of reversibly oxidizable components present in the mitochondria to which electrons are supplied from the TCA cycle. These components reduce molecular oxygen and drive the reactions of oxidative phosphorylation.

restriction The selective vulnerability of foreign DNA to attack by bacterial endonucleases.

restriction endonucleases Enzymes that cleave DNA only at specific base sequences recognized by the enzymes.

retinal A derivative of vitamin A_1, which may exist in *cis* or *trans* forms, that is essential for vision.

retinal isomerase An enzyme that catalyzes the conversion of *trans*-vitamin A_1 to 11-*cis*-vitamin A_1.

retinal reductase An enzyme that catalyzes the interconversion of *trans*-retinal and 11-*cis*-retinal.

reversible inhibition The reduction of enzymic activity as a result of the reversible binding of a small molecule.

rhodopsin A protein of the retina that consists of opsin plus 11-*cis*-retinal.

riboflavin Vitamin B_2, an isoalloxazine derivative, which is a component of the flavin coenzymes.

riboflavin kinase An enzyme that forms FMN from riboflavin and ATP.

ribonuclease An enzyme that hydrolyzes phosphodiester bonds of RNA.

ribonucleic acid (RNA) A class of nucleic acid containing ribose that mediates protein synthesis.

ribonucleoside diphosphate reductase An enzyme that catalyzes the reduction of ribonucleotides to deoxyribonucleotides by thioredoxin.

ribosephosphate isomerase An enzyme that catalyzes the interconversion of D-ribulose-5-phosphate and D-ribose-5-phosphate.

ribosephosphate pyrophosphatase An enzyme that forms 5-phosphoribose-1-pyrophosphate.

ribosomes Cytoplasmic particles composed of RNA and protein that are the sites of protein synthesis.

ribulosediphosphate carboxylase An enzyme that catalyzes the carboxylation of ribulose-1,5-diphosphate to form 3-phosphoglycerate.

rifamycin An antibiotic that blocks transcription.

RNA polymerase The enzyme that catalyzes the transcription of RNA from DNA.

rod cells Cells of the mammalian retina that are concerned with vision at low light levels.

salt bridge An electrostatic bond between two oppositely charged groups.

saponification The alkaline hydrolysis of triacylglycerols to form free fatty acids.

sarcolemma The membrane enclosing a muscle fiber.

sarcomere The repeating structural unit of myofibrils.

sarcoplasm The intracellular fluid of muscle cells.

Schiff's base A compound formed by an amino group and an aldehyde.

secondary structure The conformation of the polypeptide backbone of a protein.

sedimentation coefficient A parameter characterizing the rate of movement of a solute in a centrifugal field.

sedimentation equilibrium An ultracentrifugal technique for molecular weight determination that depends upon the equilibrium distribution of solute in a centrifugal field.

sedimentation velocity An ultracentrifugal technique in which the rates of sedimentation of solutes are measured.

segment long spacing fibrils A form of regenerated collagen fibrils consisting of short polarized segments of length 260 nm.

semiquinone A free radical derived from hydroquinone.

sequenator An automated apparatus for the determination of amino acid sequence.

serine hydratase An enzyme that converts serine to pyruvate.

serine hydroxymethyltransferase An enzyme that catalyzes the cleavage of threonine to acetaldehyde and glycine.

serotonin A vasoconstrictor derived from tryptophan.

serum The watery portion of blood plasma remaining after removal of the fibrin clot.

serum albumin A plasma protein that contributes to the maintenance of osmotic pressure and binds and transports various small molecules.

SF_1 fragments Two equivalent fragments formed from the globular head of myosin by papain treatment.

shadow-casting A technique used to enhance contrast in electron microscopy by evaporating a metal of high density, such as gold, onto the sample.

sickle-cell anemia A hereditary blood disease resulting from a mutationally altered hemoglobin.

side-chain That portion of an amino acid that varies from one amino acid to the next.

sigmoidal curve An S-shaped curve, as for O_2 binding by hemoglobin.

single displacement reactions Enzymic reactions in which two substrates are bound simultaneously to form a ternary complex with the enzyme.

skeletal muscle A striated, multinucleate form of muscle subject to voluntary control, occurring in the arms, legs, face, and so on. It is the most abundant tissue found in vertebrates.

smooth muscle A nonstriated, uninucleate, involuntary form of muscle found in blood vessel and digestive tract walls, skin, and uterus.

somatic cells The ordinary functional cells of an organism that are not involved in reproduction.

somatotropin A protein hormone secreted by the pituitary that stimulates protein synthesis and bone growth.

somatotropin release inhibitor hormone (SRIH) A peptide hormone secreted by the hypothalamus that inhibits the release of thyrotropin and somatotropin.

specific rotation The rotation, in degrees, of light passing through one decimeter of a solution containing a g/ml of optically active solute.

spermidine A polyamine of wide biological occurrence.

spermine A polyamine of wide biological occurrence. It differs from spermidine in containing an additional $—(CH_2)_3NH_2$ group.

sphingolipid A phospholipid containing sphingosine.

sphingomyelin A sphingolipid in which a fatty acid is in amide linkage to the amino group of sphingosine.

sphingosine A nitrogenous alcohol containing an extended hydrocarbon chain.

sphingosine acyltransferase An enzyme that catalyzes the formation of ceramide from fatty acyl CoA and sphingosine.

spindle A system of cellular microtubules appearing during mitosis.

squalene epoxide lanosterol cyclase An enzyme that converts squalene 2, 3-epoxide to lanosterol.

squalene monooxygenase An enzyme that converts squalene to squalene 2, 3-epoxide.

squalene synthase An enzyme that forms squalene by reduction of presqualene.

staining The treatment of a protein with an electron-opaque material to enhance its visibility in electron microscopy.

starch The principal reserve carbohydrate of plants.

steady state The period of constant velocity in an enzymic reaction.

sterane The parent hydrocarbon of the steroids.

stereospecific numbering A convention of numbering the glycerol carbon atoms in glycerophosphatides.

steroids A class of compounds containing four fused hydrocarbon rings.

stop codon A trinucleotide in mRNA that signals the end of polypeptide synthesis.

striated muscle Skeletal muscle.

structural gene A gene that codes for a specific protein molecule (other than a repressor).

substrate A substance that is chemically transformed as the result of the action of an enzyme.

succinate dehydrogenase An enzyme of the TCA cycle that catalyzes the oxidation of succinate to fumarate.

succinyl CoA synthetase An enzyme of the TCA cycle that catalyzes the conversion of succinyl CoA to succinic acid.

sucrose A disaccharide composed of glucose and fructose.

sucrose phosphatase An enzyme that hydrolyzes sucrose-6-phosphate to sucrose.

sucrose phosphate synthase An enzyme that catalyzes the synthesis of sucrose-6′-phosphate from UDP-glucose and fructose-6-phosphate.

sugars The basic monomeric carbohydrates, such as glucose and fructose.

sulfatide A cerebroside that contains a sulfate ester of galactose.

supercoils The structures arising from further twisting of the double helix of circular DNA.

superhelix density The number of superhelical turns per 10 base pairs in a DNA supercoil.

suppressor gene A gene that suppresses the effects of a mutation in a second gene.

symmetry model A model for allosteric behavior that postulates a simultaneous transition in the conformation of all protomers upon the binding of substrate.

synapse The end plate of a nerve that is the juncture between two nerve cells.

target tissue The tissue acted upon by a particular hormone.

Tay-Sachs disease A hereditary disease characterized by a high level of ganglioside in brain tissue.

telophase The final phase of mitosis during which the chromosomes resume their resting appearance and the nuclear membranes are reformed.

template A DNA molecule that guides the biosynthesis of a new DNA molecule or of mRNA by specific base pairing.

terpenes A class of compounds whose structures are based upon the isoprene unit.

tertiary structure The mode of folding of the polypeptide backbone of a protein.

T-even bacteriophages A form of bacterial viruses that infect *E. coli* cells.

thioesterase An enzyme that cleaves a fatty acyl ACP to form the free fatty acid.

thioester bond An esterlike linkage formed by a sulfhydryl and a carboxyl group.

thioether bond An etherlike linkage in which oxygen is replaced by sulfur.

thiohemiacetal linkage A hemiacetal-like bond formed by a sulfhydryl and an aldehyde group.

thiolase An enzyme that catalyzes the reaction with CoA of a β-keto derivative of a fatty acyl CoA to form the fatty acyl CoA with two fewer carbon atoms.

thioredoxin An enzyme that is involved in the formation of deoxyribonucleotides, removing an oxygen from the ribonucleotide.

thioredoxin reductase An enzyme that catalyzes the conversion of thioredoxin to the form with two free sulfhydryls.

thoracic duct A lymphatic vessel draining the intestine.

thrombin The enzyme whose action upon fibrinogen causes the polymerization of the latter and results in blood clotting.

thromboplastin A lipoprotein that is an element of the blood clotting system.

thymidylate synthetase An enzyme that catalyzes the formation of deoxythymidylate from deoxyuridylate.

thyroglobulin A thyroid protein whose tyrosine residues are precursors of thyroxine.

thyrotropin A protein hormone secreted by the pituitary that stimulates the thyroid gland.

thyrotropin releasing hormone (TRH) A peptide hormone secreted by the hypothalamus that stimulates the release of thyrotropin and prolactin.

thyroxine binding globulin A plasma protein that binds and transports thyroxine.

tobacco mosaic virus The infective agent of a disease of tobacco plants.

transaldolase An enzyme that catalyzes the transfer of three carbon atoms from sedoheptulose-7-phosphate to glyceraldehyde-3-phosphate to form fructose-6-phosphate.

transaminase An enzyme that catalyzes the transfer of an amino group from an amino acid to an α-keto acid.

transduction The transfer of hereditary characteristics between different bacterial cells by a virus.

transfer RNA (tRNA) A form of RNA that functions as an adaptor in arranging amino acids along a strand of mRNA.

transferrin A plasma protein that functions in iron transport.

transforming principle DNA that carries hereditary characteristics between bacterial cells from different strains.

transketolase An enzyme that catalyzes the transfer of two carbon atoms from xylulose-5-phosphate to ribose-5-phosphate to form sedoheptulose-7-phosphate.

transglycosylation The exchange of a monosaccharide between two glycosides.

transient phase The initial period of an enzymic reaction prior to attainment of the steady state.

translation The expression of mRNA base sequences as amino acid sequences of proteins.

translocation The process whereby a tRNA molecule carrying a growing polypeptide chain is shifted from the A to the P ribosomal site.

triacylglycerols Triesters of glycerol with fatty acids.

tricarboxylic acid (TCA) cycle A metabolic pathway that oxidizes an acetyl group from acetyl CoA to CO_2 and H_2O and supplies electrons to the respiratory chain.

triose phosphate isomerase An enzyme that catalyzes the interconversion of dihydroxyacetone phosphate and glyceraldehyde-3-phosphate.

tropocollagen The monomeric subunit of collagen.

tropomyosin A muscle protein located in the grooves of the helical F-actin strands.

troponin A globular muscle protein with a regulatory function in contraction.

trypsin An enzyme formed in the pancreas that catalyzes the hydrolysis of peptide bonds involving basic amino acids.

tryptophan pyrrolase An enzyme that opens the indole ring of tryptophan by reaction with molecular oxygen to form N-formylkynurenine.

tyrosine hydroxylase An enzyme that converts tyrosine to dopa.

ubiquinone A coenzyme, also known as coenzyme Q, that is important in biological oxidation-reduction processes and whose structure consists of a quinone derivative plus a long unsaturated hydrocarbon side-chain.

UDP-glucose 4-epimerase An enzyme that catalyzes the interconversion of UDP-glucose and UDP-galactose.

ultracentrifuge An apparatus designed to produce very high speeds of rotation and used to study biopolymers.

uncoupling reagents Compounds that block the phosphorylation of ADP by oxidative phosphorylation but do not interfere with electron transport down the respiratory chain.

unwinding protein A protein involved in DNA replication whose function is to mediate the separation of the strands of the parental DNA.

urate oxidase An enzyme that catalyzes the oxidation of uric acid to allantoin.

ureidosuccinic acid (N-carbamoyl aspartic acid) An intermediate in the biosynthesis of orotic acid that is formed by the combination of aspartic acid and carbamoyl phosphate.

uric acid A degradation product of purines in mammals, in which the ring structure is preserved.

uronic acid A sugar acid formed by oxidation of the terminal alcohol group.

vacuoles Fluid-filled cellular cavities.

valinomycin A peptide antibiotic that specifically binds K^+ and accelerates its diffusion through a membrane.

vasopressin A peptide hormone secreted by the pituitary that promotes reabsorption of water by the kidney.

verdoglobin A degradation product of hemoglobin in which the heme ring is opened up.

vesicles Membrane-bounded cellular compartments.

viruses Obligatorily parasitic noncellular microorganisms that cannot replicate or display any biological activity outside of a host cell.

vitamin An essential dietary component that is usually a precursor of a coenzyme.

vitamin A_1 A fat-soluble vitamin important for vision.

water-soluble vitamins Vitamins soluble in aqueous media, including B and C.

weak electrolyte An acid whose extent of ionization depends upon conditions.

Wilson's disease A disease characterized by a deficiency of ceruloplasmin.

wobble base An anticodon base less specific than the other two in its interaction with its partner in the codon.

xanthine A purine into which guanine and adenine are converted in the initial step of their degradation.

xanthine oxidase An enzyme that catalyzes the oxidation of xanthine to uric acid.

xanthophylls A class of photosynthetic pigments.

zonal sedimentation An ultracentrifugal technique in which the sedimenting solute is separated into discrete regions.

zwitterion An ion containing positive and negative charges.

zygote A cell formed by the fusion of two gametes, one of which is contributed by each parent.

zymogen An inactive precursor of an enzyme.

Selected Answers to Review Questions

Chapter 2 **2-1** (c) **2-2** **(a)**-(2); **(b)**-(1); **(c)**-(1); **(d)**-(4) **2-3** **(a)** (2), (5); **(b)** (1), (4) **2-5** 0.67, 0.998, 0.07

Chapter 3 **3-1** **(a)** 1 **3-2** (a) **3-3** **(a)** Val-Ala; **(b)** Glu-Arg; **(c)** Ser-Asp-NH_2 **3-4** **(a)** disulfide; **(b)** salt bridge, hydrogen bond; **(c)** salt bridge, hydrogen bond; **(d)** hydrophobic bond

Chapter 4 **4-1** **(a)** spot 1, residues 1–4; 2, 5–9; 3, 10; 4, 11–13;
(b) **(1)** 1 replaced by new spot;
(2) 3 and 4 replaced by new spot containing residues 10–13.

4-2

```
Leu-Cys-Arg-Cys-Lys-Cys
     |       |_______|
Ala-Cys-Lys
```

4-3 **(a)** 10.8; **(b)** 5.0; **(c)** 9.2

4-4 (b) **4-6** 5 **4-8** **(a)**

```
NH3+-Gly-Glu - Lys-COO-
          |     |
         COO-  NH3+
```

(c) 5.9

Chapter 5 **5-1** **(a)** +; **(b)** 0; **(c)** −; **(d)** −; **(e)** +; **(f)** −; **(g)** − **5-2** **(b)** **5-3** α:(a), (b); β :(a), (b)
5-4 **(a)** yes; **(b)** no; **(c)** no; **(d)** no; **(e)** no; **(f)** yes
5-5 III, B; none; I, A; II, C, B and D (in that order)
5-6 **(a)** Electrostatic repulsion of side-chains destabilizes structure; **(b)** α-helix or β-structure **5-8** ~150 nm (assuming complete extension)

Chapter 6 **6-1 (a)** Ser, 100%; **(b)** His, 50%; **(c)** Leu, 100% **6-2** 0.005 mole
6-3 $V/3$ **6-4 (a)** 70 μmole/min; **(b)** $1.5 \times 10^{-5} M$; **(c)** 70 μmole/min
6-5 (a) (3); **(b)** (5) **6-11 (a)** $E + S \rightleftarrows ES \rightarrow P$;
(b) (1) no; **(2)** no; **(3)** yes **6-12 (a)** entropy; **(b)** free energy **6-13** (d)

Chapter 7 **7-1 (a)** -0.34; **(b)** -0.29; **(c)** -0.32 **7-2** (a), (c), (d), (b)
7-3 14.7 kcal **7-5** It would become 0.06 V more negative.
7-6 0.22 V; 2×10^7

Chapter 8 **8-1** (b) and (c) are true. **8-2** (e) **8-3** 57 moles
8-4 -8.7 kcal, -7.3 kcal

Chapter 9 **9-5** (c)

Chapter 10 **10-1** Fructose-1, 6-diphosphate and glyceraldehyde-3-phosphate.
10-2 fructose + 2ADP + $2P_i \rightleftarrows$ 2 lactate + 2ATP + $2H_2O$
10-3 -26.2 kcal per mole ($\Delta G^{\circ\prime} = -32.4$ kcal) **10-4** $7 \times 10^{-5}\ M$
10-5 4×10^{-3} **10-6** 10 mM **10-7** No

Chapter 11 **11-1** 0.16 V **11-2** No, -0.32 V
11-3 The photoionization of chlorophyll.
11-4 To increase the efficiency of photosynthesis by reducing the rate of photorespiration.
11-5 Absorption of an exciton raises an electron to a higher energy level.
11-6 Absorbed radiant energy is diverted from the synthesis of carbohydrate.

Chapter 12 **12-1** (c), (e) **12-2** (a) **12-3 (1)**-(c); **(2)**-(d); **(3)**-(e); **(4)**-(a) **12-4** (c)
12-5 15 **12-6** (c) **12-7** (a) **12-8** (b) **12-9** (d)
12-10 citrate + $\frac{9}{2}O_2$ + 26ADP + 2GDP + $28P_i$
$\rightarrow 6CO_2$ + 26ATP + 2GTP + $32H_2O$
12-11 citrate + $\frac{9}{2}O_2$ + 2GDP + $2P_i \rightarrow 6CO_2$ + 2GTP + $6H_2O$

Chapter 13 **13-3** (e) **13-4** (d) **13-5** (c) **13-7** 405, 27
13-8 $7CH_3COSCoA$ + 12NADPH + $12H^+$ + 6ATP + H_2O
$\rightarrow CH_3(CH_2)_{12}COOH$ + 7CoA + $12NADP^+$ + 6ADP + $6P_i$
13-9 NADPH is generated.

Chapter 14 **14-1** (a) 70:30; (b) 46:54; (c) 53:47; (d) 79:21
14-2 anode: (a), (d); stationary: (b), (c)
14-3 (a) phosphate, palmitate, oleate, choline, glycerol;
(b) phosphate, stearate, oleate, serine, glycerol **14-4** 6

Chapter 15 **15-1** (a), (b), (c), (d) **15-2** 82 **15-3** simple diffusion
15-4 facilitated diffusion

Chapter 16 **16-1** 18 acetyl CoA + $4O_2$ + 12NADPH + $12H^+$ + 18ATP + $9H_2O$
$\rightarrow$ cholesterol + $9CO_2$ + $12NADP^+$ + 18ADP + $18P_i$ + 18CoA
16-3 (b) **16-4** All carbon atoms would be labeled.
16-5 Carbon atoms 1, 3, 5, 7, 9, 13, 15, 17, 18, 19, 21, 22, 24, 26, 27.

Chapter 17 **17-1** (d) **17-4** 70%, proline
17-7 No, molecular O_2 is bound without oxidation of Fe(II).

Chapter 18 **18-1** Both strands must be broken at the same place for a drop in molecular weight to occur.
18-3 There is a greater likelihood of nonunique sequences.
18-10 (a)-(4), (b)-(3), (b)-(4), (c)-(1), (c)-(2)
18-11 RNA is double stranded.
18-12 **(a)** Structure is single stranded with hairpin-type helices.
(b) Hyperchromicity is less; profile is broader. **(c)** A_{260}'s are similar.
(d) Renaturation is faster than for *E. coli* DNA.

Chapter 19 **19-1** **(a)** Alternating Ser-Leu copolymer. **(b)** Leu-Ile-Asp-HN_2 tripeptides.
(c) Alternating Val-Cys copolymer. **(d)** Met polymer and Asp polymer.
(e) Mixed copolymer of Lys, Asp-NH_2, His, Thr, Pro, Glu-NH_2 with random sequence.
19-7 198 **19-8** 27

Chapter 20 **20-1** (d) **20-2** (e) **20-3** (e) **20-4** (e)
20-5 2Tyr + $18O_2$ + 90ADP + $88P_i$
$\rightarrow$ $17CO_2$ + urea + 87ATP + 3AMP + $2PP_i$ + $85H_2O$
20-6 Leucine; reduction of isovaleryl CoA **20-7** 42
20-8 glucose + $4NAD^+$ + 2glutamate + $2H_2O$
$\rightarrow$ 2serine + 4NADH + 2 α-ketoglutarte + $4H^+$
20-9 4

Chapter 21 **21-1** oxaloacetate + ribose-5-phosphate + $2NH_3$ + NAD^+ + NADPH + H^+ + 4ATP $\rightarrow$ UTP + NADH + $NADP^+$ + 3ADP + AMP + $3P_i$ + H_2O
21-2 6 **21-3** N_7, C_4, C_5 **21-4** N_1, $-NH_2$ **21-5** 6

Chapter 22 **22-3** The disulfide pairing present in insulin is such as to achieve the maximum structural stability for the precursor of insulin, rather than insulin itself.
22-4 A stepwise regulatory mechanism permits amplification at each step.
22-5 insulin, glucagon, epinephrine

Chapter 23 **23-1** (e) **23-2** (c) **23-3** No precipitin reaction would occur.
23-4 Only noncovalent bonds are involved. **23-7** $10^{-7}M$

Chapter 24 **24-1** (b), (c), (f) **24-3** **(a)** 3.6×10^7; **(b)** 1.6×10^2
24-4 348 cal; 23 cal **24-5** 4×10^{-2} moles

Index

A-bands 501
Acetoacetic acid 275
Acetoacetyl CoA 232, 235–237, 266, 317
Acetyl CoA carboxylase 268
Acids 20, 21
Aconitase 238
ACP-acyltransferase 269
Actin 505, 506
Actinomycin D 382
Active acetaldehyde 236
Active acetate 232
Active site 100
Active transport 188, 295, 297, 298, 299, 300, 301
Actomyosin 506, 507
Acyl adenylate 391, 392
Acyl carrier protein 268
Acyl CoA dehydrogenase 205, 246
Acyl CoA synthase 264
Acylglycerols 257–275
Addison's disease 322
Adenine 349, 350
Adenosine diphosphate 153
Adenosine monophosphate 153
Adenosine triphosphate 153, 154
Adenosylhomocysteinase 439
S-Adenosylmethionine 135, 373
S-Adenosylmethionine decarboxylase 436
Adenylate cyclase 210, 471, 475
Adenylate kinase 264
Adenylosuccinate lyase 453
Adenylosuccinate synthetase 453
Adipose cells 258, 263
Adrenal cortex 304, 321, 322
Adrenal gland 469
Adrenal medulla 469, 471
Adrenocortical hormones 321, 322, 476
Andrenocorticotropin 322, 472
Affinity chromatography 52
Aerobic cells 152
β-alanine 459
Alanine biosynthesis 436
Alanine degradation 423
Alanine transaminase 412
Alcohol dehydrogenase 195
Alcoholic fermentation 195
Aldoses 163, 164
Aldosterone 321, 322, 324
Allantoic acid 457, 458
Allantoin 456–458
Allantoinase 458
Allolactose 381
Allosteric enzymes 121–125
Allosterism 337
Amidophosphoribosyl transferase 449, 454
Amino acids 26–47, 156
  activation 391–392
  decarboxylases 434
  degradation 419–432
  metabolism 409–441
  oxidase 411
Aminoacyl site 394, 395
Aminoacyl-tRNA synthetase 391–393
α-Amino-β-ketoadipic acid 342, 343
δ-Aminolevulinate dehydratase 342
δ-Aminolevulinic acid 342, 343
Aminopropyl transferase 436
Aminosugars 169
Amylases 181, 182
Amylopectin 173–175
Amylose 173, 174
Anaerobic cells 152
Androgens 304, 322, 323, 324
Androsterone 322, 323
Animals 1, 2
Annealing 357
Antibiotics 397
Antibodies 479, 485–490
Anticodon 393
Antidiuretic hormone 742
Antigen 485, 487–490
Antimycin A 246
Arginase 418
Arginine 29
  biosynthesis 441
  degradation 427
Argininosuccinate lyase 415
Argininosuccinate synthetase 418
Argininosuccinic acid 415
Aromatic amino acid decarboxylase 471
Asparaginase 432
Asparagine 28, 29
  biosynthesis 436
  synthetase 436

Aspartate carbamoyl transferase 444, 445, 448
Aspartate transaminase 412
Aspartic acid 28
biosynthesis 436
degradation 432
Atherosclerosis 311
ATP-citrate lyase 268
Autotrophic cells 214

Bacteriophage 358, 362, 372
Base 20, 21, 348, 349, 350
Bilayer 280
Bile 262
acids 315–317
salts 315–317
Biosynthesis of amino acids 436–441
Biotin 133, 134
Blood 22
Bohr effect 339
Bone 507, 508
Bragg equation 91
Brownian motion 73
Buffer 21, 22
Bundle-sheaf cells 227
Burk-Lineweaver plot 106
Butyryl-S-ACP 270
Butyryl-β-ketoacyl-ACP synthase 270

Calcitonin 475, 508
Calvin cycle 222–226
cAMP-dependent protein kinase 471
Carbamoyl phosphate 415, 447
synthase 415, 447
Carotenoids 305
Carbohydrate metabolism 186–211
Carbohydrates 162–184, 258
Carbonic anhydrase 336
Carboxypeptidase 121
A and B 54
Cardiac muscle 501
Carnitine 265
acyltransferase 265
Carotenes 216
Catabolite activator protein 381
Catalase 227, 342, 412
Catecholamines 469–472
CDP-diacylglycerol inositol phosphatidyl-transferase 290
CDP-ethanolamine 288
Cell 1–15
coats 187
Cellulases 182, 187
Cellulose 162, 163, 176, 177
Cell wall 8
Centrioles 7
Ceramide 286, 291
cholinephosphotransferase 291
Cerebrosides 286
Ceruloplasmin 491
Chain form of glucose 167
Chemical-coupling theory 249
Chemiosmotic hypothesis 250–252
Chemolithotrophs 152
Chemoorganotrophs 152
Chenodeoxycholic acid 315
Chirality 35
Chloramphenicol 397
Chlorella pyrenoidosa 222
Chlorophyll 216–222
Chloroplasts 8, 215–228
Cholecalciferol 313, 314
Cholestanol 309, 311
Cholesterol 304, 310, 311
biosynthesis 317–321
Cholic acid 315
Choline 279, 280
kinase 289
Chondroitin sulfate 178
Chromaffin granules 470, 471
Chromane 308
Chromosomes 5, 9–13
Chylomicrons 263
Chymotrypsin 113–119
Chymotrypsinogen 113, 114, 119
Cisternae 5
Cis-trans isomerism 261
Citrate synthase 238, 242
Citric acid 235, 238
cycle 231–252
Citrulline 415
Clotting of blood 479–484
Clupeine 375
Codon 388, 393
Coenzyme A 130, 131, 236
biosynthesis 460, 461, 462
Coenzyme Q 140, 141, 246
Coenzymes 103, 129–148
Collagen 494, 499
Competitive inhibition 106–108
Concentration gradient 73, 74
Cone cells 306
Conformation of proteins 68–95
Conjugated proteins 51
Conjugation 10, 362
Connective tissue 494–499
Constitutive mutations 381
COOH-terminal residues 54
Cooperativity 122, 337
Coprostanol 311
Corepressors 400
Corticosterone 321, 322, 324
Corticotropin 59
Cortisol 321, 322, 324, 474
C_4 plants 227
Creatine biosynthesis 432, 433
Creatine kinase 507
Cristae 6, 7, 233
Crossing over 13
Crotonyl-S-ACP 270
Crystals 88–91
CTP-synthetase 447
Cyclic AMP 209–211, 471, 475
Cyclic AMP-dependent protein kinase 210, 211
Cyclic photophosphorylation 221–222
Cystathionine 432
Cystathionine-γ-lyase 439
Cystathionine-β-synthase 439
Cysteine 29, 33
Cysteine biosynthesis 439
Cysteine degradation 423
Cysteine dioxygenase 423
Cystine 29, 33
Cytochrome a 342
Cytochrome c 61, 340, 341
Cytochromes 221, 246, 329, 340 341, 342
Cytoplasm 3
Cytosine 350

Decarboxylase 414
7-Dehydrocholesterol 311, 313, 321
Dehydro-epiandrosterone 322, 323
3-Dehydrosphinganine 290
Denaturation 94, 358

Density gradient centrifugation 360, 366, 367
Deoxycholic acid 315
2-Deoxy-D-ribose 349
Deoxyribonuclease 353
Deoxyribonucleic acid (DNA) 3, 4, 11, 247–282
Deoxyribonucleotide biosynthesis 455, 456
Deoxysugars 169, 170
Dextran 177
Diabetes 469
Diacylglycerol acytransferase 274
Dielectric constant 18
Diffusion 73–75
Diffusion coefficient 75
Digestion of carbohydrates 187, 188
Digestion of fats 262, 263
Digestion of protein 410, 411
Digitonin 249
Dihydroorotase 444
Dihydroorotate dehydrogenase 444
Dihydrosphingosine 286
Dihydroxyacetone 163
Dihydroxyacetone phosphate acyltransferase 288
1,25-Dihydroxycholecalciferol 314
Dimethylallyl pyrophosphate 317, 319, 320
Dimethylallyl transferase 317
2, 3-Diphosphoglycerate 338, 339
Diploid cells 9
Disc gel electrophoresis 47
DNA ligase 370–372
DNA polymerases 367–370
DNA replication 366–373
Dopamine 434, 469–472

Dopamine-β-hydroxylase 471
E. coli chromosome 363–365
Edman degradation 53
Elaidic acid 261
Elastase 120
Elastin 498, 499
Electron transfer 231
Elongation factor 395, 396
Elution volume 84
Enantiomorph 36
Endergonic reaction 112, 153
Endoplasmic reticulum 5
Enolase 193, 198
Enolization 170
Enoyl-ACP hydratase 266, 269
Enoyl-ACP reductase 270
Enoyl CoA hydratase 266
Enthalpy 110
Entropy 110
Enzyme inhibition 106–108
Enzyme kinetics 104–109
Enzymes 3, 100–125
Epimerization 204, 205
Epimers 164
Epinephrine 210, 434, 469–472, 476
Equilibrium constant 110, 111
Equivalence zone 488
Ergosterol 311, 313
Erythrocytes 329, 334, 335, 336
Escherichia coli 403–405, 406
Essential amino acids 410
Estradiol 323, 324
Estriol 323, 324
Estrogens 304, 323, 324
Estrone 323, 324
Ethanolamine 280
Ethanolamine kinase 288
Eukaryotic cell 3, 5–9
Evolutionary divergence 51
Exciton transfer 219
Excitons 219
Exergonic reaction 112
Extrinsic clotting mechanism 482
Extrinsic membrane proteins 296

Fabry's disease 286
Farnesyl pyrophosphate 317, 320
Fatty acids 100, 257–263
  biosynthesis 268–273
  degradation 264–267
  synthetase 268
Feedback 401
Ferredoxin 220, 226
Ferredoxin-NADP reductase 220
Ferredoxin reducing substance 220
Ferrihemes 329
Ferrochelatase 344
Ferrohemes 329
F-factor 363
Fibrin 309, 480
Fibrinogen 309, 480–484
Fibrin stabilizing factor 484
Fibrous long spacing form 497
Fibrous proteins 69
Fick's law 74
Fingerprinting 45
Fischer formulas 166, 167
Flavin adenine dinucleotide 138, 139
Flavin mononucleotide 138, 139
Fluid mosaic model 296
1-Fluoro-2,4-dinitrobenzene 34, 53
FMN adenylyltransferase 459
Folic acid 131–133
Follicle stimulating hormone 472, 473
Frame-shift mutations 405
Free energy 109–112
Freeze etching 296–297
Frictional coefficient 75
Frictional ratio 75
Fructokinase 190
Fructose diphosphatase 243
Fructose diphosphate aldolase 191, 197, 203
Fumarase 241
Fumaric acid 235, 240
Furanose 167

Galactokinase 205
Galactosamine 169
Galactosemia 205
Galactosyl transferase 206
Gametes 9
Gangliosides 286, 287
Gaucher's disease 286
Gel electrophoresis 45
Gel filtration 40, 84, 85
Genes 9-13
Genetic code 397-399
Genetic engineering 373
Genetic map 9, 364
Genome 5
Geranyl pyrophosphate 317, 320
Glands 464
Globin 329
Globular proteins 69
Glucagon 467, 469, 476
Glucokinase 190, 196, 197
Gluconeogenesis 231, 242–244, 409
Glucosamine 169
Glucose 476
  metabolism 188–204

Glucose-6-phosphate
dehydrogenase 200
Glucosephosphate isomerase 191
Glucose-1-phosphate
uridylyltransferase 206
α,1,6-Glucosidase 181
Glucosides 168
Glutamate dehydrogenase 411
Glutamate kinase 437
Glutamate synthase 428
Glutamic acid 28
biosynthesis 436
degradation 427
Glutaminase 428
Glutamine 28, 29
biosynthesis 436
degradation 427, 428
synthetase 436
γ-Glutamylcysteine synthetase
434
Glutathione biosynthesis 433,
434
Glutathione synthetase 434
Glyceraldehyde 163
Glyceraldehydephosphate
dehydrogenase 192, 197, 198
Glycerol 280
Glycerol kinase 264, 274
L-Glycerol-3-phosphate 274, 279,
280
Glycerolphosphate acyltransferase
274
Glycerol-3-phosphate
dehydrogenase 264, 274
Glycosyl-1-phosphate
nucleotidyl-transferases 204
Glycerophosphatides 279–285,
283–285
Glycine 38
biosynthesis 438, 439
degradation 423, 426
synthase 426, 439
Glycine amidinotransferase 432
Glycocholic acid 316
Glycogen 163, 176, 476
metabolism 206–211
particles 209
phosphorylase 207–211
synthase 209
Glycolate oxidase 227
Glycolysis 100, 187–199, 244
Glycoproteins 181
Glycosides 168
Glycosidic bond 168
Glyoxylate cycle 244
Glyoxysomes 245
Golgi apparatus 5, 467
Gout 458
Grana 216
Guanine 349, 350
Guanylic acid 455

H-Band 501
Hageman factor 482
Haploid cells 9
Haptens 489, 490
Haptoglobin 491
Haworth formulas 166, 167
α-Helix 86, 87, 93
Heme 328, 329
metabolism 342, 343, 344
synthetase 344
Hemiacetal 166, 167
Hemoglobin 63, 329–339
synthesis 389–391
Hemolysis 335
Henderson-Hasselbalch equation
20
Heparin 178, 179, 263
Heterotrophic cells 214
Heterozygous organism 9
Hexokinase 188–190, 196
Hexose diphosphatase 203
Hexose monophosphate shunt
200–203
Hexose-1-phosphate
uridylyltransferase 205
High energy compounds 152
Histamine 434
Histidine 29, 39
decarboxylase 434
degradation 427, 429
Histones 375
Homology 359
Homozygous organism 9
Hormones 465–476
Hyaluronic acid 177, 178
Hydrogen bonds 17, 69, 70, 71
Hydrogen electrode 19
Hydrolysis reactions 102
Hydronium ion 18
Hydrophobic bonds 18, 71, 72
3-Hydroxy acyl CoA
dehydrogenase 266
Hydroxyapatite 507
D-β-Hydroxybutyric acid 275
D-β-Hydroxybutyric acid
dehydrogenase 275
25-Hydroxycholecalciferol 314
5-Hydroxymethylcytosine 350,
405
β-Hydroxy-β-methylglutaryl CoA
275, 317, 318, 421
Hydroxymethylglutaryl-CoA
reductase 317
Hydroxymethylglutaryl-CoA
synthase 317, 318
Hydroxyproline 33
biosynthesis 437
Hyperglycemia 469
Hypochromism 356, 357
Hypothalamus 322, 472, 474
Hypothyroidism 466

I-Bands 501
Ichthyocol 499
Immunoglobulins 485–490
IMP-cyclohydrolase 453
IMP-dehydrogenase 453
Inducer 380
Inducible enzymes 376
Initiation factor 394, 395
Inosinic acid 450, 453
Inositol 280
Inositol phosphatides 285
Insulin 57, 58, 59, 60, 467–469,
476
Intercalation 355
Intragenic suppression 405
Intrinsic clotting mechanism 481
Intrinsic membrane proteins 296
Inversion of sucrose 172
Ion exchange 41
Ionophores 248
Islet cells 467
Isocitrate dehydrogenase 238,
242
Isocitrate lyase 244
Isocitric acid 235, 238
Isoelectric point 39
Isoleucine degradation 428, 430,
432
Isomerization reactions 102
Isopentenyl pyrophosphate 317,
319
Isopentenyl pyrophosphate
isomerase 317
Isoprene 304, 305
Isozymes 199

Jamaican vomiting sickness 421

β-Ketoacyl-ACP reductase 269
β-Ketoacyl-ACP synthase 269, 270, 271
α-Ketoglutarate dehydrogenase 240
α-Ketoglutaric acid 238, 239
Ketone bodies 275
Ketoses 163, 165
β-Ketostearyl CoA 272
Kinases 204

Lac operon 377, 380, 381, 399
Lac repressor 381, 399
α-Lactalbumin 206
Lactate dehydrogenase 194, 199
Lactose 173
Lactose synthase 206
Lamellae 216
Lanosterol 312, 318, 320, 321
Leucine degradation 419–421
Lignoceric acid 286
Limonene 305
Lincomycin 397
Linoleic acid 262, 273
Linolenic acid 262, 272
Lipases 259, 263
Lipids 257–291
Lipoamide dehydrogenase 236
Lipoate acetyltransferase 236
Lipoic acid 139, 140, 236
Lipoprotein lipase 263
Lipoproteins 263, 491
Liposomes 280, 283
Lithocholic acid 315
Luteinizing hormone 472, 473
Lymphocytes 485
Lysine 29
  degradation 421, 422
Lysosomes 5
Lysozyme 58, 92, 93

Malate dehydrogenase 196, 227, 244
Malate synthase 245
Malic acid 241
Malic enzyme 196, 227
Malonyl CoA 270
Malonyl-S-ACP 270
Malonyl transferase 269
Maltose 172, 173
Maple syrup disease 421
Matrix 233, 265
Meiosis 11, 12, 13
Melanin 434
Melanocyte stimulating hormone 472
Melting point 357
Membrane potential 298
Membranes 3, 100, 295–301
Mercaptalbumin 485
Meselson-Stahl experiment 366, 367
Mesophyll cells 227
Messenger RNA 348, 376, 377, 388, 389, 393
Metabolic convergence 231
Metabolism 2, 152, 153
Methemoglobin 329
Methionine 29
  degradation 428, 431, 432
Methionine adenosyltransferase 432, 439
Methylation of sugars 168
Methylmalonyl CoA 267
Methylmalonyl CoA mutase 267
Methylmalonyl CoA racemase 267
Mevalonate kinase 317
Mevalonic acid 317, 318
Micelles 262
Michaelis constant 105
Michaelis-Menten model 104–109
Missense mutation 405
Mitochondria 6, 231, 233, 245–252
Mitosis 7, 11
Mobility 39
Monosaccharides 162–171
Mucins 181
Mucopolysaccharides 162, 177–180
Muramic acid 170, 171
Muropeptide 179, 180
Mutases 204
Mutation 9, 51
Myofibril 499–507
Myoglobin 329, 339, 340
Myosin 504–507

Na^+-K^+ activated ATP-ase 299, 300, 301
NADH dehydrogenase 245, 246
Nervonic acid 286
Neuraminic acid 170, 171
NH_2-terminal residues 53, 54
Nicotinamide adenine dinucleotide 136–138, 155
Nicotinamide adenine dinucleotide biosynthesis 459, 460
Nicotinamide adenine dinucleotide phosphate 136–138, 155
Nicotinic acid mononucleotide 459
Niemann-Pick disease 286
Ninhydrin 34, 35
p-Nitrophenyl acetate 115, 116
Noncompetitive inhibition 106–108
Nonessential amino acids 410
Nonsense mutation 405
Norepinephrine 434, 469–472
Nuclear membrane 5
Nucleic acids 3, 347–383
Nucleolus 5
Nucleoproteins 375
Nucleoside 350
Nucleosomes 375
Nucleotides 347–353
Nucleus 3, 5
Nutrition 151–159

Okazaki fragments 372
Oleic acid 261, 272, 273
Oligosaccharides 171–173
Oligosaccharases 187
Operator 380
Operon model 380, 381
Opsin 307
Ornithine carbamoyl transferase 415
Ornithine decarboxylase 436
Orotate phosphoribosyl transferase 445
Orotic acid 444–446
Orotidine-5′-phosphate 445, 446
Ouabain 299
Oxaloacetic acid 235, 238, 243, 244
Oxalosuccinic acid 235, 238
Oxidation-reduction 141–148
Oxidative deamination 411, 412
Oxidative phosphorylation 153, 232, 233, 245–252
Oxytocin 472, 474, 474

Palindromes 360
Palmitic acid 260, 261, 271, 272
Palmitoleic acid 261, 272, 273
Palmitoleyl CoA 273
Palmitoyl CoA 266, 272, 273

Pantothenic acid 460, 461
Papain 121
Paper chromatography 42, 43
Paper electrophoresis 44, 45
Parathyroid hormone 474, 508
Partial specific volume 77
Partition chromatography 42
Passive facilitated diffusion 297
Passive facilitated transport 298, 299
Pasteur effect 195
Pectins 177
Pepsin 120
Peptide bond 27
Peptidoglycan 179
Peptidyl site 394, 395
Perimysium 501
Peroxidases 329, 342
pH scale 19
Phenylalanine 28
  degradation 421, 423
Phenylalanine-4-monooxygenase 421, 436
Phenylethanolamine-N-methyltransferase 471
Phenylketonuria 423
Phosphatidal ethanolamine 285
Phosphatidate phosphatase 274
Phosphatidic acid 273
  biosynthesis 288
L-α-Phosphatidic acid 283, 284
Phosphatidyl choline 281, 284
  biosynthesis 289
Phosphatidyl ethanolamine 284
  biosynthesis 288
Phosphatidyl ethanolamine methyltransferase 289
Phosphatidyl inositol biosynthesis 290
Phosphatidyl serine 282, 284
Phosphatidyl serine biosynthesis 290
5-Phospho-β-D-ribosylamine 448–450
Phosphocholine cytidylyltransferase 289
Phosphocholine transferase 289
Phosphocreatine 507
Phosphodiester bonds 352
Phosphoenolpyruvate carboxykinase 242
Phosphoenolpyruvate carboxylase 226
Phosphoethanolamine 288
Phosphoethanolamine cytidylyltransferase 288
Phosphoethanolamine transferase 288
Phosphofructokinase 191, 197, 199
Phosphoglucomutase 204
Phosphogluconate pathway 200–204
6-Phosphogluconic acid 190
Phosphoglucokinase 204
Phosphoglycerate dehydrogenase 438
Phosphoglycerate kinase 192, 223
Phosphoglyceromutase 193
Phospholipids 257, 279–291, 295
Phosphomevalonate kinase 317
4-Phosphopantetheine 268, 460, 461
Phosphopentokinase 223
5-Phosphoribose-1-pyrophosphate 445, 446, 459
Phosphoribosylaminoimidazole carboxylase 452
Phosphoribosylaminoimidazole synthetase 451
Phosphoribosylformylglycinamide synthetase 451
Phosphoribosylglycinamide formyltransferase 451
Phosphoribosylglycinamide synthetase 449
Phosphorylase a 207–211
Phosphorylase b 207–211
Phosphorylase kinase 207, 209–211
Phosphorylase phosphatase 207, 209–211
Phosphoserine phosphatase 438
Phosphoserine transferase 438
Photolithotrophs 152
Photoorganotrophs 152
Photorespiration 227–228
Photosynthesis 2, 186, 214–228
Photosynthetic organisms 151
Photosystems I and II 219–222
Pituitary 322
Pituitary hormones 472–475
Plants 1, 2, 7, 8
Plasma 479-491
Plasma albumin 480, 484, 485
Plasma cells 485
Plasmalogens 282, 284, 285
Plastids 8
Plastocyanin 221
Plastoquinone 221
Platelets 480
β-Pleated sheet 87
Pneumococcus 361
Polarized light 35
Polyenoic fatty acids 273
Polypeptides 27
Polyribonucleotide phosphorylase 397
Polysaccharides 163, 173–182
Polysome 393
Porphin 327
Porphobilinogen 342, 343, 344
Porphyrins 327, 328
Portal vein 263
Precipitin reaction 488
Pregnenolone 323
Preproinsulin 467
Presqualene synthase 317
Primary structure 50–63
Primers 103, 104, 370
Progesterone 323, 324
Progestins 323
Proinsulin 467
Prokaryotic cell 3, 4, 5
Prolactin 473
Proline 33, 93
  biosynthesis 437
  degradation 427, 428
Proline-4-monooxygenase 438
Promoter 380
Propionyl CoA carboxylase 267
Prostaglandins 262
Prosthetic group 51, 52, 103, 129
Protamines 375
Proteins 3, 26, 27
Protein synthesis 387–406
Proteolytic enzymes 56, 113–121
Prothrombin 308, 309, 480–483
Protists 2
Protomers 122
Protoporphyrin 328, 344
Proximal histidine 330
Purines 348, 349, 350
  degradation 456–458
  nucleotide biosynthesis 448–455
Puromycin 397
Putrescine 434
Pyranose 167
Pyridoxal phosphate 134, 413

Pyrimidines 348–350
  degradation 458, 459
  nucleotide biosynthesis 444–448
Pyrophosphomevalonate kinase 317
Pyruvate carboxylase 196, 242
Pyruvate decarboxylase 195
Pyruvate dehydrogenase 236, 237, 241
Pyruvate dehydrogenase kinase 237
Pyruvate dehydrogenase phosphatase 237
Pyruvate kinase 193, 198

Quaternary structure 69

Redox potential 143–148
Regulatory gene 380
Replication fork 370, 372
Repressor 380
Respiration 152, 186
Respiratory chain 232, 245–252
Restriction endonucleases 373
Restriction of DNA 373
Reticulocytes 389, 390
Retina 306, 307
Retinal isomerase 307
Retinal reductase 307
Rho factor 381
Rhodopsin 307
Riboflavin biosynthesis 459
Riboflavin kinase 459
Ribonuclease 58, 353
Ribonucleic acid (RNA) 3, 348, 387–406
Ribonucleoside diphosphate reductase 456
D-Ribose 349
Ribosephosphate isomerase 202
Ribosephosphate pyrophosphatase 445
Ribosomal RNA 375, 376
Ribosomes 3, 375, 376, 393, 394
Ribulosediphosphate carboxylase 223
Ribulosephosphate-3-epimerase 200
Rifamycin 382
RNA polymerase 374, 379–382, 388
Rod cells 306
Rolling circle model 374
Rotenone 246

Saponification 259
Sarcolemma 501
Sarcomeres 501
Satellite DNA 360
Schiff's base 413
Schlieren optical system 77, 78
SDS gel electrophoresis 85, 86
Secondary structure 69
Sedimentation coefficient 79
Sedimentation equilibrium 81, 82
Sedimentation velocity 75–80
Segment long spacing form 498
Self-association of proteins 72
Sequenator 54
Sequence determination of polypeptides 55, 56
Sequential model 125
Serine 28, 280
  biosynthesis 438, 439
  degradation 423, 426
  hydratase 425
  hydroxymethyltransferase 419, 426, 439
Serotonin 434
Serum albumin 264
Sex hormones 304
SF-fragments 505
Shortening 262
Sialic acid 171
Sickle-cell anemia 51, 61, 62
Skeletal muscle 501–507
Smooth muscle 499
Soaps 262
Somatic cells 9, 11
Somatomedin 473
Somatotropin 473
Somatotropin release inhibitor hormone 473, 474
Soret band 329
Species variations in primary structure 58–61
Spermidine 434
Spermine 434
Sphinganine 290
Sphingolipids 279, 285, 286, 295
Sphingomyelin 282, 285, 286
  biosynthesis 290, 291
Sphingosine 279, 280, 290
Sphingosine acyltransferase 291
Squalene 304, 317, 320
Squalene 2,3-epoxide 318
Squalene epoxide lanosterol cyclase 318
Squalene monooxygenase 318
Squalene synthase 317
Stacking of bases 355
Starch 163, 173–175
Steady state 104
Stearic acid 260
Stearyl CoA 272
Sterane 309
Stereospecific numbering 280
Steroids 304
Streptomycin 397
Striated muscle 501–507
Subtilisin 121
Succinate dehydrogenase 240, 241
Succinic acid 235, 239, 240, 343
Succinyl CoA 342, 431, 432
Succinyl CoA synthetase 240
Sucrose 172, 173
Sucrose gradient centrifugation 81–84
Sucrose phosphatase 205
Sucrose phosphate synthase 205
Sugar 162–171
Sugar phosphates 169
Sulfatides 286
Supercoils 359
Superhelix density 359
Suppressor genes 405, 406
Suppressor mutations 405
Supramolecular structures 100
Symmetry model 124, 125

Target tissue 464
Taurine 315
Tay-Sachs disease 286
TCA cycle 100, 153
Termination factor 396
Termination signals 381
Terpenes 304–309
Tertiary structure 69
Testosterone 322, 323, 324
T-Even bacteriophages 403
Thermodynamics 100, 109–112
Thiamine pyrophosphate 134, 135
Thin layer chromatography 44
Thioesterase 270
Thioether bonds 340
Thiolase 266, 275
Thioredoxin 455, 456

Thioredoxin reductase 455
Thoracic duct 263
Threonine 28
  degradation 419–421
Thrombin 309, 480–483
Thromboplastin 482
Thymidylate synthetase 456
Thymine 350
Thyroglobulin 467
Thyrotropin 472, 473
Thyrotropin-releasing factor 473, 474
Thyroxine 466
Thyroxine-binding globulin 467
Tobacco mosaic virus 387, 388, 401–403
Tocopherol 307, 308
Transaldolase 202
Transaminase 412
Transaminase-amidase 414
Transamination 226
Transcription 375, 379–383
Transduction 363
Transfer reactions 102
Transfer RNA 377–379, 388
Transferrin 491
Transformation 360–362
Transforming principle 361
Transglycosylation 206
Transketolase 202, 203
Translation 388–406
Translocation 396
Triacylglycerols 257–275
  biosynthesis 273
Tricarboxylic acid (TCA) cycle 231–253
Triosephosphate dehydrogenase 223
Triose phosphate isomerase 192, 203, 223
Tropocollagen 495, 496
Tropomyosin 505, 506
Troponin 505, 506, 507
Trypsin 113
Trypsinogen 113
Tryptophan 28
  degradation 421, 423
Tryptophan pyrrolase 423
Tyrosine 28
  biosynthesis 436
  degradation 421, 423
Tyrosine hydroxylase 471

Ubiquinone 140, 141
UDP-glucose-4-epimerase 205
Ultracentrifuge 76–84
Uncoupling reagents 248
Unwinding protein 370, 372
Uracil 350
Urate oxidase 458
Ureidosuccinic acid 447
Uric acid 456–458
Uridine diphosphate glucose 204, 205, 205

Vacuoles 9
Valine degradation 428, 430, 432
Valinomycin 248, 300, 301
Vasopressin 474, 475
Vesicles 249
  X 174 virus 374
Viruses 2
Vitamins 129, 130, 156–159
  A 305–307
  D 312, 313, 314, 315
  E 307
  K 308, 309

Water 15, 16, 17
Watson-Crick structure 354–356
Wobble base 393

Xanthine 457, 458
  oxidase, 458
Xanthophylls 216
Xanthylic acid 455
X-ray diffraction 68, 87–92
Xylanas 179

Z-band 501
Zonal technique 84
Zwitterions 27
Zygote 13
Zymogen 113